ANNUAL REVIEW OF
PLANT BIOLOGY

ANNUAL REVIEW OF PLANT BIOLOGY

VOLUME 53, 2002

DEBORAH P. DELMER, *Editor*
The Rockefeller Foundation

HANS J. BOHNERT, *Associate Editor*
University of Illinois, Urbana

SABEEHA MERCHANT, *Associate Editor*
University of California, Los Angeles

www.annualreviews.org   science@annualreviews.org   650-493-4400

ANNUAL REVIEWS
4139 El Camino Way • P.O. BOX 10139 • Palo Alto, California 94303-0139

ANNUAL REVIEWS
Palo Alto, California, USA

International Standard Serial Number: 1040-2519
International Standard Book Number: 0-8243-0653-8
Library of Congress Catalog Card Number: 50-13143

Typeset by Techbooks, Fairfax, VA
Printed and Bound in the United States of America

PREFACE

As many of you know, the original title for this Annual Reviews series was the *Annual Review of Plant Physiology*. With the rising importance of molecular biology in the field of plant science, a decision was made in 1988 to change the name to the *Annual Review of Plant Physiology and Plant Molecular Biology*. Although this change did help convey our inclusion of molecular biology, the name was cumbersome and, in fact, still did not reflect the wide range of topics covered by this series; topics ranging from discussions of plant structure, biochemical processes, genetics, evolution, and responses of plants to the environment have always been routinely covered in this series. Therefore, the Editorial Board this year made the decision that "simpler is better" and adopted a new name, *Annual Review of Plant Biology*. We are sure you'll be happy to have a shorter title to quote in the future, and we also very much hope you will agree with us that this title also more clearly indicates to the readership that we provide reviews that cover a broad scope of topics in plant biology.

Changes have also occurred in the composition of the Editorial Board. I am pleased and honored to take over as Editor. For the past eight years, Russell Jones has guided this series with both wit and wisdom, as evidenced by the fact that this series has consistently had the highest impact factor ratings for citations of any plant series, ranking first out of the 137 journals tracked by ISI, a number that ranks competitively with many of the very best journals in all of science. And this number has held up in spite of the rise of many new review series for the plant sciences. We believe this is so because we are the source you come to for truly comprehensive reviews—those that include a historical perspective and that allow the space to discuss issues in depth. Such reviews are particularly valuable, not only for the experts in the field, but also for those entering a new field or having to teach it for the first time.

I also note with pleasure that Sabeeha Merchant has agreed to become one of our Associate Editors. Sabeeha has served for several years on our Board, and I have been most impressed with her energy, enthusiasm, and breadth of knowledge. Our thanks are also given for the excellent contributions of outgoing members Ray Chollet and Malcolm Drew, and our welcome is given to new board members Don Ort and Steve Huber.

So ... some changes have occurred this year, but one thing has not—our commitment to provide you with the very best comprehensive reviews that cover an expansive range of topics of importance to the field of plant biology.

Deborah Delmer, Editor

Annual Review of Plant Biology
Volume 53, 2002

CONTENTS

ERRATA
An online log of corrections to *Annual Review of Plant Biology* chapters (if any, 1997 to the present) may be found at http://plant.annualreviews.org/

RELATED ARTICLES

From the *Annual Review of Biochemistry*, Volume 71 (2002)

Biochemistry of Na,K-ATPase, Jack H. Kaplan

DNA Replication in Eukaryotic Cells, Stephen P. Bell and Anindya Dutta

Great Metalloclusters in Enzymology, Douglas C. Rees

Error-Prone Repair DNA Polymerases in Prokaryotes and Eukaryotes, Myron F. Goodman

Nuclear Actin and Actin-Related Proteins in Chromatin Remodeling, Ivan A. Olave, Samara L. Reck-Peterson, and Gerald R. Crabtree

Metabolism and the Control of Circadian Rhythms, Jared Rutter, Martin Reick, and Steven L. McKnight

Eukaryotic DNA Polymerases, Ulrich Hübscher, Giovanni Maga, and Silvio Spadari

Mechanisms of Fast Protein Folding, Jeffrey K. Myers and Terrence G. Oas

From the *Annual Review of Biophysics and Biomolecular Structure*, Volume 30 (2001)

Structures and Proton-Pumping Strategies of Mitochondrial Respiratory Enzymes, Brian E. Schultz and Sunney I. Chan

A Structural View of Cre-loxP Site-Specific Recombination, Gregory D. Van Duyne

Ab Initio Protein Structure Prediction: Progress and Prospects, Richard Bonneau and David Baker

Chaperonin-Mediated Protein Folding, D. Thirumalai and George H. Lorimer

Photosystem II: The Solid Structural Era, Kyong-Hi Rhee

From the *Annual Review of Cell and Developmental Biology*, Volume 17 (2001)

Polarized Cell Growth In Higher Plants, Peter Hepler, L. Vidali, and A. Cheung

Molecular Bases of Circadian Rhythms, Stacey L. Harmer, Satchidananda Panda, and Steve A. Kay

Getting the Message Across: The Intracellular Localization of mRNAs in Higher Eukaryotes, Isabel M. Palacios and Daniel St. Johnston

Control of Early Seed Development, Adbul M. Chaudhury,
Anna Koltunow, Thomas Payne, Ming Luo, Mathew R. Tucker,
E. S. Dennis, and W. J. Peacock

Peroxisome Biogenesis, P. Edward Purdue and Paul B. Lazarow

Left-Right Asymmetry Determination in Vertebrates, Mark Mercola and
Michael Levin

From the *Annual Review of Genetics*, Volume 35 (2001)

*Hypoviruses and Chestnut Blight: Exploiting Viruses to Understand and
Modulate Fungal Pathogenesis*, Angus L. Dawe and Donald L. Nuss

Genetics and The Fitness of Hybrids, John M. Burke and Michael L. Arnold

Building a Multicellular Organism, Dale Kaiser

*The Inheritance of Genes in Mitochondria and Chloroplasts: Laws,
Mechanisms, and Models*, C. William Birky, Jr.

Models and Data on Plant-Enemy Coevolution, Joy Bergelson,
Greg Dwyer, and J. J. Emerson

From the *Annual Review of Phytopathology*, Volume 40 (2002)

Viral Sequences Integrated into Plant Genomes, Glyn Harper, Roger Hull,
Ben Lockhart, and Neil Olszewski

Molecular Basis of Recognition Between Phytophthora *Pathogens and
Their Hosts*, Brett M. Tyler

Antibiotic Use in Plant Agriculture, Patricia S. McManus, Virginia
O. Stockwell, George W. Sundin, and Alan L. Jones

Risk Assessment of Virus-Resistant Transgenic Plants, Mark Tepfer

A. A. Benson

Annu. Rev. Plant Biol. 2002. 53:1–25
DOI: 10.1146/annurev.arplant.53.091201.142547

PAVING THE PATH

A. A. Benson

Scripps Institution of Oceanography, La Jolla, California 92093-0202;
e-mail: abenson@ucsd.edu

Key Words   Sam Ruben, Rat House, ORL, radiobiochemical exploration, neutron
activation

■ **Abstract**   Fortuitous preparation and experiences led to the opportunity to use
radioactive carbon dioxides to discern the path of carbon in photosynthesis. The
search for the CO_2 acceptor led to recognition of the growth stimulatory effect of
methanol and its derivatives. With the techniques developed, radiochromatographic exploration led to discovery of major membrane lipids containing phosphorus, sulfur, and
arsenic.

CONTENTS

INTRODUCTION

Confluence of events and experience led to successes in discerning the path of carbon in photosynthesis. Enjoying the excitement of understanding how Nature works, coupled with serendipitous good fortune, I find myself describing some of these adventures in the following pages.

Managed from Above

At Caltech, on May 12, 1942, I had just finished describing my synthesis of fluoro-iodo-thyronines, sphingosine analogs, neurobiology of the sea scallop and discussion of my 10 Propositions before my Doctoral Committee, when Linus Pauling asked one more question: "Andy, can you write on the board the differential equation for decay of a radioactive isotope?" Well, I managed to get it down all right, but it was a surprise. I had no idea why he asked that question; it had absolutely nothing to do with my thesis! But, knowing his approach after several years in his classes, I wasn't surprised and proceeded to forget all about that incident. He had embarrassed me one evening in the lab, finding me balancing a 12-liter flask of hot methanol on the edge of my sink and suggested that it might not be necessary for me to go further in his quantum mechanics courses. I had not done at all well in the final exam. So, why did Pauling ask me the question? Sixty years later it dawned on me that Linus Pauling and Wendell Latimer, Dean of the College of Chemistry at U.C. Berkeley, had contrived to provide support in organic chemistry for a young and rising star of the Berkeley faculty, Chemistry Instructor Sam Ruben, who was on the verge of being advanced to Assistant Professor.

Within a week or so, after my exam and the Linus Pauling episode, I received an invitation from Chemistry Department Chairman Joel Hildebrand in U.C. Berkeley to join the faculty of the Department as Instructor. The salary would be $2000.00 per annum. That was nearly $167 a month! It was a stellar opportunity and the beginning of my life within the path of carbon in photosynthesis, eventually leading toward the honor of preparing this chapter.

BACK ON THE FARM

My father delivered me on September 24, 1917, in the wood-shingled Sisters Hospital in Modesto, not far from a similar event, the birth of Osamu Hayaishi in Stockton. Two conflagrations were going on, the war in Europe (World War I) and a distressing fire of the horse barn across the street, not a comforting event for my wonderful mother. Home was in nearby Riverbank, overlooking the Stanislaus River and superb farm land. Carl Bennett Benson had started medicine at the University of Chicago, impressed by lectures of the great physiologist, A.J. Carlson. It was cut short by tuberculosis contracted from a roommate. He chose to recover in Los Angeles at the University of Southern California School of Medicine where a life-long friendship with classmate Montague Cleeves developed. Graduating in 1915 he accepted a position in Riverbank with the Santa Fe Railroad. You can imagine my impressions at the age of five, being in the center of a circle of steam-belching screaming locomotives in the great roundhouse. My father's patient Mrs. Foster phoned my father when her incubators were hatching, asking him to drive me to her farm. There, I was excited to watch the emerging turkey chicks, which she sold to local farmers. Californians will recognize the name, Foster Farms, managed now by her grandchildren who were delivered by my father. They now process two million chickens each week.

Coming from first-generation families of Swedish immigrants, farming in Minnesota, my father soon found a Minnesosta-like site near Modesto on the tree-lined bank of the Tuolumne River with a parallel verdant slough screaming with redwing blackbirds, roving coots, turtles, and, later, huge French frogs (we called them French frogs, though I don't know how French they were as they came from the nearby town of French Camp). He managed to establish an adjacent farm for his parents and brothers and sisters. No place on earth could have been finer for a young boy: pristine Nature and fine uncles for help and advice.

We moved from Riverbank to Escalon where my father established a small hospital and outfitted me with a donkey and small plough to work the vacant lot. Later, when we moved to Modesto, I rode my donkey, a two-day trip, which young reporter Max Foster noted in the Modesto paper as Another Ass Comes to Town. High school there was exciting. The band was highly ranked in the country, finishing second in national competition for two consecutive years. Amos Alonzo Stagg came to cheer our football team. Willie Brown, the chemistry teacher, had been on the Stanford football team. He was big, and everyone enjoyed his contagious enthusiasm. Students feared and loved him. At home a chemistry set enhanced my scientific enthusiasm, and the fifth of July, even more. Collecting unspent fireworks bombs early in the morning, I rebuilt them with new fuses and enjoyed creating mayhem at odd hours. It was hazardous sport like my experiments with my father's portable X-ray apparatus and fluoroscopic screen. I spent evenings studying and grinding and figuring a telescope mirror. Later at Caltech I visited the shop weekly to watch the grinding of the 200-inch mirror. Now, I carry my Caltech Athenaeum membership card signed by Edwin P. Hubble who was Secretary at the time.

UNDERGRADUATE DAYS AT UNIVERSITY
OF CALIFORNIA, BERKELEY

As I finished high school in 1935, my father took me to Berkeley for an interview with Wendell Latimer, Chairman of the Department of Chemistry. He was encouraging, and I enrolled. There were only 20 students in chemistry lab sections with Professors Hildebrand and Latimer. Those two semesters were a true privilege. Chemistry classes were exciting, and with this select group of top students, the lab camaraderie was enriching. Daily, we were shooed out of the lab at five o'clock by Carlos, a unique janitor who would attract little attention except for his attitude. His kindly gruff forcefulness rarely failed to clear the lab. Only later did I realize that his forcefulness stemmed from his close relationship with the Dean, who I understand would sometimes need to borrow money from Carlos. With years of collecting postage stamps from the waste baskets of the Dean and Chairman, Carlos Abascal had become a resourceful stamp dealer. We maintained a delightful friendship and exchanged letters for several decades.

Hot summers were spent with Chick Cleeves in the dry yard of our apricot farm, making trays and managing the fruit cutting, sulfuring and then spreading the stacked trays from their wagons on the long rail track. It was hot but productive—nearly five tons of dried fruit at 14¢ a pound. Consequently, I am immune to sulfur dioxide. Tending irrigation levees, I know how to manage a shovel. I enjoyed building a brick house with a master brick mason from Hull, England, and developed sincere respect for master craftsmen as well as some skill at slacking lime and mixing "mud."

In my spare time I enjoyed some nonchemistry courses. I took an optical physics course from Luis Alvarez (later, Nobel Laureate) and remember his arrival for class one morning almost pale with fevered excitement. The word that Lise Meitner and Otto Hahn (Nobel Laureate) had reported nuclear fission was a cataclysmic shock in the Berkeley physics department. The optics lecture was dispensed with. Of course, the fact that Europe was on the verge of conflict enhanced his apprehension. I had been taking a superb course in European history from Professor Kerner who brought day-to-day tension to reality as Hitler took the Sudetenland and Austria. In the other corner of the campus I was enjoying courses in civil, mechanical, chemical, and electrical engineering as well as fluid mechanics. Having survived courses in physics, I found the engineering courses interesting, easy, ultimately useful, and just fun. A few of our lab mates often snuck under the cloud of smoke from the Alhambra Casino Philippine cigar of G.N. Lewis into his Thursday afternoon seminars. At the time I was making trinitrotriphenyl methane for Glenn Seaborg's research with Lewis on the color of molecules.

CLIMBING AT CALTECH: 1939–1942

With help from above, my applications to several schools were accepted. I chose Caltech because of its proximity to home and to the Cleeves family. Arriving together, Dick Noyes and I developed a lasting friendship. We took the first of several qualifying exams in Pasadena, September, 103°F temperatures. That first evening, Dick broke his ankle at table tennis. He recovered well, and we enjoyed countless expeditions and climbs in the San Gabriels and the Sierra Nevada as well as scaling the wall of Crellin Laboratory early Sunday mornings.

In the Niemann laboratory, directly above Linus Pauling's office, I enjoyed fellowship and advice from superb colleagues. Jim Mead, a year ahead of me, guided the technical aspects of my research on synthesis of fluorinated thyroxine analogs, a 20-step synthesis that started with a gallon of anethole (anise oil) oxidized, in flames, within a flask of hot fuming nitric acid and ended with two grams of difluorodiiodothyronine. Jim was working on the structure of sphingosine, which later became part of my thesis work on periodate and lead tetraacetate degradation of its vicinal amino glycol. This experience led directly to my subsequent work with Al Bassham on sedoheptulose and ribulose degradations. I enjoyed rapport with Dick Lewis, son of Gilbert Newton Lewis; Chuck Wagner; John Hays; Ernie Redeman and Kurt Mislow, postdoctoral associates. Jim Mead led beach expeditions to a rented house in South Laguna where we fished with spear and snorkel for hours till we were blue. My handmade dive mask is now an antique. Mead was a close associate and good friend; I served as best man at his wedding. This association proved invaluable in 1960 when our family moved West and spent a year in Jim's Laboratory of Nuclear Medicine and Radiation Biology of UCLA's School of Medicine before moving to La Jolla and Scripps Institution of Oceanography.

The war in Europe was raging, and I recall clearly, as does everyone, the news of December 7, 1941. Conscription had begun and decisions were considered. Our family supported conscientious objectors and sympathized with the tenets of the Society of Friends. A small group of us at Caltech met frequently at the noon hour with Bob Emerson, descendent of the distinguished Quaker family. I registered as a Conscientious Objector. This later became a problem for my position in the Berkeley chemistry department, which was to become increasingly involved in research for the war effort.

BACK TO BERKELEY

Life in the Rat House

So when I arrived in Berkeley, Latimer directed me to the Rat House where Sam welcomed me to a small office/laboratory and a mass of dirty glassware freshly deserted by Henry Taube (Nobel Laureate, 1983). Sam introduced me to the Warburg Apparatus in the dingy lab upstairs lighted by bare clear light bulbs. He handed me his copy of Manometric Methods by Burris, Stauffer, and Umbreit and explained that he needed to know more about the ratios of photosynthesis to respiration in his *Chlorella*. Needless to say, I did not realize at the time that Bob Burris was but a few years my senior, nor that his son John would later be earning his doctorate in my laboratory, nor that his student Nathan Edward Tolbert (20, 23, 29) would become a close friend and colleague who later would convince the world that photorespiration exists (in spite of my failure to consider it very remarkable).

Ruben & Kamen (28) had observed "reversible fixation" of $C^{11}O_2$[1] forming a water-extractable product possessing a carboxyl group. They had deduced from their early fixation experiments that the product(s) of dark fixation could result from the reversibility of the reaction—$RH + CO_2 \leftrightarrow RCOOH$ (28, 34). van Niel pointed out that in bacterial photosynthesis no O_2 is produced and that bacteria must have access to a reducing agent to provide hydrogen for the reduction of CO_2. (No oxidizing agents present in living things are powerful enough to dehydrogenate water except for the photochemical reaction centers of photosynthetic organisms.)

At each step of these developments Sam Ruben and Martin Kamen proposed cyclic processes for the regeneration of the CO_2 acceptor, R-H. With the imposing influence of the phosphorylation processes being developed by Herman Kalckar and by Fritz Lipmann, Sam Ruben increasingly leaned toward the concept of the "high energy phosphate bond" participating in the process. At this point, I arrived in the Rat House and was immediately influenced by Sam's enthusiasm, as expressed in "Photosynthesis and Phosphorylation" (29). Sam was also influenced by the exciting developments from Otto Warburg and handed me copies of the latest papers from Berlin. I had no idea that in a few years I would be with Otto Warburg in his laboratory in Dahlem or that he would send me and Clint Fuller checks for $1000.00 to attend his lecture in Strassbourg in 1963 (17).

I joined Sam and his students in the many late-night experiments with $C^{11}O_2$ (14). They always began at approximately 8 p.m. because Martin needed the time for the bombardment of his boron target after the physicists on the "37 inch" cyclotron left for supper. On maybe three occasions, Martin called just before the experiment, "Cyc's sick, Sam": a disappointment, but an opportunity for us to go home early. With a bombardment completed though, Martin connected his evacuated "aspirator chamber" to the target and collected its gaseous $C^{11}O_2$ and $C^{11}O$. Owing to the short half-life of C^{11}, there then followed the 100-yard dash from the cyclotron to the Rat House and Sam's waiting arms and his demand to "leave at once" (19). Martin was far too radioactive. The aspirator was coupled to a copper oxide–filled tube within a hot combustion furnace for conversion of the gas mixture to pure $C^{11}O_2$ for the photosynthesis experiments. At first I was a helper while more-experienced Charlie Rice and Mary Belle Allen performed their preplanned duties. Each experiment was meticulously planned and orchestrated by Sam. I recall no experiment that failed because of errors in planning. I attended some planning discussions with Martin and Sam in the empty classroom. I was horrified at the vigor of their arguments, but they always reached agreement on the experimental plan unscathed. Sam managed the stopcocks and transfers from the liquid air-cooled spiral trap for the $C^{11}O_2$ to the waiting algae. Never was there any friction involved in working with Sam Ruben. He was a gentle human who drove himself but did not demand the same of his colleagues. He certainly appreciated

---

[1] I use the "atomic weight superscript" terminology of the time. I consider it didactically superior to the current practice, which I use in a following page.

their efforts with good humor. Charlie Rice was "Rice Crispies," and Mary Belle Allen was "Madam Curie." Sam revealed little of his family life in the lab (18). With little children to manage it was not at all easy for him or for his talented chemist wife, Helena. Sam was very interested in the Menschutkin Reaction mechanism studies of Denham Harman; I remember he repeatedly called out "Denham!" for some reason or other. Now I enjoy frequent contact with Denham, the father of our widespread concern for free radical involvement in human aging. Those late-night experiments in the Rat House were the last of a long series (28, 34). Though long-lived radiocarbon (C^{14}) was "invented" by Ruben and Kamen (27), they had not applied it for photosynthesis study; its specific activity was too low and its measurement too tedious.

Soon it was 1943 (and World War II), and Sam's efforts were directed toward a meteorological effort related to movement of gas clouds. Martin Kamen, beset by the House on Un-American Activities Committee and the State Department, was engaged by the local shipyard (19). At this point Sam gave me all of the $BaC^{14}O_3$ in the world to follow the path of carbon. Coupled with an appointment in the finest chemistry department in the country, such a golden opportunity now seems nearly unbelievable. At the time I was too embedded in the problems at hand to realize where I trod in the path of science. The path of carbon had been paved, but its end still lay beyond the horizon.

With the $C^{14}O_2$ I carried out many dark fixations with *Chlorella*, following Sam and Martin's concept of reversible reaction with R-H to yield R-COOH. To reduce the respiratory exchange, they were done in nitrogen—not a good idea. Hence, the product of such fixation was a massive accumulation of succinate. (Science is more about failure than success.) Counting C^{14} was a chore, and it was difficult to achieve reproducibility. Sam had some "Libby Screen Wall Counters" in which the weak C^{14} betas could fly through the screen wall into the G-M counting region. Because they required repeated assembly with deKhotinsky sealing compound each time, I made new ones with standard taper joints, which could easily be opened and reused. I prepared my samples on the interior surfaces of glass cylinders rotated on rollers and dried with hot air blown from a small burner. They could slide over, or away from, the screen wall of the G-M counter. Each was filled with counting gas on the high-vacuum line, and its voltage plateau determined. With the low activities, the work was tedious.

I have been asked about the evolution of our "photosynthetic algae illumination vessel," which Melvin Calvin and others called a "Lollipop." It was a result of Sam Ruben's effort to illuminate a dense algal suspension as uniformly and compactly as possible in thin flat circular Warburg flasks having one or two stopcocks and standard taper joints. I merely turned it on its side so that it could be illuminated from both sides of the thin round vessel and mixed by shaking or by an inserted gas stream. With a large bore stopcock on the bottom for rapid collection of experimental samples, it had assumed the outline of a flat lollipop.

To follow the chemical reactivity of the unknown product(s) I converted the fixed activity to derivatives, using diazomethane to methylate carboxyl groups, and acetic anhyhdride in pyridine to acetylate hydroxyl groups. Evidence for reaction was estimated from behavior of their radioactivity's partition between water and ether or water and ethyl acetate. I could confirm the conclusion of Ruben and Kamen that the product possessed a carboxyl group; the partition coefficient between water and ethyl acetate was 0.14. (Three years later, in ORL, with the daily helpful collaboration of Ed McMillan (Nobel Laureate, 1951), I crystallized the dark fixation product and co-crystallized it with succinic acid to constant specific activity. Little did I realize that he was also involved in crystallizing salts of Neptunium.)

Rat House (Berkeley) Continued

My thesis research under Carl Niemann at Caltech had involved study of the structure of the vicinal aminodihydroxy analogs of sphingosine. I was intent on establishing the location of its amino group in relation to that of its two hydroxyl groups. Clearly, periodate oxidation was a viable approach. During my three years at CalTech I had followed all the carbohydrate chemisty courses delivered by Carl Niemann, former sudent of Karl P. Link in the University of Wisconsin, Madison. All of Niemann's many publications with Link were on the subject of carbohydrate chemistry. I was probably more familiar with carbohydrate chemistry than any other faculty member in the Department of Chemistry when I arrived at U.C. Berkeley as Instructor in July, 1942. My undergraduate classmates, Dick Powell and Bob Connick, came later. The true carbohydrate scholar, of course, was the lovable Zev Hassid with whom I had become well acquainted while working with Sam Ruben in the Rat House. When I was asked by G.N. Lewis to present the Thursday afternoon seminar, I chose the then-novel periodate oxidation as my subject. I had done some of the critical experiments in the Rat House before the seminar, and the sphingosine problem was finally solved. As usual, the seminar began with G.N. striking a big match and waving its flame in front of his famous Philippine cigar. When I was finished, he asked several penetrating questions, as usual. Melvin Calvin was there, but I doubt if he recalled any of the seminar's content; carbohydrate chemistry and periodate oxidation were beyond his interests at that time.

Phosgene and Ruben's Accident

Sam had visited the army's Dugway Proving Ground where tests with goats indicated that phosgene exposure induced lung edema and that the fluid was similarly antigenic to other goats. Having been exposed to the immunological studies of Dan Campbell and Linus Pauling, my theory was that phosgene, a double acid chloride, could couple two proteins or so alter the conformation of one protein to yield a novel antigenic protein. So I developed rapid phosgene synthesis from the 20-minute half-life of $C^{11}O_2$ from Martin Kamen and the 37-inch cyclotron (27a). Sam's technique with the phosgene-exposed rat was to drop the victim into the Waring Blender to produce a protein preparation for measurement of its C-11. For Sam's own phosgene experiments I had fitted small steel bombs with valves so I could fill them with phosgene from the ancient Kahlbaum ampoules lying in sawdust in back of the chemical store room. Not having a chemical hood I did this carefully with ice-cooled ampoules and vacuum transfer to the steel bombs. I still insist that all students should recognize the odor of phosgene.

Because of my tenuous "conscientious objector" draft status, Dean Latimer was unable to extend my teaching contract, and I left for Civilian Public Service in the Nevada mountains, fighting forest fires, building dams, and logging for Forest Service construction projects. A group of technically skilled members was selected for a photogrammetry group in Reno, producing maps from aerial photographs. This led to field work for the maps in the hot, fire-prone foothills of northern California. Again with help from above, I was transferred to the Stanford Chemistry Department under Professor Francis Bergstrom to make antimalarial drugs. This, again, fostered new friendships, with Murray Luck (founder of Annual Reviews), Dick Ogg, Carl Noller, and Ted R. Norton who returned to Dow after the war. Ted later headed the Agricultural branch of Dow Chemical Company and hired Melvin Calvin as consultant for Dow. The Stanford experience was valuable and delightful. It ended with transfer to the related project at Caltech where I enjoyed exciting visits with Sam Wildman and Art Pardee.

A few weeks after I had departed in July 1943, Sam's phosgene supply became exhausted, and with his broken arm in a sling, he tried to transfer liquid phosgene (b.p. 8°C). Too impatient to cool the ampoule in ice salt, he immersed it in liquid air. The aged soft glass cracked, releasing the deadly liquid into the boiling liquid air, splattering it all over his wool sweater from which it was impossible to escape. It is said that Sam carried the boiling cauldron out of the building to avoid further dangerous exposures, then lay down on the grass by Strawberry Creek. One of his student assistants was hospitalized, and Sam succumbed in a day as his lungs filled with fluid. Only a week before, I had received a thoughtful letter from him. This great tragedy of science left the path of carbon without its real leader and Sam's family without the future it deserved (18).

May of 1946 brought the invitation from Melvin Calvin to join his bio-organic group as Associate Director of the not-yet existent photosynthesis laboratory. This would be a separate branch of his Bio-organic Group, which was developing C-14 techniques and compounds for medical applications in collaboration with John Lawrence, M.D., the brother of Ernest Orlando Lawrence, who had pioneered treatments of leukemia with radioactive phosphate.

ORL

The Old Radiation Laboratory (ORL) was the birthplace of the nuclear age and home of Ernest Lawrence's 37-inch cyclotron in the center of the old maple-floored well-built wood building. Though still functioning, the cyclotron was scheduled for transfer to UCLA's physics department. We occupied the long room just west of the central cyclotron area. Its floor was orange-yellow with uranium residues and somewhat radioactive. This problem was easily solved by a cover of low-cost brown linoleum.

I was trained in the excellent organic laboratories at Caltech, which contained some features born in European laboratories. So, the chemical hoods, lab benches, and facilities I designed for our space in ORL worked out very well. Full fluorescent fixtures were placed directly over the black hardened plate glass bench tops where tiny vials would not tip over, and if they did, a spill could be quantitatively recovered. The largest white vitreous porcelain sinks with high swing faucets with long flexible extensions were a great success with large equipment (and, later, for my controlling an explosion and fire in our laboratory at Penn State.) For viewing the radioautograms made from the two-dimensional paper chromatograms I designed a large white Formica table where they could be compared and discussed. I designed a vacuum evaporation system, based on one from Caltech, which evaporated solutions faster than the popular Büchi Rotavapor apparatus that came later. Each $C^{14}O_2$ experiment involved evaporation of the algae or leaf extract. There must have been a thousand of them. I always claimed that I could boil water faster than anyone.

That experience was reinforced by my delightful interactions with Hans Schmid from the Chemistry Department of the University of Zürich. I worked with Hans for several months on the vac line, preparing $C^{14}O_2$ and making the necessary measurements of the C^{14} activity that I had developed with Ruben in 1942–1943. After Schmid, came Roger Boissonnas' brief but delightful visit for the same purpose. Soon I had the good fortune of meeting Roger's Geneva classmate Gérard Milhaud and his charming wife, Vera, in Paris in 1952, the beginning of permanent scientific, cultural, and adventurous experiences on many research ships and river boats.

The stream of distinguished visitors passing through ORL included Ernest Lawrence and Philo Farnsworth, working on their three-color gun for television, later the SONY Trinitron; Irène Joliot Curie; Emilio Segré; Ed McMillan; Georg von Hevesy; George Burr; Hiroshi Tamiya; Fredrick C. Bersworth; and countless others.

Hiroshi Tamiya and Kazuo Shibata

Hiroshi and Nobuko Tamiya came to Berkeley from Japan during the months of September to December 1952, staying with Mrs. Kami on University Avenue. They provided a wonderful new insight into Japanese culture for all of us. He had been trained in the laboratory of René Wurmser during its leadership in development of biochemical redox systems. In Melvin Calvin's home he was always the star of the evening. Nobuko was teaching Japanese cooking, though her training at the Cordon Bleu was in French cuisine. She captivated the hearts of everyone. At the end of the 1930s Hiroshi was to begin research with C-11 from the RIKEN cyclotron, one of the four in the world. He often told me of the sad demise of the RIKEN cyclotrons destroyed during the American occupation. With his friend Harry Kelly, scientist with the Supreme Command during the American occupation, Hiroshi, physicist Seiji Kaya, physicist Ryokichi Sagane, and nuclear physicist Yoshio Nishina had led the reorganization of RIKEN (33).

I felt fortunate to be the focus of Hiroshi's interest in photosynthesis. He was anxious to confirm conclusions from his kinetic observations of the effects of oxygen on photosynthesis. We did a number of $C^{14}O_2$ fixation experiments in the hood at the east wall of our first room in ORL, with and without oxygen, but their meaning seemed to be obscure. I regret that we could not reach a definitive conclusion from the experiments. My many experiments showing the great accumulation of glycolic acid during photosynthesis in air as compared to that in air with less than normal oxygen had verified Hiroshi's predictions.

Hiroshi's further contribution to our science was his arrangement for his student Kazuo Shibata, son of Seiho Takeuchi, leading Japanese painter of the turn of the century, to join our group in ORL. That was an important step in the enhancement of our appreciation of Japanese culture and science. He joined us later aboard R.V. Alpha Helix at the Great Barrier Reef where he discovered algae with unique UV-absorbing properties. Kazuo became a friend of everyone and my best friend and adviser until his untimely death.

Paper Chromatography: The Solvents

Bill Stepka, a student in plant nutrition, brought paper chromatography from Rochester where it had been established in F.C. Steward's laboratory by C.E. Dent. Two-dimensional paper chromatography effectively separated amino acids, sugars, and other groups of compounds. Their solvents included noxious and sickening lutidine and collidine as well as phenol; each required separation of the organic phase from the "water phase" before use. The physicists in offices near our Chromatography Room on the second floor of ORL were so sensitive to such odors that several were taken to Cowell Hospital for treatment.

I formulated a phenol-water solvent by using distilled phenol with 40% of its weight of water. This gave a water-saturated phenol solution at room temperature. I selected propionic acid for addition to n-butanol for acidification and for enhancing the water content of the "organic phase" of our solvent. Knowing the necessary amounts of water, butanol, and propionic acid, I prepared two solutions, one of water and butanol and the other of water and propionic acid such that equal volumes of the two solutions would yield a solution identical with the organic phase of that system. This was simple, avoided esterification, and allowed chromatographic separation of a host of components, from lipids to polysaccharides and sugar diphosphates.

Paper chromatographic separations are, in effect, gradient elutions. The solvent loses water to the paper as it travels and becomes more "organic" in composition. Thus the sugar phosphates separated with the water-rich solvent and later the phospholipids and pigments separated in the more organic, less aqueous, solvent. I demonstrated this by analyzing the solvent composition as it traveled on the paper.

Paper chromatographic separations are the result of "partition" between the moving organic phase and a stationary aqueous phase. The partition coefficient for each substance and solvent system is unique; this was the basis of my 1943 study of the products of dark $C^{14}O_2$. I recall two of Glenn Seaborg's seminars on his work defining the actinide series. There it was clear that partition between two immiscible solvents can provide valid information, even when only a few atoms are involved.

The primary attribute of our application of two-dimensional paper chromatograpy (6) was the result of applying an aliquot portion of the "total extract" of the plants labeled in the experiment. Others had complicated interpretation of their work by chromatographing several kinds of extracts separately. It is absolutely essential that the whole assembly of products be examined in the chromatogram. Nothing can escape. The insoluble proteins and polymers then remain on the origin. They could be measured and treated chemically or enzymatically for further resolution and identification. My students may recall that I urged them to carry a sharp pocket knife for cutting out radioactive spots from the chromatogram for elution. As my father had said, "Without his pocket knife, a man is naked."

"Standardyes" for chromatographic Rf orientation were selected from Mr. Ray's fabulous chemical storeroom, on the basis of their chemical structures and estimated relative water/solvent solubilities, for co-chromatography with labeled compounds. The mobilities of Tropeolin (Orange II) and Ponceau-4R (red) were reasonably reliable for comparison with C14 compounds, the Rf of Tropeolin being the most useful. Hence one could judge the position of a labeled compound by virtue of its mobility relative to those of the standardyes, even after the solvent front had disappeared.

Pattern Recognition

Following his doctorate in ORL, Alex Wilson, as Dean in New Zealand's University of Waikato, hoped to measure "genius" in students. He realized that there are at least two types of genius—those capable in mathematical concepts and those capable in "pattern recognition." The first category includes the peoples from Asia and the Middle East, strong in numerical and business transaction capabilities. The second includes the artists with their strengths in recognition of color or structural relationships. The real geniuses in pattern recognition, Alex said, are the Australian Aboriginal "Black Trackers" whose capabilities for tracing animals or criminals over a boulder patch are well documented. Their powers for observing, storing, and detecting relationships in visual information are phenomenal. Such was the recognition of novel information revealed in the hundreds of two-dimensional paper chromatograms accumulating in our studies of ^3H_2O, $^{14}CO_2$, ^{35}S, ^{32}P, ^{74}As, ^{75}Se, and ^{125}Sb metabolism.

Norway 1951–1952

Melvin Calvin, his wife, Genevieve, and her Norwegian mother visited Norway after his 1949 coronary infarct and recovery, during which I wrote our review for the *Annual Review of Plant Physiology*. At the agricultural college, Professor Lindemann induced Calvin to send a colleague to Norway to establish a laboratory for radioisotope applications in agriculture. This resulted in my appointment as Fulbright visiting professor and in delightful experiences for my wife, Ruth, and our four children. It also began our friendship with the Helge Larsen family and with pharmacognosy Professor Arnold Nordal and his family. Nordal had generously provided me with his pure sedoheptulose and knowledge of its chemical properties. Later we collaborated in ORL and at Scripps Institution of Oceanography. Several excellent students came from Norway (25).

Invitation from Otto Warburg

While in Europe I was invited to present a paper (4), a comprehensive review of the path of carbon, including the carboxylation of a C_2 from ribulose diphosphate to yield PGA, at the meeting of the Bunsengesellschaft für physikalische Chemie in Lindau, a distinguished group of photochemists, which included Otto Warburg. Sam Ruben had introduced me to Warburg's works, his algae, and his manometry. He was very interested in our results and in an opportunity to recruit our support for his heretic contentions that four quanta were required for fixation of CO_2 and production of O_2. He invited me to Berlin to observe how he grew his special *Chlorella* and made his measurements. It was a delightfully impressive experience to have lunch with him and Herr Heiss in their home. After lunch I walked just down the street to the quiet Dahlem Museum and into a small room where I found myself alone—with *Nefertiti*. That thrilling experience still brings out the goose bumps. I appreciated Warburg's arguments, though they seemed easily interpretable on the basis of my own experiments with preilluminated algae. Later, with the Linderstrom-Langs at the Carlsberg Laboratory in Copenhagen, I presented another seminar, again with Otto Warburg in the audience. (He was visiting Denmark to see his allergy physician.) On a beautiful afternoon I drove him and Herman Kalckar to Hamlet's Castle at Helsingör. Warburg peered through an iron grate into the darkness below, "Ach, it's a perfect place for that Midwest Gang." This, of course, followed Warburg's stay in Urbana where the polemic over quantum requirement of photosynthesis had become heated.

Fraction 1 Protein: Exciting Months in the Laboratory with Jacques Mayaudon, 1954

Jacques Mayaudon came to our ORL with a Fellowship from the Belgium Foundation, I.R.S.I.A., 1954–1955. The project that Melvin Calvin asked Jacques to develop proved less than stimulating, and he came to me in ORL hoping to work on photosynthesis. Our laboratory had discovered most of the important aspects of the path of carbon in photosynthesis (3) by that time. Melvin and the whole laboratory were intently concerned with his exciting theory of the role of thioctic acid (1, 15) in the quantum conversion process of photosynthesis (17). I was anxious to follow the carboxylation process by demonstrating and, hopefully, isolating the enzyme responsible for the process that we called carboxydismutase. Jacques Mayaudon was the ideal collaborator.

Very fortunately I had frequently visited Sam Wildman at Caltech in Pasadena where my wife's family resided. I followed Wildman's (35) progress in isolating and characterizing the major protein of leaves that he named Fraction I Protein. But neither Wildman nor his associates at that time considered its possible function as the critical enzyme in the path of carbon in photosynthesis.

I had developed a procedure for isolating unlabeled ribulose diphosphate from *Scenedesmus* algae using C^{14}-ribulose diphosphate as a marker for selecting the proper stripe area of my one-dimensional paper chromatograms for elution of the unlabled compound. By that time I knew how to enhance the concentration of ribulose diphosphate by withholding CO_2 prior to extraction of the algae and then adding $C^{14}O_2$ to ribulose diphosphate in a buffered solution of the enzyme produced carboxyl-labeled phosphoglycerate, which was a measure of the activity of the enzyme.

Soon it became clear that our carboxylase activity was being concentrated by the same ammonium sulfate precipitation isolation procedures as Wildman had followed with his Fraction 1. It was exciting, and we worked feverishly up to the time I had to leave for Penn State. I consider our discovery that the enzyme catalyzing carboxylation of ribulose diphosphate was the same predominant protein isolated by Sam Wildman one of my most exciting revelations (33). Forty-six years later, Sam Wildman still recalls my phone call with the news. I typed a manuscript and submitted it through Radiation Laboratory channels. Melvin reported the isolation a few weeks later at the Brussels Biochemistry Congress, but the real discovery of its identity with Fraction I protein only appeared in a paper by Mayaudon two years later (23).

LIFE WITH LIPIDS, COPEPODS, SALMON, AND THE SEA

Radiochromatographic Exploration

Several discoveries resulted from recognition of novel products or relationships revealed on film in the radioautographs of the chromatograms, their radiograms. Among the first were the recogntion of glycolate as a product of photorespiration and of ribulose and sedoheptulose, which appeared as a result of phosphatase activity in the *Rhodospirillum rubrum* chromatograms I had prepared with Clint Fuller's photosynthetic bacteria. Later, in chromatograms of P^{32}-labeled spinach I noticed a major novel product that proved to be glycerophosphorylglycerol, GPG, the deacylated derivative of the new major phospholipid, phosphatidylglycerol. In my search for thioctic (lipoic) acid in S^{35}-labeled algae the huge radioactive spot on the chromatogram, which for a time thrilled Melvin, later proved to be the plant sulfolipid, Nature's finest surfactant molecule. Later, novel compounds appeared in radioautographs of algae labeled with sulfur, arsenic, or antimony. A review (8) of these adventures was presented upon receipt of the 1987 Supelco-AOCS Research Award.

Choline Phosphate

For several years our former colleague Nathan Tolbert had reported a fast-moving compound in the xylem of barley seedlings. At Penn State I chromatographed P^{32}-labeled barley and noticed a novel neutral product. I asked Jake Maizel to have a look at it. Jake came back in a week, having recognized it as the zwitterionic choline phosphate. We wrote a manuscript for *Plant Physiology*, added Tolbert's name, and sent it off (22). Feeling that Jake would benefit from a move to Caltech, I arranged for him to move. Thus Jake then went on to Caltech and a successful career. Later, with Tolbert's acumen and energy, the important growth regulator, chlorocholine chloride (CCC) (30) was developed. A similar compound, choline sulfate (25), now recognized as an osmoregulator, appeared in chromatograms of S^{35}-labeled salt-excreting mangrove leaves.

Phosphatidylglycerol

A tantalizing unknown phosphorus compound appeared in the middle of the chromatogram of an extract of P^{32}-labled spinach leaf. I labeled it "Up," for lack of a better name. It had appeared in similar extracts of algae. Often its concentration appeared to exceed those of other phosphate esters identified earlier by explorers like Neuberg, Fischer, Meyerhof, Lohmann, and Leloir. It seemed there could hardly be any more, especially one that was a major phosphorus compound in spinach, algae, and many other plants. It certainly had to be a well-known substance, but all attempts to show that failed. Futile search for the probable solutions to the problem of identity consumed an embarrassing year or more. Only by my good fortune and the wisdom of Hiroshi Tamiya, Assistant Professor Bunji Maruo from the Institute of Applied Microbiology arrived at State College to collaborate in 1956. I had hydrolyzed Up to produce a smaller derivative (greater Rf value on the chromatograms), yet another "unknown." Soon, Bunji had identified it as the well-known glycerophosphate. With C^{14}-labeled Up the hydrolysis product was glycerol-C^{14} and glycerophosphate-C^{14}. What could be simpler and, at the same time, more perplexing? Both were too obvious. It dawned on us that the unknown was GPG or glycerophosphorylglycerol, a frighteningly simple but completely novel combination of glycerol and phosphate. Such glycerophosphoryl esters were well known as skeletons of the glycerolphosphatides. Could GPG be such a skeleton of a phospholipid of plants? We turned our attention to the P^{32}-labeled lipids of algae. With the fatty acids removed, GPG appeared in every instance. It sometimes even exceeded the amounts of the known esters of choline, ethanolamine, and inositol. Phosphate, glycerol, and fatty acids—how could such a simple lipid be a major component of plants without it having been known long ago?

It was more likely that Nature was playing a trick on us than that we could be looking at a new and important member of membranes. Even as I carried the manuscript to the mailbox, it was frightening to consider that Nature could have withheld such a simple secret for so long. Now we know that phosphatidylglycerol may be a major membrane lipid of living things. Not only plants but bacteria and fungi chose to use it in their membranes. Such simplicity was hard to imagine. Years later, Laurens van Deenen and Eugene Kennedy honored our audacity by republishing our paper in Volume 1000 of *Biochimica et Biophysica Acta* (13). In a scholarly series of publications our student Isao Shibuya has delineated the factors regulating phosphatidylglycerol of the *Escherichia coli* cell membrane.

The Plant Sulfolipid: Nature's Finest Surfactant Molecule

Deacylation of the S^{35}-labeled lipid produced a water-soluble product (11). Acid hydrolysis further yielded a smaller molecule. C^{14}-labeled sulfolipid was isolated from Chlorella and acid hydrolyzed to yield glycerol and a sulfosugar that with the skillful collaboration of Helmut Daniel and Ralph Mumma was identified as the 6-sulfonic acid of 6-deoxyglucose (quinovose). Thus, the sulfolipid was sulfo-quinovosyl diglyceride, probably the strongest amphipathic lipid in Nature. This discovery opened a study of sulfocarbohydrate metabolism, a system analogous to the known phosphorylated fermentative metabolism. Having saturated this field, we dropped this research in 1963. For nearly 40 years the field languished for the lack of brilliant ideas or novel approaches. Recently, with the works of Ernst Heinz and Christoph Benning, progress has been made. Benning's work appears to be revealing a biological mechanism for insertion of the $-SO_3H$ group, which closely resembles the free radical addition of SO_2 to a 5,6-glucoseenide as carried out by Jochen Lehmann (21). Using modern genetic tools, Benning has closely approached the true mechanism. Possible utilization of the sulfosugar in novel biodegradable commercial surfactants becomes a possibility.

Neutron Activation Chromatography

Having worked for years in ORL in the slow neutron cloud generated by the 60-inch Crocker Cyclotron, I felt no compunction about utilizing the excellent and convenient slow neutron source of the Penn State nuclear reactor facility. Few others were using the reactor in 1958, and formalities were minimal (Not so anymore, as now one would be burdened by regulations, forms, record keeping, etc.). Our society fails to recognize the possibly beneficial effects of low levels of radiation and radioisotope exposure, which may induce our enzyme systems toward DNA repair. My own early exposures to X-ray radiation from experimenting with my father's portable X-ray Tesla system, X-ray tubes, and fluorescent screen may actually have been helpful in inducing my DNA (thymine dimerization, etc.) repair enzymes. Similarly Martin Kamen's cyclotron experience could have induced his successful radiation damage repair systems. Regions of low radiation exposure from the Chornobyl disaster now appear to develop unexpectedly low cancer incidence.

Neutron Activation Analysis had come to imply the detection and assay of radionuclides produced within a sample upon slow neutron exposure. As a result of neutron capture, the radioactive nuclide ejects a number of gamma rays. The recoil from such emissions drives the victim's nucleus out of any previous chemical bonds, thus rendering molecular chemical identification impossible. With the Penn State Reactor I succeeded in overcoming this inadequacy and developed the techniques for identifying and assaying the phosphorus compounds and some other groups of organic carbon compounds separable by our methods of two-dimensional paper chromatography. It was a simple, convenient, and reasonably precise analytical method. I think it was a brilliant methodology, but it seems neither known nor used today. It was sometimes criticized as being time consuming. Hardly so, not much time is really consumed, though the time period between the extraction and chromatography and the final result may be a week or two, depending on the analytical precision desired. This was often the criticism suffered by paper chromatography and the final radiogram on the X-ray film. However, this was never a problem; there were always plenty of other projects going on at the same time. The result was worth the wait. In fact, I often exposed two films to the paper chromatogram so that eager folks like Melvin Calvin might have something to think about while the second film was exposed properly for documentation.

Neutron activation chromatography was nicely applied for assay of phosphorylated esters derived from large animals whose size precluded normal radioisotope labeling. We determined relative concentration of the several mitochondrial membrane phospholipids of bovine and sheep liver. The lipids were deacylated and the phosphodiesters chromatographed on paper having 1-μg samples of phosphorus (orthophosphate) as control standards. The paper, rolled and sealed in a polyethylene tube, was lowered to the swimming-pool reactor and exposed to the neutron flux for six hours. Short-lived isotopes were allowed to decay a few days and the paper exposed to X-ray film to reveal the characteristic pattern of deacylated lipid products. Comparing their induced radioactivity with that induced in adjacent 1-μg standards provided for quantitative determination. The recoil from neutron capture destroys the initial bonds of the diesters but hardly distorts their location on the paper. Tatsuhiko Yagi and I followed this with an episode of hot-atom chemistry to understand the vagaries and fate of the recoiling P^{32} atom in various media (32).

Cell Membrane Model

Jim Danielli, a father of the lipid bilayer model for cell membranes, opened the first of the international conferences on Membrane Structure and Function in Frascati, Italy, with, "Ladies and Gentlemen, and—Dr. Benson." I was to present a heretic model for the membrane, completely different from his widely accepted Davson-Danielli bilayer model. With collaboration of Elliot Weier and Tae Hwa Ji in our study of chloroplast lamellae and many bacterial membranes, I was convinced that the requirements for specificity of fatty acyl chains in many systems must indicate specific interaction of membrane proteins with the hydrophobic saturated or unsaturated C_{16} to C_{22} chains in such lipids. Clearly, there were hydrophobic pockets within proteins that specifically accommodate such chains. The electron micrographs of stained membranes appeared to represent layers of such lipoproteins. The work of Tae Ji corroborated observed interactions of carotenoids, chlorophyll, and lipids consistent with that concept. Hence, I entitled our paper, The Cell Membrane, a Lipoprotein Monolayer (7). That was 34 years ago.

In 1986 Bunji Maruo alerted me to a superb publication in *Science* on the structure of oxidized cytochrome c oxidase, which included the positions of eight phospholipid molecules, three of them phosphatidylglycerol (31). That paper was the first to reveal the positions of the fatty chains of the phospholipids within a membrane protein molecule. I feel that the advent of the intense light sources now available to crystallographers, particularly the powerful SPRING-8 collector ring near Osaka, will produce further evidence supporting my 1967 contention.

The countless publications relating to the position of fatty chains in the lipid bilayer model appeared naïve to me. The two-dimensional representations of such chains seemed unrelated to reality. One of the questions is the disparity of chain lengths in such membranes and the effects of multiple unsaturation. I assembled models of such lipids and concluded that the least uncomfortable arrangements of carbons in the n-3 chains of the α-linolenic (18-3) and docosahexaenoic acid (22:6) must involve helical positioning of the several double bonds. This reduces the length of the chain to nearly that of the C-16 and C-18 saturated fatty acids. This, then, can explain the apparent symmetry of the two lipid layers of the bilayer involving the long polyunsaturated chains. It could also provide for more specific interaction of the polyunsaturated chains with hydrophobic pockets within proteins.

Wax Ester: World's Major Nutritional Energy Source

My long-time colleague Dr. Judd C. Nevenzel had collaborated for years with James F. Mead in the Laboratory of Nuclear Medicine and Radiation Biology at UCLA. Judd is a master gas chromatographer and a guru of lipidometric information. He and Mead gravitated to an interest in lipids of fish, and Judd became intensely interested in the lantern fish, which daily migrates vertically approximately 1000 m. Judd discerned that these fish produced unusual amounts of wax esters, which he investigated in my laboratory at Scripps. Interaction with my graduate student Richard F. Lee and his interest in copepod metabolism led to Nevenzel's recognition of their wax ester content, up to 70% of the dry weight of the animal. As a result we spent considerable time at sea studying the *Calanus plumchrus* population of the Strait of Georgia between Vancouver and Victoria. We hoped to discern relationships between vertical migration capabilities and the compressibility of their liquid wax ester.

Soon we recognized that wax ester is the stored energy source for copepods and most of the pelagic animals of the ocean. We found that there were no specific wax ester lipases and that other lipases could hydrolyze wax ester only one tenth as rapidly as they could hydrolyze ordinary fats. Thus the deep sea animals are forced to conserve their energy supply, allowing them time to find a mate or to capture uncertain food in a dark nutrition void. Wax ester in marine animals, then, serves as Nature's Starvation Insurance.

Later I found that corals exude some wax (cetyl palmitate) in their mucus, which is collected by small reef fish. Lee and I published an article on The Role of Wax in Marine Food Chains (12). The orange roughy, a widely marketed deep sea fish from New Zealand and Australia, became popular in the 1980s. If an orange roughy were broiled whole, the result would be immediate diarrhea because of the high wax ester content of its skin and bones. The commercial fillets, on the other hand, are free of wax ester.

Salmon Research

The wax ester of the British Columbia copepods is avidly consumed by small salmon, arriving from their rivers rich in nutrients from the melting snow and glaciers. The newly nutritious seawater gives rise to a great diatom bloom, the food source for *Calanus plumchrus* and other copepods that conserve the highly unsaturated fatty acids of these phytoplankton in their wax esters, providing energy for the young salmon. Beginning in 1968 I led expeditions to study the mechanisms of the rapid aging of spawned Pacific salmon, which die, totally emaciated, a week after spawning, The 15 medical scientists aboard our RV Alpha Helix each discerned an important process leading to demise of the salmon. A major effort developed from the interests of Gérard Milhaud from the Institut Pasteur in Paris. Having developed the thyroid hormone, calcitonin, for treating osteoporosis in man, he proceeded to study its function in the salmon. Today salmon calcitonin, a 32 amino acid peptide, is an important therapeutic agent for treating osteoporosis. In 1971 our site of research moved to the more accessible village of Alert Bay with its vigorous fishing industry. I designed and equipped four successive laboratories around which our friend and benefactor Bob Peterson, founder of the Jack-in-the-Box chain of drive-in restaurants, built research ships and provided continuous enthusiastic support for our work. Canadian fishery officer Ray Scheck recruited assistance of a skillful salmon fishing family, Lily and Porgie Jolliffe. Members of the native Kwakiutl Tribe and its Nimpkish Band of Alert Bay, the Hunt family has a long and distinguished history. Lily (Kakasolas) Jolliffe, a gorgeous strong-willed lady with great cultural depth, is of the Hunt family.

Franz Boas

I gave a talk on salmon aging to the local Sons of Norway one evening and met a very interested member who was familiar with the salmon fishing industry. Later he presented me with his classic copy of Kaptein Jacobsen's Reiser til Nordamerikas Nordvestkyst. It had engendered the salmon fishery of Alaska. Jacobsen had traveled the coast of British Columbia in 1881, collecting Native American artefacts for Hagenbeck museums in Hamburg, Berlin, and Prague. At a "showing" of a group of coastal peoples in Berlin, young Franz Boas and his friend Sigmund Freud became interested in these unique humans from the Pacific Coast. Boas was excited at the prospect and obtained from Jacobsen a letter of introduction to George Hunt, Chief of the Kwakiutl nation at Fort Rupert on Vancouver Island. Boas arrived there in 1886 and spent 15 years with Chief Hunt, documenting the language and culture of the Kwakiutl. It was the beginning of scientific anthropology.

Needless to say, reading that Franz Boas of La Jolla High School was a winner in the Westlinghouse Science Talent Search (now called the Intel Science Talent Search) in California was exciting. I called his mother and learned that he was indeed a descendent of Franz Boas. I couldn't resist taking him to Alert Bay to meet Lily and her sister Laura and to experience a night of fishing on a salmon seine boat. He had never met the 65 members of the Boas family who had enjoyed a reunion with the 300-member Hunt family in Alert Bay a few years earlier. With his visit a great success, young Franz Edward went off to Harvard with happy memories.

When I was a freshman student at Berkeley, anthropology Professor Ronald Olson was assigned to be my Adviser. Subsequently I took some of his courses, which included the works of Franz Boas and the culture of the coastal Indians. Thirty years later, I found myself in the midst of these very people whose culture Boas had documented. It has been a great privilege.

Arsenate in the Sea

When it became apparent that the analyses for phosphate in seawater also included arsenate, Johnson and Pilson discovered that tropical surface waters of the Sargasso Sea and the Caribbean could include up to four times as much arsenate as phosphate, both absorbed by the same transport system. Our algae when fed radioarsenate, As-74, produced much arsenolipid and several of its hydrolysis products (16). It was an exciting adventure. Edmonds and Francesconi in Perth had identified arsenobetaine in rock lobster tails of Western Australia, so I suspected the arsenolipid could be the result of production of the arseno analog of phosphatidylcholine. Bob Cooney and Ralph Mumma developed a neat paper electrophoretic method for identification of arsenicals, based on their acid dissociation constants. Later, on the Great Barrier Reef, Roger Summons and I found the highest reported arsenic content, 1000 ppm, in the kidney of the giant clam, *Tridacna maxima*. Cooney and I had been misled by our tracer identifications based on chromatography in various solvents and electrophoresis in several pH buffers. Edmonds and Francesconi successfully identified the arsenical compounds as 5-dimethylarsenoribose derivatives, the lipid being a 3'-arsenoribosylphosphatidylglycerol. We found relatively small amounts of methanearsonate and dimethylarsinate, which many laboratories had identified by chromatography and mass spectrometry as major arsenicals. We interpret this discrepancy as the result of slow accumulation of methylarsonate and dimethylarsinate, whereas the inital products of arsenate metabolism are the arsenosugar and arsenolipid as revealed in our chromatograms.

The long history of toxicity of arsenite and the reputed ability of the "arsenic eaters of Styria" to tolerate up to five lethal doses with lunch suggest careful metabolic adaptions by their gut flora. Such adaptations could as easily develop in communities with excessive arsenic in drinking water. Protective mechanisms may develop, just as with low radiation exposures.

Methanol and Plant Productivity

In 1970 when fuel oil prices soared Arthur Nonomura had isolated an alga that exuded oil droplets. With Fred Wolf in the Calvin Laboratory he had isolated the most productive strain of *Botryococcus braunii*. Trying to improve its growth rate, Arthur asked me for suggestions. "Give them some methanol, Arthur, they'll love it." Of course I wasn't at all confident about this, but Arthur tried methanol and the alga grew twice as fast. That was the beginning of our methanol episode. Thirty years earlier I had investigated the metabolism of C^{14}-methanol in *Scenedesmus* and *Chlorella* while in search of CO_2 fixation intermediates. It was probably the first synthetic C^{14}-methanol and was made by my dear friend Bert Tolbert, brother of Nathan Edward Tolbert.

After his research tenure Arthur started a cotton ranch in Arizona and observed that foliar methanol application prevented afternoon wilting. Without midday depression of photosynthesis associated with wilt, the methanol-treated cotton showed more rapid development and an earlier harvest time than controls. After the boost in growth was verified in a DuPont-supported project, an optimistic publication appeared in the *Proceedings of the National Academy of Sciences* (26). Not surprisingly, some attempts to reproduce the observations without long duration high photorespiratory rates were inconclusive. Arthur continued his efforts, as several convincing confirmations were reported. Finally, with Roland Douce, Richard Bligny, and their capable collaborators, we examined the fate of ^{13}C-methanol in living plant cells as revealed by nuclear magnetic resonance (NMR). Dramatic development of the absorption peak for ^{13}C-methyl glucoside appeared on the screen. Though there was no a priori reason for concluding that it was an important growth factor, Arthur tried applying methyl glucoside solutions to the roots of his test plants. Success was immediate and reproducible. After developing a penetrating foliar solution with methyl glucoside, the metabolite proved far more consistent in activity over a wider range of parameters than simple foliar methanol treatment. Even C-4 plants responded to methyl glucoside applications. Field tests with hybrid corn and canola demonstrated 10% and 15% enhancement by methyl glucoside. Methanol and methyl glucoside reduce the effects of the afternoon drop in photosynthetic activity of field crops. Though there is no clear explanation yet for the observed effects, the treatment is effective and convincing.

Le Metre

The Paris Academy, in their infinitesimal wisdom, decreed the ten-millionth of the Earth's quadrant should be the standard of measure, a "Gift from God" to the countries of Europe that had found their many units of weights and measures complicating trade and communication. Those wealthy "scientists" in their crimson robes had never built a house, baked bread, or prepared meals for a family; nor had they measured tire pressures or the temperatures of concern to humans in Europe as well as the tropics.

The names of the units that people must use should be one syllable, as in inch, foot, pound, mile, and quart. A unit's name should not have to repeatedly inflict on the writer, speaker, or listener the fact that the unit has one thousand meters in it or that it is a hundredth or a thousandth of some larger unit.

Now, if one plans to devise a measuring system with accuracy sufficient for the majority of the earth's humans, the scale of such measurement should range from 00 to 99, providing one per cent precision. That scale should encompass most of the measurements of interest to most of the people involved, a principal of a "democratic" system of measurement.

Though not planned by Herr Gabriel Daniel Fahrenheit, his temperature scale, from $00°$ to $99°$, includes most of the temperatures to which humans are exposed and which they measure most frequently. It avoids the negative temperatures and imprecision of the Celsius scale.

For our gas, water, and air pressure measurement needs, the scale from 00 to 99 psi provides a range useful for most applications: 3 psi for household gas pressure, 70 psi for household water pressure, 24 to 95 psi for tire pressures. These applications encompass most of the needs of our families, plumbers, and auto mechanics.

The problem is not the decimal nature of the metric system. Thomas Jefferson was hoping to decimalize the foot and the pound; he had had great success in creating our decimalized currency. The head start and European adoption of the Paris Academy's Metric System was the downfall of his plan. This philippic contends that contrivance of the Metre was the greatest disaster of the eighteenth century that plagues less perspicacious societies to this day.

EPILOGUE

I have taken the liberty of the privilege provided by Annual Reviews to describe some of my experiences that provided the skills and background for following the path of carbon in photosynthesis (5, 9, 10) as well as to relate some personal concerns and events of interest beyond the path. Even though the Editors may have expected me to just write about the path of carbon and our many adventures within ORL, the wooden building that was the birthplace of the cyclotron and Carbon-14 and the source of the short-lived Carbon-11 used by Sam Ruben and Martin Kamen in their history-making exploration with radioactive carbon dioxide (1938–1942), I am sure that the photosynthesis story is familiar to most plant physiologists. It has been repeatedly described in elegantly competent works by Dr. J.A. Bassham (2) and most recently by our colleague Dr. R.C. Fuller (17).

Though such autobiography might only be as important as yesterday's newspaper, something only good for wrapping fish, I was enheartened by the words of Joseph Priestly, whose experiments inaugurated the modern era of photosynthesis research. He began his memoirs, published in 1806, with the following modest declaration: "Having thought it right to leave behind me some account of my friends and benefactors, it is in a manner necessary that I also give an account of myself and as the like has been done by many persons and for reasons which posterity has approved I make no apology for following their example."

I have tried to follow my path through fortuitous and serendipitous events, many of which resulted from guidance by my parents, teachers, and and broadly cultured colleagues. Though my own level of culture pales in the shadows of those of my mentors, good fortune and health have been blessings indeed.

Mentors

Father, Carl Bennett Benson, M.D.; Mother, Emma Carolina Alm; Irina Barsegova; Dee Dorgan Benson; Michael D. Berry; Montague Cleeves, M.D.; Helmut Daniel; Robert Emerson; Fritz Goreau; A. Baird Hastings; R. Barry Holtz; Benjamin F. Howell, Jr.; George N. Jolliffe, Jr.; Francis C. Knowles; Mutsuyuki Kochi, M.D.; Helge Larsen; Bunji Maruo; Marie Mathers; James F. Mead; James E. Merlin; Gérard Milhaud, M.D.; Stanley L. Miller; Ralph O. Mumma; Judd C. Nevenzel; Arnold Nordal; Ted R. Norton; Margaret Painter; Stuart Patton; Elizabeth Baker Pelz; Robert Oscar Peterson; Harry Rosenberg; Sam Ruben; Kenji Sakaguchi; Paul D. Saltman; Pete Scholander; Michio Seki; Kazuo Shibata; J. Rudi Strickler; Roger E. Summons; Hiroshi Tamiya; Eduard A. Titlyanov; Bert M. Tolbert; Eberhard G. Trams; Viktor E. Vaskovsky; Ellen C. Weaver; and A.A. Yayanos.

Memorable Experiences

(*a*) Charles Lindbergh at Sacramento, 1927. (*b*) Fireworks, Gagarin Celebration, Moscow, 1961. (*c*) Surtsey eruption, 1964. (*d*) Isla Socorro Sea Bridge, 1979. (*e*) Solar eclipse, 5 min, La Paz, Baja California, 1991. (*f*) Guest at presentation of Enrico Fermi Award to Martin Kamen, 1997.

ACKNOWLEDGMENTS

I am indebted to skillful proofreading of my manuscript by John F. Kern and Carole Mayo.

DEDICATION

In memory of Chris, Bonnie, Kazuo, Juan, and Big Paul.

Visit the Annual Reviews home page at www.annualreviews.org

LITERATURE CITED

1. Barltrop JA, Hayes PM, Calvin M. 1954. The chemistry of 1,2-Dithiolane (trimethylene disulfide) as a model for the primary quantum conversion act in photosynthesis. *J. Am. Chem. Soc.* 76:4348–67

2. Bassham JA, Calvin M. 1957. *The Path of Carbon in Photosynthesis.* Englewood Cliffs, NJ: Prentice-Hall. 104 pp.

3. Bassham JA, Benson AA, Kay LD, Harris AZ, Wilson AT, Calvin M. 1954. The path of carbon in photosynthesis. XXI.

The cyclic regeneration of carbon dioxide acceptor. *J. Am. Chem. Soc.* 76:1760–70

4. Benson AA. 1952. Mechanism of biochemical photosynthesis. *Zeitschr. Elektrochem.* 56:848–54

5. Benson AA. 1968. The cell membrane, a lipoprotein monolayer. 1967. In *Membrane Models and the Formation of Biological Membranes*, ed.L Bolis, BA Pethica, Frascati, pp. 190–202. Amsterdam: North Holland Publ.

6. Benson AA. 1987. Radiochromatographic exploration. *J. Am. Oil Chem. Soc.* 64:1309–14

7. Benson AA. 1996. Naïve steps along the path. *CR Acad. Sci. Paris Life Sci.* 319:843–47

8. Benson AA. 1998. The path of carbon in photosynthesis: 1942–1955. In *Discoveries in Plant Biology*, ed S-D Kung, S-F Yang, 1:197–213. Singapore: World Sci. Publ. 371 pp.

9. Benson AA. 2002. Following the path of carbon in photosynthesis: a personal story. History of photosynthesis. In *Photosynthesis Research*. Dordrecht, The Neth.: Kluwer Acad. Publ. Spec. Issue. In press

10. Benson AA, Bassham JA, Calvin M, Goodale TC, Hass VA, Stepka W. 1950. Paper chromatography and radioautography of the products. *J. Am. Chem. Soc.* 72:1710–18

11. Benson AA, Daniel H, Wiser R. 1959. A sulfolipid in plants. *Proc. Natl. Acad. Sci. USA* 45:1582–87

12. Benson AA, Lee RF. 1975. The role of wax in oceanic food chains. *Sci. Am.* 232:77–86

13. Benson AA, Maruo B. 1989. A "nova" in phosphate metabolism, GPG, and discovery of phosphatidylglycerol: a commentary. *Biochim. Biophys. Acta* 1000:447–58

14. Benson AA, Miyachi S. 1992. Interview. *Cell Sci.* (Jpn.) 8:55–69

15. Calvin M. 1954. Chemical and photochemical reactions of thioctic acid and related disulfides. *Fed. Proc.* 13:697–711

16. Cooney RV, Mumma RO, Benson AA. 1978. Arsonium phospholipid in algae. *Proc. Natl. Acad. Sci. USA* 75:4262–64

17. Fuller RC. 1999. Forty years of microbial photosynthesis research: where it came from and what it led to. *Photosynth. Res.* 62:1–29

18. Johnston HS. 2003. *A Bridge Not Attacked.* New Jersey: World Sci. Publ.

19. Kamen MD. 1985. *Radiant Science, Dark Politics: A Memoir of the Nuclear Age.* Berkeley, CA: Univ. Calif. Press. 348 pp.

20. Knowles FC, Benson AA. 1983. The biochemistry of arsenic. *Trends Biochem. Sci.* 8:178–89

21. Lehmann J, Benson AA. 1964. The plant sulfolipid. IX. Sulfosugar syntheses from methyl hexoseenides. *J. Am. Chem. Soc.* 86:4469–72

22. Maizel JV, Benson AA, Tolbert NE. 1956. Identification of phosphoryl choline as an important constituent of plant saps. *Plant Physiol.* 31:407–8

23. Mayaudon J. 1957. Study of association between the main nucleoprotein of green leaves and carboxydismutase. *Enzymologia* 18:345–54

24. Mayaudon J, Benson AA, Calvin M. 1957. Ribulose-1,5-diphosphate from and CO_2 fixation by Tetragonia expansa leaves extract. *Biochim. Biophys. Acta* 23:342–51

25. Nissen P, Benson AA. 1961. Choline sulfate in higher plants. *Science* 134:1759

26. Nonomura AM, Benson AA. 1992. The path of carbon in photosynthesis. XXIV. Improved crop yields with methanol. *Proc. Natl. Acad. Sci. USA* 89:9794–98

27. Ruben S. 1943. Photosynthesis and phosphorylation. *J. Am. Chem. Soc.* 65:279–82

27a. Ruben S, Benson AA. 1945. The physiological action of phosgene. In *Committee of Gas Casualties, "Fasiculus on Chemical Warfare Medicine,"* pp. 327, 641. Washington, DC: Natl. Res. Counc. Comm. Treat. Gas Casualties

28. Ruben S, Kamen MD. 1941. Long-lived radioactive carbon: C^{14}. *Phys. Rev.* 59:349–54

29. Ruben S, Kamen MD, Hassid WZ, DeVault DC. 1939. Photosynthesis with radio-carbon. *Science* 90:570–71

30. Tolbert NE. 1997. The C_2 oxidative photosynthetic carbon cycle. *Annu. Rev. Plant Physiol.* 48:1–25

31. Tsukihara T, Aoyama H, Yamashita E, Tomizaki T, Yamaguchi H, et al. 1996. The whole structure of the 13-subunit oxidized cytochrome c oxidase at 2.8 Å. *Science* 272:1136–44

32. van Niel CB, Ruben S, Carson SF, Kamen MD, Foster JW. 1942. Radioactive carbon as an indicator of carbon dioxide utilization. VIII. The role of carbon dioxide in cellular metabolism. *Proc. Natl. Acad. Sci. USA* 28:8–15

33. Wildman SG. 1998. Discovery of Rubisco. In *Discoveries in Plant Biology*, ed. S-D Kung, S-F Yang, pp. 163–73. Singapore: World Sci. Publ.

34. Yagi T, El-Kinawy SA, Benson AA. 1963. Phosphoryations and phosphonations of glycerol by recoil atoms. *J. Am. Chem. Soc.* 85:3462–65

35. Yoshikawa H, Kauffman J. 1994. *Science Has No National Borders. Harry C. Kelly and the Reconstruction of Science and Technology in Postwar Japan.* Cambridge/Tokyo: MIT Press/Mita Press. 137 pp.

Annu. Rev. Plant Biol. 2002. 53:27–43
DOI: 10.1146/annurev.arplant.53.091401.110929

NEW INSIGHTS INTO THE REGULATION AND FUNCTIONAL SIGNIFICANCE OF LYSINE METABOLISM IN PLANTS

Gad Galili

*Department of Plant Sciences, The Weizmann Institute of Science, Rehovot 76100 Israel;
e-mail: gad.galili@weizmann.ac.il*

Key Words essential amino acids, stress, glutamate

■ **Abstract** Lysine is one of the most limiting essential amino acids in vegetative foods consumed by humans and livestock. In addition to serving as a building block of proteins, lysine is also a precursor for glutamate, an important signaling amino acid that regulates plant growth and responses to the environment. Recent genetic, molecular, and biochemical evidence suggests that lysine synthesis and catabolism are regulated by novel concerted mechanisms. These include intracellular compartmentalization of enzymes and metabolites, complex transcriptional and posttranscriptional controls of genes encoding enzymes in lysine metabolism during plant growth and development, as well as interactions between different metabolic fluxes. The recent advances in our understanding of the regulation of lysine metabolism in plants may also prove valuable for future production of high-lysine crops.

CONTENTS

INTRODUCTION

Human and monogastric mammals cannot synthesize 10 of the 20 most common amino acids found in proteins and therefore need to obtain them from the diet. Among these "essential" amino acids, lysine is most limiting in the major cereal grains, and it is therefore considered as a nutritionally significant amino acid (6). The nutritional importance of lysine has motivated extensive studies on the biochemical regulation of its metabolism, and recent studies demonstrated that lysine metabolism also represents a valuable model system for elucidating novel mechanisms of metabolic regulation. These recent studies also suggest that, in addition to being an important essential amino acid, lysine is also a precursor for a metabolic pathway that functions in plant development and response to stress (2, 20).

Studies on the regulation of lysine synthesis in plants date back to the middle of the twentieth century, when the plant enzymes responsible for production of lysine from aspartate were identified and characterized based on information derived from bacteria. In 1984, the first lysine-overproducing plant mutant was isolated, and the mutation was identified in a gene encoding one of the enzymes of lysine biosynthesis, namely, dihydrodipicolinate synthase (DHPS), reducing its sensitivity to feedback inhibition by lysine (38). Research from the 1990s was characterized by studies expressing key recombinant bacterial enzymes of the lysine biosynthetic pathway, either constitutively or in a seed-specific manner, in transgenic plants (4, 7, 16, 23, 44, 50, 51). These studies, together with additional research dissecting the pattern of expression of plant genes encoding enzymes in lysine biosynthesis (57, 62, 63), allowed, for the first time, functional dissection of fundamental regulatory networks of lysine biosynthesis. These regulatory networks included (*a*) intracellular localization of the enzymes of lysine biosynthesis; (*b*) developmental, physiological, and environmental signals; (*c*) tissue and organ specificity; and (*d*) interactions between different metabolic fluxes. These functional studies also led to the discovery that some plant organs possess not only an active pathway of lysine biosynthesis, but also an active, highly regulated, and multifunctional pathway of lysine catabolism (20, 23). In this review, I describe new insights into the regulation and functional significance of lysine metabolism in plants, which have mostly emerged from recent molecular and genetic studies.

REGULATION OF LYSINE BIOSYNTHESIS IN PLANTS

Lysine Regulates Its Own Synthesis by Feedback Inhibition Loops

In plants, like in many bacterial species, lysine is synthesized from aspartate through a branch of the aspartate-family pathway (Figure 1). Another branch of this

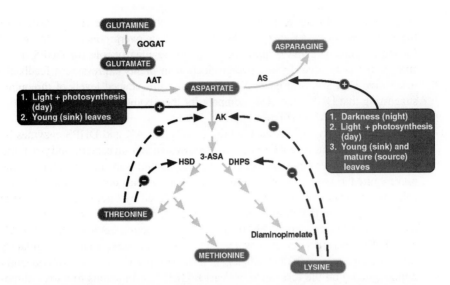

Figure 1   Schematic diagram of the metabolism of glutamine, glutamate, aspartate, asparagine, and the aspartate-family amino acids: lysine, threonin, and methionine. Factors controlling the channeling of aspartate either toward lysine, threonine, and methionine or toward asparagine are indicated on the left and right, respectively. GOGAT, glutamate synthase; AAT, aspartate aminotransferase; AS, asparagine synthetase; AK, aspartate kinase; DHPS, dihydrodipicolinate synthase; HSD, homoserine dehydrogenase; 3-ASA, 3-aspartic semialdehyde. Dashed arrows with a minus sign represent end-product feedback inhibition loops. Solid arrows with a plus sign represent positively affecting factors.

pathway leads to the synthesis of two additional essential amino acids, threonine and methionine (Figure 1). The entire aspartate-family pathway, except the last step of methionine synthesis (methionine synthase), occurs in the plastid and is regulated by several feedback inhibition loops (see Ref. 19 for details). Aspartate kinase (AK) consists of several isozymes, which are feedback inhibited either by lysine or by threonine. Lysine also feedback inhibits the activity of DHPS, the first enzyme committed to its own synthesis, whereas threonine partially inhibits the activity of homoserine dehydrogenase (HSD), the first enzyme committed to the synthesis of threonine and methionine. The AK isozymes exist either as monofunctional polypeptides that contain the lysine-sensitive AK activity or bifunctional AK/HSD enzymes that contain both the threonine-sensitive AK and HSD isozymes linked on a single polypeptide (see Ref. 19 for more detailed review).

Although both the monofunctional AK and DHPS activities are feedback inhibited by lysine, DHPS is the major limiting enzyme for lysine biosynthesis. The in vitro activities of plant DHPS enzymes are much more sensitive to lysine inhibition than those of the lysine-sensitive AK enzymes (Ki of ~5–50 μM and ~200–600 μM lysine for plant DHPS and AK enzymes, respectively) (see Ref. 19 for more detailed review). Moreover, expression of feedback-insensitive DHPS enzymes in plant cells results in lysine overproduction, whereas expression of feedback-insensitive AK enzymes leads to threonine overproduction (see Ref. 19 for more detailed review), supporting a major regulatory role for DHPS in lysine synthesis in vivo. Lysine overproduction in plants expressing a feedback-insensitive DHPS is also generally associated with reduced levels of threonine, suggesting that DHPS and HSD compete for their common substrate 3-aspartic-semialdehyde (3-ASA) (Figure 1) (4, 18, 51).

The mechanisms of feedback inhibition of plant AK and DHPS enzymes by lysine and threonine are not known. However, evidence from bacteria and yeast suggest that this process may not occur by simple enzyme-substrate interactions and may require the assistance of molecular chaperones. Yeast possesses a single cytosolic monofunctional AK enzyme, which is inhibited by threonine (45). Genetic analysis suggested that inhibition of the yeast AK by threonine may be mediated by the molecular chaperone FKBP12 (1). An interaction between the yeast AK and FKBP12 was first detected by a yeast "two-hybrid" screen, and the regulatory role of this interaction in the feedback inhibition property of AK was subsequently implied, based on the resistance of yeast FKBP12 null mutants to a toxic threonine analog (1). It will be of interest to test whether yeast FKBP12 null mutants also overproduce threonine, similarly to transgenic yeast expressing a recombinant threonine-insensitive AK (17). Moreover, it will be interesting to elucidate what domain of the yeast AK interacts with FKBP12. The sensitivities of the bacterial and yeast AK enzymes to inhibition by lysine or threonine are determined by specific domains that apparently bind these amino acids. These domains are located either in the C-terminal part of the monofunctional AK enzymes or in the linker region between the AK and HSD domains of the bifunctional AK/HSD enzymes (40, 41, 58). Because FKBP12 modulates the sensitivity of yeast AK to threonine inhibition rather than the catalytic activity of this enzyme, per se, it is likely that this molecular chaperone interacts with the C-terminal regulatory domain of the yeast AK.

Is DHPS activity also regulated by molecular chaperones? In *Escherichia coli*, cell wall synthesis is strongly dependent on the molecular chaperone GroE, and defective production of GroE is associated with cell lysis (34). Diaminopimelate, the last intermediate of the lysine branch of the aspartate-family pathway (Figure 1), is also a major precursor for *E. coli* cell wall biosynthesis (11), and depletion of GroE is associated with significant reduction in the level of diaminopimelate (34). Notably, this defect can be partially overcome by overexpression of the *DapA* gene encoding DHPS but not any of the genes encoding other enzymes of the lysine biosynthesis pathway (34). Moreover, immediately before lysis, the level of DHPS in the GroE-depleted cells is reduced to 16% of its normal level (34). In fact, DHPS was the first protein demonstrated in vivo to be both influenced by GroE and essential for cell growth. Whether GroE directly affects DHPS activity in bacteria or that its effect is indirect via the modulation of *DapA* gene expression has yet to be elucidated.

The regulation of feedback inhibition of plant AK and DHPS enzymes is not understood. Because GroE is an abundant and important chaperone in plant plastids, it may be possible that this enzyme regulates the activity of plant DHPS enzymes in a similar manner to its function in bacteria. The possible role of FKBP12 in regulating plant AK activities is more questionable because a plastid-localized FKBP12 homologue has not yet been identified in plants. Computer analysis predicts that both the monofunctional AK and bifunctional AK/HSD isozymes of plants contain domains that recognize lysine and threonine, respectively, in similar locations to those found in their microbial counterparts (B. Gakiere, L. Song, & G. Galili, unpublished information). However, genetic studies suggest that the mechanisms of feedback inhibition may be distinct between the plant monofunctional AK and bifunctional AK/HSD isozymes. Attempts to select for plant mutants possessing feedback-insensitive AK activities by in vivo mutagenesis always resulted in mutations within the lysine-sensitive AK isozymes but not in the bifunctional AK/HSD isozymes. It may be possible that altering the threonine sensitivity of AK activity in plant AK/HSD requires several amino acid changes within the protein, whose frequency of occurrence is much lower than that obtained in standard genetic screens. Based on amino acid sequence analysis, the AK/HSD enzymes of maize were suggested to contain elements resembling those of the "two component" regulatory mechanism, which controls the expression of various genes in prokaryotes and eukaryotes (37). The potential significance of these putative elements in the regulation of the plant AK/HSD isozymes, and whether such putative elements may interact with the domains that recognize threonine for feedback inhibition, still has to be elucidated.

Are the Enzymes and Metabolites of the Aspartate-Family Pathway Randomly Distributed Inside Plastids?

The current paradigm for the biochemical regulation of lysine synthesis is based mostly on interpretation of in vitro biochemical properties of the AK and DHPS enzymes. However, a deeper look into this issue identifies an important discrepancy between in vitro and in vivo observations. As discussed above, the in vitro activities of plant DHPS enzymes are much more sensitive to feedback inhibition by lysine than the lysine-sensitive AK isozymes. Simple interpretation of these results suggests that the lysine-sensitive AK would barely be inhibited in vivo because DHPS will prevent accumulation of a sufficient lysine level to inhibit AK activity. However, mutations that reduce the feedback-inhibition sensitivity of the lysine-sensitive AK cause significant threonine, but not lysine, overproduction (see Ref. 19 for more detailed review). Thus, the lysine-sensitive AK activity is apparently inhibited in vivo by lysine, but this is more important for regulating the flux toward the threonine branch, rather than the lysine branch of the aspartate-family pathway (Figure 1).

Another interesting discrepancy between the in vivo and in vitro results is related to the concentrations of lysine and threonine that are available for the DHPS and AK enzymes within plastids. Assuming that the in vitro properties of the DHPS and AK enzymes reflect their in vivo properties, one would expect that the concentration of lysine in the plastid stroma will be in the range of tens (for DHPS) to few hundreds (for AK) μM in order to enable fine regulation via enzyme feedback inhibition.

However, data from nonaqueous subcellular fractionation experiments suggest that the concentration of lysine in the plastid is near 1 mM (59), which is much lower than the Ki values of the AK and DHPS enzymes. Assuming that the amino acid concentrations measured by nonaqueous subcellular fractionation techniques reflect their actual in vivo concentrations, then all AK and DHPS enzymes are expected to be nearly inactive in vivo, but this expectation is not supported by the in vivo studies.

How can the above discrepancies between the in vitro and in vivo results be reconciled? It is possible that the in vivo enzymatic properties of AK and/or DHPS may be regulated by posttranlational modifications, which are not considered in vitro. Alternatively, the enzymes and end-product amino acids of the aspartate-family pathway may not be uniformly distributed inside the plastids, so that the enzymes are not always situated in an environment of inhibitory concentrations of amino acids. Such an hypothesis requires some level of intraorganelle compartmentalization of the enzymes and amino acids within the stroma of the plastid, which could be achieved either by physical separation or by substrate channeling via multienzyme complexes. Intraorganelle compartmentalization of metabolites in plant cells is not a new concept. For instance, a number of processes that are simultaneously subjected to differential regulation by Ca^{2+} are often explained by differential distribution of Ca^{2+} within a given organelle. In addition, results obtained by biochemical experiments and protein-protein interactions imply that enzymes of the flavonoid biosynthetic pathway in *Arabidopsis* assemble as a macromolecular complex with contacts between multiple proteins (8, 9).

Lysine Biosynthesis is Regulated by Developmental, Environmental, and Physiological Signals

Evidence of the developmental regulation of lysine biosynthesis in plants was obtained in early studies, based on enzyme activity measurements in different plant tissues (see Ref. 19 for review). Subsequent attempts to functionally dissect cellular, developmental, physiological, and environmental signals that regulate lysine biosynthesis used lysine-overproducing transgenic plants that constitutively expressed a bacterial feedback-insensitive DHPS. The suitability of such transgenic plants for functional dissection of lysine biosynthesis is based on the following: Although DHPS is the major limiting enzyme for lysine biosynthesis, the synthesis of this amino acid also depends on a battery of additional enzymes (Figure 1). Thus, significant lysine overproduction was expected to occur in these transgenic plants only in the specific tissues or growth conditions where the genes encoding the entire set of lysine biosynthetic enzymes were abundantly expressed. Indeed, the lysine level in the transgenic plants constitutively expressing the bacterial DHPS fluctuated considerably under different growth conditions, being higher in young leaves and floral organs than in old leaves and positively responding to light intensity (50, 62). Moreover, when shoots from the lysine-overproducing transgenic tobacco plants were grafted onto a wild-type stock, lysine level in the developing grafted transgenic shoots decreased significantly (R. Amir, B. Avidan,

O. Shaul, & G. Galili, unpublished information). This result suggests that lysine biosynthesis in plants is regulated by specific compound(s) (likely hormones) that are transported by the vascular system.

The regulation of lysine biosynthesis was studied further by analyzing the expression patterns of two Arabidopsis genes encoding AK/HSD and DHPS enzymes, using Northern blot analyses and promoter fusions to the β-glucuronidase (GUS) reporter gene. The developmental expression pattern of both genes was very similar, i.e., they were highly expressed in germinating seedlings, actively dividing and growing young shoot and root tissues, various organs of the developing flowers, as well as the developing embryos (57, 63). Exposure of etiolated seedlings to light results in an altered pattern of GUS staining in the hypocotyls and cotyledons, suggesting that expression of the AK/HSD and DHPS genes is also regulated by light (57, 63). This was supported by additional studies showing that the levels and activities of the barley AK isozymes are increased by light and phytochrome (46). The similarities in the developmental and light-regulated patterns of expression of the AK and DHPS genes suggest that many (but not all; see below in this section) of the genes encoding enzymes of the aspartate-family pathway may be subject to coordinated regulation of expression.

Expression of the Arabidopsis AK/HSD gene is also regulated by the photosynthesis-related metabolites sucrose and inorganic phosphate (62). Its expression is strongly enhanced by sucrose and repressed by phosphate. The antagonistic effect of phosphate indicates that the response to sucrose may be attained, at least in part, through fluctuation of cellular phosphate levels in response to photosynthesis-dependent sugar phosphorylation (28, 48, 56).

Recently, the regulation of expression of two Arabidopsis genes encoding lysine-sensitive AK isozymes was also studied using promoter-GUS constructs (21). Notably, the two genes showed significantly altered patterns and levels of expression. Expression of one isogene was much more predominant than the other in vegetative tissues. Both isogenes were highly expressed at the reproductive stage, but only one of these genes was expressed in fruit tissue (21). The response of these genes to light and metabolic signals was not reported.

Metabolic Interactions Between the Aspartate-Family Amino Acids and Amide Amino Acid Metabolism

Aspartate, the substrate of AK, serves not only as the precursor for the aspartate-family pathway, but it is also the immediate precursor for the amide amino acid asparagine via the activity of asparagine synthetase (Figure 1). Asparagine possesses several important functions in plants. Its synthesis from aspartate is important to support massive protein synthesis during the daytime (29, 31). In addition, asparagine serves as an important molecule for nitrogen storage during the night, when photosynthesis is inactive, as well as for nitrogen transport between source and sink tissues (29, 31). How is the metabolic channeling of aspartate into asparagine or the aspartate-family amino acids regulated? Molecular analyses suggest that this channeling may be regulated by the expression of the genes encoding asparagine synthetase and AK. Plants possess two forms of asparagine synthetase genes. Expression of one is induced by light and sucrose (similar to the gene encoding AK/HSD) to enable asparagine synthesis during the day, whereas expression of the other is repressed by light and sucrose and is induced during the night (29, 30). Thus, during the day aspartate is apparently channeled both into asparagine and into the aspartate-family pathway to allow synthesis of all of its end-product amino acids. During the night, the aspartate-family pathway is apparently blocked, mainly owing to repression of expression of the AK genes (62), and aspartate channels preferentially into asparagine. Indeed, asparagine levels are much higher while lysine levels are lower during the night than the daytime (29).

Channeling of aspartate into the aspartate-family pathway may not only be regulated by photosynthesis and "day/night" cycles. An unexpected observation supporting such a possibility was recently reported by analysis of an Arabidopsis DHPS knockout mutant (12, 49). In 20-day-old wild-type Arabidopsis plants grown in culture, free lysine levels range between ~50 and ~350 nanomoles per gram fresh weight, whereas free threonine levels range between ~200 and ~1,000 nanomoles per gram fresh weight [calculation from Table 1 of Ref. (12)]. From this estimation, it is expected that complete redirection of the lysine branch into the threonine branch in the Arabidopsis DHPS knockout mutant should enhance free threonine content by no more than two- to threefold. Yet, free threonine levels in the DHPS knockout plants were markedly elevated between 10- and 80-fold, depending on growth conditions, even though lysine levels were generally not reduced more than 50% (the DHPS knockout mutant is not lethal because Arabidopsis possesses two DHPS genes) (12, 49). What is the reason for excess threonine production in the DHPS knockout mutant? It is likely that the reduction in DHPS activity triggers an enhanced conversion of amide amino acids to threonine via aspartate (Figure 1). This apparently occurs by a process that either increases the expression of AK genes, and perhaps also other genes of the threonine branch of the aspartate-family pathway, and/or suppresses the feedback inhibition properties of the threonine-sensitive AK isozymes. Such a process may be regulated by novel mechanism(s) that can sense the reduced level of free lysine or its intermediate metabolites in the plant cells.

REGULATION OF LYSINE CATABOLISM IN PLANTS

Regulation of Lysine Metabolism During Seed Development: The Emerging Story of Lysine Catabolism

The regulation of lysine production during seed development was studied in transgenic plants expressing bacterial feedback-insensitive DHPS enzymes under the control of a seed-specific promoter derived from a gene encoding a seed storage protein (23). The choice of a storage protein gene promoter was based on the assumption that lysine biosynthesis is spatially and temporally coordinated with storage protein production during seed development. Hence, the bacterial DHPS would be expressed in storage cells where other enzymes of the lysine biosynthetic pathway are also produced. Seed-specific expression of the bacterial DHPS in transgenic tobacco plants resulted in increased lysine synthesis, but the free lysine level in the mature seeds was not higher than in wild-type seeds (23). Developing seeds of these transgenic plants also possessed over 10-fold higher activity of lysine-ketoglutarate reductase (LKR), the first enzyme in the pathway of lysine catabolism (Figure 2), suggesting that the low lysine level in mature seeds of the transgenic tobacco plants resulted from enhanced lysine catabolism (23). This hypothesis was later confirmed by showing that knockout mutants in the Arabidopsis lysine catabolism pathway possess higher seed lysine levels than wild type (61). The bacterial DHPS was expressed in a seed-specific manner in two additional

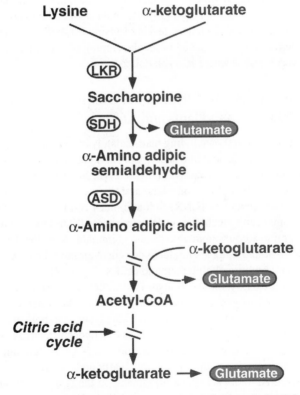

Figure 2 The lysine catabolism pathway and metabolites derived from it. LKR, lysine-ketoglutarate reductase; SDH, saccharopine dehydrogenase; ASD, aminoadipic semialdehyde dehydrogenase. Broken arrows represent several nonspecified enzymatic reactions. Glutamate residues are situated inside gray ovals. The three glutamate residues that can be produced from a single lysine molecule are illustrated in these ovals.

transgenic dicotyledonous crop plants, namely soybean and rapeseed (16, 33). Seeds of these transgenic plants overproduced lysine, but they also contained significantly higher levels of lysine catabolic products than their wild-type parents.

In contrast to dicotyledonous plants in which storage protein synthesis generally takes place in the developing embryo, the synthesis of storage proteins in cereal seeds occurs mainly in the endosperm (52). Likewise, based on in situ analysis, the lysine catabolism pathway was suggested to function mostly in the outer layers of the cereal endosperm (25). It is thus expected that expression of a bacterial DHPS, under control of an endosperm-specific storage protein gene promoter, will result in enhanced lysine production and perhaps also accumulation of lysine catabolites. This expectation was found to be incorrect because lysine overproduction in transgenic maize seeds was observed only when the bacterial DHPS was expressed under an embryo-specific, but not an endosperm-specific, promoter (33). Notwithstanding, in spite of the fact that lysine catabolism may not be active in the developing maize embryo (25), the embryo-specific overproduction of lysine was associated with increased levels of lysine catabolic products (33). It may be possible that lysine produced in the developing maize embryo is transported into endosperm tissues where it is catabolized.

The α-Amino Adipic Acid Pathway: A Ubiquitous Pathway of Lysine Catabolism in Plants and Animals

In plants and animals (including mammals), lysine is catabolized via saccharopine and α-amino adipic acid into acetyl-CoA and glutamate by a pathway often called the α-amino adipic acid pathway (Figure 2) (2, 20). Two enzymes linked on a single bifunctional polypeptide control the first two steps of this pathway (2, 20). Lysine-ketoglutarate reductase (LKR) first combines lysine and α-ketoglutarate to form saccharopine, and saccharopine dehydrogenase (SDH) then converts saccharopine into α-amino adipic semialdehyde and glutamate. Additional enzymatic reactions convert α-amino adipic semialdehyde, via α-amino adipic acid, into acetyl-CoA and several additional glutamate molecules (Figure 2).

In mammals, lysine catabolism is stimulated in brain tissues (47) and is apparently used to generate glutamate, which regulates nerve signal transmission via glutamate receptors (43). Defects in the lysine catabolism pathway in humans cause a severe genetic disorder called familial hyperlysinemias, which is associated in some patients with mental retardation (32, 60). The functional significance of lysine catabolism in plants is still not clearly understood. However, recent results employing Arabidopsis knockout mutants in the lysine catabolism pathway suggest that this pathway functions in controlling lysine homeostasis in plants. Transgenic Arabidopsis plants expressing a bacterial DHPS in a seed-specific manner overproduce lysine and have a normal phenotype. However, crossing these transgenic plants to a homozygous knockout mutant in the lysine catabolism pathway results in a severe arrest of seed germination (X. Zhu & G. Galili, unpublished information). It is likely that the arrest of germination results from the inability of the germinating embryos to catabolize the excess lysine contributed from seed reserves.

In addition to controlling lysine homeostasis, the pathway of lysine catabolism may also be important for plant development and responses to abiotic stress. As described below, production of the first two enzymes of this pathway is subject to complex regulation during plant development and in response to stress conditions.

Lysine Catabolism: An Exquisitely Regulated Metabolic Pathway

Although the α-amino adipic acid pathway consists of many enzymatic steps (Figure 2), its flux is apparently regulated mainly by the first two enzymes, LKR and SDH. Both in plants and animals, LKR and SDH are linked on a single, bifunctional LKR/SDH polypeptide encoded by the bifunctional *LKR/SDH* gene. Detailed discussion of the biochemical properties of the LKR/SDH polypeptide, regulation of the *LKR/SDH* gene, and the functional significance of the lysine catabolic pathway can also be found in recent reviews (2, 20).

Expression of the *LKR/SDH* gene in plants is subject to complex transcriptional (and/or posttranscriptional) regulation as well as posttranslational control. Lysine accumulation in developing tobacco seeds, either by feeding external lysine or by expressing a bacterial DHPS, stimulates LKR activity via an intracellular signaling cascade requiring Ca^{2+} and protein phosphorylation/dephosphorylation (22, 23). In vitro studies suggest that the lysine-mediated stimulation of LKR activity occurs by phosphorylation of the LKR/SDH polypeptide by casein kinase-II (2, 35). However, it may be possible that casein kinase-II also regulates transcription of the *LKR/SDH* gene. Expression of the maize *LKR/SDH* gene is regulated by the transcription factor Opaque2 (25) whose transcriptional activity is regulated by casein kinase-II (10).

Under normal growth conditions, the *LKR/SDH* gene is most highly expressed in floral organs and developing seeds (25, 54). However, as mentioned above, the expression pattern of the *LKR/SDH* gene in developing seeds may differ between dicots and monocots. In situ mRNA hybridization suggests that in Arabidopsis the *LKR/SDH* gene is expressed in the developing embryo and the outer layers of the endosperm (54). In contrast, in situ analysis of SDH activity suggests that in maize the *LKR/SDH* gene is expressed predominantly in the outer endosperm layers.

Expression of the *LKR/SDH* gene is not only subject to developmental regulation. Its expression is stimulated upon exposure to osmotic and salt stresses, as well as in response to abscisic acid (13, 36; A. Stepansky & G. Galili, unpublished information). In addition, *LKR/SDH* gene expression is also apparently stimulated in abscission zones because *LKR/SDH*-related sequences are present at a relatively high frequency in an EST database derived from cotton boll abscission zones (3 out of ~1800 ESTs) (http://www.genome.clemson.edu/projects/cotton/).

The bifunctional LKR/SDH polypeptide is not the only gene product of the *LKR/SDH* gene. The Arabidopsis *LKR/SDH* gene encodes two additional smaller polypeptides containing monofunctional LKR and monofunctional SDH activities (55; G. Tang, X. Zhu, & B. Gakiere, unpublished information). Also, DNA sequence analysis suggests that two of the three *LKR/SDH*-related ESTs in the cotton boll abscission zone database actually encode a monofunctional LKR, whereas the third encodes a bifunctional LKR/SDH (G. Tang, X. Zhu, & G. Galili, unpublished information). The monofunctional SDH is regulated by a promoter that is located inside the coding region of the *LKR/SDH* gene and also serves as a part of the LKR/SDH open reading frame (55). The monofunctional LKR is encoded by two different mechanisms. One involves a polyadenylation site inside an intron within the linker region between the LKR and SDH coding sequences of the *LKR/SDH* gene. The second is consistent with trans-splicing where the 3′ region of the monofunctional LKR mRNA is donated by a different gene (G. Tang, X. Zhu, & G. Galili, unpublished information). Thus, the plant *LKR/SDH* gene is an unusual composite locus encoding three related, but distinct, enzymes of lysine catabolism.

Why are LKR and SDH linked on a single bifunctional polypeptide, and why do plants also produce monofunctional LKR and SDH isozymes? Enzymatic studies suggest that in the bifunctional LKR/SDH enzyme the SDH domain functionally interacts with the LKR domain to downregulate LKR activity. Under in vitro conditions, the V_{max} of the Arabidopsis monofunctional LKR is significantly higher, whereas its K_m for lysine is significantly lower than the values of its counterpart LKR that is linked to SDH (20; G. Tang, X. Zhu, & G. Galili, unpublished information). In addition, the in vitro LKR activity of LKR/SDH, but not of the monofunctional LKR, is modulated by the ionic strength of the incubation assay, and the linker region between LKR and SDH plays an essential role in this regulation (20, 24; X. Zhu & G. Galili, unpublished information). It can therefore be hypothesized that LKR/SDH is a highly regulated enzyme in which LKR activity can be negatively regulated by the SDH domain by a mechanism that is controlled by lysine and operates via phosphorylation/dephosphorylation of LKR by casein kinase-II and a putative protein phosphatase. In contrast, the monofunctional LKR is a highly efficient enzyme (20). Furthermore, it is possible that plants possess two different types of lysine catabolism fluxes, a highly regulated flux operating in tissues where the bifunctional LKR/SDH predominates and a very efficient flux in tissues, like abscission zones, where the monofunctional LKR predominates.

Do plants require two different fluxes of lysine catabolism? This could be important for the possible dual functions of the lysine catabolic pathway, i.e., to control lysine homeostasis and also to efficiently regenerate glutamate. It is possible that production of the bifunctional LKR/SDH predominates in specific tissues, e.g., developing flowers and seeds, where lysine levels need to be tightly regulated to prevent toxicity while preventing depletion of free lysine. In terminal tissues, like abscission zones, and perhaps also under acute abiotic stresses, production of a monofunctional LKR may enable rapid catabolism of lysine into glutamate. Besides being a building block of proteins, glutamate also serves as a signaling and stress-related molecule in plants (see Ref. 20 for a more detailed review). Glutamate is the immediate precursor for proline, which accumulates under osmotic and salt stresses and functions as a protective osmolyte (39). Glutamate is also the immediate precursor of the stress-related molecule γ-amino butyric acid (3) as well as for arginine, which is the direct precursor of stress-related molecules, polyamines, and nitric oxide (27, 53). Because plants possess functional homologues of animal glutamate receptors (5, 14, 26), it may be possible that, similar to its function in animal cells, the plant lysine catabolic pathway is also important for generating the ligand for activation of glutamate receptors.

Is Lysine Catabolism via LKR/SDH Regulated Similarly Between Plants and Animals?

Although the α-amino adipic acid pathway is an important pathway for brain function in humans, the regulation of *LKR/SDH* gene expression represents a relatively unique situation where the knowledge from plant research is more advanced than in animals. In fact, the *LKR/SDH* cDNA was first cloned from plants (15, 54), and only later was it identified in mammals, based on DNA sequence homology to the plant counterpart (25, 42). Interestingly, the linker region between the LKR and SDH domains of LKR/SDH, which serves a number of important functions in the expression and function of the plant *LKR/SDH* locus (see discussion above), is missing in all of the animal LKR/SDH sequences identified to date (25). Although the significance of this fundamental difference is still not clear, perhaps the linker region evolved in plants to serve specific functions that are unique to them.

CONCLUSIONS AND FUTURE PROSPECTS

Early studies of the regulation of lysine biosynthesis in plants were mainly directed at improving the content of this essential amino acid. However, more recent studies demonstrated that the pathway of lysine biosynthesis may also serve as a valuable model system for elucidating regulatory mechanisms of amino acid metabolism. Expression of bacterial feedback-insensitive DHPS enzymes in transgenic plants proved to be an effective approach to elucidate various factors that regulate lysine biosynthesis. The mechanisms through which these factors operate were studied subsequently by analyzing the expression patterns of the plant genes encoding AK and DHPS. These studies, together with the more recent analysis of DHPS knockout mutants, showed that lysine biosynthesis is regulated by a complex network of interacting processes and by interpathway communication with amide amino acid metabolism. These results provide a challenge for future elucidation of mechanisms responsible for the complex control of lysine metabolism in plants.

Transgenic DHPS experiments paved the way for the elucidation of the significance of the pathway of lysine catabolism. This pathway appears to be complex and highly regulated, yielding novel information about the control of gene expression and amino acid metabolism. The molecular regulation and functional significance of lysine catabolism in plant growth and response to abiotic stress has still to be elucidated.

Studies on lysine catabolism also contribute to breeding efforts to improve lysine levels in plants and show that this must be done by manipulating lysine synthesis and catabolism. Future studies should elucidate whether a reduction in lysine catabolism can be obtained by a gene knockout approach or whether lysine catabolism can be transiently reduced in seeds in order to obtain significant increases in lysine levels with minimal phenotypic perturbations and yield loss.

Finally, modern genomic metabolic profiling and nuclear magnetic resonance approaches will be required for an in-depth analysis of flux control of lysine synthesis and catabolism, as well as for elucidating novel interactions between metabolic fluxes.

ACKNOWLEDGMENTS

I thank Dr. Brian A. Larkins for critical reading of the manuscript. The research in my laboratory was supported by grants from the FrameWork Program of the Commission of the European Communities, the Israel Academy of Sciences and Humanities, National Council for Research and Development, Israel, Grant No. BIO4-CT97-2182, and the MINERVA Foundation, Germany. The author is an incumbent of the Bronfman Chair of Plant Sciences.

Visit the Annual Reviews home page at www.annualreviews.org

LITERATURE CITED

1. Alarcon CM, Heitman J. 1997. FKBP12 physically and functionally interacts with aspartokinase in *Saccharomyces cerevisiae. Mol. Cell. Biol.* 17:5968–75
2. Arruda P, Kemper EL, Papes F, Leite A. 2000. Regulation of lysine catabolism in higher plants. *Trends Plant Sci.* 5:324–30
3. Baum G, Lev-Yadun S, Fridmann Y, Arazi T, Katsenelson H, et al. 1996. Calmodulin binding to glutamate decarboxylase is required for regulation of glutamate and GABA metabolism and normal development in plants. *EMBO J.* 15:2988–96
4. Ben Tzvi-Tzchori I, Perl A, Galili G. 1996. Lysine and threonine metabolism are subject to complex patterns of regulation in *Arabidopsis. Plant Mol. Biol.* 32:727–34
5. Brenner ED, Martinez-Baboza N, Clark AP, Liang QS, Stevenson D, Coruzzi GM. 2000. Arabidopsis mutants resistant to BMAA, a cycad-derived glutamate receptor agonist. *Plant Physiol.* 124:1615–24
6. Bright SWJ, Shewry PR. 1983. Improvement of protein quality in cereals. *CRC Crit. Rev. Plant Sci.* 1:49–93
7. Brinch-Pedersen H, Galili G, Knudsen S, Holm PB. 1996. Engineering of the aspartate family biosynthetic pathway in barley (*Hordeum vulgare* L.) by transformation with heterologous genes encoding feedback-insensitive aspartate kinase and dihydrodipicolinate synthase. *Plant Mol. Biol.* 32:611–20
8. Burbulis IE, Pelletier MK, Cain CC, Shirley BW. 1996. Are flavonoids synthesized by a multi-enzyme complex? *SAAS Bull. Biochem. Biotechnol.* 9:29–36
9. Burbulis IE, Winkel-Shirley B. 1999. Interactions among enzymes of the *Arabidopsis* flavonoid biosynthetic pathway. *Proc. Natl. Acad. Sci. USA* 96:12929–34
10. Ciceri P, Gianazza E, Lazzari B, Lippoli G, Genga A, et al. 1997. Phosphorylation of Opaque2 changes diurnally and impacts

its DNA binding activity. *Plant Cell* 9:97–108

11. Cohen GN, Saint-Girons I. 1987. Biosynthesis of threonine, lysine and methionine. In *Escherichia coli and Salmonella typhimurium: Cellular and Molecular Biology*, ed. FC Neidhardt, pp. 429–44. Washington, DC: Am. Soc. Microbiol.

12. Craciun A, Jacobs M, Vauterin M. 2000. *Arabidopsis* loss-of-function mutant in the lysine pathway points out complex regulation mechanisms. *FEBS Lett.* 487:234–38

13. Deleu C, Coustaut M, Niogert M-F, Larher F. 1999. Three new osmotic stress-regulated cDNAs identified by differential display polymerase chain reaction in rapeseed leaf discs. *Plant Cell Environ.* 22:979–88

14. Dennison KL, Spalding EP. 2000. Glutamate-gated calcium fluxes in *Arabidopsis*. *Plant Physiol.* 124:1511–14

15. Epelbaum S, McDevitt R, Falco SC. 1997. Lysine-ketoglutarate reductase and saccharopine dehydrogenase from *Arabidopsis thaliana*: nucleotide sequence and characterization. *Plant Mol. Biol.* 35:735–48

16. Falco SC, Guida T, Locke M, Mauvais J, Sandres C, et al. 1995. Transgenic canola and soybean seeds with increased lysine. *Bio-Technology* 13:577–82

17. Farfan MJ, Aparicio L, Calderon IL. 1999. Threonine overproduction in yeast strains carrying the *HOM3-R2* mutant allele under the control of different inducible promoters. *Appl. Environ. Microbiol.* 65:110–16

18. Frankard V, Ghislain M, Jacobs M. 1992. Two feedback-insensitive enzymes of the aspartate pathway in *Nicotiana sylvestris*. *Plant Physiol.* 99:1285–93

19. Galili G. 1995. Regulation of lysine and threonine synthesis. *Plant Cell* 7:899–906

20. Galili G, Tang G, Zhu X, Gakiere B. 2001. Lysine catabolism: a stress and development super-regulated metabolic pathway. *Curr. Opin. Plant Biol.* 4:261–66

21. Jacobs M, Vauterin M, De Waele E, Craciun A. 2001. Manipulating plant biochemical pathways for improved nutritional quality. In *Plant Biotechnology and Transgenic Plants*. New York: Marcel Dekker

22. Karchi H, Miron D, Ben-Yaacov S, Galili G. 1995. The lysine-dependent stimulation of lysine catabolism in tobacco seeds requires calcium and protein phosphorylation. *Plant Cell* 7:1963–70

23. Karchi H, Shaul O, Galili G. 1994. Lysine synthesis and catabolism are coordinately regulated during tobacco seed development. *Proc. Natl. Acad. Sci. USA* 91:2577–81

24. Kemper EL, Cord-Neto G, Capella AN, Goncalves-Butruile M, Azevedo RA, Arruda P. 1998. Structure and regulation of the bifunctional enzyme lysine-oxoglutarate reductase-saccharopine dehydrogenase in maize. *Eur. J. Biochem.* 253:720–29

25. Kemper EL, Neto GC, Papes F, Moraes KC, Leite A, Arruda P. 1999. The role of Opaque2 in the control of lysine-degrading activities in developing maize endosperm. *Plant Cell* 11:1981–94

26. Kim SA, Kwak JM, Jae SK, Wang MH, Nam HG. 2001. Overexpression of the *AtGluR2* gene encoding an Arabidopsis homolog of mammalian glutamate receptors impairs calcium utilization and sensitivity to ionic stress in transgenic plants. *Plant Cell Physiol.* 15:74–84

27. Klessig DF, Durner J, Noad R, Navarre DA, Wendehenne D, et al. 2000. Nitric oxide and salicylic acid signaling in plant defense. *Proc. Natl. Acad. Sci. USA* 97:8849–55

28. Krapp A, Hofmann B, Schafer C, Stitt M. 1993. Regulation of the expression of rbcS and other photosynthetic genes by carbohydrates: a mechanism for the "sink regulation" of photosynthesis? *Plant Physiol.* 99:627–31

29. Lam H-M, Coschigano K, Schultz C, Melo-Oliveira R, Tjagen G, et al. 1995. Use of Arabidopsis mutants and genes to study amide amino acid biosynthesis. *Plant Cell* 7:887–98

30. Lam HM, Chiu J, Hsieh MH, Meisel L,

Oliveira IC, et al. 1998. Glutamate-receptor genes in plants. *Nature* 396:125–26

31. Lam HM, Hsieh MH, Coruzzi G. 1998. Reciprocal regulation of distinct asparagine synthetase genes by light and metabolites in *Arabidopsis thaliana*. *Plant J.* 16:345–53

32. Markovitz PJ, Chuang DT, Cox RP. 1984. Familial hyperlysinemias: purification and characterization of the bifunctional aminoadipic semialdehyde synthase with lysine-ketoglutarate reductase and saccharopine dehydrogenase activities. *J. Biol. Chem.* 259:11643–46

33. Mazur B, Krebbers E, Tingey S. 1999. Gene discovery and product development for grain quality traits. *Science* 285:372–75

34. McLennan N, Masters M. 1998. GroE is vital for cell-wall synthesis. *Nature* (London) 392:139

35. Miron D, Ben-Yaacov S, Karchi H, Galili G. 1997. In vitro dephosphorylation inhibits the activity of soybean lysine-ketoglutarate reductase in a lysine-regulated manner. *Plant J.* 12:1453–58

36. Moulin M, Deleu C, Larher F. 2000. L-lysine catabolism is osmo-regulated at the level of lysine-ketoglutarate reductase and saccharopine dehydrogenase in rapeseed leaf discs. *Plant Physiol. Biochem.* 38:577–85

37. Muehlbauer GJ, Somers DA, Matthews BF, Gengenbach BG. 1994. Molecular genetics of the maize (*Zea mays* L.) aspartate kinase-homoserine dehydrogenase gene family. *Plant Physiol.* 106:1303–12

38. Negrutiu I, Cattoir-Reynearts A, Verbruggen I, Jacobs M. 1984. Lysine overproducer mutants with an altered dihydrodipicolinate synthase from protoplast culture of *Nicotiana sylvestris* (Spegazzini and Comes). *Theor. Appl. Genet.* 68:11–20

39. Nuccio ML, Rhodes D, McNeil SD, Hanson AD. 1999. Metabolic engineering of plants for osmotic stress resistance. *Curr. Opin. Plant Sci.* 2:128–34

40. Omori K, Imai Y, Suzuki S, Komatsubara S. 1993. Nucleotide sequence of the *Serratia marcescens* threonine operon and analysis of the threonine operon mutations which alter feedback inhibition of both aspartokinase I and homoserine dehydrogenase I. *J. Bacteriol.* 175:785–94

41. Omori K, Komatsubara S. 1993. Role of serine 352 in the allosteric response of *Serratia marcescens* aspartokinase I–homoserine dehydrogenase I analyzed by using site-directed mutagenesis. *J. Bacteriol.* 175:959–65

42. Papes F, Kemper EL, Cord-Neto G, Langone F, Arruda P. 1999. Lysine degradation through the saccharopine pathway in mammals: involvement of both bifunctional and monofunctional lysine-degrading enzymes in mouse. *Biochem. J.* 344:555–63

43. Papes F, Surpili MJ, Langone F, Trigo JR, Arruda P. 2001. The essential amino acid lysine acts as precursor of glutamate in the mammalian central nervous system. *FEBS Lett.* 488:34–38

44. Perl A, Shaul O, Galili G. 1992. Regulation of lysine synthesis in transgenic potato plants expressing a bacterial dihydrodipicolinate synthase in their chloroplasts. *Plant Mol. Biol.* 19:815–23

45. Rafalski JA, Falco SC. 1988. Structure of the yeast *Hom3* gene which encodes aspartokinase. *J. Biol. Chem.* 263:2146–51

46. Rao S, Kochhar S, Kochhar V. 1999. Analysis of photocontrol of aspartate kinase in barley (*Hordeum vulgare* L.) seedlings. *Biochem. Mol. Biol. Int.* 47:347–60

47. Rao VV, Pan X, Chang YF. 1992. Developmental changes in L-lysine-ketoglutarate reductase in rat brain and liver. *Comp. Biochem. Physiol.* 103B:221–24

48. Sadka A, DeWald DB, May GD, Park WD, Mullet JE. 1994. Phosphate modulates transcription of soybean *VspB* and other sugar-inducible genes. *Plant Cell* 6:737–49

49. Sarrobert C, Thibaud MC, Contard-David P, Gineste S, Bechtold N, et al. 2000. Identification of an *Arabidopsis thaliana* mutant accumulating threonine resulting from

mutation in a new dihydrodipicolinate synthase gene. *Plant J.* 24:357–67

50. Shaul O, Galili G. 1992. Increased lysine synthesis in transgenic tobacco plants expressing a bacterial dihydrodipicolinate synthase in their chloroplasts. *Plant J.* 2: 203–9

51. Shaul O, Galili G. 1993. Concerted regulation of lysine and threonine synthesis in tobacco plants expressing bacterial feedback-insensitive aspartate kinase and dihydrodipicolinate synthase. *Plant Mol. Biol.* 23:759–68

52. Shotwell MA, Larkins BA. 1989. The biochemistry and molecular biology of seed storage proteins. In *The Biochemistry of Plants*, ed. A Marcus, pp. 297–345. San Diego, CA: Academic

53. Smith TA. 1985. Polyamines. *Annu. Rev. Plant Physiol.* 36:117–43

54. Tang G, Miron D, Zhu-Shimoni JX, Galili G. 1997. Regulation of lysine catabolism through lysine-ketoglutarate reductase and saccharopine dehydrogenase in *Arabidopsis*. *Plant Cell* 9:1305–16

55. Tang G, Zhu X, Tang X, Galili G. 2000. A novel composite locus of *Arabidopsis* encoding simultaneously two polypeptides with metabolically related but distinct functions in lysine catabolism. *Plant J.* 23:195–203

56. Thorne JH, Giaquinata RT. 1984. Pathways and mechanisms associated with carbohydrate translocation in plants. In *Storage Carbohydrates in Vascular Plants: Distribution, Physiology and Metabolism*, ed.

DH Lewis, pp. 75–96. Cambridge, UK: Cambridge Univ. Press

57. Vauterin M, Frankard V, Jacobs M. 1998. The *Arabidopsis thaliana dhdps* gene encoding dihydrodipicolinate synthase, key enzyme of lysine biosynthesis, is expressed in a cell-specific manner. *Plant Mol. Biol.* 39:695–708

58. Viola RE. 2001. The central enzymes of the aspartate family of amino acid biosynthesis. *Acc. Chem. Res.* 34:339–49

59. Winter H, Robinson DG, Heldt HW. 1993. Subcellular volumes and metabolite concentrations in barley leaves. *Planta* 191: 180–90

60. Woody NC. 1964. Hyperlysinemia. *Am. J. Dis. Child.* 108:543

61. Zhu X, Tang G, Granier F, Bouchez D, Galili D. 2001. A T-DNA insertion knockout of the bifunctional lysine-ketoglutarate reductase/saccharopine dehydrogenase gene elevates lysine levels in Arabidopsis seeds. *Plant Physiol.* 126:1539–45

62. Zhu-Shimoni XJ, Galili G. 1998. Expression of an Arabidopsis aspartate kinase/ homoserine dehydrogenase gene is metabolically regulated by photosynthesis-related signals, but not by nitrogenous compounds. *Plant Physiol.* 116:1023–28

63. Zhu-Shimoni XJ, Lev-Yadun S, Matthews BF, Galili G. 1997. Expression of an aspartate kinase homoserine dehydrogenase gene is subject to specific spatial and temporal regulation in vegetative tissues, flowers and developing seeds. *Plant Physiol.* 113:695–706

Annu. Rev. Plant Biol. 2002. 53:45–66
DOI: 10.1146/annurev.arplant.53.092701.143332

SHOOT AND FLORAL MERISTEM MAINTENANCE IN ARABIDOPSIS

Jennifer C. Fletcher

*Plant and Microbial Biology Department, University of California Berkeley,
USDA Plant Gene Expression Center, Albany, California 94710;
e-mail: fletcher@nature.berkeley.edu*

Key Words stem cell, signal transduction, CLAVATA, receptor kinase, feedback loop

■ **Abstract** The shoot apical meristem (SAM) of higher plants functions as a site of continuous organogenesis within which a small pool of pluripotent stem cells replenishes the cells incorporated into lateral organs. This article summarizes recent results demonstrating that the fate of stem cells in Arabidopsis shoot and floral meristems is controlled by overlapping spatial and temporal signaling systems. Stem cell maintenance is an active process requiring constant communication between neighboring groups of SAM cells. Information flows via a ligand-receptor signal transduction pathway, resulting in the formation of a spatial feedback loop that stabilizes the size of the stem cell population. Termination of stem cell activity during flower development is achieved by a temporal feedback loop involving both stem cell maintenance genes and flower patterning genes. These investigations are providing exciting insights into the components and activities of the stem cell regulatory pathway and into the interaction of this pathway with molecular mechanisms that control floral patterning.

CONTENTS

1040-2519/02/0601-0045$14.00

INTRODUCTION

Plants are the products of meristems. During embryogenesis, angiosperm plants generate two distinct apical meristems, the root apical meristem and the shoot apical meristem (SAM), that act throughout the life of the plant as continuous sources of new cells for organogenesis. These meristems consist of small populations of morphologically undifferentiated, pluripotent stem cells located at the tips of roots and shoots, respectively. The root apical meristem produces the cells of the primary and lateral root system, whereas the shoot apical meristem produces the leaves, stems, and flowers that compose the above-ground architecture of the plant. Thus the proper function of root and shoot apical meristems is critical for normal growth and development. This article focuses on the problem of how a stable stem cell population is maintained in the SAM of the model plant *Arabidopsis thaliana*.

Leaves, stems, and flowers are initiated sequentially from the Arabidopsis SAM. During vegetative development, the SAM generates leaf primordia directly from its flanks in a stereotypical spatial arrangement. At the end of the vegetative phase, environmental and endogenous factors cue the plant to undergo the transition to flowering and reproductive development. During this phase the stem elongates, secondary SAMs are formed in the axils of leaves, and floral meristems are generated on the flanks of both the primary and secondary SAMs. The primary reproductive SAM is also referred to as the inflorescence, or flower-bearing, meristem. Floral meristems are small, spherically shaped mounds of cells, which produce four types of lateral organs in concentric rings called whorls. Sepals are initiated first in the outermost whorl, followed by petals in the second whorl, and stamens in the third whorl. The floral meristem is then consumed in the formation of the central carpels, which form the gynoecium that ultimately encloses the seeds of the next generation. Thus the floral meristem eventually terminates, whereas the SAM grows indefinitely.

The main activities of the angiosperm SAM throughout development are maintenance of the pluripotent stem cell population, organ initiation, and stem production to generate the architecture characteristic of each plant species. Stem formation occurs in the deeper layers of the meristem, adding breadth and girth to the shoot. Organs are produced on the flanks of the meristem, while a pool of pluripotent stem cells is preserved at the apex that replenishes the cells that have been incorporated into organ primordia or stem. To function as a site of continuous organogenesis, the SAM must maintain a constant balance between loss of stem cells through differentiation and their replacement through cell division. Mutational analysis has revealed the importance of this balanced state for proper development: Mutant Arabidopsis plants with reduced meristem activity fail to form the full complement of lateral organs and often terminate growth prematurely, whereas mutant plants with hyperactive meristems have greatly enlarged stems and produce supernumerary organs. In shoot apical meristems, perpetuation of the balanced state throughout development is mediated by a complex spatial signaling network of both positive and negative factors. In floral meristems, termination of meristem activity requires a temporal feedback loop involving both a stem cell regulator and a phase-specific floral patterning factor.

SHOOT AND FLORAL MERISTEM STRUCTURE

Shoot Apical Meristem Organization

The organization of the SAM explains how plants are able to grow while still producing organs. The SAM of flowering plants is established in the embryo and has specific structural characteristics (Figure 1a). Cytological and histological studies show that the angiosperm SAM consists of three distinct radial domains. A small cluster of enlarged, highly vacuolated cells lie at the apex of the SAM. This cluster of cells, which comprises the reservoir of pluripotent stem cells, is termed the central zone (61). These cells divide infrequently relative to the other cells in the SAM (45, 61). The cells in the central zone are surrounded by a peripheral zone of small, densely staining cells that divide more frequently than the cells in the central zone. Beneath the central zone, in the deeper layers of the meristem, are columns of large vacuolated cells referred to as the rib zone. These cells constitute the meristem pith and contribute to the bulk of the stem (61).

As stem cells in the central zone divide, their progeny cells are gradually displaced toward the flanks of the meristem into the peripheral zone (61). Cells in the peripheral zone then become incorporated either into lateral organ primordia or into the internodal region of stem between the organs (19, 43, 61). Mosaic analysis of Arabidopsis embryonic SAMs revealed that the cells on the meristem periphery are more restricted in their developmental potential than those at the apex (27). The peripheral zone therefore represents a transitional region of the SAM, where the descendents of the pluripotent central zone cells begin to acquire more specified fates. As the shoot tip grows upward owing to cell divisions in the rib zone, cells in the peripheral zone are left behind to proliferate and undergo differentiation and are replenished by the division of cells in the central zone. In this way the plant is able to maintain a reservoir of stem cells at the apex of the SAM, while simultaneously generating lateral structures on the flanks.

An additional level of organization is the arrangement of the SAM into distinct cell layers called the tunica and corpus (53, 57). In Arabidopsis and most other dicots (22), the tunica consists of an overlying epidermal L1 layer and a subepidermal L2 layer. Each of these layers is a single cell thick, and each remains clonally distinct from the others because the cells within them only divide in an anticlinal orientation, perpendicular to the plane of the meristem (66). The corpus, or L3, is not a single cell layer but rather a group of cells that lies beneath the tunica and divides in all planes. The L1 layer cells are the precursors of the epidermis of shoots, leaves, and flowers, whereas L2 cell layer derivatives provide the mesodermal cells and the germ cells of pollen grains and ovules. The L3 cells generate the stem vasculature and pith as well as the innermost cells of leaves and floral organs. The tunica-corpus organization of the angiosperm SAM therefore reflects specific

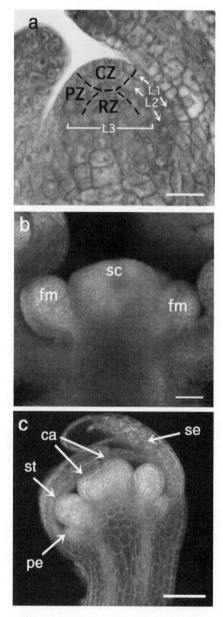

Figure 1 Structural features of Arabidopsis shoot apical and floral meristems. (*a*)
Section through a vegetative SAM showing the cell layers (L1, L2, and L3) and the
histologically defined domains. CZ, central zone; PZ, peripheral zone; RZ, rib zone.
(*b*) Confocal micrograph of an indeterminate SAM. The stem cell reservoir (sc) is at
the apex of the meristem and floral meristems (fm) arise from the flanks. (*c*) Confocal
micrograph of a determinate flower, after the floral meristem has produced sepals (se),
petals (pe), stamens (st), and two carpels (ca) in the center of the flower.

patterns of cell divisions. Surface growth occurs through anticlinal divisions of the tunica cells, whereas internal growth is achieved through the variable division orientations of the corpus cells.

Although highly regular patterns of cell division are detected in the SAMs of most plants, these patterns do not reflect a lineage-based specification of cell fate. Mosaic analysis reveals that there is no fixed pattern of SAM cell lineage beyond the general preservation of the clonal layers (21, 27, 54), and cells in one layer occasionally invade another layer without causing detectable developmental defects (66). Thus the fate of a cell in the SAM is determined by position rather than by clonal origin (62). Because cells in the both the tunica and the corpus of the SAM participate throughout development in meristem maintenance and organ formation (55, 56), these functions must be coordinated between all the cell layers. It is therefore critical that SAM cells be in continuous communication with neighboring cells within their own layer and in adjoining layers in order to assess their relative positions in the meristem and behave accordingly.

Determinant Versus Indeterminant Meristems

The organization of floral meristems is similar to that of SAMs. Floral meristems consist of tunica and corpus cell layers and have a central zone of stem cells that provides progeny cells for organogenesis in the peripheral zone (61). Genetic studies support a long-held view that flowers are modified shoots and floral organs are modified leaves (6, 25, 51, 70), which indicates that shoots and flowers are homologous structures. Although they share a common structural template, these two types of meristems vary in several important ways. One major difference between them is the type and arrangement of the lateral structures that they produce. SAMs form leaves and their associated meristems, generally in a spiral pattern, whereas floral meristems generate sepals, petals, stamens, and carpels in a whorled pattern. A second critical difference between Arabidopsis shoot and floral meristems is that the SAM is indeterminate and grows indefinitely (Figure 1*b*), whereas the floral meristem is determinate and terminates once the four whorls of organs have been produced (61) (Figure 1*c*). The stem cell reservoir in floral meristems is therefore transient, and its activity must cease at the correct stage of development to allow carpel formation in the center of the flower. Like maintenance of the stem cell pool in the SAM, the termination of stem cell activity in the floral meristem is critical for proper plant development and is under strict genetic control.

MAINTENANCE OF THE SHOOT APICAL MERISTEMATIC STATE

CLV Genes Restrict Meristem Cell Accumulation

In Arabidopsis, three *CLAVATA* genes (*CLV1*, *2*, and *3*) are required to regulate the size of the stem cell reservoir in the SAM. Recessive loss-of-function *clv1*, *clv2*, or *clv3* mutants form enlarged SAMs beginning during embryogenesis

(14, 15, 32, 35). The SAMs of *clv* mutant plants enlarge progressively throughout development, such that by the time the plant makes the transition to flowering the SAM has often undergone fasciation and grows as a mound or a strap instead of a point (Figure 2*a,b*, see color insert). The term fasciation is derived from the Latin fasces or fasciculum, meaning a bundle or packet. The production of such fasciated SAMs by *clv* mutant plants indicates that the wild-type function of *CLV1*, *CLV2*, and *CLV3* is to restrict shoot apical meristem activity. Mutants with fasciated SAM phenotypes resembling those of the *clv* mutants have also been reported in tomato (46, 65), soybean (73), and maize (62a).

SAM enlargement can be caused by the presence of either more cells than normal or larger cells than normal. Confocal laser scanning microscopy revealed that the *clv* mutant SAMs are larger because they contain many more cells than wild-type SAMs. Because the *CLV* genes do not affect cell size, they must instead control either the rate of cell division in the SAM central zone or the rate at which cells exit the central zone. A study comparing cell division rates between wild-type Arabidopsis plants and those carrying the *clv3-2* null allele found that the mitotic index of stem cells in the central zone is actually slightly lower, not higher, in mature *clv3* inflorescence apices than in the wild type. Thus it appears that *CLV* gene activity does not limit cell division rates in the center of the SAM. Rather, the CLV genes appear to control stem cell accumulation by regulating the rate at which cells in the central zone make the transition from the meristem into organ primordia.

clv mutants generate enlarged floral meristems as well as enlarged shoot apices. By the stage at which the sepal primordia arise in the outer whorl, *clv* floral meristems are already much taller than wild-type meristems and consist of many more cells (14, 15, 32). The increase in floral meristem size is closely correlated with an increase in floral organ number in all whorls, suggesting that the extra cells are allocated into additional floral organ primordia. The floral organ number increase in *clv* mutant plants is most extreme in the center of the flower. Floral meristems produced by the most severely affected *clv* mutants can generate up to 7 or 8 carpels (four times the normal number) that fuse to form a club-shaped fruit. This phenotype gives the *clavata* mutants their name, from the Latin word clavatus meaning club-like. In addition to generating extra organs, floral meristems of mutants carrying strong *clv* alleles often fail to stop proliferating upon carpel formation and accumulate undifferentiated cells in the center of the mature flower gynoecium. This loss of floral determinacy is further evidence that the *CLV* genes act to regulate the balance between meristem cell accumulation and differentiation and that their activity is required to prevent the buildup of undifferentiated cells at the center of the floral meristem.

Mutations in any one of the three *CLV* loci cause nearly identical shoot and floral meristem phenotypes, and genetic studies show that *CLV1*, *CLV2*, and *CLV3* interact to control meristem size during development. Double mutants generated between weak *clv1* and *clv3* alleles display severe *clv* mutant phenotypes, whereas those generated between strong *clv1* and *clv3* alleles have phenotypes indistinguishable from either single mutant (15). In addition, doubly heterozygous *clv1/+*

clv3/+ plants display an intermediate *clv* phenotype. This combination of nonallelic noncomplementation, mutual epistasis and the sensitivity of each gene product to a reduction in the level of the other provides strong genetic evidence that *CLV1* and *CLV3* act in the same developmental pathway. *clv2* mutants display shoot and floral meristem phenotypes that are similar to but slightly weaker than those of *clv1* and *clv3* mutants, and strong *clv1* and *clv3* alleles are epistatic to *clv2* with regard to these traits (32). However, although the *clv1* and *clv3* mutant phenotypes are restricted to the above-ground meristems, *clv2* mutants display more pleiotropic phenotypes such as elongated pedicels and reduced anther locule number. Thus *CLV1*, *CLV2*, and *CLV3* act in the same pathway in shoot and floral meristems, but *CLV2* also functions more broadly to regulate other aspects of development.

The cloning of the three *CLV* genes revealed why their mutant phenotypes are so similar. The *CLV1* gene encodes a protein consisting of 21 extracellular leucine-rich repeats (LRRs), a transmembrane domain, and an intracellular serine/threonine kinase domain (16). The *CLV2* gene encodes a receptor-like protein that consists of extracellular LRRs, a transmembrane domain, and an 11 amino acid cytoplasmic tail (29). Both proteins contain paired cysteines in their extracellular domains that potentially allow for homo- or heterodimerization through the formation of disulphide bridges. The *CLV3* gene encodes a 96 amino acid, predicted extracellular protein (20). Clonal analysis of an unstable *clv3* allele revealed that wild-type CLV3 function in either the L1 or L3 cell layer alone could confer a wild-type phenotype on the whole meristem (20). Thus CLV3 can function in a cell nonautonomous fashion, as is characteristic of diffusible signaling molecules, suggesting that CLV3 might act as the ligand for the CLV1 and/or CLV2 receptors.

CLV1 and CLV2 are members of large families of receptor proteins found in both plants and animals. Over 150 putative LRR receptor-like kinases (LRR-RLKs) that resemble CLV1 and at least 30 LRR receptor-like proteins that resemble CLV2 have been identified in the Arabidopsis genome (65a). Several Arabidopsis LRR-RLKs regulate various aspects of development (30, 67), one is involved in defense responses (23) and another is the brassinosteroid receptor (38). However, the functions of the vast majority of the Arabidopsis LRR receptor-like proteins are unknown. The LRR is a common motif found in protein binding domains of both animal and plant proteins (9), suggesting that the extracellular LRR domains of the CLV1 and CLV2 receptors may bind a protein or peptide ligand. By analogy with animal receptor kinases, many of which function in signaling cascades (2), binding of the ligand to the extracellular domains of the plant receptor(s) is predicted to cause a conformational change in the protein, resulting in the activation of the cytoplasmic domain to elicit a cellular response.

Like *CLV1* and *CLV2*, *CLV3* is also a member of a gene family. The CLV3 protein contains a 14 amino acid region near the carboxyl terminus that is highly conserved among three maize embryo surrounding region (ESR) proteins as well as the putative products of 24 small open reading frames in the Arabidopsis genome and ESTs from several other plants (17). These genes have been grouped into a family of *CLV3/ESR-like* (*CLE*) genes. The maize *ESR* genes are expressed in developing endosperm in a restricted zone surrounding the embryo, and like *CLV3* they encode putative secreted polypeptides (4, 50). Based on these data, the *ESR* proteins are proposed to mediate interactions between the endosperm and embryo. Although few of the Arabidopsis *CLE* genes are represented in EST databases, all of them are expressed in tissues based on RT-PCR analysis (V.K. Sharma & J.C. Fletcher, unpublished data). Each Arabidopsis CLE protein contains a putative signal peptide or signal anchor, suggesting that all of these proteins may act as extracellular signaling molecules. Given the very large number of CLV1 and CLV2-like LRR receptor proteins in the Arabidopsis genome, it is tempting to speculate that the *CLE* genes encode ligands involved in diverse signal transduction events effected via different combinations of LRR-RLKs. Members of the CLE family have not been identified in animals or fungi, which suggests that *CLE* gene functions may be restricted to plants.

CLV Receptor Complex

The three *CLV* genes encode components of a signal transduction pathway that communicates cell fate information between neighboring groups of shoot and floral meristem cells. Biochemical evidence confirms that CLV1, 2, and 3 are bound together in a plasma membrane-bound signaling complex (Figure 3). The CLV1 protein is detected in two complexes in plant extracts, a 185-kD complex and a more abundant 450-kD heteromeric complex (69). CLV1 is present in both complexes as an ∼185-kD disulphide-linked multimer, with approximately twice as much CLV1 in the larger complex as in the smaller one. The 450-kD complex does not form in *clv1* mutants lacking an active kinase domain, indicating that the larger form represents the active complex and that the 185-kD complex is inactive. The 450-kD complex also does not form in *clv3* mutant plants, so the presence of CLV3 is required for proper assembly of an active signaling complex. When all three components are expressed in yeast cells, the CLV3 protein binds specifically to the active form of CLV1 and associates with CLV2 and kinase-active CLV1 at the plasma membrane (68). These experiments convincingly demonstrate that CLV3 is the ligand for the CLV1 receptor complex. Interestingly, a kinase-inactive form of CLV1 expressed in yeast fails to bind CLV3 protein (68), indicating that the CLV1 cytoplasmic domain is necessary for facilitating ligand binding as well as for downstream signal transduction.

Based on genetic evidence, the CLV1 active complex is also likely to include CLV2, although antibodies specific to CLV2 are not yet available to confirm this directly. Mutations in *CLV2* reduce CLV1 protein levels by more than 90% and affect the accumulation of both the 185-kD and the 450-kD CLV1 complexes (29). The remaining CLV1 protein in extracts from *clv2* mutant plants is found in a novel ∼600-kD complex. If this complex retains some function in vivo, it would explain why all known *clv2* mutations confer weaker shoot and floral meristem phenotypes than do the *clv1* and *clv3* mutations. For instance, in the absence of CLV2, the CLV1 protein may have the ability to form homomultimers or interact with other

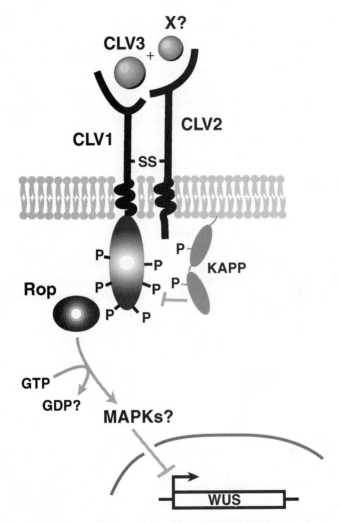

Figure 3 The CLV signaling complex. The CLV1 LRR receptor kinase forms a heteromeric complex with the CLV2 LRR receptor-like protein at the plasma membrane of interior SAM cells. Binding of the CLV3 ligand, possibly in association with another protein (X), stimulates assembly of an active signaling complex that also contains a phosphatase (KAPP) and a Rho-like GTPase (Rop). The signal is relayed from the cytosol to the nucleus, potentially via a MAP kinase cascade, to limit *WUS* expression.

receptor-like proteins to bind CLV3 and establish a basal level of signaling that slightly restricts meristem cell accumulation.

The question of whether any of the components of the complex are rate limiting has also been addressed biochemically. Approximately 75% of the CLV3 protein appears to be bound to the CLV1/CLV2 receptor complex in cauliflower extracts

(68). The high proportion of receptor-bound CLV3 suggests that the CLV3 ligand availability is rate limiting for meristem signaling and that most or all of the CLV3 protein that reaches the CLV1/CLV2 receptor complex will be bound and cause activation of the pathway. The other 25% of CLV3 protein is not receptor associated and is detected as a multimer of approximately 25 kD. It is not known whether this complex consists of a CLV3 homomultimer or if other proteins are present. Such putative CLV3-associated factors might act as co-ligands or mediate CLV3 binding to the receptor complex, or possibly sequester CLV3 from the receptor complex for release under the appropriate conditions.

Signaling via cell surface receptors generally requires cytosolic factors that relay, amplify, and/or attenuate the signal. Several proteins have been identified that interact with the active form of the CLV1 complex on the cytoplasmic side of the plasma membrane. The CLV1 kinase domain associates with a kinase-associated protein phosphatase (KAPP) that is expressed in meristems in a region encompassing the *CLV1*-expressing cells (63, 64, 72). Co-immunoprecipitation experiments have shown that KAPP is part of the 450-kD active complex, but in vitro association of CLV1 and KAPP requires CLV1 kinase activity and KAPP phosphorylation (69). The KAPP kinase interaction domain contains a phosphoserine/phosphothreonine binding forkhead-associated (FHA) domain (39), suggesting that KAPP binding to the CLV1 kinase domain may occur via binding to phosphoserine residues. Overexpression of the KAPP phosphatase in wild-type plants causes a weak *clv* mutant phenotype; thus KAPP is thought to dephosphorylate CLV1 and act as a negative regulator of the meristem growth control pathway (64, 72). KAPP is a shared component of multiple signal transduction pathways, as it also interacts with a number of other receptor-like kinases (8, 63). KAPP involvement in multiple signaling processes is one possible reason why loss-of-function KAPP mutations have not been identified, as they may be lethal.

The CLV active complex also co-precipitates with one or more members of the Rop subfamily of plant Rho/Rac small GTPase-related proteins (37). There are at least 10 Rop family members in Arabidopsis, many of which are expressed in shoots (37), but it is not yet clear which Rop(s) participates in CLV signaling. Rho/Rac GTPases are members of the Ras GTPase superfamily of cytosolic proteins. In animals and fungi, Ras GTPases mediate many receptor tyrosine kinase signaling events by switching on intracellular protein kinase cascades that control cellular processes such as polarized cell growth, actin cytoskeletal reorganization, and cell polarity establishment (12, 49). Ras GTPases are not found in the Arabidopsis genome (65a) and have yet to be isolated from other plants, so Rho/Rac GTPases such as the Rop proteins may be functionally analogous to Ras GTPases in plant signal transduction. Arabidopsis Rop proteins participate in developmental processes such as embryogenesis, lateral organ morphogenesis, and shoot apical dominance and growth (36). Based on similar roles for Ras GTPases in animals, it has been proposed that Rop GTPases in the SAM may participate in meristem signaling by activating a mitogen-activated protein kinase (MAPK)-like cascade in response to CLV1 kinase activation (69).

CLV Signal Transduction

The mRNA expression patterns of the *CLV1* and *CLV3* genes provided important clues to how the CLV signal transduction pathway controls shoot and floral meristem size. The *CLV1* and *CLV3* mRNAs are initially expressed at the heart stage of embryogenesis, approximately midway through embryonic development, in a small group of cells between the embryonic seed leaves. After germination, *CLV3* mRNA accumulates in a few cells at the apex of the SAM, predominantly in the L1 and L2 tunica cells of the region corresponding to the central zone (20). *CLV3* transcripts are not detected in initiating lateral organ primordia but reappear in the central zone of the floral meristem shortly after its separation from the SAM. *CLV3* continues to be expressed at the floral meristem apex until the carpel primordia initiate in the center of the flower. *CLV3* mRNA is therefore associated with the pluripotent stem cell population throughout development. *CLV1* mRNA is also expressed in shoot and floral meristems throughout development, but the transcripts are localized in the deeper regions of the SAM, mainly in the L3 corpus layers. *CLV1* mRNA is not detected in the L1 cell layer and in the SAM is also absent from the L2 layer (16). Thus the *CLV3* expression domain overlies the *CLV1* expression domain, indicating that in shoot and floral meristems the stem cells and the underlying cells communicate with one another via the CLV signaling pathway. RNA blot analysis detects *CLV2* transcripts in shoots and developing flowers, as well as in a number of other tissues (29).

The *CLV3* expression domain is greatly enlarged in the fasciated shoot apical meristems of *clv1* and *clv3* mutant plants, whereas the size of the peripheral zone of non-*CLV3* expressing cells is not increased relative to wild-type plants (20). Because *CLV3* is expressed exclusively in stem cells, this result confirms that *clv1* and *clv3* mutants accumulate excess stem cells in their shoot apices. Further, *CLV3* mRNA can be detected in the center of mature *clv1* mutant flowers, between the carpels, long after *CLV3* is normally downregulated in wild-type plants. Thus *clv1* mutants accumulate extra stem cells in both shoot and floral meristems. Comparable enlargement of the *CLV3* expression domain also occurs in *clv2* and *clv3* mutant meristems, indicating that *CLV1*, *CLV2*, and *CLV3* all act to limit the number of *CLV3*-expressing stem cells.

The mRNA expression domains of both *CLV1* and *CLV3* enlarge coordinately in *clv1, 2,* or *3* mutant plants (20, 29). These observations were interpreted to mean that the expansion of the *CLV* expression domains is coordinated by a positive, stem cell-promoting pathway, which in turn is negatively regulated by the stem cell-restricting *CLV* pathway. Overexpression of *CLV3* in transgenic Arabidopsis plants has allowed those two pathways to be distinguished (7). Stem cells are not correctly maintained when *CLV3* is constitutively expressed at high levels, as the transgenic plants germinated normally but ceased to initiate organs after the production of the first leaves. Some transgenic lines expressed lower levels of constitutive *CLV3* and retained some meristem function. However, these plants failed to generate the full complement of flowers and floral organs, again demonstrating an inability to replenish stem cells. The abundance of the CLV3 ligand is therefore the critical factor that determines the size of the stem cell reservoir and, as a result, the number of lateral organs that can be produced by the meristem. When the CLV3 transgene was introduced into a *clv1* or *clv2* mutant background, the transgenic plants exhibited the typical *clv* mutant phenotype despite expressing high levels of *CLV3*. Thus CLV3 signaling requires functional CLV1 and CLV2, and the terminal meristem phenotypes observed in the transgenic CLV3 plants are due to enhanced CLV3 signaling through the CLV1/CLV2 receptor complex.

Feedback Regulation of Stem Cell Fate

Achieving a balanced meristematic state requires input both from genes that restrict SAM activity and from genes that promote SAM activity. In Arabidopsis, one gene that is critical for SAM activity is the *KNOTTED1*-like homeobox gene *SHOOTMERISTEMLESS* (*STM*) (41). Plants homozygous for loss-of-function *stm* mutations fail to establish and maintain a functional SAM and do not undergo postembryonic development (1) (Figure 2c). *STM* is expressed throughout the shoot apex and is required to prevent the specification of stem cells as organ cells by repressing the expression of organ-specific Myb genes in the SAM (11). *STM* interacts genetically with the *CLV* loci. *stm clv* double mutants develop some vegetative and floral organs, indicating that *clv* mutations can partially suppress the *stm* mutant phenotypes and restore postembryonic growth (13). The *clv* mutations suppress the *stm* mutant phenotypes in a dominant fashion, and vice versa, revealing that the *stm* phenotype is sensitive to the level of *CLV* activity and that the *clv* phenotypes are likewise sensitive to the level of *STM* activity (13, 32). Thus *STM* and the *CLV* loci act in parallel pathways to competitively regulate SAM function.

The Arabidopsis *WUSCHEL* (*WUS*) gene also promotes meristem function, and *WUS* has been identified as a key target of CLV meristem signal transduction. Loss-of-function *wus* mutants fail to organize a functional SAM in the embryo (58). The cells at the *wus* embryonic apex appear to be incorrectly specified because instead of establishing themselves as a stem cell population they differentiate without producing organ primordia (33). After germination, *wus* mutants sporadically develop multiple transient meristems that form only a few organs before terminating prematurely in aberrant flat structures. Unlike wild-type plants, which initiate organs from the flanks of the SAM, *wus* mutants initiate lateral organ primordia randomly across the entire shoot apex. Reiterative generation and premature termination of *wus* meristems eventually produces bushy plants with multiple rosettes. It is from this phenotype that the mutants received their name, as wuschel is a German word meaning tousled. Some *wus* adventitious meristems eventually form abnormal inflorescences that develop aerial rosettes of leaves and terminate prematurely after producing a reduced number of flowers. *wus* floral meristems produce near-normal numbers of sepals and petals in the outer whorls but fail to generate the full complement of inner whorl organs and usually terminate in a solitary stamen (Figure 2*d*). These phenotypes reveal that *WUS* activity is necessary to sustain the pluripotent stem cell pools in shoot and floral meristems after they are established (33).

WUS is a component of the CLV signal transduction pathway, as *wus clv1* and *wus clv3* double mutant plants resemble *wus* single mutant plants. *WUS* encodes a novel subtype of homeodomain proteins, and the WUS protein is localized to the nucleus (44). From these observations, we infer that *WUS* is likely to function at the transcriptional level to promote stem cell fate and that this activity is downregulated by CLV signaling. During embryogenesis, *WUS* is expressed in apical cells prior to the initial appearance of *CLV1* and *CLV3*. As the plants mature, *WUS* transcripts become restricted to a small group of cells in the internal layers of the shoot and floral meristems. These cells lie beneath the *CLV3* expression domain and overlap the *CLV1* domain in the L3. Maintenance of the *WUS* expression domain requires the activity of *FASCIATED1* (*FAS1*) and *FAS2*, which encode components of Arabidopsis chromatin assembly factor-1 (31). FAS1 and FAS2 are therefore likely to promote stable *WUS* gene transcription by facilitating chromatin assembly. In *clv3* mutant meristems, the *WUS* expression domain expands laterally toward the meristem flanks and also upward into the subepidermal layer (7, 58). Conversely, the arrested meristems of *CLV3* overexpressing plants, which phenocopy the *wus* loss-of-function mutant phenotype, do not express *WUS* mRNA (7). Thus signaling through the CLV pathway leads to negative regulation of stem cell activity by restricting the size of the *WUS* expression domain.

The negative, stem cell–restricting *CLV* pathway therefore targets *WUS*, which is sufficient to specify stem cell fate. When *WUS* is expressed under the control of the *AINTEGUMENTA* (*ANT*) promoter, which drives *WUS* transcription in all initiating organ primordia on the flanks of the SAM (58), the resulting *pANT::WUS* transgenic seedlings do not form any lateral organs. Instead, the shoot apex consists entirely of undifferentiated meristematic cells, a graphic demonstration that *WUS* activity is sufficient to confer stem cell fate. In wild-type plants, *CLV3* mRNA is only found at the very apex of the SAM, but in *pANT::WUS* seedlings, *CLV3* transcripts can be detected on the periphery of the meristematic cell mass. Thus *WUS* activity is also sufficient to induce *CLV3* transcription, indicating that *WUS* is a critical component of a stem cell–promoting pathway that counterbalances the negative, stem cell–restricting pathway by preserving the *CLV3*-expressing stem cell reservoir at the SAM apex.

Despite its central role in specifying stem cells, *WUS* does not appear to be the only target of CLV signal transduction. A genetic screen for second-site modifiers of a weak *clv* mutant phenotype resulted in the isolation of several recessive *poltergeist* (*pol*) suppressor mutants (75). Although *pol* single mutants are nearly indistinguishable from wild-type plants, *POL* appears to promote stem cell activity because *pol clv* double mutants accumulate fewer stem cells in their shoot and floral meristems than *clv* single mutants. *pol* mutations suppress the phenotypes of many *clv* alleles in a semidominant fashion but have no effect on CLV1 receptor activation. *POL* is therefore likely to function downstream of the *CLV* loci and to be a target of downregulation in response to CLV signaling. *pol* enhances the *wus* shoot and floral meristem phenotypes, and dominant interactions are observed between the *pol* and *wus* mutations. These data suggest that *POL* acts redundantly with *WUS* to specify stem cells in shoot and floral meristems. They also indicate that although *WUS* can almost completely compensate for the lack of *POL* function *POL* is not able to effectively compensate for *WUS*.

Thus stem cell activity in Arabidopsis shoot and floral meristems is mediated by mutual regulation and signaling across the layers of the meristem, involving both positive and negative interactions. The CLV3 signal originates from the stem cells at the apex of shoot and floral meristems and is perceived by the underlying CLV1-expressing cells (Figure 4a). Binding of the CLV3 ligand to the receptor

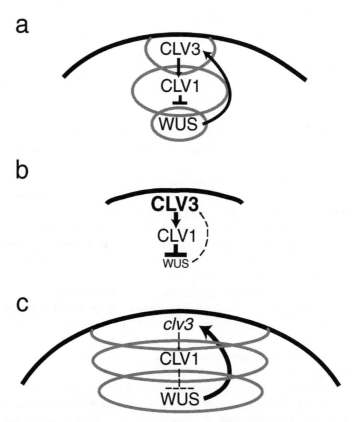

Figure 4   Spatial feedback loop controlling stem cell maintenance in indeterminate meristems. (*a*) A wild-type shoot apical meristem. *CLV3*, *CLV1*, and *WUS* are expressed in overlapping domains (*grey circles*). *WUS* is expressed in a small group of cells in the interior of the SAM where it signals to the overlying cells to maintain the *CLV3*-expressing stem cell population in the superficial layers. CLV3 signaling through the CLV1 receptor complex restricts *WUS* to its narrow domain in the L3. (*b*) Constitutive CLV3 signaling (*bold type*) downregulates *WUS* expression, extinguishing the WUS-mediated signal that preserves the stem cells and causing SAM termination. (*c*) Disruption of the CLV pathway causes enlargement of the *WUS* expression domain, leading to excess stem cell accumulation and coordinated expansion of the meristem.

complex in the underlying cells is likely to occur via the extracellular LRR domains of CLV1 and CLV2. Ligand binding in the presence of the active CLV1 kinase domain induces assembly of the active receptor complex, permitting downstream signal transduction. Signaling through the *CLV* pathway targets *POL* and *WUS*, limiting the scope of *WUS* activity by restricting its expression to a small group of cells in deeper regions of the meristem. When *CLV3* is overexpressed, constitutive signaling enhances this negative pathway, abolishing *WUS* transcription and resulting in the complete loss of stem cells (Figure 4*b*). Activity of the positive pathway mediated by *WUS* promotes the expression of *CLV3* and the persistence of the *CLV3*-expressing stem cell pool. *WUS* mRNA is not detected in the L1 or L2 layers of wild-type meristems, so stem cell activity in these superficial cell layers is likely to be maintained by an inductive signal mediated by *WUS*. Disruption of the negative pathway in *clv* mutants leads to increased *WUS* activity, causing excess stem cell accumulation and expansion, both lateral and upward, of the *WUS* expression domain (Figure 4*c*). Conversely, disruption of the positive pathway in *wus* mutants leads to the specification of insufficient numbers of stem cells, causing premature meristem termination. These regulatory mechanisms provide a stable feedback system that tends toward equilibrium, enabling the SAM to sustain organogenesis throughout the life of the plant.

REGULATION OF FLORAL MERISTEM CELL FATE

Floral meristems arise on the flanks of the SAM and are organized in a similar fashion. Floral meristems express *CLV3* and *WUS* in adjacent domains at the meristem apex and maintain a pool of stem cells during the formation of the sepals, petals, and stamens on the meristem flanks. But as previously noted, floral meristems are not equivalent to SAMs because they generate different types of organs, in different spatial arrangements, and because they ultimately terminate in the formation of the female reproductive organs. Their separate identity is conferred by floral meristem identity genes such as *LEAFY (LFY)* and *APETALA1 (AP1)*, which are transcribed in initiating floral meristem primordia but not in the SAM (3, 24, 70). These genes direct the expression of the floral homeotic genes in overlapping spatial domains. The floral homeotic genes in turn specify the identity of the four types of floral organs from the outer whorls to the inner whorls, according to the well-characterized ABC model (18, 28).

The homeotic gene *AGAMOUS (AG)*, which encodes a flower-specific MADS domain transcription factor (5, 74), is unique in that it is required for both floral meristem termination and organ identity specification. In *ag* mutants, petals instead of stamens arise in the third whorl, and a new flower is formed in the fourth whorl in place of carpels. As a result, *ag* mutant flowers consist entirely of whorls of sepals and petals that are reiterated indefinitely. These flowers are like shoots in that they remain indeterminate and continue to produce organs. Conversely, transgenic plants that constitutively express *AG* form inflorescence meristems that terminate in a solitary flower (48). Thus *AG* is required for floral meristem determinacy and is sufficient to convert an indeterminate meristem into a determinate meristem.

AG expression is regulated by several factors. *AG* is initially expressed in the center of floral meristems in the cells that will generate the stamens and carpels, and transcription persists until the late stages of organ development (74). *AG* is a direct target of activation by *LFY* (10), but because LFY protein is present throughout the floral meristem (59) at least one other factor must contribute regional specificity to *AG* induction. *LFY*-dependent, spatially restricted *AG* induction in floral meristems is conferred by *WUS* (34, 40) (Figure 5). *WUS* is expressed in the center of the floral meristem in a subset of the cells that eventually express *AG*. The flowers

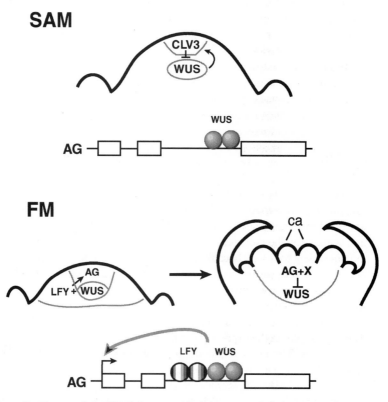

Figure 5 Temporal feedback loop regulating stem cell termination in determinate floral meristems. (*top*) Schematic of an indeterminate SAM, showing the interaction between *CLV3* and *WUS* in their respective domains (*grey circles*). In the SAM, LFY is absent and *AG* expression is not induced. (*bottom*) Schematic of a determinate floral meristem over time. LFY is present throughout the young floral meristem. Both LFY and WUS bind to *AG* enhancer sequences and cooperate to induce *AG* transcription in the center of the developing flower. At the time of carpel (ca) initiation, AG and an additional factor (X) repress *WUS* expression to terminate stem cell activity.

Figure 2 Arabidopsis shoot apical meristem mutants. (*a*) Wild-type Landsberg *erecta* SAM growing outward as a point and initiating lateral organs on the flanks. (*b*) Enlarged *clv3-2* SAM growing as a mound (*arrow*) and producing many extra flowers. (*c*) An *stm-11* seedling that lacks a functional SAM and forms cotyledons but no other lateral organs. (*d*) A *wus-1* adventitious SAM that has formed a reduced number of flowers. These flowers lack the inner organs, often terminating in a single stamen (*arrowhead*).

produced by rare *wus* mutant floral meristems lack all carpels and most stamens, the organ types that are specified by *AG* (33). *WUS* misexpression in the second whorl of developing flowers causes ectopic stamen and carpel formation, which results from ectopic *AG* induction (34, 40). Thus *WUS* expression, in the presence of *LFY*, is sufficient to activate *AG* in floral meristems.

The LFY and WUS proteins cooperate to activate *AG*. WUS binds directly to homeodomain protein consensus sequences in the second intron of *AG* (40). These sites are adjacent to binding sites for LFY, but LFY and WUS seem to bind independently to the *AG* enhancer sequence in the second intron. The WUS binding sites, like the LFY binding sites, are important for activity of the *AG* enhancer in vivo. However, neither LFY nor WUS appears to be sufficient to activate *AG*, as neither protein can activate an *AG* reporter construct in vitro or when expressed individually in yeast cells (40). Furthermore, the requirement for *WUS* to induce *AG* transcription is not absolute, as one stamen is usually formed in *wus* mutant flowers (33). *POL*, which acts redundantly with *WUS* to promote stem cell activity, is one candidate for a factor that might stimulate a sufficient level of *AG* transcription in the absence of *WUS* to achieve limited stamen specification without rescue of the determinacy function. This hypothesis is consistent with the observation that floral organ identity requires lower levels of *AG* activity than does meristem determinacy (47).

AG and *WUS* function as key regulators of determinate and indeterminate growth, respectively, suggesting that they have antagonistic roles in the flower. Expression analysis revealed that *WUS* transcripts persist in the center of *ag* mutant flowers after the production of many whorls of organs (34, 40), and moderate *WUS* overexpression at the floral apex causes a partial indeterminacy phenotype akin to that observed in plants with reduced *AG* function (47, 58, 60). Determinacy is restored in *ag wus* double mutant flowers, which resemble *wus* flowers, indicating that *ag* indeterminacy is dependent on the ectopic activity of *WUS*. Thus prolonged *WUS* expression is sufficient to make floral meristems indeterminant, and one role of *AG* is to downregulate *WUS* and terminate stem cell activity prior to carpel formation. *AG* appears to act at least partially independently of the *CLV* pathway to repress *WUS*, as the size of the *WUS* expression domain is larger in *ag clv1* flowers than in *ag* flowers (40).

Thus stem cell termination in floral meristems involves a temporal negative feedback loop whereby *WUS* activity in the floral meristem induces its own repressor. The LFY and WUS proteins bind independently to *AG* regulatory sequences in the second intron and cooperate to direct *AG* expression in the center of the floral meristem. Once activated, *AG* represses *WUS* and prevents further renewal of the floral stem cell reservoir. Although *WUS* is present in the SAM, *AG* induction does not occur there, presumably because *LFY* is absent. Thus, as is observed (71), the meristems of 35S::*LFY* plants that express *LFY* constitutively should terminate prematurely in a flower. Meristem termination does not occur immediately following germination of 35S::*LFY* plants, however, so additional factors are required to make the meristem competent to respond to *LFY* and/or

AG (71). In addition, the *AG* expression domain is larger than the *WUS* expression domain, indicating either that other factors are also involved in *AG* induction outside the *WUS* domain or that the WUS protein itself moves from cell to cell as has been reported for the transcription factors DEFICIENS, LFY, and KN1 (42, 52, 59).

CONCLUDING REMARKS

A renewable stem cell reservoir is absolutely critical for plant growth and development. But far from being a default state, stem cell maintenance at the apex of shoot and floral meristems requires constant signaling between *CLV3*-expressing and *WUS*-expressing cells. This signaling establishes a stable feedback loop between stem cell–promoting and stem cell–restricting pathways to preserve stem cell number at equilibrium. Termination of stem cell activity under the appropriate circumstances also depends on feedback regulation, via a pathway that functions over time rather than across space. Molecular genetic and biochemical studies are rapidly advancing our knowledge of the basic components and mechanisms that govern meristem maintenance and termination, but there are many more lines of inquiry to pursue. Various aspects of the CLV ligand-receptor interaction need to be clarified, including the role of CLV2 in the complex and the composition of the 25-kD CLV3 multimer. Additional intracellular components of the CLV pathway also remain to be identified, including those that relay the signal from the CLV receptor complex into the nucleus and the molecule(s) that represses *WUS* transcription. At the cellular level, it will be important to discover the targets of the *WUS* transcription factor and to determine where *POL* fits into the picture. Another mystery that remains to be addressed is the nature of the WUS-mediated inductive signal. One possibility is that the WUS protein itself moves through plasmodesmata to directly induce *CLV3* expression in overlying meristem cells. Alternatively, the signal may be one of the CLE proteins or another small secreted molecule. Finally, we still do not understand how the stem cell pathway and the floral patterning pathway interact to terminate the floral meristem. How does *WUS*, which is expressed in a few interior meristem cells, activate *AG* in a much larger domain in the center of the floral meristem? What is the identity of the protein that acts with AG to downregulate *WUS*? With many of the tools in hand to address these issues, the investigation of shoot and floral meristem biology will be rewarding for years to come.

ACKNOWLEDGMENTS

I am grateful to members of my laboratory for insightful discussions and to Dr. Sheila McCormick for critical review of the manuscript. My laboratory's work on plant development is supported by the U.S. Department of Agriculture (CRIS 5335-21000-013-D) and the National Science Foundation (IBN 0110667).

Visit the Annual Reviews home page at www.annualreviews.org

LITERATURE CITED

1. Barton MK, Poethig RS. 1993. Formation of the shoot apical meristem in *Arabidopsis thaliana*: an analysis of development in the wild type and in the *shoot meristemless* mutant. *Development* 119:823–31

2. Becraft P. 1998. Receptor kinases in plant development. *Trends Plant Sci.* 3:384–88

3. Blazquez MA, Soowal LN, Lee I, Weigel D. 1997. *LEAFY* expression and flower initiation in Arabidopsis. *Development* 124:3835–44

4. Bonello J-F, Opsahl-Ferstad H-G, Perez P, Dumas C, Rogowsky PM. 2000. *Esr* genes show different levels of expression in the same region of maize endosperm. *Gene* 246:219–27

5. Bowman JL, Smyth DR, Meyerowitz EM. 1989. Genes directing flower development in *Arabidopsis. Plant Cell* 1:37–52

6. Bowman JL, Smyth DR, Meyerowitz EM. 1991. Genetic interactions among floral homeotic genes of Arabidopsis. *Development* 112:1–20

7. Brand U, Fletcher JC, Hobe M, Meyerowitz EM, Simon R. 2000. Dependence of stem cell fate in *Arabidopsis* on a feedback loop regulated by *CLV3* activity. *Science* 289:617–19

8. Braun DM, Stone JM, Walker JC. 1997. Interaction of the maize and Arabidopsis kinase interaction domains with a subset of receptor-like kinases: implications for transmembrane signaling in plants. *Plant J.* 12:83–95

9. Buchanan SGSC, Gay NJ. 1996. Structural and functional diversity in the leucine-rich repeat family of proteins. *Prog. Biophys. Mol. Biol.* 65:1–12

10. Busch MA, Bomblies K, Weigel D. 2000. Activation of a floral homeotic gene in Arabidopsis. *Science* 285:585–87

11. Byrne ME, Barley R, Curtis M, Arroyo JM, Dunham M, et al. 2000. *Asymmetric leaves1* mediates leaf patterning and stem cell function in *Arabidopsis. Nature* 408:967–71

12. Chant J, Stowers L. 1995. GTPase cascades choreographing cellular behavior: movement, morphogenesis, and more. *Cell* 81:1–4

13. Clark SE, Jacobsen SE, Levin JZ, Meyerowitz EM. 1996. The CLAVATA and SHOOT MERISTEMLESS loci competitively regulate meristem activity in Arabidopsis. *Development* 122:1567–75

14. Clark SE, Running MP, Meyerowitz EM. 1993. *CLAVATA1*, a regulator of meristem and flower development in Arabidopsis. *Development* 119:397–418

15. Clark SE, Running MP, Meyerowitz EM. 1995. *CLAVATA3* is a specific regulator of shoot and floral meristem development affecting the same processes as *CLAVATA1. Development* 121:2057–67

16. Clark SE, Williams RW, Meyerowitz EM. 1997. The *CLAVATA1* gene encodes a putative receptor kinase that controls shoot and floral meristem size in Arabidopsis. *Cell* 89:575–85

17. Cock JM, McCormick S. 2001. A large family of genes that share homology with *CLAVATA3. Plant Physiol.* 126:939–42

18. Coen ES, Meyerowitz EM. 1991. The war of the whorls—genetic interactions controlling flower development. *Nature* 353:31–37

19. Esau K. 1977. *Anatomy of Seed Plants.* New York: Wiley

20. Fletcher JC, Brand U, Running MP, Simon R, Meyerowitz EM. 1999. Signaling of cell fate decisions by *CLAVATA3* in *Arabidopsis* shoot meristems. *Science* 283:1911–14

21. Furner IJ, Pumfrey JE. 1992. Cell fate in the shoot apical meristem of *Arabidopsis thaliana. Development* 115:755–64

22. Gifford EM. 1954. The shoot apex in angiosperms. *Bot. Rev.* 20:429–47

23. Gomez-Gomez L, Boller T. 2000. FLS2: a LRR receptor-like kinase involved in recognition of the flagellin elicitor in Arabidopsis. *Mol. Cell* 5:1–20

24. Gustafson-Brown C, Savidge B, Yanofsky MF. 1994. Regulation of the Arabidopsis floral homeotic gene *APETALA1*. *Cell* 76:131–43

25. Honma T, Goto K. 2001. Complexes of MADS-box proteins are sufficient to convert leaves into floral organs. *Nature* 409:525–29

26. Deleted in proof

27. Irish VF, Sussex IM. 1992. A fate map of the *Arabidopsis* embryonic shoot apical meristem. *Development* 115:745–53

28. Jack T. 2001. Relearning our ABCs: new twists on an old model. *Trends Plant Sci.* 6:310–16

29. Jeong S, Trotochaud AE, Clark SE. 1999. The Arabidopsis *CLAVATA2* gene encodes a receptor-like protein required for the stability of the *CLAVATA1* receptor-like kinase. *Plant Cell* 11:1925–33

30. Jinn T-L, Stone JM, Walker JC. 2000. *HAESA*, an *Arabidopsis* leucine-rich repeat receptor kinase, controls floral organ abcission. *Genes Dev.* 14:108–17

31. Kaya H, Shibahara K-I, Taoka K-I, Iwabuchi M, Stillman B, Araki T. 2000. *FASCIATA* genes for chromatin assembly factor-1 in *Arabidopsis* maintain the cellular organization of apical meristems. *Cell* 104:131–42

32. Kayes JM, Clark SE. 1998. *CLAVATA2*, a regulator of meristem and organ development in *Arabidopsis*. *Development* 125:3843–51

33. Laux T, Mayer KFX, Berger J, Jurgens G. 1996. The *WUSCHEL* gene is required for shoot and floral meristem integrity in *Arabidopsis*. *Development* 122:87–96

34. Lenhard M, Bohnert A, Jurgens G, Laux T. 2001. Termination of stem cell maintenance in *Arabidopsis* floral meristems by interactions between *WUSCHEL* and *AGAMOUS*. *Cell* 105:805–14

35. Leyser HMO, Furner IJ. 1992. Characterisation of three shoot apical meristem mutants of *Arabidopsis thaliana*. *Development* 116:397–403

36. Li H, Shen J-J, Zheng Z-L, Lin Y, Yang Z. 2001. The Rop GTPase switch controls multiple developmental processes in Arabidopsis. *Plant Physiol.* 126:670–84

37. Li H, Wu G, Ware D, Davis KR, Yang Z. 1998. Arabidopsis Rho-related GTPases: differential gene expression in pollen and polar localization in fission yeast. *Plant Physiol.* 118:407–17

38. Li J, Chory J. 1997. A putative leucine-rich repeat receptor kinase involved in brassinosteroid signal transduction. *Cell* 90:929–38

39. Li J, Smith GP, Walker JC. 1999. Kinase interaction domain of kinase-associated protein phosphatase, a phosphoprotein-binding domain. *Proc. Natl. Acad. Sci. USA* 96:7821–26

40. Lohmann JU, Hong RL, Hobe M, Busch MA, Parcy F, et al. 2001. A molecular link between stem cell regulation and floral patterning in *Arabidopsis*. *Cell* 105:793–803

41. Long JA, Moan EI, Medford JI, Barton MK. 1996. A member of the KNOTTED class of homeodomain proteins encoded by the *STM* gene of *Arabidopsis*. *Nature* 379:66–69

42. Lucas WJ, Bouche-Pillon S, Jackson DP, Nguyen L, Baker L, et al. 1995. Selective trafficking of KNOTTED1 homeodomain protein and its mRNA through plasmodesmata. *Science* 270:1980–83

43. Lyndon RF. 1990. *Plant Development: The Cellular Basis*. London: Unwin Hyman

44. Mayer KFX, Schoof H, Haecker A, Lenhard M, Jurgens G, Laux T. 1998. Role of *WUSCHEL* in regulating stem cell fate in the *Arabidopsis* shoot meristem. *Cell* 95:805–15

45. Medford JI, Behringer FJ, Callos JD,

Feldmann KA. 1992. Normal and abnormal development in the *Arabidopsis* vegetative shoot apex. *Plant Cell* 4:631–43

46. Merton TR, Burdick AB. 1954. The morphology, anatomy and genetics of a stem fasciation in *Lycopersicon esculentum*. *Am. J. Bot.* 41:726–32

47. Mizukami Y, Ma H. 1995. Separation of *AG* function in floral meristem determinacy from that in reproductive organ identity by expressing antisense *AG* RNA. *Plant Mol. Biol.* 28:767–84

48. Mizukami Y, Ma H. 1997. Determination of *Arabidopsis* floral meristem identity by *AGAMOUS*. *Plant Cell* 9:393–408

49. Nagata K-I, Hall A. 1996. The Rho-GTPase regulates protein kinase activity. *BioEssays* 18:529–31

50. Opsahl-Ferstad H-G, Le Deunff E, Dumas C, Rogowsky PM. 1997. *ZmEsr*, a novel endosperm-specific gene expressed in a restricted region around the maize embryo. *Plant J.* 12:235–46

51. Pelaz S, Ditta GS, Baumann E, Wisman E, Yanofsky MF. 2000. B and C floral organ identity functions require *SEPALLATA* MADS-box genes. *Nature* 405:200–3

52. Perbal M-C, Haughn G, Saedler H, Schwarz-Sommer Z. 1996. Non–cell-autonomous function of the *Antirrhinum* floral homeotic proteins *DEFICIENS* and *GLOBOSA* is exerted by their polar cell-to-cell trafficking. *Development* 122:3433–41

53. Poethig RS. 1987. Clonal analysis of cell lineage patterns in plant development. *Am. J. Bot.* 74:581–94

54. Poethig RS, Coe EHJ, Johri MM. 1986. Cell lineage patterns in maize *Zea mays* embryogenesis: a clonal analysis. *Dev. Biol.* 117:392–404

55. Poethig RS, Sussex IM. 1985. The cellular parameters of leaf development in tobacco: a clonal analysis. *Planta* 165:170–84

56. Poethig RS, Sussex IM. 1985. The developmental morphology and growth dynamics of the tobacco leaf. *Planta* 165:158–69

57. Satina S, Blakeslee AF, Avery AG. 1940. Demonstration of the three germ layers in the shoot apex of *Datura* by means of induced polyploidy in periclinal chimeras. *Am. J. Bot.* 27:895–905

58. Schoof H, Lenhard M, Haecker A, Mayer KFX, Jurgens G, Laux T. 2000. The stem cell population of *Arabidopsis* shoot meristems is maintained by a regulatory loop between the *CLAVATA* and *WUSCHEL* genes. *Cell* 100:635–44

59. Sessions A, Yanofsky MF, Weigel D. 2000. Cell-cell signaling and movement by the floral transcription factors LEAFY and APETALA1. *Science* 289:779–81

60. Sieburth LE, Running MP, Meyerowitz EM. 1995. Genetic separation of third and fourth whorl functions of *AGAMOUS*. *Plant Cell* 7:1249–58

61. Steeves TA, Sussex IM. 1989. *Patterns in Plant Development*. New York: Cambridge Univ. Press

62. Stewart RN. 1978. Ontogeny of the primary body in chimeral forms of higher plants. In *The Clonal Basis of Development*, ed. S Subtelny, IM Sussex. New York: Academic

62a. Taguchi-Shiubara F, Yuan Z, Hake S, Jackson D. 2001. The *fasciated ear2* gene encodes a leucine-rich repeat receptor-like protein that regulates shoot meristem proliferation in maize. *Genes Dev.* 15:2755–66

63. Stone JM, Collinge MA, Smith RD, Horn MA, Walker JC. 1994. Interaction of a protein phosphatase with an Arabidopsis serine-threonine receptor kinase. *Science* 266:793–95

64. Stone JM, Trotochaud AE, Walker JC, Clark SE. 1998. Control of meristem development by CLAVATA1 receptor kinase and kinase-associated phosphatase interactions. *Plant Physiol.* 117:1217–25

65. Szymkowiak EJ, Sussex IM. 1992. The

internal meristem layer (L3) determines floral meristem size and carpel number in tomato periclinal chimeras. *Plant Cell* 4:1089–100

65a. The Arabidopsis Genome Initiative. 2000. Analysis of the genome sequence of the flowering plant *Arabidopsis thaliana*. *Nature* 408:796–814

66. Tilney-Bassett RAE. 1986. *Plant Chimeras*. London: E Arnold

67. Torii KU, Mitsukawa N, Oosumi T, Matsuura Y, Yokoyama R, et al. 1996. The Arabidopsis *ERECTA* gene encodes a putative receptor protein kinase with extracellular leucine-rich repeats. *Plant Cell* 8:735–46

68. Trotochaud A, Jeong S, Clark SE. 2000. CLAVATA3, a multimeric ligand for the CLAVATA1 receptor-kinase. *Science* 289:613–17

69. Trotochaud AE, Hao T, Wu G, Yang Z, Clark SE. 1999. The CLAVATA1 receptor-like kinase requires CLAVATA3 for its assembly into a signaling complex that includes KAPP and a Rho-related protein. *Plant Cell* 11:393–405

70. Weigel D, Alvarez J, Smyth DR, Yanofsky MF, Meyerowitz EM. 1992. *LEAFY* controls floral meristem identity in *Arabidopsis*. *Cell* 69:843–59

71. Weigel D, Nilsson O. 1995. A developmental switch sufficient for flower initiation in diverse plants. *Nature* 377:495–500

72. Williams RW, Wilson JM, Meyerowitz EM. 1997. A possible role for kinase-associated protein phosphatase in the *Arabidopsis* CLAVATA1 signaling pathway. *Proc. Natl. Acad. Sci. USA* 94:10467–72

73. Yamamoto E, Karakaya HC, Knap HT. 2000. Molecular characterization of two soybean homologs of *Arabidopsis thaliana CLAVATA1* from the wild type and fasciation mutant. *Biochim. Biophys. Acta* 1491:333–40

74. Yanofsky MF, Ma H, Bowman JL, Drews GN, Feldmann KA, Meyerowitz EM. 1990. The protein encoded by the *Arabidopsis* homeotic gene *agamous* resembles transcription factors. *Nature* 346:35–39

75. Yu LP, Simon EJ, Trotochaud AE, Clark SE. 2000. *POLTERGEIST* functions to regulate meristem development downstream of the *CLAVATA* loci. *Development* 127:1661–70

Annu. Rev. Plant Biol. 2002. 53:67–107
DOI: 10.1146/annurev.arplant.53.091901.161540

NONSELECTIVE CATION CHANNELS IN PLANTS

Vadim Demidchik, Romola Jane Davenport, and Mark Tester

Department of Plant Sciences, University of Cambridge, Downing Street, Cambridge, CB2 3EA, United Kingdom; e-mail: vd211@cam.ac.uk, rjd23@cam.ac.uk, mat10@cam.ac.uk

Key Words   ion transport, ion channels, electrophysiology, glutamate receptors, cyclic nucleotide–gated channels

■ **Abstract**   Nonselective cation channels are a diverse group of ion channels characterized by their low discrimination between many essential and toxic cations. They are ubiquitous in plant tissues and are active in the plasma membrane, tonoplast, and other endomembranes. Members of this group are likely to function in low-affinity nutrient uptake, in distribution of cations within and between cells, and as plant Ca^{2+} channels. They are gated by diverse mechanisms, which can include voltage, cyclic nucleotides, glutamate, reactive oxygen species, and stretch. These channels dominate tonoplast cation transport, and the selectivity and gating mechanisms of tonoplast nonselective cation channels are comprehensively reviewed here. This review presents the first classification of plant nonselective cation channels and the first full description of nonselective cation channel candidate sequences in the Arabidopsis genome.

CONTENTS

1040-2519/02/0601-0067$14.00

INTRODUCTION

Cation channels are integral membrane proteins that form transmembrane pores to catalyze passive movements of cations across membranes. They play important roles in most plant processes, including nutrient acquisition and long-distance transport, osmotic regulation, intracellular transport, signaling, and development. K^+ channels are a major group of channels that are highly selective for K^+ over other ions and have been characterized in a wide variety of plant cells (72, 136, 154). However, over the past decade or so, increasing evidence has emerged for the existence of plant cation channels that have much lower selectivity among cations— these are termed nonselective cation channels (NSCCs). Their low selectivity poses a challenge for electrophysiological studies, as identification of the ionic species responsible for the observed currents can be difficult. In this article, the electrophysiological properties of this group of ion channels in higher plants are reviewed for the first time, as are their possible physiological roles and molecular identities.

DEFINITION OF NONSELECTIVE CATION CHANNELS

NSCCs in both plants and animals are a large, heterogeneous group of channels, precluding a simple comprehensive definition. Generally, they show a high selectivity for cations over anions, but a low selectivity among monovalent cations under a wide range of ionic conditions. However, they are not strictly nonselective—the use of this term for such channels in plant cells follows the terminology established within the animal literature for similar channels (144). These channels usually have a similar permeability to a wide range of monovalent cations, many of which are impermeable or poorly permeable through K^+- and Na^+-selective channels. They show $K^+:Na^+$ selectivity ratios between 0.3 and 3, although this is not exclusive. Nevertheless, NSCCs with a greater discrimination between K^+ and Na^+ discriminate poorly between other alkali metal cations. Some NSCCs are permeable to large cations such as $Tris^+$, TEA^+, and $choline^+$.

Some NSCCs are also permeable to inorganic divalent cations, but unlike highly selective Ca^{2+} channels in, for example, neurons, these NSCCs allow measurable permeation of monovalent cations even in the presence of millimolar concentrations of Ca^{2+}. NSCCs may also display low selectivity between different divalent cations, and some allow permeation of Mg^{2+}, an ion with a high dehydration energy that rarely permeates Ca^{2+}-selective channels.

Owing to the diversity of NSCCs, features such as unitary conductance, voltage dependence, agonist activation, Ca^{2+} inhibition, and sensitivity to organic inhibitors are not diagnostic, although these may prove useful for classification of different NSCCs. Nevertheless, a significant group of NSCCs reveals insensitivity to organic inhibitors of Ca^{2+}-selective and K^+-selective channels such as verapamil, nifedipine, and TEA^+.

Classification within the nonselective cation channels is problematic, owing to our current inability to assign particular currents to particular genes. Thus, a phenomenological taxonomy based on the currently described electrophysiological features may well be soon surpassed by a molecular taxonomy. Therefore, in this review, NSCCs are divided into the following categories, based on current electrophysiological measurements:

1. plasma membrane,
2. tonoplast,
3. mitochondria, plastids, and other endomembranes.

NSCCs in these membranes can be classified on the basis of features such as

1. voltage sensitivity,
2. activation by cytosolic Ca^{2+},
3. activation by amino acids, notably glutamate,
4. mechanosensitivity.

These categories are not exclusive but should be nested, using further features as necessary—for example, plasma membrane time-dependent hyperpolarization-activated channels may be divided on the basis of their sensitivity to reactive oxygen species. It remains to be seen whether the differences between these groups are due to small numbers of base changes or reflect evolutionarily distinct lineages of ion channels.

Although the molecular identities of the NSCCs considered in this review currently remain obscure, we only consider transporters containing α-helices and not consider porin-like channels. In eukaryotes, porin-type channels containing β-pleated sheets are restricted to the outer membranes of mitochondria and chloroplasts [(152); although one report of porins in the plasma membrane of human B lymphocytes should be noted (57)]. We also do not consider plasmodesmata (58, 111), annexins (82), and channels formed by heat shock proteins (11), thionins, and defensins (18).

EXPERIMENTAL CHARACTERIZATION OF NONSELECTIVE CATION CHANNELS

NSCCs can be studied using standard methods for studying ion channels, such as electrophysiological techniques; radioactive tracer flux measurements; changes in intracellular cation activities; and molecular biological, and biochemical techniques (155, 170).

Electrophysiological techniques allow direct examination of currents passing through the channel with high time resolution, while also allowing various levels of control of the solution composition and voltage. These techniques provide information on conductance, selectivity, voltage dependence, kinetics, pharmacology, and regulation of ion channels at whole-cell and single channel levels. Modeling based on electrophysiological data can predict some features of the channel structure and essential principles of channel functioning.

To facilitate electrophysiological study of NSCCs at the level of the whole cell or vacuole, the activity of other transporters should be minimized. This can be done in various ways. In patch-clamp experiments (where H^+-ATPase activity is often negligible), K^+ and anion channel conductances can be minimized by substitution of permeant inorganic anions (e.g., Cl^-, SO_4^{2-}, NO_3^-) for poorly permeant organic anions (e.g., gluconate$^-$, citrate^{2-}) and by application of inhibitors that neither inhibit NSCCs nor permeate the channel being blocked (e.g., millimolar Cs^+ or TEA^+ for most plant K^+ channels). However, caution should be used when applying some inhibitors—for example, channel-like nonselective ion conductances can be induced in plant plasma membranes by application of verapamil (12) and quinine (see Figure 2 in 153). There are no known specific blocking agents for NSCCs, nor are there likely to be any, given the diversity of this class of channel.

Exclusion of ATP from the pipette can also be an effective method for reducing selective membrane conductances (such as that due to the activity of K^+ and anion channels) to facilitate the study of NSCCs (6, 43). Of course, the effect of ATP on the NSCC being studied should also be monitored because cytosolic ATP inhibits some classes of NSCCs in animal cells (144). F. Lemtiri-Chlieh (unpublished results) found that cytosolic Mg^{2+} inhibited NSCCs in guard cell plasma membrane, so exclusion of this may also be necessary to observe NSCCs.

Separation of NSCC currents from "leaks" between the electrode glass and membrane can also be problematic, especially if the NSCC currents activate rapidly (e.g., see Voltage-Insensitive NSCCs). Besides ensuring that whole-cell characteristics are correlated with those of single channels, a useful device is to ensure the currents are protein mediated. For example, micromolar concentrations of the histidine modifier, diethylpyrocarbonate (DEPC), is an effective inhibitor of NSCCs (43; P. Essah, R. J. Davenport & M. Tester, unpublished results). Incorporation of channels into planar lipid bilayers has also proven useful for studying rapidly activating NSCCs.

Standard electrophysiological approaches (e.g., measuring changes in reversal potentials of currents with changes in ion composition) can provide data on selectivity that are difficult to interpret (e.g., see Slow-Activating NSCCs). The conductance of particular ions in complex solutions cannot easily be predicted but is critical to the physiological role of NSCCs. To try to address these problems, novel techniques are required to try to measure ion permeation in complex solutions. For example, to measure Ca^{2+} permeation through NSCCs, a series of channel-aequorin fusions are being used to report rises in Ca^{2+} adjacent to the channel pore that are due to Ca^{2+} entry through the channel pore (R. O'Mahoney & M. Tester, unpublished results). To measure simultaneously the permeation of several ions through NSCCs, vibrating extracellular multiple-barrelled ion-selective electrodes could be used to measure gradients due to net fluxes of the ions across the membrane (115, 159) of cells overexpressing particular channel subunits. Such approaches should enable identification of the physiologically relevant permeation of nonselective channels, thus enabling attribution of function (e.g., Ca^{2+} channel, Na^+ channel) to specific gene products.

Cation fluxes can be measured directly using radioactive tracers and can be deduced by measuring changes in ion activities using ion-selective electrodes and luminescent and fluorescent techniques (63). Correlation of channel properties measured electrophysiologically with properties in less-perturbed systems can provide indications of the significance of the role of a particular channel in the intact plant (e.g., 38, 107).

These cellular techniques must be complemented by molecular approaches. In particular, analysis of the Arabidopsis genome has revealed at least two classes of ion channels (cyclic nucleotide–gated channels and glutamate receptors) that had not been predicted by electrophysiological analyses. By analogy with animal cells, these channels could be NSCCs (see Cyclic Nucleotide–Gated NSCCs and Glutamate-Activated NSCCs below). Use of both heterologous expression systems, such as yeast, Xenopus oocytes, and insect cells, as well as homologous misexpression, will, when combined with cellular techniques, enable in-depth studies of the selectivity, control, and physiological role of channels encoded by a wide range of plant genes.

EARLY OBSERVATIONS

It is notable that many early electrophysiological studies of ion channels identified nonselective cation channels in both animals (62) and plants (14, 148), but their unprepossessing characteristics such as "leaks" or "short-circuits" often caused them to be sidelined by more "attractive" channels showing high selectivity and/or time-dependent gating. Even when some channels were shown to be clearly nonselective, the desire of plant scientists to "force" channels into one of the three classical classes (K^+, Ca^{2+}, Cl^-) led to physiological assignments of channels on the basis of rather sketchy evidence (154).

The first paper explicitly acknowledging the existence of nonselective cation channels in plants was by Stoeckel & Takeda (150), who patch clamped the plasma membrane of triploid endosperm cells. However it has only been in the past five years or so that nonselective cation channels have been recognized as a group of ion channels in plants distinct from K^+-selective or Ca^{2+}-selective channels and their importance in plant physiology recognized (7, 38, 146, 160, 174, 185). Nevertheless, many measurements of their activity have been made, some of which are presented here, with a view to illustrating their abundance, diversity, and potential physiological roles.

PLASMA MEMBRANE NSCCs

Depolarization-Activated NSCCs

Currents exhibiting properties consistent with the activity of depolarization-activated NSCCs have been measured several times (see Table 1). In a single-channel study on protoplasts derived from *Arabidopsis thaliana* leaf mesophyll cells, Spalding et al. (149) described outwardly rectifying currents that selected weakly between monovalent cations. Slowly activating outwardly rectifying monovalent cation currents were also observed in Arabidopsis tissue-cultured cells (25).

By incorporating plasma membrane–enriched vesicles from roots of rye (*Secale cereale*) into planar lipid bilayers, White (180–182, 184) and White & Ridout (190) measured two depolarization-activated NSCC types with high permeability to Ca^{2+} ($P_K:P_{Ca} \sim 2.0$). Voltage dependence of the lower conductance channel (Table 1) shifted in parallel with changes in the reversal potential of the channel, and this channel was sensitive to micromolar concentrations of various inhibitors of Ca^{2+}-selective channels. A similar channel from wheat roots was studied using PLB techniques by Piñeros & Tester and coworkers (122–124, 188a). Both NSCC types have been proposed to function as depolarization-activated Ca^{2+} channels, underlying whole-cell depolarization-activated Ca^{2+} currents (156) and $^{45}Ca^{2+}$ fluxes into plasma membrane–enriched vesicles (79, 108). These channels would allow simultaneous efflux of K^+ and thus prevent excessive depolarization of E_m during Ca^{2+} influx (180). However, the physiological importance of Ca^{2+} influx through depolarization-activated channels remains obscure, although their control by cytoskeletal organization (155b) remains a tantalizing observation.

Slowly activating depolarization-activated channels with weak selectivity for cations over anions have been observed in xylem parenchyma (40, 174, 176) and Phaseolus seed coat (W. H. Zhang, M. Skerrett, N. A. Walker, J. W. Patrick & S. D. Tyerman, unpublished results). Fast-activating currents of low selectivity were also observed in seed coats (196; W. H. Zhang, M. Skerrett, N. A. Walker, J. W. Patrick & S. D. Tyerman, unpublished results), and single channels permeable to Ca^{2+}, K^+, and Cl^- were recorded at very positive voltages in Arabidopsis guard cells (120). The low selectivity and cell specificity of these channels suggests that they function in high volume release of solutes during turgor adjustment or nutrient transport. However their anomalous anion permeability makes it unlikely that they are NSCCs (which are characterized by high cation selectivity).

TABLE 1 Properties of NSCCs in the plant plasma membrane

Preparation	Selectivity	Pharmacology and regulation[a]	Unitary conductance[b]	Kinetics and other properties[c]	References
Depolarization-activated					
A. thaliana leaf mesophyll	P_K (1.00) ≥ P_{Na} (0.84)	I: cytosolic Ca^{2+} S: SH-reduction	109 pS (250 mM K^+); 92 pS (250 mM Na^+)	Ø: occurred in 20% of patches	(149)
A. thaliana guard cells	P_{Cs} ≈ P_{Cl} ≈ P_{Ca}	Ø	Ø	Ø; single-channel activity in whole-cell recordings	(120)
A. thaliana cell culture	P_K (1.00) = P_{NH4} (1.00) > P_{Na} (0.43) > P_{Li} (0.36) > P_{Cs} (0.17) P_{Li} < P_{Cl}	Ø	Ø	SA	(25)
Hordeum vulgare root xylem	P_K ≈ P_{Na} ≈ P_{Rb} ≈ P_{Cs}>	Ø	14, 26, 36, and 97 pS (120 mM K^+)	SA; ins. to TEA^+	(40, 174, 176)
Phaseolus vulgaris seed coats	P_K (1.00) > P_{NH4} (0.75) ≈ P_{Na} (0.72) ≈ P_{Cs} > P_{TEA} (0.61) ≈ $P_{choline}$ < P_{Cl}	I: TEA^+, Gd^{3+}, La^{3+}	Ø	RA and SA; ins. to TEA^+, Ca^{2+}, Cs^+, flufenamate	(196; Zhang et al., unpubl.)
Secale cereale root	P_{Ba} > P_{Sr} > P_{Mn} > P_{Mg} > P_{Ca} (2.01) > P_{Co} > P_K (1.00) > P_{Rb} > P_{Cs} > P_{Na} (0.78) > P_{Li}	I: TEA^+, quinine, ruthenium red, diltiazem, verapamil. S: ATP	451 pS (100 mM K^+); 278 pS (100 mM Na^+); 135 pS (100 mM Ca^{2+})	PLB; ins. to nifedipine	(180, 182, 184, 190)
Secale cereale root	P_{Ca} (2.60) > P_{Ba} (1.66) > P_K (1.00)	Ø	174 pS (100 mM K^+); 98 pS (100 mM Na^+); 40 pS (100 mM Ca^{2+})	PLB	(181)
Triticum aestivum root	P_{Cu} (6.67) > P_{Cu} > P_{Co} > P_{Mn} > P_{Ni} ≈ P_{Mg} > P_{Zn} > P_{Ca} (1.00) > P_{Sr} > P_{Ba} (0.61) ≫ P_K (0.06–0.02)	I: La^{3+}, Al^{3+}, Gd^{3+}, ruthenium red, diltiazem, verapamil. S: ATP	30.5 pS (10 mM Ca^{2+})	PLB; ins. to bepridil, 1,4 dihydropyridines, ABA, IP_3	(122–124, 188a)
Hyperpolarization-activated					
A. thaliana guard cells	P_K ≈ P_{Ca} ≈ P_{Ba} ≈ P_{Mg} ≫ P_{TEA}	I: verapamil, La^{3+} S: H_2O_2, ABA	Ø	Ø	(119)
A. thaliana root	P_{Ba} > P_{Cu} (1.00) ≈ P_{Mg} > P_{Mn} (0.10) ≫ P_K(0.07)	I: La^{3+}, Al^{3+}, Gd^{3+}, nifedipine S: cytosolic Ca^{2+}	22 pS (10 mM Ca^{2+})	SA; ins. to increasing cytosolic Mg^{2+}	(167)

(Continued)

TABLE 1 (*Continued*)

Preparation	Selectivity	Pharmacology and regulation[a]	Unitary conductance[b]	Kinetics and other properties[c]	References
Hordeum vulgare root xylem	P_K (1.00) > P_{Cs} (0.43) ≈ P_{Rb} (0.41) > P_{Li} (0.23) ≈ P_{Na} (0.20)	I: TEA$^+$, Ca^{2+}, Ba^{2+}, La^{3+}; S: G protein	8.3 pS (100 mM K$^+$) or 2 pS (120 mM K$^+$)	SA; ins. to increasing cytosolic Ca^{2+}	(40,174–176)
Phaseolus vulgaris seed coats	P_{NH4} (1.2) > P_K (1.00) > P_{Na} (0.82) ≈ P_{Cs} (0.80) > P_{TEA} (0.53) ≈ $P_{choline}$ (0.51)	I: Ca^{2+}, Gd^{3+}, La^{3+}	Ø	RA; insensitive to TEA$^+$, Ba^{2+}, Cs$^+$, flufenamate	(Zhang et al., unpubl.)
Triticum aestivum root	P_K ≈ P_{Na} ≈ P_{Cs} Ca^{2+}-permeable	I: TEA$^+$, choline$^+$	Ø	Ø; appeared as "spiky" currents at V < −140 mV	(161)
Vicia faba guard cells	P_K/P_{Ca} = 0.5 or 3.3 at apoplastic 1 or 10 mM Ca^{2+} respectively	Ø	Ø	SA; Na$^+$ and Mg^{2+} are impermeable	(61)
Vicia faba guard cells	P_{Cs} ≈ P_{Ca}		54 pS (30 mM Cs$^+$)	Ø	(140)
Voltage-insensitive					
A. thaliana root	P_K (1.49) > P_{NH4} (1.24) > P_{Rb} (1.15) ≈ P_{Cs} (1.10) ≈ P_{Na} (1.00) > P_{Li} (0.73) > P_{TEA} (0.47)	I: H$^+$, Ca^{2+}, Ba^{2+}, Zn^{2+}, La^{3+}, Gd^{3+}, quinine, DEPC	10.5 pS (50 mM Na$^+$)	RA; ins. to TEA$^+$, nifedipine, verapamil, flufenamate, amiloride	(43)
Aster guard cells	P_K ≈ P_{Na} ≈ P_{Cs}	I: Ca^{2+} S: internal NaCl	Ø	RA; ins. to Cs$^+$, single-channel activity in whole-cell recordings	(168)
Nitella flexilis internodes	P_K (1.0) > P_{Na} (0.7) ≈ P_{Cs} (0.7) ≈ P_{Li} (0.7) > P_{Ca} (0.3)	I: excess NaCl and temperature stress S: Cu^{2+}	Ø	RA; ins. to H$^+$, Ca^{2+}, Ba^{2+}, TEA$^+$, Cs$^+$, nifedipine, verapamil	(42, 44, 146)
Petroselinum crispum culture	P_{Cs}/P_K = 0.16	I: La^{3+}, Gd^{3+} S: elicitors	245 pS (50 mM Ca^{2+})	Ø; single-channel activities in whole-cell recordings	(198)
Secale cereale root	P_{NH4} > P_K (1.00) = P_{Rb} > P_{Cs} > P_{Na} (0.73) > P_{Li} ≫ P_{TEA} (0.30)	I: quinine	38.5 pS (100 mM K$^+$) 23.0 pS (100 mM Na$^+$)	PLB; ins. to TEA$^+$, Ba^{2+}, Ca^{2+}, charybdotoxin	(183, 191)
Secale cereale root	P_{Na}/P_K = 3.8–9.9 P_{Ca}/P_K = 2.2	I: Ca^{2+}, Ba^{2+}, TEA$^+$, Cs$^+$, quinine	Ø	RA	(188)

	Selectivity[b]	Modulators[a]	Conductance[b]	Notes[c]	Ref.
Secale cereale callus	$P_K (1.0) > P_{Na} (0.8)$	S: cytosolic Mg-ATP	8 pS (100 mM Na$^+$); 7 pS (100 mM K$^+$)	RA; cytosolic Mg-ATP induced rapid run-down	(6)
Triticum aestivum root	$P_{NH4} (2.06) > P_{Rb} > P_K (1.23) > P_{Cs} > P_{Na} (1.00) > P_{Li} \gg P_{TEA} > P_{Ca} (0.21)$	I: Ca^{2+}, Mg^{2+}, Gd^{3+}, flufenamate	44 pS (100 mM Na$^+$)	PLB; ins. to TEA$^+$, verapamil, MgATP, cAMP, cGMP, amiloride, glutamate, spermine	(38)
Triticum aestivum root	$P_K (2.6) > P_{Na} (1.0) > P_{Cs} (0.12)$	I: Ca^{2+}, adequate K$^+$ supplement	Ø	RA; ins. to Cs$^+$	(22)
Triticum aestivum root cortex	$P_K (1.00) > P_{Na} (0.80)$	I: Ca^{2+}	~30pS (100 mM Na$^+$)	RA	(161)
Zea mays root cortex Ca^{2+}-activated	$P_{Na} (2.1) > P_K (1.00)$	I: Ca^{2+}	15 pS (102 mM Na$^+$)	RA; ins. to TTX, TEA$^+$, Cs$^+$	(131)
Haemanthus, Clivia endosperm	$P_K \approx P_{Li} \approx P_{Na} \approx P_{Cs}$	S: Ba^{2+}	34.4 pS (100 mM K$^+$)	SA; depolarization-activated; ins. to TEA$^+$	(150)
Pisum sativum leaf epidermis	$P_{Na} (5.12) > P_{Li} (2.03) > P_K (1.00)$	Ø	35 pS (100 mM KCl)	Ø; depolarization-activated; ins. to Ba^{2+}, TEA$^+$	(59)
Glutamate-activated *A. thaliana* root	$P_K \approx P_{Na} \approx P_{Cs}$ permeable for Ca^{2+}	I: La^{3+}, Gd^{3+}, quinine	Ø	RA	(Demidchik et al., submitted)
Mechanosensitive *Allium cepa* bulb epidermis	$P_{Ca} > P_K$	I: La^{3+}, Gd^{3+} S: herbicide	Ø	Ø	(46, 47)

[a] I, inhibition; S, stimulation; Ø, no details in cited references.

[b] Cation concentration of "extracellular" solution in parentheses.

[c] RA, rapid activation; SA, slow activation; PLB, characterized in PLB; ins, insensitive.

We have not included outward-rectifying channels with relatively high K:Na selectivity (e.g., 130, 132, 174). However some KORs are Ca^{2+}-permeable and may bear a close molecular relationship to more nonselective channels (see Molecular Characterization of NSCCs in Plants).

Hyperpolarization-Activated NSCCs

These channels form a heterogeneous group with varying properties (Table 1). As with the depolarization-activated NSCCs, the hyperpolarization-activated group can show either fast or slow activation kinetics, over timescales of milliseconds to seconds. Reports of cation selectivity vary, and monovalent cation selectivity has not always been tested. The single channels underlying these currents show a wide range of conductances. It remains to be tested whether all plant hyperpolarization-activated cation channels are NSCCs.

Hyperpolarization-activated Ca^{2+} currents have been recorded in guard cells of several species (61, 71, 119, 140). The selectivity of these currents varies. In *Vicia faba*, slowly activating currents were Mg^{2+}-impermeable but relatively K^+-permeable (61), whereas single channels of low conductance (12.8 pS) were reported with high selectivity for Ca^{2+} over K^+ (71) as well as channels of higher conductance (54 pS) permeable to both Cs^+ and Ca^{2+} (120, 140). In Arabidopsis guard cells, hyperpolarization-activated currents were equally selective for Ca^{2+} and Mg^{2+} (119). Hyperpolarization-activated Ca^{2+} currents are activated by ABA, possibly via a direct interaction of channel and ABA (71) and by reactive oxygen species (119). The role of these channel types in guard cell signal transduction is becoming well established and has been reviewed recently in this series (138).

Hyperpolarization-activated divalent cation-selective (Mg^{2+}-permeable) channels have been reported in Arabidopsis root hair plasma membrane accessed by laser ablation of the cell wall (166, 167) as well as in GFP-labeled protoplasts isolated from the elongation zone of Arabidopsis roots (86a). Ca^{2+}-permeable, instantaneous, hyperpolarization-activated NSCCs were described in wheat root protoplasts and attributed to high-conductance channels (161; S. D. Tyerman & R. J. Davenport, unpublished). In Phaseolus seed coats, rapidly activating inward currents were nonselective for monovalent cations and were attributed to a channel of 22 pS (in 10 mM KCl) (W. H. Zhang, M. Skerrett, N. A. Walker, J. W. Patrick & S. D. Tyerman, unpublished). In barley xylem parenchyma, a time-dependent monovalent cation–permeable conductance was reported to be activated by cytosolic GTP and GMP (40, 174–176). Some of these currents may be involved in Ca^{2+} influxes related to signaling (167), but they have also been proposed to play a role in the uptake of nutrients and, perhaps also, toxic radioactive isotopes such as $^{137}Cs^+$ and $^{90}Sr^{2+}$ (147, 187).

Hyperpolarization-activated NSCCs may also be involved in transducing plant responses to elicitors, stimuli that trigger any plant response to pathogens (56, 80). Addition of partially purified intracellular fluid derived from tissue of tomato leaf infected with the avirulent (i.e., hypersensitive response inducing) race of the fungus, *Cladosporium fulvum*, activates hyperpolarization-activated Ca^{2+}-permeable channels in the plasma membrane of protoplasts of *Lycopersicon esculentum* (69, 70). Although a detailed study of the channel selectivity was not made, the channel characteristics are reminiscent of those of the slowly activating hyperpolarization-activated NSCCs described above. Activation was slow, and the response to elicitors may be mediated by a heterotrimeric G-protein-dependent phosphorylation of the channel protein (70).

Voltage-Insensitive NSCCs

Voltage-insensitive (or, at most, weakly voltage-sensitive) NSCCs appear to form a significant pathway for Na^+ influx from the soil solution into roots and have therefore been studied in the context of salinity tolerance. Such channels have been measured in several species, including the roots of rye (183, 189, 191), maize (131), wheat (22, 38, 113, 161), and Arabidopsis (43) as well as barley suspension cells (6) (Table 1).

These channels are partially inhibited by extracellular Ca^{2+}, and it is likely that this reflects at least some of the Ca^{2+}-sensitive component of Na^+ influx observed in intact roots (6, 38, 160, 185). These channels also allow large movements of NH_4^+, and they may play a role in the intact plant for the uptake of NH_4^+ (38, 183). The wheat NSCC channel observed in PLBs was also permeable to Ca^{2+} (38), and it has been suggested that this channel type could also play a role in store-operated Ca^{2+} influx to replenish depleted intracellular Ca^{2+} stores (P. J. White, unpublished).

In *Aster* guard cells, Véry et al. (168) described a rapidly activated, voltage-independent current and a 70-pS Na^+-permeable channel. This channel was approximately 80% blocked when external Ca^{2+} was increased from 0.05 to 1 mM but was not blocked by Cs^+. In the halophytic *Aster tripolium*, whole-cell Na^+ conductance was 4–5 times larger than that in the nonhalophytic *Aster amellus*, suggesting a role for these channels in sensitivity of guard cells to salinity.

In the charophyte alga, *Nitella flexilis*, fast-activating, voltage-insensitive cation currents were also found (44, 146). These currents showed equal permeability to a range of monovalent cations and $P_{Ca}:P_K \sim 0.3$. Measurements of fluxes using radioactive isotopes showed that NSCCs were responsible for a major proportion of total influx of both monovalent and divalent cations. The *Nitella* NSCC is probably activated by reactive oxygen species (42, 44) and is inhibited by long-term exposure of cells to salinity or elevated temperatures (43a).

Voltage-insensitive NSCCs may also be involved in the response of some cells to elicitors (see Hyperpolarization-Activated NSCCs). Exposure of protoplasts from parsley cells to nanomolar concentrations of the oligopeptide elicitor P-13 derived from a cell wall protein of *Phytophthora soja* reversibly activated NSCCs in 70% of protoplasts tested (198). These channels appeared to be voltage-independent, at least for the voltage range tested in this work. Single-channel events could be resolved in whole-cell mode, with openings often lasting for some hundred milliseconds or even seconds. The channels had a $P_{Ca}:P_K$ of 0.16 with 240 mM external $CaCl_2$, increasing at decreasing extracellular Ca^{2+} concentrations. However no channel activation was found in outside-out configuration, and there was a delay in channel activation of 2–5 min after addition of elicitor, suggesting that the elicitor did not activate NSCCs directly but through components mediating signal transfer between the elicitor receptor and the ion channel. Results with intact cells, obtained using ion-selective electrodes and measurements of $^{45}Ca^{2+}$ influx, are consistent with those obtained by the more intrusive electrophysiological techniques (84).

Ca^{2+}-Activated NSCCs

Ca^{2+}-activated NSCCs constitute a very large class of animal NSCCs, although the molecular identity of these channels is unknown. There have been several reports of Ca^{2+} activation of plasma membrane NSCCs in a range of plant tissues (Table 1), however we include in this section only NSCCs for which a direct effect of Ca^{2+} on the channel was claimed. Ca^{2+}-activated NSCCs that activated upon depolarization were measured in endosperm cells of species of *Haemanthus* and *Clivia* (150) and in epidermal cells of expanding pea leaves (59). In other cases, the effect of cytosolic Ca^{2+} was only tested in whole-cell configuration (e.g., 139) or was shown to be absent at the single-channel level (e.g., 174). The physiological role of Ca^{2+}-activated channels remains to be investigated.

Cyclic Nucleotide–Gated NSCCs

There are 20 CNGCs-like genes in the Arabidopsis genome (see Cyclic Nucleotide–Gated NSCCs), but until recently there was only indirect evidence for the existence of cyclic nucleotide–sensitive transport. Cyclic nucleotides appeared to stimulate Ca^{2+} influx in cultured carrot cells (97) and inhibit Na^+ influx in Arabidopsis roots (107a; P. A. Essah, R. J. Davenport & M. Tester, unpublished data). In protoplasts isolated from the root tips of an Al^{3+}-tolerant line of wheat (but not in an Al^{3+}-sensitive near isogenic line), Al^{3+} stimulated activity of a slow-activating K^+-selective cation channel only in the presence of cytosolic cAMP (195). The mechanism of interaction between cAMP and the channels carrying the current was not determined. A direct interaction could reflect the function of putative cyclic nucleotide–binding domains in plant K^+ channels or indicate a contribution of CNGC gene family members to time-dependent K^+ currents. A direct effect of cyclic nucleotides on ion channels has been demonstrated recently in Arabidopsis (107a). In protoplasts isolated from Arabidopsis roots, addition of micromolar concentrations of cAMP or cGMP to the cytosolic face of plasma membrane patches caused a rapid (frequently within seconds) decrease (by at least 10%) in the open probability of Na^+-permeable voltage-independent NSCCs. In about one third of protoplasts patch clamped in whole-cell mode, addition of membrane-permeable dibutyryl cyclic nucleotides inhibited NSCC currents. These results are consistent with at least some of the NSCC-mediated Na^+ influx into Arabidopsis roots being catalyzed by one or more CNGCs.

Glutamate-Activated NSCCs

The activation of NSCCs (*iGLRs*) by glutamate and other amino acids is a key event underlying excitatory synaptic transmission and a number of other complex physiological phenomena in animal cells (48). Addition of glutamate to intact Arabidopsis seedlings caused increases in cytosolic Ca^{2+} activity, as measured by cytosolically targeted aequorin (45). Evidence was consistent with Ca^{2+} having entered across the plasma membrane, and the kinetics of the Ca^{2+} increases were consistent with a direct activation of cation channels by glutamate.

Similar results have also been recently obtained with protoplasts from Arabidopsis roots, work complemented by patch-clamp electrophysiological measurements of glutamate activation of cation currents (V. Demidchik, P. A. Essah & M. Tester, submitted) (Table 1). Low millimolar concentrations of Na-glutamate activated voltage-insensitive NSCCs, which displayed similar permeability to K^+, Na^+, and Cs^+ as well as allowed the permeation of significant amounts of Ca^{2+}. The probability of observing glutamate-activated currents increased with increasing glutamate concentration (up to 29% at 3 mM); half-maximal activation was seen at 0.2 to 0.5 mM glutamate. Glutamate-activated NSCC was inhibited by millimolar quinine and micromolar La^{3+} or Gd^{3+}. The probability and the amplitude of the glutamate-activated currents positively correlated with the magnitude of the current occurring before addition of glutamate, i.e., cells having larger inward Na^+ currents before glutamate addition had a greater probability of being activated by glutamate and had a larger glutamate-activated current. This suggests that glutamate-activated currents can also be responsible for some background current occurring prior to addition of glutamate. Glutamate also stimulated unidirectional influx of $^{22}Na^+$ into intact roots (by up to 25%).

Mechanosensitive NSCCs

Given the central importance of turgor and touch to the development and physiology of plants, it is remarkable that there has not been more work on the control of ion channels by alterations of cell volume and/or membrane tension. Changes in turgor are known to rapidly alter cytosolic Ca^{2+} activities, some of which are known to come from extracellular stores (93), and this effect can be cell-specific and induce oscillations in cytosolic Ca^{2+} (87). However, there are very few reports on mechanosensitive ion channels (Table 1). Cosgrove & Hedrich (33) reported a range of channels in guard cells, including two cation-selective channels. However, these channels were not studied in enough detail to enable further comment on their possible function in turgor sensing in guard cells. Ding & Pickard (46, 47) found a stretch-activated current in onion bulb epidermal cells that was permeable to both Ca^{2+} and K^+. The potential role of these channels in transduction of gravitropic, mechanical, or, even, low-temperature signals also remains untested.

TONOPLAST NSCCs

The vacuole is a compartment occupying most of the volume of most plant cells and having many important physiological functions, such as turgor maintenance, storage of useful and toxic substances, regulation of cytoplasmic ionic homeostasis, and signaling. Communication between vacuole and cytoplasm occurs via the vacuolar membrane—the tonoplast. Several classes of NSCCs have been described in the tonoplast, and in contrast to the plasma membrane, these are the best-studied group of ion channels in the tonoplast. There are two well-established NSCC classes in the tonoplast of terrestrial plants: so-called slow-activating (SV) and fast-activating (FV) channels. FV channels are active within 1 ms of a change in voltage at physiological cytosolic Ca^{2+} activities ($<0.3–1$ μM) and blocked by higher cytosolic Ca^{2+}, whereas SV channels activate slowly upon changes in voltage at much higher cytosolic Ca^{2+} ($>0.5–5$ μM).

Slow-Activating NSCCs

Since the first observations of these types of channel in barley leaf mesophyll protoplasts by Hedrich et al. in 1986 (73), SV channels have been found in a large number of plant cells. They appear to be a relatively homogenous group with similar properties in different tissues and species.

SELECTIVITY   In early studies, a selectivity sequence in vacuoles from storage cells of sugar beet taproots was measured as P_K (1.00) $\approx$ P_{Na} (1.00) $>$ $P_{acetate}$ (0.37) $\approx$ P_{NO3} (0.31) $\approx$ P_{malate} (0.20) $\approx$ P_{Cl} (0.17) (34, 75). Similar selectivities (in particular, $P_K \approx P_{Na}$) were measured in *Acer pseudoplatanus* culture (32), *Plantago media* and *Plantago maritima* roots (104), and seagrass, *Posidonia oceanica* (23). However, early reports of significant anion permeability appear to be in error (2, 112, 125, 126).

By studying permeation through single channels, Pottosin et al. (125, 126) have shown unambiguously that, when the monovalent cation to Ca^{2+} ratio is 1000 or 10,000 on both sides of the membrane, the channel is highly selective for K^+ over Cl^- (by about two orders of magnitude, as determined by reversal potentials), but as the monovalent cation to Ca^{2+} ratio is reduced, P_{Cl}:P_K for the channel increases. In the same study, Ca^{2+} or Mg^{2+} currents were shown to saturate at approximately 0.5 to 1 mM, whereas Na^+ or K^+ currents tended to saturate at approximately 200 to 300 mM. Divalent cations appear to move through the channel more slowly than monovalent cations; moreover, they also limit a rate of ion movement across the channel, blocking monovalent cation currents. In this respect, ion activities used in experiments must be taken into account when considering the physiological significance of previous measurements of selectivities.

Cytosolic TMA^+ did not permeate SV channels in sugar beet vacuoles, suggesting the diameter of the pore of the SV channel is less than 0.6 nM (65). Dobrovinskaya et al. (49, 50) found that millimolar $TRIS^+$, putrescine and quaternary ammonium ions, and high micromolar spermidine and spermine asymmetrically blocked SV channels from *Beta vulgaris* taproot vacuoles both from cytosolic and vacuolar sides. TEA^+ did not pass the selectivity filter, whereas other substances acted as permeable blockers. When substances were applied at the vacuolar side, block revealed a single binding site, whereas when they were applied at the cytoplasmic side, block developed owing to multiple binding of blockers inside the pore (49). The sensitivity of SV channels to quaternary ammonium ions (substances that are synthesized in plant cells under some stress conditions) suggests the potential involvement of SV channels in response to stress.

UNITARY CONDUCTANCE   In most cases, unitary conductances were in the range of 60–80 pS with 50 mM symmetrical K^+ salts (34, 73–75, 105). In symmetrical 100 mM KCl, SV channels from *Vigna unguiculata* revealed two unitary conductances of 98 and 25 pS, respectively (106). Under the same conditions using *Chenopodium rubrum* suspension cell tonoplast, Reifarthet et al. (129) measured a conductance of 83 pS, similar to the larger channel seen by Maathuis & Prins (106). However, at higher K^+ concentrations, channel conductance did not increase significantly [103 pS in 150 mM (65), 90 pS in 200 mM (2), 106 pS in 400 mM (23)], presumably owing to saturation of the channel binding sites. Differences in unitary conductances between workers are likely to be due to use of different concentrations of Ca^{2+}, Mg^{2+}, and $TRIS^+$ on both the vacuolar and the cytosolic sides of the tonoplast (see below). In a fully "nonblocked" state (with saturating cytosolic 50 μM Ca^{2+} and saturating K^+ or Na^+ of 0.1–0.3 M), Pottosin et al. (125) measured maximal unitary K^+ and Na^+ conductances of 385 and 305 pS, respectively. In the same study, unitary Mg^{2+} and Ca^{2+} conductances were measured as 18 and 13 pS, respectively (at saturating cytosolic Mg^{2+} or Ca^{2+} activities of 1–10 mM).

GATING AND KINETICS   It was originally found that all SV channels started to activate at membrane potentials positive of approximately +10 to +30 mV (cytosolic side positive) and that, at membrane potentials at which the tonoplast is generally thought to rest (approximately −30 to −50 mV), channel activity is very low. Furthermore, activation of the channel is slow, occurring over a time course of 0.5 to 5 s (2, 23, 34, 74, 75, 104, 105). However, it has been recently found that, at physiological cytosolic Mg^{2+} (submillimolar and low millimolar range), the activation curve shifts to hyperpolarized (physiological) potentials (24, 121, 125) and accelerates the speed of opening (24). Cytosolic Mg^{2+} also decreases the Ca^{2+} activity required for activation of the SV channel (4, 16, 24, 39, 121). Thus, the channel activity in the resting cell may be greater than originally thought.

Given the permeability of the channel to Ca^{2+}, it is possible that the channel acts as a pathway for Ca^{2+}-activated Ca^{2+} release (173, but see 128), similar to mechanisms generally accepted to occur in many animal endomembranes. Evidence for Mg^{2+}-induced sensitization of Ca^{2+}-induced Ca^{2+} release at positive potentials has also been obtained in experiments measuring $^{45}Ca^{2+}$ efflux from tonoplast vesicles (16).

Calmodulin appears to mediate Ca^{2+} activation of SV channels; although at physiological cytosolic Ca^{2+} activities, calmodulin has no effect on SV channels (3, 15, 141, 179). Interestingly, in *Vicia faba* guard cells, SV channels were modulated by Ca^{2+}-dependent, calmodulin-stimulated protein phosphatase 2B (calcineurin), which at low concentrations activated these channels, however at high concentrations it inhibited them (3). More studies in physiological conditions are needed to investigate further the role of calmodulin in Ca^{2+}-induced SV channel activation.

Another potentially important modulator of SV channel activity is pH, on both the cytosolic and vacuolar sides of the channel. In physiological conditions (submicromolar Ca^{2+} and 0.4 mM Mg^{2+}), alkalinization of the cytoplasm from pH 7.3 to pH 8.0 caused an approximately 10-fold increase in SV channel activity in *Beta vulgaris* taproot vacuoles (39). SV channels in guard cells also opened as the pH on both sides of the tonoplast became more alkaline (141). An increase in cytosolic pH accompanies ABA-induced stomatal closure (81), and it can be speculated that this increased pH could activate SV channels to provide Ca^{2+} influx from the vacuole and thus increase cytosolic Ca^{2+} activity.

As the vacuole contains the largest store of ions in the plant cell, the regulation of channel-mediated transtonoplast ion fluxes is important in ionic stress conditions (salinity, cold, heat, drought, and other abiotic stresses). Unfortunately, little is known about this regulation. Maathuis & Prins (104) compared the effects of salinity on open probability of SV channels from the roots of the salt-sensitive species, *Plantago media*, and its salt-tolerant relative, *Plantago maritima*. They found the probability of channel opening halved when plants were grown in 25 mM NaCl, whereas growth at 100 mM NaCl caused an 11-fold decrease in opening in *Plantago media* and complete suppression of the channel in *Plantago maritima*. At the same time, selectivity and amplitude of unitary conductance remained unchanged. The implication is that this downregulation of the channel is an adaptive response to high salinity, perhaps to minimize movement of Na^+ from the vacuole back into the cytosol. This raises the question of why the SV channel is in the tonoplast at such high densities in the first place.

The action of cytosolic $H_2PO_4^-$ and 14-3-3 proteins may also be physiologically relevant. Phosphate appeared to slow the rate of activation of the SV channel in vacuoles from *Beta vulgaris* taproots (51); and a 14-3-3 protein was shown to reduce SV channel–mediated currents in barley mesophyll cells (165). In whole-vacuole configuration, 14-3-3 protein (added to the cytosolic face) decreased outward currents by approximately 80% within 5 s after application. However, more work is required before the physiological significance of either of these observations becomes clear.

PHARMACOLOGY   In vacuoles from storage cells of sugar beet taproots, the SV channel was inhibited by a wide range of inhibitors of K^+-, Ca^{2+}-, and Cl^--selective channels (74, 104) (see above), including charybdotoxin, an inhibitor of animal Ca^{2+}-activated large-conductance K^+ channel (with an EC_{50} of ~20 nM) (177, 178).

Fast-Activating NSCCs

Fast-activating vacuolar (FV) channels are widespread (4, 19, 39, 75, 157), and as with SV channels, they are nonselective for cations. However, in contrast to SV channels, they are rapidly activated and are blocked by elevated cytosolic Ca^{2+}. Moreover, the unitary conductances of FV channels appear to be approximately five to ten times smaller than those of SV channels.

SELECTIVITY   In the first report on FV channels, Hedrich & Neher (75) showed the nonselectivity of these channels to monovalent cations. These findings were confirmed in later reports, such as in barley leaf mesophyll, where NH_4^+ was found to be ~25% more permeable than K^+, Rb^+, and Cs^+, which in turn are 20% to 30% more permeable than Na^+ and Li^+ (21). However, more recent reports suggest that the cation:anion selectivity could be much higher than originally reported (75), with recently measured values for P_K:P_{Cl} being 30 (157) and 150 (2). FV channels are impermeable for divalent cations (157), with micromolar Ca^{2+} and Mg^{2+} inhibiting permeation of monovalent cations (20).

UNITARY CONDUCTANCE   Being approximately 20 pS in symmetrical 100 mM K^+ in vacuoles from barley mesophyll cells and red beet storage tissue (50, 157), unitary conductances of FV channels are much smaller than unitary conductances of SV channels. In *Vicia faba* guard cell vacuoles, conductances were even smaller, being 6.4 pS in symmetrical 200 mM K^+ (2).

GATING AND KINETICS   The opening of FV channels changes rapidly in response to changes in voltage, over a time course of less than 1 ms (I.I. Pottosin, personal communication). In most preparations, both in whole-cell and single-channel recordings, FV channels show complex voltage dependence. Over physiological voltages, open probability is relatively constant and increases sharply at both very negative and very positive voltages (20, 21, 75, 157). However, in some studies, FV channels revealed linear I/V relations (39, 65). This discrepancy probably can be explained by different ionic conditions used in these studies (mostly by different applied Ca^{2+}, Mg^{2+}, and Cl^- activities in cytosol and vacuole). The key feature is that FV channels have a significant probability of being open in the intact resting cell.

Cytosolic and vacuolar Ca^{2+} and Mg^{2+} are probably the most important physiological regulators of FV channels. Increasing cytosolic Ca^{2+} blocks FV channels, inhibiting both probability of opening and current amplitude of FV channel with a half-maximal inhibition varying with different preparations [e.g., K_d of 6 μM in barley mesophyll (157) and 80 μM in red beet taproot (49)]. Vacuolar Ca^{2+} also inhibits inward currents (i.e., into the cytosol) through FV channels with a K_d of 14 μM in barley mesophyll cell vacuoles (157). Millimolar Ca^{2+} almost completely inhibited the inward current, apparently by reducing single-channel conductance but not open probability. In the same preparation, outward current was only slightly decreased (37%) by 2 mM vacuolar Ca^{2+}. This suggests that, in vivo, FV channels would mainly function to move ions into vacuoles and would not be involved in release of ions.

Brüggemann et al. (20) have found a voltage-dependent inhibition of FV channels by cytosolic Mg^{2+}, with a K_d of 10 μM at −60 mV. At physiological vacuolar potentials (from −20 to −80 mV) and physiological Mg^{2+} concentrations [from 0.1 to 1 mM (194)], FV channel–mediated currents were inhibited by over 95% (20). [Interestingly, cytosolic Mg^{2+} of 0.1 to 1 mM sensitizes SV channels (see above).] It is probable that, at physiological cytosolic Mg^{2+} (approximately 0.1 to 1 mM) and Ca^{2+} (approximately 100 to 500 nM), both FV and SV channels have low activity, consistent with the large differences in ion concentrations between the cytosol and vacuole. Significant changes in cytosolic Ca^{2+} and Mg^{2+} activities can probably activate them.

Many other parameters can also modulate FV channel activity. As observed for SV channels, increasing cytosolic pH (from 7.3 to 8.0) can activate FV channels (2, 39), although acidification (from 7.5 to 6.5) had no effect (20). FV channels can be activated by cytosolic ATP and ATPγS, although at acidic vacuolar pHs this effect was prevented (39). Micromolar concentrations of cytosolic polyamines (putrescine, spermidine, and spermine) inhibited FV channels in barley mesophyll cells (19), as observed for SV channels. As discussed above for SV channels, this sensitivity to polyamines may imply a role for FV channels in stress responses.

FV channels are significantly permeable to physiologically important NH_4^+ and K^+ as well as to the toxic cations, Na^+ and Cs^+. As inward currents into the cytosol are blocked by vacuolar Ca^{2+} (157), FV channels could provide a pathway for loading of monovalent cations into the vacuole. This efflux could be important for osmoregulation (including regulation of stomatal opening), maintenance of the tonoplast electrical potential, and storage of NH_4^+ and Na^+, although whether the electrochemical potential difference favors such passive movements for many of these ions in various conditions must be doubtful.

NSCCs IN OTHER ENDOMEMBRANES

There is a paucity of selectivity data on ion channels in other plant endomembranes, so it is difficult to identify whether particular channels are NSCCs. Nevertheless, investigators have characterized weakly voltage-dependent NSCCs in chloroplast thylakoids that upon illumination are likely to be involved in the efflux of cations, such as K^+ and Mg^{2+}, to balance charge build-up due to the pumping into the stromal lamella of H^+ (60, 127, 155a).

Endoplasmic Reticulum (ER)

The ER can act as a Ca^{2+} store, releasing Ca^{2+} as part of intracellular signaling pathways. Several studies using the planar lipid bilayer technique have revealed the existence of Ca^{2+}-permeable ion channels in the ER membranes of cells that respond to mechanical stimuli. Klüsener et al. (90, 91) characterized a channel from ER of the touch-sensitive tendrils of *Bryonia dioica*. A second channel was characterized from endomembranes (most likely ER) of root tips of garden cress (*Lepidium sativum*) (92). In each case, the Ca:K selectivity of the channels was low (6.6 and 9.4, respectively), suggesting that these channels are NSCCs. Gating was voltage dependent (inwardly rectifying) and controlled by the Ca^{2+} gradient across the ER membrane. The Bryonia channel was inhibited by verapamil and lanthanides, as well as H_2O_2. The Lepidium channel was sensitive to lanthanides, but not verapamil, and was strongly blocked by the Ca^{2+}-ATPase inhibitor erythrosin B, leading the authors to suggest that the channel could consist of the uncoupled transmembrane domain of a Ca^{2+}-ATPase (92).

Symbiosome Membranes

In nodulated legumes, the symbiotic N_2-fixing bacteria are contained within the plant cell cytosol in a symbiosome bounded by a plant-derived membrane. Exchange of fixed nitrogen (NH_3 and NH_4^+) and reduced carbon between the plant cytosol and the peribacteroid space occurs across this membrane. Patch-clamp studies of soybean symbiosomes indicate that the major route of NH_4^+ influx to the cytosol is via an NSCC of subpicosiemen conductance (162, 192). The channel appears to be intrinsically voltage insensitive but gated by divalent cations. Millimolar concentrations of Mg^{2+} or Ca^{2+} on either side of the membrane eliminated cation efflux from that side and caused influx to that side to become time dependent, suggesting that NH_4^+ currents are inwardly rectified in vivo by cytosolic Mg^{2+} blockade (192). Verapamil added to the bacteroid side inhibited cation influx to the plant cytosol with a K_d of 2.6 μM, but there was no evidence of divalent cation permeation of the channel (192).

MOLECULAR CHARACTERIZATION OF NSCCs IN PLANTS

Very few plant NSCC channels have been characterized at the molecular level. NSCCs are difficult to identify by complementation or mutant screens because few physiological functions have been ascribed and the low selectivity of the channels makes design of a specific screen difficult. Several NSCCs were identified by a functional assay for calmodulin binding (see Cyclic Nucleotide–Gated NSCCs), however the physiological characterization of most plant NSCCs is inadequate for this type of screen. Moreover, there is evidence that at least some NSCC gene candidates are toxic in *E. coli* (R. J. Davenport, unpublished data), which could preclude identification by expression cloning in some heterologous systems and vitiate the use of most cDNA libraries for cloning.

Identification of genes on the basis of similarity to animal NSCC sequences has been successful in identifying existing Arabidopsis ESTs as potential NSCC genes. However, animal NSCCs constitute a large and heterogeneous group, and most types remain relatively poorly characterized at both the electrophysiological and molecular levels, making design of homology-based gene isolation strategies difficult. There is also a lack of close fit between the electrophysiological characteristics of animal and plant NSCCs.

The sequencing of the complete Arabidopsis genome has revealed for the first time the full complement of *NSCC* gene candidates for a plant species. The genome is predicted to contain 30 members of the voltage-gated ion channel superfamily (including 20 cyclic nucleotide–gated channel-like sequences), 6 "double pore" putative K^+ channels, a putative NSCC-2-like channel, and 20 glutamate receptor-like genes (Figure 1, see color insert), as well as over 100 nonchannel cation transporters (110). Over 2000 other putative membrane protein sequences await classification (171, 172). The extent of the cyclic nucleotide–gated channel (*CNGC*)-like and glutamate receptor (*iGLR*)-like gene families is particularly surprising, given the relative paucity of electrophysiological evidence for ligand-gated ion channel activity. The discussion below focuses on cation channel-like sequences in Arabidopsis, with reference to other plant species and transporters where these have been characterized.

A Cautionary Note

Many animal ion channel subunits function in heteromeric assembly with like subunits and require additional nonhomologous subunits for native function. Therefore subunits within a family may show no transport function when expressed as homomeric proteins or may exhibit quite different selectivity, conductance, and gating from that observed in vivo. Subunits that usually function in highly selective ion channel complexes may also function as nonselective ion channels in other tissues or oligomeric associations, raising the possibility that ion channel families previously identified as highly selective channel types may contain NSCC members [some examples are the ENaC Na^+ channel alpha subunit (85) and the GIRK4 subunit of animal G-protein-coupled KIRs (145)]. There are also examples from the animal and fungal literature of functional channels assembling from association of channel-like subunits from different gene families. In *Saccharomyces cerevisiae*, CCH1, a subunit homologous to voltage-gated Ca^{2+} channel alpha subunits, and MID1, with a weaker similarity to other members of the voltage-gated superfamily, associate to form a Ca^{2+}-permeable channel (103, 117). In other cases, subunits assumed not to be channel forming on the basis of structure may function as ion channels and may even interact with more conventional ion channel subunits [e.g., the cystic fibrosis transmembrane conductance regulator (CFTR) (89)]. The possibility of channel assembly from subunits from different gene families both greatly increases the diversity to be expected and the difficulty of identifying those genes involved in NSCC activity. Hence we include K^+ channel sequences in this discussion of NSCC candidates and also refer briefly to transporter sequences that are not predicted to form ion channels.

The Voltage-Gated Ion Channel Superfamily

The voltage-gated ion channel superfamily comprises a number of cation channel families with a gradation from highly selective voltage-gated channels to ligand-gated NSCCs that can be insensitive to voltage (the cyclic nucleotide–gated channels, CNGCs). All members of the superfamily contain single or multiple repeats of a motif consisting of six transmembrane domains (S1-S6: 6TM), with a pore region between S5 and S6 (P or H5). Members include depolarization-activated K^+, Na^+, and Ca^{2+} channels (e.g., Shaker and the neuronal Na^+ and Ca^{2+} channels); depolarization-activated K^+ channels with cyclic nucleotide binding (CNB) motifs (ether-a-go-go/ERG/ELK family); plant inwardly rectifying (KAT and AKT families) and outwardly rectifying (SKOR, GORK) K^+ channels with CNB motifs; hyperpolarization-activated cyclic nucleotide-binding (HCN) cation channels of intermediate selectivity ($P_{Na}:P_K$ of 0.2 to 0.4); and the CNGCs (see Figure 1). In animal cells, members with a single 6TM domain (all the K^+ and CNGC channels) assemble as tetramers, whereas the depolarization-activated Ca^{2+} and Na^+ channels contain four repeats of the six transmembrane domain motif and a single subunit is sufficient for channel formation. Recently several genes have been identified encoding two repeats of the 6TM motif, providing an evolutionary link between the $1 \times 6TM$ and $4 \times 6TM$ families (see AtCCH1 below). Many of the ion channels also have additional nonhomologous subunits that modulate gating. Interestingly, all plant members of this superfamily except one, AtCCH1, have a $1 \times 6TM$ structure and a C-terminal CNB domain.

Domains affecting gating and selectivity have been characterized for a number of these cation channel families. Highly selective K^+ channels are characterized by a GYGD motif in the pore region, as well as conserved W and T residues (Figure 2, see color figure). Mutations of these residues can be sufficient to alter selectivity and to render the mutant channel cation nonselective or completely nonfunctional (26, 76). The first glycine (G) residue is generally conserved in cation channel pore domains (including Ca^{2+}-selective channels and the animal iGLRs and CNGCs). The S4 domain is implicated in voltage sensitivity and contains a regular repeat of positively charged residues in depolarization-activated channels. In most animal 6TM K^+ channels, the N terminus is involved in subunit coassembly, whereas in plant K^+ channels (and animal EAG-types) the C terminus is implicated (37).

CYCLIC NUCLEOTIDE–GATED CHANNELS   The plant CNGCs are similar in sequence to animal CNGCs and also to the EAG/ERG/ELK and HCN channels (Figure 2). CNGCs function in animals as Ca^{2+} and monovalent cation–permeable NSCCs in the visual and olfactory systems and are also expressed in other tissue types.

Cyclic nucleotide binding increases probability of opening, without desensitization. In rod photoreceptors, CNGCs mediate Na^+ influx and K^+ efflux in darkness to maintain a depolarized E_m. Extracellular Ca^{2+} both permeates and partially blocks these channels (as occurs in Na^+-permeable NSCCs in plant roots; see Voltage-Insensitive NSCCs). These channels retain several conserved residues of K^+ channel pores but have a GET motif instead of GYGD (Figure 2). The S4 region lacks the regular arrangement of cationic residues thought to confer voltage sensitivity in depolarization-activated channels. EAG/ERG/ELK K^+ channels have GFGN pore motifs as well as an S4 motif of regular cationic repeats more typical of the Shaker-type channels. The channels are highly selective for K^+ over Na^+ and are activated by depolarization, but some may also function as inward rectifiers [owing to tail currents or relief of depolarization-induced inactivation at negative voltages (197)]. HCN channels act as "pacemakers" in specific nerve, heart, and sensory cells and act in other cell types to control E_m and membrane conductance, by mediating mainly Na^+ influx upon hyperpolarization (86). They lack the stringent K^+ selectivity of depolarization-activated K^+ channels and have a GYG(A/Q/R/K) motif instead of GYGD in the pore region. The S4 domain contains a disrupted repeat of cationic residues that have been shown to dictate voltage sensitivity (28). The plant KAT and AKT channels have a similar S4 motif [although the N and C termini of these channels also affect voltage sensitivity (109)].

The plant *CNGC* genes are predicted to encode proteins with 6 TM and a pore region between S5 and S6. The pore region retains the conserved W and the first G of the GYGD motif but lacks the negatively charged residues (E, D) thought to interact with permeating cations. The exception to this is AtCNGC2, the best characterized of the proteins, which has an unusual ANDL pore motif (Figure 2). The pore and S6 domain constitute a single exon, with splice site positions similar to those of EAG/ERG/ELK channels (Figure 2). The putative S4 region of plant CNGCs contains an irregular sequence of cationic residues suggesting that, as in animal CNGCs, this region may not function as a voltage sensor (Figure 2). The C terminal portion contains a conserved cyclic nucleotide-binding domain motif, and this is partially overlapped by a CaM-binding domain. In animal CNGCs, CaM antagonizes the agonist effect of cyclic nucleotides on channel gating. However the animal channels have a CaM-binding domain near the N terminus on the alpha subunit, which is purported to act via conformational changes on the C terminal CNB domain.

The first *CNGC*-like genes in plants were isolated using calmodulin-binding assays of expressed cDNA libraries derived from barley aleurone tissue [HvCBT1 (142)] and tobacco leaf [NtCBP4 (10)]. Twenty *CNGC*-like genes have been identified in the Arabidopsis genome, of which eight have been identified as full-length cDNAs [designated AtCNGC1-6,10,20 (94, 110)]. Partial ESTs of *CNGC*-like genes exist for a number of plant species. The high number of *CNGC* genes in Arabidopsis, and the likelihood that these gene products may associate in various heteromeric arrangements, suggests that this gene family encodes a huge functional diversity of ion channels.

Of the plant *CNGC*-like genes, three have been demonstrated to encode proteins with ion transport properties. Xenopus oocytes expressing *AtCNGC1, 2* and *NtCBP4* were reported to demonstrate novel inwardly rectifying currents when pretreated with membrane-permeant cAMP and cGMP (101, 102). These channels appeared to be nonselective for monovalent cations, with the curious exception of Na^+, and were partially blocked by extracellular Ca^{2+} (101, 102). Cyclic AMP and cGMP were equally effective in stimulating inward current in cells expressing *AtCNGC2* (102), and cAMP was reported to interact directly with the channel (78). AtCNGC2 may also transport Ca^{2+} because AtCNGC2-transformed HEK cells loaded with fura-2 showed an increase in cytosolic Ca^{2+} after pretreatment with membrane-permeant cAMP and cGMP. Ca^{2+} increase could be blocked by Mg^{2+}, indicating that Ca^{2+} was entering the cytosol from the external medium (102).

NtCBP4 is implicated in heavy metal transport because overexpression increased Pb^{2+} sensitivity and accumulation, although it reduced Ni^{2+} accumulation and toxicity (10). Disruption of function of NtCPB4 (by overexpression of a truncated subunit in tobacco) and AtCNGC1 (in an Arabidopsis T-DNA insertion mutant) reduced Pb^{2+} accumulation and sensitivity (151). Both HvCBT1 and NtCBP4 are plasma membrane localized, supporting a role for CNGCs in cation influx (10, 142).

Other studies have failed to demonstrate a direct role for *CNGC*-like gene products in ion transport. When expressed in yeast, *HvCBT1* did not complement the yeast K^+ transport mutant CY162, and mutation of the pore motif from GQNL to GYGD actually inhibited yeast growth, suggesting that the gene could be interacting with and disrupting the function of endogenous transporter subunits (142). *AtCNGC1* and *AtCNGC2* partially complemented CY162 (94), but neither Kohler et al. nor Shuurink et al. used membrane-permeant cyclic nucleotides to activate the gene products. *AtCNGC2* did permit growth of CY162 at low (2 mM) K^+ in the presence of membrane permeant cAMP (102). The possibility remains that some plant CNGCs may be inhibited by cyclic nucleotides (107a; see Cyclic Nucleotide–Gated NSSCs) and activated by CaM/Ca^{2+} (9). Alternatively, some of the plant CNGCs may function like CNGC 6TM beta subunits in animals, some of which do not form functional homomeric channels but form functional heteromers with native channel alpha subunits (143).

Calmodulin binding has been demonstrated for several plant CNGC proteins. The CaM-binding regions of NtCBP4, AtCNGC1, and AtCNGC2 have been localized to a part of the putative CNB domain, suggesting that these channels may integrate messages from several signaling pathways via a direct competitive interaction between Ca^{2+}-CaM and cyclic nucleotides (8, 96). AtCNGC1 and AtCNGC2 show differing affinities for the different Arabidopsis CaM isoforms, and AtCNGC1 interacts only weakly with Ca^{2+}-CaM, providing evidence for functional diversity between the gene products (96).

Plant CNGCs are implicated in programmed cell death (PCD) during defense and senescence. Where demonstrated, tissue localization is consistent with PCD during senescence. *AtCNGC2* is expressed in mature cotyledons and in senescing floral and seed tissues (95). *HvCBT1* was isolated from a cDNA library prepared from barley aleurone (a tissue that also undergoes PCD during germination) (142), but additional tissue localization has not been demonstrated. AtCNGC2 is also involved in defense-related PCD because the *dnd1* (defense, no death) mutation, which results in failure to initiate hypersensitive cell death in response to pathogen attack, maps to a truncated *AtCNGC2* (31). Despite sequence similarity to voltage-insensitive Na^+-permeable animal CNGCs, the inward rectification and Ca^{2+} permeability associated with AtCNGC2, and the abundance of evidence implicating these genes in PCD, suggest that these genes could underlie the hyperpolarization-activated Ca^{2+} currents recently shown to be triggered in response to various stress signals, including the PCD trigger oxidative stress (see Hyperpolarization-Activated NSCCs). The novel selectivity of these channels and their high sequence similarity to plant 6TM K^+ channels suggest that they may be more closely related to K^+ than CNGC channels. These channels share many structural and sequence similarities with EAG/ERG/ELK-type K^+ channels and may have diverged from a 6TM CNB precursor by mutation of the pore region and modification of the CNB domain to accommodate competitive CaM binding.

AKT AND KAT FAMILIES   Two families of 6TM channels have been cloned from several plant species and shown to form highly selective monomeric K^+ channels in heterologous systems, with some sensitivity to inhibition or modulation by cyclic nucleotides (67, 77). The putative pore domain contains a GYGD motif (Figure 2). The channels are inwardly rectifying, the S4 domain contains a disrupted cationic repeat region similar to hyperpolarization-activated HCN channels in animal cells, and mutation of these residues reduces voltage sensitivity. KAT and AKT subunits assemble as heteromers in oocytes (13, 52), although not in cultured insect cells or yeast (164). Both channel types share C-terminal CNB domains with other plant 6TM channels (except AtCCH1), and this domain has been implicated in coassembly of subunits into heteromeric channels (37). Therefore there is a possibility of coassembly of K^+ channel subunits with other channel subunit types to form novel ion channels. The function of the six C-terminal ankyrin repeats in AKT proteins has not been demonstrated but raises the possibility that these proteins could function in mechanosensing.

SKOR AND GORK FAMILY   SKOR and GORK from Arabidopsis act as outwardly rectifying (OR) K^+-selective channels when expressed as monomers in heterologous systems (1, 68). They contain GYGD pore motifs, putative ankyrin-binding domains, and a disrupted series of cationic residues in the putative S4 domain almost identical to that of KAT and AKT proteins, despite the opposite voltage sensitivity of the two channel types (OR versus IR) (Figure 2). *SKOR* is expressed in root stelar cells and is Ca^{2+} permeable when expressed in oocytes. Therefore it is likely to mediate the OR currents of intermediate selectivity measured in protoplasts from stelar cells (68). Although these proteins seem unlikely to contribute to OR NSCC currents, it remains possible that SKOR and GORK associate in vivo with other subunits to create plasma membrane OR NSCCs.

AtCCH1   The Arabidopsis genome contains a single gene with similarity to depolarization-activated highly selective Ca^{2+} channels. *AtCCH1* (At4g03560, also known as AtTPC1) is predicted to encode a 2 × 6TM channel with strong similarity to the first two 6TM repeats of N-type neuronal Ca^{2+} channels. It shows greatest similarity to a 2 × 6TM channel-like gene recently isolated from rat tissue (83). Neither the rat gene nor *AtCCH1* has the pore motifs characteristic of neuronal Ca^{2+} channels, although a key tryptophan (W) is conserved (Figure 2). The rat gene has not been functionally characterized.

AtCCH1 (AtTPC1) shows Ca^{2+} transport activity. Expression in a yeast CCH1 mutant restored Ca^{2+} influx to wild-type levels and rescued growth on Ca^{2+}-deficient medium (64a). Sucrose-induced elevation of cytosolic Ca^{2+} was higher in leaves transiently overexpressing *AtCCH1* (measured by aequorin luminescence) and reduced in leaves transformed with antisense *AtCCH1*. Gene transcription was detected in root, shoot, flowers, and siliques. The authors speculated that AtCCH1 functions as a depolarization-activated Ca^{2+} channel involved in sugar signaling.

The predicted amino acid sequence of AtCCH1 shows only weak similarity to the *S. cerevisiae* CCH1 sequence, which is 4 × 6TM and forms a Ca^{2+}-permeable channel in association with the MID1 subunit (103, 117). The yeast CCH1 contains the cationic pore residue (E) implicated in Ca^{2+} binding in animal Ca^{2+} channels, as well as the cationic residues associated with voltage sensitivity in the S4 transmembrane domains (117). It has been hypothesized that animal 4 × 6TM Na^{+} and Ca^{2+} channels evolved from successive duplication of a K^{+} channel 1 × 6TM domain, with a Ca^{2+} or Na^{+} selective 2 × 6TM intermediate. The discovery of a 2 × 6TM-type protein in plants, a bacterial 1 × 6TM protein with similarity to Ca^{2+} channels (54), and a 4 × 6TM eukaryotic protein without Na^{+} or Ca^{2+} channel motifs (100) suggests that this channel family may be older and more diverse than previously hypothesized. It is not clear yet whether the *AtCCH1* gene encodes a Ca^{2+}-selective or cation-nonselective channel. Nor is it known whether the AtCCH1 subunit functions as a homomer or associates with a dissimilar pore-forming subunit (as in the case of the yeast CCH1) to form a cation channel.

Ionotropic Glutamate Receptors

Ionotropic glutamate receptors are NSCCs that function as Na^{+}- or Ca^{2+}-permeable channels in the nervous system of animals. In mammals, iGLRs are found in the postsynaptic membranes of neural junctions. Na^{+}-permeable iGLRs (AMPA and kainate receptors) open upon binding of the neurotransmitter glutamate to depolarize the postsynaptic membrane and initiate the action potential. These receptors desensitize rapidly. If the depolarization is large enough, then Ca^{2+}-permeable iGLRs (NMDA receptors) open upon glutamate and glycine binding to release Ca^{2+} into the cell and initiate signaling processes leading to long-term potentiation associated with learning. NMDA receptors activate and desensitize with slower kinetics than the AMPA/KA types. AMPA receptors can also act as Ca^{2+} channels, their selectivity for monovalent over divalent cations depending on post-transcriptional RNA editing of a single base in the pore region (from Q to R).

Channels are formed from the heteromeric assembly of subunits into pentamers or, more likely, tetramers, and the huge diversity of iGLR activity observed in vivo derives largely from the heteromeric combinations generated. Some iGLR subunits do not form functional homomers in heterologous systems.

In the current model of iGLR structure, each subunit (of 800 to 1500 residues) comprises a long extracellular N-terminal domain, two extracellular domains, S1 and S2, three membrane-spanning domains and a putative pore region, and an intracellular C-terminal domain (Figure 1). The N-terminal domain resembles bacterial periplasmic binding proteins (PBPs) specific for neutral amino acids, but it has no known function. The S1 and S2 domains are homologous to the lobes of bacterial amino acid–binding PBPs and are thought to interact with glutamate via a "Venus fly-trap" mechanism. The first two transmembrane regions (M1, M2) and the intervening pore region (P) lie between S1 and S2 and are postulated to have arisen via insertion of an inverted K^+ channel domain into the hinge region of a PBP (193) (Figure 1). Evidence for this evolutionary sequence was provided by the recent discovery of the only bacterial iGLR homologue, GluR0 from the cyanobacterium, *Synechocystis* (Figure 1) (27). GluR0 comprises only the S1, S2, M1, M2, and P regions. The pore region contains the GYGD sequence typical of K^+ channels, and the channel is selective for K^+ over other monovalent cations. When expressed in oocytes, GluR0 was activated by glutamate and glutamine and demonstrated relatively slow activation and desensitization.

The plant iGLRs closely resemble vertebrate iGLRs in predicted structure and in the sequences of the glutamate-binding and first transmembrane regions. The splice sites dividing the various domains are also preserved between plants and animals. However the pore region differs greatly from that of animal iGLRs, and indeed of cation channels generally, having a cluster of positively charged residues within the putative selectivity filter (Figure 2). Nevertheless, the region retains a channel-like hydropathy profile. An intrapore splice site is retained in plant and animal iGLRs as well as some K^+ channels (e.g., Shaker, the plant SKOR and GORK channels, and some animal and yeast double-barrelled K channels) and may provide a mechanism for rapid evolution of channel properties by intron shuffling. The N-terminal domain is also distinctive in plants, with similarity to metabotropic glutamate receptors, $GABA_B$ receptors, and Ca^{2+} sensors and has been suggested to mediate GABA sensitivity in plants (158).

The existence of amino acid receptors in plants was predicted by some researchers (29), but the evidence for glutamate as a signaling molecule in plants remains sparse (see Glutamate-Activated NSCCs). Some antagonists (DNQX) and agonists (BMAA) of animal iGLRs have been shown to alter plant responses to light (17, 99).

Characterization of plant *iGLR* genes has progressed slowly. There are 20 *iGLRs* predicted in the Arabidopsis genome, of which only seven have been cloned as cDNAs (98). Overexpression of (genomic) *AtGluR2* (renamed *AtGLR3.2*) produced symptoms of Ca^{2+} deficiency in transgenic Arabidopsis, although total Ca^{2+} content of the plants was unaltered, indicating a perturbation of Ca^{2+} distribution or

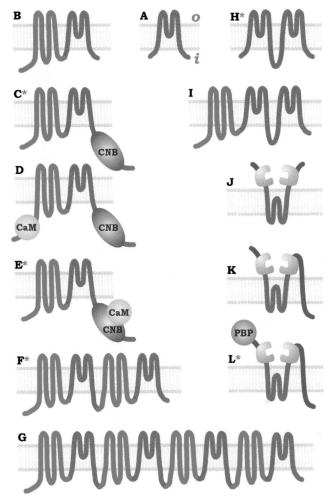

Figure 1 Structural domains of some families of cation transporters. Single alpha subunits are shown for each type. Asterisk indicates presence in plants. CaM, calmodulin-binding domain; CNB, cyclic nucleotide-binding domain; PBP, periplasmic amino acid binding protein-like domain. (*A*) 2TM IP, e.g., KscA from *Streptomyces lividans*. (*B*) 6TM IP, e.g., Shaker-type depolarization-activated K$^+$ channels. (*C*) 6TM 1P, e.g., HCN, EAG/ERG/ELK, and plant AKT and KAT channels. (*D*) 6TM 1P animal CNGC channels. (*E*) 6TM 1P plant putative CNGC channels. (*F*) 2x6TM IP, e.g., AtCCH1. (*G*) 4x6TM 1P, e.g., depolarization-activated neuronal Na$^+$ and Ca^{2+} channels. (*H*) 2x2TM 1P, e.g., KCO family. (*I*) 8TM 1P channels, e.g., TOK K$^+$ channels. (*J*) Bacterial iGLR, GluR0. (*K*) Eukaryotic low molecular weight iGLR-like proteins. (*L*) Plant and animal iGLRs with N-terminal regions homologous to PBPs, metabotropic glutamate receptors, and GABA$_B$ receptors.

A

P

S6

AtCNGC1:	KFFYCFWWGL	QNLISLGQNL	KTSTYIWEIC	FAVFISIAGL	VLFSF----LIGNM	QTYL	407
AtCNGC2:	KILYPIFWGL	MTLITFANDL	EPTSNWLEVI	FSIVMLSGL	LLFTL----LIGNI	QVFL	452
AtCNGC3:	KFFYCFWWGL	RNLIALGQNL	KTSAFEGEII	FAIVICISGL	LLFVM----LIGNI	QKYL	398
AtCNGC4:	KILFPIFWGL	MTLISFGN-L	ESTTENSEVV	FNIIVLTSGL	LLVTM----LIGNI	VFL	417
AtCNGC5:	KYCYCLWWGL	QNLITLGQGL	ETSTYPMEII	FSISLAISGL	LLFAI----LIGNM	QTYL	412
AtCNGC6:	KYFFCLWWGL	QNLITLGQGL	ETSTYPGEVI	FSITLAIAGL	LLFAI----LIGNM	QTYL	435
AtCNGC10:	KFFYCFWWGL	RNLIALGQNL	QTSKFVGEII	FAISICISGL	VLFAI----LIGNM	QKYL	389
AtCNGC20:	RYSYSLFWGF	QISTIAGNQ	VPSYFLGEVF	FTMGIIGIGL	LLFAI----LIGNM	QNFLQALGKI	521
NtCBT4:	KFFYCFWWGL	QNLISLGQNL	QTSTFIWEMC	FAVFISIAGL	VLFAF----LIGNM	QTCL	400
NtCBT7:	KLSYCFWWGL	RNLSSLGQGL	NTSDFLGEII	FAVFICIIGL	ILFSL----LIGNM	QEYL	397
HvCBT1:	KLFYCVWWGL	QNLSSLGQNL	KTSTYAWENL	FAVFVSISGL	VLFSI----LIGNM	QTYL	394
HsCNG3:	KYIYSLYWST	LTLITIGETP	PPVKDE-EYL	FVVVDFLVGV	LIFAT----IVGNV	GSMI	403
OcCNG2:	EYIICLYWST	LTLITIGETP	PPVKDE-EYL	FVIFDFLIGV	LIFAT----IVGNV	GSMI	375
DmEAG:	MVTALYFTM	TCMTSVGFGN	VAAETDNEKV	FTICMMIIAA	LLVAT----IFGHV	TTII	492
HCN2:	MVINHSWSE LYSFALFKAM	SHMLCIGYGR	QAPESMTDIW	LTMLSMIVGA	TCYAM----FIGHA	TALIQSLDSS RQYQEHY	480
SHAKER:	SIFDAFWWAV	VTMTTVGYGD	MIPVGVWGKI	VGSLCAIAGV	LTIALPVPVIVSNF	NYFY	485
AKT1:	RYTSMYWSI	TTLITTGYGD	LHPVNTKEMI	FDIFYMLFNL	GLTAY----LIGNM	TNLV	292
KAT1:	RYVTALYWSI	TTLTTTGYGD	FHAENPREML	FDIFFMMFNL	GLTAY----LIGNM	TNLV	300
SKOR:	RYTTSMYFAV	VTMATVYGD	IHAVNMREMI	FAMVYISFDM	GLTAY----LIGNM	TALI	327
KCO1:	GVVDALYFCI	VTMTTVGYGD	IVPNSSASRL	LACAFVFSGM	VLVCH----LLSRA	ADYL	161

B

AtCNGC1	FFQYIPRFIR	IYPLYKEVTR	TSGILTETA	255
AtCNGC2	LFQFLPKIYH	CICLMRRMQK	VTGYIFGTI	289
AtCNGC3	FCQYVPRIAR	IYPLFKEVTR	TSGLVTETA	243
AtCNGC4	LFQYLPKIYH	SIRHLRRNAT	LSGYIFGTV	257
AtCNGC5	LVQYIPRFLR	VLPLTSELKR	TAGVFAETA	252
AtCNGC6	LVQYIPRFLR	MYPLSSELKR	TAGVFAETA	279
AtCNGC10	IAQYVPRILR	MYPLYTEVTR	TSGIVTETA	234
AtCNGC20	LFQYIPKLYR	LLPFLAGQTP	TGFIFESAW	364
NtCBT4	FTQYIPRVLR	VYPLYREVTR	TSGILTETA	249
NtCBT7	FTQYVPIFR	IFPLYREVTR	TTGFFTETA	244
HvCBT1	ICQYVPRVIR	IRPLYLQITR	SAGITETA	241
HsCNG3	TNYPEVRFNR	LLKFSRLFEF	FDRTETRTN	296
OcCNG2	IHNPELRFNR	LLHFARMEF	FDRTETRTS	268
DmEAG	SALKVRLLR	LGRVVRKLDR	YLEYGAAML	372
HCN2	KTARALRIVR	FTKILS-LLR	LLRLSRLIR	345
SHAKER	NQAMSLAILR	VIRIVR-VFR	IFKLSHSK	368
AKT1	GLFNMLRLWR	LRRVGALFAR	LEKDRNFNY	186
KAT1	RIISMLRLWR	LRRVSSLFAR	LEKDIRFNY	193
SKOR	RYLLLIRLYR	VHRVILFFHK	MEKDIRINY	213

C

AtGLR1.1	SMMLWFGFST	IVFAH-REKL	QK	656
AtGLR2.8	GTSFWFSFST	MVFAH-REKV	VS	610
AtGLR3.1	ITILWTFSFST	MFFSH-RETT	VS	675
AtGLR3.2	VTILWFSFST	MFFSH-RENT	VS	645
AtGLR3.4	ITIFWFSFST	MFFSH-RENT	VS	624
AtGLR3.5	ITVFWFSFST	MFFSH-RENT	VS	671
AtGLR3.7	STMLLFSFST	LFKRN-EDT	IS	636
GluR6	LNSFWFGVGA	LMQQ-SELM	PK	629
NMDAR1	SSAMWFSWGV	LLNSGIGEGA	PR	625
GluR0	QNGMWFALVT	LTTVGYGDRS	PR	217
SHAKER	PDAFWWAVVT	MTTVGYGDMY	PV	451
HsCNGC3	IYSILWSTLI	LTTIG--ETP	PP	372
CaC2.1	NVLWALLTLF	TVSTG--EGW	PM	1369
RyR	CYMFHVSTGL	RAGGGIGDEI	ED	4833
IP3R	CIVIVLSHGL	RSGGGVGDVL	RK	2554

Figure 2 Alignments of plant ion channel sequences with other cation channels. (*A*) Alignments of putative pore (P) and S6 domains, predicted for plant CNGC full-length cDNAs, rabbit olfactory receptor OcCNG3, human cone photoreceptor HsCNG3, Drosophila DmEAG, human HCN2, Drosophila Shaker K$^+$ channel, and Arabidopsis K$^+$ channels KAT1, AKT1, SKOR, and KCO1 (first pore region + M2). (*B*) Alignments of putative S4 voltage-sensing domains of the same channel sequences, showing positively charged residues including histidine (pK$_a$ = 6.5). (*C*) Alignment of the putative pore regions predicted for plant iGLR full-length cDNAs, Synechocystis GluR0, rat GluR6, rat NMDAR1, human N-type Ca channel alpha 1B subunit pore motif III (CaC2.1), mouse type 1 IP$_3$ receptor (IP$_3$R), and mouse ryanodine receptor (RyR). Chemical properties of amino acid side groups are represented in color: *red*, basic; *purple*, acidic; *blue*, hydrophilic; *yellow*, aliphatic; *orange*, aromatic; *pink*, proline and glycine; *green*, cysteine. Splice sites, where known, are indicated by *vertical bars*.

efficacy rather than uptake (88). The plants were also hypersensitive to Na^+ and K^+ stress, but no measurements of plant uptake or content of these ions were made. The promoter region of *AtGluR2* is active in vascular tissues of root and shoot, especially in cells adjacent to xylem vessels. Very recently two reports of transport activity of Arabidopsis iGLRs have been made. AtGLR3.7 transported monovalent cations and Ca^{2+} in Xenopus oocytes (26a). AtGLR3.4 transported Ca^{2+} in oocytes, whereas AtGLR2.8 did not appear to mediate currents (98a). In the case of both AtGLR3.7 and AtGLR3.4, the currents were constitutive and showed no response to glutamate or other agonists of animal iGLRs. This may be an artifact of homomeric characterization. Alternatively, plant iGLRs may correspond to the voltage-insensitive, constitutively active NSCCs so commonly observed in electrophysiological studies. The function of the highly conserved putative amino acid–binding domains (S1, S2) remains to be determined.

One problem with at least some of these genes is their toxicity in bacteria and possibly in other heterologous systems. Even low-level expression of these genes, driven by adventitious promoters in common cloning vectors, is sufficient to eliminate correctly spliced cDNAs of some iGLRs from cDNA populations (R. J. Davenport, unpublished data). This may account partly for the low proportion of these genes that have been cloned as cDNAs. Other causes may be the low frequency or unusual expression patterns of some iGLRs. *AtGluR2* transcript could not be detected by Northern blotting (88), and analysis of putative promoter regions of *iGLRs* suggests that transcription of some *iGLRs* could be specific to seeds (R. J. Davenport, unpublished data). Interestingly, of the seven cDNAs cloned, five belong to the third subgroup of AtGLRs, suggesting that the subgroups are functionally specialized.

Most of the cloned and predicted *iGLRs* in the Arabidopsis genome (including *AtGLR3.2*) are predicted by membrane targeting prediction programs to be plasma membrane localized. However several are predicted to be endomembrane targeted. Glutamate plays several critical intracellular roles and is transported between organelles and the cytosol during amino acid synthesis and catabolism and photorespiration. It is possible that glutamate plays a key role in intracellular coordination of intracellular process of photosynthesis and nitrogen metabolism, via modulation of ion fluxes between compartments through iGLRs in endomembranes.

Other Channel Sequences

In addition to the 6TM IR K^+ channel genes, the Arabidopsis genome contains six *KCO* genes predicted to encode OR K^+ channels. These genes belong to the family of "double-barrelled" ion channel genes, comprising 2×2TM domains homologous to the pore-containing S5-S6 region of the 6TM channels. This family includes the ORK, TASK, and TREK proteins, some of which encode mechanosensitive K^+ channels. In contrast to most two-pore channels, both pore regions of the plant KCO proteins contain the GYGD motif. Despite this, KCO1 is only moderately selective for K^+: K^+ (1.00) > NH_4^+ (0.3) > Na^+ (0.15) > Li (0.1). KCO1 is tonoplast targeted (K. Czempinski, J-M. Frachisse, C. Maurel, H. Barbier-Brygoo & B. Müller-Röber, submitted), outwardly rectifying, and activated by cytosolic Ca^{2+} (36), raising the possibility that the KCO gene family could encode subunits of the vacuolar SV channel.

A single Arabidopsis gene has been identified with some similarity to the SEC62 gene family (At3g20920). SEC62 proteins are thought to function in protein import into the ER in yeast, but they have been also suggested to function as NSCCs (the NSCC-2 family) (134) in the plasma membrane of mammalian cells, although there is little molecular evidence for this. Other types of cation channel-like sequences have not yet been identified in plant genomes. Plants apparently lack several major evolutionary groups of ion channels, including NSCCs such as the CAN (Ca^{2+}-activated NSCC), transient receptor potential (TRP) NSCCs, sarcoplasmic reticulum (SR) Ca^{2+} channels, and IP_3 receptor groups [despite physiological evidence for the latter two channel types (114)]. However some channel types are harder to identify [e.g., the yeast vacuolar TRP homologue (118)], and plants may contain novel cation channel types not recognized by sequence similarity searches. The Saccharomyces genome contains very few channel sequences compared with Arabidopsis but does contain a TRP homologue, as well as MID1, which has weak homology to 6TM channels. It is interesting that the main groups of NSCC gene candidates in plants, the *CNGCs* and *iGLRs*, are absent from yeast but present in animals. Both families are represented by single members in the cyanobacterium Synechocystis, suggesting that these proteins may have ancient functions associated with photosynthesis or nitrogen metabolism (27, 116). Further sequencing of fungal genomes will reveal whether the absence of these gene families is specific to yeast (perhaps as a result of elimination of genes for multicellular functions from the yeast genome) or represents a more general evolutionary divergence of the fungal kingdom.

Cation Transporters of Nonchannel Topology

Low-capacity transporters such as the HKT, KUP, and HAK families are capable of passive ion transport at high substrate concentrations (so-called dual-affinity transport) (64, 135). These K^+ transporters appear to consist of four repeats of a 2TM + P domain like that of the prokaryotic K^+ channel KcsA, with elements of the conventional K^+ channel pore signature, and it has been suggested that they may form functional channel-like complexes (53, 55). In heterologous systems, HvHAK1 mediates high-affinity K^+ uptake and low-affinity Na^+ transport (135), whereas TaHKT1 from wheat is selective for K^+ and Na^+ (in symport) at micromolar concentrations but transports other monovalent cations passively at millimolar concentrations (66, 133). However, the Arabidopsis HKT1 homologue appears to be a selective Na^+ transporter (163). It remains to be determined what contribution these transporters make to low-affinity nonselective cation uptake.

LCT1 is a unique gene cloned from wheat by complementation of a K^+ transport-defective *S. cerevisiae* mutant. Overexpression in yeast moderately increases uptake of monovalent and divalent cations (5, 30, 137), but the mechanism of transport has not been determined. The gene is predicted to encode a 10TM protein, with no similarity to other transporters. LCT1 is thought to be localized to the yeast plasma membrane but has not been characterized in plant systems.

Molecular Conclusions: Why NSCCs?

The main families of NSCCs in plants, the iGLRs and CNGCs, as well as the putative Ca channel AtCCH1, were originally identified as cation channels by sequence homology with animal cation channels. However the plant sequences differ from other ion channels in key residues within the putative pore region lining the ion channel and conferring ion selectivity in K^+-, Ca^{2+}-, and Na^+-selective channels as well as divalent:monovalent cation selectivity in animal iGLRs and CNGCs. Mutations of pore residues can convert highly selective channels to NSCCs, suggesting that a number of alternative sequences could be selected for in NSCC channel evolution. It is curious that both the CNGCs and more particularly the iGLRs in plants show high homology with animal sequences in the ligand-binding and transmembrane domains around the putative pore region but low homology in the pore itself.

The unconserved region of the plant genes is flanked by an intron in the middle of the pore itself and a second intron shortly downstream in the end of the pore region (iGLRs) or the end of S6 (CNGCs), forming a small complete exon in all the *AtCNGCs* and the third subfamily of *iGLRs* (Figure 2). The pore region has been identified as an "evolutionary hotspot" of gene recombination in K^+ channel genes (169), although most modern K^+ channel genes exist as a single exon. The retention of ancient splice sites in plant channel genes suggests the evolutionary maintenance of common ligand sensitivity and overall channel structure, in parallel with shuffling and rearrangement of the pore to give novel ion selectivity. Such rearrangements may have been necessary to adapt NSCCs to the relatively dilute, low-Na^+ solutions to which plants are usually exposed or may reflect the early divergence of animal and plant ligand-gated ion channels from common K^+-selective, glutamate, or cyclic nucleotide–gated ion channel ancestors.

Most of the gene products in Arabidopsis do not correspond closely to measured NSCCs in planta. This probably reflects the paucity of data yet available for these genes but may also be due to the limitations of homomeric characterization. The identification of channel subunits contributing to NSCC channels in vivo will require extensive single cell type sampling, colocalization, and coexpression studies to determine the interactions of the various channel subunits expressed in particular cell types. More information is also needed on the membrane localization of the gene products to account for the full repertoire of observed NSC currents.

There is no evidence in plants for the evolution of any highly selective cation channels in plants except K^+ channels (although AtCCH1 remains to be characterized electrophysiologically). However plants have a high diversity of NSCCs. An obvious question is why any organism should evolve NSCCs in place of highly selective cation channels, given the physical possibility of high selectivity. One possible answer is that although plants possessed the raw genetic material to evolve highly selective cation channels (e.g., 6TM channels) they lacked the evolutionary pressure. However this type of explanation cannot account for the ubiquity of NSCCs in animals, where highly selective and nonselective ion channels coexist. Indeed, the existence of NSCCs in animal cells remains largely unexplained, except in cases where the bidirectional movement of different ions (K^+ out, Na^+ in) serves to fix E_m at an intermediate voltage (e.g., photoreceptor CNGCs; see The Voltage-Gated Ion Channel Superfamily). NSCCs have several advantages over highly selective ion channels. One advantage is that they can transport different ions in both directions, allowing flexibility in setting E_m with a single channel. For instance, the depolarization caused by Ca^{2+} entry through NSCCs could be offset by K^+ efflux via the same channel. NSCCs interact less strongly with permeant cations and so may be able to maintain uptake of the target cation in the presence of cations that block more selective channels. This would be an advantage in fluctuating soil solutions, where K^+ uptake must be maintained in the presence of Cs^+, for instance, or Ca^{2+} in the presence of other divalent cations. These channels also provide a passive route for uptake of cations for which high selectivity may not be possible (for instance, NH_4^+ or Mg^{2+}). Relatively indiscriminate cation transport may be a positive advantage for turgor-controlled ion channels, which must mediate rapid exchange of solutes.

CONCLUSIONS

There is a range of NSCCs in several different membranes in plants, controlled by parameters such as voltage, Ca^{2+}, and glutamate. They can allow movements of monovalent or divalent cations and appear to constitute the primary pathway of Ca^{2+} uptake and signaling in plants (although the full repertoire of plant channels and genes is not yet known). NSCCs are involved in a wide range of plant processes, including low-affinity nutrient acquisition and allocation, turgor control, intracellular transport, and signaling. The low selectivity of these channels also implicates them in uptake of toxic cations, notably Na^+. Some of the genes encoding plant NSCCs have been identified, but most remain uncharacterized. The unexpected evolutionary relationships of some of these genes has prompted further investigation of the control mechanisms of native NSCCs. Misexpression studies of native and mutated *NSCC* gene candidates should help to elucidate the roles of NSCCs in planta and the physiological importance of their low selectivity, as well as establishing the scope for manipulating ion uptake and distribution within plants.

ACKNOWLEDGMENTS

We thank many colleagues for sharing unpublished findings and the many members of the transport groups in Cambridge for their intellectual input and stimulation. Special thanks are due to Julia Davies. Research in the authors' laboratories is supported by the BBSRC, The Leverhulme Foundation, a Royal Society/NATO Fellowship (V.D.), and a Royal Society Dorothy Hodgkin Fellowship (R.J.D.).

Visit the Annual Reviews home page at www.annualreviews.org

LITERATURE CITED

1. Ache P, Becker D, Ivashikina N, Dietrich P, Roelfsema MRG, Hedrich R. 2000. GORK, a delayed outward rectifier expressed in guard cells of *Arabidopsis thaliana*, is a K(+)-selective, K(+)-sensing ion channel. *FEBS Lett.* 486:93–98

2. Allen GJ, Amtmann A, Sanders D. 1998. Calcium-dependent and calcium-independent K^+ mobilization channels in *Vicia faba* guard cell vacuoles. *J. Exp. Bot.* 49:305–18

3. Allen GJ, Sanders D. 1995. Calcineurin, a type 2B protein phosphatase, modulates the Ca^{2+}-permeable slow vacuolar ion channel of stomatal guard cells. *Plant Cell* 7:1473–83

4. Allen GJ, Sanders D. 1996. Control of ionic currents in guard cell vacuoles by cytosolic and luminal calcium. *Plant J.* 10:1055–69

5. Amtmann A, Fischer M, Marsh EL, Stefanovic A, Sanders D, Schachtman DP. 2001. The wheat cDNA *LCT1* generates hypersensitivity to sodium in a salt-sensitive yeast strain. *Plant Physiol.* 126:1061–71

6. Amtmann A, Laurie S, Leigh RA, Sanders D. 1997. Multiple inward channels provide flexibility on Na^+/K^+ discrimination at the plasma membrane of barley suspension culture cells. *J. Exp. Bot.* 48:481–97

7. Amtmann A, Sanders D. 1999. Mechanisms of Na^+ uptake by plant cells. *Adv. Bot. Res.* 29:75–112

8. Arazi T, Kaplan B, Fromm H. 2000. A high-affinity calmodulin-binding site in a tobacco plasma-membrane channel protein coincides with a characteristic element of cyclic nucleotide-binding domains. *Plant Mol. Biol.* 42:591–601

9. Arazi T, Kaplan B, Sunkar R, Fromm H. 2000. Moving signals and molecules through membranes. *Biochem. Soc. Trans.* 28:471–75

10. Arazi T, Sunkar R, Kaplan B, Fromm H. 1999. A tobacco plasma membrane calmodulin-binding transporter confers Ni^{2+} tolerance and Pb^{2+} hypersensitivity in transgenic plants. *Plant J.* 20:171–82

11. Arispe N, De Maio A. 2000. ATP and ADP modulate a cation channel formed by Hsc70 in acidic phospholipid membranes. *J. Biol. Chem.* 275:30839–43

12. Babourina O, Shabala S, Newman I. 2000. Verapamil-induced kinetics of ion flux in oat seedlings. *Aust. J. Plant Physiol.* 27:1031–40

13. Baizabal-Aguirre VM, Clemens S, Uozomi N, Schroeder JI. 1999. Suppression of inward-rectifying K^+ channels KAT1 and AKT2 by dominant negative point mutations in the KAT1 alpha subunit. *J. Membr. Biol.* 167:119–25

14. Beilby MJ. 1986. Potassium channels and different states of *Chara plasmalemma*. *J. Membr. Biol.* 89:241–49

15. Bethke PC, Jones RL. 1994. Ca^{2+}-calmodulin modulates ion channel activity in storage protein vacuoles of barley aleurone cells. *Plant Cell* 6:277–85

16. Bewell MA, Maathuis FJM, Allen GJ, Sanders D. 1999 Calcium-induced calcium release mediated by a voltage-activated cation channel in vacuolar vesicles from red beet. *FEBS Lett.* 458:41–44

17. Brenner ED, Martinez-Barboza N, Clark AP, Liang QS, Stevenson DW, Coruzzi GM. 2000. Arabidopsis mutants resistant to S(+)-β-methyl-α,β-diaminopropionic acid, a cycad-derived glutamate receptor agonist. *Plant Physiol.* 124:1615–24

18. Broekaert WF, Cammue BPA, De Bolle MFC, Thevissen K, De Samblanx GW, Osborn RW. 1997. Antimicrobial peptides from plants. *Crit. Rev. Plant Sci.* 16:297–323

19. Brüggemann LI, Pottosin II, Schönknecht G. 1998. Cytoplasmic polyamines block the fast-activating vacuolar cation channel. *Plant J.* 16:101–5

20. Brüggemann LI, Pottosin II, Schönknecht G. 1999. Cytoplasmic magnesium regulates the fast activating vacuolar cation channel. *J. Exp. Bot.* 50:1547–52

21. Brüggemann LI, Pottosin II, Schönknecht G. 1999. Selectivity of the fast activating vacuolar cation channel. *J. Exp. Bot.* 50:837–76

22. Buschmann PH, Vaidynathan R, Gassmann W, Schroeder JI. 2000. Enhancement of Na^+ uptake currents, time-dependent inward-rectifying K^+ channel currents, and K^+ channel transcripts by K^+ starvation in wheat root cells. *Plant Physiol.* 122:1387–97

23. Carpaneto A, Cantù AM, Busch H, Gambale F. 1997. Ion channels in the vacuoles of the seagrass *Posidonia oceanica. FEBS Lett.* 412:236–40

24. Carpaneto A, Cantù AM, Gambale F. 2001. Effects of cytoplasmic Mg^{2+} on slowly activating channels in isolated vacuoles of *Beta vulgaris. Planta* 213:457–68

25. Cerana R, Colombo R. 1992. K^+ and Cl^- conductance of Arabidopsis thaliana plasma membrane at depolarised voltages. *Bot. Acta* 105:273–77

26. Chapman ML, Krovetz HS, VanDongen AMJ. 2001. GYGD pore motifs in neighbouring potassium channel subunits interact to determine ion selectivity. *J. Physiol.* 530:21–33

26a. Cheffings CM. 2001. Calcium channel activity of a plant glutamate receptor homologue. *Int. Workshop Plant Membr. Biol., 12th, Madison*

27. Chen G-Q, Cui C, Mayer ML, Gouax E. 1999. Functional characterisation of a potassium-selective prokaryotic glutamate receptor. *Nature* 402:817–21

28. Chen J, Mitcheson JS, Lin M, Sanguinetti MC. 2000. Functional roles of charge residues in the putative voltage sensor of the HCN2 pacemaker channel. *J. Biol. Chem.* 275:36465–71

29. Chiu J, DeSalle R, Lam H-M, Maisel L, Coruzzi G. 1999. Molecular evolution of glutamate receptors: a primitive signaling mechanism that existed before plants and animals diverged. *Mol. Biol. Evol.* 16:826–38

30. Clemens S, Antosiewicz DM, Ward JM, Schachtman DP, Schroeder JI. 1998. The plant cDNA LCT1 mediates the uptake of calcium and cadmium in yeast. *Proc. Natl. Acad. Sci. USA* 95:12043–48

31. Clough SJ, Fengler KA, Yu I-C, Lippok B, Smith RK, Bent AF. 2000. The *Arabidopsis dnd1* 'defense no death' gene encodes a mutated cyclic nucleotide–gated ion channel. *Proc. Natl. Acad. Sci. USA* 97:9323–28

32. Colombo R, Cerana R, Lado P, Peres A. 1988. Voltage-dependent channels permeable to K^+ and Na^+ in the membrane of *Acer pseudoplatanus* vacuoles. *J. Membr. Biol.* 103:227–36

33. Cosgrove DJ, Hedrich R. 1991. Stretch-activated chloride, potassium, and calcium channels coexisting in plasma membranes of guard cells of *Vicia faba* L. *Planta* 186:143–53

34. Coyaud L, Kurkdjian A, Kado R, Hedrich R. 1987. Ion channels and ATP-driven pumps involved in ion transport across the tonoplast of sugarbeet vacuoles. *Biophys. Biochim. Acta* 902:263–68

35. Deleted in proof

36. Czempinski K, Zimmermann S, Ehrhardt T, Müller-Röber B. 1997. New structure and function in plant K^+ channels: KCO1, an outward rectifier with a steep Ca^{2+} dependency. *EMBO J.* 16:2565–75

37. Daram P, Urbach S, Gaymard F, Sentenac H, Chérel I. 1997. Tetramerisation of the AKT1 plant potassium channel involves its C terminal cytoplasmic domain. *EMBO J.* 16:3455–63

38. Davenport RJ, Tester M. 2000. A weakly voltage-dependent, nonselective cation

82. Isas JM, Cartailler JP, Sokolov Y, Patel DR, Langen R, et al. 2000. Annexins V and XII insert into bilayers at mildly acidic pH and form ion channels. *Biochemistry* 39:3015–22

83. Ishibashi K, Suzuki M, Imai M. 2000. Molecular cloning of a novel form (two-repeat) protein related to voltage-gated sodium and calcium channels. *Biochem. Biophys. Res. Commun.* 270:370–76

84. Jabs T, Tschope M, Colling C, Hahlbrock K, Scheel D. 1997. Elicitor-stimulated ion fluxes and O_2^- from the oxidative burst are essential components in triggering defense gene activation and phytoalexin synthesis in parsley. *Proc. Natl. Acad. Sci. USA* 94:4800–5

85. Jain L, Chen XJ, Malik B, Al-Khalili O, Eaton DC. 1999. Antisense oligonucleotides against the α-subunit of ENaC decrease lung epithelial cation-channel activity. *Am. J. Physiol.* 276:L1046–51

86. Kaupp UB, Seifert R. 2001. Molecular diversity of pacemaker ion channels. *Annu. Rev. Physiol.* 63:325–57

86a. Kiegle E, Gilliham M, Haseloff J, Tester M. 2000. Hyperpolarisation-activated calcium currents found only in cells from the elongation zone of *Arabidopsis thaliana* roots. *Plant J.* 21:225–29

87. Kiegle E, Moore C, Haseloff J, Tester M, Knight M. 2000. Cell-type specific calcium responses to drought, NaCl, and cold in *Arabidopsis* root: a role for endodermis and pericycle in stress signal transduction. *Plant J.* 23:267–78

88. Kim SA, Kwak JM, Jae SK, Wang MH, Nam HG. 2001. Overexpression of the At-GluR2 gene encoding an Arabidopsis homolog of mammalian glutamate receptors impairs calcium utilization and sensitivity to ionic stress in transgenic plants. *Plant Cell Physiol.* 42:74–84

89. Kirk KL. 2000. New paradigms of CFTR chloride channel regulation. *Cell. Mol. Life Sci.* 57:623–34

90. Klüsener B, Boheim G, Liss H, Engelberth J, Weiler EW. 1995. Gadolinium-sensitive, voltage-dependent calcium release channels in the endoplasmic reticulum of a higher plant mechanoreceptor organ. *EMBO J.* 14:2708–14

91. Klüsener B, Boheim G, Weiler EW. 1997. Modulation of the ER Ca^{2+} channel BCC1 from tendrils of *Bryonia dioica* by divalent cations, protons and H_2O_2. *FEBS Lett.* 407:230–34

92. Klüsener B, Weiler EW. 1999. A calcium-selective channel from root tip endomembranes of cress. *Plant Physiol.* 119:1399–405

93. Knight MR, Smith SM, Trewavas AJ. 1992. Wind-induced plant motion immediately increases cytosolic calcium. *Proc. Natl. Acad. Sci. USA* 89:4967–71

94. Kohler C, Merkle T, Neuhaus G. 1999. Characterisation of a novel gene family of putative cyclic nucleotide- and calmodulin-regulated ion channels in *Arabidopsis thaliana*. *Plant J.* 18:97–104

95. Kohler C, Merkle T, Roby D, Neuhaus G. 2001. Developmentally regulated expression of a cyclic nucleotide–gated ion channel from *Arabidopsis* indicates its involvement in programmed cell death. *Planta* 213:327–32

96. Kohler C, Neuhaus G. 2000. Characterisation of calmodulin binding to cyclic nucleotide–gated ion channels from *Arabidopsis thaliana*. *FEBS Lett.* 4710:133–36

97. Kurosaki F, Kaburaki H, Nishi A. 1994. Involvement of plasma membrane–located calmodulin in the response decay of cyclic nucleotide–gated cation channel of cultured carrot cells. *FEBS Lett.* 340:193–96

98. Lacombe B, Becker D, Hedrich R, Chiu J, DeSalle R, et al. 2001. On the identity of plant glutamate receptors. *Science* 292:1486–87

98a. Lacombe B, Meyerhoff O, Steinmeyer R, Becker D, Hedrich R. 2001. *Role of Arabidopsis ionotropic glutamate receptors*. Presented at Assoc. Canaux Ionoques, 12th, La Londe les Maures

99. Lam H-M, Chiu J, Hsieh M-H, Meisel L,

Oliviera IC, et al. 1998. Glutamate receptor genes in plants. *Nature* 396:125–26

100. Lee JH, Cribbs LL, Perez-Reyes E. 1999. Cloning of a novel four repeat protein related to voltage-gated sodium and calcium channels. *FEBS Lett.* 445:231–36

101. Leng Q, Mercier RW, Kaplan B, Hillel F, Berkowitz GA. 2000. *Voltage clamp analysis of cloned plant cyclic nucleotide gated ion channels.* Presented at Plant Biol., San Diego

102. Leng Q, Mercier RW, Yao W, Berkowitz GA. 1999. Cloning and first functional characterisation of a plant cyclic nucleotide–gated cation channel. *Plant Physiol.* 121:753–61

103. Locke EG, Bonilla M, Liang L, Takita Y, Cunningham KW. 2000. A homolog of voltage-gated Ca^{2+} channels stimulated by depletion of secretory Ca^{2+} in yeast. *Mol. Cell. Biol.* 20:6686–94

104. Maathuis FJM, Prins HBA. 1990. Patch clamp studies on root cell vacuoles of salt-tolerant and salt-sensitive *Plantago* species. *Plant Physiol.* 92:23–28

105. Maathuis FJM, Prins HBA. 1991. Inhibition of inward rectifying tonoplast channels by a vacuolar factor: physiological and kinetic implications. *J. Membr. Biol.* 122:251–58

106. Maathuis FJM, Prins HBA. 1991. Outward current conducting ion channels in tonoplasts of *Vigna unguiculata. J. Plant Physiol.* 139:63–69

107. Maathuis FJM, Sanders D. 1995. Contrasting roles in ion transport of two K^+-channel types in root cells of *Arabidopsis thaliana. Planta* 197:456–64

107a. Maathuis FJM, Sanders D. 2001. Sodium uptake in *Arabidopsis thaliana* roots is regulated by cyclic nucleotides. *Plant Physiol.* 127:1617–25

108. Marshall J, Corzo A, Sanders D. 1994. Membrane potential-dependent calcium transport in right-side-out plasma membrane vesicles from *Zea mays* L. roots. *Plant J.* 5:683–94

109. Marten I, Hoshi T. 1997. Voltage-dependent gating characteristics of the K^+ channel KAT1 depend on the N and C termini. *Proc. Natl. Acad. Sci. USA* 94: 3448–53

110. Mäser P, Thomine S, Schroeder JI, Ward JM, Hirschi K, et al. 2001. Phylogenetic relationships within cation transporter families of Arabidopsis. *Plant Physiol.* 126:1646–87

111. Meiners S, Xu A, Schindler M. 1991. Gap junctions protein homologue from Arabidopsis thaliana: evidence for connexins in plants. *Proc. Natl. Acad. Sci. USA* 88:4119–22

112. Miedema H, Pantoja O. 2001. Anion modulation of the slow-vacuolar channel. *J. Membr. Biol.* 183:137–45

113. Moran N, Ehrenstein G, Iwasa K, Bare C, Mishke C. 1984. Ion channels in plasmalemma of wheat protoplasts. *Science* 226:835–38

114. Muir SR, Bewell MA, Sanders D, Allen GJ. 1997. Ligand-gated Ca^{2+} channels and Ca^{2+} signalling in higher plants. *J. Exp. Bot.* 48:589–97

115. Newman IA. 2001. Ion transport in roots: measurement of fluxes using ion-selective microelectrodes to characterize transporter function. *Plant Cell Environ.* 24:1–14

116. Ochoa de Alda JA, Houmard J. 2000. Genomic survey of cAMP and cGMP signalling components in the cyanobacterium *Synechocystis* PCC 6803. *Microbiology* 146:3183–94

117. Paidhungat M, Garrett S. 1997. A homolog of mammalian, voltage-gated calcium channels mediates yeast pheromone-stimulated Ca^{2+} uptake and exacerbates the *cdc1* (Ts) defect. *Mol. Cell. Biol.* 17:6339–47

118. Palmer CP, Zhou X-L, Lin J, Loukin SH, Kung C, Saimi Y. 2001. A TRP homolog in *Saccharomyces cerevisiae* forms an intracellular Ca^{2+}-permeable channel in the yeast vacuolar membrane. *Proc. Natl. Acad. Sci. USA* 98:7801–5

119. Pei ZM, Murata Y, Benning G, Thomine

S, Klusener B, et al. 2000. Calcium channels activated by hydrogen peroxide mediate abscisic acid signalling in guard cells. *Nature* 406:731–34

120. Pei ZM, Schroeder JI, Schwarz M. 1998. Background ion channel activities in *Arabidopsis* guard cells and review of ion channel regulation by protein phosphorylation events. *J. Exp. Bot.* 49:319–28

121. Pei ZM, Ward JM, Schroeder JI. 1999. Magnesium sensitizes slow vacuolar channels to physiological cytosolic calcium and inhibits fast vacuolar channels in fava bean guard cell vacuoles. *Plant Physiol.* 121:977–86

122. Piñeros M, Tester M. 1995. Characterisation of a voltage-dependent Ca^{2+}-selective channel from wheat roots. *Planta* 195:478–88

123. Piñeros M, Tester M. 1997. Calcium channels in plant cells: selectivity, regulation and pharmacology. *J. Exp. Bot.* 48(Spec. Issue):551–77

124. Piñeros M, Tester M. 1997. Characterisation of the high-affinity verapamil-binding site in a plant plasma membrane Ca^{2+}-selective channel. *J. Membr. Biol.* 157:139–45

125. Pottosin II, Dobrovinskaya OR, Muñiz J. 2000. The slow vacuolar channel: permeability, block, and regulation by permeable cations. *Mater. Primer Congr. Responsab. Proy. Investig. Cienc. Nat., Veracruz, 8–11 Oct.* http://www.conacyt.mx/daic/proyectos/congresos/naturales/pdf/29473–N.pdf

126. Pottosin II, Dobrovinskaya OR, Muñiz J. 2001. Conduction of monovalent and divalent cations in the slow vacuolar channel. *J. Membr. Biol.* 181:55–65

127. Pottosin II, Schönknecht G. 1996. Ion channel permeable for divalent and monovalent cations in native spinach thylakoid membranes. *J. Membr. Biol.* 152:223–33

128. Pottosin II, Tikhonova LI, Hedrich R, Schonknecht G. 1997. Slowly activating vacuolar channels can not mediate Ca^{2+}-induced Ca^{2+} release. *Plant J.* 12:1387–98

129. Reifarth FW, Weiser T, Bentrup FW. 1994. Voltage- and Ca^{2+}-dependence of the K^+ channel in the vacuolar membrane of *Chenopodium rubrum* L. suspension cells. *Biochim. Biophys. Acta* 1192:79–87

130. Roberts SK, Tester M. 1997. Permeation of Ca^{2+} and monovalent cations through an outwardly rectifying channel in maize root stelar cells. *J. Exp. Bot.* 48:839–46

131. Roberts SK, Tester M. 1997. Patch clamp study of Na^+ transport in maize roots. *J. Exp. Bot.* 48:431–40

132. Romano LA, Miedema H, Assmann SM. 1998. Ca^{2+}-permeable, outwardly-rectifying K^+ channels in mesophyll cells of *Arabidopsis thaliana*. *Plant Cell Physiol.* 39:1133–44

133. Rubio F, Gassmann W, Schroeder JI. 1995. Sodium-driven potassium uptake by the plant potassium transporter HKT1 and mutations conferring salt tolerance. *Science* 270:1660–63

134. Saier MH. 2001. The nonselective cation channel family, NSCC-2. http://www.biology.ucsd.edu/~msaier/transport/1_A_15.html

135. Santa-María GE, Rubio F, Dubcovsky J, Rodríguez-Navarro A. 1997. The *HAK1* gene of barley is a member of a large gene family and encodes a high-affinity potassium transporter. *Plant Cell* 9:2281–89

136. Schachtman DP. 2000. Molecular insights into the structure and function of plant K^+ transport mechanisms. *Biochim. Biophys. Acta* 1465:127–39

137. Schachtman DP, Kumar R, Schroeder JI, Marsh EL. 1997. Molecular and functional characterisation of a novel low-affinity cation transporter (LCT1) in higher plants. *Proc. Natl. Acad. Sci. USA* 94:11079–84

138. Schroeder JI, Allen GJ, Hugouvieux V, Kwak JM, Waner D. 2001. Guard cell

signal transduction. *Annu. Rev. Plant Physiol. Plant Mol. Biol.* 52:627–58

139. Schroeder JI, Hagiwara S. 1989. Cytosolic calcium regulates ion channels in the plasma membrane of *Vicia faba* guard cells. *Nature* 338:427–30

140. Schroeder JI, Hagiwara S. 1990. Repetitive increases in cytosolic Ca^{2+} of guard cells by abscisic acid activation of nonselective Ca^{2+} permeable channels. *Proc. Natl. Acad. Sci. USA* 87:9305–9

141. Schulz-Lessdorf B, Hedrich R. 1995. Protons and calcium modulate SV-type channels in the vacuolar-lysosomal compartment-channel interaction with calmodulin inhibitors. *Planta* 197:655–71

142. Schuurink RC, Shartzer SF, Fath A, Jones RL. 1998. Characterisation of a calmodulin-binding transporter from the plasma membrane of barley aleurone. *Proc. Natl. Acad. Sci. USA* 95:1944–49

143. Shapiro MS, Zagotta WN. 2000. Structural basis for ligand selectivity of heteromeric olfactory nucleotide-gated channels. *Biophys. J.* 78:2307–20

144. Siemen D. 1993. Nonselective cation channels. In *Nonselective Cation Channels: Pharmacology, Physiology, and Biophysics*, ed. D Siemen, J Hescheler, pp. 3–27. Basel: Birkhauser Verlag

145. Silverman SK, Lester HA, Dougherty DA. 1998. Asymmetrical contributions of subunit pore regions to ion selectivity in an inward rectifier K^+ channel. *Biophys. J.* 75:1330–39

146. Sokolik AI. 1999. Nonselective ion conductivity of plasmalemma as an important component of the system of membrane transport of ions in plants. *Dokl. Acad. Nauk. Belar.* 43:77–80 (In Russian)

147. Sokolik AI, Demko G, Gorobchenko N, Yurin VM. 1997. Basic mechanisms of ^{137}Cs uptake by root system of plants. *Radiat. Biol. Radioecol.* 37:787–96 (In Russian)

148. Sokolik AI, Yurin VM. 1981. Transport properties of potassium channels of the plasmalemma in *Nitella* cells at rest. *Sov. Plant Physiol.* 28:206–12

149. Spalding EP, Slayman CL, Goldsmith MHM, Gradmann D, Bertl A. 1992. Ion channels in Arabidopsis plasma membrane—transport characteristics and involvement in light-induced voltage changes. *Plant Physiol.* 99:96–102

150. Stoeckel H, Takeda K. 1989. Calcium-activated, voltage-dependent, non-selective cation currents in endosperm plasma membrane from higher plants. *Proc. R. Soc. London Ser. B* 237:213–31

151. Sunkar R, Kaplan B, Bouche N, Arazi T, Dolev D, et al. 2000. Expression of a truncated tobacco NtCBP4 channel in transgenic plants and disruption of the homologous Arabidopsis CNGC1 gene confer Pb^{2+} tolerance. *Plant J.* 24:533–42

152. Tamm LK, Arora A, Kleinschmidt JH. 2001. Structure and assembly of β-barrel membrane proteins. *J. Biol. Chem.* 276:32399–402

153. Tester M. 1988. Pharmacology of K^+ channels in the plasmalemma of the green alga *Chara corallina*. *J. Membr. Biol.* 103:159–69

154. Tester M. 1990. Plant ion channels: whole-cell and single channel studies. *New Phytol.* 114:305–40

155. Tester M. 1997. Techniques for studying ion channels: an introduction. *J. Exp. Bot.* 48:353–59

155a. Tester M, Blatt MR. 1989. Direct measurement of K^+ channels in thylakoid membranes by incorporation of vesicles into planar lipid bilayers. *Plant Physiol.* 91:249–52

155b. Thion L, Mazars C, Nacry P, Bouchez D, Moreau M, et al. 1998. Plasma membrane depolarization-activated calcium channels, stimulated by microtubule-depolymerizing drugs in wild-type *Arabidopsis thaliana* protoplasts, display constitutively large activities and a longer half-life in *ton* 2 mutant cells affected in the organization of cortical microtubules. *Plant J.* 13:603–10

156. Thuleau P, Ward JM, Ranjeva R, Schroeder JI. 1994. Voltage-dependent calcium-permeable channels in the plasma membrane of a higher plant cell. *EMBO J.* 13:2970–75

157. Tikhonova LI, Pottosin II, Dietz KJ, Schönknecht G. 1997. Fast-activating cation channels in barley mesophyll vacuoles. Inhibition by calcium. *Plant J.* 11:1059–70

158. Turano F, Panta GR, Allard MW, van Berkum P. 2001. The putative glutamate receptors from plants are related to two superfamilies of animal neurotransmitter receptors via distinct evolutionary mechanisms. *Mol. Biol. Evol.* 18:1417–20

159. Tyerman SD, Beilby M, Whittington J, Juswono U, Newman I, Shabala S. 2001. Oscillations in proton transport revealed from simultaneous measurements of net current and net proton fluxes from isolated root protoplasts: MIFE meets patch-clamp. *Aust. J. Plant Physiol.* 28:591–604

160. Tyerman SD, Skerrett M. 1999. Root ion channels and salinity. *Sci. Hortic.* 78:175–235

161. Tyerman SD, Skerrett M, Garrill A, Findlay GP, Leigh RA. 1997. Pathways for the permeation of Na^+ and Cl^- into protoplasts derived from the cortex of wheat roots. *J. Exp. Bot.* 48:459–80

162. Tyerman SD, Whitehead LF, Day DA. 1995. A channel-like transporter for NH_4^+ on the symbiotic interface of N_2-fixing plants. *Nature* 378:629–32

163. Uozumi N, Kim EJ, Rubio F, Yamaguchi T, Muto S, et al. 2000. The Arabidopsis *HKT1* gene homolog mediates inward Na^+ currents in *Xenopus laevis* oocytes and Na^+ uptake in *Saccharomyces cerevisiae*. *Plant Physiol.* 122:1249–59

164. Urbach S, Cherel I, Sentenac H, Gaymard F. 2000. Biochemical characterization of the Arabidopsis K^+ channels KAT1 and AKT1 expressed or co-expressed in insect cells. *Plant J.* 23:527–38

165. van den Wijngaard PWJ, Bunney TD, Roobeek I, Schönknecht G, de Boer AH. 2001. Slow vacuolar channels from barley mesophyll cells are regulated by 14-3-3 proteins. *FEBS Lett.* 488:100–4

166. Véry A-A, Davies JM. 1998. Laser microsurgery permits fungal plasma membrane single-ion-channel resolution at the hyphal tip. *Appl. Environ. Microbiol.* 64:1569–72

167. Véry A-A, Davies JM. 2000. Hyperpolarization-activated calcium channels at the tip of *Arabidopsis* root hairs. *Proc. Natl. Acad. Sci. USA* 97:9801–6

168. Véry A-A, Robinson MF, Mansfield TA, Sanders D. 1998. Guard cell cation channels are involved in Na^+-induced stomatal closure in a halophyte. *Plant J.* 14:509–21

169. Wang ZW, Kunkel MT, Wei A, Butler A, Salkoff L. 1999. Genomic organization of nematode 4TM K^+ channels. *Ann. NY Acad. Sci.* 868:286–303

170. Ward JM. 1997. Patch-clamping and other molecular approaches for the study of plasma membrane transporters demystified. *Plant Physiol.* 114:1151–59

171. Ward JM. 2001. *Arabidopsis Membrane Protein Library.* http://www.cbs.umn.edu/arabidopsis

172. Ward JM. 2001. Identification of novel families of membrane proteins from the model plant *Arabidopsis thaliana. Bioinformatics* 17:560–63

173. Ward JM, Schroeder JI. 1994. Calcium-activated K^+ channels and calcium-induced calcium release by slow vacuolar ion channels in guard cell vacuoles implicated in the control of stomatal closure. *Plant Cell* 6:669–83

174. Wegner LH, de Boer AH. 1997. Properties of two outward-rectifying channels in root xylem parenchyma cells suggest a role in K^+ homeostasis and

long-distance signaling. *Plant Physiol.* 115:1707–19

175. Wegner LH, de Boer AH. 1997. Two inward K⁺ channels in the xylem parenchyma cells of barley roots are regulated by G-protein modulators through a membrane-delimited pathway. *Planta* 203:506–16

176. Wegner LH, Raschke K. 1994. Ion channels in the xylem parenchyma of barley roots. A procedure to isolate protoplasts from this tissue and patch-clamp exploration of salt passageways into xylem vessels. *Plant Physiol.* 105:799–813

177. Weiser R, Benturp FW. 1990. (+)Tubocurarine is a potent inhibitor of cation channels in the vacuolar membrane of *Chenopodium rubrum* L. *FEBS Lett.* 277:220–22

178. Weiser R, Bentrup FW. 1991. Charybdotoxin blocks cation-channels in the vacuolar membrane of suspension cells of *Chenopodium rubrum* L. *Biochim. Biophys. Acta* 1066:109–10

179. Weiser R, Blum W, Bentrup FW. 1991. Calmodulin regulates the Ca²⁺-dependent slow-vacuolar ion channel in the tonoplast of *Chenopodium rubrum* suspension cells. *Planta* 185:440–42

180. White PJ. 1993. Characterization of high-conductance, voltage-dependent cation channel from the plasma membrane of rye roots in planar lipid bilayers. *Planta* 191:541–51

181. White PJ. 1994. Characterization of a voltage-dependent cation-channel from the plasma membrane of rye (*Secale cereale* L.) roots in planar lipid bilayers. *Planta* 193:186–93

182. White PJ. 1996. Specificity of ion channel inhibitors for the maxi cation channel in rye root plasma membranes. *J. Exp. Bot.* 47:713–16

183. White PJ. 1996. The permeation of ammonium through a voltage-independent K⁺ channel in the plasma membrane of rye roots. *J. Membr. Biol.* 152:89–99

184. White PJ. 1998. The kinetics of quinine

blockade of the maxi cation channel in the plasma membrane of rye roots. *J. Membr. Biol.* 164:275–81

185. White PJ. 1999. The molecular mechanism of sodium influx to root cells. *Trends Plant Sci.* 4:245–46

186. Deleted in proof

187. White PJ, Broadley MR. 2000. Mechanisms of caesium uptake by plants. *New Phytol.* 147:241–56

188. White PJ, Lemtiri-Chlieh F. 1995. Potassium currents across the plasma membrane of protoplasts derived from rye roots: a patch-clamp study. *J. Exp. Bot.* 46:497–511

188a. White PJ, Piñeros M, Tester M, Ridout MS. 2000. Cation permeability and selectivity of a root plasma membrane calcium channel. *J. Membr. Biol.* 174:71–83

189. White PJ, Ridout MS. 1995. The K⁺ channel in the plasma membrane of rye roots has a multiple ion residency pore. *J. Membr. Biol.* 143:37–49

190. White PJ, Ridout MS. 1999. An energy-barrier model for the permeation of monovalent and divalent cations through the maxi cation channel in the plasma membrane of rye roots. *J. Membr. Biol.* 168:63–75

191. White PJ, Tester MA. 1992. Potassium channels from the plasma membrane of rye roots characterized into planar lipid bilayers. *Planta* 186:188–202

192. Whitehead LF, Day DA, Tyerman SD. 1998. Divalent cation gating of an ammonium permeable channel in the symbiotic membrane from soybean nodules. *Plant J.* 16:313–24

193. Wo ZG, Oswald RE. 1995. Unravelling the modular design of glutamate-gated ion channels. *Trends Neurosci.* 18:161–68

194. Yazaki Y, Asukagawa N, Ishikaway Y, Eiji O, Sakata M. 1988. Estimation of cytosolic free Mg²⁺ levels and phosphorylation potentials in mung bean root tips by in vivo ³¹P-NMR spectroscopy. *Plant Cell Physiol.* 29:919–24

195. Zhang WH, Ryan PR, Tyerman SD. 2001. Malate-permeable channels and cation channels activated by aluminum in the apical cells of wheat roots. *Plant Physiol.* 125:1459–72

196. Zhang WH, Walker NA, Tyerman SD, Patrick JW. 2000. Fast activation of a time-dependent outward current in protoplasts derived from coats of developing *Phaseolus vulgaris* seeds. *Planta* 211:894–98

197. Zhou Z, Gong Q, Ye B, Fan Z, Makielski JC, et al. 1998. Properties of HERG channels stably expressed in HEK 293 cells studied at physiological temperature. *Biophys. J.* 74:230–41

198. Zimmermann S, Nurnberger T, Frachisse JM, Wirtz W, Guern J, et al. 1997. Receptor-mediated activation of a plant Ca^{2+}-permeable ion channel involved in pathogen defense. *Proc. Natl. Acad. Sci. USA* 94:2751–55

Annu. Rev. Plant Biol. 2002. 53:109–30
DOI: 10.1146/annurev.arplant.53.091701.153921

REVEALING THE MOLECULAR SECRETS OF MARINE DIATOMS

Angela Falciatore and Chris Bowler

*Laboratory of Molecular Plant Biology, Stazione Zoologica "A. Dohrn," Villa Comunale,
I-80121 Naples, Italy; e-mail: chris@alpha.szn.it*

Key Words   iron, marine algae, photoperception, quorum sensing, silica

■ **Abstract**   Diatoms are unicellular photosynthetic eukaryotes that contribute close
to one quarter of global primary productivity. In spite of their ecological success in the
world's oceans, very little information is available at the molecular level about their biology. Their most well-known characteristic is the ability to generate a highly ornamented silica cell wall, which made them very popular study organisms for microscopists in
the last century. Recent advances, such as the development of a range of molecular tools,
are now allowing the dissection of diatom biology, e.g., for understanding the molecular
and cellular basis of bioinorganic pattern formation of their cell walls and for elucidating key aspects of diatom ecophysiology. Making diatoms accessible to genomics
technologies will potentiate greatly these efforts and may lead to the use of diatoms to
construct submicrometer-scale silica structures for the nanotechnology industry.

CONTENTS

DIATOMS AND ECOSYSTEM EARTH

The origin of life and the subsequent changes in oceanic productivity that followed
the evolution of photosynthesis have profoundly influenced the geochemistry
of the Earth for the past three billion years (29). In the contemporary ocean, marine

phytoplankton, made up of photosynthetic bacteria such as prochlorophytes and cyanobacteria, and eukaryotic microalgae, such as chromophytes (brown algae), rhodophytes (red algae), and chlorophytes (green algae), represent the major contributors of marine carbon fixation. In some regions of the ocean, these organisms can fix approximately the same amount of carbon, a few grams per square meter per day, as a terrestrial forest (82). Today, the oceans cover 70% of the Earth's surface, and on a global scale they are thought to contribute approximately one half of the total primary productivity of the planet.

Diatoms are a group of unicellular chromophyte algae that colonize the oceans down to depths to which photosynthetically available radiation can penetrate. They are thought to be the most important group of eukaryotic phytoplankton, responsible for approximately 40% of marine primary productivity (29). They also play a key role in the biogeochemical cycling of silica (84) owing to their requirement for this mineral for cell wall biogenesis.

There are well over 250 genera of extant diatoms, with perhaps as many as 100,000 species (71, 88) ranging across three orders of magnitude in size (about as many as land plants) and exhibiting a remarkable variety of shapes. The well-studied small-celled species (5–50 μm) tend to be most abundant at the beginning of spring and autumn, when nutrients are not limiting and when light intensity and day length are optimal for diatom photosynthesis. When nutrients run out they will often aggregate into flocs that sink quickly out of the photic zone. The giant diatoms (which can reach 2–5 mm in size) are ubiquitous in all oceans, and their abundance shows less seasonal variability. Their silica cell walls predominate in the sediments of the ocean floor, thus making them serious players in ocean biogeochemistry over geologically significant timescales (45). Besides planktonic diatoms, which are found in all open water masses, there are many benthic forms, growing on sediments or attached to rocks or macroalgae, and some species can also be found in soil (59). Diatoms also constitute a large proportion of the algae associated with sea ice in the Antarctic and Arctic. Furthermore, in warm oligotrophic seas it is possible to find symbioses between nitrogen-fixing bacteria and cyanobacteria and diatoms (91). In these areas the fixation of nitrogen by these endosymbionts contributes a significant amount of nitrogen to the ecosystem.

In spite of their ecological relevance, very little is known about the basic biology of diatoms (78). What are the molecular secrets behind their success? One possibility is that they have an extraordinary capacity for finding different adaptive solutions (e.g., physiological, biochemical, behavioral) to different environments. It has also been proposed that the major factor behind ecological success is their siliceous cell wall (see example in Figure 1). Smetacek (81, 82) has argued that the many different shapes and sizes of diatoms evolved to provide a robust first line of defense against various type of grazers, therefore being the functional equivalents of the waxy cuticles, trichomes, and spines of higher plants. Plankton defense systems are poorly studied, but an emerging idea is that protection against grazers may be an important factor in determining the composition and succession

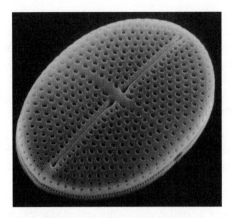

Figure 1   Electron micrograph of the diatom *Mastogloia binotata* (Grun.) cl. at 5800 times magnification. Photo courtesy of Keigo Osada.

of phytoplankton. If this is the case plankton evolution may be ruled as much by protection as by short-term adaptations or competition (82).

DIATOM BIOLOGY

General Characteristics

Diatoms are Bacillariophycea within the division Heterokontophyta (also known as Stramenopiles). Their most characteristic feature is the ability to generate a highly patterned external wall composed of amorphous silica [(SiO$_2$)n(H$_2$O)], known as the frustule (Figure 1; Figure 2, see color insert). This is constructed of two almost equal halves, with the smaller fitting into the larger like a Petri dish (Figure 2). The larger of the two halves is denoted the epitheca, and the inner one is denoted the hypotheca. Each theca is typically composed of two parts: the valve (which forms the larger outer surface) and a girdle (circular bands of silica attached to the edge of the valve) (Figure 2). The siliceous material of the frustule is laid down by largely unknown mechanisms in highly regular patterns that leave the wall beautifully ornamented. Pattern design is faithfully reproduced from generation to generation, implicating a strict genetic control of the process. The precision of this nanoscale architecture far exceeds the capabilities of present-day materials science engineering, indicating that the understanding of the process will one day be exploitable in nanotechnological applications (15a, 63, 70, 73).

Diatoms are generally classified into two major groups depending on the symmetry of their frustules (88). Centric diatoms are radially symmetrical, whereas pennate diatoms are elongated and bilaterally symmetrical. The former group tends to be planktonic, whereas the latter are benthic, living on sediments or other surfaces. Some of the pennate diatoms are able to glide along surfaces, owing to the presence of a crevice (known as raphe) within one or both of the frustules through which mucilage is secreted to aid movement (Figure 2).

Cell Division and Cell Wall Biogenesis

As far as is known, vegetative diatom cells are diploid. The normal asexual method of reproduction is by division of one cell into two, with each valve of the parent cell becoming an epitheca of the daughter cell (75). Each daughter cell must therefore generate a new hypotheca (Figure 3, see color insert). The process of frustule formation is only partially understood and is largely based on microscopical observations. Prior to cell division, the cell elongates, pushing the epitheca away slightly from the hypotheca, and the nucleus divides by an "open" mitosis. After the protoplast has divided in two by invagination of the plasma membrane, each daughter cell must generate a new hypotheca. Remarkably, this structure, which must cover one half of the cell, is commonly generated by the polarized generation of a huge vesicle known as the silica deposition vesicle (SDV). Hypotheca biogenesis involves the laying down of a precise silica lattice work followed by its coating with an organic matrix that prevents its dissolution. Once generated, the entire structure is then exocytosed, after which the two daughter cells can separate. This whole process is not well understood but has been reviewed recently (96).

Analysis of frustule composition has revealed the presence of specific organic components in addition to silica. Modified peptides known as silaffins together with putrescine-derived polyamines bind the silica scaffold extremely tightly and can only be removed following complete solubilization of silica with anhydrous hydrogen fluoride. Both the silaffins and polyamines can promote silica precipitation in vitro, generating a network of nanospheres with diameters between 100 nm and 1 μm, depending on the molecules used (51, 52). It is therefore possible that such components generate the basic building blocks for silica cell wall formation. *Cylindrotheca fusiformis* contains two major types of silaffins, denoted silaffin-1A (4 kDa) and silaffin-1B (8 kDa). Both are proteolytically derived from the product of a single gene, denoted *Sil1*.

Hydrogen fluoride-extractable material also contains other proteinaceous fractions of high molecular mass, denoted pleuralins (pleuralin-1 was formerly known as HEP200) (53, 54). Immunolocalization experiments have revealed that pleuralin-1 is specifically localized to the terminal girdle band (known as pleural band) of the epitheca but that it is also targeted to the pleural band of the hypotheca of a cell that has mitotically divided but not yet generated a hypotheca (54). It is therefore a terminal differentiation marker for the epitheca. A small gene family encodes pleuralin proteins in the *C. fusiformis* genome, which are localized together (53). All encode proteins with characteristic repeat domains.

Calcium-binding glycoproteins known as frustulins (49, 50) have been localized to the outer coating of diatom cell walls (87). To date, five different types of frustulins have been described, based on their different molecular weights: α-frustulin (75 kDa), β-frustulin (105 kDa), γ-frustulin (140 kDa), δ-frustulin (200 kDa), and ε-frustulin (35 kDa), all of which contain characteristic acidic cysteine-rich domains (ACR domains). The function of this domain is not yet known. Immunological studies have demonstrated that, unlike pleuralin-1, the frustulins are localized ubiquitously over the external surface of the cell wall (53, 87), although it is possible that individual frustulins have specific localization profiles. Because they are not an integral component of the siliceous cell wall, they are not thought to participate in the silica biomineralization process.

The chemical form of Si available to marine diatoms is mainly undissociated silicic acid, $Si(OH)_4$ (65). It is transported into diatom cells via novel membrane-localized silica transporters. Five different silicic acid transporter (SIT) genes have been isolated from *C. fusiformis* (40). These genes encode an integral membrane protein with 10 membrane-spanning domains and a long hydrophyllic carboxyl-terminal region containing coiled-coil domains, possibly involved in mediating interactions with other proteins. Heterologous hybridization experiments have revealed that such genes are likely to be present in other diatom species. This new class of transporter, which has no known homologs beyond the diatoms, has been functionally characterized as a silicic acid transporter by overexpressing the cDNA in *Xenopus laevis* oocytes (40, 40a). Analysis of *SIT* gene regulation indicated that they are differentially expressed during cell division and cell wall biogenesis (65). To explain these variations it has been hypothesized that different SIT proteins could have distinct transport characteristics or subcellular localizations.

The deposition of new siliceous valves between mitosis and daughter cell separation necessitates a precise coupling between silicon metabolism and the cell cycle (65). In several species, it has been observed that silica uptake precedes cell division (88). A series of silica-dependent steps have been identified during diatom cell division. Two arrest points appear universal, one at the G1/S boundary and another during G2/M associated with the construction of new valves (16). The arrest point at G1/S has been hypothesized to be indicative of a silica dependency for DNA synthesis, and the data of Vaulot and coworkers (89, 89a) suggest that this block serves to determine whether sufficient silica is available to allow completion of frustule biogenesis.

Diatom Sex

For the vast majority of diatom species, the Petri-dish nature of the frustule and its unusual mode of biogenesis lead to a reduction in size during successive mitotic divisions in one of the daughter cells (Figure 3). Mitotically dividing diatom populations therefore decrease in size over time. Regeneration of the original size typically occurs via sexual reproduction, followed by auxospore formation. Gametogenesis occurs once cells decrease in size to approximately 30–40% of the maximum diameter. This is known as the critical size threshold. The resulting male and female gametes combine to create a diploid auxopore that is larger than either parent. This newly created cell then proceeds along the asexual pathway until an appropriate trigger once again elicits gametogenesis. Sexual reproduction in diatoms involves a range of mechanisms (reviewed in 62). In centric diatoms, sex is almost universally oogamous, with flagellated male gametes. Within the pennate diatoms, there is much more variety, including anisogamy, isogamy, and automixis. Only fragmentary information is available because almost all studies are based on microscopic observation of what is a very rare event. Diatom sexuality is in fact limited to brief periods (minutes or hours) that may occur less than once a year in some species and that involve only a small number of vegetative cells within a population (62).

Sex can sometimes be induced when vegetative cells are exposed to unfavorable growth conditions. Factors such as light, nutrients, salinity, and temperature shifts (5, 89, 89a) have all been reported to induce the switch from asexual to sexual reproduction in some diatoms. Armbrust has identified a new gene family expressed during the onset of sexual reproduction in the centric diatom *Thalassiosira weissflogii*, by exploiting the fact that sexuality within this species can be induced by a shift in the diurnal light regime (3). Three of the sexually induced genes, *Sig1*, *Sig2*, and *Sig3*, are members of a novel gene family that encode proteins containing epidermal growth factor (EGF)-like domains present in extracellular proteins that promote cell-cell interactions during different stages of development in animals. The strong sequence similarity to components of the extracellular matrix (ECM) of mammalian cells is interesting in that the ECM is not present in higher plants. It has been suggested that the SIG polypeptides are involved in sperm-egg recognition (3).

Diatom Photosynthesis

Like in other photosynthetic eukaryotes, the photosynthetic apparatus of diatoms is housed within plastids inside the cell. The thylakoid membranes within the plastid have the typical structure of the Heterokontophyta, grouped into stacks (lamellae) of three, all enclosed by a girdle lamella (Figure 2) (88). Diatoms are brown in color owing to the presence of the accessory carotenoid pigment fucoxanthin, which is located together with chlorophyll *a* and *c* in their plastids. For this reason, diatom and other brown algal plastids have been called phaeoplasts, to distinguish them from the rhodoplasts and chloroplasts of red and green algae, respectively. Centric diatoms generally have large numbers of small discoid plastids, whereas pennate diatoms tend to have fewer plastids, sometimes only one (88). Fucoxanthin and chlorophylls are bound within the light-harvesting antenna complexes by fucoxanthin, chlorophyll *a/c*–binding proteins (FCP).

The FCP proteins are integral membrane proteins localized on the thylakoid membranes within the plastid, and their primary function is to target light energy to chlorophyll *a* within the photosynthetic reaction centers (72). In the pennate diatom *Phaeodactylum tricornutum*, two *FCP* gene clusters have been identified containing, respectively, four and two individual *FCP* genes (12). They show sequence similarity to the chlorophyll *a/b*–binding protein genes (*CAB*) of plants and green algae. Like CAB proteins, diatom FCP proteins are encoded in the nucleus. Although they are functionally and structurally strongly related to higher plant CAB proteins, the transport mechanisms that target them to the diatom plastid are very different, owing to the fact that diatom plastids are enclosed within four membranes rather than two membranes as in land plants (see Diatom Phylogeny below). The N-terminal translocation sequences of immature FCP proteins are in fact bipartite. One of these is likely to be utilized for translocation through the outer membrane, whereas the other (a more conventional plastid transit peptide) is utilized for transport through the inner two membranes (11). The former sequence resembles an endoplasmic reticulum signal peptide and, indeed, is capable of cotranslational import and processing by microsomal membranes. The process of plastid protein targeting and import has been more thoroughly studied for a related presequence from the γ-subunit of the plastid ATP synthase from the centric diatom *Odontella sinensis* (55).

In *T. weissflogii*, semiquantitative RT-PCR analysis of *FCP* gene expression revealed that transcript levels decrease during prolonged darkness and are highly induced following a subsequent shift to white light (58). In *Dunaliella tertiolecta*, a marine green alga, a shift from high to low irradiance leads to a rapid induction of *CAB* gene expression (57). It has been proposed that this regulation allows increased production of light-harvesting pigment complexes in conditions of low light intensity and that redox signaling from the chloroplast to the nucleus controls the process (25). It will be interesting to determine whether such mechanisms also exist in diatoms and whether they contribute to the ability of diatoms to photosynthesize at optimal levels over a wide range of light intensities (30).

One potentially significant observation that has recently been made is that diatoms may be capable of C_4 photosynthesis (77). This is a specialized form of photosynthesis that is restricted to a few land plants, such as sugar cane and maize, and that allows a more efficient utilization of available CO_2. The report by Reinfelder and colleagues is the first description of C_4 photosynthesis in a marine microalga, and the data suggest that C_4 carbon metabolism may be confined to the cytoplasm, away from the RUBISCO-dependent reactions within the plastid. If shown to be a universal feature of diatoms, C_4 photosynthesis may provide a further explanation for their ecological success in the world's oceans, in that it provides a means of storing carbon for use in less favorable conditions. Identification of the genes encoding key enzymes of C_4 metabolism will provide the molecular tools for exploring the universality of this type of photosynthesis in diatoms.

PERCEPTION OF ENVIRONMENTAL SIGNALS

Living organisms are constantly bombarded with information to which they must react. In plant cells, both external and internal signals are amplified and communicated by complex signal transduction pathways, most of which are initiated by the activation of receptor proteins (85). The marine environment potentially contains a great number of physicochemical signals that can be utilized to control organismal adaptive responses to changing local conditions. The importance of molecular sensing of environmental signals in diatoms has recently been examined by studying calcium-dependent signal transduction in *P. tricornutum* (28). Transgenic diatom cells containing the calcium-sensitive photoprotein aequorin were generated to follow changes in calcium homeostasis in response to a range of relevant environmental stimuli. The results revealed sensing systems for detecting and responding to fluid motion, osmotic stress, and nutrient limitation. In particular, an exquisitely sensitive calcium-dependent signaling mechanism was induced by iron, a key nutrient controlling diatom abundance in the ocean (see below). Based on our knowledge of calcium signaling in other organisms, the physiological responses of diatoms to environmental changes are therefore likely to be regulated by sense-process-respond chains involving specific receptors and feedback mechanisms, whose activity is determined by the previous history of the cell. The further characterization of signal transduction pathways that tune cellular metabolism to ambient light and nutrient levels will be an important step for understanding the molecular mechanisms that contribute to the ecological success of diatoms.

Nutrients

External nutrient concentrations are key regulators of phytoplankton growth. In the marine environment, nutrients such as nitrate, silicate, and phosphate are extremely important, and strong evidence also implicates dissolved iron as being a limiting resource for phytoplankton growth in many regions of the oceans (see 29 and references therein). The "iron hypothesis" has been tested and supported through in situ iron fertilization experiments in the equatorial Pacific Ocean and elsewhere (15, 18, 64), which resulted in a clear and unambiguous physiological response to the addition of iron—a massive phytoplankton bloom of predominantly diatoms. Subsequently, some effort has been dedicated to developing methods to determine iron deficiency in phytoplankton in situ. Flavodoxin is the first such example of a molecular marker for studying diatom cell physiology in their natural context (56). Because the appearance of this protein correlates with iron deficiency, flavodoxin has been used as an immunological probe to map the degree of iron stress in natural populations.

To better understand some of the nutrient-sensing mechanisms in marine algae, it will be necessary to clone genes encoding nutrient transporters. Besides the previously described silicic acid transporters (SIT), Hildebrand & Dahlin recently reported the cloning and initial characterization of the first genes encoding nitrate transporters (NAT) in a marine organism, from the diatom *C. fusiformis* (39). The NAT proteins are predicted to have 12 membrane-spanning domains with hydrophyllic amino- and carboxy-termini that are located in the cytoplasm. The proteins show significant homology to the nitrate transporter family NRT2 of *Aspergillus nidulans*, but no similarity to the NTR1 class of *Arabidopsis*. *NAT* gene expression in *C. fusiformis* was sensitive to both the level and the type of

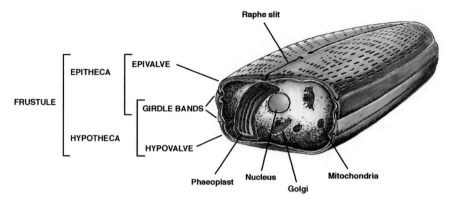

Figure 2 Schematic overview of the general structural features of a pennate diatom. See text for further details.

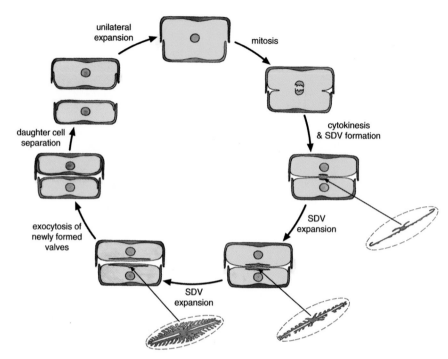

Figure 3 Schematic overview of mitotic cell division and hypovalve formation in a pennate diatom. Following mitosis and cytokinesis, a specialized vesicle known as the silica deposition vesicle (SDV) forms between the nucleus and the plasma membrane, at a position where the new hypovalve will be generated. The SDV elongates into a tube and then spreads out perpendicularly to eventually form a huge vesicle along one side of the cell. A new valve is formed within the SDV by the transport of silica, proteins, and polysaccharides into it, and once complete, it is exocytosed from the cell. The two daughter cells can then separate and grow unidirectionally along the cell division axis by the biogenesis of girdle bands, which are also formed within SDVs. However, one daughter cell is always smaller than the other, owing to the different sizes of the parental thecae from which they are derived. For more details see text and Reference (96). Cell division in centric diatoms is similar, although the timing of events can be different (75).

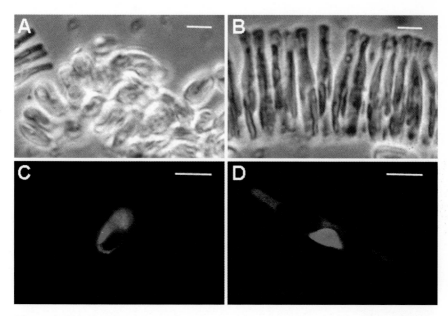

Figure 4 The diatom *P. tricornutum*. Oval (*A, C*) and fusiform (*B, D*) morphotypes. The *top panels* show transmitted light images of cells. Oval cells produce a large amount of mucilage and tend to form aggregates (*A*). Fusiform cells grown on agar plates can form chains (*B*). The *lower panels* show fluorescence of oval (*C*) and fusiform (*D*) cells genetically transformed with a cytosolic GFP construct. *Green*, GFP; *red*, chlorophyll fluorescence; *blue*, nuclei. Scale bars are 5 μm. Images courtesy of O. Malakhova & M. Mutarelli.

nitrogen source added: Transcript levels were high in the presence of nitrate, were lower in nitrite-grown cultures, and were highly repressed when ammonium was used as the nitrogen source. Southern hybridization experiments indicated that at least four copies of *NAT* genes are present in *C. fusiformis* and that multiple copies are present in other diatom species, thus raising the exciting possibility of developing "universal" probes for estimating nitrate transport capacity in marine samples.

Interestingly, we have recently found expressed sequence tags (ESTs) of *P. tricornutum* encoding proteins with high homology to several other channel and transporter proteins (e.g., copper transporter, ammonium transporter, ABC transporter, glycine betaine transporter, inorganic phyrophosphate transporter) (79). The generation of ESTs therefore provides a powerful new approach for elucidating nutrient transport systems in diatoms.

Owing to the important role played by iron in controlling phytoplankton growth, it will be important to clone iron transporter genes. Recently a gene encoding a membrane protein directly involved in Fe(III) uptake was cloned from maize (19) by complementation of a yeast iron uptake mutant. Expression of the maize gene in the mutant restored growth on Fe(III)-phytosiderophore media. Similar complementation experiments in yeast could be useful for identifying such proteins in diatoms.

Light

Light is another essential factor regulating the abundance of photosynthetic organisms in the oceans. In addition to constituting their principal energy source, it provides microalgae with positional information from their local environment. As previously mentioned, it is known that marine phytoplanktonic organisms are extremely efficient at photoacclimation and that they are highly sensitive to changes in spectral quality and light intensity (30). Although much is known about the mechanisms of photoperception and signal transduction in plants living on land, the marine environment imposes different constraints. The spectral distribution of solar irradiance is identical only on land and at the surface of water. Light fluence and quality change drastically with depth, and in oceanic waters, blue light predominates at greater depths (46). Furthermore, land plants are immobile, whereas many algae live suspended along the water column and can experience dramatic changes in the light field. Consequently, phytoplanktonic organisms are likely to have developed different ecological strategies to adapt to the variability of the light conditions. For example, complementary chromatic adaptation is a process by which cyanobacteria alter the composition of their photosynthetic light-harvesting apparatus (the phycobilisomes) in response to the spectral quality of the available light (36). Kehoe & Grossman reported that the chromatic adaptation sensor is similar to plant phytochrome photoreceptors (44).

Recently, Béjà et al. (7, 8) described a new type of bacterial rhodopsin, discovered through genomic analyses of naturally occuring marine bacterioplankton, that can mediate light-driven proton-pumping activity. These data indicate that an unsuspected phototrophic pathway of bacterially mediated light-driven energy generation may occur in the oceans and influence significantly the fluxes of carbon and energy.

Another method that phytoplankton utilize to respond to light fluctuations is movement. Many photosynthetic flagellate algae actively search for optimal light conditions by means of phototaxis (80). The green alga *Chlamydomonas reinhardtii* has developed a rudimentary light-sensing system that enables it to measure the light intensity and direction in an eyespot, a highly pigmented organelle in an almost equatorial position of the cell. The photoexcitation of a rhodopsin-type photoreceptor in the eyespot generates a photoreceptor current across the eyespot region of the cell membrane that can induce flagellar movement.

Planktonic diatoms are not able to move actively because they do not have flagella, thus they are subject to passive movements such as sinking and water turbulence. The diatom *Ditylum brightwellii* can move up and down the water column by changing the ionic composition of its large vacuole (32). Other species can control their buoyancy by making colonies, which slows settling in the water column because of the increase in surface to volume ratio (59). Some benthic diatoms, on the other hand, move by a mechanism involving the secretion of mucilage through their raphe slit (59). Indeed, some diatoms living in sediments have been observed to migrate up to the surface before dawn and down again during the first hours of darkness or to move up and down according to the tides (38). In some locations, this phenomenon is so dramatic that the sand can be observed to change color. Some, albeit limited, action spectra have been performed to study diatom phototaxis (38).

Light-mediated plastid reorientation has been reported in different diatom species. The action spectrum for plastid orientation in *Pleurosira laevis* revealed that two different photoreceptors could be involved in this response: a green light-sensitive photoreceptor that controls plastid dispersion and a UV-A/blue light-sensitive system that controls plastid assemblage (33). It is logical to propose that a phototropin-type photoreceptor, homologous to *Arabidopsis* NPL1, is responsible for the latter phenomenon (43).

To further understand photoperception mechanisms in diatoms, Leblanc et al. (58) performed action spectra of *FCP* gene expression using the Okazaki large spectrograph. These experiments revealed responses to UV-A, blue, green, red, and far-red light wavelengths, suggesting the presence of cryptochrome-, rhodopsin-, and phytochrome-like receptors in marine diatoms. The apparent presence of phytochrome-mediated responses is of particular interest given the low fluences of red and far-red light wavelengths in marine environments. In fact, because water selectively absorbs light of longer wavelengths, most of the light below 15 m is confined to the blue-green (400–550 nm) region of the spectrum (46). However, recent estimates of irradiance at red wavelengths, produced by the natural fluorescence of chlorophyll in marine phytoplankton, support the idea that these signals could also be perceived and utilized as information even below the depth of maximal penetration of solar-derived red light (M. Ragni & M. Ribera, manuscript in preparation). Clearly then, isolation of phytochrome photoreceptors and the subsequent elucidation of their function in diatoms, whose phylogeny is separated from that of higher green plants and from other algae (see below) (79), will provide exciting information about the photoperception mechanisms utilized in the aquatic environment and promises to reveal new insights into photoreceptor evolution.

Neighbor Perception

In addition to the physico-chemical signals described above, marine organisms are also likely to be able to sense and respond to the presence of other organisms, both friend and foe. Interestingly, marine diatoms are able to control the population size of the zooplankton that eat them. Miralto et al. (69) reported that diatoms synthesize specific aldehyde molecules that can arrest mitosis during egg development of their principal grazers, the copepods. These molecules also display antimitotic activity in diatoms (17). In addition to being defined as "defense molecules," these aldehydes may therefore have a "signaling" role for controlling diatom population size, e.g., by acting as a suicide trigger for bloom termination. Such a proposal is a radical alternative to the traditional dogmas about bloom decline based on nutrient depletion.

A related phenomenon that has been well studied in bacteria is quorum sensing (74). Quorum sensing is a cell-density-dependent regulation of specific gene expression in response to extracellular chemical signals, produced by the bacteria themselves. The most well-studied signaling system is the acyl-homoserine lactone (Acyl-HSL) system, used by a large number of Gram-negative bacteria that interact with plant and animal hosts. When the population reaches a critical density within a host, the Acyl-HLS signal, produced by specific enzymes and detected by specific receptors, activates the expression of specific genes necessary for continued success in the host (74). Signals such as these enable individual bacteria to function in a more multicellular manner by coupling gene expression to the attainment of an optimal population size.

One example of intercellular communication in diatoms can be inferred from studies of iron uptake mechanisms in algal communities (41). In this work, it was demonstrated that different algae utilize different complexed forms of iron, e.g., cyanobacteria can transport Fe-complexed siderophores, whereas diatoms preferentially utilize iron complexed in porphyrins. Because siderophores are important molecules in cyanobacterial metabolism, as are porphyrins in diatom metabolism, such specifically differentiated uptake mechanisms provide a means whereby algal cells of the same type can "altruistically" preserve the well being of their own community, but not of others. This could also be used for quorum sensing.

Similarly, the natural fluorescence of chlorophyll in marine phytoplankton could be utilized by individual cells to detect the presence of neighboring organisms (M. Ragni & M. Ribera, manuscript in preparation). Confirmation of the existence of such a fluorescence-based system of neighbor perception would be of great interest and would have far-reaching implications for phytochrome photoreceptor evolution.

DIATOM PHYLOGENY

The plastids of all photosynthetic organisms are likely to have arisen at least 1.5 billion years ago from the engulfment of a photosynthetic bacterium by a unicellular eukaryotic heterotroph (86, 88). Analysis of the sequences of plastid genomes from a range of eukaryotic algae and from higher and lower plants clearly supports the hypothesis that all photosynthetic eukaryotes are derived from a single endosymbiotic event involving a photosynthetic bacterium highly similar to extant cyanobacteria (9, 21, 48, 76). Functional analysis of the photosynthetic apparatus from different algae also supports this view (e.g., 92). Hence, all plastids appear to have a monophyletic origin.

Current knowledge suggests that the initial endosymbiotic event gave rise eventually to two major plastid lineages: chloroplasts and rhodoplasts. Green algae, as well as their descendants, the higher plants, contain chloroplasts that are characterized by the presence of stacked thylakoid membranes and the use of chlorophyll *a* and *b* for light harvesting. Red algae, on the other hand, contain rhodoplasts, which utilize chlorophyll *a* and phycobilisomes for the capture of light energy.

Chromophyte algae, such as diatoms, differ fundamentally from the majority of photosynthetic eukaryotes: Whereas the plastids of red algae, green algae, and plants are normally surrounded by two membranes, diatom plastids have four membranes. It is therefore believed that diatoms and related chromophyte algae arose following a secondary endosymbiotic event in which a eukaryotic alga was engulfed by a second eukaryotic heterotroph (21, 34). In such a scenario, a flagellate host contributed the nucleus, endomembranes, and mitochondria to the chimera, whereas the photosynthetic endosymbiont provided its plastid and plasma membrane (that perhaps became the inner three membranes of the plastid). Such a hypothesis is supported by the finding of a second nucleus (the nucleomorph) between the outer and inner two membranes in some chromophytes, presumably derived from the nucleus of the ancestral algal endosymbiont (21). The nucleomorph of the chromophyte *Guillardia theca* was recently sequenced (22). The genome is highly reduced (0.55 Mb) and lacks almost all genes for metabolic functions, which have been transferred to the nucleus of the secondary host. There are, however, 30 genes for plastid-localized proteins, and all these genes are essential for plastid function, presumably explaining why the nucleomorph has persisted during evolution.

Analysis of plastid ribosomal RNAs as well as comparisons of plastid genomes strongly support the hypothesis that diatom plastids were derived from a secondary endosymbiosis involving an alga from the rhodoplast lineage (10, 21, 48). However, diatoms have not retained phycobilins for light harvesting and instead use chlorophyll *a* and *c* together with the brown carotenoid fucoxanthin (chlorophyll *b* is never found in diatoms).

The ancestral heterotrophic host that paired with the red alga to give rise to the phaeoplast lineage is believed to have been a small unicellular flagellate belonging to the Oomycetes (lower fungi) and similar to *Cafeteria roenbergensis* (66, 88).

Guillou et al. (37) have subsequently identified a new class of biflagellate algae, the Bolidophyceae, that appears to be a sister group of the diatoms. Phylogenetic analysis based on SSU rDNA sequences together with the discovery of fucoxanthin as a major carotenoid strongly suggest that these newly identified *Bolidomonas* species could be similar to the ancestral heterokonts that gave rise to the diatom lineage (66). This hypothesis is also consistent with the most recent eukaryotic phylogenetic trees (6).

The fossil record suggests that the secondary endosymbiotic event is unlikely to have occurred much before the Permian-Triassic boundary, between 259 and 285 million years ago, just prior to the end-Permian mass extinction (66, 67). It has been proposed that the combination of autotrophy with the ability to form resting spores in the heterokont algae may have allowed them to survive the mass extinction and to subsequently diversify into the many heterokont classes that are currently in existence (67). The oldest diatom fossils clearly represent centric species, whereas pennate diatoms appear to have evolved much later (88).

In phylogenetic terms, diatoms therefore appear to have long since lost contact with the green algal lineage from which higher plants are derived and have followed a parallel evolutionary path for at least 650 million years (9). Analysis of several thousand ESTs generated by random sequencing of *P. tricornutum* cDNAs indeed confirms that many diatom genes are more homologous to animal rather than plant counterparts (79).

TOWARD A DIATOM MODEL SPECIES

The arrival of the new millenium has brought a paradigm shift in biological research. No longer are we confined to the study of single genes and single isolated phenomena, as was the case for the past 30 years or so. The completion of several genome projects and the sheer volume of sequence data available in the public databases now allow the simultaneous study of several thousand genes as well as the possibility of finely dissecting highly complex processes at the whole-genome level.

In the year 2002, where does diatom research stand? Perhaps up until now, the most significant success stories have been a realization of the importance of diatoms for aquatic and marine ecosystems and a convincing theory of their phylogenetic origins (10, 21, 48, 66, 88). But their unique origins and ecological success imply that they possess many novel cellular characteristics.

Unfortunately, many of the molecular secrets of diatoms still await discovery, mainly as a result of a complete lack of the necessary methodologies. However, in the last few years, a serious effort has been made to establish the molecular-based techniques required to understand the basic biology of these organisms (78). The recent development of genetic transformation systems for a limited number of diatoms has removed one of the major bottlenecks for such studies (1, 24, 26, 31). These systems are based on helium-accelerated particle bombardment of exogenous DNA, followed by selection of transfected cells using antibiotics. An important factor behind the success of these experiments was the realization that endogenous promoter and regulatory sequences are necessary to drive the expression of foreign genes.

Genetic transformation technologies are most advanced for the pennate diatom *P. tricornutum* in which a range of antibiotic resistance genes can be used to select for transgenic clones, including phleomycin, kanamycin, and nourseothricin (1, 26, 94). Even though transformation efficiencies are low, on the order of 10^{-6} per μg plasmid DNA, it is possible to co-transform plasmids without selectable markers at an efficiency of approximately 60%. Reporter genes commonly used in other systems such as chloramphenicol acetyl transferase, luciferase, β-glucuronidase, and green fluorescent protein (GFP) (27) can now be easily introduced and expressed in *P. tricornutum*. These reporter genes can provide the molecular tools for many experimental applications, such as identifying sequences important for controlling the transcription of genes and for defining the subcellular localization of specific proteins.

Genetic transformation is also an essential tool for enhancing the applied aspects of diatom research. Dunahay et al. (23, 24) attempted to enhance lipid biosynthesis in *Cyclotella cryptica* by introducing multiple copies of the acetyl CoA carboxylase gene. Most significantly, Zaslavskaia et al. (95) recently reported the trophic conversion of *P. tricornutum*, an obligate photoautotroph, into a heterotroph by metabolic engineering. In this work, genes encoding glucose transporters from human erythrocytes (*glut1*) or from the microalga *Chlorella kessleri* (*hup1*) were expressed in *P. tricornutum*, and the transgenic cells exhibited glucose uptake and were able to grow in the absence of light. The trophic conversion of *P. tricornutum* and other diatoms may increase the use of fermentation technology for large-scale commercial exploitation of diatoms by overcoming the financial and logistical limitations associated with light-dependent growth. Moreover, this conversion will facilitate the generation of photosynthetic mutants, which will help researchers to understand diatom photosynthesis, as has proved so successful in the green alga *C. reinhardtii* (35, 38a).

What should be the major future objectives in diatom research? First and foremost, research should concentrate on the novel aspects of diatom biology. In our opinion the most important research objectives can be classified into three groups:

1. ecological: elucidation of the molecular mechanisms underlying the ecological success of diatoms;

2. cell biological: elucidation of novel protein targeting mechanisms; diatom life cycles, including the regulation of critical size thresholds and diatom sex; silica-based bioinorganic pattern formation and its genetic basis;

3. nanotechnological: harnessing the novel mechanisms responsible for the fabrication of micrometer-scale silica structures for new industrial applications.

For all of these aspects it is essential that multiparallel molecular genetic approaches be developed to underpin research efforts. First and foremost, more information about the characteristics of diatom genomes is required: e.g., What is the range in genome size from species to species? What is the corresponding range in chromosome number? Can diatom sex be controlled to allow genetic crosses to be performed under defined laboratory conditions? Are there significant differences in GC content, codon usage, and methylation?

Although there are some answers to these questions in the literature (e.g., 20, 42, 47, 79, 90), the available information is fragmentary and incomplete. A clear priority is simply to catalog more diatom genes, e.g., through EST programs, as has recently been reported in *P. tricornutum* (79). In this regard, it is incredible that up until last year the public databases of DNA sequences contained less than 50 accessions for nuclear-encoded protein coding sequences from diatoms. EST programs in other diatom species should be initiated to allow interspecies differences to be determined.

With increasing interest in understanding the ecological importance of diatoms, it has become clear that more molecular tools must be developed. Moreover, almost all studies have been performed on pennate diatom species such as *P. tricornutum* and *C. fusiformis*, neither of which possesses highly ornamented cell walls nor has significant ecological importance. However, it is difficult to identify a diatom species that is of universal ecological importance. Perhaps the closest is *Skeletonema costatum*, which is more or less ubiquitous in all coastal waters (59, 88). Genetic transformation has not yet been reported in this species.

In addition to extending technologies to ecologically relevant diatom species, it is important to improve existing transformation efficiencies. If several hundreds or thousands of independent transformants could be obtained from a single transformation experiment, it would be possible to generate whole libraries containing different insertions. This could be useful to generate insertionally mutagenized libraries in which every single diatom gene had been randomly inactivated. It would also permit the rational utilization of reverse genetics approaches to identify novel promoters or protein targeting sequences, e.g., by using promoterless or ATG-less GFP constructs. Establishing other technologies, e.g., for inactivating gene expression, is also important. Because the limited information available suggests that homologous recombination is unlikely in diatoms (26), alternative systems must be developed to allow the specific inactivation of specific genes, such as antisense and sense suppression, and RNA interference (83, 93). Owing to the universality of the basic cellular mechanisms exploited in these approaches, there is a good likelihood that they will be effective in diatoms.

It is reasonable to expect that within the next five years a diatom genome will be sequenced. Before this happens, the diatom research community will have to decide which species to use. The situation is in some ways analogous to that of the early 1990s when the plant research community decided to focus efforts on *Arabidopsis thaliana*. At the time, there was considerable controversy over the decision—tobacco had been used much more extensively in molecular research because of the ease with which it could be transformed, and the agricultural lobby was much more in favor of sequencing the genome of a plant of agricultural interest. Nonetheless, the choice to use *Arabidopsis* prevailed and now, with the availability of its complete genome sequence (2) as well as powerful functional genomics platforms in several research centers worldwide, it is clear that the right decision was made.

Which diatom species should be chosen for genome-level research? Essential requirements include:

1) ease of growth under defined laboratory conditions,
2) ease of genetic transformation,
3) small genome size,
4) availability of genetic-based methodologies (isolation of mutants, a genome map, sexual crosses),
5) knowledge of basic biochemistry.

Currently, no single diatom species can fulfill these criteria and represent, at the same time, a good model from an ecological point of view. Those that come closest are *P. tricornutum* and *C. fusiformis*, at least for criteria 1, 2, and 5. Both have also been reasonably well studied in terms of ecology and physiology. However, sex has not been described in either of them. The genome size of *P. tricornutum* was measured as between 55 and 100 Mb (20, 90), the smallest of any diatom examined in previous work (90). Subsequent measurements by us (79) and by M. Hildebrand (personal communication) have found the *P. tricornutum* genome to be more on the order of 15 Mb, slightly higher than that of the yeast *Saccharomyces cerevisiae*, whose sequence was completed in 1997 (68), whereas the *C. fusiformis* genome is approximately 60 Mb. It is important to note that several thousand ESTs have been generated from *P. tricornutum*, and its genome properties have been characterized in much more detail than any other diatom (79). Its genome displays very low levels of methylation, normally associated with transcriptionally inactive DNA, suggesting that its genome is gene dense with only low amounts of noncoding "junk" DNA (42). It would therefore appear to be the best choice for a sequencing initiative. Furthermore, recent research in our laboratory using transgenic marker genes suggests that it is possible to perform sexual crosses with this species (O. Malakhova & C. Bowler, unpublished observations).

However, *P. tricornutum* is a rather atypical diatom in that it is polymorphic. It exists as three different morphotypes (oval, fusiform, and triradiate), which are only partially silicified (Figure 4, see color insert) (13, 14, 60, 61). Nonetheless, phylogenetic analysis performed on 18S rRNA places it in the middle of the pennate diatom lineage (D. Vaulot, personal communication). It would therefore appear that *P. tricornutum* is the most appropriate diatom for a genome sequencing project, at least for pennate diatoms.

A parallel project could also be considered for a centric diatom, although in this case it is very difficult to choose a representative model. Even less molecular work has been conducted with centric diatoms, in spite of their enormous ecological relevance. One of the more commonly studied is *T. weissflogii*, which could be transiently transfected with a *GUS* reporter gene (26). Furthermore, competitive semiquantitative RT-PCR techniques have been optimized for gene expression studies with this organism (58). At this time, the only centric diatom that can be stably transformed is *C. cryptica* (24). Armbrust recently characterized the structural features of the eight known nuclear genes of *T. weissflogii*, with the goal of furthering the potential of this diatom for molecular studies (4). However, the development of molecular tools for this diatom is just beginning, and the time might be premature to determine if it could be the best centric diatom for a genome sequence project. As an alternative, some researchers have proposed *Thalassiosira pseudonana*, which appears to have a genome size similar to that of *P. tricornutum* (M. Hildebrand, personal comunication). At press time, this diatom was on the list of organisms to be sequenced by the U.S. Department of Energy (http//www.er.doc.gov/production/ober/EPR/mig_cont.html).

In conclusion, although diatom research has made some important advances in recent years, it is clear that radical measures are required to make it accessible to the enormously powerful genomic and postgenomic research platforms. Given the novelty and potential applicability of certain aspects of diatom ecology and cell biology, this is surprising. However, now that a range of molecular technologies are in place, we hope that more researchers will be attracted to this field so that progress can be accelerated.

ACKNOWLEDGMENTS

We thank all the numerous colleagues who have helped us to understand diatom biology over the past years. We are especially grateful to Mark Hildebrand and Daniel Vaulot for sharing unpublished information and to Ian Nettleton for the illustrations. We apologize that, owing to size restrictions, it has not been possible to discuss all the work of our colleagues. Finally, this review is dedicated to the memory of Gaetano Salvatore, without whom our research on diatoms would never have begun.

Visit the Annual Reviews home page at www.annualreviews.org

LITERATURE CITED

1. Apt KE, Kroth-Pancic PG, Grossman AR. 1997. Stable nuclear transformation of the diatom *Phaeodactylum tricornutum*. *Mol. Gen. Genet.* 252:572–79

2. *Arabidopsis* Genome Initiative. 2000. Analysis of the genome sequence of the flowering plant *Arabidopsis thaliana*. *Nature* 408:796–815

3. Armbrust EV. 1999. Identification of a new gene family expressed during the onset of sexual reproduction in the centric diatom *Thalassiosira weissflogii*. *Appl. Env. Microbiol.* 65:3121–28

4. Armbrust EV. 2000. Structural features of nuclear genes in the centric diatom *Thalassiosira weissflogii* (Bacillariophyceae). *J. Phycol.* 36:942–46

5. Armbrust EV, Chisholm SW. 1990. Role of light and the cell cycle on the induction of spermatogenesis in a centric diatom. *J. Phycol.* 26:470–78

6. Baldauf SL, Roger AJ, Wenk-Siefert I,

Doolittle WF. 2000. A kingdom-level phylogeny of eukaryotes based on combined protein data. *Science* 290:972–77

7. Béjà O, Aravind L, Koonin EV, Suzuki MT, Hadd A, et al. 2000. Bacterial rhodopsin: evidence for a new type of phototrophy in the sea. *Science* 289:1902–6

8. Béjà O, Spudich EN, Spudich JL, Leclerc M, DeLong EF. 2001. Proteorhodopsin phototrophy in the ocean. *Nature* 411: 786–89

9. Besendahl A, Qiu YL, Lee J, Palmer JD, Bhattacharya D. 2000. The cyanobacterial origin and vertical transmission of the plastid tRNA(Leu) group-I intron. *Curr. Genet.* 37:12–23

10. Bhattacharya D, Medlin L. 1995. The phylogeny of plastids: a review based on comparisons of small-subunit ribosomal RNA coding regions. *J. Phycol.* 31:489–98

11. Bhaya D, Grossman A. 1991. Targeting proteins to diatom plastids involves transport through an endoplasmic reticulum. *Mol. Gen. Genet.* 229:400–4

12. Bhaya D, Grossman AR. 1993. Characterization of gene clusters encoding the fucoxanthin chlorophyll proteins of the diatom *Phaeodactylum tricornutum. Nucl. Acids Res.* 21:4458–66

13. Borowitzka MA, Chiappino ML, Volcani BE. 1977. Ultrastructure of a chain-forming diatom *Phaeodactylum tricornutum. J. Phycol.* 13:162–70

14. Borowitzka MA, Volcani BE. 1978. The polymorphic diatom *Phaeodactylum tricornutum*: ultrastructure of its morphotypes. *J. Phycol.* 14:10–21

15. Boyd PW, Watson AJ, Law CS, Abraham ER, Trull T. 2000. A mesoscale phytoplankton bloom in the polar Southern Ocean stimulated by iron fertilization. *Nature* 407:695–702

15a. Brott LL, Nait RR, Pikas DJ, Kirkpatrick SM, Tomiln DW, et al. 2001. Ultrafast holographic nanopatterning of biocatalytically formed silica. *Nature* 413:29–93

16. Brzezinski MA, Olson RJ, Chisholm SW. 1990. Silicon availability and cell-cycle progression in marine diatoms. *Mar. Ecol. Prog. Ser.* 67:83–96

17. Casotti R, Mazza S, Ianora A, Miralto A. 2001. *Strategies to reduce mortality in marine and freshwater phytoplankton.* ASLO Aquat. Sci. Meet. Spec. Sess., Albuquerque, New Mex.

18. Coale KH, Johnson KS, Fitzwater SE, Gordon RM, Tanner S, et al. 1996. A massive phytoplankton bloom induced by an ecosystem-scale iron fertilization experiment in the equatorial Pacific Ocean. *Nature* 383:495–501

19. Curie C, Panaviene Z, Loulergue C, Dellaporta SL, Briat JF, et al. 2001. Maize *yellow stripe1* encodes a membrane protein directly involved in Fe(III) uptake. *Nature* 409:346–49

20. Darley WM. 1968. Deoxyribonucleic acid content of the three cell types of *Phaeodactylum tricornutum* Bohlin. *J. Phycol.* 4:219–20

21. Delwiche CF, Palmer JD. 1997. The origin of plastids and their spread via secondary symbiosis. In *Origins of Algae and Their Plastids*, ed. D Bhattacharya, pp. 53–86. Vienna/New York: Springer Verlag

22. Douglas S, Zauner S, Fraunholz M, Beaton M, Penny S, et al. 2001. The highly reduced genome of an enslaved algal nucleus. *Nature* 410:1091–96

23. Dunahay TG, Jarvis EE, Dais SS, Roessler PG. 1996. Manipulation of microalgal lipid production using genetic engineering. *Appl. Biochem. Biotech.* 57/58:223–31

24. Dunahay TG, Jarvis EE, Roessler PG. 1995. Genetic transformation of the diatoms *Cyclotella cryptica* and *Navicula saprophila. J. Phycol.* 31:1004–12

25. Escoubas J-M, Lomas M, LaRoche J, Falkowski PG. 1995. Light intensity regulation of *cab* gene transcription is signaled by the redox state of plastoquinone pool. *Proc. Natl. Acad. Sci. USA* 92:10237–41

26. Falciatore A, Casotti R, Leblanc C, Abrescia C, Bowler C. 1999. Transformation of nonselectable reporter genes in marine diatoms. *Mar. Biotech.* 1:239–51

27. Falciatore A, Formiggini F, Bowler C. 2001. Reporter genes and in vivo imaging. In *Molecular Plant Biology: A Practical Approach*, ed. P Gilmartin, C Bowler, 2:265–83. Oxford, UK: Oxford Univ. Press

28. Falciatore A, Ribera D'Alcalà M, Croot P, Bowler C. 2000. Perception of environmental signals by a marine diatom. *Science* 288:2363–66

29. Falkowski PG, Barber RT, Smetacek V. 1998. Biogeochemical controls and feedbacks on ocean primary production. *Science* 281:200–6

30. Falkowski PG, LaRoche J. 1991. Acclimation to spectral irradiance in algae. *J. Phycol.* 27:8–14

31. Fischer H, Robl I, Sumper M, Kröger N. 1999. Targeting and covalent modification of cell wall and membrane proteins heterologously expressed in the diatom *Cylindrotheca fusiformis* (Bacillariophyceae). *J. Phycol.* 35:113–20

32. Fisher AE, Harrison PJ. 1996. Does carbohydrate content affect the sinking rate of marine diatoms? *J. Phycol.* 32:360–65

33. Furukawa T, Watanabe M, Shihira-Ishikawa I. 1998. Green- and blue-light-mediated chloroplast migration in the centric diatom *Pleurosira laevis*. *Protoplasma* 203:214–20

34. Gibbs SP. 1981. The chloroplasts of some algal groups may have evolved from endosymbiotic eukaryotic algae. *Ann. New York Acad. Sci.* 361:193–208

35. Grossman AR. 2000. *Chlamydomonas reinhardtii* and photosynthesis: genetics to genomics. *Curr. Opin. Plant Biol.* 3:132–37

36. Grossman AR, Schaefer MR, Chiang GG, Collier JL. 1993. Environmental effects on the light-harvesting complex of cyanobacteria. *J. Bacteriol.* 175:575–82

37. Guillou L, Chretiennot-Dinet MJ, Medlin LK, Claustre H, Loiseaux–de Goer S, et al. 1999. *Bolidomonas*: a new genus with two species belonging to a new algal class, the Bolidophyceae (Heterokonta). *J. Phycol.* 35:368–81

38. Harper MA. 1977. Movements. In *The Biology of Diatoms*, ed. D Werner, pp. 224–49. Berkeley, CA: Univ. Calif. Press

38a. Harris EH. 2000. *Chlamydomonas* as a model organism. *Annu. Rev. Plant Physiol. Plant Mol. Biol.* 52:363–406

39. Hildebrand M, Dahlin K. 2000. Nitrate transporter genes from the diatom *Cylindrotheca fusiformis* (Bacillariophyceae): mRNA levels controlled by nitrogen source and by the cell cycle. *J. Phycol.* 36:702–13

40. Hildebrand M, Dahlin K, Volcani BE. 1998. Characterization of a silicon transporter gene family in *Cylindrotheca fusiformis*: sequences, expression analysis, and identification of homologs in other diatoms. *Mol. Gen. Genet.* 260:480–86

40a. Hildebrand M, Volcani BE, Gassmann W, Schroeder JI. 1997. A gene family of silicon transporters. *Nature* 385:68–89

41. Hutchins DA, Witter AE, Butler A, Luther GW III. 1999. Competition among marine phytoplankton for different chelated iron species. *Nature* 400:858–61

42. Jarvis EE, Dunahay TG, Brown LM. 1992. DNA nucleoside composition and methylation in several species of microalgae. *J. Phycol.* 28:356–62

43. Kagawa T, Sakai T, Suetsugu N, Oikawa K, Ishiguro S, et al. 2001. *Arabidopsis* NPL1: a phototropin homolog controlling the chloroplast high-light avoidance response. *Science* 291:2138–41

44. Kehoe DM, Grossman AR. 1996. Similarity of a chromatic adaptation sensor to phytochrome and ethylene receptors. *Science* 273:1409–12

45. Kemp AES, Pike J, Pearce RB, Lange CB. 2000. The "fall dump"—a new perspective on the role of a "shade flora" in the annual cycle of diatom production and export flux. *Deep-Sea Res.* 47:2129–54

46. Kirk JTO. 1983. *Light and Photosynthesis in Aquatic Ecosystems.* Cambridge, UK: Cambridge Univ. Press

47. Kociolek JP, Stoermer EF. 1989. Chromosome numbers in diatoms: a review. *Diatom Res.* 4:47–54

48. Kowallik KV. 1992. *Origin and Evolution of Plastids from Chlorophyll-a + c-Containing Algae: Suggested Ancestral Relationships to Red and Green Algal Plastids,* ed. RA Lewin. New York/London: Chapman Hall

49. Kröger N, Bergsdorf C, Sumper M. 1994. A new calcium-binding glycoprotein family constitutes a major diatom cell wall component. *EMBO J.* 13:4676–83

50. Kröger N, Bergsdorf C, Sumper M. 1996. Frustulins: domain conservation in a protein family associated with diatom cell walls. *Eur. J. Biochem.* 239:259–64

51. Kröger N, Deutzmann R, Bergsdorf C, Sumper M. 2000. Species-specific polyamines from diatoms control silica morphology. *Proc. Natl. Acad. Sci. USA* 97:14133–38

52. Kröger N, Deutzmann R, Sumper M. 1999. Polycationic peptides from diatom biosilica that direct silica nanosphere formation. *Science* 286:1129–32

53. Kröger N, Lehmann G, Rachel R, Sumper M. 1997. Characterization of a 200-kDa diatom protein that is specifically associated with a silica-based substructure of the cell wall. *Eur. J. Biochem.* 250:99–105

54. Kröger N, Wetherbee R. 2000. Pleuralins are involved in theca differentiation in the diatom *Cylindrotheca fusiformis. Protist* 151:263–73

55. Lang M, Apt KE, Kroth PG. 1998. Protein transport into "complex" diatom plastids utilizes two different targeting signals. *J. Biol. Chem.* 273:30973–78

56. LaRoche J, Boyd PW, McKay RML, Geider RJ. 1996. Flavodoxin as an in situ marker for iron stress in phytoplankton. *Nature* 382:802–5

57. LaRoche J, Mortain-Bertrand A, Falkowski PG. 1991. Light intensity–induced changes in *cab* mRNA and light-harvesting complex II apoprotein levels in the unicellular chlorophyte *Dunaliella tertiolecta. Plant Physiol.* 97:147–53

58. Leblanc C, Falciatore A, Bowler C. 1999. Semi-quantitative RT-PCR analysis of photoregulated gene expression in marine diatoms. *Plant Mol. Biol.* 40:1031–44

59. Lee RE. 1999. Heterokontophyta, Bacillariophyceae. In *Phycology,* ed. RE Lee, pp. 415–58. Cambridge, UK: Cambridge Univ. Press

60. Lewin JC. 1958. The taxonomic position of *Phaeodactylum tricornutum. J. Gen. Microbiol.* 18:427–32

61. Lewin JC, Lewin RA, Philpott DE. 1958. Observations on *Phaeodactylum tricornutum. J. Gen. Microbiol.* 18:418–26

62. Mann DG. 1993. Patterns of sexual reproduction in diatoms. *Hydrobiologia* 269/270:11–20

63. Mann S, Ozin GA. 1996. Synthesis of inorganic materials with complex form. *Nature* 382:313–18

64. Martin JH, Coale KH, Johnson KS, Fitzwater SE, Gordon RM, et al. 1994. Testing the iron hypothesis in ecosystems of the equatorial Pacific Ocean. *Nature* 371:123–29

65. Martin-Jezequel V, Hildebrand M, Brzezinski MA. 2000. Silicon metabolism in diatoms: implications for growth. *J. Phycol.* 36:821–40

66. Medlin LK, Kooistra WCH, Schmid A-MM. 2000. A review of the evolution of the diatoms—a total approach using molecules, morphology and geology. In *The Origin and Early Evolution of the Diatoms: Fossil, Molecular and Biogeographical Approaches,* ed. A Witkowski, J Sieminska, pp. 13–35. Cracow: W Szafer Inst. Bot., Pol. Acad. Sci.

67. Medlin LK, Kooistra WHCF, Gersonde R, Sims PA, Wellbrock U. 1997. Is the origin of the diatoms related to the end-Permian mass extinction? *Nova Hedwigia* 65:1–11

68. Mewes HW, Albermann K, Bahr M, Frishman D, Gleissner A, et al. 1997. Overview of the yeast genome. *Nature* 387 (Suppl.):7–65

69. Miralto A, Barone G, Romano G, Poulet SA, Ianora A, et al. 1999. The insidious effect of diatoms on copepod reproduction. *Nature* 402:173–76

70. Morse DE. 1999. Silicon biotechnology: harnessing biological silica production to construct new materials. *Trends Biotechnol.* 17:230–32

71. Norton TA, Melkonian M, Andersen RA. 1996. Algal biodiversity. *Phycologia* 35: 308–26

72. Owens TG. 1986. Light-harvesting function in the diatom *Phaeodactylum tricornutum. Plant Physiol.* 80:732–38

73. Parkinson J, Gordon R. 1999. Beyond micromachining: the potential of diatoms. *Trends Biotechnol.* 17:190–96

74. Parsek M, Greenberg EP. 2000. Acyl-homoserine lactone quorum sensing in Gram-negative bacteria: a signaling mechanism involved in associations with higher organisms. *Proc. Natl. Acad. Sci. USA* 97:8789–93

75. Pickett-Heaps J, Schmid A-MM, Edgar LA. 1990. The cell biology of diatom valve formation. In *Progress in Phycological Research*, ed. FE Round, DJ Chapman. 7:1–168. Bristol, UK: BioPress Ltd.

76. Qiu Y-L, Palmer JD. 1999. Phylogeny of early land plants: insights from genes and genomes. *Trends Plant Sci.* 4:26–30

77. Reinfelder JR, Kraepiel AML, Morel FMM. 2000. Unicellular C_4 photosynthesis in a marine diatom. *Nature* 407:996–99

78. Scala S, Bowler C. 2001. Molecular insights into the novel aspects of diatom biology. *Cell. Mol. Life Sci.* 58:1666–73

79. Scala S, Carels N, Falciatore A, Chiusano ML, Bowler C. 2002. Genome properties of the diatom *Phaeodactylum tricornutum* and comparison with the genomes of other eukaryotes. *Plant Physiol.* In press

80. Sineshchekov OA, Govorunova EV. 1999. Rhodopsin-mediated photosensing in green flagellated algae. *Trends Plant Sci.* 4:58–63

81. Smetacek V. 1999. Diatoms and the ocean carbon cycle. *Protist* 150:25–32

82. Smetacek V. 2001. A watery arms race. *Nature* 411:745

83. Smith NA, Singh SP, Wang MB, Stoutjesdijk PA, Green AG, et al. 2000. Total silencing by intron-spliced hairpin RNAs. *Nature* 407:319–20

84. Tréguer P, Nelson DM, Van Bennekom AJ, DeMaster DJ, Leynaert A, et al. 1995. The silica balance in the world ocean: a reestimate. *Science* 268:375–79

85. Trewavas A. 2000. Signal perception and transduction. In *Biochemistry and Molecular Biology of Plants*, ed. BB Buchanan, W Gruissem, RL Jones, pp. 930–87. Rockville, MA: Am. Soc. Plant Physiol.

86. Valentin K, Cattolico RA, Zetsche K. 1992. *Phylogenetic Origin of the Plastids.* New York/London: Chapman Hall

87. van de Poll WH, Vrieling EG, Gieskes WWC. 1999. Location and expression of frustulins in the pennate diatoms *Cylindrotheca fusiformis, Navicula pelliculosa,* and *Navicula salinarum* (Bacillariophyceae). *J. Phycol.* 35:1044–53

88. Van Den Hoek C, Mann DG, Johns HM. 1997. *Algae. An Introduction to Phycology.* Cambridge, UK: Cambridge Univ. Press

89. Vaulot D, Olson RJ, Chisholm SW. 1986. Light and dark control of the cell cycle in two phytoplankton species. *Exp. Cell Res.* 167:38–52

89a. Vaulot D, Olson RJ, Merkel SM, Chisholm SW. 1987. Cell cycle response to nutrient starvation in two marine phytoplankton species. *Mar. Biol.* 95:625–30

90. Veldhuis MJW, Cucci TL, Sieracki ME. 1997. Cellular DNA content of marine phytoplankton using two new fluorochromes: taxonomic and ecological implications. *J. Phycol.* 33:527–41

91. Villareal TA. 1989. Division cycles in

the nitrogen-fixing *Rhizosolenia* (Bacillariophyceae)–*Richelia* (Nostocaceae) symbiosis. *Br. Phycol. J.* 24:357–65

92. Wolfe GR, Cunningham FX, Durnford D, Green BR, Gantt E. 1994. Evidence for a common origin of chloroplasts with light-harvesting complexes of different pigmentation. *Nature* 367:566–68

93. Zamore PD, Tuschl T, Sharp PA, Bartel DP. 2000. RNAi: double-stranded RNA directs the ATP-dependent cleavage of mRNA at 21 to 23 nucleotide intervals. *Cell* 101:25–33

94. Zaslavskaia LA, Lippmeier JC, Kroth PG, Grossman AR, Apt KE. 2000. Transformation of the diatom *Phaeodactylum tricornutum* (Bacillariophyceae) with a variety of selectable marker and reporter genes. *J. Phycol.* 36:379–86

95. Zaslavskaia LA, Lippmeier JC, Shih C, Ehrhardt D, Grossman AR, et al. 2001. Trophic conversion of an obligate photoautotrophic organism through metabolic engineering. *Science* 292:2073–75

96. Zurzolo C, Bowler C. 2001. Exploring bioinorganic pattern formation in diatoms. A story of polarized trafficking. *Plant Physiol.* 127:1339–45

Annu. Rev. Plant Biol. 2002. 53:131–58
DOI: 10.1146/annurev.arplant.53.092701.180236

ABSCISSION, DEHISCENCE, AND OTHER CELL SEPARATION PROCESSES

Jeremy A. Roberts, Katherine A. Elliott, and Zinnia H. Gonzalez-Carranza

Division of Plant Science, School of Biosciences, Sutton Bonington Campus, University of Nottingham, Loughborough, Leics LE12 5RD United Kingdom; e-mail: jeremy.roberts@nottingham.ac.uk, katherine.elliot@nottingham.ac.uk, zinnia.gonzalez@nottingham.ac.uk

Key Words biotechnology, cell wall degradation, ethylene, gene expression, hydrolytic enzymes

■ **Abstract** Cell separation is a critical process that takes place throughout the life cycle of a plant. It enables roots to emerge from germinating seeds, cotyledons, and leaves to expand, anthers to dehisce, fruit to ripen, and organs to be shed. The focus of this review is to examine how processes such as abscission and dehiscence are regulated and the ways new research strategies are helping us to understand the mechanisms involved in bringing about a reduction in cell-to-cell adhesion. The opportunities for using this information to manipulate cell separation for the benefit of agriculture and horticulture are evaluated.

CONTENTS

INTRODUCTION

In the 20 years since abscission was last reviewed in this series (128), there has been an explosion in the number of new techniques, particularly in the fields of molecular genetics and cell biology, that have had a profound impact on our understanding of plant development and morphological features such as cell wall structure and function (26, 109). The aim of this review is to assess how far our knowledge of abscission, and other developmental phenomena where cell separation takes place, has advanced over the last two decades and point the way that future studies might take.

CELL SEPARATION PROCESSES IN PLANTS

Plant cells are joined together by an adhesive matrix composed primarily of pectin. This feature provides a plant with rigidity and strength, thus enabling the aerial parts to optimally intercept solar radiation, the roots to explore different soil horizons, and structures such as fruits and pods to nurture and protect developing seeds. The links between cells also provide a framework for synchronizing events such as lateral and longitudinal expansion; however, their very nature imposes a restriction on the activities and autonomy of individual cells, and elaborate mechanisms are in place to loosen the bonds or break them entirely. Incidences of the precisely coordinated breakdown of cell-to-cell adhesion play a key role throughout the life cycle of a plant (108). Cell separation facilitates (*a*) radicles to appear from germinating seeds, (*b*) primary roots to penetrate the soil and laterals to emerge, (*c*) cotyledons and leaves to expand and gaseous exchange to take place, (*d*) pollen to be released from anthers, (*e*) fruit to soften, and (*f*) pods to dehisce and organs to be shed. A common feature of all these processes is that the cell wall is degraded, and we are steadily accumulating information to suggest that common mechanisms may bring this about. Indeed recent evidence indicates that processes such as anther dehiscence, pod shatter, and the shedding of seeds may involve a similar sequence of events (69, 119). Although the biochemical processes that lead to wall breakdown may be comparable, the nature of the signals that induce these changes are likely to be different so that the process only occurs at critical spatial and temporal locations. Moreover, there are instances where wall breakdown may only be required to dissolve the bonds surrounding an individual cell or, even, one face of the wall.

Although the primary focus of this review is on abscission and dehiscence, our objective is also to highlight those events that take place at other sites where cell separation has been documented. It is not our intention to consider all processes in a comprehensive way, and phenomena such as fruit softening have recently received an entire article devoted to the topic (45a).

ABSCISSION

Morphogenesis of the Abscission Zone

The shedding of plant organs takes place at predetermined positions called abscission zones (AZs) (49, 128, 137). Anatomical studies have revealed that prior to cell separation the site where wall breakdown is destined to occur can frequently be morphologically identified as comprising a layer (or layers) of isodiametrically flattened cells. This observation has led to the hypothesis that the AZ is composed of positionally differentiated target cells (98). The concept that an AZ is a predetermined site for specific inter- and intracellular signaling events is well established, and there is convincing morphological and biochemical evidence that cells that constitute this zone respond in different ways from their neighbors to the same hormonal cues (12, 49, 108, 137).

The differentiation of the zone would appear to occur many months before organ separation is destined to take place, and leaves of deciduous trees may be induced to shed soon after they have reached full expansion in the spring (98). By isolating and characterizing mutants showing aberrant organ shedding, it has been possible to begin the process of unraveling the morphogenetic events that contribute to the differentiation process. In tomato, the entire flower can be shed if cell separation in the knuckle region of the pedicel is triggered. This site not only leads to the loss of flowers but is also activated at the time of fruit fall. Two mutants of tomato, *jointless* and *lateral suppressor*, do not develop a pedicel AZ, and the mutated genes responsible for these phenotypes have been identified. The wild-type *JOINTLESS* gene has been identified by map-based cloning and was found to encode a MADS-box protein (85). This family of transcription factors contributes to the establishment of specific sites of cell differentiation (1). Interestingly, the *MADS-box* genes, *SHP1* and *SHP2* (for further details see below), play a key role in regulating the formation of the dehiscence zone (DZ) in *Arabidopsis* siliques (81). Whether these transcription factors define the position of a zone where cell separation takes place or just initiate its formation remains to be clarified (85). The *LATERAL SUPRESSOR* gene has also been characterized and was found to encode a member of a different group of putative transcriptional activators (124).

The LS peptide exhibits homology to a group of VHID proteins that play a role in transducing signals associated with gibberellins (100). The mechanism by which this protein brings about its effects is unclear as it induces a range of pleiotropic features, the principal one being an absence of lateral buds. As the leaves and floral organs of both *jointless* and *lateral supressor* plants abscind, it seems likely that the morphogenetic signals that regulate the formation of AZs at different tissue sites are not the same. Not all nonshedding mutants appear to be the consequence of a failure of the AZ to differentiate, and in the lupin mutant *abs1*, events subsequent to zone formation appear to be blocked (24).

As yet only a handful of abscission mutants have been documented and the genes that they encode characterized. *Arabidopsis* has proved to be an excellent model system for the generation and isolation of mutants although the inability of this species to shed its leaves requires floral organ abscission to be the primary screen and the fragility of some of these tissues after desiccation may make a rapid and reliable assessment of cell separation a challenge. However, progress in this area is being made, and a number of delayed abscission mutants have now been isolated (99).

An innovative approach that has been taken to study the signals involved in zone differentiation has been to use periclinal chimeras generated from the *jointless* mutant and wild-type plant material (133). These studies showed that the genotypic origin of L3 (the layer that originates the vascular tissues) dictated the formation of a functional pedicel AZ. Chimeras comprising L1 and L2 from *jointless* formed normally differentiated AZs within the pedicel tissues. Further work is necessary to determine whether the signal responsible for zone differentiation is the JOINT-LESS protein itself and whether this protein has the capacity to pass from the L3 layer into L2 and L1.

Formation of Secondary AZs

Although the main sites where abscission occurs are predetermined, under certain circumstances an analogous process might take place at positions where shedding would not normally occur (108). Such secondary AZs have been well documented although relatively few studies have focused on the mechanisms by which they might be triggered. In *Impatiens sultani*, secondary zones can be routinely induced in internodal segments, and the position of abscission is influenced by applying auxin to the base of the tissue sections (145, 146). A more detailed examination of secondary abscission in *Phaseolus vulgaris* petioles (90) has revealed that cross talk between ethylene and auxin plays an important role in dictating the site of cell separation. Ethylene seemed to induce the formation of the zone, whereas applied IAA dictated where it would be sited. This study is the first to compare and contrast the events associated with primary and secondary abscission and to demonstrate that many morphological and biochemical features are shared between the two systems. In particular, there is an increase in dictyosome activity and the accumulation of a β-1,4-glucanase with a pI of 9.5. An important distinction that may exist between the formation of secondary zones in bean and in other species is that differentiation does not involve cell division. It is not clear whether the separation layer that develops in bean petioles is strictly an AZ and, if so, whether its formation must be preceded by the expression of abscission-related transcription factors to define its position. In *Brassica napus*, we have obtained evidence to indicate that mRNAs encoding cell wall degrading enzymes such as polygalacturonase can accumulate in ethylene-treated petioles as long as the AZ is attached (Z.H. Gonzalez-Carranza, unpublished data). Whether this is the consequence of mRNA migrating from the zone or of a signal moving from the site of cell separation to induce selective gene expression is not clear; however, it does raise the possibility that primary and secondary AZs may not be analogous structures.

REGULATION OF THE TIMING OF ABSCISSION

The differentiation of the site where cell separation will take place is just the first phase in the abscission process (Figure 1, see color insert). The second step is to trigger the cells that make up the AZ so that cell wall hydrolysis occurs. Although the two processes may be closely linked temporally in organs such as flowers, in leaves many months may separate the formation from the activation of the zone. The critical questions are what factors regulate the timing of the cell separation process, are they comparable at different sites where abscission takes place, and perhaps most importantly, is there a single trigger that is responsible. The shedding of plant organs is normally associated with the senescence of the distal organ; however, a spectrum of environmental factors can prematurely precipitate leaf, flower, or fruit fall. Invariably these are associated with stress brought about as a consequence of a deficit or surplus of water, extremes of temperature, or pest and pathogen attack (49, 137). Tissue senescence and cell stress are commonly associated with elevations in ethylene production, and this has led to the hypothesis that the gas is a natural regulator of abscission (12, 66). Certainly ethylene is most effective at accelerating the process, particularly in excised pieces of tissue, and these "explants" are routinely used to study the biochemical and molecular changes associated with abscission as excision and exposing to ethylene synchronizes and accelerates the cell separation process (128).

The availability of mutants that exhibit insensitivity to ethylene provides material by which the role of this hormone in the regulation of abscission can be dissected in more detail. The ethylene insensitive mutant of *Arabidopsis Etr1-1* exhibits a delay in the shedding of floral parts (7, 17, 99), but abscission does eventually take place. Although no detailed analysis has been carried out on this mutant to confirm that "conventional" cell separation has taken place, these observations place ethylene in the chain of events that regulate the timing of abscission but indicate that its contribution may not be an absolute requirement. There is good reason to believe that a balance between different hormones may be the key factor that regulates the timing of the process (137) with ethylene acting as a natural accelerator of abscission and auxin (IAA) operating as a brake. Such a mechanism would enable the AZ to be protected from elevated ethylene production at a time when abscission would be inopportune, for instance, during the ripening of climacteric fruit. Although the capacity of auxins to delay abscission has been extensively documented, our understanding of how IAA prevents the cells within the AZ from responding to ethylene is limited and is one of the major challenges for the next phase of abscission research.

As the name suggests, a role for abscisic acid (ABA) in regulating abscission was once broadly accepted (128). The contemporary view is that ABA probably stimulates the abscission process of leaves and flowers through its ability to bring forward tissue senescence and the associated ethylene climacteric particularly of excised tissues. However, there is some evidence to suggest that ABA could play a more direct role in abscission of organs such as seeds (122) and possibly some flowers (2). This observation raises the question of whether the shedding of different tissues is under the influence of the same hormonal stimulus and whether the key regulating ligand has yet to be identified.

ANATOMICAL CHANGES ASSOCIATED WITH ABSCISSION

Attempts to elucidate the mechanisms responsible for abscission have been hampered by the limited number of cells that constitute the site of cell separation with the most well-documented model systems being *P. vulgaris*, *Lycopersicon esculentum*, and *Sambucus nigra* (108). The latter species has proved to be particularly productive as up to 30 layers of cells constitute the leaflet AZ facilitating the analysis of the biochemical and molecular events that take place prior to wall breakdown (136, 147).

Cells that make up the abscission zone can be largely distinguished prior to cell separation. They comprise a few layers of cells that are normally smaller than adjacent nonseparating cells and are more densely cytoplasmic (128). As separation proceeds dilation of the golgi vesicles can be observed, and activation of the endomembrane system takes place. Finally the middle lamella shows signs of degradation, and dissociating cellulose microfibrils can frequently be viewed. At this time the separating cells round up, and it has been proposed that this provides the hydraulic mechanism that severs the vascular trace, which is the only tissue that does not appear to undergo wall breakdown (127). These changes in the integrity of the cell wall are often highly restricted; however, it is not yet clear whether this is the consequence of focal autolysis or the existence of target substrates within the walls of the separating cells. Certainly in *S. nigra* leaflets there is evidence for a qualitative difference between the protein complement of the AZ cell wall and adjacent tissues (89), and analyses of the wall matrix at the site of separation of oilpalm fruit has revealed elevated levels of unmethylated pectin (61).

BIOCHEMICAL AND MOLECULAR CHANGES ASSOCIATED WITH ABSCISSION

It is evident that shedding is accompanied by a spectrum of changes in gene expression. Strategies to study this have either focused on specific genes predicted to be upregulated during abscission, for instance, those encoding cell wall degrading enzymes, or have made use of the fact that nonseparating tissues can be used as a control tissue for the analysis of differentially expressed genes.

β-1,4-glucanase

The first enzyme proposed to contribute to wall loosening at the site of abscission was β-1,4-glucanase, or cellulase (see 128). Activity was found to increase substantially in the petiole AZ during shedding of *P. vulgaris* leaves, and the gene encoding the enzyme was subsequently cloned using a cDNA probe from ripening avocado fruit (142). Expression is upregulated specifically in the AZ after treatment with ethylene, but this can be repressed by exposure to IAA. Increases in β-1,4-glucanase expression have since been reported during the abscission of *S. nigra* leaflets (134), tomato flowers (80), and flowers and leaves of pepper (41).

β-1,4-glucanases belong to a large gene family, and in tomato seven different isozymes (Cel1 to Cel7) have been cloned (13, 14, 16, 35, 80). During cell separation in the flower pedicel an increase in expression of Cel1, Cel2, and Cel5 has been detected (35, 48, 71). Using an antisense RNA strategy, expression of either Cel1 or Cel2 has been downregulated individually, which has resulted in the need for a greater force to bring about pedicel abscission (15, 79). In contrast, tomato fruit from such transgenic plants do not exhibit delayed or reduced softening even though both β-1,4-glucanase transcripts have been reported to be upregulated during ripening of wild-type fruit. These observations suggest that cell separation during abscission and ripening may involve different wall loosening events. In *Arabidopsis* there are over 20 putative β-1,4-glucanases that can be identified within the genome (34), and a specific abscission-related β-1,4-glucanase has yet to be identified. Preliminary analysis of gene expression in *B. napus* petioles has identified more than six different β-1,4-glucanases although it is not yet clear which, if any, of these isoforms may contribute to the cell separation process (Z.H. Gonzalez-Carranza, unpublished data).

Polygalacturonase

A correlation between increasing polygalacturonase (PG) activity and cell separation was first detected in ripening fruit (65). These studies provoked an upsurge of interest in the role of this enzyme in abscission, and increases in enzyme activity were reported during the shedding of leaves (136), flowers (141), and fruit (9, 61). In an examination of transgenic tomato material, where the fruit specific PG had been downregulated by antisense mRNA, it was revealed that the abscission-related enzyme was significantly different from that associated with fruit softening (135). This was subsequently confirmed when a PG that was upregulated during tomato flower abscission was characterized (72). Three PG isoforms (TAPG1, TAPG2, and TAPG4) are associated with abscission in tomato, and, although these are highly homologous (80–90% at the nucleotide level), their transcripts (1.5 kB) are substantially shorter than the fruit enzyme (1.9 kB) and share a sequence identity of only 50% (73). The time course of expression of the abscission-related PGs differs with TAPG2 and TAPG4 accumulating in the pedicel zone prior to the onset of cell separation (at 6 h), whereas TAPG1 can be detected 6 h later at the time of shedding. The temporal and spatial patterns of *TAPG1* and *TAPG4* expression have been further studied by translationally fusing their promoters to the reporter gene *β-glucuronidase (GUS)* and using these constructs to transform tomato plants. Expression of GUS was localized specifically within the AZ of leaf petioles, flower and fruit pedicels, petal corolla, and stigma. Expression was promoted by ethylene and inhibited by IAA. The expression of *TAPG1* and of *TAPG4* were subtly different. When driven by the promoter of the former gene, GUS accumulation occurs throughout the leaf and flower AZs, but when driven by the promoter of the latter, it occurs most notably in the vascular tissue (63). The authors concluded that the stele may be the origin of a signal that coordinates the cell separation process, a hypothesis originally proposed by Thompson & Osborne (138) after carrying out surgical manipulation of the vascular cylinder of *P. vulgaris* AZs. Further analysis of the *TAPG1* promoter has revealed that fusion of 320 bp upstream from the ATG start codon to *GUS* is sufficient to drive expression within the AZ tissues.

The expression of an abscission-related PG (*PGAZAT*) has recently been characterized in floral zones of *Arabidopsis* (50). The gene was identified by isolating its homologue in *B. napus* using an RT-PCR strategy. Expression in *B. napus* can be seen during leaf and floral organ abscission, and mRNA accumulation can be detected prior to leaf shedding. Expression of *PGAZAT* has been studied by fusing the promoter to the reporter genes *GUS* or *Green Fluorescence Protein (GFP)*. Reporter gene activity during floral organ shedding in *Arabidopsis* is initially detected at the base of the anther filaments, and this occurs both during "natural" and ethylene-promoted abscission. It is intriguing that expression, as determined by GUS or GFP accumulation, appears to be restricted to certain cells within the separation layer with a ring of cells within the cortex first showing signs of activity (Figures 2*I,K*, see color insert). In contrast to the observation in tomato, no evidence exists that expression commences in the vascular tissue. The expression profile raises the possibility that not all cells within the separation layer are responding to the coordinating signals with the same temporal pattern. In the floral zone of *Arabidopsis* it could be particularly important for expression of hydrolytic enzymes to be precisely coordinated as uncontrolled wall degradation might bring about shedding of the developing silique.

In the *Arabidopsis* genome over 75 putative PG genes can be identified (108). One approach to identify the role of these genes in cell separation processes is to analyze the phenotype of specific gene knockouts. Recently a null of *PGAZAT* was identified, and the time course of ethylene-promoted flower abscission was found to be slightly delayed in mutant compared to wild-type plants (K.A. Elliott, unpublished data). This observation provides further evidence that the activity of PGAZAT contributes to the timing of organ abscission; however as separation continues to take place in its absence, additional members of the PG gene family might also be upregulated during abscission (119) or PG might be only one member of a cocktail of hydrolytic enzymes that bring about wall dissolution.

Expansin

Expansins are associated with cell wall loosening during expansion growth and ripening (25). Circumstantial evidence to support a role for this class of wall protein in abscission came from the report that expression of *AtEXP10* (a member of the expansin family in *Arabidopsis*) could be observed specifically at the base of the leaf petiole (23). This proposal was strengthened by the demonstration that downregulation of the gene resulted in an increase in force necessary to separate the petiole from the body of the plant. However, the interpretation of these observations should be viewed with caution as *Arabidopsis* does not naturally shed its leaves and the authors could find no evidence of expression of *AtEXP10* at the base of separating sepals, petals, or flowers. More convincing evidence of a role for expansins in organ shedding has come from studies on *S. nigra* where an increase in activity has been found in the AZ of ethylene-treated leaflets (S.J. McQueen-Mason, unpublished). Two expansin genes have now been cloned from these tissues and have been found to be specifically upregulated during cell separation (E.J. Belfield, unpublished). These preliminary observations should act as a catalyst in the search for abscission-related expansins, and the impact of downregulating the expression of these genes on organ shedding is awaited with interest.

It is clear from the preceding discussion that a cocktail of proteins (116) may contribute to the wall loosening that accompanies shedding of plant organs. What is not yet clear is whether different isoforms of these proteins contribute to separation at different sites and whether the trans-acting factors that regulate expression are the same. It is likely that many of the answers will come from studies on *Arabidopsis* where the generation of targeted gene knockouts is becoming routine, and it is relatively straightforward to generate mutants where the expression of reporter genes has been manipulated and mutated genes characterized.

Pathogenesis-Related Proteins

The shedding of plant organs provides an ideal site for invasion by pathogens. It is therefore no great surprise to discover that associated with cell separation is an increase in the accumulation of pathogenesis-related (PR) proteins. In the *P. vulgaris* petiole AZ chitinases (82a) and PR-1, -2, and -5 type proteins (36) accumulate during leaf fall, whereas in ethylene-treated *S. nigra* leaflets genes encoding a spectrum of PR proteins are expressed as cell separation proceeds (30). Upregulation preempts infection and in some cases seems to be promoted by explant generation rather than by exposure to ethylene (30). How expression of these proteins is coordinated is unclear. However, reactive oxygen intermediates might be involved (78) and metallothionein (MT)-like proteins might act to scavenge the free radicals that would otherwise accumulate within the AZ tissues (29).

The promoters of the genes encoding PR proteins show strong specificity (38), and if the bean chitinase promoter is fused to *GUS* and is used to transform *Arabidopsis* then expression is observed at the base of the floral parts at the time of shedding (7). In *Arabidopsis* a number of genes, such as β1,3 glucanases (144) and jasmonic acid biosynthetic enzymes (77), associated with plant defense are upregulated specifically in the floral AZ under conditions that promote organ loss.

Other Genes and Proteins

It is now a common strategy to use reporters such as GUS to identify spatial and temporal patterns of gene expression, and this strategy has enabled the identification of additional abscission-related genes. These include such genes as a tryptophan synthase β (106) and the abscisic acid insensitive gene *ABI3* (111). Interpretation of such data must be made cautiously in the absence of information relating to transcript or protein accumulation. Intriguingly, a novel gene encoding a leucine-rich repeat is upregulated during floral organ abscission and at the base of the leaf petioles in *Arabidopsis* (70). Downregulation of this HAESA gene results in a reduced capacity of the transgenic plants to shed floral parts, with the degree of inhibition being closely correlated with the reduction in HAESA protein. As yet the role of this protein in abscission is unknown, and the explanation for why it may be expressed in the absence of leaf petiole shedding is unclear.

Over the last few years a number of genes that are upregulated during abscission have been identified. Some of these contribute to the process of cell separation, others act to protect the exposed fracture surface from pathogenic attack, and the role of the remainder is not yet clear. What also remains to be resolved is whether every cell within an AZ is induced to express the same genes or whether the zone is a chimera of cell types contributing different proteins to the process (107). The use of reporter genes and the in situ localization of abscission-related mRNAs should help to resolve this question. Genechip array analysis has already proved to be a sensitive and reproducible means of studying gene expression (55a), and this strategy should generate additional information about the changes in transcription that take place specifically in AZ during cell separation.

POD DEHISCENCE

Pod dehiscence results in the premature shedding of seeds from siliques prior to and during harvest. This phenomenon, also known as pod shatter, can cause substantial reductions in seed yield from a wide range of crop species including oilseed rape (84), soybean (103), sesame (33), and the legume birdsfoot trefoil (52). Most studies on the process of dehiscence have been centered on oilseed rape (*B. napus*), where premature seed shedding is an economically significant problem. Pod shatter can cause a loss of up to 50% of the potential seed yield if harvesting is delayed by adverse conditions (84). Moreover, seeds that are shed and persist in the soil give rise to weed oilseed rape plants contaminating crops that are subsequently grown (see below for examples).

Examination of *B. napus* pods shows the fruit is a bivalve silique, i.e., the fruit wall (carpel) encloses two seed-containing valves separated by a pseudoseptum and a replar region (104). Each DZ is located at the carpel margins adjacent to the septum and runs the whole length of the silique. Pod development in *B. napus* can be divided into three stages. The first stage occurs 0–20 days after anthesis (DAA), during which the siliques reach their maximum length. The seeds do not begin to grow until the pods are almost full size (62). The site at which dehiscence will take place cannot be distinguished in pods at 10 DAA. However, the progressive differentiation of cells in the replum by 20 DAA results in the formation of a distinct region 1–3 cells wide that separates the vascular tissue from the valve edges (91). This narrow strip of cells constitutes the DZ.

The second stage of pod development occurs between 20 and 50 DAA. From 20 DAA, once pod elongation has ceased, secondary cell wall material is deposited in the walls of valve edge cells, and the replum becomes lignified (91). The DZ cells do not exhibit wall thickening at any stage of pod development. Once the lignification process is complete (approximately 35 DAA) the DZ is enclosed by thickened tissues, and the cells exhibit a progressive reduction in both volume and organellar content (91). From 40 DAA onward cell wall degradation occurs in the DZ, resulting in the breakdown of the middle lamella and a loss in cellular cohesion (91, 102).

During the final stages of pod development, which occur 50–70 DAA, the lignified cells undergo senescence. As the pod becomes desiccated, tensions in the fruit wall are created, and the weakened DZ cell walls eventually give way, resulting in the shattering of the pod and release of the seed (129). The tension in the pod is believed to be caused primarily by the lignification of the endocarp cells surrounding the DZ. In *Brassica juncea* lines that have a reduced tendency to shatter the endocarp layer *enb* is not completely lignified (129).

The general pattern of pod development in *Arabidopsis thaliana* is similar to that described for *B. napus*, although the whole process occurs at a much faster rate (40, 129). Interestingly, although the many different ecotypes of *Arabidopsis* exhibit wide genetic diversity and natural variations within phenotypes, no differences in fruit dehiscence have been recorded. A similar process where tensions in the pod and a zone of weakness contribute to pod dehiscence has also been described for other species including sesame (33) and soybean (139). By reducing the tensions created in the pod wall, or by strengthening the zone of weakness, pod dehiscence can be reduced. In soybean, varieties showing an increase in the length or width of the bundle cap that traverses the DZ have reduced seed loss (139).

Morphogenesis of the Pod DZ

Recently, a number of genes that contribute to the formation of the DZ within *Arabidopsis* siliques have been identified (40, 53, 81). Two of these, *SHATTERPROOF 1* and *2* (*SHP1* and *SHP2*), are members of the extended MADS-box gene family in *Arabidopsis* (83). *SHP1* and *SHP2* have 87% identity at the amino acid level and show almost identical patterns of expression in the developing *Arabidopsis* pod (43, 83, 123). *shp1* and *shp2* single mutants exhibit no detectable difference in phenotype from the wild type. However siliques of the double mutant fail to dehisce, and seeds can be accessed only if the fruit is opened manually (81). Further analysis of *shp1 shp2* pods indicates that these two genes are required for the correct development of the *Arabidopsis* fruit valve margin, including the differentiation of DZ cells and the lignification of the valve margin cells (81).

Another MADS-box gene *FRUITFULL* (*FUL*) is involved in the development of *Arabidopsis* pods (53). The valve cells of *ful* mutant siliques fail to elongate and differentiate after fertilization, leading to short pods full of overcrowded seeds. Mature *ful* pods fail to dehisce normally, probably as a consequence of a highly abnormal valve-replum boundary, and frequently, the valve ruptures prematurely because of the internal pressure caused by the growing seeds (53). The *FRUIT-FULL* gene also appears to play an important role in the dehiscence process as the DZ fails to form in 35S:*FUL* pods resulting in indehiscent fruit (82). Recently, the *FRUITFULL* gene has been shown to act as a negative regulator of *SHATTER-PROOF* gene expression (39).

The *SPATULA* (*SPT*) gene encodes a basic-helix-loop-helix transcription factor, and studies on mutants of this gene indicate that it promotes the growth of carpel margins and of pollen tract tissues derived from them (60). In addition, *SPT* expression was seen in the stomium of the anther, in the AZ within the funiculus, and in the DZ of the silique, all being regions where cell separation takes place. In the last case *SPT* expression becomes restricted to the valve edges in the maturing silique, in a pattern resembling that observed for the *SHATTERPROOF* genes. However, valve dehiscence is not affected in *spt* mutants, and SPT may play a redundant role downstream of the SHP proteins.

In sesame the only varieties with high seed retention are those homozygous for the recessive indehiscence gene (*id*) (33). This again is the result of abnormal differentiation occurring within the carpel, as mutants exhibit an increased number of mesocarp cell layers at the region of separation and/or an increase in the number of endocarp cell layers at the tip.

REGULATION OF THE TIMING OF POD DEHISCENCE

During the process of dehiscence all of the cells of the DZ undergo cell wall breakdown at approximately the same time. For this to occur there must be some trigger that starts the process of separation, and the same or a second signal may coordinate the events that result in pod shatter.

To date little work has been carried out to determine which hormonal ligands or environmental factors may trigger dehiscence. Generally, pod shatter occurs at the same time as seed abscission, a process developed by plants to disperse their seeds. However, dehiscence can be induced by infection with the pod midge *Dasineura brassicae* (93) in immature *B. napus* pods as young as 20 DAA. This fly deposits eggs into young siliques and upon emergence the larvae, while remaining trapped inside the pod, feed on the developing seeds. Prior to pupation, the larvae make their escape by triggering the process of cell separation causing premature dehiscence. Although this phenomenon is similar to that seen during normal developmental shatter, with a loss of cellular cohesion occurring in the DZ, a few differences do occur. For example, separation is commonly localized to the site where the larva is trapped, and the separated DZ cells from infected pods contain apparently functional cytoplasm, whereas this is not seen in comparable cells during developmental shatter (91, 93). Such differences may indicate that dehiscence is regulated differently in the two processes.

Unlike abscission, there is little evidence to suggest that ethylene acts as a regulator of dehiscence. *Arabidopsis* mutants that have nonfunctional ethylene receptors, for example, *Etr1*, exhibit a normal time course of silique opening (99). Although a transitory climacteric of ethylene produced by the seeds is detectable prior to senescence in *B. napus* pods (21, 92), elevated β-1,4-glucanase activity in the DZ occurs at a time when the ethylene level in the pods is falling. Moreover, in parthenocarpic siliques the peak in ethylene occurs 20 days later than that observed in seeded pods (21), whereas dehiscence is only delayed in the former compared to the latter by a few days. Similarly suppression of ethylene production by treatment of pods with aminoethoxyvinylglycine (AVG) treatment has only a small impact on the timing of pod shatter (21). From these results it has been proposed that a low threshold level of ethylene is needed for cell separation and that the appearance of the ethylene climacteric in seeded pods may accelerate rather than initiate the process. Recently, work (110) on *Brassica rapa* siliques detected two peaks in ethylene production, first when the seeds and silique had a high growth rate and second when the silique reached its maximum fresh weight, before the pod begins to undergo senescence and desiccation. After this point (45 DAA) there is a rapid decrease in ethylene production to relatively low levels in the phase prior to dehiscence (55–60 DAA).

Auxins have also been proposed to play a role in the regulation and timing of pod dehiscence. A decrease in indole-3-acetic acid (IAA) was observed specifically in the DZ prior to moisture loss in *B. napus* pods, and this was correlated with an increase in β-1,4-glucanase activity (19). Furthermore, the treatment of pods with an auxin mimic, 2-methyl-4-chlorophenoxyacetic acid (4-CPA), delayed β-1,4-glucanase activity and cell separation in the DZ by approximately 10 days. More force was needed to open 4-CPA pods when compared to control pods of the same developmental stage and this was due to incomplete separation in the DZ (19). It was proposed that a decline in auxin concentrations in the DZ may be the trigger that elevates cell wall degrading enzyme activity and hence promotes pod dehiscence. Studies on parthenocarpic pods indicate that seeds may be the source of the IAA (19). IAA levels in the DZ of these pods remained low throughout pod development, and dehiscence was delayed by approximately 4 days, thus indicating that a decrease in auxin level may be one trigger, but not the only prerequisite for pod dehiscence.

BIOCHEMICAL AND MOLECULAR CHANGES ASSOCIATED WITH POD DEHISCENCE

β-1,4-glucanase

A number of groups have observed an increase in β-1,4-glucanase activity specifically in the DZ tissue prior to *B. napus* pod shatter. Furthermore, when *B. napus* pods are treated with auxin, a delay in both increased β-1,4-glucanase activity and cell separation in the DZ was observed (19). Although several fragments of β-1,4-glucanases mRNAs have been amplified from the DZ of *B. napus* pods, their expression patterns have yet to be resolved (J.A. Roberts, unpublished data).

Polygalacturonase

The primary site of breakdown in the pod DZ is the middle lamella (91), and an increase in PG activity has been detected in *B. napus* pods toward the end of development (102). Extracts of wall-bound proteins prepared from DZ tissue 6 WAA (weeks after anthesis) were able to degrade polygalacturonic acid, and the results indicated that an endo-PG was responsible for the depolymerization of the substrate. Further evidence to support a role for both PG and β-1,4-glucanase in pod dehiscence is provided by work performed on pod midge infected *B. napus* plants, where the larvae induce pod dehiscence to facilitate their escape (93). In infected plants an increase in the activity of both PG and β-1,4-glucanase immediately adjacent to the site of premature splitting was observed. However, it is possible that this increase in enzyme activity is derived from the insect, rather than originating from the plant itself.

By performing RT-PCR on DZ RNA extracted from *B. napus* pods, PG mRNA has been cloned by two separate groups (68, 102). Full-length cDNA clones have been obtained and were termed SAC66 and RDPG1, respectively. They both encode what appears to a homologous transcript of 1.7 kb (97% similarity at the nucleotide level) with similarity to fruit endo-PGs. Northern analysis indicates that these transcripts are expressed specifically in the DZ of *B. napus* pods, with SAC66 increasing at 30 DAA and reaching a plateau at 45 DAA (68), whereas *RDPG1* was expressed at low levels during the first 5 WAA but became abundantly expressed 6 WAA (102).

The homologue of SAC66 has been isolated from *Arabidopsis*, and this gene, called *SAC70*, has 87% identity and 91% similarity at the amino acid level to SAC66 (69). The promoter of *SAC70* also exhibits close sequence homology to that of the *B. napus RDPG1*, particularly in the first 500 bp upstream of the initiation codon. A fragment of *SAC70* promoter (1.408 kb) was translationally fused to the *GUS* gene and transformed into *B. napus* (69). GUS activity was seen at three sites within these plants, all of which suggest that this PG is involved in cell separation. Strong expression was seen in the DZ of the pod from 35 DAA onward, the timing of which agrees with the northern results observed for *SAC66* (68). GUS activity was also seen in the junction between the seed and funiculus, which suggests that the same PG may play a role in pod dehiscence and seed abscission. This would be a sensible strategy by the plant, to ensure that pod opening and seed abscission are coordinated to occur simultaneously. The third site at which the *SAC70::GUS* is expressed is in the stomium of the anthers, the site where cell dissociation occurs during anther dehiscence. The same promoter was fused to the barnase gene, which encodes a ribonuclease, and this construct was transformed into *B. napus* (69). The resultant plants were male sterile, and anatomical studies revealed that their anthers failed to undergo dehiscence. The pollen in the anthers of these plants remained functional, and these plants could be pollinated, suggesting that it was the failure of anther dehiscence that caused them to be infertile. If these plants were pollinated with either their own pollen or nontransgenic pollen the resulting pods exhibited a reduced susceptibility to shatter, thus providing evidence that the PG is involved in cell separation at the DZ (69). Transgenic *Arabidopsis* plants containing the same *SAC70* constructs have since been generated and found to give comparable results, demonstrating that analogous transcriptional control systems are present in *Arabidopsis* and *B. napus* (108).

The promoter of the *B. napus RDPG1* gene has also been fused to *GUS* and transformed in *Arabidopsis* (119). Results indicate that expression is primarily associated with anther and silique dehiscence but that low levels of expression can also be detected during floral organ abscission and at other sites where cell separation may take place.

Other Genes and Proteins

Differential screening of a cDNA library generated from the DZ of *B. napus* pods identified two genes that are upregulated specifically in the DZ during pod dehiscence (27, 28). One of these (SAC51) encodes a proline-rich protein that has sequence homology with a number of peptides whose functions are unknown. These proteins, many of which appear to be secreted, are found in a range of plant species and different plant organs, and they are upregulated in response to a number of environmental and developmental stimuli (51, 150). The second clone, SAC25, encodes a peptide that has some homology with dehydrogenases. As yet, the precise role of either of these two genes during pod dehiscence is unknown.

Recently, *AtSUC3*, a gene encoding an *Arabidopsis* sucrose transporter, has been localized to the subepidermal cell layers of carpels. There may be an AtSUC3-driven import of sucrose into this layer of cells, which become lignified during cell maturation, and this may be necessary for pod shatter (94). Alternatively, Barker et al. (4) suggested a function for AtSUC3 in sucrose sensing as there is evidence that sucrose can act as a signal molecule in plants (22, 112, 148). A similar protein, AtSUC1, which is localized in the anthers, may generate an osmotic gradient that results in the dehydration of the endothecium and anther dehiscence (130).

BREEDING OF PLANTS RESISTANT
TO POD DEHISCENCE

Several strategies for breeding crops with a reduced capacity to shatter are currently underway. In *B. napus* there is little variation for resistance to shattering between current cultivars; however, resistant lines have been found within the crops' diploid parents (*B. oleracea* and *B. rapa*) as well as within other members of the Brassicae. Synthetic *B. napus* lines have been generated from different lines of *B. oleracea* and

B. rapa, and these show increased variation in pod shattering susceptibility (95). Resistance to shattering in these lines appears to be associated with the presence of extra vascular tissue within the DZ and with the failure of the walls of the DZ cells to degrade. These synthetic lines, however, contained many agronomically deleterious traits that made them unsuitable as cultivars (95). Intergeneric crosses between *B. napus* and *Sinapis alba* (commonly known as yellow mustard) are also being generated, in an attempt to transfer a number of beneficial agronomic traits to oilseed rape, including resistance to drought stress and pod shatter (11).

As with *B. napus* there is little variation in pod shattering resistance within birdsfoot trefoil (*Lotus corniculatus*), and breeding to reduce shatter through recurrent selection has been unsuccessful (52). At present, attempts are being made to transfer the indehiscent seed pod trait from distantly related species (e.g., by interspecific somatic hybridization) (reviewed in 52). In soybean, studies are currently underway to identify quantitative trait loci (QTL) that condition resistance to pod dehiscence (3). Five putatively independent RFLP markers were associated with pod dehiscence, one of which accounted for 44% of the variation in shatter.

ANTHER DEHISCENCE

Dehiscence is not restricted to pods and also takes place during the final stages of anther development, resulting in the release of pollen. Anther dehiscence, therefore, has an important role in crop production, being necessary for the generation of seeds that may be either harvested or used to produce the next generation.

Anther development can be divided into two stages. During the first stage the morphology of the anther is determined, differentiation of cells and tissues occurs, and the microspore mother cells undergo meiosis. In stage two the filament extends, the anther enlarges, and eventually, tissue degeneration, dehiscence, and the release of pollen grains occur (46). The cellular processes that regulate differentiation of the anther and cause it to switch to a program of cell degeneration and dehiscence are not yet known.

The process of anther dehiscence is essentially quite similar to that described for pods. The endothecium cell walls become thickened, and there is enzymatic breakdown of the cell walls in the stomium and circular cell cluster. Just prior to dehiscence, the epidermal and endothecial cells become swollen and turgid, rupturing the weakened stomium. Subsequent desiccation of the endothecium causes differential shrinkage of the thickened and unthickened regions of the cell wall. This causes an outward bending force that leads to retraction of the anther wall and full opening of the stomium, thus allowing release of the pollen (10, 46, 75). Using a cell ablation strategy to destroy specific anther cell types, Beals & Goldberg (5) demonstrated that a set of functional stomium cells is required for anther dehiscence in tobacco.

Male sterile mutants have been identified in many plants including *Arabidopsis* (74). Although many of these male sterility genes have not been cloned and studied, phenotypic analyses suggest that some of these mutations affect the differentiation and/or function of various anther cell types, including the stomium, tapetum, and endothecium (18, 74). The *ms35* mutation in *Arabidopsis* (previously known as *msH*) causes male sterility by disrupting the development of the lignified secondary thickening in the anther endothecium, thus preventing the anthers from dehiscing (32). Another *Arabidopsis* mutant, designated as *late dehiscence*, has stamens that are morphologically similar to those of wild type. However, the *late dehiscence* anthers fail to dehisce on time and release their pollen grains after the stigma is no longer receptive to pollination (46).

Little is known about the signal that regulates the onset of anther dehiscence. However, male-sterile tobacco anthers that lack tapetal cells and pollen grains undergo a normal dehiscence process (76, 86) suggesting that dehiscence is not triggered by signals originating in either of these two cell types. A T-DNA mutagenesis screen for *Arabidopsis* male-fertility mutants resulted in the isolation of *delayed dehiscence1* plants, which exhibit a delay in stomium degeneration and, hence, anther dehiscence (120). The DELAYED DEHISCENCE1 gene encodes 12-oxophytodienoate reductase, an enzyme in the jasmonic acid biosynthesis pathway (121). In *delayed dehiscence1* plants there is no DELAYED DEHISCENCE1 mRNA, nor is there any 12-oxophytodienoate reductase activity. The defect in anther dehiscence can be rescued, and seed set can be obtained in this mutant by the exogenous application of jasmonic acid, suggesting that jasmonic acid signaling plays a role in the regulation of anther dehiscence (121). Other jasmonic acid mutants exhibit problems in anther dehiscence. The *coi1* mutant, identified by root-growth insensitivity to jasmonic acid, and the triple *fad* mutant (*fad3-2, fad7-2, fad8*), created to be deficient in trienoic acid production, are both defective in anther dehiscence and pollen development (42, 87, 149). Triple *fad*, which leads to a defect in the precursors of jasmonic acid biosynthesis, can be rescued by exogenous jasmonic acid; *coi1* however cannot be rescued in this way, and the mutation may cause a defect in jasmonic acid perception. It must be noted, however, that other *Arabidopsis* jasmonic acid mutants such as *jin* (6), *jar* (131), and a tomato mutant *def1*, which is blocked in the octadecanoid pathway (64), are all fertile.

A number of anther-specific mRNAs have been isolated and shown to encode proteins, such as pectate lyases, PGs, thiol endopeptidases, and glycine-rich and proline-rich cell wall proteins (76, 88, 126), that may be involved in the dehiscence process. Also, at the time of dehiscence in the anthers of *Lathyrus odoratus* L., high levels of β-1,4-glucanase activity have been recorded (97).

A sucrose carrier, AtSUC1, which is expressed in anther connective tissue during the final stages of *Arabidopsis* anther development and immediately prior to anther dehiscence, has been isolated (130). AtSUC1 might play a role in anther dehiscence by causing sucrose uptake in a ring of parenchyma cells surrounding the connective tissue, which causes these cells to import water from the adjacent cells of the anther walls. The water flow from the anther walls to the connective tissue will result in the endothecial cells becoming dehydrated and may cause the tensions that eventually lead to anther opening (130).

MANIPULATION OF ABSCISSION
AND DEHISCENCE PROCESSES

The processes of abscission and dehiscence are critical to the life cycle of a plant. In addition, they are phenomena that have significance to agriculture and horticulture, and as a consequence, major benefits could be gained by manipulating either the timing or the site at which they occur (108).

By attenuating the expression of genes involved in AZ differentiation, ethylene production, or cell wall breakdown it may be possible to alter the timing of organ shedding or even prevent it from taking place. Delaying the shedding of leaves, flowers, fruit, or seeds may make it possible to enhance photosynthetic assimilation, extend the longevity of a horticultural species, or elevate the yield of a crop. Alternatively loosening ripe fruit could facilitate their harvest, and toward the end of a growing season accelerating the fall of immature flowers and fruit might prevent them from acting as a site for attack by pathogens.

Conventional breeding methods to improve pod shatter are limited, as there is generally very little variation in this phenomenon between the different cultivars available. An alternative approach to increase resistance to pod shatter would be to inhibit the process of dehiscence itself. This may be by generating crops that do not have a DZ in the pods, as seen in the SHATTERPROOF mutants in *Arabidopsis*. However, whether seeds can then be recovered from the pods by conventional threshing methods remains to be seen. Another strategy to inhibit shatter may be to stop the breakdown of the cell walls in the dehiscence layer, either by manipulating the enzymes responsible or by strengthening the region of the DZ itself. The antisense expression of cell wall degrading enzymes such as PGs and β-1,4-glucanases may prevent cell wall breakdown and dehiscence, although as the process probably involves a cocktail of different enzymes it may only delay it. A better method of controlling dehiscence may be to identify the genes and/or ligand that act upstream of these enzymes, as switching this off would prevent the activation of all the downstream events.

OTHER CELL SEPARATION PROCESSES

Abscission and dehiscence are similar phenomena in that the overall objective of the event is to shed specific tissues or organs. However, cell separation also occurs at a spectrum of other times during the life cycle of a plant (see Figure 3, see color insert), and the following section summarizes what we know about the regulation of some of these other processes.

Seeds

The primary role of the seed is to protect and nurture the developing embryo. This role necessarily also constrains the process of seedling emergence, and evidence is accumulating that cell wall degradation takes place in a coordinated way to facilitate germination. At least three expansins are associated with endosperm

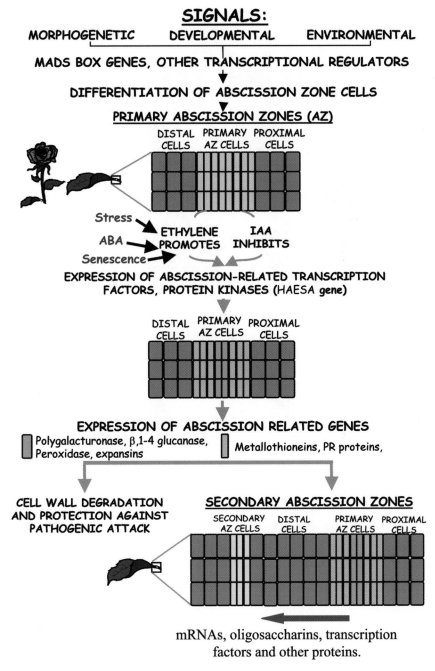

Figure 1 A model outlining some of the events that take place during the formation and response of primary and secondary abscission zones (AZs).

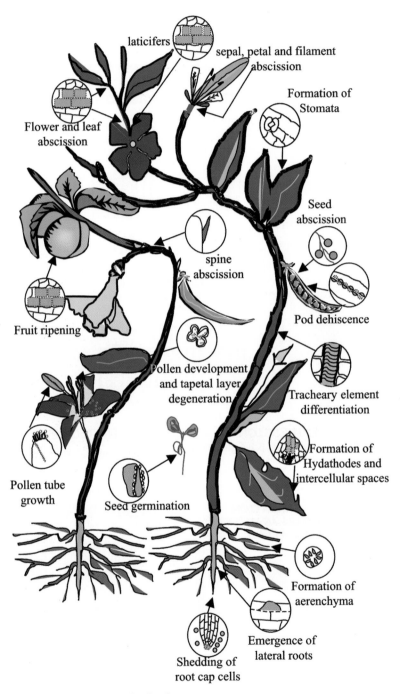

Figure 2 Sites of cell separation in plants.

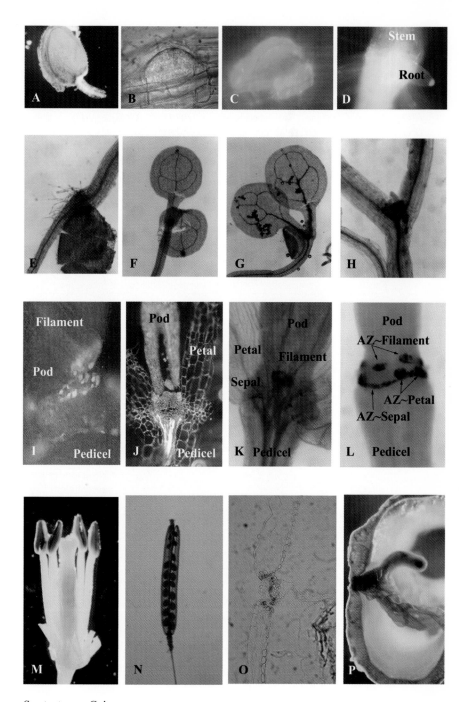

See text page C-4

Figure 3 (page C-3) GUS or GFP expression patterns seen in transgenic *Arabidopsis* seedlings driven by polygalacturonase promoters from either *Arabidopsis* (*PGAZAT*, *SAC70*) or *Brassica napus* (*PGAZBRAN*). *PGAZAT* & *PGAZBRAN* are promoters of abscission-related PGs, and *SAC70* is the promoter of a dehiscence-related PG. (*A*) seed testa (*PGAZAT::GUS*), (*B*) lateral root primordia (*PGAZAT::GFP*) bright field, (*C*) lateral root primordia (*PGAZAT::GFP*), (*D*) root cap (*PGAZAT::GFP*), (*E*) root:shoot junction (*PGAZBRAN::GUS*), (*F*) cotyledon (*PGAZAT::GUS*), (*G*) cotyledon (*PGAZBRAN::GUS*), (*H*) leaf primordia (*PGAZBRAN::GUS*), (*I*) abscinding flower (*PGAZAT::GFP*), (*J*) section though abscinding flower (*PGAZAT::GUS*) dark field, (*K*) abscinding flower (*PGAZAT::GUS*), (*L*) silique base (*PGAZAT::GUS*), (*M*) anthers (*SAC70::GUS*), (*N*) pod (*SAC70::GUS*), (*O*) section through DZ of pod (*SAC70::GUS*), (*P*) seed AZ (*SAC70::GUS*).

weakening during tomato seed germination. In situ hybridization analysis has revealed that the localization of the mRNAs encoding these enzymes is at the micropylar endosperm cap region, and so they may be involved in radicle emergence (20).

Roots

The mucilaginous sheath that surrounds a root as it penetrates the soil is primarily composed of cells that are shed from the cap region. Although the precise role of these cells is uncertain it is evident that they are of high viability and capable of being independently cultured (57). In many ways the cells resemble those that are observed at the exposed fracture of an abscinding leaf or flower (143), and the demonstration that release is correlated with elevated pectolytic enzymes (56) and β-1,4-glucanase (143) activity supports this. The mechanism responsible for bringing about their separation is unknown, and a role for ethylene in the process has not been examined.

The loss of adhesion between root cells is also a feature associated with the outgrowth of lateral roots. These organs are initiated within the pericycle, and localized cell separation enables them to push their way through the cortical and epidermal cells while causing minimal tissue damage. In *Allium porrum* lateral root development is associated with an increase in the activity of PG, and at least two isoforms of the enzyme may be responsible (101). In situ localization revealed that the PG accumulated between the primordium apex and the cortical tissues. Whether the PG originates from the developing lateral or the cells with which it comes into contact is unclear; however, our evidence from reporter gene studies suggests that it may be the latter (see Figures 2*B*,*C*).

Cell separation can occur as part of the natural developmental program of a growing root, and it can also be initiated by phenomena such as stress. Roots of species such as rice and maize respond to anaerobic conditions by generating aerenchyma. This phenomenon involves the collapse of targeted cells in the roots and stems to produce continuous gas spaces that are used to transport oxygen from the aerial tissues (37). Aerenchyma development involves both protoplast disintegration and breakdown of the middle lamella (125), and there is convincing evidence that formation is regulated to an extent by ethylene (59, 117).

The molecular mechanisms that bring about aerenchyma development remain unknown. However, Saab & Sachs (117) have identified a xyloglucan endo-transglycosylase (XET) gene from maize that may be involved in wall dissolution. Expression of the XET was upregulated in root and mesocotyl tissues within 12 h of flooding, and its accumulation could be prevented by treatment with the ethylene synthesis inhibitor aminooxyacetic acid (AOA). Increased β-1,4-glucanase activity has also been reported to accompany aerenchyma formation (58).

Leaves and Stems

During leaf development intercellular air spaces develop to enable gaseous diffusion to take place. The formation of these spaces occurs at discontinuous locations sited at the junctions between cells (31) and is associated with an increase in activity of cell-wall degrading enzymes (67). Other places within the leaf where cell separation is likely to take place include the formation of structures such as hydathodes (see Figure 2F) where guttation from the xylem system may occur. Cell to cell adhesion may also be lost within the epidermal cells of the leaf, for instance, at the site of stoma formation where pore formation is associated with the specific breakdown of the wall joining the guard cells together (118, 132).

Laticifers are specialized cell types that are present in some groups of flowering plants. Natural products such as thebaine, heroine, morphine, codeine, and rubber are derived from laticifer exudates, commonly referred to as "latex." The development of this particular cell type begins with the gradual disappearance of common walls between differentiating laticifer elements throughout the plant. Evidence shows that in *Papaver somniferum* genes encoding cell wall degrading enzymes such as pectin methylesterase (PME), pectin acetylesterase (PAE), and pectate lyase (PL) are expressed in a latex-specific manner (105). Although in this study the expression of genes encoding β-1,4-glucanase, PG, and XET did not appear to be limited to laticifers, it is likely that cell wall modifications in these cells take place in a similar way to wall degradation found at other sites in plants.

Fruit

An extensive body of research has been carried out to determine how fruit softening takes place. In many fruit, cell wall degradation is involved, and a spectrum of hydrolytic enzymes including β-1,4-glucanases (8, 41, 140), PGs (44, 54, 55), pectinmethylesterases (96), expansins (113–115), and β-glucosidase (45) accompany ripening. In climacteric fruit there is convincing evidence that ethylene plays a central role in coordinating the softening process, whereas in nonclimacteric fruit the regulatory stimuli are not yet defined (45a).

FUTURE STRATEGIES FOR STUDYING CELL SEPARATION PROCESSES

It is likely that the sites identified above represent only a fraction of those where wall breakdown takes place either routinely as part of a developmental sequence or in response to an environmental stimulus. If some of the events that bring about wall dissolution are common, then further information about sites where cell to cell adhesion is lost may come about by analysing expression patterns of the large families of genes that encode wall degrading enzymes. We have begun this analysis by studying the expression of different PGs in *Arabidopsis*, and we have identified a spectrum of sites where this enzyme may accumulate and perhaps contribute to wall loosening. These include the testa, root cap, site of lateral root emergence, hydathodes, vascular tissue, leaf primordia, floral and seed AZs, and anther and pod DZs (see Figure 2). The use of specific gene knockouts will also prove invaluable to ascertain which peptides play critical roles in defining the sites of cell wall disassembly and bring about wall dissolution. If gene knockouts reveal that certain isoforms of wall hydrolases play specific roles in causing wall breakdown, then the expression of different isoforms at these sites may enable us to determine whether substrate specificity plays an important role in regulating the site of cell separation. This is an important area for future studies and will be assisted by functional studies on the different domains of enzymes such as PG (33a) and by detailed biochemical analyses of the cell wall matrix of separating cells (147a). The search for the trans-acting factors that regulate cell separation will be aided by the identification of domains within the promoters of hydrolytic enzymes that are critical for expression during such processes as abscission and dehiscence and with the isolation and characterization of mutants that exhibit a delay in undergoing cell separation or fail to exhibit the phenomenon.

The past 20 years has seen major advances in our understanding of how cell separation is regulated in plants. In particular, genes involved in the differentiation of abscission and dehiscence zones have been characterized, and some of the events associated with wall degradation identified. In this review we highlight some key questions that remain to be addressed, and we predict that over the next decade many of the answers should be forthcoming.

ACKNOWLEDGMENTS

The authors would like to thank colleagues for access to prepublication information and the BBSRC and BIOGEMMA U.K. Ltd. for funding their work.

Visit the Annual Reviews home page at www.annualreviews.org

LITERATURE CITED

1. Alvarez-Buylla ER, Liljegren SJ, Pelaz S, Gold SE, Burgeff C, et al. 2000. MADS-box gene evolution beyond flowers: expression in pollen, endosperm, guard cells, roots and trichomes. *Plant J.* 24: 457–66
2. Aneja M, Gianfagna T, Ng E. 1999. The roles of abscisic acid and ethylene in the abscission and senescence of cocoa flowers. *Plant Growth Regul.* 27:149–55
3. Bailey MA, Mian MAR, Carter TE, Ashley DA, Boerma HR. 1997. Pod dehiscence of soybean: identification of quantitative trait loci. *J. Hered.* 88:152–54
4. Barker L, Kuhn C, Weise A, Schulz A, Gebhardt C, et al. 2000. SUT2, a putative sucrose sensor in sieve elements. *Plant Cell* 12:1153–64
5. Beals TP, Goldberg RB. 1997. A novel

6. cell ablation strategy blocks tobacco anther dehiscence. *Plant Cell* 9:1527–45
6. Berger S, Bell E, Mullet JE. 1996. Two methyl jasmonate-insensitive mutants show altered expression on *AtVsp* in response to methyl jasmonate and wounding. *Plant Physiol.* 111:525–31
7. Bleecker AB, Patterson SE. 1997. Last exit: senescence, abscission and meristem arrest in *Arabidopsis*. *Plant Cell* 9:1169–79
8. Bonghi CN, Ferrarese L, Ruperti B, Tonutti P, Ramina A. 1998. Endo-1,4-β-glucanases are involved in peach fruit growth and ripening and regulated by ethylene. *Physiol. Plant* 102:346–52
9. Bonghi CN, Rascio A, Ramina A, Casadoro G. 1992. Cellulase and polygalacturonase involvement in the abscission of

leaf and fruit explants of peach. *Plant Mol. Biol.* 20:839–48

10. Bonner LL, Dickinson HG. 1989. Anther dehiscence in *Lycopersicon esculentum*. I. Structural aspects. *New Phytol.* 113:97–115

11. Brown J, Brown AP, Davis JB, Erickson D. 1997. Intergeneric hybridization between *Sinapis alba* and *Brassica napus*. *Euphytica* 93:163–68

12. Brown KM. 1997. Ethylene and abscission. *Physiol. Plant* 100:567–76

13. Brummell DA, Bird CR, Schuch W, Bennett AB. 1997. An endo-1,4-β-glucanase expressed at high levels in rapidly expanding tissues. *Plant Mol. Biol.* 33:87–95

14. Brummell DA, Catala C, Lashbrook CC, Bennett AB. 1997. A membrane-anchored E-type endo-1,4-β-glucanase is localized on Golgi and plasma membranes of higher plants. *Proc. Natl. Acad. Sci. USA* 94:4794–99

15. Brummell DA, Hall BD, Bennett AB. 1999. Antisense suppression of tomato endo-1,4-β-glucanase Cel2 mRNA accumulation increases the force required to break fruit abscission zones but does not affect fruit softening. *Plant Mol. Biol.* 40:615–22

15a. Bryant JA, Hughes SG, Garland JM, eds. 2000. *Programmed Cell Death in Animals and Plants*. Oxford: BIOS Sci.

16. Catala C, Rose JKC, Bennett AB. 1997. Auxin regulation and spatial localization of an endo-1,4-β-D-glucanase and a xyloglucan endotransglycosylase in expanding tomato hypocotyls. *Plant J.* 12:417–26

17. Chao Q, Rothenberg M, Solano R, Roman G, Terzaghi W, Ecker JR. 1997. Activation of ethylene gas response pathway in *Arabidopsis* by the nuclear protein ethylene insensitive 3 and related proteins. *Cell* 89:1133–44

18. Chaudhury AM. 1993. Nuclear genes controlling male fertility. *Plant Cell* 5:1277–83

19. Chauvaux N, Child R, John K, Ulvskov P,

Borkhardt B, et al. 1997. The role of auxin in cell separation in the dehiscence zone of oilseed rape pods. *J. Exp. Bot.* 48:1423–29

20. Chen F, Bradford KJ. 2000. Expression of an expansin is associated with endosperm weakening during tomato seed germination. *Plant Physiol.* 124:1265–74

21. Child R, Chauvaux N, John K, Ulvskov P, Van Onckelen HA. 1998. Ethylene biosynthesis in oilseed rape pods in relation to pod shatter. *J. Exp. Bot.* 49:829–38

22. Chiou TJ, Bush DR. 1998. Sucrose is a signal molecule in assimilate partitioning. *Proc. Natl. Acad. Sci. USA* 95:4784–88

23. Cho H-T, Cosgrove DJ. 2000. Altered expression of expansin modulates leaf growth and pedicel abscission in *Arabidopsis thaliana*. *Proc. Natl. Acad. Sci. USA* 97:9783–88

24. Clement J, Atkins C. 2001. Characterization of a non-abscission mutant of *Lupinus angustifolius*. I. Genetic and structural aspects. *Am. J. Bot.* 88:31–42

25. Cosgrove DJ. 2000. New genes and new biological roles for expansins. *Curr. Opin. Plant Biol.* 3:73–78

26. Cosgrove DJ. 2001. Wall structure and wall loosening. A look backwards and forwards. *Plant Physiol.* 125:131–34

27. Coupe SA, Taylor JE, Isaac PG, Roberts JA. 1993. Identification and characterization of a proline-rich mRNA that accumulates during pod development in oilseed rape (*Brassica napus* L.). *Plant Mol. Biol.* 23:1223–32

28. Coupe SA, Taylor JE, Isaac PG, Roberts JA. 1994. Characterization of a mRNA that accumulates during development of oilseed rape pods. *Plant Mol. Biol.* 24:223–27

29. Coupe SA, Taylor JE, Roberts JA. 1995. Characterization of an mRNA encoding a metallothionein-like protein that accumulates during ethylene-promoted abscission of *Sambucus nigra* L. leaflets. *Planta* 197:442–47

30. Coupe SA, Taylor JE, Roberts JA. 1997. Temporal and spatial expression of

mRNAs encoding pathogenesis-related proteins during ethylene-promoted leaflet abscission in *Sambucus nigra*. *Plant Cell Environ.* 20:1517–24

31. Dale JE, Milthorpe FL. 1981. General features of the production and growth of leaves. In *The Growth and Functioning of Leaves*, ed. JE Dale, FL Milthorpe, pp. 151–78. Cambridge, UK: Cambridge Univ. Press

32. Dawson J, Sozen E, Vizir I, Van Waeyenberge S, Wilson ZA, Mulligan BJ. 1999. Characterization and genetic mapping of a mutation (ms35) which prevents anther dehiscence in *Arabidopsis thaliana* by affecting secondary wall thickening in the endothecium. *New Phytol.* 144:213–22

33. Day JS. 2000. Anatomy of capsule dehiscence in sesame varieties. *J. Agric. Sci.* 134:45–53

33a. Degan FD, Child R, Svendsen I, Ulvskov P. 2001. The cleavable N-terminal domain of plant endopolygalacturonases from clade B may be involved in a regulated secretion mechanism. *J. Biol.Chem.* 276: 35297–304

34. del Campillo E. 1999. Multiple endo-1,4-β-D-glucanase (cellulase) genes in *Arabidopsis*. *Curr. Top. Dev. Biol.* 46:39–61

35. del Campillo E, Bennett AB. 1996. Pedicel breakstrength and cellulase gene expression during tomato flower abscission. *Plant Physiol.* 111:813–20

36. del Campillo E, Lewis LN. 1992. Identification and kinetics of accumulation of proteins induced by ethylene in bean abscission zones. *Plant Physiol.* 98:955–61

37. Drew MC, He C-J, Morgan PW. 2000. Ethylene-triggered cell death during aerenchyma formation in roots. See Ref. 15a, pp. 183–92

38. Eyal Y, Meller Y, Levyadun S, Fluhr R. 1993. A basic-type PR-1 promoter directs ethylene responsiveness, vascular and abscission zone-specific expression. *Plant J.* 4:225–34

39. Ferrandiz C, Liljegren SJ, Yanofsky MF. 2000. Negative regulation of the SHAT-TERPROOF genes by FRUITFULL during *Arabidopsis* fruit development. *Science* 289:436–38

40. Ferrandiz C, Pelaz S, Yanofsky MF. 1999. Control of carpel development. *Annu. Rev. Biochem.* 68:321–54

41. Ferrarese L, Trainotti L, Moretto P, Polverino de Laureto P, Rascio N, Casadoro G. 1995. Differential ethylene-inducible expression of cellulase in pepper plants. *Plant Mol. Biol.* 29:735–47

42. Feys BJ, Benedetti CE, Penfold CN, Turner JG. 1994. *Arabidopsis* mutants selected for resistance to the phytotoxin coronatine are male sterile, insensitive to methyl jasmonate, and resistant to a bacterial pathogen. *Plant Cell* 6:751–59

43. Flanagan CA, Hu Y, Ma H. 1996. Specific expression of *AGL1* MADS-box gene suggests regulatory functions in *Arabidopsis* gynoecium and ovule development. *Plant J.* 10:343–53

44. Gallego PP, Zarra I. 1998. Cell wall autolysis during kiwifruit development. *Ann. Bot.* 81:91–96

45. Gerardi C, Blando F, Santino A, Zacheo G. 2001. Purification and characterization of a beta glucosidase abundantly expressed in ripe sweet cherry (*Prunus avium* L) fruit. *Plant Sci.* 160:795–805

45a. Giovannoni J. 2001. Molecular biology of fruit maturation and ripening. *Annu. Rev. Plant Physiol. Plant Mol. Biol.* 52:725–49

46. Goldberg RB, Beals TP, Sanders PM. 1993. Anther development: basic principles and practical applications. *Plant Cell* 5:1217–29

47. Deleted in proof

48. Gonzalez-Bosch C, del Campillo E, Bennett AB. 1997. Immunodetection and characterization of tomato endo-1,4-β-glucanase cell protein in flower abscission zones. *Plant Physiol.* 114:1541–46

49. Gonzalez-Carranza ZH, Lozoya-Gloria E, Roberts JA. 1998. Recent developments in abscission: shedding light on the shedding process. *Trends Plant Sci.* 3:10–14

50. Gonzalez-Carranza ZH, Whitelaw CA,

Swarup R, Roberts JA. 2002. Temporal and spatial expression of a polygalacturonase during leaf and flower abscission in *Brassica napus* and *Arabidopsis thaliana*. *Plant. Physiol.* In press

51. Goodwin W, Palls JA, Jenkins GI. 1996. Transcripts of a gene encoding a putative cell wall-plasma membrane linker protein are specifically cold-induced in *Brassica napus*. *Plant Mol. Biol.* 31:771–81

52. Grant WF. 1996. Seed pod shattering in the genus Lotus (*Fabaceae*): a synthesis of diverse evidence. *Can. J. Plant Sci.* 76:447–56

53. Gu Q, Ferrandiz C, Yanofsky MF, Martienssen R. 1998. The FRUITFULL MADS-box gene mediates cell differentiation during *Arabidopsis* fruit development. *Development* 125:1509–17

54. Hadfield KA, Bennett AB. 1998. Polygalacturonases: many genes in search of a function. *Plant Physiol.* 117:337–43

55. Hadfield KA, Rose JKC, Yaver DS, Berka RM, Bennett AB. 1998. Polygalacturonase gene expression in ripe melon fruit supports a role for polygalacturonase in ripening-associated pectin disassembly. *Plant Physiol.* 117:363–73

55a. Harmer SL, Hogenesch JB, Straume M, Chang H-S, Hans B, et al. 2000. Orchestrated transcription of key pathways in *Arabidopsis* by the circadian clock. *Science* 290:2110–13

56. Hawes MC, Lin H-J. 1990. Correlation of pectolytic enzyme activity with the programmed release of cells from root caps of pea (*Pisum sativum*). *Plant Physiol.* 94:1855–59

57. Hawes MC, Pueppke SG. 1986. Isolated peripheral root cap cells: yield from different plants, and callus formation from single cells. *Am. J. Bot.* 73:1466–73

58. He C-J, Drew MC, Morgan PW. 1994. Induction of enzymes associated with lysigenous aerenchyma formation in roots of *Zea mays* during hypoxia or nitrogen starvation. *Plant Physiol.* 105:861–65

59. He C-J, Morgan PW, Drew MC. 1996. Transduction of an ethylene signal is required for cell death and lysis in the root cortex of maize during aerenchyma formation induced by hypoxia. *Plant Physiol.* 112:463–72

60. Heisler MGB, Atkinson A, Bylstra YH, Walsh R, Smyth DR. 2001. *SPATULA*, a gene that controls development of carpel margin tissues in *Arabidopsis*, encodes a bHLH protein. *Development* 128:1089–98

61. Henderson J, Davies HA, Heyes SJ, Osborne DJ. 2001. The study of a monocotyledon abscission zone using microscopic, chemical, enzymatic and solid state ^{13}C CP/MAS NMR analyses. *Phytochemistry* 56:131–39

62. Hocking PJ, Mason L. 1993. Accumulation, distribution and redistribution of dry matter and mineral nutrients in fruits of canola (oilseed rape) and the effect of nitrogen fertiliser and windrowing. *Aust. J. Agric. Res.* 44:1377–88

63. Hong S-B, Sexton R, Tucker ML. 2000. Analysis of gene promoters for two tomato polygalacturonases expressed in abscission zones and the stigma. *Plant Physiol.* 123:869–81

64. Howe GA, Lightner J, Browse J, Ryan CA. 1996. An octadecanoid pathway mutant (JL5) of tomato is compromised in signaling for defense against insect attack. *Plant Cell* 8:2067–77

65. Huber DJ. 1983. The role of cell wall hydrolases in fruit softening. *Hortic. Rev.* 5:169–215

66. Jackson MB, Osborne DJ. 1970. Ethylene, the natural regulator of leaf abscission. *Nature* 225:1019–22

67. Jeffree CE, Dale JE, Fry SC. 1986. The genesis of intercellular spaces in developing leaves of *Phaseolus vulgaris* L. *Protoplasma* 132:90–98

68. Jenkins ES, Paul W, Coupe SA, Bell SJ, Davies EC, Roberts JA. 1996. Characterization of an mRNA encoding a polygalacturonase expressed during pod

development in oilseed rape (*Brassica napus* L.). *J. Exp. Bot.* 47:111–15

69. Jenkins ES, Paul W, Craze M, Whitelaw CA, Weigand A, Roberts JA. 1999. Dehiscence-related expression of an *Arabidopsis thaliana* gene encoding a polygalacturonase in transgenic plants of *Brassica napus*. *Plant Cell Environ.* 22:159–67

70. Jinn T-L, Stone JM, Walker JC. 2000. *HAESA*, an Arabidopsis leucine-rich repeat receptor kinase, controls floral organ abscission. *Genes Dev.* 14:108–17

71. Kalaitzis P, Hong S-B, Solomos T, Tucker ML. 1999. Molecular characterization of a tomato endo-β-1,4-glucanase gene expressed in mature pistils, abscission zones and fruit. *Plant Cell Physiol.* 40:905–8

72. Kalaitzis P, Koehler SM, Tucker ML. 1995. Cloning of tomato polygalacturonase expressed in abscission. *Plant Mol. Biol.* 28:647–56

73. Kalaitzis P, Solomos T, Tucker ML. 1997. Three different polygalacturonases are expressed in tomato leaf and flower abscission, each with a different temporal expression pattern. *Plant Physiol.* 113:1303–8

74. Kaul MLH. 1988. *Male Sterility in Higher Plants*. Berlin: Springer-Verlag

75. Keijzer CJ. 1987. The process of anther dehiscence and pollen dispersal. I. The opening mechanism of longitudinally dehiscing anthers. *New Phytol.* 105:487–98

76. Koltunow AM, Truettner J, Cox KH, Wallroth M, Goldberg RB. 1990. Different temporal and spatial gene expression patterns occur during anther development. *Plant Cell* 2:1201–24

77. Kubigsteltig I, Laudert D, Weiler EW. 1999. Structure and regulation of the *Arabidopsis thaliana* allene oxide synthase gene. *Planta* 208:63–71

78. Lamb CJ. 1994. Plant disease resistance genes in signal perception and transduction. *Cell* 76:419–22

79. Lashbrook CC, Giovannoni JJ, Hall BD, Fischer RL, Bennett AB. 1998. Transgenic analysis of tomato endo-β-1,4-glu-

canase gene function. Role of *cel1* in floral abscission. *Plant J.* 13:303–10

80. Lashbrook CC, Gonzalez-Boch C, Bennett AB. 1994. Two divergent endo-β-1,4-glucanase genes exhibit overlapping expression in ripening fruit and abscising flowers. *Plant Cell* 6:1485–93

81. Liljegren SJ, Ditta GS, Eshed Y, Savidge B, Bowman JL, Yanofsky MF. 2000. *SHATTERPROOF* MADS-box genes control seed dispersal in *Arabidopsis*. *Nature* 404:766–70

82. Liljegren SJ, Ferrandiz C, Alvarez-Buylla ER, Pelaz S, Yanofsky MF. 1998. *Arabidopsis* MADS-box genes involved in fruit dehiscence. *Flower. Newsl.* 25:9–19

82a. Lim MAG, Kelly P, Sexton R, Trewavas AJ. 1987. Identification of chitinase mRNA in abscission zones from bean. *Plant Cell Environ.* 10:741–46

83. Ma H, Yanofsky MF, Meyerowitz EM. 1991. *AGL1–AGL6*, an *Arabidopsis* gene family with similarity to floral homeotic and transcription factor genes. *Genes Dev.* 5:484–95

84. Macleod J. 1981. Harvesting. In *Oilseed Rape Book*, pp. 107–19. Cambridge, UK: Agric. Publ.

85. Mao L, Begum D, Chuang H, Budiman MA, Szymkowiak EJ, et al. 2000. *JOINTLESS* is a MADS-box gene controlling tomato flower abscission zone development. *Nature* 406:910–13

86. Mariani C, De Beuckeleer M, Truettner J, Leemans J, Goldberg RB. 1990. Induction of male sterility in plants by a chimaeric ribonuclease gene. *Nature* 347:737–41

87. McConn M, Browse J. 1996. The critical requirement for linolenic acid is pollen development, not photosynthesis, in an *Arabidopsis* mutant. *Plant Cell* 8:403–16

88. McCormick S. 1991. Molecular analysis of male gametogenesis in plants. *Trends Genet.* 7:298–303

89. McManus MT, Osborne DJ. 1991. Identification and characterization of an ionically-bound cell wall glycoprotein expressed preferentially in the leaf rachis

abscission zone of *Sambucus nigra* L. *J. Plant Physiol.* 138:63–67

90. McManus MT, Thompson DS, Merriman C, Lyne L, Osborne DJ. 1998. Transdifferentiation of mature cortical cells to functional abscission cells in bean. *Plant Physiol.* 116:891–99

91. Meakin PJ, Roberts JA. 1990. Dehiscence of fruit in oilseed rape. 1. Anatomy of pod dehiscence. *J. Exp. Bot.* 41:995–1002

92. Meakin PJ, Roberts JA. 1990. Dehiscence of fruit in oilseed rape. 2. The role of cell wall degrading enzymes. *J. Exp. Bot.* 41:1003–11

93. Meakin PJ, Roberts JA. 1991. Induction of oilseed rape pod dehiscence by *Dasineura brassica. Ann. Bot.* 67:193–97

94. Meyer S, Melzer M, Truernit E, Hummer C, Besenbeck R, et al. 2000. AtSUC3, a gene encoding a new Arabidopsis sucrose transporter, is expressed in cell layers adjacent to the vascular tissue and in a carpel cell layer. *Plant J.* 24:869–82

95. Morgan CL, Bruce DM, Child R, Ladbrooke ZL, Arthur AE. 1998. Genetic variation for pod shatter resistance among lines of oilseed rape developed from synthetic *B. napus. Field Crops Res.* 58:153–65

96. Nairn JC, Lewendowski DJ, Burns JK. 1998. Genetics and expression of two pectinesterase genes in Valencia orange. *Physiol. Plant* 102:226–35

97. Neelam A, Sexton R. 1995. Cellulase (endo beta-1,4 glucanase) and cell wall breakdown during anther development in the sweet pea (*Lathyrus odoratus* L)—isolation and characterisation of partial cDNA clones. *J. Plant Physiol.* 146:622–28

98. Osborne DJ. 1989. Abscission. *Crit. Rev. Plant Sci.* 8:103–29

99. Patterson SE. 2001. Cutting loose. Abscission and dehiscence in Arabidopsis. *Plant Physiol.* 126:494–500

100. Peng J, Richards DE, Hartley NM, Murphy GP, Devos KM, et al. 1999. "Green revolution" genes encode mutant gibberellin response modulators. *Nature* 400:256–61

101. Peretto R, Favaron F, Bettini V, De Lorenzo G, Marini S, et al. 1992. Expression and localization of polygalacturonase during the outgrowth of lateral roots in *Allium porrum* L. *Planta* 188:164–72

102. Petersen M, Sander L, Child R, Van Onckelen H, Ulvskov P, Borkhardt B. 1996. Isolation and characterization of a pod dehiscence zone-specific polygalacturonase from *Brassica napus. Plant Mol. Biol.* 31:517–27

103. Philbrook B, Oplinger ES. 1989. Soybean field losses as influenced by harvest delays. *Agron. J.* 81:251–58

104. Picart JA, Morgan DG. 1984. Pod development in relation to pod shattering. *Asp. Appl. Biol.* 6:101–10

105. Pilatzke-Wunderlich I, Nessler CL. 2001. Expression and activity of cell-wall degrading enzymes in the latex of opium poppy, *Papaver somniferum* L. *Plant Mol. Biol.* 45:567–76

106. Pruitt KD, Last RL. 1993. Expression patterns of duplicate tryptophan synthase β genes in *Arabidopsis thaliana. Plant Physiol.* 102:1019–26

107. Roberts JA. 2000. Abscission and dehiscence. See Ref. 15a, pp. 203–11

108. Roberts JA, Whitelaw CA, Gonzalez-Carranza ZH, McManus M. 2000. Cell separation processes in plants—models, mechanisms and manipulation. *Ann. Bot.* 86:223–35

109. Roberts K. 2001. How the cell wall acquired a cellular context. *Plant Physiol.* 125:127–30

110. Rodriguez-Gacio MD, Matilla AJ. 2001. The last step of the ethylene biosynthesis pathway in turnip tops (*Brassica rapa*) seeds: alterations related to development and germination and its inhibition during desiccation. *Physiol. Plant* 112:273–79

111. Rohde A, Van Montagu MV, Boerjan W. 1999. The *ABSCISIC ACID-INSENSITIVE* (*ABI3*) gene is expressed during vegetative quiescence processes in

Arabidopsis. Plant Cell Environ. 22:261–70

112. Rook F, Gerrits N, Kortstee A, van Kampen M, Borrias M, et al. 1998. Sucrose-specific signalling represses translation of the *Arabidopsis ATB2* bZIP transcription factor gene. *Plant J.* 15:253–63

113. Rose JKC, Bennett AB. 1999. Cooperative disassembly of the cellulose-xyloglucan network of plant cell walls: parallels between cell expansion and fruit ripening. *Trends Plant Sci.* 4:176–83

114. Rose JKC, Cosgrove SJ, Albersheim P, Darvill AG, Bennett AB. 2000. Detection of expansin proteins and activity during tomato fruit ontogeny. *Plant Physiol.* 123:1583–92

115. Rose JKC, Lee HH, Bennett AB. 1997. Expression of a divergent expansin gene is fruit-specific and ripening related. *Proc. Natl. Acad. Sci. USA* 94:5955–60

116. Ruperti B, Whitelaw CA, Roberts JA. 1999. Isolation and expression of an allergen-like mRNA from ethylene-treated *Sambucus nigra* leaflet abscission zones. *J. Exp. Bot.* 50:733–34

117. Saab IN, Sachs MM. 1996. A flooding-induced xyloglucan *endo*-transglycosylase homolog in maize is responsive to ethylene and associated with aerenchyma. *Plant Physiol.* 112:385–91

118. Sack FD. 1987. The development and structure of stomata. In *Stomatal Function*, ed. E Zieger, GD Farquhar, IR Cowan, pp. 59–89. Stanford, CA: Stanford Univ. Press

119. Sander L, Child R, Ulvskov P, Albrechtsen M, Borkhardt B. 2001. Analysis of a dehiscence zone endo-polygalacturonase in oilseed rape (*Brassica napus*) and *Arabidopsis thaliana*: evidence for roles in cell separation in dehiscence and abscission zones, and in stylar tissues during pollen tube growth. *Plant Mol. Biol.* 46:469–79

120. Sanders PM, Bui AQ, Weterings K, McIntire KN, Hsu YC, et al. 1999. Anther developmental defects in *Arabidopsis thaliana* male-sterile mutants. *Sex. Plant Reprod.* 11:297–322

121. Sanders PM, Lee PY, Biesgen C, Boone JD, Beals TP, et al. 2000. The *Arabidopsis* DELAYED DEHISCENCE1 gene encodes an enzyme in the jasmonic acid synthesis pathway. *Plant Cell* 12:1041–61

122. Sargent JA, Osborne DJ, Dunford SM. 1984. Cell separation and its hormonal control in the Gramineae. *J. Exp. Bot.* 35:1663–67

123. Savidge B, Rounsley SD, Yanofsky MF. 1995. Temporal relationship between the transcription of two *Arabidopsis* MADS box genes and the floral organ identity genes. *Plant Cell* 7:721–33

124. Schumacher K, Schmitt T, Rossberg M, Schmitz G, Theres K. 1999. The *Lateral suppressor* (*Ls*) gene of tomato encodes a new member of the VHIID protein family. *Proc. Natl. Acad. Sci. USA* 96:290–95

125. Schussler EE, Longstreth DJ. 2000. Changes in the cell structure during the formation of root aerenchyma in *Sagittaria lancifolia* (Alismataceae). *Am. J. Bot.* 87:12–19

126. Scott R, Hodge R, Paul W, Draper J. 1991. The molecular biology of anther differentiation. *Plant Sci.* 80:167–91

127. Sexton R, Redshaw AJ. 1981. The role of cell expansion in the abscission of *Impatiens* leaves. *Ann. Bot.* 48:745–57

128. Sexton R, Roberts JA. 1982. Cell biology of abscission. *Annu. Rev. Plant Physiol.* 33:133–62

129. Spence J, Vercher Y, Gates P, Harris N. 1996. Pod shatter in *Arabidopsis thaliana*, *Brassica napus* and *B. juncea*. *J. Microsc.* 181:195–203

130. Stadler R, Truernit E, Gahrtz M, Sauer N. 1999. The AtSUC1 sucrose carrier may represent the osmotic driving force for anther dehiscence and pollen tube growth in *Arabidopsis. Plant J.* 19:269–78

131. Staswick PE, Su WP, Howell SH. 1992. Methyl jasmonate inhibition of root-growth and induction of a leaf protein are decreased in an *Arabidopsis thaliana*

mutant. *Proc. Natl. Acad. Sci. USA* 89: 6837–40

132. Stevens RA, Martin ES. 1978. Structural and functional aspects of stomata. 1. Developmental studies in *Polypodium vulgare*. *Planta* 142:307–16

133. Szymkowiak EJ, Irish EE. 1999. Interactions between *jointless* and wild-type tomato tissues during development of the pedicel abscission zone and the inflorescence meristem. *Plant Cell* 11:159–75

134. Taylor JE, Coupe SA, Picton SJ, Roberts JA. 1994. Isolation and expression of a mRNA encoding an abscission-related β1,4 glucanase from *Sambucus nigra*. *Plant Mol. Biol.* 24:961–64

135. Taylor JE, Tucker GA, Lasslett Y, Smith CJS, Arnold CM, et al. 1990. Polygalacturonase expression during leaf abscission of normal and transgenic tomato plants. *Planta* 183:133–38

136. Taylor JE, Webb STJ, Coupe SA, Tucker GA, Roberts JA. 1993. Changes in polygalacturonase activity and solubility of polyuronides during ethylene-stimulated leaf abscission in *Sambucus nigra*. *J. Exp. Bot.* 258:93–98

137. Taylor JE, Whitelaw CA. 2001. Signals in abscission. *New Phytol.* 151:323–39

138. Thompson DS, Osborne DJ. 1994. A role for the stele in intertissue signaling in the initiation of abscission in bean leaves (*Phaseolus vulgaris* L.). *Plant Physiol.* 105:341–47

139. Tiwari SP, Bhatia VS. 1995. Characters of pod anatomy associated with resistance to pod-shattering in soybean. *Ann. Bot.* 76:483–85

140. Trainotti L, Spolaore S, Pavanello A, Baldan B, Casadoro G. 1999. A novel E-type endo-β-1,4-glucanase with a putative cellulose-binding domain is highly expressed in ripening strawberry fruits. *Plant Mol. Biol.* 40:323–32

141. Tucker GA, Schindler CB, Roberts JA. 1984. Flower abscission in mutant tomato plants. *Planta* 160:164–67

142. Tucker ML, Sexton R, del Campillo E, Lewis LN. 1988. Bean abscission cellulase: characterization of a cDNA clone and regulation of gene expression by ethylene and auxin. *Plant Physiol.* 88:1257–62

143. Uheda E, Akasaka Y, Daimon H. 1997. Morphological aspects of the shedding of surface layers from peanut roots. *Can. J. Bot.* 75:607–11

144. Volko SM, Boller T, Ausubel FM. 1998. Isolation of new Arabidopsis mutants with enhanced disease susceptibility to *Pseudomonas syringe* by direct screening. *Genetics* 149:537–48

145. Warren Wilson J, Warren Wilson PM, Walker ES. 1987. Abscission sites in nodal explants of *Impatiens sultani*. *Ann. Bot.* 60:693–704

146. Warren Wilson PM, Warren Wilson J, Addicott FT, McKenzie RH. 1986. Induced abscission sites in internodal explants of *Impatiens sultani*: a new system for studying positional control. *Ann. Bot.* 57:511–30

147. Webb STJ, Taylor JE, Coupe SA, Ferrarese L, Roberts JA. 1993. Purification of β-1,4 glucanase from ethylene-treated abscission zones of *Sambucus nigra*. *Plant Cell Environ.* 16:329–33

147a. Willats WGT, Orfila C, Limberg G, Buchholt HC, van Alebeek G-JWM, et al. 2001. Modulation of the degree and pattern of methyl-esterification of pectic homogalacturonan in plant cell walls. *J. Biol. Chem.* 276:19404–13

148. Wobus U, Weber H. 1999. Sugars as signal molecules in plant seed development. *Biol. Chem.* 380:937–44

149. Xie D-X, Feys BF, James S, Nieto-Rostro M, Turner JG. 1998. *COI1*: an *Arabidopsis* gene required for jasmonate-regulated defense and fertility. *Science* 280:1091–94

150. Yasuda E, Ebinuma H, Wabiko H. 1997. A novel glycine-rich/hydrophobic 16kDa polypeptide gene from tobacco: similarity to proline-rich protein genes and its wound-inducible and developmentally regulated expression. *Plant Mol. Biol.* 33:667–78

Annu. Rev. Plant Biol. 2002. 53:159–82
DOI: 10.1146/annurev.arplant.53.100301.135154

PHYTOCHELATINS AND METALLOTHIONEINS: Roles in Heavy Metal Detoxification and Homeostasis

Christopher Cobbett

Department of Genetics, University of Melbourne, Parkville, Australia 3052;
e-mail: ccobbett@unimelb.edu.au

Peter Goldsbrough

Department of Horticulture and Landscape Architecture, Purdue University,
West Lafayette, Indiana 47907-1165; e-mail: goldsbrough@hort.purdue.edu

Key Words heavy metals, detoxification, homeostasis, phytochelatin synthase

■ **Abstract** Among the heavy metal-binding ligands in plant cells the phytochelatins (PCs) and metallothioneins (MTs) are the best characterized. PCs and MTs are different classes of cysteine-rich, heavy metal-binding protein molecules. PCs are enzymatically synthesized peptides, whereas MTs are gene-encoded polypeptides. Recently, genes encoding the enzyme PC synthase have been identified in plants and other species while the completion of the Arabidopsis genome sequence has allowed the identification of the entire suite of MT genes in a higher plant. Recent advances in understanding the regulation of PC biosynthesis and MT gene expression and the possible roles of PCs and MTs in heavy metal detoxification and homeostasis are reviewed.

CONTENTS

INTRODUCTION

Some heavy metals, particularly copper and zinc, are essential micronutrients for a range of plant physiological processes via the action of Cu- and Zn-dependent enzymes. These and other nonessential heavy metal ions, such as cadmium, lead, and mercury, are highly reactive and consequently can be toxic to living cells. Thus plants, like all living organisms, have evolved a suite of mechanisms that control and respond to the uptake and accumulation of both essential and nonessential heavy metals. These mechanisms include the chelation and sequestration of heavy metals by particular ligands. The two best-characterized heavy metal-binding ligands in plant cells are the phytochelatins (PCs) and metallothioneins (MTs).

MTs are cysteine-rich polypeptides encoded by a family of genes. In contrast, PCs are a family of enzymatically synthesized cysteine-rich peptides. The history of studies of PCs and MTs in plants and other species provides some salutary examples of the development of unsubstantiated dogmas that have been overturned by recent studies. In the search for MTs similar to those that had been characterized in animal species, early studies in plants repeatedly identified PCs. Like MTs in animals, PCs in plants are heavy metal-inducible, heavy metal-binding, cysteine-rich polypeptides, and in the absence of evidence for MTs in plants, it was initially suggested that PCs might be functionally analogous to MTs. The dichotomy that MTs were animal-specific ligands and PCs were plant specific became entrenched over time. Even when MT genes (repeatedly referred to as MT-like genes) were described in plants and it became clear that some microorganisms expressed both MTs and PCs, the notion that if PCs were not plant-specific ligands then at least they were "nonanimal" was maintained. Only with the isolation of PC synthase genes from plants and the demonstration that functional homologues exist in at least some animal species has this artificial dichotomy been discarded. It is also satisfying to describe an example where studies, largely in plant systems, have informed our understanding of extensively studied mechanisms in animal species.

The now apparent breadth of PC function across the plant and animal kingdoms leads to questions about nomenclature. PCs have been given various alternative names (including Class III MTs) over the years, but "phytochelatins" is the name that has been most widely adopted. The term phytochelatins has, however, never been truly accurate, particularly as PCs were first discovered in the yeast *S. pombe*, and is even less so because PC synthase genes have now been identified in animals. Nonetheless, it serves to distinguish one broad class of heavy metal-binding compounds from another and has become so entrenched in the literature that there seems little reason to change it. Its use also encompasses something of the history of PC and MT research.

This review describes recent advances in our understanding of the expression and function of both PCs and MTs in plants, with reference to other organisms where appropriate. PCs and MTs are discussed separately because at present there is no evidence that their spheres of function in plant cells even overlap, although it is likely that this would be the case. Much of the recent work has involved molecular genetic studies in Arabidopsis, which also provides a focus for this review. The isolation and characterization of PC-deficient mutants of Arabidopsis has provided considerable impetus to research into PC biosynthesis and function in plants. A similar set of mutants is required to illuminate the function of MTs.

PHYTOCHELATINS

Structure and Biosynthetic Pathway

PCs form a family of structures with increasing repetitions of the γ-GluCys dipeptide followed by a terminal Gly; $(\gamma$-GluCys$)_n$-Gly, where n is generally in the range of 2 to 5. PCs have been identified in a wide variety of plant species and in some microorganisms. In addition, a number of structural variants, for example, $(\gamma$-GluCys$)_n$-β-Ala, $(\gamma$-GluCys$)_n$-Ser and $(\gamma$-GluCys$)_n$-Glu, have been identified in some plant species. The reader is directed to previous reviews for a more detailed coverage of these early studies (14, 58, 59, 84).

PCs are structurally related to glutathione (GSH; γ-GluCysGly), and numerous physiological, biochemical, and genetic studies have confirmed that GSH (or, in some cases, related compounds) is the substrate for PC biosynthesis (14, 58, 59, 84). In particular, genetic studies have confirmed that GSH-deficient mutants of *S. pombe* as well as *Arabidopsis* are PC deficient and hypersensitive to Cd. A list of mutants that identify a role for particular genes in PC biosynthesis or function is shown in Table 1. In addition, a schematic including the PC biosynthetic pathway is illustrated in Figure 1.

Identification of PC Synthase Genes

The enzyme catalyzing the biosynthesis of PCs from GSH, phytochelatin synthase, was first characterized by Grill et al. (24) in 1989. However, it was not until 1999 that the cloning of PC synthase genes was described. The PC synthase gene was first identified genetically in Arabidopsis. Cd-sensitive, *cad1*, mutants are PC-deficient but have wild-type levels of GSH. They also lack PC synthase activity, suggesting a defect in the PC synthase gene (28).

PC synthase genes were isolated simultaneously by three research groups using different approaches. Two groups used expression of Arabidopsis and wheat cDNA libraries in *S. cerevisiae* to identify genes [*AtPCS1* (78) and *TaPCS1* (11), respectively] conferring increased Cd resistance. The third group identified *AtPCS1* through the positional cloning of the *CAD1* gene of Arabidopsis (25). A similar sequence was identified in the genome of *S. pombe*, and targeted deletion mutants of that gene are, like Arabidopsis *cad1* mutants, Cd sensitive and PC deficient, confirming the analogous function of the two genes in the different organisms. Heterologous expression of the *CAD1/AtPCS1* and *SpPCS* genes (25) or purification of epitope-tagged derivatives of SpPCS (11) and AtPCS1 (78) was used to demonstrate both were necessary and sufficient for GSH-dependent PC

TABLE 1   Mutants affected in phytochelatin biosynthesis and function

Organism[a]	Gene/locus	Activity/function	Reference
PC biosynthesis			
Sp	Gsh1	γ-glutamylcysteine synthetase/ GSH biosynthesis	(23, 51)
At	CAD2/RML1	γ-glutamylcysteine synthetase/ GSH biosynthesis	(14, 80)
Sp	Gsh2	glutathione synthetase/ GSH biosynthesis	(23, 51)
At	CAD1	PC synthase/PC biosynthesis	(28)
Sp	Pcs1	PC synthase/PC biosynthesis	(11, 25)
Ce	Pcs1	PC synthase/PC biosynthesis	(79)
PC function			
Sp	Hmt1	PC-Cd vacuolar membrane ABC-type transporter	(53, 54)
Sp	Ade2, 6, 7, 8	Metabolism of cysteine sulfinate to products involved in sulphide biosynthesis; also required for adenine biosynthesis	(32, 71)
Sp	Hmt2	Mitochondrial sulfide: quinone oxidoreductase/detoxification of sulphide	(74)
Ca	Hem2	Porphobilinogen synthase/ siroheme biosynthesis (cofactor for sulfite reductase)	(31)

[a]At, A. thaliana; Sp. S. pombe; Ca, Candida albicans; Ce; C. elegans.

biosynthesis in vitro. This combination of genetic, molecular, and biochemical data was a conclusive demonstration that these genes encode PC synthase.

There is a second PC synthase gene, *AtPCS2*, in Arabidopsis with significant identity to *CAD1/AtPCS1* (25). This was an unexpected finding because PCs were not detected in a *cad1* mutant after prolonged exposure to Cd, suggesting the presence of only a single active PC synthase in wildtype (28). *AtPCS2* is transcribed, and expression experiments have demonstrated it encodes a functional PC synthase enzyme (C. Cobbett & A. Savage, unpublished data). The physiological function of this gene remains to be determined. In most tissues *AtPCS2* is expressed at a relatively low level compared with *AtPCS1*. However, because AtPCS2 has been preserved as a functional PC synthase through evolution, it must presumably be the predominant PC synthase in some tissue(s) or environmental conditions, thereby conferring a selective advantage. Full-length or partial cDNA clones encoding presumptive PC synthases have also been isolated from other plant species, including *Brassica juncea* and rice (Table 2).

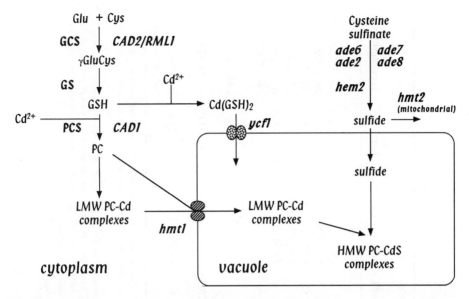

Figure 1 Genes and functions contributing to Cd detoxification in plants and fungi as a composite of various functions identified through the isolation of Cd-sensitive mutants of different organisms that express PCs. Refer to text and Table 1 for a more detailed description of the various functions. Gene loci are shown in italics: *CAD1* and *CAD2/RML1* are in *Arabidopsis*; *hmt1*, *hmt2*, *ade2*, *ade6*, *ade7*, and *ade8* are in *S. pombe*; *ycf1* is in *S. cerevisiae*; and *hem2* is in *C. glabrata*. Enzyme abbreviations: GCS, γ-glutamylcysteine synthetase; GS, glutathione synthetase; PCS, phytochelatin synthase.

Some Animals Express a PC Synthase

Through the history of studies of heavy metal detoxification in animals there has been no evidence for the presence of PCs. Thus, it came as a surprise when database searches identified similar genes in the nematodes *Caenorhabditis elegans* and *C. briggsae* and in the slime mould *Dictyostelium discoideum*. In addition, using polymerase chain reaction, investigators have identified partial sequences with homology to the plant and yeast PC synthase genes from the aquatic midge, *Chironomus*, and earthworm species (C. Cobbett & W. Dietrich, unpublished data). Recent work has demonstrated that the *C. elegans* and *D. discoideum* genes encode PC synthase activity. *CePCS1* was able to rescue either a Cd-sensitive *ycf1* mutant of *S. cerevisiae* or a PC synthase-deficient mutant of *S. pombe* and catalysed PC biosynthesis in vivo in both heterologous hosts and in vitro (12, 77). Similarly, expression of the *D. discoidium* PC synthase in *S. cerevisiae* is also able to catalyze PC biosynthesis in vivo and confer increased Cd-resistance (C. Cobbett & A. Savage, unpublished data). Significantly, the suppression of *CePCS1* expression by using the double-stranded RNA interference technique resulted in Cd sensitivity, thereby demonstrating an essential role for PCs in heavy metal detoxification

TABLE 2 PC synthase enzymes predicted from DNA sequences

| Organism | Protein | Predicted protein | | Accession No.[c] |
		N + Con + Var = aa (kD)[a]	Arrangement of Cys residues in Var (number)[b]	
Plants				
Arabidopsis thaliana	AtPCS1	4 + 208 + 263 = 485 (54.4)	C CC CC3C2C C1C C (10)	AF135155
	AtPCS2	4 + 207 + 241 = 452 (51.5)	C C CC3C2C C (7)	AC003027
Brassica juncea	BjPCS1	4 + 208 + 263 = 485 (54.3)	C CC C CC3C2C C1C C (11)	AJ278627
Wheat	TaPCS1	4 + 208 + 288 = 500 (55.0)	C C C C2C CC3C2C C5C (14)	AF093752
Rice	OsPCS1	2 + 208 + 289 = 499 (55.6)	C C C2C CC3C2C C (12)	(C. Cobbett, unpublished)
Others				
Schizosaccharomyces pombe	SpPCS1	40 + 207 + 167 = 414 (46.7)	C CC4CC3CC (6)	Z68144
Caenorhabditis elegans	CePCS1	6 + 209 + 156 = 371 (42.1)	C CC4CC6CC C2C7C3C (11)	AF299332
Dictyostelium discoideum	DdPCS1	133 + 208 + 285 = 626 (70.5)	[C3C1C C2C] C C CC3C C C1C7C [5](9)	(C. Cobbett, unpublished)

[a]From an alignment, the amino acid sequences have been divided into a conserved domain (Con), a variable (Var) C-terminal domain, and an N-terminal (N) extension. The conserved domain corresponds to amino acids 5 to 212 of the AtPCS1 sequence and is arbitrarily based on the level of conservation accross the plant and nonplant sequences. The total number of amino acid residues (aa) and predicted MW (kD) are indicated.

[b]The arrangement of Cys (C) residues in the variable C-terminal domain is indicated. The number of amino acids separating Cys residues is indicated where that number is less than 8. Otherwise, a gap indicates an unspecified number of amino acid residues. Because the C-terminal regions of plant PCS proteins can be aligned, corresponding Cys residues are vertically aligned in the table. The C-terminal Cys residues in the nonplant sequences cannot be aligned, and this is not implied in the table. The DdPCS1 sequence has an extended N-terminal region that also contains Cys residues indicated by brackets. Total number of Cys residues in the variable C-terminal domains is indicated in parentheses.

[c]A single accession number is shown for each sequence, although there may be multiple entries in the sequence databases.

in *C. elegans* (77). This shows clearly that PCs play a wider role in heavy metal detoxification in biology than previously expected. In contrast, it appears that some organisms do not (or probably do not) express a PC synthase. There is, for example, no evidence for PC synthase–homologous sequences in the *S. cerevisiae*, *Drosophila melanogaster*, or mouse and human genomes. One view of the limited selection of species in which such sequences have been identified is that organisms with an aquatic or soil habitat are more likely to express PCs. However, the recent report of partial PC synthase–homologous ESTs in, for example, the mosquito-borne parasitic nematode *Brugia malayi* (77) undermines this simplistic categorization.

PC Synthase Enzymes and Their Regulation

The predicted molecular weights of various PC synthase enzymes deduced from DNA sequences are given in Table 2 and range from 42 kD to 70 kD. A comparison of the deduced amino acid sequences shows that the N-terminal regions of the plant, yeast, and animal PC synthases are very similar (40–50% identical), whereas the C-terminal sequences show little apparent conservation of amino acid sequence. The most apparent common feature of the C-terminal regions is the occurrence of multiple Cys residues, often as adjacent pairs or near pairs (Table 2). The arrangements of Cys residues is reminiscent of those found in MTs (see below). The C-terminal regions of the *Arabidopsis* and *S. pombe* PC synthase proteins, for example, have 10 and 7 Cys residues, respectively, of which 4 and 6, respectively, are as adjacent pairs. However, there is no apparent conservation of the positions of these Cys residues relative to each other. In contrast, monocot (TaPCS1) and dicot (AtPCS1) plant PC synthase sequences can be aligned across their entire length (55% identity) (11). The former contains 14 Cys residues, including two pairs, in the C-terminal domain. The *S. pombe* and *D. discoideum* sequences also contain N-terminal extensions, which in the latter also contains clusters of Cys residues that may play a role similar to the C-terminal Cys clusters.

When a PC synthase activity was first identified (from cultured cells of *Silene cucubalis*) it was characterized as a γ-GluCys dipeptididyl transpeptidase (EC 2.3.2.15) (24). It catalyzed the transpeptidation of the γ-GluCys moiety of GSH onto a second GSH molecule to form PC_2 or onto a PC molecule to produce an $n + 1$ oligomer. The enzyme was described as a tetramer of Mr 95,000 with a Km for GSH of 6.7 mM. The MW of this purified enzyme seems inconsistent with the MWs of the PC synthase sequences deduced from both dicot and monocot plant genes (Table 2). Furthermore, there is no evidence that cloned plant PC synthase enzymes are multimeric. This suggests a protein mixture may have been purified from *S. cucubalis* and that the PC synthase activity was not the major component detected in MW determinations. PC synthase activities have also been detected in pea (36), tomato (8), and *Arabidopsis* (28).

In vivo studies have shown that PC synthesis can be induced by a range of metal ions in *S. pombe* and in both intact plants and plant cell cultures (see 58). Kinetic studies using plant cell cultures demonstrated that PC biosynthesis occurs within minutes of exposure to Cd and is independent of de novo protein synthesis. The enzyme appears to be expressed independently of heavy metal exposure. It has been detected in extracts of plant cell cultures or tissues grown in the presence of only trace levels of essential heavy metals. Together, these observations indicate that PC synthase is primarily regulated by activation of the enzyme in the presence of heavy metals. In vitro, the partially purified enzyme from *S. cucubalis* was active only in the presence of a range of metal ions. The best activator tested was Cd, followed by Ag, Bi, Pb, Zn, Cu, Hg, and Au cations. This result has been mirrored by in vitro studies of PC synthase expressed in *E. coli* or in *S. cereviseae*, where the enzyme was activated to varying extents by Cd, Cu, Ag, Hg, Zn, and Pb ions (11, 25, 78).

Early models for the activation of PC synthase assumed a direct interaction between metal ions and the enzyme but raised the question of how the enzyme might be activated by such a wide range of metals. A significant recent study has provided evidence for an alternative model that provides a solution to this dilemma (79). With the cloning of PC synthase genes, the expression and purification of tagged recombinant derivatives of the enzyme has led to a more comprehensive understanding of the mechanisms of enzyme activation and catalysis. Vatamaniuk et al. (79) demonstrated that, in contrast to earlier models of activation, metal binding to the enzyme, per se, is not responsible for catalytic activation. Although AtPCS1 binds Cd ions at high affinity (Kd $= 0.54 \pm 0.20 \ \mu$M) and high capacity (stoichiometric ratio $= 7.09 \pm 0.94$) (78), it has a much lower affinity for other metal ions, such as Cu, which are equally effective activators.

The kinetics of PC synthesis are consistent with a mechanism in which heavy metal glutathione thiolate (e.g., Cd.GS$_2$) and free GSH act as γ-Glu-Cys acceptor and donor. First, modeling using the known binding constants of GSH and Cd showed that, in the presence of physiological concentrations of GSH and μM concentrations of Cd, essentially all of the Cd would be in the form of a GSH thiolate. Second, S-alkylglutathiones can participate in PC biosynthesis in the absence of heavy metals. These observations are consistent with a model in which blocked glutathione molecules (metal thiolates or alkyl substituted) are the substrates for PC biosynthesis. Thus the role of metal ions in enzyme activation is as an integral part of the substrate, rather than interacting directly with the enzyme itself. In this way, any metal ions that form thiolate bonds with GSH may have the capacity to activate PC biosynthesis, subject to possible steric constraints in binding at the active site of the enzyme. Early work suggested that PC biosynthesis in vitro was ultimately terminated by the PC products chelating the activating metal ions (or could be prematurely terminated by the addition of a metal chelator such as EDTA) (43), which provides a mechanism to autoregulate the biosynthesis of PCs. Viewed from a perspective where the metal ion forms part of the substrate, termination of the reaction results simply from exhaustion of substrate.

The conserved N-terminal domain of PC synthase is presumed to be the catalytic domain. The *cad1-5* mutation of Arabidopsis is a nonsense mutation that would result in premature termination of translation downstream of the conserved domain (25). The truncated polypeptide would lack 9 of the 10 Cys residues in the C-terminal domain. Of all the *cad1* mutants characterized, *cad1-5* is the least sensitive to Cd and makes the highest residual level of PCs on exposure to Cd (28). This suggests that the C-terminal domain is not absolutely required for catalysis. Because the work of Vatamaniuk et al. (79) suggests that heavy metal "activation" of PC synthase is in fact an integral component of catalysis, what then is the role of the multiple Cys residues in the variable C-terminal domain? Because the truncation of the *cad1-5* mutant polypeptide produces a mutant phenotype, the C-terminal domain clearly has some role in activity. This domain probably acts to enhance activity by binding metal glutathione complexes, bringing them into closer proximity to the catalytic domain.

Studies indicating PC synthase is expressed constitutively and levels of enzyme are generally unaffected by exposure of cell cultures or intact plants to Cd suggest the induction of PC synthase gene expression is unlikely to play a significant role in regulating PC biosynthesis. This is supported by analysis of the expression of *AtPCS1/CAD1* that showed that levels of mRNA were not influenced by exposure of plants to Cd and other metals, thus suggesting an absence of regulation at the level of transcription (25, 79). In contrast, analysis of *TaPCS1* expression in roots indicated increased levels of mRNA on exposure to Cd (11). This suggests that, in some species, PC synthase activity may be regulated at both the transcriptional and posttranslational levels. Little is known about the tissue specificity of PC synthase expression and/or PC biosynthesis. In a study in tomato, activity was detected in the roots and stems of tomato plants but not in leaves or fruits (8).

Sequestration to the Vacuole

In both plant and yeast, PC-Cd complexes are sequestered to the vacuole. In *S. pombe*, this process has been most clearly demonstrated through studies of the Cd-sensitive mutant *hmt1*. In extracts of *S. pombe*, two PC-Cd complexes (referred to as HMW and LMW) can be clearly resolved using gel-filtration chromatography. The *hmt1* mutant is unable to form the HMW complexes. The *Hmt1* gene encodes a member of the family of ATP-binding cassette (ABC) membrane transport proteins that is located in the vacuolar membrane (53). Both HMT1 and ATP are required for the transport of LMW PC-Cd complexes into vacuolar membrane vesicles (54) (Figure 1). In *S. cerevisiae*, which appears not to express a PC synthase, YCF1 is also a member of the ABC family of transporters and carries $(GSH)_2Cd$ complexes to the vacuole (41). Interestingly, in *C. elegans*, various mutations affecting ABC transporter proteins also confer heavy metal sensitivity (4). It is possible that these transporters are involved in the sequestration of PC-metal complexes in *C. elegans*. The site of such sequestration is still unidentified.

In plants, sequestration of PCs to the vacuole has also been observed. In mesophyll protoplasts derived from tobacco plants exposed to Cd, almost all of both the Cd and PCs accumulated was confined to the vacuole (81). An ATP-dependent, proton gradient-independent activity, similar to that of HMT1, capable of transporting both PCs and PC-Cd complexes has been identified in oat roots (65). Nonetheless, a plant gene encoding this function has not yet been identified. A recent inventory (66) of the ABC transporter protein genes in the Arabidopsis genome has not revealed a clearly identifiable homologue of HMT1.

Sulfide Ions and PC Function

In some plants and in the yeasts *S. pombe* and *Candida glabrata*, sulphide ions play an important role in the efficacy of Cd detoxification by PCs. HMW PC-Cd complexes contain both Cd and acid-labile sulfide. The incorporation of sulfide into the HMW complexes increases both the amount of Cd per molecule and the stability of the complex. Some complexes with a comparatively high ratio of S^{2-}:Cd consist of aggregates of 20Å-diameter particles, which themselves consist of a CdS crystallite core coated with PCs (18, 60).

The analysis of Cd-sensitive mutants of *S. pombe* deficient in PC-Cd complexes has provided evidence for the importance of sulfide in the function of PCs. These include some mutants affected in steps in the adenine biosynthetic pathway (71). Juang et al. (32) have shown that these enzymes, in addition to catalyzing the conversion of aspartate to intermediates in adenine biosynthesis, could also utilize cysteine sulfinate, a sulfur-containing analog of aspartate, to form other sulfur-containing compounds, which may be intermediates or carriers in the pathway of sulfide incorporation into HMW complexes (Figure 1). Whether sulfide is involved in the detoxification of other metal ions by PCs is unknown.

Using other Cd-sensitive mutants of *S. pombe* and *Candida glabrata*, investigators have identified additional functions important in sulfide metabolism. In *C. glabrata*, the *hem2* mutant is deficient in porphobilinogen synthase, which is involved in siroheme biosynthesis (31). Siroheme is a cofactor for sulfite reductase required for sulfide biosynthesis (Figure 1). This deficiency may contribute to the Cd-sensitive phenotype. However, additional studies are required to establish the precise influence of this pathway on PC function. In *S. pombe*, the *hmt2* mutant hyperaccumulates sulfide in both the presence and absence of Cd (74). The *HMT2* gene encodes a mitochondrial sulfide:quinone oxidoreductase, which was suggested to function in the detoxification of endogenous sulfide. The role of HMT2 in Cd tolerance is uncertain, but one possibility is that it detoxifies excess sulfide generated during the formation of HMW PC-Cd complexes after Cd exposure (Figure 1).

Metals Other than Cd

Although both induction of PCs in vivo and activation of PC synthase in vitro is conferred by a range of metal ions, there is little evidence supporting a role for PCs in the detoxification of such a wide range of metal ions. For metals other than Cd, there are few studies demonstrating the formation of PC-metal complexes either in vitro or in vivo. PCs can form complexes with Pb, Ag, and Hg in vitro (for example, see 47, 59). Maitani et al. (45) used inductively coupled plasma-atomic emission spectroscopy in combination with HPLC separation of native PC-metal complexes in the roots of *Rubia tinctorum*. PCs were induced to varying levels by a wide range of metal ions tested. The most effective appeared to be Ag, arsenate, Cd, Cu, Hg, and Pb ions. However, the only PC complexes identified in vivo were with Cd, Ag, and Cu ions. PC complexes formed in response to Pb and arsenate, but these complexes contained copper ions and not the metal ion used for induction of synthesis. This seems to conflict with the model for PC synthase activity whereby a metal-GSH thiolate is the substrate for PC-metal biosynthesis. It may indicate that some metals in complexes with PCs can be exchanged for others. In contrast, however, Schmöger et al. (68) have clearly demonstrated the formation of PC-As complexes in vivo and in vitro.

The clearest evidence for the role of PCs in heavy metal detoxification comes from characterization of the PC synthase–deficient mutants of *Arabidopsis* and *S. pombe*. A comparison of the relative sensitivity of the *Arabidopsis* and *S. pombe* mutants to different heavy metals revealed a similar but not identical pattern (25). In both organisms, PC-deficient mutants are highly sensitive to Cd and arsenate. For other metals, including Cu, Hg, Ag, Zn, Ni, and selenite ions, the mutants showed little or no sensitivity. Suppression of PC synthase in *C. elegans* also resulted in a Cd-sensitive phenotype, but the effect on responses to other metals has not been reported (77). Thus, PCs play a clear role in Cd and arsenate detoxification. Cu, for example, is a strong activator of PC biosynthesis both in vivo and in vitro, yet PC-deficient mutants show relatively little sensitivity to Cu. PCs also form complexes with Cu in vivo. It is possible, nonetheless, that PC-Cu complexes are relatively poorly sequestered to the vacuole, that they are comparatively transient, or that there is an alternative, more effective, mechanism for Cu detoxification.

The Roles of PCs

Although PCs clearly can have an important role in metal detoxification, alternative primary roles of PCs in plant physiology have also been proposed. These have included roles in essential metal ion homeostasis and in Fe or sulphur metabolism (see 67, 84). However, there is currently no direct evidence that PCs have functions outside of metal detoxification. These proposals stem from the expectation that the levels of Cd and As, for example, to which most organisms would be exposed in the natural, nonpolluted environment would not be sufficiently high to select for such a detoxification mechanism. Most experimental studies in plants have used Cd concentrations above 1 μM (67). In contrast, it has been estimated that solutions of nonpolluted soils contain Cd concentrations ranging up to 0.3 μM (82). Wagner (82) has argued that only at high levels of Cd exposure (not generally found in natural environments) might PCs play a role. Counter to this is the observation that a PC-deficient mutant of Arabidopsis is highly sensitive to concentrations of Cd as low as 0.6 μM in agar medium (28). Even at concentrations of Cd where the mutant is not obviously sensitive, the wild type may nonetheless have a selective advantage. This suggests that PCs may have a role in heavy metal detoxification in an unpolluted environment. The existence of a PC-deficient mutant of Arabidopsis and the isolation of, for example, a PC-deficient insertion mutant of *C. elegans* may allow the role of PCs in organisms in unpolluted environments to be assessed.

METALLOTHIONEINS

Structure

Metallothionein proteins, products of mRNA translation, are characterized as low molecular weight, cysteine-rich, metal-binding proteins (33). Although PCs conform to many of these characteristics, the enzymatic synthesis of PCs distinguishes them from MT proteins. Since their discovery more than 40 years ago as Cd-binding proteins present in horse kidney, MT proteins and genes have been found throughout the animal and plant kingdoms as well as in the prokaryote Synechococcus. The large number of cysteine residues in MTs bind a variety of metals by mercaptide bonds. MTs typically contain two metal-binding, cysteine-rich domains that give these metalloproteins a dumbbell conformation. MT proteins are classified based on the arrangement of Cys residues (9). Class I MTs contain 20 highly conserved Cys residues based on mammalian MTs and are widespread in vertebrates. MTs without this strict arrangement of cysteines are referred to as Class II MTs and include all those from plants and fungi as well as nonvertebrate animals. In this MT classification system, PCs are, somewhat confusingly, described as Class III MTs.

Shortly after the discovery of PCs as an important metal ligand required for tolerance of plants to Cd, a MT protein was identified in wheat (39), and a number of MT genes were isolated from plants. The plant Class II MT proteins can be further classified based on amino acid sequence. Robinson et al. (62) first identified two plant MT types based on the position of cysteine residues in the predicted proteins. Since then, the number of characterized plant MT genes has increased dramatically, and because many do not conform to these two groups, additional categories have been added. Other classifications have been proposed (2, 42, 59). However, the system presented below builds on the one developed by Robinson et al. (62). and is able to place almost all of the known plant MT genes into four categories based on amino acid sequence (Figure 2). Type 1 and Type 2 MTs follow the classification of Robinson et al. (62), Type 3 includes many MTs expressed during fruit ripening, and Type 4 is exemplified by the wheat Ec protein, the first characterized plant MT protein.

Type 1 MTs contain a total of six Cys-Xaa-Cys motifs (where Xaa represents another amino acid) that are distributed equally among two domains. In the majority of Type 1 MTs, the two domains are separated by approximately 40 amino acids that include aromatic amino acids. This large spacer is a common feature of plant MTs and contrasts with most other MTs in which cysteine-rich domains are separated by a spacer of less than 10 amino acids that do not include aromatic residues. Within the Type 1 MTs, those from various Brassicaceae (Arabidopsis and Brassica species) have a number of distinguishing features, including a much shorter spacer between the cysteine-rich domains and an additional Cys residue (5, 85).

```
Type 1
          *        *   *          *        *         *
AtMT1a   MADSNCGCGS SCKCGDSCSC EKNY......  ..........  ..........  ......NKEC DNCSCGSNCS CGSNCNC
AtMT1c   MAGSNCGCGS SCKCGDSCSC EKNY......  ..........  ..........  ......NKEC DNCSCGSNCS CGSSCNC
BnMT1    MAGSNCGCGS GCKCGDSCSC EKNY......  ..........  ..........  ......NTEC DSCSCGSNCS CGDSCSC
OsMT1a   MS...CSCGS SCSCGSNCSC GKKYPDLEEK SSSTKATVVL GVAPEKKQQF EAAAESGETA HGCSCGSSCR CNP.CNC
PsMT1    MSG..CGCGS SCNCGDSCKC NKRSSGLSYS EMETTETVIL GVGPAKIQFE GAEMSAASED GGCKCGDNCT CDP.CNCK
MsMT1    MSG..CNCGS SCNCGDNCKC NSRSSGLGYL EGETTETVIL GVGPAKIHFE GAEMGVAAED GGCKCGDSCT CDP.CNCK

Type 2
          **         *                                                        *            *      *
AtMT2a   MSCCGGNCGC GSGCKCGNGC GGCKMYPDLG FSGETTTTET FVLGVAPAMK NQYEASGESN NAENDACKCG SDCKCDPCTC K
BoMT2    MSCCGGNCGC GSGCKKCGNGC GGCKMYPDLG FSGELTTTET FVFGVAPTMK NQHEASGEGV .AENDACKCG SDCKCDPCTC E
AtMT2b   MSCCGGSCGC GSACKCGNGC GGCKRYPDL. ...ENTATET LVLGVAPAMN SQYEASGETF VAENDACKCG SDCKCNPCTC K
PhMT2    MSCCGGNCGC         GGCKMYPDFS YT.ESTTTET LILGVGPEKT SFGSMEMGES PAEN.GCKCG SDCKCDPCTC SK
SvMT2    MSCCNGNCGC GSACKCGSGC GGCKMFPDFA E..GSSGSAS LVLGVAP.MA SYFDAREMEMG VATENGCKCG DNCQCNPCTC K
OsMT2    MSCCGGNCGC GSSCQCGNGC GGCK.YSEVB PTTTTTPLAD ATNKGSGAAS GGSEMGAENG SCGCNTCKCG TSCGCSCCNC N

Type 3
          *                                                        *     *    *
AtMT3    MSSNCGSCDC ADKTQCVKKG TSYTFDIVET QESYKEAMIM DVGAEENNAN CKCKCGSSCS CVNCTCCPN
MaMT3    MS.TCGNCDC VDKSQCVKKG NSYGIDIVET EKSYVDEVIV AAEAAEHDG. .KCKCGAACA CTDCKCGN
AdMT3    MSDKCGNCDC ADSSQCVKKG NS..IDIVET DKSYIEDVVM GVPAAESGG. .KCKCGTSCP CVNCTCD
OsMT3    MSDKCGNCDC ADKSQCVKKG TSYGVIVBA EKSHFBEV.. .AAGBENGG. .CKCGTSCS CTDCKCGK
GhMT3    MSDRGRNCDC ADRSQCTK.G NSNTM.IIET EKSYINTAVM DAPAENDG.. .KCKCGTGCS CTDCTCGH
PgMT3    MSSDCGNCDC ADKSQCTKKG FQID.GIVET SYEMGHGGD. ..VSLEND.. .CKCGPNCQ CGTCTCHT

Type 4
          *        *          *         *                   *            *          *    *   *   *
AtMT4a   MADTGKGSSV AGCNDSCGCP SPCPGGNSCR CRM..R.EAS AGDQHMVCP CGEHCGCNPC NCPKTQTQTS AKG....CTC GEGCTCASCA T
AtMT4b   MADTGKGSAS ASCNDRCGCP SPCPGGESCR CRM..MSEAS GGDQEHNTCP CGEHCGCNPC NCPKTQTQTS AKG....CTC GEGCTCATCA A
PhMT4    MADL.RGSS. AICDERCGCP SPCPGGVACR CASGGAATAG GGDMEHKKCP CGEHCGCNPC TCPKSEGTTA GSGK.AHCKC GPGCTCVQCA S
ZmMT4    MG........ ..DDKCGCA VPCPGGKDCR CTS...G..S GGQREHTTCG CGEHCECSPC TCGRATMPSG RENRRANCSC GASCNCASCA SA
TaMT4    MG........ .CDDKCGCA VPCPGGTGCR CTS...ARSG AAAGEHTTCG CGEHCGCNPC ACGREGTPSG RANRRANCSC GAACNCASCG SATA
OsMT4    MG........ ..CDDKCGCA VPCPGGTGCR CAS...S.AR SGGGDHTTCS CGDHCGCNPC RCGRESOPTG RENRRAGCSC GDSCTCASCG STTTTAPAAT T
```

Figure 2 Alignment of plant MT amino acid sequences. Examples of the four types of plant MTs are shown. Cysteine residues are in bold, and conserved cysteines in each type are indicated by a star. The protein sequences are predicted from gene sequences in Arabidopsis (At), Brassica napus (Bn), rice (Os), pea (Ps), alfalfa (Ms), Brassica oleracea (Bo), petunia (Ph), Silene vulgaris (Sv), banana (Ma), kiwifruit (Ad), cotton (Gh), Picea glauca (Pg), maize (Zm), and wheat (Ta).

Type 2 MTs also contain two cysteine-rich domains separated by a spacer of approximately 40 amino acid residues. However, the first pair of cysteines is present as a Cys-Cys motif in amino acid positions 3 and 4 of these proteins. A Cys-Gly-Gly-Cys motif is present at the end of the N-terminal cysteine-rich domain. Overall, the sequences of the N-terminal domain of Type 2 MTs are highly conserved (MSCCGGNCGCS). The C-terminal domain contains three Cys-Xaa-Cys motifs. By contrast, the spacer region separating these domains in Type 2 MTs is much more variable between species.

Type 3 MTs contain only four Cys residues in the N-terminal domain. The consensus sequence for the first three is Cys-Gly-Asn-Cys-Asp-Cys. The fourth cysteine is not part of a pair of cysteines but is contained within a highly conserved motif, Gln-Cys-Xaa-Lys-Lys-Gly. The six Cys residues in the C-terminal cysteine-rich domain are arranged in Cys-Xaa-Cys motifs. As with the majority of Type 1 and Type 2 plant MTs, the two domains are separated from each other by approximately 40 amino acid residues.

Type 4 MTs differ from other plant MTs by having three cysteine-rich domains, each containing 5 or 6 conserved cysteine residues, which are separated by 10 to 15 residues. Most of the cysteines are present as Cys-Xaa-Cys motifs. Although a large number of Type 4 MTs have not been identified, compared to those from monocots, Type 4 MTs from dicots contain an additional 8 to 10 amino acids in the N-terminal domain before the first cysteine residue.

The vast majority of plant MT genes have been identified in the angiosperms. A number of species, including Arabidopsis, rice, and sugarcane (A. Figueira, personal communication), contain genes encoding all four types of MTs. This indicates that evolution of the four plant MT types predates the separation of monocots and dicots, and it is likely that the majority of flowering plants also contain the four different MT types. The presence of four types of MTs in plants with distinct arrangements of cysteines contrasts with the situation in animals. For example, the four mouse MTs all contain the same conserved cysteines, although they do differ in tissue expression (35). The diversity of the plant MT gene family suggests that these may differ not only in sequence but also in function. There is little information about MT genes in nonflowering plant species. However, genes encoding Type 3 MTs have been cloned from several gymnosperms (7). A MT-encoding gene has also been isolated from *Fucus vesiculocus*, a brown alga (48). This MT does not fit readily into any of the four plant types described above but, primarily on the basis of the cysteine residues, is equally similar to Arabidopsis MT1a and an oyster MT. Further studies are needed to determine if the diverse MT gene family present in angiosperms is also found in other divisions of the plant kingdom.

Gene Structure

Genomic DNA sequences have been determined for a small number of MT genes, and these provide some additional support for the classification system described above. Almost all plant MT genes contain an intron located close to the end of the N-terminal cysteine-rich domain. However, the position of this intron varies in genes encoding different MT types. The single intron in Type 1 MT genes from monocots disrupts the codon after the last cysteine codon in the N-terminal cysteine-rich domain. All Type 3 MT genes that have been characterized contain two introns, and the first lies in the same relative position after the end of the N-terminal cysteine-rich domain. However, the first intron in genes encoding Type 1 MTs from dicots and Type 4 MTs lies in the codon preceding the last cysteine codon of the first domain. This is the same position as the single intron in Arabidopsis Type 2 MT genes and the first of two introns in Type 2 MT genes in rice. Interestingly, another classification of MTs has identified the Brassica MTs as variant forms of Type 2 plant MTs (2). Overall, apart from the difference between Type 1 MT genes from monocots and dicots, the position of the first intron in plant MT genes supports the classification of plant MTs into four types based on amino acid sequence.

The Arabidopsis genome sequence has provided information on how the seven members of the MT gene family are organized. The *MT1a* and *MT1c* genes lie within 4 kb as an inverted repeat on chromosome 1; *MT2a* and *MT3* are at distinct positions on chromosome 3; both *MT4a* and *MT4b* lie on chromosome 2 but are not closely linked; finally, *MT2b* is positioned on chromosome 5. One pseudo-gene, *MT1b*, has been identified in Arabidopsis and is also found on chromosome 5. Mapping and genome sequencing have demonstrated that MT genes are also distributed across different chromosomes in the tomato (22) and rice genomes. However, evidence of MT gene clustering has been found in cotton where three MT genes were identified within a 10-kb fragment of genomic DNA (30).

MT Proteins

The wheat Ec Type 4 MT protein was purified from embryos as a zinc binding protein and provided the first evidence that plants contained not only PCs but also MTs as cysteine-based metal ligands (39). The ensuing flood of information about plant genes and cDNAs encoding MT proteins has not been accompanied by a corresponding increase in knowledge about the expression or distribution of MT proteins. Consequently, there has been a trend to describe these as "metallothionein-like genes," for fear that they were in fact not translated into bona fide metal-binding proteins. However, it would seem quite perverse if these genes were not translated into proteins in plants, given that many are highly expressed and encode proteins with known metal-binding motifs and similarity to proteins required for specific functions in animals and fungi. Indeed, evidence for the occurrence of several MT proteins has been obtained in Arabidopsis (50). Peptide fragments derived from MT1a, MT2a, MT2b, and MT3 were identified after purification of the proteins under anaerobic conditions using several chromatographic separations, including Zn affinity chromatography. Immunoblot analysis also demonstrated that expression of MT1 and MT2 proteins corresponded to observed RNA levels in terms of tissue specificity and induction by copper treatment. Difficulties in identifying MTs in plants may arise from instability of these proteins in the presence of oxygen. There is, however, a critical need for more information about the distribution and form of MTs in plants, including the metals that are bound to these proteins in vivo. This search would be assisted by the development of simpler purification procedures and isoform-specific antibodies for MTs as well as the application of protein-tagging procedures (e.g. myc or GFP) for in situ localization of these proteins.

Although it has been difficult to study MT proteins in planta, several plant MTs have been expressed in microbial hosts to examine the metal-binding properties of these proteins and their ability to provide metal tolerance. When expressed in *E. coli*, the pea Type 1 MT, PsMTa, bound Cu, Cd, and Zn, with the highest affinity for Cu (73). Similarly, a recombinant Fucus MT fusion protein showed a greater affinity for Cu than Cd, and the pH required for dissociation of Cd from Fucus MT was approximately 2 pH units higher than for a recombinant human MT. Arabidopsis MTs have been expressed in MT-deficient strains of yeast (85) and Synechococcus (63) and were able to complement these mutations in terms of restoring tolerance to copper and zinc, respectively. These studies provide important evidence that plant MTs are capable of providing a biological function—metal tolerance—albeit in nonplant systems. A number of studies have examined the effects of ectopic expression of MT proteins from various sources on metal tolerance in plants. Although some of these studies have resulted in increased metal tolerance or altered distribution of metals in plants, they have not been informative regarding the function of endogenous plant MTs.

MT Gene Expression

In attempting to shed light on their function, investigators have relied primarily on RNA blot hybridization to study the expression of MT genes during development and in response to various environmental factors. More detailed localization of MT mRNAs or MT gene promoter activity has been obtained in a small number of cases through in situ hybridization and reporter gene expression studies. Many MT genes are expressed at very high levels in plant tissues, at least in terms of transcript abundance. Direct evidence comes from gene profiling experiments in rice using the serial analysis of gene expression (SAGE) protocol (46). Transcripts from four MT genes comprised almost 3% of the transcripts in two-week-old seedlings. A Type 2 MT gene contributed 1.26% of all transcripts, the single most abundantly expressed gene in this tissue. Transcripts of two Type 3 MT genes accounted for an additional 1.25% of the mRNAs. Furthermore, ESTs for MT genes are among the most prevalent in randomly sequenced cDNA libraries from a number of plants. For example, a Type 2 MT gene accounted for 0.4% of the tomato ESTs, and 0.5% of maize ESTs were derived from a Type 1 MT gene (72). The large number of MT genes that have been identified by differential screening of cDNA libraries also indicates that RNAs encoding MTs are abundant in many other plant species.

Expression of Type 4 MTs, such as the wheat Ec MT, is restricted to developing seeds. Type 4 MT genes contain promoter sequences with homology to ABA-response elements, and their expression is regulated by ABA (83). These genes follow the same regulatory program as a large number of other genes that are expressed during the maturation of embryos and whose RNAs persist until imbibition and germination. Kawashima et al. (34) proposed that this embryo-specific

MT provides a mechanism for storing zinc that is required during germination. Expression of MT genes from other types has also been observed in developing seeds (A. Figueira, personal communication). If these MTs are involved in accumulation and storage of metals in seeds, they may play an important role in determining the concentrations of metals in grains. As more attention is paid to the nutritional composition of foods as opposed to simple calorific value, MTs may provide one mechanism to manipulate metal concentration in seeds (44).

Expression of other plant MT genes is not restricted to a single organ and cannot be categorized simply as that of Type 4 MTs (59). However, a number of general observations can be made about the expression of these genes. RNA expression of Type 1 MT genes tends to be higher in roots than shoots, whereas the reverse is observed generally for Type 2 MTs. In plants that produce fleshy fruits, e.g., banana (13), apple (61), and kiwi (40), Type 3 MT RNAs are highly expressed in the fruits as they ripen. Type 3 MTs are also expressed at high levels in leaves of plants that do not produce fleshy fruits, such as Arabidopsis (W. Bundithya & P. Goldsbrough, unpublished observation).

A small number of studies have examined the expression of MT genes at a more detailed level, using in situ hybridization and reporter gene expression. In both Arabidopsis and *Vicia faba*, MT RNA expression in leaves was shown by in situ hybridization to be highest in trichomes (19, 20). GUS reporter genes driven by the Arabidopsis *MT1a* and *MT2a* promoters also direct GUS expression preferentially in trichomes under some conditions (W. J. Guo, W. Bundithya, & P. Goldsbrough, unpublished observations). There are a number of possible explanations for high levels of MT gene expression in trichomes. Toxic metals such as Cd accumulate in trichomes (64). Although the metals that bind to MTs in most plant tissues are not known, MTs may be required for detoxification of metals that are deposited in trichomes. Foley & Singh (19) have suggested that specific metal-binding enzymes are highly expressed in trichomes and that MT expression may be involved in the delivery of metals into these specialized cells. Expression of Arabidopsis MT genes has also been localized to the phloem in a number of tissues (20; W. Bundithya, W. J. Guo, & P. Goldsbrough, unpublished observations), raising the possibility that MTs could play a role in metal ion transport. However, as yet, MTs have not been identified among the phloem exudate proteins characterized from various plants, but the MT proteins have been difficult to purify by standard methods.

The expression of some MT genes changes during development. One interesting example is the dramatic increase in MT RNA levels in senescing leaves. This was first reported for a Type 1 MT gene in *Brassica napus* (5) and has been confirmed in Arabidopsis and rice (20, 29). MT RNA expression in senescing leaves appears to be localized primarily to phloem tissue (6). At least two other genes that are specifically involved in copper homeostasis, one encoding a metal chaperone (AtCCH) and the other a copper transporter (AtRAN1), are also expressed in senescing leaves (26, 27). The homologous proteins in yeast, Atx1 and Ccc2, are involved in the delivery of copper to the trans-Golgi network for incorporation into copper-requiring proteins. They are components of an integrated system for the regulated uptake and distribution of copper (56). This system is able to maintain the level of free copper ions below one ion per yeast cell, thereby preventing the damage that reactive copper ions can cause through the production of reactive oxygen species to membranes, proteins, and nucleic acids (57). Why are plant homologues of these genes expressed in senescing leaves? One possibility is that MTs are required to chelate copper released from metalloproteins that are being catabolized in senescing leaves. In the absence of MTs, or another ligand, free copper ions would precipitate a cascade of oxidative damage and disrupt the controlled senescence program. Expression of MTs in phloem during leaf senescence also suggests that MTs might serve as a chaperone for long-distance transport of copper. The other plant homologs of this copper homeostasis system, AtCCH and AtRAN1, may play a role in the export of copper from leaves to sinks such as developing seeds. The human homolog of AtCCH is thought to deliver copper to a transporter that is responsible for copper efflux from intestinal epithelial cells (56). This copper transporter is defective in Menkes disease patients, resulting in accumulation of copper in the intestine and consequent copper deficiencies in other tissues. AtRAN1, the Arabidopsis homolog of the Menkes copper transporter, may therefore participate in copper efflux from senescing leaves, and this could require the partner copper chaperone protein, AtCCH. It is of interest that MT gene expression has also been observed in other processes that involve apoptosis, including leaf abscission (17) and the hypersensitive response to pathogens (6).

A large number of reports have described the effects of various environmental factors on MT RNA expression in plants, and these have been tabulated by Rauser (59). Overall, these experiments show little in the way of consistent trends for conditions that modulate expression of specific MT types across species. For example, various metals including copper had either no effect or repressed MT gene expression in many species. However, copper induced expression of a Type 1 MT gene in Arabidopsis, rice, wheat, and tobacco (10, 29, 70, 85) as well as MT genes in Fucus (48) and *P. oceanica* (21). Type 1 MT genes are also induced by a variety of other stresses, including aluminum, cadmium, nutrient deprivation, and heat shock, in wheat and rice (29, 70), suggesting that MTs may be expressed as part of a general stress response, although an indirect connection to metal ion status could exist. It has been proposed that iron deficiency, which induces MT gene expression in barley and pea, mediates this response by increasing copper uptake (52, 62). That expression of MT genes in animals is also affected by a tremendous variety of conditions (33) is worth noting. Currently there is no information about the mechanisms that regulate transcription of plant MT genes in vegetative tissues, in contrast to the detailed knowledge of metal regulation of MT gene expression in yeast and mammals. The yeast transcription factor Ace1 is activated by copper and binds to elements in the CUP1 promoter to stimulate transcription of this MT gene (38). In mammals, MTF1 activates transcription of MT genes in response to metals by binding to metal response elements in MT gene promoters (1). This transcription factor is essential because an MTF1-null mutation is lethal in mice, even though MT-deficient mice are normal unless they are subjected to cadmium or zinc toxicity.

Function of Metallothioneins

What are the functions of MT genes in plants? In animals, MTs protect against cadmium toxicity (35), but this function in plants is clearly provided by PCs. Reconciling all the available data on plant MTs into a simple model may be impossible and may also be unrealistic given the diverse family of MT genes in plants. However, there is evidence to support the hypothesis that MTs are involved in copper tolerance and homeostasis in plants: Some plant MTs are functional copper-binding proteins; expression of some MT genes is induced by copper; MT gene expression in senescing leaves is coordinated with a set of genes involved in copper homeostasis; the level of expression of a Type 2 MT gene correlated closely with copper tolerance in a group of Arabidopsis ecotypes (49); expression of a Type 2 MT gene is elevated in a copper-sensitive mutant that accumulates copper (76); more recently, copper tolerant populations of *S. vulgaris* have been shown to have higher RNA expression and gene copy number of a Type 2 MT gene (75). In addition, PCs do not provide tolerance to copper in Arabidopsis, indicating that another mechanism, perhaps involving MTs, must be involved. While supporting a role for MTs in copper tolerance, this evidence is not conclusive.

The most direct approach to answering this and other questions about the function of MTs is to identify and analyze plants with defined MT-null genotypes. T-DNA insertional mutagenesis is well developed for this objective (37). However, MT genes present very small targets for this approach (less than 1 kb), and the probability of finding insertions even in populations of 50,000 lines is not high. Targeted gene disruption strategies using transposable element "launch pads" inserted close to specific MT loci may be more efficient (69), whereas RNA interference may provide an alternative that is not dependent on identifying DNA insertions into MT genes (16). It may be necessary to combine null mutants for more than one MT gene and test these plants under a variety of conditions in order to observe any phenotype (3). Use of this approach in a variety of model plants is necessary and overdue in order to provide definitive answers about MT function.

In spite of the availability of such experimental tools to study animal MTs, the function of these proteins remains somewhat of an enigma (55). Mammalian MTs have a highly conserved sequence, are expressed in many tissues, and respond to a wide variety of regulatory factors. Although these observations hint at an important function for MTs in mammals, the only role that has been established unequivocally is in protection against cadmium and zinc toxicity (55). Therefore, although MTs are expressed ubiquitously and conserved in plants, determining their function remains a future challenge.

FUTURE PROSPECTS

The use of model systems to study the biosynthesis, expression, regulation, and function of both PCs and MTs has offered significant advances in recent years. For PCs, the characterization of Cd-sensitive mutants of *S. pombe*, the organism in which PCs were first recognized, has identified, with the remarkable exception of the PC synthase gene, various genes involved in PC biosynthesis or function. Thus far, in Arabidopsis, only mutants and genes in the PC biosynthetic pathway have been isolated. The parallel studies in *S. pombe* point to a number of additional functions still not discovered. There is clear evidence that Cd, in plants as in *S. pombe*, is sequestered to the vacuole in complexes with sulfide. Yet, for example, there is no apparent homologue of the vacuolar transporter, HMT1, in Arabidopsis. Clearly, further studies in plants are required to identify these additional functions. Also, as yet, no MT-deficient mutants in Arabidopsis have been characterized, and in view of the possibility of redundancy among the members of the MT gene family, the full suite of MT-deficient mutants is likely to be required to adequately determine the functions of the various genes.

The potential for the use of plants for the detoxification or "phytoremediation" of polluted environments is being increasingly examined. The manipulation of PC expression is one potential mechanism for increasing the capacity of plants for phytoremediation. Understanding the effect of the overexpression, possibly in a tissue-specific manner, of the genes of the GSH/PC biosynthetic pathway on metal tolerance and accumulation will soon lead to indications as to their usefulness in this endeavor. Here too, genes controlling other aspects of PC function may be required.

ACKNOWLEDGMENTS

The authors wish to thank present and former members of their laboratories for their contributions to aspects of the work reviewed here. Thanks to Metha Meetam and Woei-Jiun Guo for helpful comments. Research in the authors' laboratories is supported by Australian Research Council (C.C.) and USDA-NRI (P.G.).

Visit the Annual Reviews home page at www.annualreviews.org

LITERATURE CITED

1. Andrews GK. 2000. Regulation of metallothionein gene expression by oxidative stress and metal ions. *Biochem. Pharmacol.* 59:95–104

2. Binz PA, Kägi JHR. 2001. *Metallothionein.* http://www.unizh.ch/~mtpage/MT.html

3. Bouche N, Bouchez D. 2001. Arabidopsis gene knockout: phenotypes wanted. *Curr. Opin. Plant Biol.* 4:111–17

4. Broeks A, Gerrard B, Allikmets R, Dean M, Plasterk RH. 1996. Homologues of the human multidrug resistance genes MRP and MDR contribute to heavy metal resistance in the soil nematode *Caenorhabditis elegans. EMBO J.* 15:6132–43

5. Buchanan-Wollaston V. 1994. Isolation of cDNA clones for genes that are expressed during leaf senescence in *Brassica napus.* Identification of a gene encoding a senescence-specific metallothionein-like protein. *Plant Physiol.* 105:839–46

6. Butt A, Mousley C, Morris K, Beynon J, Can C, et al. 1998. Differential expression of a senescence-enhanced metallothionein gene in *Arabidopsis* in response

to isolates of *Peronospora parasitica* and *Pseudomonas syringae*. *Plant J.* 16:209–21

7. Chatthai M, Kaukinen KH, Tranbarger TJ, Gupta PK, Misra S. 1997. The isolation of a novel metallothionein-related cDNA expressed in somatic and zygotic embryos of Douglas fir: regulation by ABA, osmoticum, and metal ions. *Plant Mol. Biol.* 34:243–54

8. Chen JJ, Zhou JM, Goldsbrough PB. 1997. Characterization of phytochelatin synthase from tomato. *Physiol. Plant.* 101:165–72

9. Cherian GM, Chan HM. 1993. Biological functions of metallothioneins—a review. See Ref. 71a, pp. 87–109

10. Choi D, Kim HM, Yun HK, Park JA, Kim WT, Bok SH. 1996. Molecular cloning of a metallothionein-like gene from *Nicotiana glutinosa* L. and its induction by wounding and tobacco mosaic virus infection. *Plant Physiol.* 112:353–59

11. Clemens S, Kim EJ, Neumann D, Schroeder JI. 1999. Tolerance to toxic metals by a gene family of phytochelatin synthases from plants and yeast. *EMBO J.* 18:3325–33

12. Clemens S, Schroeder JI, Degenkolb T. 2001. *Caenorhabditis* expresses a functional phytochelatin synthase. *Eur. J. Biochem.* 268:3640–43

13. Clendennen SK May GD. 1997. Differential gene expression in ripening banana fruit. *Plant Physiol.* 115:463–69

14. Cobbett CS. 2000. Phytochelatins and their role in heavy metal detoxification. *Plant Physiol.* 123:825–33

15. Cobbett CS, May MJ, Howden R, Rolls B. 1998. The glutathione-deficient, cadmium-sensitive mutant, *cad2-1*, of *Arabidopsis thaliana* is deficient in γ-glutamylcysteine synthetase. *Plant J.* 16:73–78

16. Cogoni C, Macino G. 2000. Post-transcriptional gene silencing across kingdoms. *Genes Dev.* 10:638–43

17. Coupe SA, Taylor JE, Roberts JA. 1995. Charactersiation of an mRNA encoding a metallothionein-like protein that accumulates during ethylene-promoted abscission of *Sambucus nigra* L. leaflets. *Planta* 97:442–47

18. Dameron CT, Reese RN, Mehra RK, Kortan AR, Carroll PJ, et al. 1989. Biosynthesis of cadmium sulfide quantum semiconductor crystallites. *Nature* 338:596–97

19. Foley RC, Singh KB. 1994. Isolation of a *Vicia faba* metallothionein-like gene: expression in foliar trichomes. *Plant Mol. Biol.* 26:435–44

20. Garcia-Hernandez M, Murphy A, Taiz L. 1998. Metallothioneins 1 and 2 have distinct but overlapping expression patterns in Arabidopsis. *Plant Physiol.* 118:387–97

21. Giordani T, Natali L, Maserti BE, Taddei S, Cavallini A. 2000. Characterization and expression of DNA sequences encoding putative Type-II metallothioneins in the seagrass *Posidonia oceanica*. *Plant Physiol.* 123:1571–81

22. Giritch A, Ganal M, Stephan UW, Baumlein H. 1998. Structure, expression and chromosomal localisation of the metallothionein-like gene family of tomato. *Plant Mol. Biol.* 37:701–14

23. Glaeser H, Coblenz A, Kruczek R, Ruttke I, Ebert-Jung A, Wolf K. 1991. Glutathione metabolism and heavy metal detoxification in *Schizosaccharomyces pombe*. Isolation and characterization of glutathione-deficient, cadmium-sensitive mutants. *Curr. Genet.* 19:207–13

24. Grill E, Loffler S, Winnacker E-L, Zenk MH. 1989. Phytochelatins, the heavy-metal-binding peptides of plants, are synthesized from glutathione by a specific γ-glutamylcysteine dipeptidyl transpeptidase (phytochelatin synthase). *Proc. Natl. Acad. Sci. USA* 86:6838–42

25. Ha S-B, Smith AP, Howden R, Dietrich WM, Bugg S, et al. 1999. Phytochelatin synthase genes from *Arabidopsis* and the yeast, *Schizosaccharomyces pombe*. *Plant Cell* 11:1153–64

26. Himelblau E, Amasino RM. 2000. Delivering copper within plant cells. *Curr. Opin. Plant Biol.* 3:205–10

27. Himelblau E, Mira H, Lin SJ, Culotta V, Penarrubia L, Amasino RM. 1998. Identification of a functional homolog of the yeast copper homeostasis gene *ATX1* from Arabidopsis. *Plant Physiol.* 117:1227–34

28. Howden R, Goldsbrough PB, Andersen CR, Cobbett CS. 1995. Cadmium-sensitive, *cad1*, mutants of *Arabidopsis thaliana* are phytochelatin deficient. *Plant Physiol.* 107:1059–66

29. Hsieh HM, Liu WK, Huang PC. 1995. A novel stress-inducible metallothionein-like gene from rice. *Plant Mol. Biol.* 28:381–89

30. Hudspeth RL, Hobbs SL, Anderson DM, Rajasekaran K, Grula JW. 1996. Characterization and expression of metallothionein-like genes in cotton. *Plant Mol. Biol.* 31:701–5

31. Hunter TC, Mehra RK. 1998. A role for *HEM2* in cadmium tolerance. *J. Inorg. Biochem.* 69:293–303

32. Juang R-H, MacCue KF, Ow DW. 1993. Two purine biosynthetic enzymes that are required for cadmium tolerance in *Schizosaccharomyces pombe* utilize cysteine sulfinate *in vitro*. *Arch. Biochem. Biophys.* 304:392–401

33. Kagi JHR. 1993. Evolution, structure and chemical activity of class I metallothioneins: an overview. See Ref. 71a, pp. 29–56

34. Kawashima I, Kennedy TD, Chino M, Lane BG. 1992. Wheat E_c metallothionein genes: like mammalian Zn^{2+} metallothionein genes, wheat Zn^{2+} metallothionein genes are conspicuously expressed during embryogenesis. *Eur. J. Biochem.* 209:971–76

35. Klaassen CD, Liu J, Choudhuri S. 1999. Metallothionein: an intracellular protein to protect against cadmium toxicity. *Annu. Rev. Pharmacol. Toxicol.* 39:267–94

36. Klapheck S, Schlunz S, Bergmann L. 1995. Synthesis of phytochelatins and homo-phytochelatins in *Pisum sativum* L. *Plant Physiol.* 107:515–21

37. Krysan PJ, Young JK, Sussman MR. 1999. T-DNA as an insertional mutagen in Arabidopsis. *Plant Cell* 11:2283–90

38. Labbe S, Thiele DJ. 1999. Pipes and wiring: the regulation of copper uptake and distribution in yeast. *Trends Microbiol.* 7:500–5

39. Lane BG, Kajioka R, Kennedy TD. 1987. The wheat germ Ec protein is a zinc-containing metallothionein. *Biochem. Cell. Biol.* 65:1001–5

40. Ledger SE, Gardner RC. 1994. Cloning and characterization of five cDNAs for genes differentially expressed during fruit developement of kiwifruit (*Actinidia deliciosa* var. *deliciosa*). *Plant Mol. Biol.* 25:877–86

41. Li Z-S, Lu Y-P, Zhen R-G, Szczypka M, Thiele DJ, Rea PA. 1997. A new pathway for vacuolar cadmium sequestration in *Saccharomyces cerevisiae*: YCF1-catalyzed transport of *bis*(glutathionato) cadmium. *Proc. Natl. Acad. Sci. USA* 94:42–47

42. Liu JY, Lu T, Zhao NM. 2000. Classification and nomenclature of plant metallothionein-like proteins based on their cysteine arrangement patterns. *Acta Bot. Sin.* 42:649–52

43. Loeffler S, Hochberger A, Grill E, Winnacker E-L, Zenk M-H. 1989. Termination of the phytochelatin synthase reaction through sequestration of heavy metals by the reaction product. *FEBS Lett.* 258:42–46

44. Lucca P, Hurrell R, Potrykus I. 2001. Genetic engineering approaches to improve the bioavailability and the level of iron in rice grains. *Theor. Appl. Genet.* 102:392–97

45. Maitani T, Kubota H, Sato K, Yamada T. 1996. The composition of metals bound to class III metallothionein (phytochelatin and its desglycyl peptide) induced by various metals in root cultures of *Rubia tinctorum*. *Plant Physiol.* 110:1145–50

46. Matsumura H, Nirasawa S, Terauchi R. 1999. Transcript profiling in rice (*Oryza sativa* L.) seedlings using serial analysis of gene expression (SAGE). *Plant J.* 20:719–26

47. Mehra RK, Tran K, Scott GW, Mulchandani P, Sani SS. 1996. Ag(I)-binding to phytochelatins. *J. Inorg. Biochem.* 61: 125–42

48. Morris CA, Nicolaus B, Sampson V, Harwood JL, Kille P. 1999. Identification and characterization of a recombinant metallothionein protein from a marine alga, *Fucus vesiculosus*. *Biochem. J.* 338:553–60

49. Murphy A, Taiz L. 1995. Comparison of metallothionein gene expression and nonprotein thiols in ten Arabidopsis ecotypes. *Plant Physiol.* 109:945–54

50. Murphy A, Zhou J, Goldsbrough P, Taiz L. 1997. Purification and immunological identification of metallothioneins 1 and 2 from *Arabidopsis thaliana*. *Plant Physiol.* 113:1293–301

51. Mutoh N, Hayashi Y. 1988. Isolation of mutants of *Schizosaccharomyces pombe* unable to synthesize cadystin, small cadmium-binding peptides. *Biochem. Biophys. Res. Commun.* 151:32–39

52. Okumura N, Nishizawa NK, Umehara Y, Mori S. 1991. An iron deficiency-specific cDNA from barley roots having two homologous cysteine-rich MT domains. *Plant Mol. Biol.* 12:531–33

53. Ortiz DF, Kreppel L, Speiser DM, Scheel G, McDonald G, Ow DW. 1992. Heavy-metal tolerance in the fission yeast requires an ATP-binding cassette-type vacuolar membrane transporter. *EMBO J.* 11:3491–99

54. Ortiz DF, Ruscitti T, MacCue KF, Ow DW. 1995. Transport of metal-binding peptides by HMT1, a fission yeast ABC-type vacuolar membrane protein. *J. Biol. Chem.* 270:4721–28

55. Palmiter RD. 1998. The elusive function of metallothioneins. *Proc. Natl. Acad. Sci. USA* 95:8428–30

56. Pena MMO, Lee J, Thiele DJ. 1999. A delicate balance: homeostatic control of copper uptake and distribution. *J. Nutr.* 129:1251–60

57. Rae TD, Schmidt PJ, Pufhal RA, Culotta VC, O'Halloran TV. 1999. Undetectable intracellular free copper: the requirement of a copper chaperone for superoxide dismutase. *Science* 284:805–8

58. Rauser WE. 1995. Phytochelatins and related peptides: structure, biosynthesis, and function. *Plant Physiol.* 109:1141–49

59. Rauser WE. 1999. Structure and function of metal chelators produced by plants: the case for organic acids, amino acids, phytin and metallothioneins. *Cell. Biochem. Biophys.* 31:19–48

60. Reese RN, White CA, Winge DR. 1992. Cadmium sulfide crystallites in Cd-(γ-EC)nG peptide complexes from tomato. *Plant Physiol.* 98:225–29

61. Reid SJ, Ross GS. 1997. Up-regulation of two cDNA clones encoding metallothionein-like proteins in apple fruit during cool storage. *Physiol. Plant* 100:183–89

62. Robinson NJ, Tommey AM, Kuske C, Jackson PJ. 1993. Plant metallothioneins. *Biochem. J.* 295:1–10

63. Robinson NJ, Wilson JR, Turner JS. 1996. Expression of the type 2 metallonthionein-like gene MT2 from *Arabidopsis thaliana* in Zn^{2+}-metallothionein-deficient *Synechococcus* PCC 7942: putative role for MT2 in Zn^{2+} metabolism. *Plant Mol. Biol.* 30:1169–79

64. Salt DE, Prince RC, Pickering IJ, Raskin I. 1995. Mechanisms of cadmium mobility and accumulation in Indian mustard. *Plant Physiol.* 109:1427–33

65. Salt DE, Rauser WE. 1995. MgATP-dependent transport of phytochelatins across the tonoplast of oat roots. *Plant Physiol.* 107:1293–301

66. Sanchez-Fernandez R, Davies TGE, Coleman JOD, Rae PA. 2001. The *Arabidopsis thaliana* ABC protein superfamily, a complete inventory. *J. Biol. Chem.* 276: 30231–44

67. Sanita di Toppi L, Gabbrielli R. 1999.

Response to cadmium in higher plants. *Environ. Exp. Bot.* 41:105–30

68. Schmöger MEV, Oven M, Grill E. 2000. Detoxification of arsenic by phytochelatins in plants. *Plant Physiol.* 122:793–802

69. Smith D, Yanai Y, Liu YG, Ishiguro S, Okada K, et al. 1996. Characterization and mapping of Ds-GUS-T-DNA lines for targeted insertional mutagenesis. *Plant J.* 10:721–32

70. Snowden KC, Richards KD, Gardner TC. 1995. Aluminum-induced genes—induction by toxic metals, low calcium and wounding and pattern of expression in root tips. *Plant Physiol.* 107:341–48

71. Speiser DM, Ortiz DF, Kreppel L, Ow DW. 1992. Purine biosynthetic genes are required for cadmium tolerance in *Schizosaccharomyces pombe*. *Mol. Cell. Biol.* 12:5301–10

71a. Suzuki KT, Imura N, Kimura M, eds. 1993. *Metallothionein III*. Basel: Birkhauser

72. TIGR Gene Indices. 2001. http://www.tigr.org/tdb/tgi.shtml

73. Tommey AM, Shi J, Lindsay WP, Urwin PE, Robinson NJ. 1991. Expression of the pea gene PsMTa in *E. coli*—metalbinding properties of the expressed protein. *FEBS Lett.* 292:48–52

74. Vande Weghe JG, Ow DW. 1999. A fission yeast gene for mitochondrial sulfide oxidation. *J. Biol. Chem.* 274:13250–57

75. van Hoof NALM, Hassinen VH, Hakvoort H, Ballintijn KF, Schat H, et al. 2001. Enhanced copper tolerance in *Silene vulgaris* (Moench) Garcke populations from copper mines is associated with increased transcript levels of a 2b-type metallothionein gene. *Plant Physiol.* 126:1519–27

76. van Vliet C, Anderson CR, Cobbett CS. 1995. Copper-sensitive mutant of *Arabidopsis thaliana*. *Plant Physiol.* 109:871–78

77. Vatamaniuk OK, Bucher EA, Ward JT, Rea PA. 2001. A new pathway for heavy metal detoxification in animals—phytochelatin synthase is required for cadmium tolerance in *Caenorhabditis elegans*. *J. Biol. Chem.* 276:20817–20

78. Vatamaniuk OK, Mari S, Lu Y-P, Rea PA. 1999. AtPCS1, a phytochelatin synthase from *Arabidopsis*: isolation and *in vitro* reconstitution. *Proc. Natl. Acad. Sci. USA* 96:7110–15

79. Vatamaniuk OK, Mari S, Lu Y-P, Rea PA. 2000. Mechanism of heavy metal ion activation of phytochelatin (PC) synthase—blocked thiols are sufficient for PC synthase-catalyzed transpeptidation of glutathione and related thiol peptides. *J. Biol. Chem.* 275:31451–59

80. Vernoux T, Wilson RC, Seeley KA, Reichheld JP, Muroy S, et al. 2001. The *ROOT MERISTEMLESS1/CADMIUM SENSITIVE2* gene defines a glutathionedependent pathway involved in initiation and maintenance of cell division during postembryonic root development. *Plant Cell* 12:97–109

81. Vogeli-Lange R, Wagner GJ. 1990. Subcellular localization of cadmium and cadmium-binding peptides in tobacco leaves. Implication of a transport function for cadmium-binding peptides. *Plant Physiol.* 92:1086–93

82. Wagner GJ. 1993. Accumulation of cadmium in crop plants and its consequences to human health. *Adv. Agron.* 51:173–212

83. White CN, Rivin CJ. 1995. Characterization and expression of a cDNA encoding a seed-specific metallothionein in maize. *Plant Physiol.* 108:831–32

84. Zenk MH. 1996. Heavy metal detoxification in higher plants—a review. *Gene* 179:21–30

85. Zhou J, Goldsbrough PB. 1994. Functional homologs of animal and fungal metallothionein genes from Arabidopsis. *Plant Cell* 6:875–84

Annu. Rev. Plant Biol. 2002. 53:183–202
DOI: 10.1146/annurev.arplant.53.100301.135245

Vascular Tissue Differentiation and Pattern Formation in Plants

Zheng-Hua Ye

*Department of Botany, University of Georgia, Athens, Georgia 30602;
e-mail: zhye@dogwood.botany.uga.edu*

Key Words auxin, procambium, positional information, venation, xylem

■ **Abstract** Vascular tissues, xylem and phloem, are differentiated from meristematic cells, procambium, and vascular cambium. Auxin and cytokinin have been considered essential for vascular tissue differentiation; this is supported by recent molecular and genetic analyses. Xylogenesis has long been used as a model for study of cell differentiation, and many genes involved in late stages of tracheary element formation have been characterized. A number of mutants affecting vascular differentiation and pattern formation have been isolated in *Arabidopsis*. Studies of some of these mutants have suggested that vascular tissue organization within the bundles and vascular pattern formation at the organ level are regulated by positional information.

CONTENTS

1040-2519/02/0601-0183$14.00

INTRODUCTION

Plant vascular tissues, xylem and phloem, evolved as early as the Silurian period some 430 million years ago. Evolution of vascular tissues solved the problem of long-distance transport of water and food, thus enabling early vascular plants to gradually colonize the land (71). In primitive vascular plants, vascular tissues are organized in a simple pattern such that xylem is located at the center and phloem surrounds xylem. With the evolution of diverse vascular plants, vascular tissues also evolved to have a variety of organizations (28). In a given cross section of primary stems and roots, the most prominent variation of anatomical structures among different species is the organization of vascular tissues. In the stems of woody plants, the vascular tissue, secondary xylem or wood, provides both mechanical strength and long-distance transport of water and nutrients, which enables shoots of some woody plants to grow up to 100 m tall. Vascular tissues have long been chosen as a model for study of cell differentiation (48, 73, 79). In this review, I first briefly describe the general anatomical features of vascular tissues that will be useful to readers who are not familiar with this subject, and then devote my discussion mainly to the latest progress and current status of the study of vascular differentiation and pattern formation. For additional information, readers are referred to several recent excellent reviews that cover additional aspects of vascular differentiation and pattern formation (9, 11–13, 23, 34, 35, 64, 74, 78).

VASCULAR TISSUES

Vascular tissues are composed of two basic units, xylem and phloem. Xylem transports and stores water and nutrients, transports plant hormones such as abscisic acid and cytokinin, and provides mechanical support to the plant body. Phloem provides paths for distribution of the photosynthetic product sucrose and for translocation of proteins and mRNAs involved in plant growth and development. Xylem is composed of conducting tracheary elements and nonconducting elements such as xylary parenchyma cells and xylary fibers. Tracheary elements in angiosperms typically are vessel elements that are perforated at both ends to form continuous hollow columns called vessels (Figure 1a, see color insert). Tracheary elements in gymnosperms are tracheids that are connected through bordered pits to form continuous columns. Phloem is composed of conducting sieve elements and nonconducting cells such as parenchyma cells and fibers. Sieve elements of nonflowering plants are sieve cells that are connected with each other through sieve areas. Sieve elements of most flowering plants are sieve tube members that are connected through sieve plates to form continuous columns (28, 58).

Vascular tissues can be formed from two different meristematic tissues, procambium and vascular cambium. During the primary growth of stems and roots, procambial initials derived from apical meristems produce primary xylem and primary phloem. Vascular cambium initials, which are originated from procambium and other parenchyma cells when plants undergo secondary growth, give rise to secondary xylem, commonly called wood, and secondary phloem. Vascular cambium is typically composed of two types of initials: fusiform initials that produce tracheary elements and xylary fibers in the longitudinal system of wood and ray initials that produce ray parenchyma cells in the transverse system of wood (28, 58).

VASCULAR PATTERNS

Vascular Tissue Organization Within a Vascular Bundle

There is great plasticity in the organization of vascular tissues within a vascular bundle as long as vascular tissues are functional for transport. The common vascular organization within a bundle is a parallel placement of xylem and phloem, a pattern called collateral vascular bundles (Figure 1c). In some families such as Cucurbitaceae and Solanaceae, xylem is placed in parallel with external phloem and internal phloem, a pattern called bicollateral vascular bundles. Several less-common vascular tissue organizations were also evolved in vascular plants. In some monocot plants such as *Acorus* and *Dracaena*, phloem is surrounded by a continuous ring of xylem, a pattern called amphivasal vascular bundles (Figure 1d). In contrast, amphicribral vascular bundles, which are found in some angiosperms and ferns, have xylem surrounded by a ring of phloem (58).

Vascular Tissue Organization at the Organ Level

Conducting elements of xylem and phloem form continuous columns, a vascular system throughout the plant body for transport of water, nutrients, and food. Similar to the diverse organizations of vascular tissues seen within vascular bundles, vascular plants have also evolved a diversity of patterns for placement of vascular bundles in the stele. In primary stems and roots, two major patterns for placement of vascular bundles are recognized. One is the protostele in which xylem forms a solid mass at the center of the stele and phloem surrounds xylem. This is considered to be a primitive type of vascular pattern that is commonly seen in shoots of many nonseed vascular plants and in the primary roots of many dicot plants. The other is the siphonostele in which individual vascular bundles are arranged in the stele. Based on the arrangement of vascular bundles in the stele, siphonostele is generally grouped into two major patterns. In one, vascular bundles are organized as a ring in the stele, a pattern called eustele, which is mainly seen in stems of dicots and in roots of monocots. In the other, vascular bundles are scattered throughout the ground tissue, a pattern called atactostele, which is commonly seen in stems of monocot plants. Siphonostele may have evolved from the protostele by gradual replacement of the solid mass of xylem at the center with parenchyma cells (58).

In leaves, vascular bundles, commonly called veins, are organized in distinct patterns among different species. Leaves of most dicot plants have a midvein and a network of minor veins. Leaves of most monocot plants typically have veins run in parallel. Ginkgo leaves have an open dichotomous venation pattern. Many subtle variations of leaf venation patterns among different species have been recorded (76).

MODEL SYSTEMS FOR STUDYING VASCULAR DEVELOPMENT

Coleus

It is apparent that the complexity of vascular tissues and their organizations presents a big challenge for studying the molecular mechanisms underlying vascular differentiation and pattern formation. At the same time, vascular tissues represent a model for understanding many aspects of fundamental biological questions regarding cell specification, cell elongation, cell wall biosynthesis, and pattern formation. To study the different aspects of vascular development, it is ideal to choose simple or genetically manipulable systems. One of the early systems used for vascular study is Coleus in which the stems were used to study roles of auxin and cytokinin in the induction of xylem and phloem formation (3, 4). The advantage of Coleus is that the stems are big enough for easy excision of vascular tissues and subsequent analysis of effects of external factors on vascular differentiation. However, this system has been limited to physiological studies.

Zinnia

Tissue culture has long been used to study the effects of hormones on xylem and phloem differentiation (3). The most remarkable in vitro system developed so far is the zinnia in vitro tracheary element induction system (34). In this system, isolated mesophyll cells from young zinnia leaves can be induced to transdifferentiate into tracheary elements in the presence of auxin and cytokinin (Figure 1b). The advantage of this system is that isolated mesophyll cells are nearly homogeneous, and the induction rate of tracheary elements can reach up to 60%. Thus the biochemical and molecular changes associated with the differentiation of a single cell type, tracheary elements, can be monitored. A number of genes associated with tracheary element formation have been isolated and characterized by using this system (34, 61). The zinnia in vitro tracheary element induction system presents an excellent source for isolation of genes essential for different aspects of tracheary element differentiation, including cell specification, patterned secondary wall thickening, and programmed cell death.

Arabidopsis

With the introduction of the model plant *Arabidopsis* as a genetic system for studying plant growth and development, *Arabidopsis* has been adopted as a powerful system for genetic dissection of vascular differentiation and pattern formation. Unlike the zinnia system, which is limited to the study of one cell-type differentiation, *Arabidopsis* can be used to study not only the differentiation of multiple cell types in the vascular tissues but also vascular differentiation and pattern formation at the organ level. Recent studies of *Arabidopsis* mutants have opened new avenues for understanding the molecular mechanisms regulating various aspects of vascular development, such as alignment of vascular strands (17, 69), formation of a network of veins in leaves (16, 18, 19, 24, 43, 52, 69), division of procambial cells (55, 81), differentiation of primary and secondary xylem (36, 105), and organization of vascular tissues within the bundles in leaves (59, 60, 95, 96, 104) and stems (104). It is apparent that the *Arabidopsis* system is still not fully exploited, and novel mutant-screening approaches should be employed to isolate more mutants affecting various aspects of vascular differentiation and pattern formation.

APPROACHES USED FOR STUDYING VASCULAR DEVELOPMENT

Vascular development has traditionally been studied using physiological, biochemical, and molecular approaches. Early physiological studies have established that the plant hormones auxin and cytokinin are important for vascular differentiation (3, 77). A number of proteins and genes involved in different stages of tracheary element formation such as secondary wall thickening and cell death have been characterized using biochemical and subtractive hybridization approaches (34).

With the recent advance of molecular tools and the introduction of the *Arabidopsis* genetic system, many new approaches have been applied to the research of vascular differentiation. One powerful approach that goes beyond *Arabidopsis* is the large-scale sequencing of the expressed sequences from cambium and secondary xylem of pine (2) and poplar (86). Categorization of the genes expressed in the vascular cambium and secondary xylem by microarray technology (42a) will provide invaluable tools for further study of proteins involved in the differentiation of different cell types in wood. A similar approach using PCR-amplified fragment length polymorphisms has been applied to the zinnia system for isolation of genes involved in the differentiation of tracheary elements (61).

Another powerful approach is to isolate mutants with defects in vascular development. Isolation of genes that regulate vascular differentiation and pattern formation is essential for the study of vascular development because these genes can be used as tools for isolation of upstream and downstream genes by molecular and genomic approaches such as direct target screening, microarray, and yeast two-hybrid analysis. Many *Arabidopsis* mutants affect vascular patterning or normal formation of vascular strands (Tables 1–4), but none completely block the vascular cell differentiation, presumably because of the potential lethality to plants. New mutant-screening approaches such as temperature-sensitive mutants and T-DNA enhancer trap lines should be exploited.

VISUALIZATION OF VASCULAR TISSUES

The most prominent feature of vascular tissues is the presence of tracheary elements with thickened secondary wall and lignin deposition. Tracheary elements can be easily visualized by histological staining with dyes such as toluidine blue and phloroglucinol-HCl (66). For large organs such as stems of *Arabidopsis*, free-hand sections stained with the dyes often give satisfactory anatomical images (92). For high resolution, thin or ultrathin sections should be sought (66). For observation of leaf venation pattern, leaves can be cleared with chloral hydrate and then observed under light microscope (16, 57). Recently, confocal microscopy, which can give high-quality images, has been applied to visualize vascular tissues in leaves and roots. After staining with basic Fuchsin, the lignified tracheary elements in leaves and roots can be readily seen under a confocal microscope (17, 25).

Molecular markers can be used to visualize the differentiation of vascular tissues. For example, the promoters of *ATHB8* (6) and phosphoinositol kinase (27) genes, which are expressed in procambial cells, can be used as early markers of vascular differentiation. The promoter of *TED3* gene, which is specifically expressed in xylem cells (44), can be used as a marker of xylogenesis.

PROCESSES OF VASCULAR DIFFERENTIATION

Owing to the existence of multiple cell types and various organizations of vascular tissues, one can imagine that the molecular mechanisms controlling the vascular differentiation are also complicated, and many genes may be involved in vascular development. Because most of the research on vascular differentiation has been focused on xylem differentiation and very few studies have been done on phloem differentiation (90), I focus my discussion on the processes of xylem formation as follows: formation of procambium and vascular cambium, initiation of xylem differentiation, cell elongation, secondary wall thickening, and cell death.

Formation of Procambium and Vascular Cambium

Vascular tissues are differentiated from meristematic cells: procambial cells during primary growth and vascular cambium cells during secondary growth. Procambial cells in roots and stems are derived from apical meristems. Procambial cells in leaves are formed during very early stages of leaf development. It is clear that the sites for procambial cell initiation determine the pattern of vascular organization and that the activity of procambial cells controls the differentiation of vascular tissues. The central question is how molecular signals mediate the initiation of procambial cells and promote their division, which continuously provides precursor cells for differentiation of xylem and phloem.

It has long been proposed that auxin, which is polarly transported from shoot apical meristem and young leaves, induces formation of procambial cells. Early physiological studies have clearly demonstrated that the signals for induction of procambial cell formation are derived from apex and that exogenous auxin could replace the function of apex in the induction of procambial cell formation (3, 77). Roles of auxin in the induction of procambial cell formation have been supported by genetic studies in *Arabidopsis* mutants. Mutation of the *MP* gene, which encodes an auxin-response factor, disrupts the normal formation of continuous vascular strands (10, 41, 69). Mutants such as *pin1* (36, 67) and *gnom* (52, 85) with defects in auxin polar transport show dramatic alterations in vascular differentiation. The *PIN1* gene encodes an auxin efflux carrier (36), and the *GNOM* gene encodes a membrane-associated guanine-nucleotide exchange factor for an ADP-ribosylation factor G protein that is required for the coordinated polar location of PIN1 protein (85). Further studies on the roles of additional auxin polar transport carriers and auxin response factors will help us understand the roles of auxin in procambial cell formation.

Cytokinin is essential for promoting the division of procambial cells (3). Mutation of the *WOL/CRE1* gene, which encodes a cytokinin receptor (47, 55), leads to differentiation of all procambial cells into protoxylem (55, 81). Crossing of *wol* with *fass*, a mutant with supernumerary cell layers, shows that WOL is not essential for phloem and metaxylem formation, indicating that WOL is involved in promotion of procambial cell division. WOL is localized in procambial cells in roots and embryos (55). Because there are several other WOL-like cytokinin receptors in the *Arabidopsis* genome (82), it will be interesting to investigate whether they are also involved in promoting procambial cell division.

Little is known at the molecular level about how auxin and cytokinin induce procambial cell formation. Because procambial cells are dividing cells, auxin and cytokinin are likely to stimulate cell proliferation by regulating the cell cycle progression (22a,b). Recently, investigators have shown that an inositol phospholipid kinase, which is involved in the synthesis of phosphoinositide signaling molecules, is predominantly expressed in procambial cells (27). Because auxin induces the formation of phosphoinositides that may be involved in cell proliferation (29), phosphoinositides might be involved in the auxin and cytokinin signal transduction pathways, leading to procambial cell formation. The auxin response factors such as MP are obvious candidates for involvement in auxin signaling, and further characterization of their functions will be essential for understanding how auxin initiates procambial cell formation. A number of other genes such as *ATHB8* (6) and *Oshox1* (80) are expressed in procambial cells, but their precise roles in procambial cell formation are not known. Overexpression of *ATHB8* leads to overproduction of vascular tissues, suggesting that ATHB8 might be involved in stimulation of procambial cell activity (7).

Vascular cambium, a lateral meristem, is derived from procambial cells and other parenchyma cells, such as interfascicular cells in stems and pericycle cells in roots, when organs initiate secondary growth. Auxin may regulate cambium activity (3). Recently, auxin has been shown to be distributed in a gradient across the cambial zone of pine stems (93, 94). In addition, reduction in auxin polar transport in the inflorescence stems of the *ifl1* mutants leads to a block of vascular cambium activity at the basal parts of stems (103, 105). The block of vascular cambium activity in the *ifl1* mutants is associated with reduced expression of auxin efflux carriers PIN3 and PIN4 (106), suggesting important roles of polar auxin flow in vascular cambium activity. Because many auxin efflux carrier homologues have been identified in the *Arabidopsis* genome, it will be important to investigate which carriers play central roles in the formation of vascular cambial cells. Induction of vascular cambial cell formation by auxin is likely mediated through the protein kinase PINOID (21) because mutation of the *PINOID* gene completely abolishes the vascular cambium formation (R. Zhong & Z-H. Ye, unpublished observations). Cytokinin is considered essential for the continuous division of vascular cambium cells, which supply precursor cells for differentiation into xylem and phloem (3). Because *Arabidopsis* stems and roots undergo secondary growth (7, 26, 102), it will be interesting to investigate whether WOL or other cytokinin receptor homologs are involved in the regulation of vascular cambium cell division in *Arabidopsis*. Dissection of the signaling transduction pathways of auxin and cytokinin that lead to vascular cambium formation is essential for our understanding of vascular cambium development.

Initiation of Xylem Differentiation

Procambium and vascular cambium are polar in terms of the final fates of their daughter cells. The daughter cells may become either xylem precursor cells or phloem precursor cells, depending on their positions. This suggests that the cambial cells at different positions receive different signals that specify different cell fates. Auxin may act as a patterning agent for differentiation of vascular tissues. Auxin is distributed in a gradient across the cambial region (93, 94), indicating that different levels of auxin together with other signaling molecules such as cytokinin are important for vascular cell differentiation (3, 4, 77, 78). Transgenic studies have shown that alterations of endogenous auxin level dramatically affect xylem formation (50, 75). Although the phenotype of the *wol* mutant suggests that cytokinin is not directly involved in xylem differentiation, in vitro studies in zinnia indicate that both auxin and cytokinin are required for induction of xylem cell formation. It is possible that other WOL-like genes play roles in xylem cell differentiation. In addition to auxin and cytokinin, other factors such as brassinosteroid (49, 98, 99) and phytosulfokine, a peptide growth factor (56), might play important roles in the stimulation of xylogenesis.

Little is known about the signal transduction pathways of auxin and cytokinin, which lead to xylem cell formation. MP, an auxin response factor, is likely involved in this process because mutation of the *MP* gene results in misaligned xylem strands (41, 69). Many other auxin response factors have been identified (39), and studies of their functions will likely help us to further understand vascular cell differentiation. Several other transcription factors such as *Arabidopsis* ATHB8

(6, 7), rice Oshox1 (80), and aspen PttHB1 (42) might also play roles in xylem cell differentiation. Auxin-insensitive mutants *axr6* (43) and *bodenlos* (40) are defective in venation pattern, and their corresponding genes are likely involved in auxin signaling pathways important for vascular differentiation. In addition, the maize *wilted* mutant causes a partial block of metaxylem cell formation (68) (Figures 2*a,b*, see color insert), and the *Arabidopsis eli1* mutant shows discontinuous xylem strands (17). Isolation and functional characterization of these genes will be important for further dissection of the molecular mechanisms underlying xylem cell differentiation (Table 1).

So far, no mutants with a complete block of xylem cell differentiation have been isolated presumably because these kinds of mutants are lethal. This greatly hinders the utilization of the genetic approach to study xylogenesis. One complementary approach is to use the zinnia in vitro tracheary element induction system to isolate genes associated with xylogenesis. Because isolated zinnia mesophyll cells can be induced to transdifferentiate into tracheary elements, this system has long been exploited to isolate genes involved in different stages of xylogenesis (34). Many genes that are induced within hours after hormonal treatment have been isolated in the zinnia system using PCR-amplified fragment length polymorphisms (61). Researchers anticipate that homologous genes will be found in *Arabidopsis* and their functions in xylogenesis studied by using T-DNA or transposon knock-out mutants.

TABLE 1   Mutants affecting vascular differentiation

Mutant	Species	Vascular phenotype	Gene product	Reference
wilted	Maize	Disrupted metaxylem differentiation	Unknown	68
wilty-dwarf	Tomato	Compound perforation plate in vessels instead of wild-type simple perforation plate	Unknown	1, 72
wol	*Arabidopsis*	Block of procambial cell division	Cytokinin receptor	47, 55, 81
mp	*Arabidopsis*	Misaligned vessel elements	Auxin response factor	10, 41
pin1	*Arabidopsis*	Increased size of vascular bundles in stems	Auxin efflux carrier	36, 67
ifl1	*Arabidopsis*	Reduced secondary xylem differentiation in stems	Homeodomain leucine zipper protein	105
eli1	*Arabidopsis*	Discontinuous xylem strands	Unknown	17
irx1	*Arabidopsis*	Reduced secondary wall formation in xylem cells	Cellulose synthase catalytic subunit	87, 92
irx2	*Arabidopsis*	Reduced secondary wall formation in xylem cells	Unknown	92
irx3	*Arabidopsis*	Reduced secondary wall formation in xylem cells	Cellulose synthase catalytic subunit	88, 92
gpx	*Arabidopsis*	Gapped xylem	Unknown	91
fra2	*Arabidopsis*	Reduced length of vascular cells	Katanin-like microtubule severing protein	15

Cell Elongation

After initiation of vascular cell differentiation, the conducting cells, tracheary elements in the xylem and sieve elements in the phloem, undergo significant elongation before the tubular conducting system is formed. Because developing conducting cells cease to elongate when the secondary cell wall starts to be laid down (1), which is typical of diffuse cell elongation, the molecular mechanisms regulating the elongation of conducting cells are likely similar to those for other nonvascular cells. A katanin-like microtubule-severing protein AtKTN1 is important for the normal elongation of both xylem and phloem cells (15), indicating that microtubules regulate cell elongation in vascular tissues. Microtubules are thought to direct the orientation of cellulose microfibril deposition, which in turn determines the axis of cell elongation. Cell elonagtion requires the loosening of the existing cellulose-hemicellulose network, a process mediated by cell wall loosening enzymes such as expansins (22). Expansin mRNA is preferentially localized at the ends of differentiating tracheary elements in zinnia, suggesting that expansins are important in the elongation of vessel cells (46). Plant hormones are clearly involved in regulation of vascular cell elongation. Mutation of genes involved in brassinosteroid biosynthesis results in a dramatic reduction in length of all cells including vascular cells (20).

Secondary Wall Thickening

After elongation, tracheary elements undergo secondary wall thickening with annular, helical, reticulate, scalariform, and pitted patterns (58). The thickened secondary wall provides mechanical strength to the vessels for withstanding the negative pressure generated through transpiration. The patterned secondary wall thickening is regulated through controlled deposition of cellulose microfibrils, a process that is apparently regulated by the patterns of cortical microtubules located underneath the plasma membrane. Pharmacological studies have shown that disruption of the cortical microtubule organization completely alters the patterns of secondary wall thickening (30–32). Little is known about how the cortical microtubules form different patterns and how they regulate the patterns of secondary wall thickening. In addition to cortical microtubules, microfilaments also appear to be important for the normal patterning of secondary wall in tracheary elements (51).

There has been a significant progress in the characterization of genes involved in the synthesis of secondary wall, including synthesis of cellulose and lignin. Several *Arabidopsis* mutants affecting secondary wall formation have been isolated (91, 92); two of which encode cellulose synthase catalytic subunits that are specifically involved in cellulose synthesis in the secondary wall (87, 88). Isolation of these genes will further expand our understanding of secondary wall biosynthesis.

Lignin impregnated in the cellulose and hemicellulose network provides additional mechanical strength to the secondary wall and also renders the secondary wall waterproof owing to its hydrophobic nature. Monolignols are synthesized through the phenylpropanoid pathway and are exported into the secondary wall where they are polymerized into lignin polymers. Most genes involved in monolignol biosynthesis have been isolated and characterized, and readers are referred to a recent review on this topic (97).

Cell Death

After fulfilling cellular activities necessary for building up a secondary wall, developing tracheary elements undergo cell death to remove their cellular contents, and in the case of vessel elements, their ends are perforated to form tubular columns called vessels. The ends of vessel elements can be perforated with a single hole, a pattern called simple perforation plate, or with more than one hole, a pattern called complex perforation plate (28). Perforation sites contain only the primary wall that is digested by cellulase during autolysis, whereas the secondary wall impregnated with lignin is resistant to cellulase attack. However, to date, no cellulase genes have been shown to be specifically expressed at the late stages of xylogenesis. Perforation plate patterns, whether simple or complex, are controlled by the patterned deposition of secondary wall on both ends of vessel elements. It is extremely interesting to note that mutation of a gene in the tomato *wilty-dwarf* mutant (1a, 72) converts the wild-type simple perforation plate in vessels into a compound perforation plate (Figures 2c,d). Isolation of the corresponding gene should shed new insight into the mechanisms controlling the patterned secondary wall deposition.

Hydrolytic enzymes including cysteine proteases (8, 45, 65, 101, 102), serine proteases (8, 38, 101, 102), and nucleases (5, 89, 100) are highly induced during xylogenesis, and they are stored in vacuoles before autolysis occurs (35). Cell death is initiated by disruption of the vacuole membrane, resulting in release of hydrolytic enzymes into the cytosol (37, 38, 53, 65). One of the biochemical markers for the cell death of tracheary elements is the degradation of nuclear DNA that can be detected by terminal deoxynucleotidyl transferase-mediated dUTP nick-end labeling (37, 63). Little is known about what signals trigger the biosynthesis of a battery of hydrolytic enzymes and the final disruption of vacuoles, except for a possible involvement of calcium influx and an extracellular serine protease in the initiation of cell death of tracheary elements (38).

VASCULAR PATTERN FORMATION

Vascular Bundles

Vascular tissues, xylem and phloem, within a vascular bundle can be organized into distinctive patterns, such as collateral, amphivasal, and amphicribral bundles. Recent genetic analysis has begun to unravel the molecular mechanisms underlying vascular pattern formation. Studies from three mutants have revealed that the vascular tissue organization within the bundles is controlled by positional

TABLE 2   Mutants affecting vascular tissue organization within vascular bundles

Mutant	Species	Vascular phenotype	Gene product	Reference
phan	*Antirrhinum*	Amphicribral vascular bundles in leaves instead of wild-type collateral vascular bundles	MYB transcription factor	95, 96
phb-1d	*Arabidopsis*	Amphivasal vascular bundles in leaves instead of wild-type collateral vascular bundles	Homeodomain leucine zipper protein	59, 60
avb1	*Arabidopsis*	Amphivasal vascular bundles in leaves and stems instead of wild-type collateral vascular bundles	Homeodomain leucine zipper protein	104

information (Table 2). In *Arabidopsis* leaves, vascular tissues within a bundle are organized as collateral, i.e., xylem is parallel to phloem. In the bundle, xylem is positioned next to the adaxial side of leaves, and phloem is positioned next to the abaxial side of leaves. The leaves of *Arabidopsis phb-1d* mutant, which exhibits a loss of abaxial characters, have amphivasal vascular bundles, i.e., xylem surrounds phloem (59). Similarly, in the leaves of the *Arabidopsis avb1* mutant, which shows a partial loss of leaf polarity (R. Zhong & Z-H. Ye, unpublished observations), the collateral vascular bundles are transformed into amphivasal bundles (104). In contrast, in the leaves of the *Antirrhinum phan* mutant, which causes a loss of adaxial cell fate, the collateral vascular bundles are transformed into amphicribral bundles, i.e., phloem surrounds xylem (95). This suggests that, when the positional information that determines the normal placement of xylem is disrupted, xylem forms a circle around phloem by default, as seen in the *phb-1d* and *avb1* mutants. Similarly, when the positional information that determines the normal placement of phloem is disrupted, phloem forms a circle around xylem by default, as seen in the *phan* mutant. This also indicates that similar positional information is utilized by plants to control leaf polarity and vascular tissue organization in leaves. The *PHB* (60) and *AVB1* (R. Zhong & Z-H. Ye, unpublished observations) genes have been cloned, and they encode proteins belonging to a family of homeodomain leucine-zipper transcription factors. The *PHAN* gene encodes a MYB transcription factor (96). With the availability of these molecular tools, it will be possible to further investigate how these transcription factors regulate the positional signals that direct various organizations of vascular tissues. Because auxin and cytokinin are inducers of vascular differentiation, it is reasonable to propose that the positional information might regulate the positions of the hormonal flow that determine the formation of various vascular tissue organizations.

In *Arabidopsis* inflorescence stems, the vascular tissues within a bundle are also organized as collateral. In the bundle, xylem is positioned next to the center of stems, and phloem is positioned next to the periphery (Figure 1c). In the stems of the *avb1* mutant, the normal collateral placement of xylem and phloem is disrupted, leading to formation of amphivasal vascular bundles with xylem surrounding phloem (104) (Figure 1*d*). This suggests that leaves and stems might share the same molecular mechanisms in controlling the organization of vascular tissues.

Vascular Patterning at the Organ Level

At the organ level, vascular tissues can be arranged in a variety of patterns. In primary stems and roots, vascular bundles can be organized as a single ring or in a scattered pattern. In stems and roots with secondary growth, vasculature can be organized as a single ring, multiple concentric rings, multiple separated rings, or multiple scattered bands (58). In leaves, vascular bundles, often called veins, form diverse patterns such as parallel and networked arrangements. The positions of the polar auxin flow may determine the pattern of vascular tissues at the organ level (77, 78). This proposal has been supported by both pharmacological and genetic studies in *Arabidopsis*. Alteration of auxin polar transport by auxin polar transport inhibitors (57, 83) and by mutation of genes affecting auxin polar transport (12, 57) dramatically alters the venation pattern in *Arabidopsis* leaves. With the identification of all putative auxin efflux carriers in the *Arabidopsis* genome, it is now possible to further investigate the roles of these carriers in determining the vascular patterns in different organs.

Genetic analysis has indicated that vascular patterns at the organ level are also regulated by positional information (Tables 3 and 4). Mutation of the *Arabidopsis* *AVB1* gene not only transforms the collateral vascular bundles into amphivasal bundles, but also disrupts the ring-like organization of vascular bundles in stems (104). In the *avb1* mutant, multiple bundles are branched into the pith, a pattern reminiscent of those seen in the monocot stems. This suggests that, when the positional information that determines the ring-like vascular organization is disrupted, additional vascular bundles are formed in pith by default, as seen in the *avb1* mutant. It will be interesting to investigate the distribution patterns of auxin efflux carriers in the *avb1* mutant. The importance of positional information in regulating vascular patterning is also demonstrated by several organ polarity mutants such as the *yabby* (84) and *ago1* (14) mutants. These mutants exhibit altered venation patterns in leaves. The *YABBY* genes encode putative transcription factors (84), and the *AGO* gene encodes a protein with unknown functions (14).

A number of other mutants affecting vascular patterning have been isolated (Tables 3 and 4). Most of these mutants were isolated based on alterations of the

TABLE 3   Mutants affecting vascular patterns in stems and roots

Mutant	Species	Vascular phenotype	Gene product	Reference
avb1	*Arabidopsis*	Disruption of the ring-like vascular bundle organization in stems	Homeodomain leucine zipper protein	104
lsn1	Maize	Disorganization of vascular bundles in roots and scutellar nodes	Unknown	54

TABLE 4   Mutants affecting leaf venation patterns

Mutant	Species	Vascular phenotype	Gene product	Reference
midribless	Pearl millet	Lack of midrib	Unknown	70
mbl	*Panicum maximum*	Lack of midrib	Unknown	33
lop1	Arabidopsis	Midvein bifurcation, discontinuous and reduced number of veins	Unknown	19
mp	Arabidopsis	Discontinuous and reduced number of veins	Auxin response factor	10, 41
bdl	Arabidopsis	Discontinuous and reduced number of veins	Unknown	40
ago	Arabidopsis	Reduced number of veins	Protein with unknown identity	14
fil-5 yab3-1	Arabidopsis	Reduced number of veins	Transcription factors	84
cvp1, cvp2	Arabidopsis	Discontinuous and reduced number of veins	Unknown	18
van1, van2 van3, van4 van5, van6	Arabidopsis	Discontinuous and reduced number of veins	Unknown	52
gnom/van7	Arabidopsis	Discontinuous and reduced number of veins	Guanine-nucleotide exchange factor	52, 85
sfc	Arabidopsis	Discontinuous and reduced number of veins	Unknown	24
axr6	Arabidopsis	Discontinuous and reduced number of veins	Unknown	43
hve	Arabidopsis	Reduced number of veins	Unknown	16
ixa	Arabidopsis	Reduced number of veins and free-ending vascular strands	Unknown	16
ehy	Arabidopsis	Exess of hydathodes	Unknown	16
pin1	Arabidopsis	Reduced number of veins	Auxin efflux carrier	12
ifl1	Arabidopsis	Reduced number of veins	Homeodomain leucine zipper protein	105

venation pattern in cotyledons or leaves. These mutations cause discontinuous, random, or reduced numbers of veins (16, 18, 24, 33, 52, 70). Recently, a maize mutant with an alteration of vascular patterns in roots and scutellar nodes has been described (54). All these vascular pattern mutants also display defects in other aspects of plant growth and development, indicating that the genes affected are involved in multiple processes important for normal plant development. Isolation of the corresponding genes in these mutants and further characterization of their functions will undoubtedly contribute to the dissection of pathways involved in vascular pattern formation.

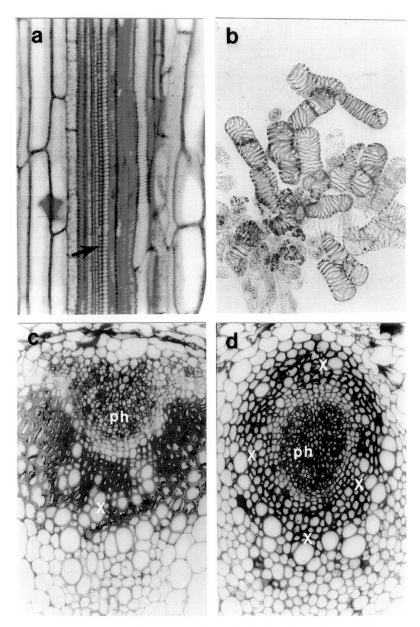

Figure 1 Anatomy of vascular tissues. (*a*) Longitudinal section of an *Arabidopsis* stem showing vessels (*arrow*). (*b*) Tracheary elements differentiated from isolated zinnia mesophyll cells. (*c*) Cross section of the wild-type *Arabidopsis* stem showing a collateral vascular bundle. (*d*) Cross section of the *Arabidopsis avb1* mutant stem showing an amphivasal vascular bundle. ph, phloem; x, xylem. Figures 1*c* and *d* were reproduced with permission from (104).

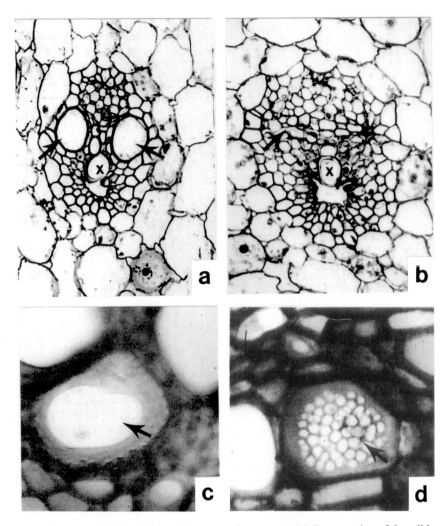

Figure 2 Anatomical phenotypes of two vascular mutants. (*a*) Cross section of the wild-type maize stem showing two prominent metaxylem cells (*arrows*) in a vascular bundle. x, protoxylem. (*b*) Cross section of the maize *wilted* mutant stem showing the absence of metaxylem cells in a vascular bundle. (*c*) Cross section of the wild-type tomato stem showing the simple perforation plate (*arrow*) in a vessel. (*d*) Cross section of the tomato *wilty-dwarf* mutant showing a compound perforation plate (*arrow*) in a vessel. Figures 2*a* and *b* were reproduced with permission from (68). Figures 2*c* and *d* were reproduced with permission from (1).

CONCLUSIONS

Significant progress has been made in our understanding of vascular differentiation and pattern formation, which lays a foundation for further dissecting these complicated processes at the molecular level. The isolation of genes involved in auxin polar transport and signaling of auxin and cytokinin will help us to investigate how auxin polar flow is spatially regulated and how auxin and cytokinin signals are transduced to induce vascular differentiation. With the availability of many vascular pattern mutants and further characterization of their corresponding genes, it will soon be possible to investigate how positional signals determine the organization of vascular tissues. Although the zinnia system will still be an important player in the search for genes specifically involved in tracheary element formation, the model plant *Arabidopsis* is undoubtedly a powerful genetic system for investigating the molecular mechanisms regulating different aspects of vascular differentiation and pattern formation. Because the inflorescence stems and roots of *Arabidopsis* undergo secondary growth, *Arabidopsis* is also a useful genetic tool for studying wood formation. It is anticipated that a combined application of molecular, genetic, genomic, cellular, and physiological tools will soon lead to many exciting discoveries regarding the molecular mechanisms underlying vascular tissue differentiation and vascular tissue patterning.

ACKNOWLEDGMENTS

Work in the author's laboratory was supported by a grant from the Cooperative State Research, Education, and Extension Service, U.S. Department of Agriculture.

Visit the Annual Reviews home page at www.annualreviews.org

LITERATURE CITED

1. Abe H, Funada R, Ohtani J, Fukazawa K. 1997. Changes in the arrangement of cellulose microfibrils associated with the cessation of cell expansion in tracheids. *Trees* 11:328–32

1a. Alldridge NA. 1964. Anomalous vessel elements in wilty-dwarf tomato. *Bot. Gaz.* 125:138–42

2. Allona I, Quinn M, Shoop E, Swope K, Cyr SS, et al. 1998. Analysis of xylem formation in pine by cDNA sequencing. *Proc. Natl. Acad. Sci. USA* 95:9693–98

3. Aloni R. 1987. Differentiation of vascular tissues. *Annu. Rev. Plant Physiol.* 38:179–204

4. Aloni R. 1993. The role of cytokinin in organised differentiation of vascular tissues. *Aust. J. Plant Physiol.* 20:601–8

5. Aoyagi S, Sugiyama M, Fukuda H. 1998. *BEN1* and *ZEN1* cDNAs encoding S1-type DNases that are associated with programmed cell death in plants. *FEBS Lett.* 429:134–38

6. Baima S, Nobili F, Sessa G, Lucchetti S, Ruberti I, Morelli G. 1995. The expression of the *Athb-8* homeobox gene is restricted to provascular cells in *Arabidopsis thaliana. Development* 12:4171–82

7. Baima S, Possenti M, Matteucci A, Wisman E, Altamura MM, et al. 2001. The

Arabidopsis ATHB-8 HD-Zip protein acts as a differentiation-promoting transcription factor of the vascular meristems. *Plant Physiol.* 126:643–55

8. Beers EP, Freeman TB. 1997. Proteinase activity during tracheary element differentiation in zinnia mesophyll cultures. *Plant Physiol.* 113:873–80

9. Beers EP, Woffenden BJ, Zhao C. 2000. Plant proteolytic enzymes: possible roles during programmed cell death. *Plant Mol. Biol.* 44:399–415

10. Berleth T, Jürgens G. 1993. The role of the monopteros gene in organising the basal body region of the *Arabidopsis* embryo. *Development* 118:575–87

11. Berleth T, Mattsson J. 2000. Vascular development: tracing signals along veins. *Curr. Opin. Plant Biol.* 3:406–11

12. Berleth T, Mattsson J, Hardtke CS. 2000. Vascular continuity and auxin signals. *Trends Plant Sci.* 5:387–93

13. Berleth T, Sachs T. 2001. Plant morphogenesis: long-distance coordination and local patterning. *Curr. Opin. Plant Biol.* 4:57–62

14. Bohmert K, Camus I, Bellini C, Bouchez D, Caboche M, Benning C. 1998. *AGO1* defines a novel locus of *Arabidopsis* controlling leaf development. *EMBO J.* 17:170–80

15. Burk DH, Liu B, Zhong R, Morrison WH, Ye Z-H. 2001. A katanin-like protein regulates normal cell wall biosynthesis and cell elongation. *Plant Cell* 13:807–27

16. Candela H, Marinez-Laborda A, Micol JL. 1999. Venation pattern formation in *Arabidopsis thaliana* vegetative leaves. *Dev. Biol.* 205:205–16

17. Cano-Delgado AI, Metzlaff K, Bevan MW. 2000. The *eli1* mutation reveals a link between cell expansion and secondary cell wall formation in *Arabidopsis thaliana*. *Development* 127:3395–405

18. Carland FM, Berg B, FitzGerald JN, Jinamornphongs S, Nelson T, Keith B. 1999. Genetic regulation of vascular tissue patterning in *Arabidopsis*. *Plant Cell* 11:2123–37

19. Carland FM, McHale NA. 1996. *LOP1*: a gene involved in auxin transport and vascular patterning in *Arabidopsis*. *Development* 122:1811–19

20. Choe S, Noguchi T, Fujioka S, Takatsuto S, Tissier CP, et al. 1999. The *Arabidopsis dwf7/ste1* mutant is defective in the sterol C-5 desaturation step leading to brassinosteroid biosynthesis. *Plant Cell* 11:207–21

21. Christensen SK, Dagenais N, Chory J, Weigel D. 2000. Regulation of auxin response by the protein kinase PINOID. *Cell* 100:469–78

22. Cosgrove DJ. 1999. Enzymes and other agents that enhance cell wall extensibility. *Annu. Rev. Plant Physiol. Plant Mol. Biol.* 50:391–417

22a. D'Agostino IB, Kieber JJ. 1999. Molecular mechanisms of cytokinin action. *Curr. Opin. Plant Biol.* 2:359–64

22b. den Boer BGW, Murray JAH. 2000. Triggering the cell cycle in plants. *Trends Cell Biol.* 10:245–50

23. Dengler N, Kang J. 2001. Vascular patterning and leaf shape. *Curr. Opin. Plant Biol.* 4:50–56

24. Deyholos MK, Cordner G, Beebe D, Sieburth LE. 2000. The *SCARFACE* gene is required for cotyledon and leaf vein patterning. *Development* 127:3206–13

25. Dharmawardhana DP, Ellis BE, Carlson JE. 1992. Characterization of vascular lignification in *Arabidopsis thaliana*. *Can. J. Bot.* 70:2238–44

26. Dolan L, Roberts K. 1995. Secondary thickening in roots of *Arabidopsis thaliana*: anatomy and cell surface changes. *New Phytol.* 131:121–28

27. Elge S, Brearley C, Xia H-J, Kehr J, Xue H-W, Mueller-Roeber B. 2001. An *Arabidopsis* inositol phospholipid kinase strongly expressed in procambial cells: synthesis of Ptdlns(4,5)P_2 and Ptdlns(3,4,5)P_3 in insect cells by 5-phophorylation of precursors. *Plant J.* 26:561–71

28. Esau K. 1977. *Anatomy of Seed Plants.* New York: Wiley. 2nd ed.

29. Ettlinger C, Lehle L. 1988. Auxin induces rapid changes in phosphatidylinositol metabolites. *Nature* 331:176–78

30. Falconer MM, Seagull RW. 1985. Xylogenesis in tissue culture: taxol effects on microtubule reorientation and lateral association in differentiating cells. *Protoplasma* 128:157–66

31. Falconer MM, Seagull RW. 1986. Xylogenesis in tissue culture II: cell shape and secondary wall patterns. *Protoplasma* 133:140–48

32. Falconer MM, Seagull RW. 1988. Xylogenesis in tissue culture III: continuing wall deposition during tracheary element development. *Protoplasma* 144:10–16

33. Fladung M, Bossinger G, Roeb GW, Salamini F. 1991. Correlated alterations in leaf and flower morphology and rate of leaf photosynthesis in a *midribless* (*mbl*) mutant of *Panicum maximum* Jacq. *Planta* 184:356–61

34. Fukuda H. 1997. Tracheary element differentiation. *Plant Cell* 9:1147–56

35. Fukuda H. 2000. Programmed cell death of tracheary elements as a paradigm in plants. *Plant Mol. Biol.* 44:245–53

36. Gälweiler L, Guan C, Müller A, Wisman E, Mendgen K, et al. 1998. Regulation of polar auxin transport by AtPIN1 in *Arabidopsis* vascular tissue. *Science* 282: 2226–30

37. Groover A, DeWitt N, Heidel A, Jones AM. 1997. Programmed cell death of plant tracheary elements differentiating *in vitro*. *Protoplasma* 196:197–211

38. Groover A, Jones AM. 1999. Tracheary element differentiation uses a novel mechanism coordinating programmed cell death and secondary cell wall synthesis. *Plant Physiol.* 119:375–84

39. Guilfoyle TJ. 1998. Aux/IAA proteins and auxin signal transduction. *Trends Plant Sci.* 3:203–4

40. Hamann T, Mayer U, Jürgens G. 1999. The auxin-insensitive *bodenlos* mutation affects primary root formation and apical-basal patterning in the *Arabidopsis* embryo. *Development* 126:1387–95

41. Hardtke CS, Berleth T. 1998. The *Arabidopsis* gene *MONOPTEROS* encodes a transcription factor mediating embryo axis formation and vascular development. *EMBO J.* 17:1405–10

42. Hertzberg M, Olsson O. 1998. Molecular characterisation of novel plant homeobox gene expressed in the maturing xylem zone of *Populus tremula X trenmuloides*. *Plant J.* 16:285–95

42a. Hertzberg M, Sievertzon M, Aspeborg H, Nilsson P, Sandberg G, Lundeberg J. 2001. cDNA microarray analysis of small plant tissue samples using a cDNA tag target amplification protocol. *Plant J.* 25: 585–91

43. Hobbie L, McGovern M, Hurwitz LR, Pierro A, Liu NY, et al. 2000. The *axr6* mutants of *Arabidopsis thaliana* define a gene involved in auxin response and early development. *Development* 127:23–32

44. Igarashi M, Demura T, Fukuda H. 1998. Expression of the *Zinnia* TED3 promoter in developing tracheary elements of transgenic *Arabidopsis*. *Plant Mol. Biol.* 36:917–27

45. Iliev I, Savidge R. 1999. Proteolytic activity in relation to seasonal cambial growth and xylogenesis in *Pinus banksiana*. *Phytochemistry* 50:953–60

46. Im K-H, Cosgrove DJ, Jones AM. 2000. Subcellular localization of expansin mRNA in xylem cells. *Plant Physiol.* 123:463–70

47. Inoue T, Higuchi M, Hashimoto Y, Seki M, Kobayashi M, et al. 2001. Identification of CRE1 as a cytokinin receptor from *Arabidopsis*. *Nature* 409:1060–63

48. Iqbal M, ed. 1990. *The Vascular Cambium.* New York: Wiley

49. Iwasaki T, Shibaoka H. 1991. Brassinosteroids act as regulators of tracheary-element differentiation in isolated *Zinnia* mesophyll cells. *Plant Cell Physiol.* 32:1007–14

50. Klee HJ, Horsch RB, Hinchee MA, Hein MB, Hoffmann NL. 1987. The effects of overproduction of two *Agrobacterium tumefaciens* T-DNA auxin biosynthetic gene products in transgenic petunia plants. *Genes Dev.* 1:86–96

51. Kobayashi H, Fukuda H, Shibaoka H. 1988. Interrelation between the spatial deposition of actin filaments and microtubules during the differentiation of tracheary elements in cultured *Zinnia* cells. *Protoplasma* 143:29–37

52. Koizumi K, Sugiyama M, Fukuda H. 2000. A series of novel mutants of *Arabidopsis thaliana* that are defective in the formation of continuous vascular network: calling the auxin signal flow canalization hypothesis into question. *Development* 127:3197–204

53. Kuriyama H. 1999. Loss of tonoplast integrity programmed in tracheary element differentiation. *Plant Physiol.* 121:763–74

54. Landoni M, Gavazzi G, Rascio N, Vecchia FD, Consonni G, Dolfini S. 2000. A maize mutant with an altered vascular pattern. *Ann. Bot.* 85:143–50

55. Mähönen AP, Bonke M, Kauppinen L, Riikonen M, Benfey P, Helariutta Y. 2000. A novel two-component hybrid molecule regulates vascular morphogenesis of the *Arabidopsis* root. *Genes Dev.* 14:2938–43

56. Matsubayashi Y, Takagi L, Omura N, Morita A, Sakagami Y. 1999. The endogenous sulfated pentapeptide phytosulfokine-alpha stimulates tracheary element differentiation of isolated mesophyll cells of *Zinnia*. *Plant Physiol.* 120:1043–48

57. Mattsson J, Sung ZR, Berleth T. 1999. Responses of plant vascular systems to auxin transport inhibition. *Development* 126:2979–91

58. Mauseth JD. 1988. *Plant Anatomy*. Menlo Park, CA: Benjamin/Cummings

59. McConnell JR, Barton MK. 1998. Leaf polarity and meristem formation in *Arabidopsis*. *Development* 125:2935–42

60. McConnell JR, Emery J, Eshed Y, Bao N, Bowman J, Barton MK. 2001. Role of *PHABULOSA* and *PHAVOLUTA* in determining radial patterning in shoots. *Nature* 411:709–13

61. Milioni D, Sado P-E, Stacey NJ, Domingo C, Roberts K, McCann MC. 2001. Differential expression of cell-wall–related genes during the formation of tracheary elements in the *Zinnia* mesophyll cell system. *Plant Mol. Biol.* 47:221–38

62. Minami A, Fukuda H. 1995. Transient and specific expression of a cysteine endopeptidase during autolysis in differentiating tracheary elements from *Zinnia* mesophyll cells. *Plant Cell Physiol.* 36:1599–606

63. Mittler R, Lam E. 1995. In situ detection of nDNA fragmentation during the differentiation of tracheary elements in higher plants. *Plant Physiol.* 108:489–93

64. Nelson T, Dengler N. 1997. Leaf venation pattern formation. *Plant Cell* 9:1121–35

65. Obara K, Kuriyama H, Fukuda H. 2001. Direct evidence of active and rapid nuclear degradation triggered by vacuole rupture during programmed cell death in *Zinnia*. *Plant Physiol.* 125:615–26

66. O'Brien TP, McCully ME. 1981. *The Study of Plant Structure. Principles and Selected Methods*. South Melb., Aust.: Termarcarphi

67. Okada K, Ueda J, Komaki MK, Bell CJ, Shimura Y. 1991. Requirement of the auxin polar transport system in early stages of *Arabidopsis* floral bud formation. *Plant Cell* 3:677–84

68. Postlethwait SN, Nelson OE. 1957. A chronically wilted mutant of maize. *Am. J. Bot.* 44:628–33

69. Przemeck GKH, Mattsson J, Hardtke CS, Sung ZR, Berleth T. 1996. Studies on the role of the *Arabidopsis* gene *MONOPTEROS* in vascular development and plant cell axialization. *Planta* 200:229–37

70. Rao SA, Mengesha MH, Rao YS, Reddy CR. 1989. Leaf anatomy of *midribless* mutants in pearl millet. *Curr. Sci.* 58:1034–36

71. Raven PH, Evert RF, Eichhorn SE. 1999. *Biology of Plants.* New York: Freeman Co. 6th ed.

72. Rick CM. 1952. The grafting relations of *wilty dwarf,* a new tomato mutant. *Am. Nat.* 86:173–84

73. Roberts LW. 1976. *Cytodifferentiation in Plants. Xylogenesis as a Model System.* London, UK: Cambridge Univ. Press

74. Roberts K, McCann MC. 2000. Xylogenesis: the birth of a corpse. *Curr. Opin. Plant Biol.* 3:517–22

75. Romano CP, Hein MB, Klee HJ. 1990. Inactivation of auxin in tobacco transformed with the indoleacetic acid-lysine synthetase gene of *Pseudomonas savastanoi. Gene Dev.* 5:438–46

76. Roth-Nebelsick A, Uhl D, Mosbrugger V, Kerp H. 2001. Evolution and function of leaf venation architecture: a review. *Ann. Bot.* 87:553–66

77. Sachs T. 1991. Cell polarity and tissue patterning in plants. *Development* S1:83–93

78. Sachs T. 2000. Integrating cellular and organismic aspects of vascular differentiation. *Plant Cell Physiol.* 41:649–56

79. Savidge R, Barnett J, Napier R, ed. 2000. *Cell and Molecular Biology of Wood Formation.* Oxford, UK: BIOS Sci. Publ.

80. Scarpella E, Rueb S, Boot KJM, Hoge JHC, Meijer AH. 2000. A role for the rice homeobox gene *Oshox1* in provascular cell fate commitment. *Development* 127:3655–69

81. Scheres B, Laurenzio LD, Willemsen V, Hauser M-T, Janmaat K, et al. 1995. Mutations affecting the radial organization of the *Arabidopsis* root display specific defects throughout the embryonic axis. *Development* 121:53–62

82. Schmülling T. 2001. CREam of cytokinin signalling: receptor identified. *Trends Plant Sci.* 7:281–84

83. Sieburth LE. 1999. Auxin is required for leaf vein pattern in *Arabidopsis. Plant Physiol.* 121:1179–90

84. Siegfried KR, Eshed Y, Baum SF, Otsuga D, Drews GN, Bowman JL. 1999. Members of the *YABBY* gene family specify abaxial cell fate in *Arabidopsis. Development* 126:4117–28

85. Steinmann T, Geldner N, Grebe M, Mangold S, Jackson CL, et al. 1999. Coordinated polar localization of auxin efflux carrier PIN1 by GNOM ARF GEF. *Science* 286:316–18

86. Sterky F, Regan S, Karlsson J, Hertzberg M, Rohde A, et al. 1998. Gene discovery in the wood-forming tissue of poplar: analysis of 5692 expressed sequence tags. *Proc. Natl. Acad. Sci. USA* 95:13330–35

87. Taylor NG, Laurie S, Turner SR. 2000. Multiple cellulose synthase catalytic subunits are required for cellulose synthesis in *Arabidopsis. Plant Cell* 12:2529–39

88. Taylor NG, Scheible W-R, Cutler S, Somerville CR, Turner SR. 1999. The *irregular xylem3* locus of *Arabidopsis* encodes a cellulose synthase required for secondary cell wall synthesis. *Plant Cell* 11:769–79

89. Thelen MP, Northcote DH. 1989. Identification and purification of a nuclease from *Zinnia elegans* L.: a potential marker for xylogenesis. *Planta* 179:181–95

90. Tornero P, Conejero V, Vera P. 1996. Phloem-specific expression of a plant homeobox gene during secondary phases of vascular development. *Plant J.* 9:639–48

91. Turner SR, Hall M. 2000. The *gapped xylem* mutant identifies a common regulatory step in secondary cell wall deposition. *Plant J.* 24:477–88

92. Turner SR, Somerville CR. 1997. Collapsed xylem phenotype of *Arabidopsis* identifies mutants deficient in cellulose deposition in the secondary cell wall. *Plant Cell* 9:689–701

93. Uggla C, Mellerowicz EJ, Sundberg B. 1998. Indole-3 acetic acid controls cambial growth in Scots pine by positional signaling. *Plant Physiol.* 117:113–21

94. Uggla C, Moritz T, Sandberg G, Sundberg B. 1996. Auxin as a positional signal

in pattern formation in plants. *Proc. Natl. Acad. Sci. USA* 93:9282–86

95. Waites R, Hudson A. 1995. *phantastica*: a gene required for dorsiventrality of leaves in *Antirrhinum majus*. *Development* 121:2143–54

96. Waites R, Selvadurai HRN, Oliver IR, Hudson A. 1998. The *PHANTASTICA* gene encodes a MYB transcription factor involved in growth and dorsiventrality of lateral organs in *Antirrhinum*. *Cell* 93: 779–89

97. Whetten RW, Mackay JL, Sederoff RR. 1998. Recent advances in understanding lignin biosynthesis. *Annu. Rev. Plant Physiol. Plant Mol. Biol.* 49:585–609

98. Yamamoto R, Demura T, Fukuda H. 1997. Brassinosteroids induce entry into the final stage of tracheary element differentiation in cultured *Zinnia* cells. *Plant Cell Physiol.* 38:980–83

99. Yamamoto R, Fujioka S, Demura T, Takatsuto S, Yoshida S, Fukuda H. 2001. Brassinosteroid levels increase drastically prior to morphogenesis of tracheary elements. *Plant Physiol.* 125:556–63

100. Ye Z-H, Droste DL. 1996. Isolation and characterization of cDNAs encoding xylogenesis-associated and wounding-induced ribonucleases in *Zinnia elegans*. *Plant Mol. Biol.* 30:697–709

101. Ye Z-H, Varner JE. 1996. Induction of cysteine and serine proteases during xylogenesis in *Zinnia elegans*. *Plant Mol. Biol.* 30:1233–46

102. Zhao C, Johnson BJ, Kositsup B, Beers EP. 2000. Exploiting secondary growth in *Arabidopsis*. Construction of xylem and bark cDNA libraries and cloning of three xylem endopeptidases. *Plant Physiol.* 123:1185–96

103. Zhong R, Taylor JJ, Ye Z-H. 1997. Disruption of interfascicular fiber differentiation in an *Arabidopsis* mutant. *Plant Cell* 9:2159–70

104. Zhong R, Taylor JJ, Ye Z-H. 1999. Transformation of the collateral vascular bundles into amphivasal vascular bundles in an *Arabidopsis* mutant. *Plant Physiol.* 120:53–64

105. Zhong R, Ye Z-H. 1999. *IFL1*, a gene regulating interfascicular fiber differentiation in *Arabidopsis*, encodes a homeodomain-leucine zipper protein. *Plant Cell* 11:2139–52

106. Zhong R, Ye Z-H. 2001. Alteration of auxin polar transport in the *Arabidopsis ifl1* mutants. *Plant Physiol.* 126:549–63

Annu. Rev. Plant Biol. 2002. 53:203–24
DOI: 10.1146/annurev.arplant.53.100301.135256

LOCAL AND LONG-RANGE SIGNALING PATHWAYS REGULATING PLANT RESPONSES TO NITRATE

Brian G. Forde

Department of Biological Sciences, Lancaster University, Lancaster, LA1 4YQ, United Kingdom; e-mail: b.g.forde@lancaster.ac.uk

Key Words   auxin, cytokinin, lateral roots, leaf expansion, nitrate transport

■ **Abstract**   Nitrate is the major source of nitrogen (N) for plants growing in aerobic soils. However, the NO_3^- ion is also used by plants as a signal to reprogram plant metabolism and to trigger changes in plant architecture. A striking example is the way that a root system can react to a localized source of NO_3^- by activating the NO_3^- uptake system and proliferating lateral roots preferentially within the NO_3^--rich zone. That roots are able to respond autonomously in this fashion implies the existence of local signaling pathways that are sensitive to local changes in the external NO_3^- concentration. On the other hand, long-range signaling pathways are also needed to modulate these responses according to the plant's N status and to coordinate the allocation of resources between the root and the shoot. This review examines these signaling mechanisms and their interactions with sugar-sensing and hormonal response pathways.

CONTENTS

INTRODUCTION

As sessile organisms, plants have developed sophisticated mechanisms that enable them to modify their phenotype in response to changing environmental conditions. One important function of this phenotypic plasticity is to allow the plant to adapt its metabolism and morphology to accommodate changes in the availability and distribution of essential resources (50). Photomorphogenesis is a prime example

1040-2519/02/0601-0203$14.00

of this (1), but plants are also highly responsive to variations in the supply of nutrients, particularly nitrogen (36, 88).

Nitrogen is the mineral nutrient that most frequently limits plant growth, and NO_3^- is the major N source available in aerobic soils (66). However, concentrations of NO_3^- in the soil solution can vary by several orders of magnitude, both seasonally and from place to place within the soil, even over short distances (87). Evidence has accumulated in recent years that the NO_3^- ion is used by plants as a signal to trigger changes in metabolism and development in response to the fluctuating availability of their primary source of N (16, 84, 109, 124). Access to mutants and transgenic lines defective in NO_3^- assimilation and the plant's ability to use alternative N sources (such as ammonium and glutamine) have been instrumental in helping experimentalists make the critical distinction between the nutritional role of NO_3^- and its actions as a signal molecule.

At the physiological level, the NO_3^- ion can induce both the NO_3^- assimilatory pathway and the reprogramming of C metabolism to provide reductants and C skeletons for this pathway (16, 84, 109). A recent microarray analysis identified over 40 NO_3^--induced genes in *Arabidopsis* seedlings, many of which encode enzymes of C and N metabolism (132). At the developmental level, NO_3^- is an important regulator of processes that include root branching (36, 63), leaf expansion (68, 130), and the allocation of resources between shoot and root growth (103).

An essential aspect of NO_3^- signaling in higher plants is the existence of both local and long-range signaling pathways. This is best seen in the response of roots to localized supplies of NO_3^-. When roots encounter a NO_3^--rich patch of soil, there is a rapid physiological response, consisting of a localized and transient increase in NO_3^- uptake activity as well as a slower developmental response in the form of higher rates of lateral root proliferation within the patch (88). Ecologically, these localized responses are important because they can determine a plant's effectiveness at competing with its neighbors for limiting and heterogeneously distributed supplies of N (48, 89, 90). However, any such localized responses must be integrated with events elsewhere in the plant. Thus the localized developmental response to a NO_3^--rich patch is most intense when the N status of the whole plant is low (29, 38), implying the existence of shoot-to-root signals that can communicate changes in the N status of the shoot.

This review focuses on what is currently known about the local and long-range signaling pathways that regulate the NO_3^- uptake system in the root and the NO_3^--dependent changes in plant growth and development. Much of the emphasis is on *Arabidopsis* because it is here that a great deal of the recent progress in understanding NO_3^- signaling pathways has been made. A number of recent reviews provide detailed coverage of various aspects of the regulation of plant metabolism and development by NO_3^- and other forms of N (14, 15, 35, 43, 69, 109, 139). The posttranslational regulation of nitrate reductase by reversible phosphorylation and interactions with 14-3-3 proteins, an exciting aspect of the regulation of the NO_3^- assimilatory pathway (65), falls outside the scope of this review.

REGULATION OF NITRATE UPTAKE

Although the whole pathway of NO_3^- assimilation is highly regulated (122), the single most important regulatory step seems to be the process of NO_3^- influx across the plasma membrane of root cortical and epidermal cells (23, 44). Nitrate influx is an active process, driven by the H^+ gradient and catalyzed by a combination of high- and low-affinity transport systems (HATS and LATS, respectively) (34, 39, 43). The NO_3^- uptake system is inducible by the external NO_3^- supply and feedback regulated by the products of N assimilation (34, 39, 43). That this regulation involves both local and long-range signaling pathways is well established from "split-root" experiments in which the supply of NO_3^- to different parts of a root system can be experimentally manipulated: If one half of a split-root system is deprived of NO_3^- (−N) while the other half remains well supplied with NO_3^- (+N), then the NO_3^- uptake activity of the +N half is upregulated, despite the absence of any change in the external NO_3^- supply to those roots (40, 57, 58). On the other hand, induction of the uptake system depends on the local presence of external NO_3^- (75).

Regulation by the External Nitrate Supply

NITRATE INDUCTION OF NITRATE TRANSPORTER GENES   Physiological studies established that one component of the HATS for NO_3^-, the iHATS (inducible high-affinity transport system), is strongly induced by NO_3^- (34, 43). The iHATS is now known to be a multicomponent system that is partly encoded by genes of the *NRT2* family (34, 43), which has seven members in *Arabidopsis* (18, 77). One of the *Arabidopsis* genes (*AtNRT2.1*) has been characterized in detail and shown to make a major contribution to the iHATS in this species (11, 33, 40, 61, 141). In addition, the *Arabidopsis AtNRT1.1* (*CHL1*) gene encodes a dual-affinity NO_3^- transporter that contributes significantly to the iHATS under certain nutritional conditions (133).

All *NRT2* genes tested thus far (and *AtNRT1.1*) are expressed at very low levels in roots in the absence of NO_3^- and are rapidly induced (within minutes) when NO_3^- is resupplied (18, 34, 77). The induction kinetics of the NO_3^--inducible *NRT2* and *NRT1* genes are very similar to those of genes known to be induced in the primary response to NO_3^-. Primary response genes are those whose transcriptional induction is independent of prior protein synthesis, and established primary response genes for NO_3^- include the *NIA* gene for NR in maize (45) and the maize genes for ferredoxin-dependent GOGAT and the plastidic form of glutamine synthetase (85). In the case of the *NIA* gene, the NO_3^- ion, and not a downstream metabolite, has been shown to provide the inductive signal (25, 81), and experiments with NR-deficient mutants of *Arabidopsis* and *Nicotiana plumbaginifolia* suggest the same is true for the induction of the NO_3^- transporter genes (55, 61).

THE SIGNAL TRANSDUCTION PATHWAY FOR NITRATE INDUCTION OF GENE EXPRES-
SION   Redinbaugh & Campbell (84) proposed the existence of a constitutively
expressed NO_3^- response pathway consisting of a NO_3^- sensor and downstream
NO_3^- regulatory proteins responsible for the transcriptional activation of the primary response genes. A decade later, there is little concrete information about any
of the components of this signaling pathway. The proposed NO_3^- sensor is thought
to be most likely located on the external face of the plasma membrane (35, 84). The
strongest evidence in favor of this is the observation that *NIA* transcript levels in
barley roots are highly responsive to changes in external NO_3^- supply (114), even
though tissue NO_3^- concentrations change only slowly and cytosolic NO_3^- concentrations are strongly buffered by the vacuolar NO_3^- pool (114, 127). Nonetheless,
it is also possible that an intracellular NO_3^- sensor could be sensitive to small
fluctuations in the cytosolic NO_3^- concentration.

In bacteria, NO_3^- signaling is via a His-to-Asp phosphorelay system, and external NO_3^- is sensed by means of a pair of transmembrane histidine kinases specified
by the *NarQ* and *NarX* genes (108). Related His-to-Asp phosphorelay systems operate in cytokinin and ethylene signaling in plants (12, 72), but although there are
data suggesting that phosphorylation reactions (and Ca^{2+} ions) may be involved
in NO_3^- induction of gene expression (97, 115), there is still no direct evidence for
a His-to-Asp phosphorelay pathway for NO_3^- sensing in plants.

Other candidates for the role of NO_3^- sensor can be envisaged. In yeast, there
is a class of membrane transporters that are used to monitor the availability of
key nutrients in the medium. For example, Mep2p, one of three ammonium transporters in yeast, acts as a sensor for environmental NH_4^+ in regulating the initiation
of the process of pseudohyphal differentiation (64). Thus NO_3^- sensing must be
considered as a possible role for one or more of the seven NRT2 proteins encoded
in the *Arabidopsis* genome (18, 77).

The plasma membrane–bound form of NR (PM-NR), which is quite distinct
from the cytosolic form and is located on the outer surface of the plasma membrane
(56, 112, 134), has also been discussed as a possible NO_3^- sensor (see Ref. 35).
Recently a novel plasma membrane–bound enzyme activity that catalyzes the reduction of NO_2^- to nitric oxide (NO) was reported in tobacco roots (113). Stöhr
et al. suggest that this nitrite:NO-reductase might act in concert with PM-NR to
convert external NO_3^- to NO, which could readily pass through the plasma membrane and act as an intermediary in NO_3^- signaling. Nitric oxide is well established
in mammals as an important signal molecule and is now recognized as having
diverse regulatory roles in plants (3). Although a possible role in NO_3^- signaling
was first suggested a number of years ago (6), as yet there is no evidence that NO
itself is able to induce the expression of NO_3^--inducible genes.

The NO_3^- sensing mechanism in fungi is similarly obscure, but other components of the N regulatory system are well characterized. In *Aspergillus nidulans* and *Neurospora crassa*, NO_3^- induction is mediated by the pathway-specific
positive-control genes, *nirA* and *nit-4*, respectively (17). N-metabolite repression
is mediated by the *areA* and *nit-2* genes, which encode Zn^{2+}-finger DNA-binding
proteins of a type that recognizes a GATA sequence motif. GATA-binding proteins

might have a role in NO_3^- regulation in higher plants (83). An in vivo footprinting technique detected an NO_3^--inducible DNA-binding activity in transgenic tobacco leaves that interacted with sequences within a region of the spinach *NII* promoter thought to be involved in NO_3^- induction and containing two adjacent GATA motifs (83). An in vivo footprinting technique detected a NO_3^--inducible DNA-binding activity in tobacco leaves that recognized sequences within a region of the promoter thought to be involved in NO_3^- induction and containing two adjacent GATA motifs (83). An alternative candidate for a "nitrate box" was identified in the *Arabidopsis NIA1* and *NIA2* promoters (51). When this 12-bp conserved sequence is mutated, the NO_3^- inducibility of the *NIA1* and *NIA2* promoters is lost. Sequences related to the nitrate box $[(a/t)_7 Ac/gTCA]$ have been identified in the promoter-proximal regions of a wide range of NO_3^--regulated genes, including one within a putative NO_3^--regulatory domain in the spinach *NII* promoter (73), in the region required for the NO_3^- response of the birch *NII* gene (135), and in the upstream sequences of three *NRT2* genes (129). In vitro studies indicated the presence of nuclear proteins in tobacco leaves and nonphotosynthetic suspension cultures that bind in a sequence-specific manner to the regions of the *NIA1* and *NIA2* promoters containing these motifs (51). This binding activity was not dependent on prior treatment of the tissues with NO_3^- (51). Constitutive expression of the binding activity is consistent with the evidence that protein synthesis is not required for NO_3^- induction of NR (45). One model proposed was that the binding factor might be modified by the NO_3^- signal in a way that does not affect its DNA-binding properties, but only its ability to activate transcription (51). A recent preliminary report suggests that the yeast one-hybrid system has successfully been used to clone the gene for a DNA-binding protein that recognizes the nitrate box motif (105). If a role for this protein in NO_3^- induction of gene expression can be demonstrated, it would provide our first real insight into the NO_3^- primary response pathway.

Long-Range Signals

FEEDBACK REGULATION BY THE N STATUS OF THE PLANT   The NO_3^- uptake system is also feedback regulated according to the internal N status of the plant, a means of coordinating uptake rates with the plant's demand for N (35, 123). The feedback regulation of *NRT2* genes by the products of NO_3^- assimilation closely parallels changes in HATS activity (40, 61, 128). The use of inhibitors of N assimilation suggested that feedback repression of *NRT2* genes in *Arabidopsis* and barley may be mediated by both NH_4^+ and glutamine (or other amino acids) (128, 141).

A recent study using split-roots demonstrated that the feedback regulation of the *AtNRT2.1* gene was dependent on long-range signals from the shoot and occurred even under moderate N limitation (40). This pattern of regulation contrasted markedly with that of another gene that is derepressed by N deficiency: The expression of the *AtAMT1.1* NH_4^+ transporter gene was only upregulated under severe N stress, and it responded to the local availability of N in the root and not to long-range signals from the shoot (40). Thus it appears that both local and long-range feedback regulatory pathways operate in *Arabidopsis* roots but that the long-range controls dominate with respect to regulation of *NRT2* genes. The ability of amino acids to inhibit NO_3^- uptake in maize suspension cultures indicates that long-distance signals do not have an exclusive role in the feedback regulation of NO_3^- uptake (78). Surprisingly, the *AtNRT1.1* gene for the dual-affinity NO_3^- transporter does not appear to be susceptible to any kind of feedback regulation by N metabolites (32, 61).

The feedback regulation we have been discussing has been at the level of *NRT2* mRNA abundance. However, regulation may also be occurring at the translational or posttranslational levels (37, 128). In transgenic tobacco plants expressing the *NpNRT2.1* gene under the control of a constitutive promoter, NO_3^- influx was still repressed by the presence of external NH_4^+, even though expression of the transgene at the mRNA level was maintained (37). Western blot analysis has provided evidence that the barley HvNRT2.1 protein is posttranslationally modified in a way that is dependent on the presence or absence of reduced forms of N (NH_4^+ or amino acids) in the external medium (M. Hansen & B.G. Forde, unpublished observations). Whether long-range signals are involved in this regulatory mechanism is unknown.

HOW ARE CHANGES IN THE NITROGEN STATUS OF THE SHOOT COMMUNICATED TO THE ROOT?   Phloem-borne amino acids have been prominent candidates for the role of shoot-derived signal for feedback regulation of NO_3^- uptake (13, 52). Rapid cycling of amino acids between the shoot and the root occurs, so that changes in the N status of the shoot should rapidly be reflected in the rate of delivery of amino acids to the root (13, 52). However, contrary to the requirements of this model, N deficiency can sometimes lead to an increase rather than a decline in amino acid cycling (e.g., 80). Furthermore, split-root studies on mung bean (*Ricinus communis*) found no correlation between rates of NO_3^- uptake and either quantitative or qualitative changes in the amino acid content of the phloem (121). Further evidence suggesting the existence of long-distance signals other than amino acids was obtained from studies with mung bean plants carrying *Agrobacterium*-induced stem tumors (70). Nitrate uptake in the tumorized plants was markedly inhibited, even though there were no qualitative differences in the amino acid composition of the phloem sap compared to control plants, and the overall amino acid concentration in the phloem sap was actually lower than in the controls (70). Although an observed parallel decline in the sucrose content of the phloem could provide one explanation for the reduction in NO_3^- uptake rates (see below), Mistrik et al. also considered the possibility that plant hormones (ethylene, abscisic acid, and auxin) could play a role (70). The latter suggestion has been given increased credence by a recent report that the *AtNRT1.1* gene is auxin-regulated in *Arabidopsis* roots (19) (see below).

The phloem is not only a conduit for metabolites and hormones; it also carries proteins and other macromolecules with the potential to act as highly specific long-range signals (95, 119). A 31-kDa protein (ΔN-CmPP36) present in the phloem sap of *Cucurbita maxima* and related to cytochrome b_5 reductases may be involved in long-distance signaling of the Fe status of the shoot (137). Thus we must keep an open mind to the possibility that novel phloem-mobile signals, such as peptides or RNA molecules, could communicate information to the root about the N status of the shoot.

A possible component of the machinery responsible for monitoring the N status of the shoot has been identified in the form of the *Arabidopsis GLB1* gene (49). *GLBI* is a nuclear gene encoding a chloroplast protein related to the bacterial PII family of regulatory proteins that play a key role in the coordination of prokaryotic N and C metabolism (74). Although there is evidence that this PII-like protein is involved in regulating anthocyanin production in response to changes in the N and C supply (49), a wider role for *GLB1* in regulating other metabolic or developmental responses to the N/C balance has yet to be demonstrated.

DIURNAL REGULATION AND SUGAR SIGNALING   Diurnal regulation of NO_3^- uptake has been reported in a number of plant species, including *Arabidopsis*, with activities generally being highest during the light period and lowest in the dark (24, 60, 61, 67, 79), and these diurnal fluctuations in uptake rate are generally correlated with changes in *NRT1* and *NRT2* transcript levels (61, 67, 76). In *Arabidopsis*, both the *AtNRT2.1* and *AtNRT1.1* genes are diurnally regulated in a similar manner to the HATS activity (61). In each case, the decline in transcript levels during the dark period could be delayed by supplying sucrose to the medium (61). The finding that *AtNRT2.1* and *AtNRT1.1* are both positively regulated by the sucrose supply but are differentially responsive to feedback regulation by reduced forms of N indicates that the effect of sucrose is not mediated through its influence on the overall N/C balance of the plant (59). A role for sucrose in regulating expression of NO_3^- transporter genes was also surmised from studies with tobacco, which showed that diurnal changes in *NRT2* expression were correlated with changes in the sugar content of the root, whereas there were no major fluctuations in the overall amino acid content of the root (67).

Sucrose acts as a signal regulating the expression of a variety of plant genes, and there is evidence for both sucrose-specific and glucose-specific signal transduction pathways operating in plants (14, 107). Thus it seems likely that diurnal fluctuations in the supply of sugars from the shoot are acting as long-range signals to regulate expression of the NO_3^- transporter genes, so helping to coordinate the processes of NO_3^- uptake and photosynthesis (111). A model for the regulation of *AtNRT2.1* and *AtNRT1.1* expression by local and long-range signaling pathways is presented in Figure 1.

REGULATION OF GROWTH AND DEVELOPMENT

Local Signals Regulating Lateral Root Growth

LOCALIZED STIMULATION OF LATERAL ROOT ELONGATION   When the root systems of many plant species are exposed to a localized supply of NO_3^-, they respond by increasing the rate of lateral root proliferation specifically within the NO_3^--rich

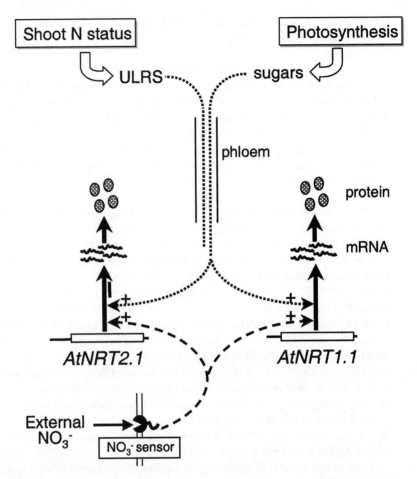

Figure 1 Local and long-range signaling pathways regulating expression of NO_3^- transporter genes in *Arabidopsis*. Translational and posttranslational controls, which are known to exist (37, 128), are not depicted because of insufficient information. ULRS, unidentified long-range signal. See text and Ref. (40) for further details.

zone (88). In barley, this response also occurred with ammonium and inorganic phosphate and consisted of an increase in both the rate of lateral initiation and the rate of lateral root elongation (27–29). When *Arabidopsis* seedlings were grown on vertical agar plates and presented with a localized NO_3^- supply, they showed a similar response, although in this case the effect was restricted to a two- to threefold increase in elongation rates with no increase in lateral root numbers (138, 140). The increased elongation rates were due to enhanced meristematic activity in the lateral root tips (140).

The signal for increased meristematic activity in the lateral roots appears to come from the NO_3^- ion, and not from a product of NO_3^- assimilation (138, 140),

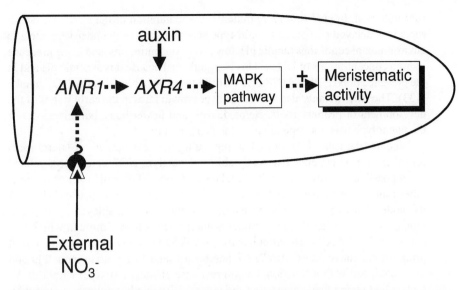

Figure 2   Local signal transduction pathway for NO_3^- stimulation of lateral root elongation. The model has been updated from one published previously (140).

implying that cells in the lateral root tips are equipped with NO_3^- sensor and a signal transduction pathway to convert the NO_3^- signal into a growth response (Figure 2). Two genes encoding possible components of this signal transduction pathway have been identified. The first (*ANR1*) is a NO_3^--regulated member of the MADS-box family of transcription factors (138). When *ANR1* expression was downregulated by antisense or cosuppression, the resulting transgenic lines no longer displayed the lateral root response to localized NO_3^- supplies (138). The second (*AXR4*) is a gene first identified as an auxin-sensitivity gene with an important role in root gravitropism (47). Unlike other auxin-resistant mutants, *axr4* mutants have no obvious shoot phenotype and no cross-resistance to other plant hormones (47). However, like the *ANR1* downregulated lines, an *axr4* mutant failed to respond to a localized NO_3^- treatment (140), suggesting an overlap between the auxin and the NO_3^- response pathways in the regulation of lateral root growth (71). Auxin signaling in *Arabidopsis* roots involves a mitogen-activated protein kinase (MAPK) signaling pathway, and the activation of this pathway by auxin is dependent on *AXR4* (71). A tentative model integrating the possible components of a signaling pathway for NO_3^- regulation of lateral root elongation is shown in Figure 2.

LATERAL ROOT OUTGROWTH AND THE AtNRT1.1 NITRATE TRANSPORTER   A recent report has unexpectedly implicated the AtNRT1.1 dual-affinity NO_3^- transporter (see above) in the regulation of lateral root development (46). Surprisingly for a NO_3^- transporter gene, *AtNRT1.1* was most strongly expressed in the primary and lateral root tips and in young leaves and developing flower buds. In the developing lateral root, its expression was detectable even in the earliest stages after lateral root initiation. In *chl1* (*atntr1.1*) mutants, the maturation of the emerging lateral roots was delayed compared to wild-type seedlings, but only under very specific environmental conditions (acidic pH, low NO_3^- concentrations, and in the presence of high concentrations of NH_4^+). This is significant because it is at acidic pH and in the presence of NH_4^+ that AtNRT1.1 contributes most to high-affinity NO_3^- uptake (133). The *chl1* mutants also showed slower growth rates in the early stages of the development of primary roots, stems, leaves, and flower buds, but this aspect of the phenotype was not dependent on the NO_3^- supply (46).

Whether the role of AtNRT1.1 in regulating the development of lateral roots (or other organs) is dependent on its ability to catalyze NO_3^- uptake or is related to its possible contribution to the regulation of intracellular pH (92) or to some other function in transport or signaling (46), has yet to be established. Whatever its mode of action, the defect in lateral root maturation displayed by the *chl1* mutants indicates that there is a hitherto unreported pathway through which low concentrations of NO_3^- can stimulate the growth and maturation of the lateral root primordium. Intriguingly, *AtNRT1.1* has been found to be auxin-inducible and to be equipped with a functional auxin response element (AuxRE) within its 5′-transcribed region (19), suggesting the potential for another point of intersection between auxin and NO_3^- response pathways.

Long-Range Signals Regulating Root Branching

SHOOT CONTROL OF LATERAL ROOT DEVELOPMENT   Drew and colleagues demonstrated the importance of the local NO_3^- supply in determining the rate of lateral root proliferation (27–29), but it was also apparent that the NO_3^- supply to the remainder of the root system had a significant influence on the intensity of the localized response (29). Experiments with split-root systems have confirmed the existence of systemic controls on NO_3^--induced root branching and provided strong evidence that the regulatory signals arise in the shoot (38, 58, 102).

The importance of NO_3^- accumulation in the shoot as a signal regulating root growth was demonstrated using tobacco transformants expressing very low levels of NR (103, 110). Under conditions where the low NR lines accumulated high concentrations of NO_3^- in the shoot, the amino acid and protein content remained low, but there was a major shift in the shoot:root ratio and a marked decrease in lateral root density. Although changes in carbon allocation to the root were observed in tobacco, these could not account for the effect on root branching (110). Similar effects have been noted in wild-type *Arabidopsis* seedlings, where external NO_3^- concentrations of 10 mM or more strongly inhibited the outgrowth (but not the initiation) of lateral roots (138–140). Confirming the importance of tissue NO_3^- concentrations for this effect, lateral root outgrowth in an NR-deficient mutant was found to be more sensitive to the high NO_3^- treatment than in the wild type (140).

POSSIBLE ROLE OF AUXIN   As a major plant growth regulator that is synthesised in shoots and transported to roots, auxin is likely to play a key role in shoot-to-root communication (4). Auxin is important for both lateral root initiation and the later stages of lateral root development (9, 10), and inhibiting auxin transport from the shoot with N-1-naphthylphthalamic acid (NPA) can markedly reduce the number of emerged lateral roots (86). A number of components of an auxin signaling pathway that regulates lateral root development have been identified (e.g., 93, 120, 136). Thus one obvious way in which the NO_3^- status of the shoot could generate a long-distance signal to regulate root branching would be through modulating the auxin supply to the root. Unfortunately the evidence that this occurs is still largely absent. In soybean, a high external NO_3^- supply (8 mM) did cause a 70% decrease in the auxin content of the root (8), but in this case, no effects on root branching were observed. Thus, although auxin is a plausible long-distance signal mediating the inhibitory effect of NO_3^- on lateral root development, this hypothesis still needs to be properly tested. Figure 3 suggests two possible models

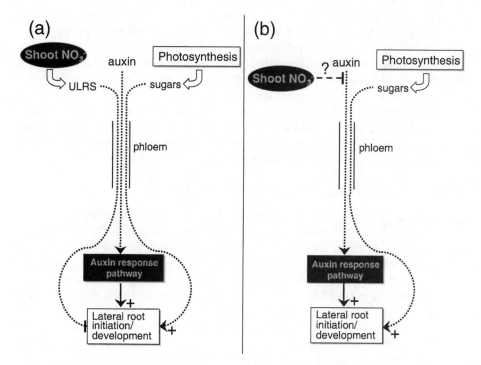

Figure 3 Long-range signaling pathways for the regulation of root branching. Two alternative models for the interaction between the auxin and NO_3^- signaling pathways are shown: (*a*) A phloem-mobile ULRS (unidentified long-range signal) generated by the NO_3^- status of the shoot inhibits lateral root initiation/development. (*b*) Auxin is the long-distance (positive) signal, and the accumulation of NO_3^- in the shoot inhibits auxin synthesis (or auxin export to the root), resulting in reduced root branching. In each case, sugars transported from the shoot act in parallel as additional signals, providing a means of integrating root branching with the carbon status of the shoot.

for the long-range regulation of root branching and the interactions between NO_3^- and auxin signaling.

INTERACTIONS WITH SUGAR SIGNALING   There is evidence in *Arabidopsis* (20) and in wheat (5) [although not in tobacco (110)] that increased supplies of carbohydrates to the roots, whether delivered through the phloem or by external supply, are correlated with increased lateral root densities. Whether these reported effects were due to the roots being carbon-limited or whether sugars were acting as signals to regulate lateral root initiation is not clear. However, in a study where the external concentrations of both NO_3^- and sucrose were varied, sucrose stimulated lateral root development most strongly at the high NO_3^- concentrations that were inhibitory to lateral root outgrowth (140). Because the high NO_3^- concentrations were specifically inhibitory to lateral root maturation, and not to the growth of mature laterals or primary roots, it seemed unlikely that the effect could be due to carbon limitation. The most plausible alternative is that sucrose (or a sucrose metabolite) has a signaling role in the control of lateral root development.

Recently, De Veylder et al. identified a sucrose-regulated member of the D-type cyclins that is expressed during the formation of lateral root primordia (26). Given the known role of D-type cyclins in regulating the onset of cell division (104), this could be the basis for another long-distance signaling pathway in which sugars delivered through the phloem modulate the rate of lateral root development (Figure 3). A role for sugar signaling in the control of biomass partitioning between roots and shoots has been proposed (31), and the regulation of root branching by such a pathway could provide one way for this control to operate.

Long-Range Signals Regulating Leaf Expansion

One of the most commonly observed responses to nutrient limitation is a shift in the allocation of resources from shoot growth to root growth (30), enabling the plant to concentrate its available resources on exploring the soil volume. Growth responses to changes in the nutrient supply can be very rapid, suggesting an efficient mechanism for root-to-shoot signaling. Leaf expansion is particularly sensitive to fluctuations in the NO_3^- supply: When the roots are deprived of NO_3^-, there is a rapid decline in the rate of leaf expansion, which can be quickly reversed by resupplying NO_3^- (68). Effects on both cell division and cell size are reported to contribute to the reduction in final leaf size in NO_3^--deprived mung bean and tobacco plants (91, 130). A number of recent developments have thrown new light on the possible nature of the signaling pathway responsible for this example of long-range growth regulation.

In tobacco, the rapid inhibition of leaf expansion that occurred in NO_3^--deprived plants could not be prevented by replacing the NO_3^- with NH_4^+ (130, 131). Physiological analysis of the NO_3^-- and NH_4^+-supplied plants detected no differences in the N or C status of the plants, nor in their osmotic potentials or rates of water uptake, that could account for the effect on leaf growth. Furthermore, there was no evidence of NH_4^+ toxicity. However, growth inhibition in the NH_4^+-fed plants was correlated with a sharp reduction in the concentration of Z-type cytokinins (zeatin and zeatin riboside) in the xylem sap (130). Similar correlations between the N supply and cytokinin biosynthesis have been reported previously for other species (e.g., 100, 101), and in maize, NO_3^- (or ammonium) treatment of N-starved plants stimulates cytokinin biosynthesis in the root and, within 4 h, increases fluxes of cytokinins to the shoot (117). Because cytokinins are important regulators of cell division and growth (22), it seems plausible that these changes in cytokinin production could provide the long-range signals that regulate leaf morphogenesis in accordance with changes in the NO_3^- supply (2, 130).

How changes in the NO_3^- supply are linked to changes in cytokinin metabolism and the identity of the key regulatory step in the cytokinin biosynthetic pathway are still unknown. However, at the other end of the signaling pathway—in the shoot—a possible signal transduction pathway for the perception and response to the cytokinin signal has emerged. In leaves of maize and *Arabidopsis*, a group of response regulator genes (*ZRR* and *ARR*, respectively) have been identified that are among the few genes known to be induced in the primary response to cytokinins (7, 21, 98). Response regulators are components of His-to-Asp phosphorelay signal transduction pathways [see above and Ref. (99)]. The cytokinin-inducible genes belong to the type-A family of response regulators that lack the C-terminal output domain characteristic of the type-B family (125), and their expression can be induced in leaves by either the direct application of cytokinins or by resupplying NO_3^- to roots of N-starved plants (118). It has been proposed that the type-A response regulator genes are part of a cytokinin-mediated signaling pathway that enables gene expression in leaves to be modulated by the inorganic N supply to the roots (99). The finding that one of the cytokinin-regulated genes (ARR5) is expressed in shoot apical meristems (21) would be consistent with a possible role in regulating leaf expansion in response to the NO_3^- supply.

Arabidopsis has a family of histidine kinases belonging to the same His-to-Asp signaling system as the response regulators (126), and one of these (CRE1) has been identified as a cytokinin receptor (53), and two others (CKI1 and AHK) are candidates for additional cytokinin receptors (54, 116). As yet, no formal connection between these histidine kinases and the type-A response regulators has been made, but potentially one or more of them could be components of a NO_3^--cytokinin signaling pathway as depicted in Figure 4.

CONCLUDING REMARKS

Many fundamental questions about the NO_3^- signaling pathways discussed here remain unanswered. For example, we still have little idea how the root senses external NO_3^-, nor how the NO_3^- signal is transduced into the transcriptional activation of the primary response genes. Progress is being made toward identifying potential transcription factors that bind to *cis*-acting elements upstream of NO_3^--inducible genes (51, 105, 106). However, these biochemical approaches will need to be complemented by genetics. As yet, no clear NO_3^- regulatory mutants have been identified, but the availability of transgenic *Arabidopsis* plants carrying reporter genes

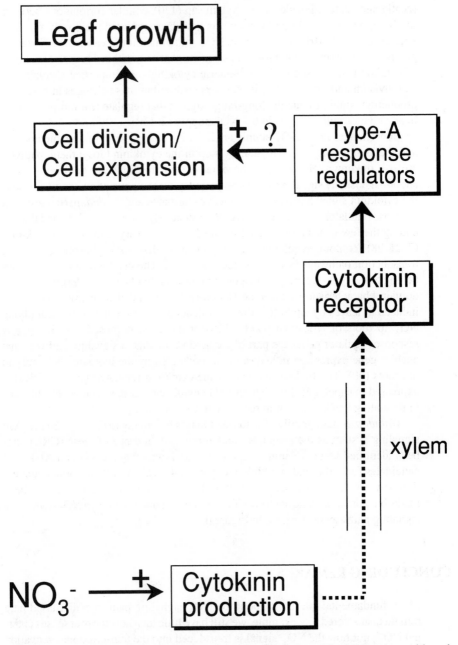

Figure 4   Long-range signaling pathway for NO_3^- regulation of leaf expansion. Although NO_3^- and (in maize) NH_4^+ stimulate the fluxes of Z-type cytokinins to the shoot and trigger expression of the cytokinin-inducible type-A response regulators (see text), that these response regulators have any role in regulating leaf expansion has not been demonstrated.

under the control of NO_3^--inducible promoters (62) opens new possibilities for mutant screens. Now that a number of NO_3^--regulated transcription factors are known (132, 138), the identification of their target genes becomes feasible (see 96), offering the potential for new insights into how NO_3^- signals are converted into physiological and developmental responses. New mutant screens based on the strong inhibition of lateral root growth in *Arabidopsis* seedlings grown on high NO_3^- (140) have the potential to identify genes involved in sensing NO_3^- accumulation in the shoot as well as those controlling the generation and perception of the long-range signal(s) (Figure 3).

As has become increasingly apparent, signal transduction cascades do not operate in isolation, but rather as parts of a network or matrix of interacting pathways (15, 41, 42). In view of the close interrelationships between C and N metabolism, and the need for coordination between them, it is not surprising that there should be extensive cross-talk between NO_3^- signaling and sugar-sensing pathways (Figures 1 and 3). The pervasiveness of the network into which NO_3^- signaling pathways are integrated is further emphasized by the extent of their interactions within auxin and cytokinin response pathways (Figures 2–4): Between them, auxins and cytokinins regulate almost every aspect of plant growth and development. There is also evidence of cross-talk between NO_3^- signaling and S and P nutrition (82, 94, 103), and it seems inevitable that as our understanding of NO_3^- signaling pathways progresses many more such interactions will be uncovered. Nevertheless, in the aftermath of the sequencing of the *Arabidopsis* genome, and with the myriad tools of molecular genetics and proteomics now at our disposal, we can be optimistic that we have entered an exciting new era where answers to some of the key questions posed in this review will soon be forthcoming.

ACKNOWLEDGMENTS

The author thanks Alistair Hetherington for stimulating discussions, Tony Miller and Sophie Filleur for their comments on earlier versions of the manuscript, and Nigel Crawford for access to unpublished data. Work in the author's laboratory is supported by the Biotechnology and Biological Research Council and by Norsk Hydro ASA.

Visit the Annual Reviews home page at www.annualreviews.org

LITERATURE CITED

1. Ballaré CL. 1999. Keeping up with the neighbours: phytochrome sensing and other signalling mechanisms. *Trends Plant Sci.* 4:97–102

2. Beck EH. 1996. Regulation of shoot/root ratio by cytokinins from roots in *Urtica dioica*: opinion. *Plant Soil* 185:3–12

3. Beligni MV, Lamattina L. 2001. Nitric oxide in plants: the history is just beginning. *Plant Cell Environ.* 24:267–78

4. Berleth T, Sachs T. 2001. Plant morphogenesis: long-distance coordination and local patterning. *Curr. Opin. Plant Biol.* 4:57–62

5. Bingham IJ, Blackwood JM, Stevenson EA. 1998. Relationship between tissue sugar content, phloem import and lateral root initiation in wheat. *Physiol. Plant* 103:107–13

6. Bloom AJ. 1997. Interactions between inorganic nitrogen nutrition and root development. *Z. Pflanz. Bodenk.* 160:253–59

7. Brandstatter I, Kieber JJ. 1998. Two genes with similarity to bacterial response regulators are rapidly and specifically induced by cytokinin in Arabidopsis. *Plant Cell* 10:1009–19

8. Caba JM, Centeno ML, Fernandez B, Gresshoff PM, Ligero F. 2000. Inoculation and nitrate alter phytohormone levels in soybean roots: differences between a supernodulating mutant and the wild type. *Planta* 211:98–104

9. Casimiro I, Marchant A, Bhalerao RP, Beeckman T, Dhooge S, et al. 2001. Auxin transport promotes Arabidopsis lateral root initiation. *Plant Cell* 13:843–52

10. Celenza JL, Grisafi PL, Fink GR. 1995. A pathway for lateral root formation in *Arabidopsis thaliana*. *Genes Dev.* 9:2131–42

11. Cerezo M, Tillard P, Filleur S, Munos S, Daniel-Vedele F, Gojon A. 2001. Alterations of the regulation of root NO_3^--uptake are associated with the mutation of *Nrt2.1* and *Nrt2.2* genes in *Arabidopsis*. *Plant Physiol.* 127:262–71

12. Chang C, Stadler R. 2001. Ethylene hormone receptor action in Arabidopsis. *BioEssays* 23:619–27

13. Cooper HD, Clarkson DT. 1989. Cycling of amino nitrogen and other nutrients between shoots and roots in cereals: a possible mechanism integrating shoot and root in the regulation of nutrient uptake. *J. Exp. Bot.* 40:753–62

14. Coruzzi G, Bush DR. 2001. Nitrogen and carbon nutrient and metabolite signaling in plants. *Plant Physiol.* 125:61–64

15. Coruzzi GM, Zhou L. 2001. Carbon and nitrogen sensing and signaling in plants: emerging "matrix effects." *Curr. Opin. Plant Biol.* 4:247–53

16. Crawford NM. 1995. Nitrate: nutrient and signal for plant growth. *Plant Cell* 7:859–68

17. Crawford NM, Arst HN. 1993. The molecular genetics of nitrate assimilation in fungi and plants. *Annu. Rev. Genet.* 27:115–46

18. Crawford NM, Forde BG. 2002. Molecular and developmental biology of inorganic nitrogen nutrition. In *The Arabidopsis Book*, ed. E Meyerowitz, C Somerville. In press (online ref.)

19. Crawford NM, Guo F-Q. 2002. Regulation of the dual-affinity nitrate transporter gene *AtNRT1.1* (*CHL1*) during vegetative and reproductive development in *Arabidopsis*. *J. Exp. Bot.* In press

20. Crookshanks M, Taylor G, Dolan L. 1998. A model system to study the effects of elevated CO_2 on the developmental physiology of roots: the use of *Arabidopsis thaliana*. *J. Exp. Bot.* 49:593–97

21. D'Agostino IB, Deruere J, Kieber JJ. 2000. Characterization of the response of the Arabidopsis response regulator gene family to cytokinin. *Plant Physiol.* 124:1706–17

22. D'Agostino IB, Kieber JJ. 1999. Molecular mechanisms of cytokinin action. *Curr. Opin. Plant Biol.* 2:359–64

23. Daniel-Vedele F, Filleur S, Caboche M. 1998. Nitrate transport: a key step in nitrate assimilation. *Curr. Opin. Plant Biol.* 1:235–39

24. Delhon P, Gojon A, Tillard P, Passama L. 1995. Diurnal regulation of NO_3^--uptake in soybean plants. I. Changes in NO_3^--influx, efflux, and N utilization in the plant during the day-night cycle. *J. Exp. Bot.* 46:1585–94

25. Deng MD, Moureaux T, Caboche M. 1989. Tungstate, a molybdate analog inactivating nitrate reductase, deregulates the expression of the nitrate reductase structural gene. *Plant Physiol.* 91:304–9

26. De Veylder L, Engler JD, Burssens S, Manevski A, Lescure B, et al. 1999. A new D-type cyclin of *Arabidopsis*

thaliana expressed during lateral root primordia formation. *Planta* 208:453–62

27. Drew MC. 1975. Comparison of the effects of a localized supply of phosphate, nitrate, ammonium and potassium on the growth of the seminal root system, and the shoot, in barley. *New Phytol.* 75:479–90

28. Drew MC, Saker LR. 1975. Nutrient supply and the growth of the seminal root system of barley. II. Localized, compensatory increases in lateral root growth and rates of nitrate uptake when nitrate supply is restricted to only part of the root system. *J. Exp. Bot.* 26:79–90

29. Drew MC, Saker LR, Ashley TW. 1973. Nutrient supply and the growth of the seminal root system in barley. I. The effect of nitrate concentration on the growth of axes and laterals. *J. Exp. Bot.* 24:1189–202

30. Ericsson T. 1995. Growth and shoot: root ratio of seedlings in relation to nutrient availability. *Plant Soil* 168–69:205–14

31. Farrar J. 1996. Regulation of root weight ratio is mediated by sucrose: opinion. *Plant Soil* 185:13–19

32. Filleur S, Daniel-Vedele F. 1999. Expression analysis of a high-affinity nitrate transporter isolated from *Arabidopsis thaliana* by differential display. *Planta* 207:461–69

33. Filleur S, Dorbe MF, Cerezo M, Orsel M, Granier F, et al. 2001. An Arabidopsis T-DNA mutant affected in *Nrt2* genes is impaired in nitrate uptake. *FEBS Lett.* 489:220–24

34. Forde BG. 2000. Nitrate transporters in plants: structure, function and regulation. *Biochim. Biophys. Acta* 1465:219–35

35. Forde BG, Clarkson DT. 1999. Nitrate and ammonium nutrition of plants: physiological and molecular perspectives. *Adv. Bot. Res.* 30:1–90

36. Forde BG, Lorenzo H. 2001. The nutritional control of root development. *Plant Soil* 232:51–68

37. Fraisier V, Gojon A, Tillard P, Daniel-Vedele F. 2000. Constitutive expression of a putative high-affinity nitrate transporter in *Nicotiana plumbaginifolia*: evidence for post-transcriptional regulation by a reduced nitrogen source. *Plant J.* 23:489–96

38. Friend AL, Eide MR, Hinckley TM. 1990. Nitrogen stress alters root proliferation in Douglas fir seedlings. *Can. J. For. Res.* 20:1524–29

39. Galván A, Fernández E. 2001. Eukaryotic nitrate and nitrite transporters. *Cell. Mol. Life Sci.* 58:225–33

40. Gansel X, Munos S, Tillard P, Gojon A. 2001. Differential regulation of the NO_3^-- and NH_4^+-transporter genes *AtNrt2.1* and *AtAmt1.1* in Arabidopsis: relation with long-distance and local controls by N status of the plant. *Plant J.* 26:143–55

41. Genoud T, Métraux J-P. 1999. Crosstalk in plant cell signaling: structure and function of the genetic network. *Trends Plant Sci.* 4:503–7

42. Gibson SI. 2000. Plant sugar-response pathways. Part of a complex regulatory web. *Plant Physiol.* 124:1532–39

43. Glass ADM, Brito DT, Kaiser BN, Kronzucker HJ, Kumar A, et al. 2001. Nitrogen transport in plants, with an emphasis on the regulation of fluxes to match plant demand. *Z. Pflanz. Bodenk.* 164:199–207

44. Gojon A, Dapoigny L, Lejay L, Tillard P, Rufty TW. 1998. Effects of genetic modification of nitrate reductase expression on $^{15}NO_3^-$- uptake and reduction in *Nicotiana* plants. *Plant Cell Environ.* 21:43–53

45. Gowri G, Kenis JD, Ingemarsson B, Redinbaugh MG, Campbell WH. 1992. Nitrate reductase transcript is expressed in the primary response of maize to environmental nitrate. *Plant Mol. Biol.* 18:55–64

46. Guo F-Q, Wang RC, Chen M, Crawford NM. 2001. The Arabidopsis dual-affinity nitrate transporter gene *AtNRT1.1 (CHL1)* is activated and functions in nascent organ development during vegetative and reproductive growth. *Plant Cell* 13:1–18

47. Hobbie L, Estelle M. 1995. The *axr4* auxin-resistant mutants of *Arabidopsis*

thaliana define a gene important for root gravitropism and lateral root initiation. *Plant J.* 7:211–20

48. Hodge A, Stewart J, Robinson D, Griffiths BS, Fitter AH. 1999. Plant, soil fauna and microbial responses to N-rich organic patches of contrasting temporal availability. *Soil Biol. Biochem.* 31:1517–30

49. Hsieh M-H, Lam H-M, Van De Loo FJ, Coruzzi G. 1998. A PII-like protein in Arabidopsis: putative role in nitrogen sensing. *Proc. Natl. Acad. Sci. USA* 95:13965–70

50. Hutchings MJ, de Kroon H. 1994. Foraging in plants: the role of morphological plasticity in resource acquisition. *Adv. Ecol. Res.* 25:159–238

51. Hwang CF, Lin Y, Dsouza T, Cheng CL. 1997. Sequences necessary for nitrate-dependent transcription of Arabidopsis nitrate reductase genes. *Plant Physiol.* 113:853–62

52. Imsande J, Touraine B. 1994. N demand and the regulation of nitrate uptake. *Plant Physiol.* 105:3–7

53. Inoue T, Higuchi M, Hashimoto Y, Seki M, Kobayashi M, et al. 2001. Identification of CRE1 as a cytokinin receptor from Arabidopsis. *Nature* 409:1060–63

54. Kakimoto T. 1996. CKI1, a histidine kinase homolog implicated in cytokinin signal transduction. *Science* 274:982–85

55. Krapp A, Fraisier V, Scheible WR, Quesada A, Gojon A, et al. 1998. Expression studies of *Nrt2:1Np*, a putative high-affinity nitrate transporter: evidence for its role in nitrate uptake. *Plant J.* 14:723–31

56. Kunze M, Riedel J, Lange U, Hurwitz R, Tischner R. 1997. Evidence for the presence of GPI-anchored PM-NR in leaves of *Beta vulgaris* and for PM-NR in barley leaves. *Plant Physiol. Biochem* 35:507–12

57. Lainé P, Ourry A, Boucaud J. 1995. Shoot control of nitrate uptake rates by roots of *Brassica napus* L. Effects of localized nitrate supply. *Planta* 196:77–83

58. Lainé P, Ourry A, Boucaud J, Salette J. 1998. Effects of a localized supply of nitrate on NO_3^--uptake rate and growth of roots in *Lolium multiflorum* Lam. *Plant Soil* 202:61–67

59. Lam HM, Coschigano KT, Oliveira IC, Melo-Oliveira R, Coruzzi GM. 1996. The molecular genetics of nitrogen assimilation into amino acids in higher plants. *Annu. Rev. Plant Physiol. Plant Mol. Biol.* 47:569–93

60. Lebot J, Kirkby EA. 1992. Diurnal uptake of nitrate and potassium during the vegetative growth of tomato plants. *J. Plant Nutr.* 15:247–64

61. Lejay L, Tillard P, Lepetit M, Olive FD, Filleur S, et al. 1999. Molecular and functional regulation of two NO_3^--uptake systems by N- and C-status of *Arabidopsis* plants. *Plant J.* 18:509–19

62. Leydecker MT, Camus I, Daniel-Vedele F, Truong HN. 2000. Screening for Arabidopsis mutants affected in the *Nii* gene expression using the GUS reporter gene. *Physiol. Plant.* 108:161–70

63. Leyser O, Fitter A. 1998. Roots are branching out in patches. *Trends Plant Sci.* 3:203–4

64. Lorenz MC, Heitman J. 1998. The MEP2 ammonium permease regulates pseudohyphal differentiaition in *Saccharomyces cerevisiae*. *EMBO J.* 17:1236–47

65. MacKintosh C, Meek SEM. 2001. Regulation of plant NR activity by reversible phosphorylation, 14-3-3 proteins and proteolysis. *Cell. Mol. Life Sci.* 58:205–14

66. Marschner H. 1995. *Mineral Nutrition of Higher Plants*. London: Academic. 889 pp.

67. Matt P, Geiger M, Walch-Liu P, Engels C, Krapp A, Stitt M. 2001. The immediate cause of the diurnal changes of nitrogen metabolism in leaves of nitrate-replete tobacco: a major imbalance between the rate of nitrate reduction and the rates of nitrate uptake and ammonium metabolism during the first part of the light period. *Plant Cell Environ.* 24:177–90

68. McDonald AJS, Davies WJ. 1996. Keeping in touch: responses of the whole plant to deficits in water and nitrogen supply. *Adv. Bot. Res.* 22:229–300

69. Meyer C, Stitt M. 2001. Nitrate reduction and signalling. In *Plant Nitrogen*, ed. PJ Lea, J-F Morot-Gaudry, pp. 37–59. Heidelberg: Springer

70. Mistrik I, Pavlovkin J, Wachter R, Pradel KS, Schwalm K, et al. 2000. Impact of *Agrobacterium tumefaciens*-induced stem tumors on NO_3^--uptake in *Ricinus communis*. *Plant Soil* 226:87–98

71. Mockaitis K, Howell SH. 2000. Auxin induces mitogenic activated protein kinase (MAPK) activation in roots of Arabidopsis seedlings. *Plant J.* 24:785–96

72. Mok DWS, Mok MC. 2001. Cytokinin metabolism and action. *Annu. Rev. Plant Physiol. Plant Mol. Biol.* 52:89–118

73. Neininger A, Back E, Bichler J, Schneiderbauer A, Mohr H. 1994. Deletion analysis of a nitrite reductase promoter from spinach in transgenic tobacco. *Planta* 194:186–92

74. Ninfa AJ, Atkinson MR. 2000. PII signal transduction proteins. *Trends Microbiol.* 8:172–79

75. Ohlen E, Larsson CM. 1992. Nitrate assimilatory properties of barley grown under long-term N-limitation: effects of local nitrate supply in split-root cultures. *Physiol. Plant* 85:9–16

76. Ono F, Frommer WB, von Wiren N. 2000. Coordinated diurnal regulation of low- and high-affinity nitrate transporters in tomato. *Plant Biol.* 2:17–23

77. Orsel M, Filleur S, Fraisier V, Daniel-Vedele F. 2002. Nitrate transport in plants: which gene and which control? *J. Exp. Bot.* In press

78. Padgett PE, Leonard RT. 1996. Free amino acid levels and the regulation of nitrate uptake in maize cell suspension cultures. *J. Exp. Bot.* 47:871–83

79. Pearson CJ, Steer BT. 1977. Daily changes in nitrate uptake and metabolism in *Capsicum annuum*. *Planta* 137:107–12

80. Peuke AD, Hartung W, Jeschke WD. 1994. The uptake and flow of C, N and ions between roots and shoots in *Ricinus communis* L. II. Grown with low or high nitrate supply. *J. Exp. Bot.* 45:733–40

81. Pouteau S, Cherel I, Vaucheret H, Caboche M. 1989. Nitrate reductase mRNA regulation in *Nicotiana plumbaginifolia* nitrate reductase-deficient mutants. *Plant Cell* 1:1111–20

82. Prosser IM, Purves JV, Saker LR, Clarkson DT. 2001. Rapid disruption of nitrogen metabolism and nitrate transport in spinach plants deprived of sulphate. *J. Exp. Bot.* 52:113–21

83. Rastogi R, Bate NJ, Sivasankar S, Rothstein SJ. 1997. Footprinting of the spinach nitrite reductase gene promoter reveals the preservation of nitrate regulatory elements between fungi and higher plants. *Plant Mol. Biol.* 34:465–76

84. Redinbaugh MG, Campbell WH. 1991. Higher plant responses to environmental nitrate. *Physiol. Plant* 82:640–50

85. Redinbaugh MG, Campbell WH. 1993. Glutamine synthetase and ferredoxin-dependent glutamate synthase expression in the maize (*Zea mays*) root primary response to nitrate: evidence for an organ-specific response. *Plant Physiol.* 101:1249–55

86. Reed RC, Brady SR, Muday GK. 1998. Inhibition of auxin movement from the shoot into the root inhibits lateral root development in arabidopsis. *Plant Physiol.* 118:1369–78

87. Reisenauer HM. 1966. Mineral nutrients in soil solution. In *Environmental Biology*, ed. PL Altman, DS Dittmer, pp. 507–8, Bethesda, MA: Fed. Am. Soc. Exp. Biol.

88. Robinson D. 1994. The responses of plants to non-uniform supplies of nutrients. *New Phytol.* 127:635–74

89. Robinson D. 2001. Root proliferation, nitrate inflow and their carbon costs during nitrogen capture by competing plants in patchy soil. *Plant Soil* 232:41–50

90. Robinson D, Hodge A, Griffiths BS,

Fitter AH. 1999. Plant root proliferation in nitrogen-rich patches confers competitive advantage. *Proc. R. Soc. Lond. Ser. B* 266:431–35

91. Roggatz U, McDonald AJS, Stadenberg I, Schurr U. 1999. Effects of nitrogen deprivation on cell division and expansion in leaves of *Ricinus communis* L. *Plant Cell Environ.* 22:81–89

92. Romani G, Beffagna N, Meraviglia G. 1996. Role for the vacuolar H^+-ATPase in regulating the cytoplasmic pH: an in vivo study carried out in *chl1*, an *Arabidopsis thaliana* mutant impaired in NO_3^--transport. *Plant Cell Physiol.* 37:285–91

93. Ruegger M, Dewey E, Gray WM, Hobbie L, Turner J, Estelle M. 1998. The TIR1 protein of Arabidopsis functions in auxin response and is related to human SKP2 and yeast Grr1p. *Genes Dev.* 12:198–207

94. Rufty TW, Siddiqi MY, Glass ADM, Ruth TJ. 1991. Altered $^{13}NO_3^-$-influx in phosphorus-limited plants. *Plant Sci.* 76:43–48

95. Ruiz-Medrano R, Xoconostle-Cazares B, Lucas WJ. 2001. The phloem as a conduit for inter-organ communication. *Curr. Opin. Plant Biol.* 4:202–9

96. Sablowski RWM, Meyerowitz EM. 1998. A homolog of NO APICAL MERISTEM is an immediate target of the floral homeotic genes APETALA3/PISTILLATA. *Cell* 92:93–103

97. Sakakibara H, Kobayashi K, Deji A, Sugiyama T. 1997. Partial characterization of the signaling pathway for the nitrate-dependent expression of genes for nitrogen-assimilatory enzymes using detached maize leaves. *Plant Cell Physiol.* 38:837–43

98. Sakakibara H, Suzuki M, Takei K, Deji A, Taniguchi M, Sugiyama T. 1998. A response-regulator homologue possibly involved in nitrogen signal transduction mediated by cytokinin in maize. *Plant J.* 14:337–44

99. Sakakibara H, Taniguchi M, Sugiyama T. 2000. His-Asp phosphorelay signaling:

a communication avenue between plants and their environment. *Plant Mol. Biol.* 42:273–78

100. Samuelson ME, Larsson CM. 1993. Nitrate regulation of zeatin riboside levels in barley roots: effects of inhibitors of N-assimilation and comparison with ammonium. *Plant Sci.* 93:77–84

101. Sattelmacher B, Marschner H. 1978. Nitrogen nutrition and cytokinin activity in *Solanum tuberosum*. *Physiol. Plant* 42:185–89

102. Scheible WR, Gonzalez-Fontes A, Lauerer M, Muller-Rober B, Caboche M, Stitt M. 1997. Nitrate acts as a signal to induce organic acid metabolism and repress starch metabolism in tobacco. *Plant Cell* 9:783–98

103. Scheible WR, Lauerer M, Schulze ED, Caboche M, Stitt M. 1997. Accumulation of nitrate in the shoot acts as a signal to regulate shoot-root allocation in tobacco. *Plant J.* 11:671–91

104. Shen WH. 2001. The plant cell cycle: G1/S regulation. *Euphytica* 118:223–32

105. Shiraishi N, Imanishi M, Endo T, Sugimoto T, Oji Y. 2001. *Investigation of transcription factor of spinach nitrate reductase*. Presented at Int. Symp. Inorg. N Assim., 6th, Reims

106. Sivasankar S, Rastogi R, Jackman L, Oaks A, Rothstein S. 1998. Analysis of cis-acting DNA elements mediating induction and repression of the spinach nitrite reductase gene. *Planta* 206:66–71

107. Smeekens S. 2000. Sugar-induced signal transduction in plants. *Annu. Rev. Plant Physiol. Plant Mol. Biol.* 51:49–81

108. Stewart V. 1994. Regulation of nitrate and nitrite reductase synthesis in enterobacteria. *Ant. Van Leeuw. Int. J. Gen. Mol. Microbiol.* 66:37–45

109. Stitt M. 1999. Nitrate regulation of metabolism and growth. *Curr. Opin. Plant Biol.* 2:178–86

110. Stitt M, Feil R. 1999. Lateral root frequency decreases when nitrate accumulates in tobacco transformants with low

nitrate reductase activity: consequences for the regulation of biomass partitioning between shoots and root. *Plant Soil* 215:143–53

111. Stitt M, Krapp A. 1999. The interaction between elevated carbon dioxide and nitrogen nutrition: the physiological and molecular background. *Plant Cell Environ.* 22:583–621

112. Stöhr C, Schuler F, Tischner R. 1995. Glycosyl-phosphatidylinositol-anchored proteins exist in the plasma membrane of *Chlorella saccharophila* (Kruger) Nadson: plasma membrane–bound nitrate reductase as an example. *Planta* 196:284–87

113. Stöhr C, Strube F, Marx G, Ullrich WR, Rockel P. 2001. A plasma membrane–bound enzyme of tobacco roots catalyses the formation of nitric oxide from nitrite. *Planta* 212:835–41

114. Sueyoshi K, Kleinhofs A, Warner RL. 1995. Expression of NADH-specific and NAD(P)H-bispecific nitrate reductase genes in response to nitrate in barley. *Plant Physiol.* 107:1303–11

115. Sueyoshi K, Mitsuyama T, Sugimoto T, Kleinhofs A, Warner RL, Oji Y. 1999. Effects of inhibitors for signaling components on the expression of the genes for nitrate reductase and nitrite reductase in excised barley leaves. *Soil Sci. Plant Nutr.* 45:1015–19

116. Suzuki T, Miwa K, Ishikawa K, Yamada H, Aiba H, Mizuno T. 2001. The Arabidopsis sensor His-kinase, AHK4, can respond to cytokinins. *Plant Cell Physiol.* 42:107–13

117. Takei K, Sakakibara H, Taniguchi M, Sugiyama T. 2001. Nitrogen-dependent accumulation of cytokinins in root and the translocation to leaf: implication of cytokinin species that induces gene expression of maize response regulator. *Plant Cell Physiol.* 42:85–93

118. Taniguchi M, Kiba T, Sakakibara H, Ueguchi C, Mizuno T, Sugiyama T. 1998. Expression of *Arabidopsis* response reg-

ulator homologs is induced by cytokinins and nitrate. *FEBS Lett.* 429:259–62

119. Thompson GA, Schulz A. 1999. Macromolecular trafficking in the phloem. *Trends Plant Sci.* 4:354–60

120. Tian Q, Reed JW. 1999. Control of auxin-regulated root development by the *Arabidopsis thaliana SHY2/IAA3* gene. *Development* 126:711–21

121. Tillard P, Passama L, Gojon A. 1998. Are phloem amino acids involved in the shoot to root control of NO_3^--uptake in *Ricinus communis* plants? *J. Exp. Bot.* 49:1371–79

122. Tischner R. 2000. Nitrate uptake and reduction in higher and lower plants. *Plant Cell Environ.* 23:1005–24

123. Touraine B, Clarkson DT, Muller B. 1994. Regulation of nitrate uptake at the whole plant level. In *A Whole Plant Perspective on Carbon-Nitrogen Interactions*, ed. J Roy, E Garnier, pp. 11–30. Den Haag, The Neth.: SPB Acad. Publ.

124. Trewavas AJ. 1983. Nitrate as a plant hormone. In *British Plant Growth Regulator Group Monograph 9*, ed. MB Jackson, pp. 97–110. Oxford, UK: Br. Plant Growth Regul. Group

125. Urao T, Miyata S, Yamaguchi-Shinozaki K, Shinozaki K. 2000. Possible His to Asp phosphorelay signaling in an Arabidopsis two-component system. *FEBS Lett.* 478:227–32

126. Urao T, Yamaguchi-Shinozaki K, Shinozaki K. 2000. Two-component systems in plant signal transduction. *Trends Plant Sci.* 5:67–74

127. van der Leij M, Smith SJ, Miller AJ. 1998. Remobilisation of vacuolar stored nitrate in barley root cells. *Planta* 205:64–72

128. Vidmar JJ, Zhuo D, Siddiqi MY, Schjoerring JK, Touraine B, Glass ADM. 2000. Regulation of high-affinity nitrate transporter genes and high-affinity nitrate influx by nitrogen pools in roots of barley. *Plant Physiol.* 123:307–18

129. Vidmar JJ, Zhuo DG, Siddiqi MY, Glass

ADM. 2000. Isolation and characterization of *HvNRT2.3* and *HvNRT2.4*, cDNAs encoding high-affinity nitrate transporters from roots of barley. *Plant Physiol.* 122:783–92

130. Walch-Liu P, Neumann G, Bangerth F, Engels C. 2000. Rapid effects of nitrogen form on leaf morphogenesis in tobacco. *J. Exp. Bot.* 51:227–37

131. Walch-Liu P, Neumann G, Engels C. 2001. Response of shoot and root growth to supply of different nitrogen forms is not related to carbohydrate and nitrogen status of tobacco plants. *Z. Pflanz. Bodenk.* 164:97–103

132. Wang RC, Guegler K, LaBrie ST, Crawford NM. 2000. Genomic analysis of a nutrient response in arabidopsis reveals diverse expression patterns and novel metabolic and potential regulatory genes induced by nitrate. *Plant Cell* 12:1491–509

133. Wang RC, Liu D, Crawford NM. 1998. The Arabidopsis CHL1 protein plays a major role in high-affinity nitrate uptake. *Proc. Natl. Acad. Sci. USA* 95:15134–39

134. Ward MR, Tischner R, Huffaker RC. 1988. Inhibition of nitrate transport by anti-nitrate reductase IgG fragments and the identification of plasma-membrane associated nitrate reductase in roots of barley seedlings. *Plant Physiol.* 88:1141–45

135. Warning HO, Hachtel W. 2000. Functional analysis of a nitrite reductase promoter from birch in transgenic tobacco. *Plant Sci.* 155:141–51

136. Xie Q, Frugis G, Colgan D, Chua NH. 2000. Arabidopsis NAC1 transduces auxin signal downstream of TIR1 to promote lateral root development. *Genes Dev.* 14:3024–36

137. Xoconostle-Cazares B, Ruiz-Medrano R, Lucas WJ. 2000. Proteolytic processing of CmPP36, a protein from the cytochrome b_5 reductase family, is required for entry into the phloem translocation pathway. *Plant J.* 24:735–47

138. Zhang HM, Forde BG. 1998. An Arabidopsis MADS box gene that controls nutrient-induced changes in root architecture. *Science* 279:407–9

139. Zhang HM, Forde BG. 2000. Regulation of Arabidopsis root development by nitrate availability. *J. Exp. Bot.* 51:51–59

140. Zhang HM, Jennings A, Barlow PW, Forde BG. 1999. Dual pathways for regulation of root branching by nitrate. *Proc. Natl. Acad. Sci. USA* 96:6529–34

141. Zhuo D, Okamoto M, Vidmar JJ, Glass ADM. 1999. Regulation of a putative high-affinity nitrate transporter (*NRT2;1At*) in roots of *Arabidopsis thaliana*. *Plant J.* 17:563–68

Annu. Rev. Plant Biol. 2002. 53:225–45
DOI: 10.1146/annurev.arplant.53.100201.160729

ACCLIMATIVE RESPONSE TO TEMPERATURE STRESS IN HIGHER PLANTS: Approaches of Gene Engineering for Temperature Tolerance

Koh Iba

*Department of Biology, Kyushu University, Fukuoka 812-8581 Japan;
e-mail: koibascb@mbox.nc.kyushu-u.ac.jp*

Key Words molecular breeding, global warming, membrane lipids, heat-shock proteins, compatible solutes

■ **Abstract** Temperature stresses experienced by plants can be classified into three types: those occurring at (*a*) temperatures below freezing, (*b*) low temperatures above freezing, and (*c*) high temperatures. This review outlines how biological substances that are deeply related to these stresses, such as heat-shock proteins, glycinebetaine as a compatible solute, membrane lipids, etc., and also detoxifiers of active oxygen species, contribute to temperature stress tolerance in plants. Also presented here are the uses of genetic engineering techniques to improve the adaptability of plants to temperature stress by altering the levels and composition of these substances in the living organism. Finally, the future prospects for molecular breeding are discussed.

CONTENTS

INTRODUCTION

Because plants lack the capability of locomotion as a means of responding to changes in their environment, they are exposed to various environmental stresses and must adapt to them in other ways. The most typical kind of stress plants receive from their surroundings is temperature stress. The range of temperatures experienced by plants varies both spatially and temporally at several different scales. For example, inland equatorial regions can experience maximum daytime temperatures as high as 60°C, whereas in Arctic eastern Siberia, temperatures of almost 30°C in the summer can fall to −70°C in severe winters, a range of 100°C. Altitude and local topography can also exert a significant influence. Each plant species has its own optimum temperature for growth, and its distribution is determined to a major extent by the temperature zone in which it can survive. Efforts in plant breeding have been made to efficiently cultivate varieties of plants suitable for food or livestock feed and to extend their natural range of growth and distribution. The long history of these classical genetic approaches has produced substantial successes, such as increasing the low-temperature tolerance of subtropical rice cultivars, enabling it to yield crops even in subarctic regions (87).

Recently, concerns have been voiced about the potentially serious effects on agriculture of radical global temperature change in the near future. By the latter half of the twenty-first century, global warming that results from increasing atmospheric concentrations of carbon dioxide and other greenhouse gasses could jeopardize agriculture, forestry, and other industries utilizing the natural environment (30). However, although much research has been conducted to evaluate the effects of global warming on these industries (10, 69, 81, 82, 93), efforts to search for specific and practical approaches to improve the adaptability of plants to their temperature environment have only recently begun (19, 91). In recent years, developments in the field of molecular biology have focused attention on molecular breeding methods in addition to those used in classical plant breeding. Rapid advances in molecular genetic approaches have enabled genes to be cloned, both from prokaryotes and directly from the plants themselves, that are thought to provide the key to the mechanism of temperature adaptation (70, 102). Techniques for efficiently and directly introducing genes to a variety of useful plants are also being developed and improved, and progress is being made on the practical application of such techniques (40, 56, 67). This review provides an overview of the genetic engineering approaches currently being used to improve plant tolerance to low- and high-temperature stresses and discusses future prospects. The major characteristics of the genetically engineered temperature stress tolerant plants produced so far together with specific information about the genetic engineering approaches used are summarized in Table 1.

HEAT-SHOCK PROTEINS

The heat-shock response is a reaction caused by exposure of an organisms tissue or cells to sudden high temperature stress, and it is characterized by a transient expression of heat-shock proteins (HSPs). The primary protein structure for HSPs is well conserved in organisms ranging from bacteria and other prokaryotes to eukaryotes such as higher animals and plants. Consequently, it is thought to be closely involved in the protection of the organism against heat stress and the maintenance of homeostasis (61).

TABLE 1 Genetically engineered alterations in the temperature stress tolerance of plants

Functions of transformed genes	Host plants	Effects on the enhancement of temperature and other stress tolerances	Remarks	References
Heat shock proteins				
Heat shock transcription factor (ATHSF1)	Arabidopsis	High temperatures (enhanced)		44
Heat shock protein (Hsp70)	Arabidopsis	High temperatures (lessened)	Heat-inducible antisense expression	45
Small heat shock protein (Hsp17.7)	Carrot	High temperatures (enhanced/lessened)	Constitutive expression, heat-inducible	49
Active oxygen species				
Chloroplast CuZn-superoxide dismutase	Tobacco	Intense light under low temperatures (enhanced)		88, 89
Mn-superoxide dismutase	Alfalfa	Low or freezing temperatures (enhanced)	Targeted to mitochondria or to chloroplast	54
Fe-superoxide dismutase	Alfalfa	Low or freezing temperatures (enhanced)	Targeted to chloroplast	55
Glutathione reductase	Tomato	Not ascertained	Gene from *E. coli*, targeted to chloroplast	14
Glutathione S-transferase/ Glutathion peroxidase	Tobacco	Low temperatures and salt (enhanced)		84
Compatible solutes				
Betaine aldehyde dehydrogenase	Rice	High and low temperatures and salt (enhanced)	Targeted to peroxisomes, betaine aldehyde applied exogenously	37
Choline oxidase	Arabidopsis	High, low, and freezing temperatures and salt (enhanced)	Gene from a soil bacterium, targeted to chloroplast	2, 3, 22, 85
Membrane lipids				
Glycerol-3-phosphate acyltransferase	Tobacco	Low temperatures (enhanced/lessened)		60, 64
Glycerol-3-phosphate acyltransferase	Rice	Low temperatures (enhanced)		112
Glycerol-3-phosphate acyltransferase from *E. coli*	Arabidopsis	Low temperatures (lessened)		106
Chloroplast ω-3 fatty acid desaturase	Tobacco	High and low temperatures (enhanced)	Overexpression (low temperatures), gene silencing (high temperatures)	38, 39, 62
Endoplasmic reticulum ω-3 fatty acid desaturase	Tobacco	Not ascertained		20

(Continued)

TABLE 1   *(Continued)*

Functions of transformed genes	Host plants	Effects on the enhancement of temperature and other stress tolerances	Remarks	References
Endoplasmic reticulum ω-3 fatty acid desaturase	Rice	Low temperatures (enhanced)		92
D9 desaturase	Tobacco	Low temperatures (enhanced)	Gene from a cyanobacterium	32
Trancriptional factors (activators)				
Transcriptional activator (CBF1)	Arabidopsis	Freezing (enhanced)	Induction of COR gene expression	33
cis-acting promoter element (DREB1A)	Arabidopsis	Freezing, drought, and salt (enhanced)	Driven by the stress inducible rd29A promoter	34
Zinc finger protein (SCOF-1)	Arabidopsis Tobacco	Low temperatures (enhanced)	Induction of COR gene expression	36
Transcriptional activator (ABI3)	Arabidopsis	Freezing (enhanced)	Enhancement of ABA-induced expression of genes for cold acclimation	100

The induction of HSPs is dependent on the temperature at which each species ordinarily grows. In higher plants, HSPs are generally induced by a short exposure to a temperature of 38–40°C. HSPs exist in various molecular sizes, all of which are characterized by binding to structurally unstable proteins. They perform important physiological functions as molecular chaperones (11). In addition to their functions of folding proteins immediately after translation and the transformation of proteins into a structure suited to membrane transport, they prevent the aggregation of denatured proteins and promote the renaturation of aggregated protein molecules. These functions of HSPs are deeply involved in resistance to temperature and various other kinds of stress (11). HSPs are classified into five classes based on their differences in molecular weight [HSP 100, HSP 90, HSP 70, HSP 60, and low-molecular weight HSPs (smHSP)] and are located in both the cytoplasm and organelles such as the nucleus, mitochondria, chloroplasts, and endoplasmic reticulum (ER).

The smHSPs, with a molecular weight of 15–30 kDa, are the most diverse HSPs (8). The pea HSP 18.1 (a cytosolic class I smHSP) works to prevent the aggregation of proteins denatured by heat and to reactivate them, and it has an ATP-independent molecular chaperone activity, as shown through in vitro experimental systems on model proteins such as citrate synthase (43).

Expression of the HSP genes is regulated mainly at the transcription level, and the heat-shock activated transcription factors (HSFs) recognize the cis-elements (heat-shock elements; HSEs) located in the upstream promoter region of the HSP genes, inducing their transcription (107). When an HSF gene was introduced into *Arabidopsis* to cause constitutive expression of HSP, increased thermotolerance compared to wild-type strains was observed (44). At least one smHSP localized in the cytoplasm was expressed in the transgenic *Arabidopsis*, under both non-stress and stress conditions, suggesting that smHSPs may contribute to the plants thermotolerance.

HSP 70 has the most-conserved primary protein structure across different species. Detailed research on it, termed DnaK in *Escherichia coli*, as well as in animal cells and yeasts, has indicated that it functions as a molecular chaperone. HSP 70 is thought to interrupt the interaction within and between protein molecules, for example, to facilitate their membrane transport, to bind to the ER, or to prevent the aggregation of denatured proteins. These functions are ATP dependent, and there are indications, even in plants, for an intrinsic ATPase activity of HSPs (57). In transgenic *Arabidopsis*, transformed with the heat-shock inducible antisense gene for HSP 70, repression of endogenous heat-induced HSP 70 resulted, and a lowering of thermotolerance was observed in the leaf tissues (45).

HSP 100 and HSP 90 are also suspected to function as chaperones, although there is little direct evidence related to temperature stress in plants. However, studies of *Arabidopsis* (24, 25) have been reported in which a mutant, *hot1*, lacking in HSP 101, showed a susceptibility to high temperatures.

Apart from heat shock, there are also HSPs induced by osmotic and salt stress, stress from low oxygen, dinitrophenol (DNP), arsenic compounds, other chemical agents, and plant hormones such as abscisic acid and ethylene. They might play a specific role in the denaturation of proteins, the prevention of that denaturation, and the repair function, as part of the physiological responses to diverse environmental stresses. It has been reported, for example, that salt and drought tolerance improves when the HSP gene is overexpressed in tobacco (97) and in Arabidpsis (98).

ACTIVE OXYGEN SPECIES DETOXIFICATION SYSTEMS

The active oxygen species (AOS) includes the superoxide anions (O_2^-) and hydrogen peroxide ($H_2O_2^-$) that are generated by the reduction of molecular oxygen (O_2) and the hydroxyl radicals (OH •) that are produced in reactions involving H_2O_2 and O_2^-. AOS causes damage to cells by oxidation of their constituents. The amount of AOS in plants increases when they are exposed to low temperatures, drought, high light intensity, and other stresses (9). Six enzymes, namely superoxide dismutase (SOD), ascorbate peroxidase (APX), catalase (CAT), monodehydroascorbate reductase (MDAR), dehydroascorbate reductase (DHAR), and glutathione reductase (GR), are involved in AOS-detoxification systems in higher plants. SOD catalyzes the dismutation of superoxide radicals to hydrogen peroxide and molecular oxygen in that detoxification process. Hydrogen peroxide is also detoxified by APX and CAT. The three enzymes MDAR, DHAR, and GR are involved in the recovery of ascorbic acid that has been oxidized by APX. CAT is localized in the microbodies of almost all plant species, but the other five enzymes are found as multiple isozymes, so that a series of these detoxifying enzymes exists in the chloroplast, cytoplasm, mitochondria, and microbodies, respectively.

In *Arabidopsis* subjected to low temperatures, H_2O_2 accumulates in the cells, and the enzyme activities of APX and GR increase (76). Conversely, in rice, APX and SOD levels rise when plants are transferred from cold to normal temperature (86). When tobacco plants are chilled under normal lighting conditions, chloroplastic Fe-SOD is induced, followed by cytosolic CuZn-SOD induction (103). Conspicuous induction of cytosolic CuZn-SOD also takes place under high-temperature treatment in both dark and light conditions. Cytosolic APX is induced by high temperatures, but the high-temperature induction of the gene (*APX1*) for cytosolic APX in *Arabidopsis* is regulated by the HSE (59, 96). Because enzymes related to such AOS detoxification are induced when plants are subjected to the above stresses, it was originally hoped that, by artificially introducing those genes into plants, their tolerance against temperature-induced stress could be increased. The attempts described below, most of which concentrated on the SOD genes, were carried out on the basis of such expectations. Transgenic tobacco, into which chloroplastic CuZn-SOD has been introduced, shows improved resistance to intense light and low temperatures (88, 89), whereas transgenic alfalfa potently expressing Mn-SOD and Fe-SOD in the mitochondria and chloroplasts showed improved resistance to low temperatures (54, 55). However, with transgenic tomato plants overexpressing the *E. coli* GR gene in order to increase the activation of GR, no improvement in chilling tolerance was observed (14). Glutathione S-transferase (GST) is a detoxifying enzyme that uses H_2O_2 and peroxides as substrate. Transgenic tobacco expressing GST genes in the cytoplasm shows resistance to low temperature and salt stress during the process of germination (84).

AOS-detoxification systems are composed of multiple enzymes, and these are thought to function in accordance with stress conditions in different compartments of the cell. Therefore, the protective system might not be able to function more efficiently simply by introducing each individual enzyme gene. On the basis of findings so far, there appear to be limits to the increase in resistance to stress that can be achieved by the introduction of enzyme genes involved in AOS detoxification. To create potent stress tolerant plants, a new strategy will be required to make the entire AOS-detoxification system work efficiently in a well-balanced manner.

GLYCINEBETAINE

When plants are exposed to salt, drought, and low-temperature stress, they accumulate highly soluble organic compounds of low molecular weight, called compatible solutes. These organic compounds exist in stable form inside cells and are not easily metabolized, but neither do they have any effect on cell functions, even when they have accumulated in high concentrations (111). The functions of these substances within the living organism are still unclear. However, because many plant stresses cause cell dehydration, the accumulation of these substances might play a part in increasing internal osmotic pressure and preventing loss of water from the cell. Typical compatible solutes include mannitol and other sugar alcohols, amino acids such as proline, and amino acid derivatives such as glycinebetaine. Some of these compatible solutes, like proline, are accumulated in practically all plant species, whereas others, like glycinebetaine, are distributed only among plants with a high tolerance to salt or cold (80).

Glycinebetaine is created from choline. In many plants, it is synthesized through a two-step oxidation of choline, by choline monooxygenase and betaine aldehyde dehydrogenase (4, 15, 109). The two enzyme genes have been cloned from plants (31, 68, 79, 104), and attempts have been made to introduce them separately into plants to generate glycinebetaine, but sufficient quantities of the choline substrate and the intermediate betaine aldehyde could not be obtained within the cell (75). For this and other reasons, generating a sufficient quantity of glycinebetaine in vivo to contribute to a significant stress tolerance has not been possible. However, when a transgenic rice plant containing the betaine aldehyde dehydrogenase gene was supplied with sufficient exogenous betaine aldehyde to artificially generate a considerable quantity of glycinebetaine, it showed improved tolerance against both salt and low temperatures (37).

In a different pathway, *E. coli* and *Arthrobacter globiformis* are both able to synthesize glycinebetaine from choline, using choline dehydrogenase and choline oxidase, respectively. When the bacterial choline oxidase gene (*CodA*) is introduced into tobacco or *Arabidopsis*, glycinebetaine is synthesized as a result, and not only is salt tolerance increased, but a slight though significant resistance to low temperature and freezing stress is also conferred (2, 22, 85). Equally, the lowered germination rate of seeds under high temperatures is reversed to some extent (3). However, the accumulation is minute (1 μmol GlyBet g^{-1} fresh weight), which equals 10- to 100-fold less than the amounts found in stressed naturally salt-tolerant plants. This suggests that if a method can be developed for accumulating high concentrations of glycinebetaine in a plant it could possibly lead to the development of a practical plant with high resistance to temperature or salt stress or, indeed, a wide range of other environmental stresses.

MEMBRANE LIPIDS

Lipids, which are among the basic constituents of biomembranes, have been a focus of attention since the 1960s as one of the factors affecting temperature sensitivity in plants (78). For example, the physiochemical characteristics displayed by lipid bilayers at different temperatures differ with the species of the lipid head group or their esterified fatty acids, and their lipid constituents and fatty acid constituents change depending on the environmental growth temperature (53, 94, 105). There has also been an interest in the relationship of low-temperature tolerance to the biosynthesis and rearrangement of biomembranes in response to temperature. Since the late 1980s, a series of mutant strains with altered fatty acid constituents in their biomembranes have been isolated in *Arabidopsis*, providing a breakthrough in the clarification of the pathways by which the polyunsaturated fatty acids contained in biomembrane lipids are generated (95). In the 1990s, cloning the mutated genes, one after the other, using genetic approaches (7, 18, 23, 29, 77) became possible, and useful information such as the structure of the enzymes in the biosynthetic pathway for membrane lipids became available (90). The creation of transgenic plants by introducing these cloned genes also allowed the artificial modification of the fatty acid constituents of the biomembrane lipids and enabled the physiological significance of biomembrane lipids in temperature acclimatization to be clarified more directly (20, 38, 39, 60, 62, 64).

Phosphatidylglycerol

Artificially created lipid bilayers undergo a phase transition from a highly fluid liquid crystalline phase to a more solid gel phase following a temperature decrease. At low temperatures, biomembranes also have reduced liquidity and phase transitions; ion leakage and deactivation of membrane proteins both take place. These physical characteristics enabled research into the relationship between the fatty acid composition of biomembrane lipids and low-temperature sensitivity in plants. Murata (63) and Murata et al. (65) found that the levels of 16:0 and t16:1 fatty acids in phosphatidylglycerol (PG) in leaf tissue were high in chilling-sensitive plants but low in chilling-resistant plants. PGs from the chilling-sensitive plants changed from the liquid crystalline state into the phase separation state at approximately 30°C, whereas PGs from the chilling-resistant plants went into the phase separation state at 15°C or lower, a significant difference (66). In general, PG makes up 8–10% of chloroplast lipids, and most of this lipid is found in thylakoid membranes. In chilling-resistant plants, the amount of PGs with saturated fatty acids such as 16:0, t16:1, and 18:0 (saturated PGs) is less than 20%. By contrast, most chilling-sensitive plants contain 40% or more saturated PGs (35, 74, 83). These observations suggest that the saturated PG content in the chloroplast membranes might be related to the phase transition temperature, which is related to the low-temperature adaptability of plants.

Chloroplast PG always contains 16:0 or t16:1 in the sn-2 position. In chilling-resistant plants, the unsaturated fatty acid oleic acid (18:1) tends to bind in the sn-1 position, whereas in chilling-sensitive plants, the saturated fatty acid palmitic acid (16:0) tends to bind there. These differences result from the substrate selectivity of glycerol-3-phosphate-acyltransferase. Consequently, it is currently hypothesized that the difference in their sensitivity to chilling is caused by the properties of single enzymes (64). Experiments have been conducted to overexpress the genes for glycerol-3-phosphate-acyltransferase in tobacco, in an attempt to clarify the relationship between chilling sensitivity in plants and the fatty acid molecular species in PG (64). In transgenic tobacco overexpressing the acyltransferase gene (pSQ) from squash, a plant with chilling sensitivity, the content of the saturated PGs rose significantly to 76%, compared to the 36% found in wild-type tobacco, whereas in transgenic tobacco overexpressing the same enzyme gene (pARA) from *Arabidopsis*, a plant with chilling resistance, the saturated PG content dropped slightly to 28%. After the plants were grown for 10 days under strong illumination, equivalent to approximately 7% sunlight, at 1°C, then returned to 25°C, the extent of chlorosis in the leaf tissue was the greatest in pSQ, followed by the wild-type strain, and with the least in the pARA strain.

Conversely, Wolter et al. (106) introduced the *E. coli* gene for glycerol-3-phosphate-acyltransferase (*plsB*), which is selective for saturated acyl-ACP as compared to unsaturated acyl-ACP, into *Arabidopsis* and expressed it in the chloroplasts. The saturated PG content in the chloroplast membranes of the transgenic *Arabidopsis* rose to over 50%, compared with that of 5% in the wild-type strain. The chilling sensitivity of the transgenic tobacco also increased. These findings indicate that the level of chloroplast PG saturation is one of the determining factors of chilling sensitivity of plants.

Light is a prerequisite for photosynthesis, but it is also harmful to the photosynthetic apparatus. Photoinhibition of photosynthesis is a major stress, which is aggravated if strong light is combined with other stresses, such as low temperature. How the increase in the saturated fatty acid species in PG affects photoinhibition under low-temperature conditions has been studied using the tobacco pSQ line, which is susceptible to low temperatures (60). However, no discernible difference was found in the extent of light-induced inactivation of photoinhibition, even when chilling treatment was applied. This suggests that the level of saturated PG content has no effect on the sensitivity of the photosynthetic machinery under stress induced by low temperatures. The primary target for photoinhibition is thought to be the D_1 protein of the photosystem II (PSII) reaction center (5, 6). When this protein is damaged by the photochemical reaction, de novo synthesis of the protein occurs and the active PSII is recovered. Thus, unsaturation of PG may infuence the efficiency with which the photoinhibition-damaged D_1 protein is removed from the PSII reaction center complex and replaced by a new D_1 protein (60).

Studies of the mutant strain *Arabidopsis fab1*, being defective in palmitoyl-ACP elongase and having, therefore, an elevated level of 16:0 in its cellular lipids, have provided an alternative view of the role of unsaturated PG. The saturated PG content of the wild-type *Arabidopsis* is approximately 9%, whereas this mutant contains approximately 43% saturated PG (108). This is a significant difference, even compared with the saturated PG content of plant strains with a chilling sensitivity. Even though the saturated PG content of this mutant strain is higher than that of chilling-sensitive plants such as cucumber, it still does not suffer marked chilling damage even when grown at a temperature (2°C) that would kill the cucumber. This suggests that the level of saturated PG is not the only determinant of the plant's capacity to tolerate low temperatures, but that there are also other factors involved. Incidentally, the *fab1* mutant strain undergoes chlorosis when it is grown for an extended period at 2°C and ultimately withers and dies. This suggests that a high content of saturated PG also has an effect on long-term growth at low temperatures.

Polyunsaturated Fatty Acids

Analyses of mutant strains of *Arabidopsis* have shown that polyunsaturated fatty acids in chloroplast membranes influence the size of the chloroplast and the formation of its membranes at low temperatures (28). The *fad5* mutant is deficient in the activity of a chloroplast ω-9 fatty acid desaturase and accumulates high levels of palmitic acid (16:0). Conversely, the *fad6* mutant is deficient in the activity of the chloroplast ω-6 fatty acid desaturase and accumulates high levels of 16:1 and 18:1 fatty acids. Both mutants show correspondingly reduced levels of polyunsaturated fatty acids in the chloroplast galactolipids. When young seedlings of these mutants and the wild type (grown for seven days at 22°C) were subjected to chilling treatment (5°C) for three weeks, the mutant strains showed noticeably more chlorosis than the wild type. The amount of chlorophyll also decreased to approximately half of that in the wild-type strain. The chloroplast membrane (thylakoid membrane) content in the mutant strains decreased, and the chloroplast was smaller. In contrast, when more mature plants were subjected to similar chilling treatment, no difference was observed between the wild type and the mutant strains in the amount of chlorophyll, chloroplast size, or membrane content. Conversely, in the *fad2* mutant, which is deficient in the activity of 18:1 fatty acid desaturase localized in the ER and thus has a marked decrease in the amount of polyunsaturated fatty acids in the extrachloroplast membrane lipids (47, 77), long-term culture (42 days) at 6°C inhibited growth, eventually causing plants to wither and die (58). These observations suggest that polyunsaturated fatty acids are required for low-temperature survival and that under chilling stress the polyunsaturated fatty acids in chloroplast membrane lipids contribute significantly more to the regular formation of chloroplast membranes than to maintaining their stability when mature.

The cyanobacterium gene with a broad specificity for the $\Delta 9$ desaturases (*des9*) was introduced into tobacco (D9-1 line) (32). The 16:1*c*7 fatty acids found in higher plants are bound to the sn-2 position of the chloroplastic lipid monogalactosyldiacylglycerol (MGD), and they exist only in minute quantities as intermediates in the production of 16:3. The 16:1*t*3 fatty acid that is bound to the sn-2 position of PG has characteristics similar to those of the saturated 16:0 fatty acids. The 16:0 fatty acids bound to the sn-1 position of PG, and all other lipid species are not desaturated. The D9-1 line produced 16:1*c*9 fatty acids that are not found naturally in higher plants, and they accounted for over 10% of all fatty acids esterified to individual membrane lipids. When the wild-type plants grown at 25°C were chilled at 1°C for 11 days, they underwent chlorosis, but this did not occur in the D9-1 line subjected to the same treatment. When the seeds of both the wild-type and the D9-1 plants were germinated at 10°C and grown for 52 days, the wild-type seedlings turned pale green in color and showed damage from chilling, but the D9-1 seedlings maintained growth, similar to their response at 25°C. Notably, the increase in chilling resistance, obtained by the introduction of the glycerol-3-phosphate-acyltransferase gene, is limited only to the process of recovery from chilling damage (64), yet the D9-1 line seedlings also displayed resistance to stress from direct exposure to low temperatures and reduced the lower temperature limit for plant growth (32).

Trienoic Fatty Acids

In general, trienoic fatty acids (TAs) are present in the highest percentage among fatty acids in plant membrane lipids. TAs are polyunsaturated fatty acids that have three cis double bonds, and their content varies considerably according to the plant species and the environmental conditions. TAs are formed from dienoic fatty acids (DAs) (having two cis double bonds) through the activity of ω-3 fatty acid desaturase. This desaturase seems to be deeply embedded within biomembranes. Owing to the difficulty of characterizing it by conventional biochemical approaches, its gene has been cloned using genetic techniques, namely map-based cloning methods that used mutant strains of *Arabidopsis* (7, 29). The ω-3 fatty acid desaturase genes cloned thus far are divided into two types, one localized in the chloroplasts (*FAD7* and *FAD8*) (18, 29) and the other localized in the ER (*FAD3*) (7).

Kodama et al. (38, 39) reduced the DAs (16:2+18:2) and increased the TAs (16:3 + 18:3) in leaf tissue by overexpressing the *Arabidopsis FAD7* gene in tobacco. Evaluation of low-temperature tolerance of transgenic and wild-type plants indicated no discernible difference in the performance of mature plants, but very young seedlings revealed differences in low-temperature tolerance. Transfer of wild-type plants grown at 25°C to 1°C for seven days without going through an acclimation process and subsequent return to the original temperature environment resulted in growth inhibition and chlorosis in the young leaves. This kind of damage was not observed in the transgenic tobacco, in which the level of TAs in the leaf tissue had been increased. However, when the levels of TAs in phospholipids, which are the main constituents of extrachloroplastic membranes, were increased in tobacco by overexpressing the ER-localized ω-3 fatty acid desaturase gene (*NtFAD3*), no significant difference was observed between wild-type and transgenic tobacco plants in their resistance to chilling and freezing (20). Conversely, alleviation of chilling injury at the seedling stage (92) and improvement of the germination rate under a low-temperature environment (M. Komaki & T. Shimada, personal communication) were observed in a transgenic line of rice with increased levels of TAs in their extrachloroplastic membranes. Although these results, obtained from the analysis of transgenic plants into which *FAD7* and *FAD3* genes had been introduced, provide direct proof that TAs enhance tolerance to low temperatures, the effect of TAs on the enhancement of low temperrature tolerance was relatively slight, and it might be limited to specific plant species, and their tissues or growth processes.

As indicated above, although the increase in TAs did not produce as great an improvement in low-temperature tolerance as was expected, the reverse concept, that a decrease in the level of TAs in the organism may increase the high-temperature tolerance of plants, was examined (62). In the study described (38, 39), the ω-3 fatty acid desaturase gene was linked to a potent expression promoter, such as the cauliflower mosaic virus 35S promoter, to increase the amount of enzymes produced within the plant. However, in parts of the transgenic tobacco plants that were created for this purpose, transgenic lines were found in which the expression of the intrinsic ω-3 fatty acid desaturase gene was cosuppressed by gene silencing (62).

The correlation between the TA content of the biomembranes and the ability of the plant to tolerate high temperatures was analyzed using the transgenic tobacco lines in which the activity of chloroplast-localized ω-3 fatty acid desaturase was suppressed by gene silencing. Although TAs in the chloroplast membrane lipids of these transgenic tobaccos were held at an extremely low level, the level of DAs increased in a manner corresponding to the decrease in the level of TAs. In addition, few changes were detected in the lipid molecular species of biomembranes, other than in the chloroplast membrane.

In plants grown at temperatures ranging between cool (15°C) and a more suitable growth temperature (25°C), there were no differences between the growth of two transgenic tobacco lines (T15 and T23) and that of the wild type. After germinating and cultivating these plants for 45 days at 25°C, the fresh weight of the aerial parts of the T15 and T23 plants was 489 $\pm$ 71 mg and 513 $\pm$ 88 mg, respectively, whereas the fresh weight of the aerial parts of the wild-type plants was 497 $\pm$ 43 mg. The fresh weight of the aerial parts of the T15, T23, and wild-type plants cultivated at 15°C for 45 days was 6.2 $\pm$ 1.4 mg, 6.9 $\pm$ 1.2 mg, and 6.6 $\pm$ 0.9 mg, respectively. These results in tobacco plants are consistent with the growth of the *Arabidopsis fad7fad8* mutant within the normal cultivation temperature range of 12°C to 28°C (53). Furthermore, at temperatures below 10°C, the growth of the two transgenic lines and the wild type were similarly suppressed.

Although there was no difference observed in growth between the transgenic tobacco lines and the wild type over the range of low temperatures up to the normal growth temperature, at high temperatures clear differences in growth were observed. For example, the fresh weight of the above-ground portion of plants, grown for 45 days following germination at 30°C, was 492 $\pm$ 81 mg, 445 $\pm$ 62 mg, and 399 $\pm$ 69 mg, respectively, for the T15 and T23 lines and the wild-type plants. At a higher temperature (36°C), marked differences in the growth of the transgenic tobacco lines and the wild type were observed. After cultivating plants at 36°C for 45 days, the fresh weight of the aerial parts of the T15 and T23 lines and the wild-type plants was 124 $\pm$ 49 mg, 123 $\pm$ 23 mg, and 13 $\pm$ 6 mg, respectively. Because growth of the transgenic lines continued to be uninhibited beyond 45 days at 36°C, the resistance to high temperature was not transient and thus unlike the protection conferred by induction of a heat shock protein (44). When the plants were exposed to a considerably higher temperature (47°C), the leaves of the wild type withered within two days, and after three days the plant bodies exhibited chlorosis that resulted in death. In contrast, although growth of the transgenic plants was inhibited, damage due to high temperature was avoided by the plant body. The plants then continued to grow when returned to a more suitable temperature (25°C).

Photosynthetic activity was measured to investigate the mechanism by which the improvement in high-temperature tolerance was obtained. Although photosynthetic activity in wild-type tobacco plants at high temperatures of 40°C and above decreased considerably, the decrease was mild in transgenic tobacco plants with decreased levels of TAs (62). When the chloroplast membrane was studied using differential scanning calorimetry, an endothermic peak in the range of 40–45°C that accompanies the thermal denaturation of proteins was observed in wild type, but not in the transgenic lines with decreased levels of TAs (Y. Murakami & K. Iba, unpublished data). Transgenic lines may have improved thermal stability in those proteins or functions that are associated with chloroplast membrane lipids. Prime candidates on which such protection could work are the proteins constituting the photosynthetic machinery.

High-temperature tolerance might also be conferred on plants by increasing the level of molecular species of fatty acids with a higher degree of saturation than that of DAs. In fact, it has been reported that the saturation of thylakoid membrane lipids by mutations in fatty acid desaturation (1, 27, 41) or by catalytic hydrogenation (101) increases the thermal stability of the membranes. Such increased saturation, however, could raise the temperature at which lipids, such as MGD, phase separate into nonbilayer structures, which disrupt membrane organization. Therefore, the sensitivity of such plants to low temperatures might be increased (28).

Regulation of Polyunsaturated Fatty Acid Levels by Temperature and Other Stresses

The growth rates of transgenic tobacco in which the TA level was altered did not differ significantly from that of wild-type plants (62). This was true at normal cultivation temperatures and even at the lowest temperature (15°C). These observations suggest that the reduction of the TA level might be able to confer high-temperature tolerance to plants without sacrificing their tolerance to low temperatures. However, within lower temperature ranges, especially near freezing, other factors such as a decrease in photosynthetic activity (52), growth inhibition, and chlorosis (28) might affect the plant's ability to withstand stress. Thus, alterations in the composition of membrane lipids might be a double-edged sword: That is, tolerance to either low or high temperatures, but not both, can be enhanced. Thus, if this technique is to be used to allow basic research on plant temperature-stress tolerance, clarifying how the TA level is regulated in conjunction with environmental change will be essential.

The expression of the chloroplast FAD8 ω-3 desaturase gene changes in response to a change in ambient temperature, whereas the expression of FAD7, a second chloroplast ω-3 desaturase, is not affected by temperature. With the *Arabidopsis fad7* mutant, deficient in FAD7 desaturase activity, the only activity of chloroplast ω-3 fatty acid desaturase that can be measured is that of the FAD8 desaturase (13). This makes it possible to clearly monitor the temperature dependency of the enzyme expression. Surprisingly, the expression of the *FAD8* desaturase is switched on and off by a difference of as little as a few degrees Celsius on either side of 25°C (18). This suggests that this temperature regulation operates via a mechanism that is quite different from that governing the expression of temperature-dependent genes, such as the HSP genes, found so far. Recently, Matsuda & Iba (50) constructed a series of chimeric genes, created from both the

FAD7 and *FAD8* genes that encode the isozymes of chloroplast ω-3 desaturase, and introduced them into the *Arabidopsis fad7fad8* double mutant. Analyses of these transgenic plants showed that the temperature-dependent expression of the *FAD8* gene was due, not to the 5′ flanking region, including the promoter region and the untranslated region, but to the exon/intron structure that is inherent to the *FAD8* gene. It therefore seems unlikely that *FAD8* gene expression is simply regulated at the transcriptional levels, as in the bacterial desaturase genes (99).

In the root tissues of wheat, the levels of 18:3 increase markedly at low temperatures. Conversely, in accordance with the increased accumulation of the ER ω-3 fatty acid desaturase protein, the mRNA level of the desaturase gene *TaFAD3* demonstrate minor change (26). This suggests that the increased level of 18:3 at low temperatures is regulated at the translational or posttranslational level.

The expressions of each of the ω-3 fatty acid desaturase genes, the chloroplast-localized (*FAD7*) and the ER-localized (*FAD3*) type, appear to be regulated in a complex way in response to changes in the environment or other stress-inducing factors. For example, environmental stimuli, such as wounding, salt stress, and pathogen invasion, lead to a rapid increase in a defense-related signal molecule, jasmonate. TAs, especially α-linolenic fatty acids, might play an important role as a precursor to jasmonate (16, 72), and the stimulus leads to a rapid induction of the expression of the chloroplast ω-3 desaturase genes (*FAD7* and *FAD8*) (21, 71, 73). In contrast, the expression of the ER ω-3 desaturase gene (*FAD3*) is regulated through the synergistic and antagonistic interaction of plant hormones such as auxin, cytokinin, and abscisic acid, and the tissue specificity of the expression of this gene is further modified in accordance with the growth phase in plant development (51, 110, 113). In view of these facts, the regulation of fatty acid desaturation of membrane lipids appears to be intimately related to the wide range of mechanisms that allow plants to adapt to their environment throughout development.

FUTURE PROSPECTS

Most approaches to molecular breeding have attempted to improve host resistance to temperature stress by introducing a single gene, thereby altering only a single character. However, the general view is that multiple characters are involved in tolerance over a wide range of temperatures. Consequently, approaches that aim to simultaneously alter multiple related characters are also being attempted in order to engineer better acclimation to freezing and chilling temperatures. *Arabidopsis* and tobacco transformed to express the transcription factors (activators) of those genes that play a role in acclimation have been generated. In such transgenic plants, intrinsic low-temperature-inducible genes are expressed selectively, and an enhancement of cold or freezing tolerance has been observed (33, 34, 36, 100).

Plants receive various stresses from their surrounding environment, which affect them in a complex manner. For example, when a plant is subjected to abiotic environmental stresses, such as high or low temperature, intense light, or drought, its sensitivity to biotic stresses such as viruses and bacteria could be also intensified (17, 42). Previous studies of plant resistance to temperature stress have regarded temperature extremes as the sole source of stress and have not paid attention to the multiple stresses that affect populations under natural conditions. Even if tolerance to drought and salt stresses or tolerance to freezing was conferred on plants (34), for example, this addition would still keep the tolerance within only a single stress category because these stresses are mainly caused by a single effect of dehydration on cells. As a result, the development of a more effective approach to creating plants that can resist a wide range of stresses remains a challenge for the future. To enable us to address this issue, it is necessary to identify the key strategies that plants use to deal with complex stresses of both biotic and abiotic origin. For instance, the degree of unsaturation of fatty acids in the membrane lipids of plant cells is closely related to the plant's temperature tolerance (1, 27, 28, 32, 38, 39, 60, 62, 64, 101), but at the same time, it also influences the replication of viruses infecting the plant (46). This implies that the degree of unsaturation of lipid fatty acids could possibly determine whether a viral infection occurs. In future research, emphasis should be placed on such cases where tolerance to seemingly unrelated physical stresses of this kind and viral propogation are regulated by a single factor.

The swift development in recent years of functional genomics, the comprehensive systematic analysis of transcriptomics and proteomics, will enable us to understand the route by which signals that activate resistance to various stresses are transmitted (12, 48). By inferring the origins or intersections of such signals, we might be able to discover intrinsic resistance factors that are vital in allowing plants to cope with a wide range of stresses, including biological as well as physical and chemical environmental stress.

ACKNOWLEDGMENTS

I thank current and past members of my laboratory for their many contributions and collaborations. Research in my laboratory was supported by grant RFTF96L00602 from the Japan Society for the Promotion of Science.

Visit the Annual Reviews home page at www.annualreviews.org

LITERATURE CITED

1. Alfonso M, Yruela I, Almárcegui S, Torrado E, Pérez MA, et al. 2001. Unusual tolerance to high temperatures in a new herbicide-resistant D1 mutnat from *Glycine max* (L.) Merr. cell cultures deficient in fatty acid desaturation. *Planta* 212:573–82
2. Alia, Hayashi H, Chen THH, Murata N. 1998. Tranformation with a gene for choline oxidase enhances the cold tolerance of Arabidopsis during germination and early growth. *Plant Cell Environ.* 21:232–39
3. Alia Hayashi H, Sakamoto A, Murata N. 1998. Enhancement of the tolerance of *Arabidopsis* to high temperatures by genetic engineering of the synthesis of glycinebetaine. *Plant J.* 16:155–61

4. Arakawa K, Takabe T, Sugiyama T, Akazawa T. 1987. Purification of betaine-aldehyde dehydrogenase from spinach leaves and preparation of its antibody. *J. Biochem.* 101:1485–88

5. Aro E-M, McCaffery S, Anderson JM. 1994. Recovery from photoinhibition in peas (*Pisum sativum* L.) acclimated to varying growth irradiances. *Plant Physiol.* 104:1033–41

6. Aro E-M, Virgin I, Andersson B. 1993. Photoinhibition of photosystem II. Inactivation, protein damage and turnover. *Biochim. Biophys. Acta* 1143:113–34

7. Arondel V, Lemieux B, Hwang I, Gibson S, Goodman HM, Somerville CR. 1992. Map-based cloning of a gene controlling omega-3 fatty acid desaturation in *Arabidopsis*. *Science* 258:1353–55

8. Arrigo AP, Laundry L. 1994. *The Biology of Heat Shock Proteins and Molecular Chaperones*, ed. RI Morimoto, A Tissieres, C Georgopoulos, pp. 353–73. New York: Cold Spring Harbor Lab. Press

9. Asada K. 1997. *Oxidative Stress and the Molecular Biology of Antioxidant Defences*, ed. JG Scandalios, pp. 715–35. New York: Cold Spring Harbor Lab. Press

10. Ayres MP, Lombardero MJ. 2000. Assessing the consequences of global change for forest disturbance from herbivores and pathogens. *Sci. Total Environ.* 262:263–86

11. Boston RS, Viitanen PV, Vierling E. 1996. Molecular chaperones and protein folding in plants. *Plant Mol. Biol.* 32:191–222

12. Bowler C, Fluhr R. 2000. The role of calcium and activated oxygens as signals for controlling cross-tolerance. *Trends Plant Sci.* 5:241–46

13. Browse J, McCourt P, Somerville C. 1986. A mutant of *Arabidopsis* deficient in $C_{18:3}$ and $C_{16:3}$ leaf lipids. *Plant Physiol.* 81:859–64

14. Brüggemann W, Beyel V, Brodka M, Poth H, Weil M, et al. 1999. Antioxidants and antioxidative enzymes in wild-type and transgenic *Lycopersicon* genotypes of different chilling tolerance. *Plant Sci.* 140:145–54

15. Burnet M, Lafontaine PJ, Hanson AD. 1995. Assay, purification, and partial characterization of choline monooxygenase from spinach. *Plant Physiol.* 108:581–88

16. Farmer EE. 1994. Fatty acid signaling in plants and their associated microorganisms. *Plant Mol. Biol.* 26:1423–37

17. Gaudet DA, Laroche A, Yoshida M. 1999. Low-temperature–wheat-fungal interactions: a carbohydrate connection. *Physiol. Plant.* 106:437–44

18. Gibson S, Arondel V, Iba K, Somerville C. 1994. Cloning of a temperature-regulated gene encoding a chloroplast ω-3 desaturase from *Arabidopsis thaliana*. *Plant Physiol.* 106:1615–21

19. Grover A, Agarwal M, Katiyar-Agarwal S, Sahi C, Agarwal S. 2000. Production of high temperature tolerant transgenic plants through manipulation of membrane lipids. *Curr. Sci.* 79:557–59

20. Hamada T, Kodama H, Takeshita K, Utsumi H, Iba K. 1998. Characterization of transgenic tobacco with an increased α-linolenic acid level. *Plant Physiol.* 118:591–98

21. Hamada T, Nishiuchi T, Kodama H, Nishimura M, Iba K. 1996. cDNA cloning of a wounding-inducible gene encoding a plastid ω-3 fatty acid desaturase from tobacco. *Plant Cell Physiol.* 37:606–11

22. Hayashi H, Alia H, Mustardy L, Deshnium P, Ida M, et al. 1997. Transformation of *Arabidopsis thaliana* with the *codA* gene for choline oxidase; accumulation of glycinebetaine and enhanced tolerance to salt and cold stress. *Plant J.* 12:133–42

23. Hitz WD, Carlson TJ, Booth R Jr, Kinney AJ, Stecca KL, et al. 1994. Cloning of a higher-plant plastid ω-6 fatty acid desaturase cDNA and its expression in a cyanobacterium. *Plant Physiol.* 105:635–41

24. Hong S-W, Vierling E. 2000. Mutants of *Arabidopsis thaliana* defective in the acquisition of tolerance to high temperature

stress. *Proc. Natl. Acad. Sci. USA* 97: 4392–97

25. Hong S-W, Vierling E. 2001. Hsp101 is necessary for heat tolerance but dispensable for development and germination in the absence of stress. *Plant J.* 27:25–35

26. Horiguchi G, Fuse T, Kawakami N, Kodama H, Iba K. 2000. Temperature-dependent translational regulation of the ER ω-3 fatty acid desaturase gene in wheat root tips. *Plant J.* 24:805–13

27. Hugly S, Kunst L, Browse J, Somerville C. 1989. Enhanced thermal tolerance of photosynthesis and altered chloroplast ultrastructure in a mutant of *Arabidopsis* deficient in lipid desaturation. *Plant Physiol.* 90:1134–42

28. Hugly S, Somerville C. 1992. A role for membrane lipid polyunsaturation in chloroplast biogenesis at low temperature. *Plant Physiol.* 99:197–202

29. Iba K, Gibson S, Nishiuchi T, Fuse T, Nishimura M, et al. 1993. A gene encoding a chloroplast ω-3 fatty acid desaturase complements alterations in fatty acid desaturation and chloroplast copy number of the *fad7* mutant of *Arabidopsis thaliana. J. Biol. Chem.* 268:24099–105

30. IPCC. 2001. *IPCC Third Assessment Report—Climate Change 2001.* http://www.ipcc.ch/

31. Ishitani M, Nakamura T, Han SY, Takabe T. 1995. Expression of the betaine aldehyde dehydrogenase gene in barley in response to osmotic stress and abscisic acid. *Plant Mol. Biol.* 27:307–15

32. Ishizaki-Nishizawa O, Fujii T, Azuma M, Sekiguchi K, Murata N, et al. 1996. Low-temperature resistance of higher plants is significantly enhanced by a nonspecific cyanobacterial desaturase. *Nat. Biotech.* 14:1003–6

33. Jaglo-Ottosen KR, Gilmour SJ, Zarka DG, Schabenberger O, Thomashow MF. 1998. *Arabidopsis CBF1* overexpression induces *COR* genes and enhances freezing tolerance. *Science* 280:104–6

34. Kasuga M, Liu Q, Miura S, Yamaguchi-Shinozaki K, Shinozaki K. 1999. Improving plant drought, salt, and freezing tolerance by gene transfer of a single stress-inducible transcription factor. *Nat. Biotech.* 17:287–91

35. Kenrick JR, Bishop DG. 1986. The fatty acid compositon of phosphatidylglycerol and sulfoquinovosyldiacylglycerol of higher plants in regulation to chilling sensitivity. *Plant Physiol.* 81:946–49

36. Kim JC, Lee SH, Cheong YH, Yoo C-M, Lee SI, et al. 2001. A novel cold-inducible zinc finger protein from soybean, SCOF-1, enhances cold tolerance in transgenic plants. *Plant J.* 25:247–59

37. Kishitani S, Takanami T, Suzuki M, Oikawa M, Yokoi S, et al. 2000. Compatibility of glycinebetaine in rice plants: evaluation using transgenic rice plants with a gene for peroxisomal betaine aldehyde dehydrogenase from barley. *Plant Cell Environ.* 23:107–14

38. Kodama H, Hamada T, Horiguchi G, Nishimura M, Iba K. 1994. Genetic enhancement of cold tolerance by expression of a gene for chloroplast ω-3 fatty acid desaturase in transgenic tobacco. *Plant Physiol.* 105:601–5

39. Kodama H, Horiguchi G, Nishiuchi T, Nishimura M, Iba K. 1995. Fatty acid desaturation during chilling acclimation is one of the factors involved in conferring low-temperature tolerance to young tobacco leaves. *Plant Physiol.* 107:1177–85

40. Komari T, Hiei Y, Ishida Y, Kumashiro T, Kubo T. 1998. Advances in cereal gene transfer. *Curr. Opin. Plant Biol.* 1:161–65

41. Kunst L, Browse J, Somerville C. 1989. Enhanced thermal tolerance in a mutant of *Arabidopsis* deficient in palmitic acid unsaturation. *Plant Physiol.* 91:401–8

42. Kuwabara C, Takezawa D, Fujikawa S, Arakawa K. 2001. Antifungal effect of WAS-3a protein to snow mold in winter wheat. *Plant Cell Physiol.* 42:438 (Abstr.)

43. Lee GJ, Roseman AM, Saibil HR, Vierling E. 1997. A small heat shock protein

stably binds heat-denatured model substrates and can maintain a substrate in a folding-competent state. *EMBO J.* 16: 659–71

44. Lee JH, Hübel A, Schöffl F. 1995. Derepression of the activity of genetically engineered heat shock factor causes constitutive synthesis of heat shock proteins and increased thermotolerance in transgenic *Arabidopsis*. *Plant J.* 8:603–12

45. Lee JH, Schöffl F. 1996. An *HSP70* antisense gene affects the expression of HSP70/HSC70, the regulation of HSF, and the acquisition of thermotolerance in transgenic *Arabidopsis thaliana*. *Mol. Gen. Genet.* 252:11–19

46. Lee W-M, Ishikawa M, Ahlquist P. 2001. Mutation of host Δ9 fatty acid desaturase inhibits brome mosaic virus RNA replication between template recognition and RNA synthesis. *J. Virol.* 75:2097–106

47. Lemieux B, Miquel M, Somerville C, Browse J. 1990. Mutants of *Arabidopsis* with alterations in seed lipid fatty acid compositon. *Theor. Appl. Genet.* 80:234–40

48. Maleck K, Levine A, Eulgem T, Morgan A, Schmid J, et al. 2000. The transcriptome of *Arabidopsis thaliana* during systemic acquired resistance. *Nat. Genet.* 26:403–10

49. Malik K, Slovin JP, Hwang CH, Zimmerman JL. 1999. Modified expression of a carrot small heat shock protein gene, *Hsp17.7*, results in increased or decreased thermotolerance. *Plant J.* 20:89–99

50. Matsuda O, Iba K. 2001. Identification of functional region of plastidial ω-3 fatty acid desaturase gene (*FAD8*) affecting temperature-regulated expression. *Plant Cell Physiol.* 42:444 (Abstr.)

51. Matsuda O, Watanabe C, Iba K. 2001. Hormonal regulation of tissue-specific ectopic expression of an *Arabidopsis* endoplasmic reticulum-type ω-3 fatty acid desaturase (*FAD3*) gene. *Planta* 213:833–40

52. McConn M, Browse J. 1996. The critical requirement for linolenic acid is pollen development, not photosynthesis, in an Arabidopsis mutant. *Plant Cell* 8:403–16

53. McConn M, Hugly S, Browse J, Somerville C. 1994. A mutation at the *fad8* locus of *Arabidopsis* identifies a second chloroplast ω-3 desaturase. *Plant Physiol.* 106:1609–14

54. McKersie BD, Bowley SR, Jones KS. 1999. Winter survival of transgenic alfalfa overexpressing superoxide dismutase. *Plant Physiol.* 119:839–47

55. McKersie BD, Murnaghan J, Jones KS, Bowley SR. 2000. Iron-superoxide dismutase expression in transgenic alfalfa increases winter survival without a detectable increase in photosynthetic oxidative stress tolerance. *Plant Physiol.* 122:1427–37

56. Merkle SA, Dean JFD. 2000. Forest tree biotechnology. *Curr. Opin. Biotech.* 11:298–302

57. Miernyk JA, Hayman GT. 1996. ATPase activity and molecular chaperone function of the stress70 proteins. *Plant Physiol.* 110:419–24

58. Miquel M, James D Jr, Dooner H, Browse J. 1993. *Arabidopsis* requires polyunsaturated lipids for low-temperature survival. *Proc. Natl. Acad. Sci. USA* 90:6208–12

59. Mittler R, Zilinskas BA. 1992. Molecular cloning and characterization of a gene encoding pea cytosolic ascorbate peroxidase. *J. Biol. Chem.* 267:21802–7

60. Moon BY, Higashi S, Gombos Z, Murata N. 1995. Unsaturation of the membrane lipids of chloroplasts stabilizes the photosynthetic machinery against low-temperature photoinhibition in transgenic tobacco plants. *Proc. Natl. Acad. Sci. USA* 92:6219–23

61. Morimoto RI, Tissieres A, Georgopoulos C, eds. 1994. *The Biology of Heat Shock Proteins and Molecular Chaperones*, pp. 1–30. New York: Cold Spring Harbor Lab. Press

62. Murakami Y, Tsuyama M, Kobayashi Y, Kodama H, Iba K. 2000. Trienoic fatty

acids and plant tolerance of high temperature. *Science* 287:476–79

63. Murata N. 1983. Molecular species composition of phosphatidylglycerols from chilling-sensitive and chilling-resistant plants. *Plant Cell Physiol.* 24:81–86

64. Murata N, Ishizaki-Nishizawa O, Higashi S, Hayashi H, Tasaka Y, et al. 1992. Genetically engineered alteration in the chilling sensitivity of plants. *Nature* 356:710–13

65. Murata N, Sato N, Takahashi N, Hamazaki Y. 1982. Compositions and positional distributions of fatty acids in phospholipids from leaves of chilling-sensitive and chilling-resistant plants. *Plant Cell Physiol.* 23:1071–79

66. Murata N, Yamaya J. 1984. Temperature-dependent phase behavior of phosphatidylglycerols form chilling-sensitive and chilling-resistant plants. *Plant Physiol.* 74:1016–24

67. Nadolska-Orczyk A, Orczyk W, Przetakiewicz A. 2000. Agrobacterium-mediated transformation of cereals—from technique development to its application. *Acta Physiol. Plant* 22:77–88

68. Nakamura T, Yokota S, Muramoto Y, Tsutsui K, Oguri Y, et al. 1997. Expression of a betaine aldehyde dehydrogenase gene in rice, a glycinebetaine nonaccumulator, and possible localization of its protein in peroxisomes. *Plant J.* 11:1115–20

69. Newman JE. 1980. Climate change impacts on the growing season of the North American "Corn Belt." *Biometeorology* 7:128–42

70. Nishida I, Murata N. 1996. Chilling sensitivity in plants and cyanobacteria: the crucial contribution of membrane lipids. *Annu. Rev. Plant Physiol. Plant Mol. Biol.* 47:541–68

71. Nishiuchi T, Hamada T, Kodama H, Iba K. 1997. Wounding changes the special expression pattern of the Arabidopsis plastid ω-3 fatty acid desaturase gene (*FAD7*) through differential signal transduction pathways. *Plant Cell* 9:1701–12

72. Nishiuchi T, Iba K. 1998. Roles of plastid ω-3 fatty acid desaturases in defense response of higher plants. *J. Plant Res.* 111:481–86

73. Nishiuchi T, Kodama H, Yanagisawa S, Iba K. 1999. Wound-induced expression of the *FAD7* gene is mediated by differential regulatory domains of its promoter in leaves/stem and roots. *Plant Physiol.* 121:1239–46

74. Norman HA, McMillan C, Thompson GA Jr. 1984. Phosphatidylglycerol molecular species in chilling-sensitive and chilling-resistant populations of *Avicennia germinans* L. *Plant Cell Physiol.* 25:1437–44

75. Nuccio ML, Russell BL, Nolte KD, Rathinasabapathi B, Gage DA, et al. 1998. The endogenous choline supply limits glycine betaine synthesis in transgenic tobacco expressing choline monooxygenase. *Plant J.* 16:487–96

76. O'Kane D, Gill V, Boyd P, Burdon R. 1996. Chilling, oxidative stress and antioxidant responses in *Arabidopsis thaliana* callus. *Planta* 198:371–77

77. Okuley J, Lightner J, Feldmann K, Yadav N, Lark E, Browse J. 1994. *Arabidopsis FAD2* gene encodes the enzyme that is essential for polyunsaturated lipid synthesis. *Plant Cell* 6:147–58

78. Quinn PJ, Williams WP. 1983. The structural role of lipids in photosynthetic membranes. *Biochim. Biophys. Acta* 737:223–66

79. Rathinasabapathi B, Burnet M, Russell BL, Gage DA, Liao P-C, et al. 1997. Choline monooxygenase, an unusual iron-sulfur enzyme catalyzing the first step of glycine betaine synthesis in plants: prosthetic group characterization and cDNA cloning. *Proc. Natl. Acad. Sci. USA* 94:3454–58

80. Robinson SP, Jones GP. 1986. Accumulation of glycinebetaine in chloroplasts provides osmotic adjustment during salt stress. *Aust. J. Plant Physiol.* 13:659–68

81. Rosenzweig C. 1985. Potential CO_2-induced climate effects on North American

wheat-producing regions. *Clim. Chang.* 7:367–89

82. Rosenzweig C, Parry ML. 1994. Potential impact of climate change on world food supply. *Nature* 367:133–38

83. Roughan PG. 1985. Phosphatidylglycerol and chilling sensitivity in plants. *Plant Physiol.* 77:740–46

84. Roxas VP, Smith RK Jr, Allen ER, Allen RD. 1997. Overexpression of glutathione S-transferase/glutathione peroxidase enhances the growth of transgenic tobacco seedlings during stress. *Nat. Biotech.* 15:988–91

85. Sakamoto A, Valverde R, Alia R, Chen THH, Murata N. 2000. Transformation of *Arabidopsis* with the *codA* gene for choline oxidase enhances freezing tolerance of plants. *Plant J.* 22:449–53

86. Saruyama H, Tanida M. 1995. Effect of chilling on activated oxygen-scavenging enzymes in low temperature-sensitive and -tolerant cultivars of rice (*Oryza sativa* L.). *Plant Sci.* 109:105–13

87. Sasaki T. 1997. *Science of the Rice Plant (Genetics)*, ed. T Matsuo, Y Futsuhara, F Kikuchi, H Yamaguchi, 3:534–49. Tokyo: Nobunkyo

88. Sen Gupta A, Heinen JL, Holaday AS, Burke JJ, Allen RD. 1993. Increased resistance to oxidative stress in transgenic plants that overexpress chloroplastic Cu/Zn superoxide dismutase. *Proc. Natl. Acad. Sci. USA* 90:1629–33

89. Sen Gupta A, Webb RP, Holaday AS, Allen RD. 1993. Overexpression of superoxide dismutase protects plants from oxidative stress. *Plant Physiol.* 103:1067–73

90. Shanklin J, Cahoon EB. 1998. Desaturation and related modifications of fatty acids. *Annu. Rev. Plant Physiol. Plant Mol. Biol.* 49:611–41

91. Sharkey TD. 2000. Some like it hot. *Science* 287:435–37

92. Shimada T, Wakita Y, Otani M, Iba K. 2000. Modification of fatty acid composition in rice plants by transformation with a tobacco microsomal ω-3 fatty acid desaturase gene (*NtFAD3*). *Plant Biotech.* 17:43–48

93. Smit B, Ludlow L, Brklacich M. 1988. Implications of a global climatic warming for agriculture: a review and appraisal. *J. Environ. Q.* 17:519–27

94. Smolenska G, Kuiper PJC. 1977. Effect of low temperature upon lipid and fatty acid composition of roots and leaves of winter rape plants. *Physiol. Plant.* 41:29–35

95. Somerville C, Browse J. 1991. Plant lipids: metabolism, mutants, and membranes. *Science* 252:80–87

96. Storozhenko S, De Pauw P, Van Montagu M, Inzé D, Kushnir S. 1998. The heat-shock element is a functional component of the Arabidopsis *APX1* gene promoter. *Plant Physiol.* 118:1005–14

97. Sugino M, Hibino T, Tanaka Y, Nii N, Takabe T, et al. 1999. Overexpression of DnaK from a halotolerant cyanobacterium *Aphanothece halophytice* acquires resistance to salt stress in transgenic tobacco plants. *Plant Sci.* 137:81–88

98. Sun W, Bernard C, von de Cotte B, Van Montagu M, Verbruggen N. 2001. *At-HSP17.6A*, encoding a small heat-shock protein in *Arabidopsis*, can enhance osmotolerance upon overexpression. *Plant J.* 27:407–15

99. Suzuki I, Los DA, Kanesaki Y, Mikami K, Murata N. 2000. The pathway for perception and transduction of low-temperature signals in *Synechocystis*. *EMBO J.* 19:1327–34

100. Tamminen I, Mükelä P, Heino P, Palva ET. 2001. Ectopic expression of *ABI3* gene enhances freezing tolerance in response to abscisic acid and low temperature in *Arabidopsis thaliana*. *Plant J.* 25:1–8

101. Thomas PG, Dominy PJ, Vigh L, Mansourian AR, Quinn PJ, et al. 1986. Increased thermal stability of pigment-protein complexes of pea thylakoids following catalytic hydrogenation of membrane lipids. *Biochim. Biophys. Acta* 849:131–40

102. Thomashow MF. 1999. Plant cold acclimation: freezing tolerance genes and regulatory mechanisms. *Annu. Rev. Plant Physiol. Plant Mol. Biol.* 50:571–99

103. Tsang EWT, Bowler C, Hérouart D, Van Camp W, Villarroel R, et al. 1991. Differential regulation of superoxide dismutases in plants exposed to environmental stress. *Plant Cell* 3:783–92

104. Weretilnyk EA, Hanson AD. 1990. Molecular cloning of a plant betaine-aldehyde dehydrogenase, an enzyme implicated in adaptation to salinity and drought. *Proc. Natl. Acad. Sci. USA* 87:2745–49

105. Willemot C, Hope HJ, Williams RJ, Michaud R. 1977. Changes in fatty acid composition of winter wheat during frost hardening. *Cryobiology* 14:87–93

106. Wolter FP, Schmidt R, Heinz E. 1992. Chilling sensitivity of *Arabidopsis thaliana* with genetically engineered membrane lipids. *EMBO J.* 11:4685–92

107. Wu C. 1995. Heat shock transcription factors: structure and regulation. *Annu. Rev. Cell Dev. Biol.* 11:441–69

108. Wu J, Browse J. 1995. Elevated levels of high-melting-point phosphatidylglycerols do not induce chilling sensitivity in an Arabidopsis mutant. *Plant Cell* 7:17–27

109. Valenzuela-Soto EM, Muñoz-Clares RA. 1994. Purification and properties of betaine aldehyde dehydrogenase extracted from detached leaves of *Amaranthus hypochondriacus* L. subjected to water deficit. *J. Plant Physiol.* 143:145–52

110. Yamamoto KT. 1994. Further characterization of auxin-regulated mRNAs in hypocotyl sections of mung bean [*Vigna radiata* (L.) Wilczek]: sequence homology to genes for fatty-acid desaturases and a typical late-embryogenesis-abundant protein, and the mode of expression of the mRNAs. *Planta* 192:359–64

111. Yancey PH. 1994. Compatible and counteracting solutes. *Cellular and Molecular Physiology of Cell Volume Regulation*, ed. K Strange, pp. 81–109. Austin, TX: CRC Press

112. Yokoi S, Higashi S, Kishitani S, Murata N, Toriyama K. 1998. Introduction of the cDNA for *Arabidopsis* glycerol-3-phosphate acyltransferase (GPAT) confers unsaturation of fatty acids and chilling tolerance of photosynthesis on rice. *Mol. Breed.* 4:269–75

113. Zou J, Abrams GD, Barton DL, Taylor DC, Pomeroy MK, et al. 1995. Induction of lipid and oleosin biosynthesis by (+)-abscisic acid and its metabolites in microspore-derived embryos of *Brassica napus* L. cv Reston. *Plant Physiol.* 108:563–71

Annu. Rev. Plant Biol. 2002. 53:247–73
DOI: 10.1146/annurev.arplant.53.091401.143329

SALT AND DROUGHT STRESS SIGNAL TRANSDUCTION IN PLANTS

Jian-Kang Zhu

Department of Plant Sciences, University of Arizona, Tucson, Arizona 85721;
e-mail: jkzhu@ag.arizona.edu

Key Words   osmotic stress, abscisic acid, SOS, protein kinase, phospholipid, calcium signaling, gene expression

■ **Abstract**   Salt and drought stress signal transduction consists of ionic and os-motic homeostasis signaling pathways, detoxification (i.e., damage control and repair) response pathways, and pathways for growth regulation. The ionic aspect of salt stress is signaled via the SOS pathway where a calcium-responsive SOS3-SOS2 protein ki-nase complex controls the expression and activity of ion transporters such as SOS1. Osmotic stress activates several protein kinases including mitogen-activated kinases, which may mediate osmotic homeostasis and/or detoxification responses. A number of phospholipid systems are activated by osmotic stress, generating a diverse array of messenger molecules, some of which may function upstream of the osmotic stress–activated protein kinases. Abscisic acid biosynthesis is regulated by osmotic stress at multiple steps. Both ABA-dependent and -independent osmotic stress signaling first modify constitutively expressed transcription factors, leading to the expression of early response transcriptional activators, which then activate downstream stress tolerance effector genes.

CONTENTS

INTRODUCTION

Modern-day plants are products of eons of evolution from primal living organisms in response to abiotic and biotic environmental changes. Among the abiotic factors that have shaped and continue shaping plant evolution, water availability is the most important. Water stress in its broadest sense encompasses both drought and salt stress. Because cell signaling controls plant responses and adaptation, it is probably not an exaggeration to state that water stress signaling has in large part shaped the flora on earth. Drought and salt stress, together with low temperature, are the major problems for agriculture because these adverse environmental factors prevent plants from realizing their full genetic potential.

Salt stress afflicts plant agriculture in many parts of the world, particularly irrigated land (16a). Compared to salt stress, the problem of drought is even more pervasive and economically damaging (5a). In this regard, drought stress signaling certainly merits separate treatment. Nevertheless, most studies on water stress signaling have focused on salt stress primarily because plant responses to salt and drought are closely related and the mechanisms overlap. From a practical point, salt stress can be imposed more easily and precisely in laboratory settings. Although the importance of salt and drought stress signaling was recognized long ago, few molecular components were known until recently. As such, salt and drought stress signaling was reviewed only as part of salt and drought stress tolerance (24, 31), and it has not been reviewed as a separate subject in this series.

This review focuses on general cell-based salt and drought stress signaling. Long-distance signaling within a plant and even interplant signaling are important for plant adaptation, but little mechanistic information is available. In drought stress responses, guard cell signaling is of critical importance because it is a key denominator within the plant water budget. Much effort has been justifiably dedicated to guard cell signaling and substantial advances have been made. Several excellent reviews are devoted to this subject, including Schroeder et al. (85) and Leung & Giraudat (51). Despite the fact that guard cells are specialized structures, most of what is learned there is probably applicable to other cells as well. Therefore, some of the knowledge of general cell signaling presented here has roots in guard cell studies.

INPUTS AND OUTPUTS: MAKING SENSE OF SALT AND DROUGHT STRESS SIGNALING PATHWAYS

Salt and drought stresses affect virtually every aspect of plant physiology and metabolism. Numerous changes that occur under these stresses have been documented. Although some of the changes are clearly adaptive, many may simply be pathological consequences of stress injury. True, there may be active signaling to activate pathological responses. Knowledge of them is just as important because they represent suitable targets for genetic suppression to improve salt and drought stress tolerance. In nature, for a plant to sacrifice a part of its structure constitutes an adaptive strategy to survive a stress episode.

For adaptive or presumed adaptive responses, it may be helpful to conceptually group them into three aspects: (*a*) homeostasis that includes ion homeostasis, which is mainly relevant to salt stress, and osmotic homeostasis or osmotic adjustment; (*b*) stress damage control and repair, or detoxification; and (*c*) growth control (119). Accordingly, salt and drought stress signaling can be divided into three functional categories: ionic and osmotic stress signaling for the reestablishment of cellular homeostasis under stress conditions, detoxification signaling to control and repair stress damages, and signaling to coordinate cell division and expansion to levels suitable for the particular stress conditions (Figure 1). Homeostasis signaling negatively regulates detoxification responses because, once cellular homeostasis is reestablished, stress injury would be reduced, and failure to reestablish homeostasis would aggravate stress injury. Homeostasis and detoxification signaling lead to stress tolerance and are expected to negatively regulate the growth inhibition response, i.e., to relieve growth inhibition.

It is not enough to know only the components or elements of signaling pathways (113). Good comprehension requires knowledge of precise inputs and outputs of the pathways. It is here that water stress signaling is most poorly understood. Just because certain changes occur upon drought stress treatment, one cannot assume that drought is the direct input signal, nor can one assume that the output is any

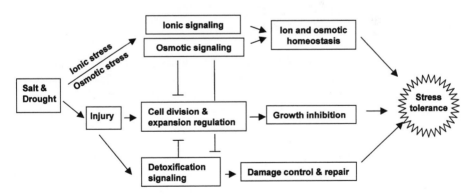

Figure 1   Functional demarcation of salt and drought stress signaling pathways. The inputs for ionic and osmotic signaling pathways are ionic (excess Na^+) and osmotic (e.g., turgor) changes. The output of ionic and osmotic signaling is cellular and plant homeostasis. Direct input signals for detoxification signaling are derived stresses (i.e., injury), and the signaling output is damage control and repair (e.g., activation of dehydration tolerance genes). Interactions between the homeostasis, growth regulation, and detoxification pathways are indicated.

of the adaptive responses. For the ionic aspect of salt stress, a signaling pathway based on the *SOS* (*Salt Overly Sensitive*) genes has been established. The input of the SOS pathway is likely excess intracellular or extracelluar Na^+, which somehow triggers a cytoplasmic Ca^{2+} signal (117). The outputs are expression and activity changes of transporters for ions such as Na^+, K^+, and H^+. The input for osmotic stress signaling is likely a change in turgor. Several osmotic stress–activated, SOS-independent protein kinases probably mediate osmotic stress signaling, but no output is known (119). Possible outputs of osmotic signaling pathways include gene expression and/or activation of osmolyte biosynthesis enzymes as well as water and osmolyte transport systems. Most of the other changes induced by salt or drought stress can be considered as part of detoxification signaling. These include (*a*) phospholipid hydrolysis; (*b*) changes in the expression of LEA/dehydrin-type genes, molecular chaperones, and proteinases that remove denatured proteins; and (*c*) activation of enzymes involved in the generation and removal of reactive oxygen species and other detoxification proteins. The input signal(s) for the detoxification pathways is mostly likely not an ionic or osmotic change, but a product of stress injury, e.g., reactive oxygen species or protein denaturation.

Water stress generally inhibits plant growth, and indirect evidence suggests active signaling to cell division and expansion machineries (119). Slower cell division under water stress is probably a result of reduced cyclin-dependent kinase (CDK) activity (87). Reduced CDK activity may be a result of combined effects of transcription suppression of cyclins and CDKs (7) and induction of CDK inhibitors (106). The direct input signal(s) for the CDK regulation is unclear but can be a product of stress injury or any of the primary or intermediary signals involved in the homeostasis and detoxification pathways.

THE SOS REGULATORY PATHWAY FOR ION HOMEOSTASIS AND SALT TOLERANCE

Although a number of possible pathways for salt, drought, or cold signaling have been proposed, none is established in terms of signaling proteins and inputs and outputs. One exception is the SOS pathway that emerged recently as a result of genetic, molecular, and biochemical analysis (117). An outline of this pathway is shown in Figure 2. Salt stress elicits a cytosolic calcium signal (42). How the calcium signal is different from that triggered by drought, cold, or other stimuli remains a mystery. A myristoylated calcium-binding protein encoded by *SOS3* presumably senses the salt-elicited calcium signal and translates it to downstream responses (32, 55). SOS3 interacts with and activates SOS2, a serine/threonine protein kinase (23, 53). SOS2 and SOS3 regulate the expression level of *SOS1*, a salt tolerance effector gene encoding a plasma membrane Na^+/H^+ antiporter (92). More importantly, SOS2 and SOS3 are required for the activation of SOS1 transport activity (78). SOS1 by itself can slightly increase the salt tolerance of a yeast mutant strain lacking all endogenous Na^+-ATPases and Na^+/H^+ antiporters (93). Coexpression of SOS3 and SOS2, together with SOS1, can dramatically

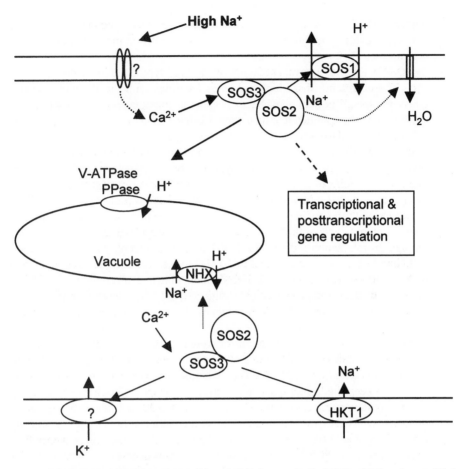

Figure 2 Regulation of ion (e.g., Na^+ and K^+) homeostasis by the SOS pathway. High Na^+ stress initiates a calcium signal that activates the SOS3-SOS2 protein kinase complex, which then stimulates the Na^+/H^+ exchange activity of SOS1 and regulates transcriptionally and posttranscriptionally the expression of some genes. SOS3-SOS2 may also stimulate or suppress the activities of other transporters involved in ion homeostasis under salt stress, such as vacuolar H^+-ATPases and pyrophosphatases (PPase), vacuolar Na^+/H^+ exchanger (NHX), and plasma membrane K^+ and Na^+ transporters.

enhance the salt tolerance of the yeast mutant (J. M. Pardo & J-K. Zhu, unpublished information). Expression of a constitutively activated SOS2 mutant could also increase the salt tolerance activity of SOS1 in the yeast mutant, implying that SOS2 kinase activity is sufficient for SOS1 activation. In a complementary study, Qiu et al. (78) showed that constitutively active SOS2 kinase could enhance a Na^+/H^+ exchange activity in purified plasma membrane vesicles from wild-type but not *sos1-1* mutant (107) plants. In *sos2-2* and *sos3-1* mutants (121), the plasma membrane Na^+/H^+ exchange activity is much lower but can be recovered to near wild-type levels by addition of activated SOS2 in vitro to the membrane vesicle preparations (78).

SOS3 belongs to a novel subfamily of EF-hand-type calcium-binding proteins (21). These proteins share high sequence identities with the B-subunit of calcineurin (type 2B protein phosphatase) and with animal neuronal calcium sensors (21, 55). They are predicted to have three EF-hands and bind to calcium with low affinity compared to calmodulin or caltractin (32; Y. Guo & J-K. Zhu, unpublished information). Only some members of the protein family contain an N-terminal myristoylation motif. SOS3 is myristoylated, which may help target SOS3 and its interacting proteins (e.g., SOS2) to membranes where their target transporters are located (e.g., SOS1). Nevertheless, not all of the SOS3 protein in plant cells is associated with membranes (32), consistent with the presence of nonmembrane-bound protein targets of the SOS3-SOS2 complex. Besides a role in salt tolerance through an interaction with SOS2, SOS3 may also interact with other proteins including SOS2-like proteins (i.e., PKS) (21) to mediate osmotic stress induction of ABA biosynthesis (L. Xiong & J-K. Zhu, unpublished information). SOS3-like proteins (SCaBPs) interact specifically with certain PKS proteins, forming distinct protein kinase complexes that likely mediate calcium signaling in response to other stimuli (21).

SOS2 also represents a novel family of proteins (PKS) so far only found in plants (21). SOS2 contains an SNF1-like catalytic domain and a unique regulatory domain that interacts with SOS3 (23, 53). The regulatory and catalytic domains of SOS2 interact to keep the kinase inactive in substrate phosphorylation, presumably by preventing substrate access to the catalytic site (21). SOS3 binding to the regulatory domain appears to disrupt the intramolecular interaction of SOS2, thus opening up the catalytic site. Serial deletion analysis identified a 21 amino acid sequence (i.e., FISL motif) in the SOS2 regulatory domain, which is necessary and sufficient for SOS3 binding. Deletion of the entire SOS2 regulatory domain (21) or simply of the FISL motif (Y. Guo & J-K. Zhu, unpublished information) results in constitutive activation of the kinase. Expression of the constitutively active SOS2 under the CaMV 35S promoter in *sos2* or *sos3* mutant plants can rescue the salt sensitive phenotype in the mutant shoot but, curiously, not in the root (Y. Guo & J-K. Zhu, unpublished information), suggesting that SOS2 kinase activity is sufficient for the SOS pathway in salt tolerance at least in the shoot.

The plasma membrane Na^+/H^+ antiporter SOS1 has a very long tail that is predicted to be on the cytoplasmic side (92). Membrane transporters with long cytoplasmic tails have been proposed to function as sensors of the solutes they transport. For example, evidence suggests that the glucose transporters Snf3 and Rgt2 in yeast can serve as glucose sensors and regulate gene expression in response to glucose starvation (75). The possibility of SOS1 being both a transporter and a sensor cannot be dismissed. If it is a Na^+ sensor, it may control SOS2 activation in plants, forming a regulatory loop because an activated SOS2 would stimulate its capacity for Na^+ efflux. Future determination of SOS2 kinase activity in response to salt treatment in wild-type and *sos1* mutants may prove or disprove this speculation. Besides being regulated by SOS2, SOS1 activity may also be regulated by SOS4. SOS4 catalyzes the formation of pyridoxal-5-phosphate, a cofactor that may serve as a ligand for SOS1 because the latter contains a putative binding sequence for this cofactor (94; H. Shi & J-K. Zhu, unpublished information).

It will not be surprising if SOS3 and SOS2 regulate Na^+ influx as well as vacuolar compartmentation systems because these systems are also vital for salt tolerance. Mutations in the *AtHKT1* gene suppressed *sos3* mutant phenotypes (81). AtHKT1 mediates Na^+ influx in oocytes and yeast (103). Ion content analysis in the *sos3hkt1* mutant supports that AtHKT1 mediates Na^+ influx in planta (81). It is possible that SOS3 and SOS2 may downregulate AtHKT1 activity under salt stress. Nonselective cation channels contribute to Na^+ influx (2, 101). Cyclic nucleotide–gated ion channels (CNGCs) are a group of nonselective cation channels that are inhibited by calcium and calmodulin. When expressed in heterologous systems (50), CNGC2 was activated by cyclic nucleotides and but did not show Na^+ permeability. However, patch clamping of Arabidopsis root cells showed that nonselective cation channels were inhibited by cAMP or cGMP in vivo (59). Arabidopsis seedling growth on 100 mM NaCl was improved by inclusion of cGMP and to a lesser extent by cAMP. Consistent with an active role of cyclic nucleotides in salt tolerance, cGMP-treated plants had less Na^+ accumulation (59). Some CNGCs may also mediate calcium influx and thus are potentially upstream of SOS3 if they can generate cytosolic calcium transients under salt stress.

PROTEIN KINASE PATHWAYS FOR OSMOTIC STRESS SIGNALING

Protein phosphorylation is such a central theme in cell signaling that its involvement in osmotic stress adaptation was predicted a while ago. Only recently, though, has direct experimental evidence been reported (118). In fact, several plant protein kinases have now been found to be activated by osmotic stress. Because of the well-known osmosensing pathway in yeast (22), much attention has been directed toward identifying a homologous pathway in plants. The yeast osmoregulatory pathways begin with either an Src-homology 3 (SH3) domain-containing membrane protein or a two-component histidine kinase, which then activates a MAP kinase cascade and leads to increased osmolyte synthesis and accumulation (22). Although plants accumulate compatible osmolytes for osmotic adjustment, whether they use similar membrane sensors and MAP kinase cascades to regulate osmolyte synthesis is unclear. Recent studies using in-gel kinase assays identified several protein kinases that are activated by osmotic stress in plants, but whether any of the pathways is directly activated by osmotic stress or by a derived stress needs to be addressed (Figure 3). In addition, a central question remains as to the outputs of the pathways, i.e., whether they regulate osmolyte biosynthesis or other stress responses.

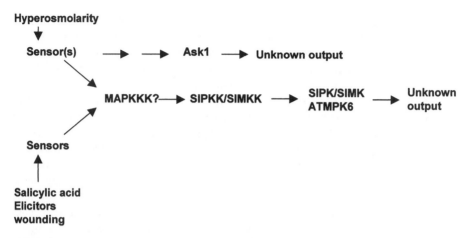

Figure 3   Activation of protein kinases by hyperosmotic stress. The MAP kinase cascade shown is also activated by other stresses. Currently, the functional significance of the kinase activation is unclear (hence the "unknown output"). SIPK, SIMK, and ATMPK6 are homologous MAP kinases from tobacco, alfalfa, and Arabidopsis, respectively.

Plants have several MAP kinases that are activated by hyperosmotic stress (Figure 3). In alfalfa cells, a 46-kD MAP kinase named SIMK became activated in response to moderate hyperosmotic stress (67). It is interesting to note that at severe hyperosmotic stress (>750 mM NaCl) SIMK was no longer activated. Instead, a smaller kinase became activated, suggesting that the two kinases function at different stress levels. In tobacco cells, an SIMK-like MAP kinase named SIPK (salicylic acid–induced protein kinase) was activated by hyperosmotic stress (64). In addition to hyperosmotic stress, hypoosmotic stress, salicylic acid, or fungal elicitors also could activate the tobacco MAP kinase (13). An Arabidopsis protein crossreacting with antibodies against the tobacco MAP kinase was found to be activated by hyperosmotic stress (29). A MAP kinase kinase that interacts with and activates the alfalfa SIMK was reported (41). A tobacco MAP kinase kinase that interacts with SIPK was also found (57). The enormous complexity of MAP kinases can be seen in a study that showed that at least three MAP kinases in Arabidopsis were enzymatically activated by salt as well as by cold, wounding, and other environmental signals (30). The picture is equally complicated in other plants. For example, other MAP kinases, e.g., WIPK in tobacco and SAMK in alfalfa, are activated by cold, drought, wounding, and biotic signals (37, 89).

In cultured tobacco cells, a 42-kD protein kinase is activated rapidly in response to hyperosmotic stress treatment (64). Partial peptide sequences from this kinase suggest that it is an ortholog of Arabidopsis ASK1, a SNF1-related kinase without a known function. In Arabidopsis seedlings, a 40-kD kinase was also identified as being rapidly activated by salt or sorbitol stress in a calcium- and ABA-independent manner (29). Although the amino acid sequence of this 40-kD Arabidopsis protein is not known, it is probably the homolog of the 42-kD kinase of tobacco. Osmotic stress activation of the Arabidopsis 40-kD protein kinase was independent of the SOS pathway because it was not affected by the *sos3* mutation. As these 40–42-kD plant kinases are clearly not of the MAP kinase type, their activation may represent a novel osmotic stress signaling pathway in plants (Figure 3).

Because osmotic stress elicits calcium signaling (42), calcium-dependent protein kinases are prime candidates that link the calcium signal to downstream responses. In a maize protoplast transient expression system, a constitutively active CDPK (calcium-dependent protein kinase) mutant activated the expression of a reporter gene that was normally responsive to osmotic stress, cold, or ABA (91). A dominant negative form of the CDPK was able to block stress or ABA induction of the reporter gene. If protein kinase specificity is maintained in the protoplast transient assay system, then the finding would be very significant in that it links calcium signaling with gene induction by osmotic stress.

Transcript levels for a number of protein kinases including a two-component histidine kinase, MAPKKK, MAPKK, and MAPK increase in response to osmotic and other stress treatments (65). It is unclear whether the protein or, more importantly, the activity levels of these kinases change upon osmotic stress treatment. For these and the osmotic stress–activated kinases discussed above, it is vital to identify the input and output of the kinases and of the pathways. The input signal could be osmotic stress (e.g., turgor changes) or derived from osmotic stress injury. The output could be osmolyte accumulation that helps establish osmotic homeostasis, stress damage protection, or repair mechanisms (e.g., induction of LEA/dehydrin-type stress genes).

OSMOTIC STRESS–ACTIVATED PHOSPHOLIPID SIGNALING

Membrane phospholipids constitute a dynamic system that generates a multitude of signal molecules (e.g., IP_3, DAG, PA, etc.) in addition to serving important structural roles during stress responses. The potential of phospholipid-based signaling in plants is still underexplored and cannot be overappreciated (68). However, it must also be pointed out that the phospholipid system, like the reactive oxygen species that is associated with it (84), is a double-edged sword: As signaling molecules at low levels, the phospholipid messengers may activate downstream adaptive responses, whereas at high levels, phospholipid-generated products may reflect stress damage or may be damaging.

Phospholipid signaling systems are typically grouped according to the phospholipases that catalyze the formation of lipid and other messengers (Figure 4). There are also novel pathways involving the formation of lipid messengers that are not the direct products of phospholipases, such as diacylglycerol pyrophosphate (DGPP) and phosphatidylinositol 3,5-bisphosphate [PI(3,5)P2] (68). The phospholipase C (PLC) pathway has been the best characterized, particularly in nonplant

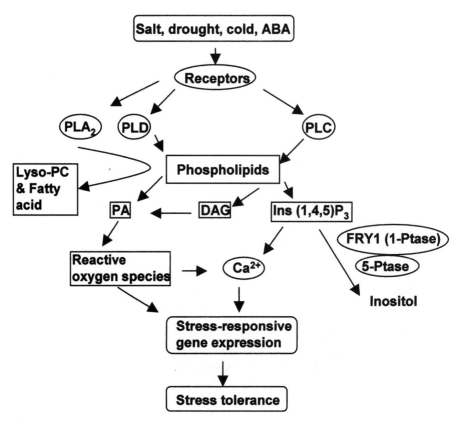

Figure 4 Phospholipid signaling under salt stress, drought, cold, or ABA. Osmotic stress, cold, and ABA activate several types of phospholipases that cleave phospholipids to generate lipid messengers (e.g., PA, DAG, and IP₃), which regulate stress tolerance partly through modulation of stress-responsive gene expression. FRY1 (a 1-phosphatase) and 5-phosphatase-mediated IP₃ degradation attenuates the stress gene regulation by helping to control cellular IP₃ levels. PLC, phospholipase C; PLD, phospholipase D; PLA₂, phospholipase A₂; PA, phosphatidic acid; DAG, diacyglycerol.

systems. PLC catalyzes the hydrolysis of phosphotidylinositol 4,5-bisphosphate (PIP₂), generating the second messengers inositol 1,4,5-trisphosphate (IP₃) and di-acylglycerol (DAG). IP₃ releases Ca^{2+} from internal stores, whereas DAG activates protein kinase C. Several studies have shown that in various plant systems IP₃ levels rapidly increase in response to hyperosmotic stress (10, 12, 25, 98). IP₃ levels also increased upon treatment with exogenous ABA in *Vicia faba* guard cell protoplasts (48) and in Arabidopsis seedlings (112). The IP₃ precursor, PIP2, is synthesized via phosphatidylinositol 4-phosphate 5-kinase. An Arabidopsis gene encoding this enzyme, *PIP5K*, is induced by osmotic stress and ABA (63). Similarly, an Arabidopsis *PLC* gene, *AtPLC1*, is also induced by salt and drought stress (26).

In guard cells, caged IP_3 induced Ca^{2+} increase in the cytoplasm and triggered stomatal closure (83). Exogenous IP_3 releases Ca^{2+} from isolated vacuoles or tonoplast vesicles (86). Microinjection as well as pharmacological experiments suggested that increases in cytoplasmic Ca^{2+} could lead to the expression of osmotic stress-responsive genes (108). Recently, a connection between phosphoinositides and stress gene expression has also been demonstrated genetically. In the Arabidopsis *fry1* mutant, superinduction of ABA- and stress-responsive gene transcription correlated with elevated IP_3 accumulation (112). In transgenic plants expressing an antisense *PLC* gene or overexpressing an $Ins(1,4,5)P_3$ 5-phosphatase gene, IP_3 levels decreased, and the plants were less sensitive to osmotic stress or ABA in germination and in stress gene induction (82). Like IP_3, other inositol phosphates such as IP_6 and $I(1,3,4)P_3$ may also function in releasing Ca^{2+} from internal stores (48, 49). Plants do not seem to have PKC genes. Thus, DAG signaling may be indirect as it can be rapidly phosphorylated to phosphatidic acid (PA), which is an important signaling molecule (66).

The cellular levels of phosphoinositide messengers are a result of both synthesis and degradation. Hence the role of phosphoinositide turnover should not be overlooked (Figure 4). Xiong et al. (112) showed that, in Arabidopsis, loss-of-function mutations in the *FRY1* gene encoding an inositol polyphosphate-1-phosphatase result in enhanced osmotic- and ABA-induction of gene transcription. In animal cells, there are two major routes for IP_3 breakdown. These are the 5-phosphatase pathway and the 3-kinase pathway, resulting in the accumulation of inositol 1, 4-bisphosphate [$Ins(1,4)P_2$] and inositol 1,3,4,5-tetraphospahte [$Ins(1,3,4,5)P_4$] intermediates, respectively (5, 60, 90). $Ins(1,3,4,5)P_4$ can be further dephosphorylated by 5-phosphatases to generate inositol 1,3,4-trisphosphate [$Ins(1,3,4)P_3$]. Animal inositol polyphosphate 1-phosphatase (IPP) hydrolyzes $Ins(1,4)P_2$ and $Ins(1,3,4)P_3$, at the 1-position (60). FRY1, also known as SAL1, was able to hydrolyze both of these two inositol polyphosphates (79). Although animal IPP isoforms can hydrolyze IP_3 directly in certain cell types, in many other cells, the 1-phosphatase does not have this ability to directly hydrolyze IP_3. Recombinant FRY1 protein was also active in directly hydrolyzing IP_3 (112). Even the 1-phosphatase activity of FRY1 toward $Ins(1,4)P_2$ and $Ins(1,3,4)P_3$ may affect the catabolism of IP_3 as the accumulation of these intermediates would slow-down IP_3 degradation.

It is interesting that, despite increased IP_3 levels and enhanced stress gene expression in *fry1*, the mutant plants were more susceptible to damage by salt, drought, or freezing stress (112). This raises the possibility that FRY1 is somehow directly involved in the homeostasis pathway of stress responses, and the *fry1* mutations disrupt cellular homeostasis and thereby lead to increased stress injury (Figure 1). Perhaps the increased stress gene expression is a compensatory mechanism to limit or repair stress injury. Similar compensatory increases in detoxification responses occur in the *sos1* mutant where high levels of proline were found (54). Although the increased stress gene expression or proline level did not correct the stress sensitive phenotypes of the respective mutants, it may be speculated that without these compensatory responses, the mutants would be even more susceptible to stress damage. In this regard, FRY1 represents an interesting point of crosstalk between the stress homeostasis and detoxification pathways. In other words, a functional FRY1 attenuates the detoxification response.

Phospholipase D cleaves membrane phospholipids to produce phosphatidic acid (PA) and free head groups. PA is a second messenger in animal cells by activating targets such as PLC and PKC (16). Osmotic stress activates PLD activity in suspension cells of *Chlamydomonas*, tomato, and alfalfa (69). In the resurrection plant *Craterostigma plantagineum* and Arabidopsis, PLD is rapidly activated by dehydration stress (18, 39). Two PLD genes were cloned from the resurrection plant; one was constitutive and the other was induced by dehydration or ABA treatment (18). When drought stress–induced PLD activity was compared between drought-resistant and -sensitive cultivars of cowpea, it was found that the activity was higher in the drought-sensitive cultivar (15). This suggests that PLD activation reflects lipolitic membrane disintegration during stress injuries. Consistent with this view, blocking PLD activity resulted in reduced stress injury and improved freezing tolerance (X-M. Wang, personal communication). In this context, it is interesting that the PLD product, PA, has evolved a signaling role to perhaps mitigate stress injury. In guard cells, PLD activity increased shortly after ABA treatment, and the application of PA mimics the effect of ABA in inducing stomatal closure (35). Sphingosine-1-phosphate (S1P), which shares the same head group with PA, has also been implicated in drought stress signaling (73). S1P concentrations increased upon drought stress, and exogenous application of S1P induced Ca^{2+} oscillations and stomatal closure.

In addition to PLC- and PLD-based lipid signaling, there are other lipid metabolizing enzymes that also respond to osmotic stress. PLA_2 cleaves phospholipids at the sn-2 position to generate lyso-phospholipids and free fatty acids. Hyperosmotic stress stimulates PLA_2 activity in algae (14, 62). The role of PLA_2 in osmotic stress adaptation and whether it is activated in higher plants is currently unclear. In yeast, phosphatidylinositol 3-phosphate is converted to $PI(3,5)P_2$ by the Fab1p enzyme when subjected to severe hyperosmotic stress (11). $PI(3,5)P_2$ levels also increase rapidly and transiently in several plants (62). This novel lipid messenger may function in osmoregulation by stimulating tonoplast H^+-ATPase activity (68).

ABA AND OSMOTIC STRESS SIGNALING

Role of ABA in Water Stress Tolerance

Although ABA has broad functions in plant growth and development, its main function is to regulate plant water balance and osmotic stress tolerance. This point is best illustrated by plant mutants that cannot produce ABA. Several ABA-deficient mutants have been reported for Arabidopsis, namely *aba1*, *aba2*, and *aba3* (43). There are also ABA-deficient mutants for tobacco, tomato, and maize (52). Without water or temperature stress, ABA-deficient mutants grow and develop relatively normally (43). The mutants, such as the Arabidopsis *aba1*, *aba2*, and *aba3*, have slightly smaller statures, which may be caused by unavoidable stress even under the best growth conditions. Additionally, the smaller stature of the *aba* mutants may also be due to ABA regulation of the cell cycle and other cellular activities. Under drought stress, ABA-deficient mutants readily wilt and die if the stress persists. Under salt stress, ABA-deficient mutants also perform poorly (110). The role of ABA in drought and salt stress is at least twofold: water balance and cellular dehydration tolerance. Whereas the role in water balance is mainly through guard cell regulation, the latter role has to do with induction of genes that encode dehydration tolerance proteins in nearly all cells. In addition, ABA is required for freezing tolerance, which also involves the induction of dehydration tolerance genes (58, 110).

Salt and Drought Stress Regulation of ABA Biosynthesis and Degradation

Osmotic stress induction of ABA accumulation is a well-known fact. Recently, some of the underlying molecular mechanisms became clear. Osmotic stress–induced ABA accumulation is a result of both activation of synthesis and inhibition of degradation. Several ABA biosynthesis genes have now been cloned (Figure 5). Zeathanxin epoxidase (known as ABA2 in tobacco and ABA1 in Arabidopsis) catalyzes the epoxidation of zeaxanthin and antheraxanthin to violaxanthin (61). The 9-*cis*-epoxycarotenoid dioxygenase (NCED) gene was first cloned from the maize *vp14* mutant (99). ABA aldehyde oxidase catalyzes the last step. ABA3, also known as LOS5, encodes a sulfurylase that generates the active form of the molybdenum cofactor required by ABA aldehyde oxidase (110).

With the cloning of these biosynthesis genes, it has been possible to determine which of them may be activated by osmotic stress. Biochemical studies suggested that the rate-limiting step is the reaction catalyzed by NCED (43). Indeed, when *VP14* and its homologous genes became available, their expression was seen to be upregulated by drought stress (34, 77, 100). It is surprising, however, that other ABA biosynthesis genes are also upregulated by osmotic stress. This is true for the Arabidopsis *ABA1* (52), for *AAO3* (88), and also for *ABA3* (110). Admittedly, the protein amount or activity has not been shown to increase in response to osmotic stress for all these genes. Nevertheless, it is evident that ABA biosynthesis is subjected to osmotic stress regulation at multiple steps (Figure 5). To date, genes responsible for ABA degradation have not been identified. Biochemical studies suggest that a cytochrome P450 monoxygenase catalyzes the first step in the oxidative degradation of ABA (45). It should not be long before this gene, which is expected to be induced by ABA but repressed by osmotic stress, is identified.

Nothing is known about the signaling between osmotic stress perception and the induction of ABA biosynthesis genes. Presumably, it involves calcium signaling and protein phosphorylation cascades.

Stress- and ABA-Regulated Gene Expression

The expression of numerous plant genes has been reported to be regulated by salt and/or drought stress (6, 95, 120). A substantial set of these genes is also responsive

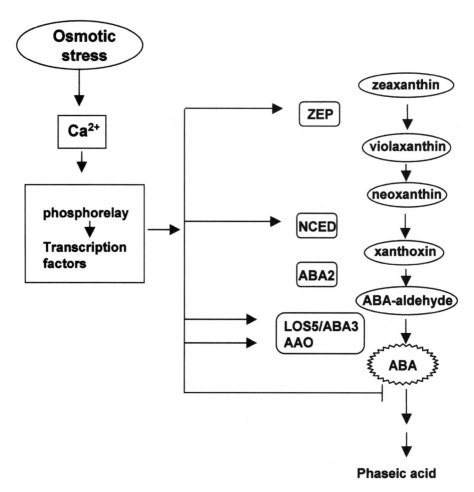

Figure 5 ABA metabolism is regulated by osmotic stress at multiple steps. The ABA biosynthesis genes ZEP, NCED, LOS5/ABA3, and AAO are upregulated by salt and drought stresses. ABA degradation is also important in controlling cellular ABA content, and biochemical evidence suggests osmotic stress inhibition of the first step of catabolism.

to ABA or cold stress. High-throughput technologies using DNA microarrays or chips are quickly replacing traditional differential screening methods in identifying new stress-regulated genes (40). Nonetheless, the reliability of microarray data is sometimes still questionable, and many resort to RNA blot analysis for confirmation. Considering the complexity of salt and drought stress responses, it is not surprising that the limited amount of DNA microarray data suggest that a substantial proportion of the genome is subjected to regulation by these stresses. For example, even in the unicellular yeast *Saccharomyces cerevisiae*, salt stress affected the expression of ∼8% of the genes in the entire genome (76, 80, 114).

One way to make sense of the large number of stress-responsive genes is to group them functionally. For salt stress, many of the induced genes function in ionic homeostasis; these include, e.g., plasma membrane Na^+/H^+ antiporters for Na^+ extrusion (92), vacuolar Na^+/H^+ antiporters for Na^+ compartmentation in the vacuole (3), and high-affinity K^+ transporters for K^+ acquisition. By increasing the ion concentration in the vacuole, the vacuolar Na^+/H^+ antiporters also function in osmotic homeostasis. Other salt- or drought-induced genes for osmotic homeostasis include, e.g., those coding for aquaporins and enzymes in osmolyte biosynthesis (6, 95, 120). Besides a potential, but still unproven, role of organic compatible osmolytes in osmotic adjustment, these osmolytes also seem to function in detoxification or damage prevention or repair (119). In fact, the majority of salt- and drought-induced genes appear to function in damage limitation or repair (119). These include the large number of osmolyte biosynthesis genes, LEA/dehydrin-type genes, detoxification enzymes, chaperones, proteases, and ubiquitination-related enzymes. Many of the LEA/dehydrin-type genes have other names that pertain to how they were initially identified; in Arabidopsis, some of them are designated as RD/COR/KIN/LTIs (118).

Evidence indicates that salt- or drought-responsive genes are under complex regulation. To study their regulation, it may be of value to consider the stress-responsive genes as either "early-response genes" or "delayed-response genes" (Figure 6). Early-response genes are induced very quickly (within minutes) and often transiently. Their induction does not require new protein synthesis because all signaling components are already in place. In contrast, delayed-response genes, which constitute the vast majority of the stress-responsive genes, are activated by stress more slowly (within hours), and their expression is often sustained. The early-response genes typically encode transcription factors that activate downstream delayed-response genes (Figure 6).

Several examples of early-response genes in salt, drought, cold, and ABA regulation have emerged. They include, e.g., the CBF/DREB gene family, RD22BP, AtMyb, and ABF/ABI5/AREB (Table 1). These genes are all rapidly induced by either ABA or one or more of the stress cues. A major research emphasis should be to define the *cis*-regulatory promoter elements in these genes that confer stress inducibility and to identify transcription factors that bind these elements and activate the early-response genes. The upstream transcription factors are typically constitutively expressed and are regulated by stress at the posttranslational level, i.e., by phosphorylation changes (Figure 6).

Major achievements in stress molecular biology in the last decade have been to define the *cis*-regulatory elements in the delayed-response genes and to clone the early-response genes coding for the element-binding proteins. Table 1 lists some of the *cis*-elements and their binding proteins known to date for salt, drought, cold, or ABA responses.

The transcriptional cascades in stress gene regulation have provided excellent opportunities for producing stress-hardy plants by regulon engineering. For example, ectopic expression of *CBF1* or *CBF3* leads to constitutive expression

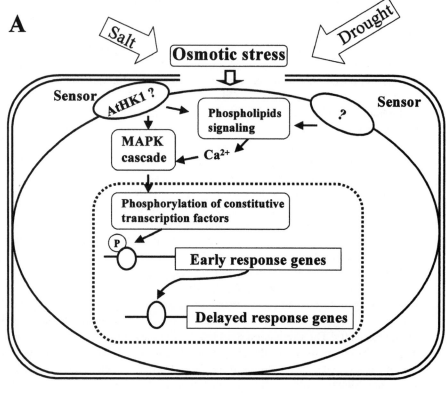

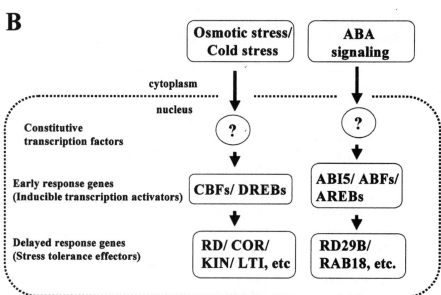

of downstream delayed-response genes such as *RD29A*, *COR15*, and *KIN1* and consequently to enhanced stress tolerance (36, 38). The use of a stress-activated promoter to drive the expression of the transcription factors avoids the pitfalls associated with constitutive expression and appears to be advantageous (38).

ABA-Dependent and ABA-Independent Signaling

Because salt and drought stress enhance ABA accumulation in plants and exogenous application of ABA can have similar effects as osmotic stress, such as in gene induction, it is reasonable to hypothesize that ABA mediates osmotic stress responses. Whether all or most of the osmotic stress responses are dependent on ABA has been of great interest. This question has often been addressed in the context of stress gene regulation, partly made possible by the availability of ABA-deficient and ABA-insensitive mutants. In particular, the Arabidopsis *aba1*, *aba2*, and *abi1* and *abi2* mutants have been extensively used for this type of study (43). One caveat of these studies is that none of the mutants are completely deficient or insensitive to ABA, often making the interpretation of ABA independence equivocal. Nevertheless, a number of studies showed that some osmotic stress responsive genes are induced completely independent of ABA, some are fully dependent on ABA, and others are only partially ABA dependent (95).

The *RD29A* gene has served as an excellent paradigm of ABA-dependent and -independent gene regulation. Early studies showed that osmotic stress induction of *RD29A* transcript accumulation is only partially blocked by *aba1* or *abi1* mutations, thus suggesting both ABA-dependent and -independent regulation (115). The ABA-independent regulation received a strong boost from a landmark paper published in 1994 (116), which reported the identification of the DRE (dehydration responsive element) sequence in the *RD29A* promoter as sufficient and necessary for osmotic stress induction. ABA cannot activate the DRE element. Whether ABA is still necessary for osmotic stress activation of DRE has not been investigated. This question could be addressed by expressing a reporter gene under a synthetic DRE-containing promoter in ABA-deficient and -insensitive mutants.

There is evidence that although ABA does not activate the DRE element, it may be required for full activation of DRE by osmotic stress. In the *los5/aba3* mutant that is deficient in ABA synthesis owing to a defective molybdenum cofactor, high salt or PEG induction of *RD29A* and other DRE-type genes is virtually abolished

←——————————————————————————

Figure 6 Model showing osmotic stress regulation of early-response and delayed-response genes. (*A*) Model integrating stress sensing, activation of phospholipid signaling and MAP kinase cascade, and transcription cascade leading to the expression of delayed-response genes. (*B*) Examples of early-response genes encoding inducible transcription activators and their downstream delayed-response genes encoding stress tolerance effector proteins. Question marks denote unknown transcription factors that activate the early-response genes.

TABLE 1 *cis*-acting promoter elements on delayed-stress response genes and transcription factors (encoded by early-response genes) that bind to them

cis element		Gene	Transcription factor name	Transcription factor type	Transcription factor expression is induced by		References
DRE	TACCGACAT	rd29A	DREB2A	AP2	Dehydration, ABA	Arabidopsis	(56, 71)
DRE		rd29A	DREB2B	AP2	Dehydration, salt	Arabidopsis	(56, 71)
DRE		rd29A	CBF1/DREB1B	AP2	Cold	Arabidopsis	(19, 96, 97)
DRE		rd29A	CBF2/DREB1C	AP2	Cold	Arabidopsis	(19, 96, 97)
DRE		rd29A	CBF3/DREB1A	AP2	Cold	Arabidopsis	(19, 96, 97)
ABRE	CACGTGGC	Em	EmBP1	bZIP		Wheat	(20)
ABRE	CCACGTGG		TAF-1	bzip		Tobacco	(74)
ABRE			OSBZ8	bZIP	ABA	Rice	(70)
ABRE			osZIP-1a	bZIP		Rice	(72)
ABRE	(T/G/C)ACGT(G/T)GC	Osem	TRAB1	bZIP	ABA	Rice	(27)
ABRE		rd29B	AREB1	bZIP	Dehydration, salt	Arabidopsis	(9, 17, 102)
ABRE	GNGGTG/GTGGNG	MsPRP2	Alfin1	zink-finger		Alfalfa	(4)
MYCRS	CANNTG	rd22	RD22BP1	myc	Dehydration, ABA	Arabidopsis	(1)
MYBRS	NyAACPyPu	rd22	AyMyb2	myb	Dehydration, ABA	Arabidopsis	(104)

(110). It is still curious why the *aba3* mutation has a stronger effect than *aba1* or *aba2* in blocking the osmotic stress induction. *LOS5/ABA3* encodes a molybdenum cofactor sulfurylase that catalyzes the synthesis of active MoCo factor required for ABA aldehyde oxidase in the last step of ABA biosynthesis. MoCo is also required for several other enzymes such as xanthine dehydrogenases (110). These latter enzymes do not appear to have a role in osmotic stress responses. Therefore, the more severe osmotic stress phenotypes of *aba3* are probably a consequence of more severe defects in ABA synthesis. Xiong et al. (110) have proposed that activation of DRE by DREB2A and related transcription factors may require ABA-dependent factor(s). Consistent with this, full activation of *RD29A* transcription depends upon the synergy between the DRE and ABRE elements (K. Shinozaki, personal communication).

The role of ABA in cold responses is still unclear. Only a few years ago, ABA was thought to have a major role in cold responses. Lang et al. (46) found transiently increased ABA accumulation in response to chilling treatment. However, other studies do not seem to find ABA accumulation under cold stress. It is clear that there is no dramatic ABA synthesis in the cold owing to the general slowdown effect of cold on cellular metabolism. Several studies found that exogenous ABA application increased the freezing tolerance of plants (8). Furthermore, cold and ABA induce a common set of genes. However, the cold stress induction appears to be completely independent of ABA. Mutations that enhance ABA induction of the *RD29A-LUC* transgene also increased osmotic stress but not cold induction (109). Nevertheless, the involvement of ABA in cold acclimation and cold-responsive gene expression cannot be ruled out. Besides blocking osmotic stress induction of genes and osmotic stress tolerance, the *los5/aba3* mutations also substantially reduce cold-responsive gene expression and freezing tolerance (110).

Although specific branches and components exist (47), the signaling pathways for salt, drought, cold, and ABA interact and even converge at multiple steps (109, 111). This was suggested by a comprehensive mutational analysis in which Arabidopsis single-gene mutations were found to affect responses to all or combinations of these signals (33). A nice example of the pathway convergence is provided by the *fry1* (*fiery1*) mutation. The mutation increases the amplitude and sensitivity of stress gene induction not only by ABA, but also by salt, drought, and cold stresses (112). An analysis of double mutants between *fry1* and *aba1* or *abi1* indicated that the cold or osmotic stress hypersensitivity in the mutant is not dependent on ABA (L. Xiong & J-K. Zhu, unpublished information). *FRY1* encodes an inositol polyphosphate 1-phosphatase that is required for IP_3 turnover. In response to ABA, wild-type plants accumulated IP_3 transiently. The IP_3 accumulation in *fry1* mutants in response to ABA was more sustained and reached higher levels. It will not be surprising if osmotic or cold stress also leads to more IP_3 accumulation in *fry1*. The results are consistent with IP_3 being a second messenger that mediates not only ABA but also salt, drought, or cold stress regulation of gene expression. Another important implication of the study is that ABA and stress signaling is desensitized by IP_3 turnover.

As discussed above, it is very intriguing that *fry1* mutant plants are less tolerant to salt, drought, or freezing stress despite enhanced expression of stress genes in the mutant (112). This observation may be explained by one or more of the following models: The first model is that too much IP_3 may compromise stress tolerance. For example, uncontrolled IP_3 accumulation could lead to unbalanced calcium signaling, which certainly would impact stress tolerance. This model can be tested by finding and analyzing other mutants with enhanced IP_3 accumulation due either to increased production or decreased degradation. Another model is that the FRY1 inositol polyphosphatase is critical for stress tolerance. Loss-of-function mutations in IP_3 biosynthesis genes such as phospholipase C are expected to suppress the IP_3 overaccumulation phenotype of *fry1*. It would be consistent with this model if the hypothetical suppressor mutations did not suppress the stress-tolerance defect of *fry1*. An interesting implication of the second model is that more stress damage leads to more IP_3 and thereby more *RD/COR/KIN*-type stress gene expression, which would support that notion of a detoxification function of RD/COR/KIN gene products.

FUTURE PROSPECTIVES

Salt and drought stress signaling has largely remained a mystery until recently. Now the molecular identities of some signaling elements have been identified. But we are still far from having a clear picture. The foremost difficulty in putting together the puzzle is not having enough pieces. Therefore, the challenge in the near future remains to identify more signaling elements. Once more components are known, signaling specificities and crosstalks can be properly addressed. Changes in gene expression or even protein amount or activity in response to water stress are not sufficient to establish whether an element is part of salt or drought signaling pathways. Any signaling component has to be established by functional necessity and functional sufficiency when possible. That is to say, plant phenotypes, be they molecular, biochemical, or physiological, are required to establish that a particular component functions in water stress signaling. In this regard, few genes believed to be involved in salt or drought signaling meet this criterion yet.

The most prominent missing elements in salt, drought, ABA, or cold signaling are the sensors or receptors. An Arabidopsis histidine kinase, AtHK1, is a candidate osmosensor because it could complement a yeast osmosensing mutant (105). As with a number of other potential regulatory genes, AtHK1 transcript level is upregulated by osmotic stress. In addition, expression of a receptor-like kinase gene was induced by abscisic acid, dehydration, high salt, and cold treatments in Arabidopsis (28). However, the functional significance, if any, of these transcript upregulations in osmotic stress responses is obscure.

Through more widespread application of forward and reverse genetic analysis in model plants and with the growing power of genomics and proteomics tools, progress in understanding water stress signaling will certainly accelerate. With a better understanding come more effective ways to improve plant salt or drought hardiness. The first phase of genetic engineering of stress hardiness has been to simply express one or several tolerance effector genes under constitutive or stress-inducible promoters. The second phase is to improve stress tolerance through engineering more effective signaling. Part of this has been achieved by overexpressing early-response transcription activators to turn on many downstream effector genes (36, 56). Components upstream of transcription factors can also be manipulated to improve stress tolerance (44). The future promises to see a much clearer picture of salt and drought signal transduction pathways and more examples of genetic improvement of water stress tolerance by fine-tuning plant sensing and signaling systems.

ACKNOWLEDGMENTS

I thank Liming Xiong, Becky Stevenson, Masaru Ohta, and Satoshi Iuchi for assistance in the preparation of this manuscript. Research in my laboratory is supported by grants from the U.S. National Institutes of Health, the U.S. National Science Foundation, and the U.S. Department of Agriculture.

Visit the Annual Reviews home page at www.annualreviews.org

LITERATURE CITED

1. Abel H, Yamaguchi-Shinozaki K, Urao T, Iwasaki T, Hosokawa D, Shinozaki K. 1997. Role of Arabidopsis MYC and MYB homologs in drought- and abscisic acid–regulated gene expression. *Plant Cell* 9:1859–68

2. Amtmann A, Sanders D. 1999. Mechanisms of Na$^+$ uptake by plant cells. *Adv. Bot. Res.* 29:75–112

3. Apse MP, Aharon GS, Snedden WA, Blumwald E. 1999. Salt tolerance conferred by overexpression of a vacuolar Na$^+$/H$^+$-antiport in *Arabidopsis. Science* 285:1256–58

4. Bastola DR, Pethe VV, Winicov I. 1998. Alfin1, a novel zinc-finger protein in alfalfa roots that binds to promoter elements in the salt-inducible MsPRP2 gene. *Plant Mol. Biol.* 38:1123–35

5. Berridge MJ, Irvine RF. 1989. Inositol phosphate and cell signaling. *Nature* 341:197–205

5a. Boyer JS. 1982. Plant productivity and environment. *Science* 218:443–48

6. Bray EA. 1993. Molecular responses to water deficit. *Plant Physiol.* 103:1035–40

7. Burssens S, Himanen K, van de Cotte B, Beeckman T, van Montagu M. 2000 Expression of cell cycle regulatory genes and morphological alterations in response to salt stress in *Arabidopsis thaliana. Planta* 211:632–40

8. Chen HH, Li PH, Brenner ML. 1983. Involvement of abscisic acid in potato cold acclimation. *Plant Physiol.* 71:362–65

9. Choi HI, Hong JH, Ha JO, Kang JY, Kim SY. 2000. ABFs, a family of ABA-responsive element binding factors. *J. Biol. Chem.* 275:1723–30

10. DeWald DB, Torabinejad J, Jones CA, Shope JC, Cangelosi AR, et al. 2001. Rapid accumulation of phosphatidylinositol 4,5-bisphosphate and inositol 1,4,

5-trisphosphate correlates with calcium mobilization in salt-stressed *Arabidopsis. Plant Physiol.* 126:759–69

11. Dove SK, Cooke FT, Douglas MR, Sayers LG, Parker PJ, Michell RH. 1997. Osmotic stress activates phosphatidylinositol-3,5-bisphosphate synthesis. *Nature* 390:187–92

12. Drobak BK, Watkins PA. 2000. Inositol (1,4,5)trisphosphate production in plant cells: an early response to salinity and hyperosmotic stress. *FEBS Lett.* 481:240–44

13. Droillard MJ, Thibivilliers S, Cazale AC, Barbier-Brygoo H, Lauriere C. 2000. Protein kinases induced by osmotic stresses and elicitor molecules in tobacco cell suspensions: two crossroad MAP kinases and one osmoregulation-specific protein kinase. *FEBS Lett.* 474:217–22

14. Einspahr KJ, Maeda M, Thompson GA Jr. 1988. Concurrent changes in *Dunaliella salina* ultrastructure and membrane phospholipid metabolism after hyperosmotic shock. *J. Cell Biol.* 107:529–38

15. El Maarouf H, Zuily-Fodil Y, Gareil M, d'Arcy-Lameta A, Pham-Thi AT. 1999. Enzymatic activity and gene expression under water stress of phospholipase D in two culitivars of *Vigna unguiculata* L. *Walp.* differing in drought tolerance. *Plant Mol. Biol.* 39:1257–65

16. English D. 1996. Phosphatidic acid: a lipid messenger involved in intracellular and extracellular signaling. *Cell. Signal.* 8:341–47

16a. Epstein E, Norlyn JD, Rush DW, Kingsbury RW, Kelly DB, et al. 1980. Saline culture of crops: a genetic approach. *Science* 210:399–404

17. Finkelstein RR, Lynch TJ. 2000. The Arabidopsis abscisic acid response gene *ABI5* encodes a basic leucine zipper transcriptional factor. *Plant Cell* 12:599–609

18. Frank W, Munnik T, Kerkmann K, Salamini F, Bartels D. 2000. Water deficit triggers phospholipase D activity in the resurrection plant *Craterostigma plantagineum. Plant Cell* 12:111–23

19. Gilmour SJ, Zarka DG, Stockinger EJ, Salazar MP, Houghton JM, Thomashow MF. 1998. Low temperature regulation of the Arabidopsis CBF family of AP2 transcriptional activators as an early step in cold-induced COR gene expression. *Plant J.* 16:433–42

20. Guiltinan MJ, Marcotte WR, Quatrano RS. 1990. A plant leucine zipper protein that recognizes an abscisic acid response element. *Science* 250:267–71

21. Guo Y, Halfter U, Ishitani M, Zhu JK. 2001. Molecular characterization of functional domains in the protein kinase SOS2 that is required for plant salt tolerance. *Plant Cell* 13:1383–400

22. Gustin MC, Albertyn J, Alexander M, Davenport K. 1998. MAP kinase pathways in the yeast *Saccharomyces cerevisiae. Microbiol. Mol. Biol. Rev.* 62:1264–300

23. Halfter U, Ishitani M, Zhu JK. 2000. The *Arabidopsis* SOS2 protein kinase physically interacts with and is activated by the calcium-binding protein SOS3. *Proc. Natl. Acad. Sci. USA* 97:3730–34

24. Hasegawa PM, Bressan RA, Zhu JK, Bohnert HJ. 2000. Plant cellular and molecular responses to high salinity. *Annu. Rev. Plant Physiol. Plant Mol. Biol.* 51:463–99

25. Heilmann I, Perera IY, Gross W, Boss WF. 1999. Changes in phosphoinositide metabolism with days in culture affect signal transduction pathways in *Galdieria suphuraria. Plant Physiol.* 119:1331–39

26. Hirayama T, Ohto C, Mizoguchi T, Shinozaki K. 1995. A gene encoding a phosphatidylinositol-specific phospholipase C is induced by dehydration and salt stress in *Arabidopsis thaliana. Proc. Natl. Acad. Sci. USA* 92:3903–7

27. Hobo T, Kowyama Y, Hattori T. 1999. A bZIP factor, TRAB1, interacts with VP1 and mediates abscisic acid-induced transcription. *Proc. Natl. Acad. Sci. USA* 96:15348–53

28. Hong SW, Jon JH, Kwak JM, Nam HG.

1997. Identification of a receptor-like protein kinase gene rapidly induced by abscisic acid, dehydration, high salt, and cold treatments in *Arabidopsis thaliana*. *Plant Physiol.* 113:1203–12

29. Hoyos ME, Zhang S. 2000. Calcium-independent activation of salicylic acid–induced protein kinase and a 40-kilodalton protein kinase by hyperosmotic stress. *Plant Physiol.* 122:1355–63

30. Ichimura K, Mizoguchi T, Yoshida R, Yuasa T, Shinozaki K. 2000. Various abiotic stresses rapidly activate Arabidopsis MAP kinases ATMPK4 and ATMPK6. *Plant J.* 24:655–65

31. Ingram J, Bartels D. 1996. The molecular basis of dehydration tolerance in plants. *Annu. Rev. Plant Physiol. Plant Mol. Biol.* 47:377–403

32. Ishitani M, Liu J, Halfter U, Kim CS, Wei M, Zhu JK. 2000. SOS3 function in plant salt tolerance requires myristoylation and calcium-binding. *Plant Cell.* 12:1667–77

33. Ishitani M, Xiong L, Stevenson B, Zhu JK. 1997. Genetic analysis of osmotic and cold stress signal transduction in *Arabidopsis thaliana*: interactions and convergence of abscisic acid–dependent and abscisic acid–independent pathways. *Plant Cell* 9:1935–49

34. Iuchi S, Kobayashi M, Taji T, Naramoto M, Seki M, et al. 2001. Regulation of drought tolerance by gene manipulation of 9-cis-epoxycarotenoid dioxygenase, a key enzyme in abscisic acid biosynthesis in Arabidopsis. *Plant J.* 27:325–33

35. Jacob T, Ritchie S, Assmann SM, Gilroy S. 1999. Abscisic acid signal transduction in guard cells is mediated by phospholipase D activity. *Proc. Natl. Acad. Sci. USA* 96:12192–97

36. Jaglo-Ottosen KR, Gilmour SJ, Zarka DG, Schabenberger O, Thomashow MF. 1998. *Arabidopsis CBF1* overexpression induces *COR* genes and enhances freezing tolerance. *Science* 280:104–6

37. Jonak C, Kiegerl S, Ligterink W, Barker PJ, Huskisson NS, Hirt H. 1996. Stress signaling in plants: a mitogen-activated protein kinase pathway is activated by cold and drought. *Proc. Natl. Acad. Sci. USA* 93:11274–79

38. Kasuga M, Liu Q, Miura S, Yamaguchi-Shinozaki K, Shinozaki K. 1999. Improving plant drought, salt, and freezing tolerance by gene transfer of a single stress-inducible transcription factor. *Nat. Biotechnol.* 17:287–91

39. Katagiri T, Takahashi S, Shinozaki K. 2001. Involvement of a novel *Arabidopsis* phospholipase D, AtPLD delta, in dehydration-inducible accumulation of phosphatidic acid in stress signaling. *Plant J.* 26:595–605

40. Kawasaki S, Borchert C, Deyholos M, Wang H, Brazille S, et al. 2001. Gene expression profiles during the initial phase of salt stress in rice. *Plant Cell* 13:889–906

41. Kiegerl S, Cardinale F, Siligan C, Gross A, Baudouin E, et al. 2000. SIMKK, a mitogen-activated protein kinase (MAPK) kinase, is a specific activator of the salt stress-induced MAPK, SIMK. *Plant Cell* 12:2247–58

42. Knight H, Trewavas AJ, Knight MR. 1997. Calcium signaling in *Arabidopsis thaliana* responding to drought and salinity. *Plant J.* 12:1067–78

43. Koornneef M, Léon-Kloosterziel KM, Schwartz SH, Zeevaart JAD. 1998. The genetic and molecular dissection of abscisic acid biosynthesis and signal transduction in *Arabidopsis*. *Plant Physiol. Biochem.* 36:83–89

44. Kovtun Y, Chiu WL, Tena G, Sheen J. 2000. Functional analysis of oxidative stress-activated mitogen-activated protein kinase cascade in plants. *Proc. Natl. Acad. Sci. USA* 97:2940–45

45. Krochko JE, Abrams GD, Loewen MK, Abrams SR, Culter AJ. 1998. (+)-abscisic acid 8′-hydroxylase is a cytochrome P450 monooxygenase. *Plant Physiol.* 118:849–60

46. Lang V, Mantyla E, Welin B, Sundberg

B, Palva ET. 1994. Alterations in water status, endogenous abscisic acid content, and expression of *rab18* gene during the development of freezing tolerance in *Arabidopsis thaliana*. *Plant Physiol.* 104:1341–49

47. Lee H, Xiong L, Gong Z, Ishitani M, Stevenson B, Zhu JK. 2001. The Arabidopsis *HOS1* gene negatively regulates cold signal transduction and encodes a RING-finger protein that displays cold-regulated nucleo-cytoplasmic partitioning. *Gene Dev.* 15:912–24

48. Lee Y, Choi YB, Suh J, Lee J, Assmann SM, et al. 1996. Abscisic acid–induced phosphoinositide turnover in guard cell protoplasts of *Vicia faba*. *Plant Physiol.* 110:987–96

49. Lemtiri-Chlieh F, MacRobbie EAC, Brearley CA. 2000. Inositol hexakisphosphate is a physiological signal regulating the K$^+$-inward rectifying conductance in guard cells. *Proc. Natl. Acad. Sci. USA* 97:8687–92

50. Leng Q, Mercier RW, Yao W, Berkowitz GA. 1999. Cloning and first functional chracterization of a plant cyclic nucleotide–gated cation channel. *Plant Physiol.* 121:753–61

51. Leung J, Giraudat J. 1998. Abscisic acid signal transduction. *Annu. Rev. Plant Physiol. Plant Mol. Biol.* 49:199–222

52. Liotenberg S, North H, Marion-Poll A. 1999. Molecular biology and regulation of abscisic acid biosynthesis in plants. *Plant Physiol. Biochem.* 37:341–50

53. Liu J, Ishitani M, Halfter U, Kim CS, Zhu JK. 2000. The *Arabidopsis thaliana SOS2* gene encodes a protein kinase that is required for salt tolerance. *Proc. Natl. Acad. Sci. USA* 97:3735–40

54. Liu J, Zhu JK. 1997. Proline accumulation and salt-stress-induced gene expression in a salt-hypersensitive mutant of Arabidopsis. *Plant Physiol.* 114:591–96

55. Liu J, Zhu JK. 1998. A calcium sensor homolog required for plant salt tolerance. *Science* 280:1943–45

56. Liu Q, Kasuga M, Sakuma Y, Abe H, Miura S, et al. 1998. Two transcription factors, DREB1 and DREB2, with an EREBP/AP2 DNA-binding domain separate two cellular signal transduction pathways in drought- and low temperature–responsive gene expression, respectively, in Arabidopsis. *Plant Cell* 10:1391–406

57. Liu Y, Zhang S, Klessig DF. 2000. Molecular cloning and characterization of a tobacco MAP kinase kinase that interacts with SIPK. *Mol. Plant Microbe Interact.* 13:118–24

58. Llorente F, Oliveros JC, Martinez-Zapater JM, Salinas J. 2000. A freezing-sensitive mutant of Arabidopsis, *frs1*, is a new *aba3* allele. *Planta* 211:648–55

59. Maathuis F, Sanders D. 2001. Sodium uptake in Arabidopsis roots is regulated by cyclic nucleotides. *Plant Physiol.* In press

60. Majerus PW. 1992. Inositol phosphate biochemistry. *Annu. Rev. Biochem.* 61: 225–50

61. Marin E, Nussaume L, Quesada A, Gonneau M, Sotta B, et al. 1996. Molecular identification of zeaxanthin epoxidase of *Nicotiana plumbaginifolia*, a gene involved in abscisic acid biosynthesis and corresponding to the ABA locus of *Arabidopsis thaliana*. *EMBO J.* 15:2331–42

62. Meijer HJG, Divecha N, van den Ende H, Musgrave A, Munnik T. 1999. Hyperosmotic stress induces rapid synthesis of phosphatidyl-D-inositol 3,5-bisphosphate in plant cells. *Planta* 208:294–98

63. Mikami K, Katagiri T, Luchi S, Yamaguchi-Shinozaki K, Shinozaki K. 1998. A gene encoding phosphatidylinositol 4-phosphate 5-kinase is induced by water stress and abscisic acid in *Arabidopsis thaliana*. *Plant J.* 15:563–68

64. Mikolajczyk M, Olubunmi SA, Muszynska G, Klessig DF, Dobrowolska G. 2000. Osmotic stress induces rapid activation of a salicylic acid–induced protein kinase and a homolog of protein kinase ASK1 in tobacco cells. *Plant Cell* 12:165–78

65. Mizoguchi T, Ichimura K, Yoshida R, Shinozaki K. 2000. MAP kinase cascades in Arabidopsis: their roles in stress and hormone responses. *Results Probl. Cell Differ.* 27:29–38

66. Munnik T, Irvine RF, Musgrave A. 1998. Phospholipid signaling in plants. *Biochim. Biophys. Acta* 1389:222–72

67. Munnik T, Ligterink W, Meskiene I, Calderini O, Beyerly J, et al. 1999. Distinct osmosensing protein kinase pathways are involved in signalling moderate and severe hyperosmotic stress. *Plant J.* 20:381–88

68. Munnik T, Meijer HJG. 2001. Osmotic stress activates distinct lipid and MAPK signaling pathways in plants. *FEBS Lett.* 498:172–78

69. Munnik T, Meijer HJG, ter Riet B, Frank W, Bartels D, Musgrave A. 2000. Hyperosmotic stress stimulates phospholipase D activity and elevates the levels of phosphatidic acid and diacylglycerol pyrophosphate. *Plant J.* 22:147–54

70. Nakagawa H, Ohmiya K, Hattori T. 1996. A rice bZIP protein, designated OSBZ8, is rapidly induced by abscisic acid. *Plant J.* 9:217–27

71. Nakashima K, Shinwari ZK, Sakuma Y, Seki M, Miura S, et al. 2000. Organization and expression of two Arabidopsis DREB2 genes encoding DRE-binding proteins involved in dehydration- and high-salinity-responsive gene expression. *Plant Mol. Biol.* 42:657–65

72. Nantel A, Quatrano RS. 1996. Characterization of three rice basic/leucine zipper factors, including two inhibitors of EmBP-1 DNA binding activity. *J. Biol. Chem.* 271:31296–305

73. Ng CK, Carr K, McAinsh MR, Powell B, Hetherington AM. 2001. Drought-induced guard cell signal transduction involves sphingosine-1-phosphate. *Nature* 410:596–99

74. Oeda K, Salinas J, Chua NH. 1991. A tobacco bZip transcription activator (TAF-1) binds to a G-box-like motif conserved in plant genes. *EMBO J.* 10:1793–802

75. Ozcan S, Dover J, Johnston M. 1998. Glucose sensing and signaling by two glucose receptors in the yeast *Saccharomyces cerevisiae. EMBO J.* 17:2566–73

76. Posas F, Chambers JR, Heyman JA, Hoeffler JP, de Nada E, Arino J. 2000. The transcriptional response of yeast to salt stress. *J. Biol. Chem.* 275:17249–55

77. Qin X, Zeevaart JAD. 1999. The 9-*cis*-epoxycarotenoid cleavage reaction is the key regulatory step of abscisic acid biosynthesis in water-stressed bean. *Proc. Natl. Acad. Sci. USA* 96:15354–61

78. Qiu Q, Guo Y, Dietrich M, Schumaker KS, Zhu JK. 2001. Characterization of the plasma membrane Na^+/H^+ exchanger in *Arabidopsis thaliana. Abstr. Int. Workshop Plant Membr. Biol., 12th, Madison, Wis.*, pp. 235

79. Quintero FJ, Garciadeblas B, Rodriguez-Navarro A. 1996. The SAL1 gene of Arabidopsis, encoding an enzyme with $3'(2')$, $5'$-bisphosphate nucleotide and inositol polyphosphate 1-phosphatase activities, increases salt tolerance in yeast. *Plant Cell* 8:529–37

80. Rep M, Krantz M, Thevelein JM, Hohmann S. 2000. The transcriptional response of *Saccharomyces cerevisiae* to osmotic shock. *J. Biol. Chem.* 275:8290–300

81. Rus A, Yokoi S, Sharkhuu A, Reddy M, Lee BH, et al. 2001. AtHKT1 is a salt tolerance determinant that controls Na^+ entry into plant roots. *Proc. Natl. Acad. Sci. USA.* 98:14150–55

82. Sanchez JP, Chua NH. 2001. Arabidopsis PLC1 is required for secondary responses to abscisic acid signals. *Plant Cell* 13:1143–54

83. Sanders D, Brownlee C, Harper JF. 1999. Communicating with calcium. *Plant Cell* 11:691–706

84. Sang Y, Cui D, Wang X. 2001. Phospholipase D and phosphatidic acid–mediated

generation of superoxide in Arabidopsis. *Plant Physiol.* 126:1449–58

85. Schroeder JI, Allen GJ, Hugouvieux V, Kwak JM, Waner D. 2001. Guard cell signal transduction. *Annu. Rev. Plant Physiol. Plant Mol. Biol.* 52:627–58

86. Schumaker KS, Sze H. 1987. Inositol 1,4,5-trisphosphate releases Ca^{2+} from vacuolar membrane vesicles of oat roots. *J. Biol. Chem.* 262:3944–46

87. Schuppler U, He PH, John PCL, Munns R. 1998. Effects of water stress on cell division and cell-division-cycle-2-like cell-cycle kinase activity in wheat leaves. *Plant Physiol.* 117:667–78

88. Seo M, Peeters AJM, Koiwai H, Oritani T, Marion-Poll A, et al. 2000. The *Arabidopsis* aldehyde oxidase 3 (*AAO3*) gene product catalyzes the final step in abscisic acid biosynthesis in leaves. *Proc. Natl. Acad. Sci. USA* 97:12908–13

89. Seo S, Okamoto M, Seto H, Ishizuka K, Sano H, Ohashi Y. 1995. Tobacco MAP kinase: a possible mediator in wound signal transduction pathways. *Science* 270:1988–92

90. Shears SB. 1992. Metabolism of inositol phosphates. *Adv. Second Messenger Phosphoprot. Res.* 26:63–92

91. Sheen J. 1996. Ca^{2+} dependent protein kinases and stress signal transduction in plants. *Science* 274:1900–2

92. Shi H, Ishitani M, Kim C, Zhu JK. 2000. The *Arabidopsis thaliana* salt tolerance gene SOS1 encodes a putative Na^+/H^+ antiporter. *Proc. Natl. Acad. Sci. USA* 97: 6896–901

93. Shi H, Quintero FJ, Pardo JM, Zhu JK. 2002. The putative plasma membrane Na^+/H^+ antiporter SOS1 controls long distance Na^+ transport in plants. *Plant Cell.* In press

94. Shi H, Xiong L, Stevenson B, Lu T, Zhu J-K. 2002. The Arabidopsis *salt overly sensitive 4* mutants uncover a critical role for vitamin B6 in plant salt tolerance. *Plant Cell.* In press

95. Shinozaki K, Yamaguchi-Shinozaki K. 1997. Gene expression and signal transduction in water-stress response. *Plant Physiol.* 115:327–34

96. Shinwari ZK, Nakashima K, Miura S, Kasuga M, Seki M, et al. 1998. An Arabidopsis gene family encoding DRE/CRT binding proteins involved in low-temperature-responsive gene expression. *Biochem. Biophys. Res. Commun.* 250:161–70

97. Stockinger EJ, Gilmour SJ, Thomashow MF. 1997. *Arabidopsis thaliana CBF1* encodes an AP2 domain-containing transcriptional activator that binds to the C-repeat/DRE, a cis-acting DNA regulatory element that stimulates transcription in response to low temperature and water deficit. *Proc. Natl. Acad. Sci. USA* 94:1035–40

98. Takahashi S, Katagiri T, Hirayama T, Yamaguchi-Shinozaki K, Shinozaki K. 2001. Hyperosmotic stress induced a rapid and transient increase in inositol 1,4,5-trisphosphate independent of abscisic acid in *Arabidopsis* cell culture. *Plant Cell Physiol.* 42:214–22

99. Tan BC, Schwartz SH, Zeevaart JAD, McCarty DR. 1997. Genetic control of abscisic acid biosynthesis in maize. *Proc. Natl. Acad. Sci. USA* 94:12235–40

100. Thompson AJ, Jackson AC, Symonds RC, Mulholland BJ, Dadswell AR, et al. 2000. Ectopic expression of a tomato 9-*cis*-epoxycarotenoid dioxygenase gene causes overproduction of abscisic acid. *Plant J.* 23:363–74

101. Tyerman SD, Skerrett IM. 1999. Root ion channels and salinity. *Sci. Hortic.* 78:175–235

102. Uno Y, Furihata T, Abe H, Yoshida R, Shinozaki K, Yamaguchi-Shinozaki K. 2000. Arabidopsis basic leucine zipper transcription factors involved in an abscisic acid-dependent signal transduction pathway under drought and high-salinity conditions. *Proc. Natl. Acad. Sci. USA* 97:11632–37

103. Uozumi N, Kim EJ, Rubio F, Yamaguchi

T, Muto S, et al. 2000. The *Arabidopsis* HKT1 gene homolog mediates inward Na⁺ currents in *Xenopus laevis* oocytes and Na⁺ uptake in *Saccharomyces cerevisiae*. *Plant Physiol.* 122:1249–60

104. Urao T, Noji M, Yamaguchi-Shinozaki K, Shinozaki K. 1996. A transcriptional activation domain of ATMYB2, a drought-inducible Arabidopsis Myb-related protein. *Plant J.* 10:1145–48

105. Urao T, Yakubov B, Satoh R, Yamaguchi-Shinozaki K, Seki B, et al. 1999. A transmembrane hybrid-type histidine kinase in Arabidopsis functions as an osmosensor. *Plant Cell* 11:1743–54

106. Wang H, Qi Q, Schorr P, Cutler AJ, Crosby WL, Fowke LC. 1998. ICK1, a cyclin-dependent protein kinase inhibitor from *Arabidopsis thaliana* interacts with both Cdc2a and CycD3, and its expression is induced by abscisic acid. *Plant J.* 15:501–10

107. Wu SJ, Lei D, Zhu JK. 1996. *SOS1*, a genetic locus essential for salt tolerance and potassium acquisition. *Plant Cell* 8:617–27

108. Wu Y, Kuzma J, Marechal E, Graeff R, Lee HC, et al. 1997. Abscisic acid signaling through cyclic ADP-ribose in plants. *Science* 278:2126–30

109. Xiong L, Ishitani M, Lee H, Zhu JK. 1999. *HOS5*—a negative regulator of osmotic stress-induced gene expression in *Arabidopsis thaliana*. *Plant J.* 19:569–78

110. Xiong L, Ishitani M, Lee H, Zhu JK. 2001. The Arabidopsis *LOS5/ABA3* locus encodes a molybdenum cofactor sulfurase and modulates cold and osmotic stress responsive gene expression. *Plant Cell.* 13:2063–83

111. Xiong L, Ishitani M, Zhu JK. 1999. Interaction of osmotic stress, ABA and low temperature in the regulation of stress gene expression in *Arabidopsis thaliana*. *Plant Physiol.* 119:205–11

112. Xiong L, Lee BH, Ishitani M, Lee H, Zhang C, Zhu JK. 2001. FIERY1 encoding an inositol polyphosphate 1-phosphatase is a negative regulator of abscisic acid and stress signaling in *Arabidopsis*. *Genes Dev.* 15:1971–84

113. Xiong L, Zhu JK. 2001. Abiotic stress signal transduction in plants: molecular and genetic perspectives. *Physiol. Plant* 112:152–66

114. Yale J, Bohnert HJ. 2001. Transcript expression in *S. cerevisiae* at high salinity. *J. Biol. Chem.* 276:15996–6007

115. Yamaguchi-Shinozaki K, Shinozaki K. 1993. Characterization of the expression of a desiccation-responsive *rd29* gene of *Arabidopsis thaliana* and analysis of its promoter in transgenic plants. *Mol. Gen. Genet.* 236:331–40

116. Yamaguchi-Shinozaki K, Shinozaki K. 1994. A novel *cis*-acting element in an Arabidopsis gene is involved in responsiveness to drought, low-temperature, or high-salt stress. *Plant Cell* 6:251–64

117. Zhu JK. 2000. Genetic analysis of plant salt tolerance using *Arabidopsis thaliana*. *Plant Physiol.* 124:941–48

118. Zhu JK. 2001. Cell signaling under salt, water and cold stresses. *Curr. Opin. Plant Biol.* 4:401–6

119. Zhu JK. 2001. Plant salt tolerance. *Trends Plant Sci.* 6:66–71

120. Zhu JK, Hasegawa PM, Bressan RA. 1996. Molecular aspects of osmotic stress in plants. *Crit. Rev. Plant Sci.* 16:253–77

121. Zhu JK, Liu J, Xiong L. 1998. Genetic analysis of salt tolerance in *Arabidopsis thaliana*: evidence of a critical role for potassium nutrition. *Plant Cell* 10:1181–92

Annu. Rev. Plant Biol. 2002. 53:275–97
DOI: 10.1146/annurev.arplant.53.100301.135248

THE LIPOXYGENASE PATHWAY

Ivo Feussner[1] and Claus Wasternack[2]

[1]Department of Molecular Cell Biology, Institute of Plant Genetics and Crop Plant Research (IPK), D-06466 Gatersleben, Germany; e-mail: feussner@ipk-gatersleben.de
[2]Department of Natural Product Biotechnology, Institute for Plant Biochemistry, D-06120 Halle, Germany; e-mail: cwastern@ipb-halle.de

Key Words lipid peroxidation, *CYP74*, oxylipins, jasmonates

■ **Abstract** Lipid peroxidation is common to all biological systems, both appearing in developmentally and environmentally regulated processes of plants. The hydroperoxy polyunsaturated fatty acids, synthesized by the action of various highly specialized forms of lipoxygenases, are substrates of at least seven different enzyme families. Signaling compounds such as jasmonates, antimicrobial and antifungal compounds such as leaf aldehydes or divinyl ethers, and a plant-specific blend of volatiles including leaf alcohols are among the numerous products. Cloning of many lipoxygenases and other key enzymes within the lipoxygenase pathway, as well as analyses by reverse genetic and metabolic profiling, revealed new reactions and the first hints of enzyme mechanisms, multiple functions, and regulation. These aspects are reviewed with respect to activation of this pathway as an initial step in the interaction of plants with pathogens, insects, or abiotic stress and at distinct stages of development.

CONTENTS

INTRODUCTION

Composition and turnover of intracellular lipids are frequently altered during plant development and are among the first targets of environmental cues. Apart from the turnover of fatty acids within lipids (12), formation of oxidized polyenoic fatty acids (PUFAs), collectively called oxylipins, is one of the main reactions in lipid alteration (28). The initial formation of hydroperoxides (7) may occur either by autooxidation (18, 109) or by the action of enzymes such as lipoxygenases (LOXs) or α-dioxygenase (α-DOX) (27). The metabolism of PUFAs via the LOX-catalyzed step and the subsequent reactions are collectively named the LOX pathway. In a number of reviews the diversity of LOX forms and LOX-derived products has been discussed (e.g., 8, 10, 33, 40, 43, 92, 104). Recent cloning, expression, and functional analysis of genes coding for LOXs and other members of the LOX pathway, as well as metabolite profiling and enzyme-mechanistic studies, shed new light on the function of LOXs and downstream enzymes acting in the different branches of the LOX pathway. Consequently, our understanding of the LOX pathway in distinct stress- and developmentally regulated processes has been substantiated. These new aspects are reviewed here. Properties and functions of 9- and 13-LOXs, allene oxide synthases (AOSs), hydroperoxide lyases (HPLs), and divinyl ether synthases (DESs) are the main focus. Finally, first conclusions regarding oxylipin profiling, one example of the new but rapidly growing field of metabolome analysis, are drawn.

THE LOX REACTION

LOXs (linoleate:oxygen oxidoreductase, EC 1.13.11.12) constitute a large gene family of nonheme iron containing fatty acid dioxygenases, which are ubiquitous in plants and animals (10). LOXs catalyze the regio- and stereo-specific dioxygenation of PUFAs containing a (1Z,4Z)-pentadiene system (26), e.g., linoleic acid (LA), α-linolenic acid (α-LeA), or arachidonic acid (Figure 1a, see color insert). Because arachidonic acid is only a minor PUFA in the plant kingdom, plant LOXs are classified with respect to their positional specificity of LA oxygenation. LA is oxygenated either at carbon atom 9 (9-LOX) or at C-13 (13-LOX) of the hydrocarbon backbone of the fatty acid leading to two groups of compounds, the (9S)-hydroperoxy- and the (13S)-hydroperoxy derivatives of PUFAs (Figure 1a). A more comprehensive classification of plant LOXs, based on comparison of their primary structure, has been proposed (102). According to their overall sequence similarity, plant LOXs can be grouped into two gene subfamilies. Those enzymes harboring no transit peptide have a high sequence similarity (>75%) to one another and are designated *type 1*-LOXs. However, another subset of LOXs carry a putative chloroplast transit peptide sequence. Based on this N-terminal extension and the fact that these enzymes show only a moderate overall sequence similarity (~35%) to one another, they have been classified as *type 2*-LOXs. To date, these LOX forms all belong to the subfamily of 13-LOXs.

Positional Specificity of LOXs

Plant LOXs are versatile catalysts because they are multifunctional enzymes, catalyzing at least three different types of reactions: (*a*) dioxygenation of lipid substrates (dioxygenase reaction) (45), (*b*) secondary conversion of hydroperoxy lipids (hydroperoxidase reaction) (67), and (*c*) formation of epoxy leukotrienes (leukotriene synthase reaction) (103). However, under physiological conditions the first reaction is most prevalent in plants. The two different regioisomers of hydroperoxy PUFAs may be determined by two independent properties of catalysis (45). (*a*) Selectivity in the initial hydrogen removal: In α-LeA this is possible either at C-11 and/or C-14, but comparison of various plant LOXs as well as site-directed mutagenesis studies revealed that only one double-allylic methylene group at C-11 in LA or α-LeA is used (26). (*b*) Selectivity in the site of oxygen insertion via rearrangement of the intermediate fatty acid radical. Consequently, when hydrogen is abstracted at C-11, molecular oxygen can be introduced at position [+2] or [−2], leading to dioxygen insertion at C-13 or C-9 (Figure 1*a*). Again, in principle, two models are used to explain the underlying reaction mechanism of positional specificity of LOXs. Based on data on mammalian LOXs, a space-related hypothesis was established (Figure 1*b*) (36, 106). Here, the fatty acid substrate penetrates the active site generally with its methyl end first. Then the depth of the substrate-binding pocket determines the site of hydrogen abstraction, and the positional specificity of molecular oxygen insertion depends on this position. However, in plant LOX reactions only one double allylic methylene group in the natural substrates LA and α-LeA seems to be accessible, rendering the space-related hypothesis unlikely. According to a second hypothesis, the substrate orientation is regarded as the key step in the determination of the position of dioxygen insertion, which then leads to varying regiospecificities of different isozymes: In the case of 13-LOXs, the active site is penetrated again by the substrate using its methyl end first. Whereas with 9-LOXs, the substrate is forced into an inverse orientation favoring a penetration with its carboxy group first (32). Consequently, a radical rearrangement at either [+2] or [−2], respectively, may be facilitated in both cases by the same mechanism within the active site. Recent data show that, at least for some plant LOXs, a combined version of both models may explain the underlying reactions because the inverse orientation of the substrate is determined by the space available in the substrate-binding pocket. This was substantiated by structural modeling and site-directed mutagenesis data for the cucumber lipid body 13-LOX (55). In the bottom area of the substrate-binding pocket, a space-filling histidine or phenylalanine residue, which occurs in nearly all plant 13-LOXs, was identified. In contrast, for all plant 9-LOXs, a valine residue was identified at this highly conserved amino acid position (26). In the cucumber lipid body 13-LOX, this histidine in the substrate-binding pocket was identified as the primary determinant for positional specificity. Its replacement by less space-filling residues, such as valine or methionine, altered the positional specificity of a 13-LOX into a 9-LOX. Structural modeling of the enzyme/substrate interaction suggests that this mutation may demask the positively charged guanidino group of an arginine residue at the bottom of the substrate pocket. As a consequence, the guanidino group may be able to form a salt bridge with the carboxylic group of the substrate, favoring an inverse head to tail substrate orientation (55). A special role of this arginine residue is also suggested by its highly conserved occurrence in all plant LOXs. In summary, the space within the active site and the orientation of the substrate are both important determinants for the positional specificity of plant LOXs and are modified by additional factors such as substrate concentration (66), the physico-chemical state of the substrate (3), pH (32), or temperature (34). However, it should be stressed that for other LOXs regiospecificity may be determined in a more complex manner (54, 57, 58). For additional details, we refer to other reviews (26, 33, 119).

Substrate Specificity of LOXs

The majority of plant LOXs strongly prefer free fatty acids as substrates (104). However, for two 13-LOXs, e.g., LOX1 from soybean seeds and a LOX from cucumber roots, activity with PUFAs esterified to phospholipids has been demonstrated (11, 76) in accordance with the suggested involvement of LOXs in membrane permeabilization. As a consequence of altering the physico-chemical properties of membranes via a modification of their fatty acid residues, fluxes of assimilates or ions may be facilitated (52, 101). For other 13-LOXs, e.g., the lipid body LOX from cucumber, the 13-LOX from barley seedlings, and the vegetative LOX d (vlx d) from soybean leaves, activity has been described with PUFAs esterified in neutral lipids such as triglycerides (20, 31, 53). For these LOXs, an involvement in triglyceride catabolism has been suggested (27).

Phylogenetic Analysis of the Plant LOX Family

The remarkable increase in sequence information allows phylogenetic tree analysis of multigene families (Figure 2, see color insert). The groups based on sequence data may help to predict at least some biochemical features and may provide suggestions regarding physiological functions. For LOXs, it is obvious that all *type 1-* and *type 2-* as well as all 9- and 13-LOXs form individual groups in separate branches of the tree. Moreover, within these subgroups, there is a distinction between monocotyledonous and dicotyledonous species. There are only a few exceptions. Some of the lipid-preferring 13-LOXs, *lox1:hv:3*, *lox1:cs:1*, and *lox1:cs:3*, group into the 9-LOX family (15, 20, 76). This discrepancy might be caused by an altered 9 to 13 ratio regarding the biochemical parameter of regiospecificity. Usually, 13-LOXs exhibit a >95% preference for 13-lipoxygenation. In the cases of these three enzymes, this preference is below 90%, down to 75%. Therefore, they might not be regarded as classical 13-LOXs (58). Interestingly, *type 1* 9-LOXs seem to be more closely related to *type 2* 13-LOXs than to *type 1* 13-LOXs. Nevertheless, prediction of regiospecificity as well as chloroplastic localization of plant LOXs is reliable with this method, whereas a prediction of substrate specificity requires more biochemical data in addition to sequence data.

Phylogenetic tree data stimulated analysis of putative functions of LOX, e.g., the existence of more than one *type 2* 13-LOX, a LOX form that can be involved in biosynthesis of JA (13, 128). The occurrence and expression patterns of several genes encoding LOX forms exhibiting nearly identical biochemical characteristics may allow the plant to respond exquisitely to the numerous environmental cues. The wound-response pathway, analyzed mainly in tomato and potato, is one such example. Both plant species carry at least two genes coding for *type 2* 13-LOXs, which are differentially expressed upon wounding of leaves and are possibly involved in biosynthesis of JA (51, 94).

Intracellular Localization of LOXs

Their intracellular localization provides clues as to the physiological functions of the different LOX forms. This has been analyzed in cotyledons, fruits, and leaves (28). A parallel occurrence of particulate, cytosolic, and vacuolar LOX forms was observed (23, 74, 116, 120, 126). In cotyledons, besides soluble LOXs, particulate LOXs were found in microsomal membranes (25), plasma membranes (82), and lipid bodies (24). Five different LOXs could be distinguished in etiolated cucumber cotyledons (25), whereas more than six different forms of LOX were detected in soybean leaves (39). In other organisms such as spinach, barley, or *Arabidopsis*, a preferential occurrence of LOXs within the envelope of leaf chloroplasts has been detected (4, 29, 122). The majority of LOXs appeared in the stroma, but at least in spinach, substantial LOX activity was detected within the envelope fraction (9). It is possible that the different LOX forms may have a specific location, which could contribute, with temporal differentiation of activity, to an orchestration of the formation of hydroperoxy PUFAs. These in turn are subsequently channeled into distinct branches of the LOX pathway. In at least two cases, the lipid body-LOXs and chloroplast LOXs, spatial and temporal expression analysis has led to the identification of physiological functions, namely involvement in storage lipid metabolism (27) and in JA formation (13).

Physiological Functions of LOXs

The involvement of *lox2:at:1* (AtLOX2) in JA biosynthesis upon wounding was shown using *antisense* plants (4). These plants lacked visible phenotype but were unable to accumulate JA upon wounding, and the expression of JA- and wound-inducible genes such as *vsp1* was reduced. However, the basal level of JA was not altered. It will be interesting to see whether the other three *type 2* 13-LOXs from *A. thaliana* are involved in JA homeostasis.

Apart from a basal compared to an induced 13-LOX activity leading to JA formation, temporal differences in LOX expression may also occur. In potato leaves, the *type 2* 13-LOXs, LOX-H3 (*lox2:st:2*) and LOX-H1 (*lox2:st:1*), are differentially induced upon wounding (94). The LOX-H3 mRNA accumulated transiently, peaking at about 30 min, whereas the LOX-H1 mRNA level was maximal at about 24 h. Depletion of LOX-H3 by antisense expression strongly reduced mRNA accumulation of the JA-responsive proteinase inhibitor II (pin2), suggesting a specific role of this LOX form in early JA formation. However, the basal level of JA was increased in these antisense plants (93), suggesting that some LOX forms can compensate, at least partially, for one another.

For certain LOXs, a role in conferring resistance against pathogens has been discussed for some time (107). For tobacco plants, this was demonstrated by *antisense* expression of a specific 9-LOX (*lox1:nt:1*) (90). In contrast to the resistant parent plants, the transgenic plants became susceptible to infection with *Phytophthora parasitica*. Such a conversion of an incompatible interaction into a compatible one suggests a crucial role for 9-LOX-derived oxylipins in conferring resistance in solanaceous plants (see below). Only 1 out of 12 9-LOX genes from potato, *lox1:st:7* (POTLX-3), is found close to *lox1:nt:1* in the phylogenetic tree analysis (Figure 2), and this specific gene is activated during plant pathogen interactions of potato (62), demonstrating again the usefulness of phylogenetic analysis. A role of a 9-LOX in plant pathogen interaction was also implicated by activation of a (9S)-hydroperoxy LA-forming enzyme upon infection of maize kernels with *Aspergillus flavus* (133).

Recently, another function for a 9-LOX has been shown in an antisense approach with a 9-LOX that specifically appears during early potato tuber development (64). The suppression of this LOX correlated with reduced tuber yield, decreased average tuber size, and a disruption of tuber formation, indicating that 9-LOX-generated metabolites are involved in the regulation of tuber enlargement (63).

At least eight different LOX forms exist in soybean. Some of these function as vegetative storage proteins and are found in paraveinal mesophyll cells (41). Others occur in another highly specialized cell layer, the mid-pericarp of pod walls (16). This cell layer of soybean pod walls may represent a novel multicellular compartment involved in defense of leguminous plants.

THE LOX PATHWAY

Data from the past five years suggest that the majority of accumulating hydro-(pero)xy fatty acids arise from action of LOXs (21, 95). A minority of PUFAs may be converted by the recently described α-DOX into α-hydro(pero)xy PUFAs (46, 96) or may give autooxidative products such as dinor isoprostanes (89). Therefore, plants predominantly contain as lipid peroxide-derived substances the (9S)-hydroperoxy and the (13S)-hydroperoxy derivatives of PUFAs (Figure 1a). These are subsequently metabolized via a number of secondary reactions. To date, four major metabolic routes for the metabolism of hydroperoxy fatty acids have been characterized (8, 27) (Figure 3): (*a*) The peroxygenase (POX) or formerly hydroperoxide isomerase pathway: Intramolecular oxygen transfer converts fatty acid hydroperoxides to epoxy- or dihydrodiol polyenoic fatty acids (7, 43). (*b*) The AOS pathway: AOS, first named hydroperoxide dehydratase (33), forms unstable allene oxides, which either undergo nonenzymatic hydrolysis leading to α- and

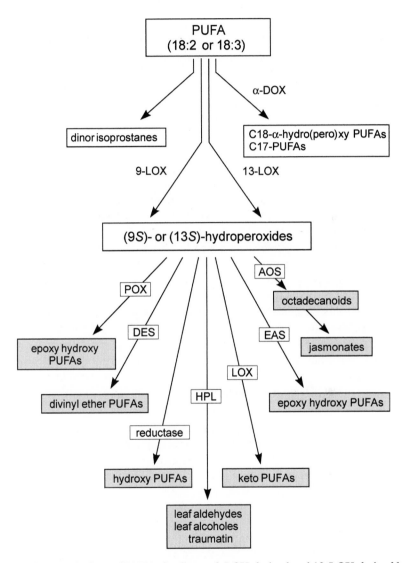

Figure 3   Metabolism of PUFAs leading to 9-LOX-derived and 13-LOX-derived hydro(pero)xy PUFAs in plants—the LOX pathway. AOS, allene oxide synthase; DES, divinyl ether synthase; α-DOX, α-dioxygenase; EAS, epoxy alcohol synthase; HPL, hydroperoxide lyase; LOX, lipoxygenase; POX, peroxygenase; PUFAs, polyunsaturated fatty acids.

γ-ketols (42) or may be metabolized to chiral (9S,13S)-12-oxo phytodienoic acid (OPDA) by an allene oxide cyclase (AOC) (Figure 4a) (136). (c) The HPL pathway: HPL, first named hydroperoxide isomerase (138), catalyzes the oxidative cleavage of the hydrocarbon backbone of fatty acid hydroperoxides. This leads to the formation of short chain aldehydes (C_6- or C_9-) and the corresponding C_{12}- or C_9-ω fatty acids (73). (d) The DES pathway: This pathway forms divinyl ethers such as colneleic acid (CA) or colnelenic acid (CnA) (40). For the POX pathway, properties of its key enzyme POX, as well as its possible physiological functions, have been summarized (8). An involvement of POX and α-DOX in pathogen defense reactions has been observed (46, 97). Recent cloning of a large number of cDNAs coding for enzymes of the three other pathways revealed that all of them belong to one P450-containing enzyme subfamily, named CYP74 (see below).

In addition to the four major pathways, there are other reactions for hydroperoxide metabolism that are less well characterized (Figure 3). (a) The LOX-catalyzed hydroperoxidase reaction (ketodiene-forming pathway): Under certain conditions, such as low oxygen pressure, LOXs are capable of catalyzing the homolytic cleavage of the O–O bond forming alkoxy radicals that may rearrange to ketodienes (67). Their endogenous occurrence has been described, but their physiological role remains to be elucidated (21, 123). (b) The epoxy alcohol synthase (EAS) pathway: Epoxy hydroxy fatty acids are formed by intramolecular rearrangement of hydro(pero)xy fatty acids catalyzed by EAS (44). Products of the EAS reaction can be regiochemically identical to the POX reaction products, but they differ with respect to their stereochemistry. To date, the EAS pathway has been found only in solanaceous species, and the EAS-generated oxylipins may be active in pathogen defense responses (38). (c) The reductase pathway: In this POX-independent reduction, hydro(pero)xy fatty acids are reduced to their corresponding hydroxy derivatives. This reduction mechanism is not fully understood but may be chemically

---

→

Figure 4 The CYP74-family. (a) Reaction mechanism. (b) Phylogenetic tree analysis of plant CYP74s. The analysis has been performed with phylip 3.5, and the proteins mentioned in the tree refer to the corresponding accession numbers in GenBank: *Arabidopsis thaliana*: AtHPL (AAC69871), AtAOS (CAA63266); *Capsicum annuum*: CaHPL (AAA97465); *Cucumis melo*: CmHPL (AAK54282); *Cucumis sativus*: CsHPL1 (AF229812), CsHPL2 (AAF64041); *Hordeum vulgare*: HvAOS1 (CAB86384), HvAOS2 (CAB63266), HvHPL (AJ318870); *Lycopersicum esculentum*: LeHPL (AAF67142), LeAOS1 (CAB88032), LeAOS2 (AAF67141), LeDES (AAG42261); *Linum usitatissimum*: LuAOS (AAA03353); *Musa acuminata*: MaHPL (CAB39331); *Medicago sativa*: MsHPL1 (CAB54847), MsHPL2 (CAB54848), MsHPL3 (CAB54849); *Medicago trunculata*: MtHPL1 (AJ316562), MtHPL2 (AJ316563), MtAOS (AJ316561); *Parthenium argentatum*: PaAOS (CAA55025); *Psidium guajava*: (AAK15070); *Physcomitrella patens*: PpAOS (AJ316566), PpDES (AJ316567); *Solanum tuberosum*: StHPL (AJ310520), StAOS (TC14225), StDES (AJ309541); *Zea maize*: ZmHPL (patent WO00/22145).

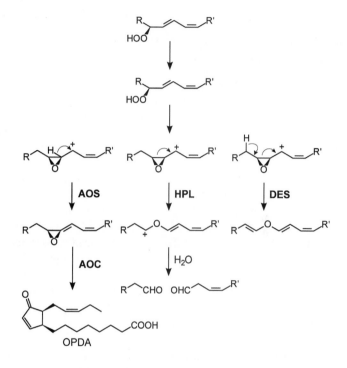

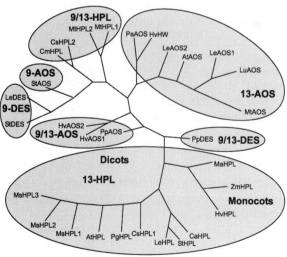

facilitated by glutathione (22). Hydroxy PUFAs are able to activate PR genes (132), and at least in seedlings, they seem to be intermediates of degradation of storage lipids (27).

The Metabolism of Hydro(pero)xy Fatty Acids—The CYP74 Family

CYP74s are members of an atypical family of P450 monooxygenases and constitute at least three different enzyme subfamilies: AOS, HPL, and DES. They do not require molecular oxygen nor NAD(P)H-dependent cytochrome P450-reductase (83, 86), and the new carbon-oxygen bonds are formed by using an acyl hydroperoxide as both the substrate and the oxygen donor (Figure 4a). Reduced affinity for CO is a common feature of the CYP74s, which are similar in catalytic properties to other P450s such as prostacyclin synthase and thromboxan synthase of the arachidonic acid cascade in animal systems (50). A common mechanistic feature of the AOS, HPL, and DES reactions is the intermediate epoxy allylic carbocation formed from the acyl hydroperoxide (35, 40, 86) (Figure 4a). AOS and DES deprotonate the carbocation at different positions leading to stable derivatives, whereas HPLs catalyze a rearrangement of the positive charge leading to a fragmentation of the molecule into two aldehydes (85). For HPL, a recent site-directed mutagenesis analysis confirmed a cysteine residue as the fifth ligand of the octahedral ligand sphere of the iron (87). EPR spectroscopy showed the high spin state of Fe(III) displaying the in vivo conformation (87).

SUBSTRATE AND PRODUCT SPECIFICITY OF CYP74s   The AOS from flax was the first member of the CYP74 family to be cloned (108). Therefore, AOS cDNAs are grouped as CYP74A. When expressed as recombinant proteins, all CYP74-cDNAs analyzed so far prefer free fatty acids as substrates, whereas fatty acid methyl esters as well as N-acyl(ethanol)amines are poor substrates (118). Recent isolation and characterization of nine different AOS-encoding cDNAs revealed different product specificities (Figure 4b). Whereas a single AOS gene has been described for A. thaliana (69), flax (108), guayule (88), and Medicago truncatula (AJ316561), two AOSs have been found thus far in tomato (56, 105) and barley (80). Within the CYP74A subfamily, most enzymes, including the enzyme from Arabidopsis and the two AOSs from tomato, are specific for 13-hydroperoxides as substrates and are called 13-AOS. Enzymes from barley (80) and Physcomitrella patens (AJ316566) show no preference between 9-hydroperoxides and 13-hydroperoxides, and as such are called 9/13-AOS. Recently, a cDNA coding for AOS with specificity for 9-hydroperoxides have been isolated (TC14225).

Two other CYP74 subfamilies contain enzymes with HPL activity. To date, 17 members have been cloned. Within the HPLs, a single cDNA has been isolated from A. thaliana (2, 79), Zea mays (patent WO00/22145), barley (AJ318870), guave (115), bell pepper (77), melon (114), potato (117), and tomato (56, 75). From M. truncatula (AJ316562, AJ 316563), potato (117), and cucumber (78), two

a

13-LOX   **9-LOX**

13-HPOTE   **9-HPOTE**

b

Space-related theory

13-LOX   9-LOX

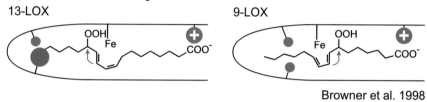

Browner et al. 1998

Orientation-dependent theory

13-LOX   9-LOX

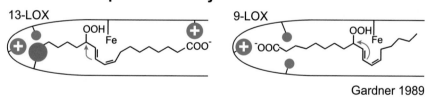

Gardner 1989

Figure 1 (*a*) LOX reaction and positional specificity. (*b*) Models detailing the underlying reaction mechanism of different regiospecificities.

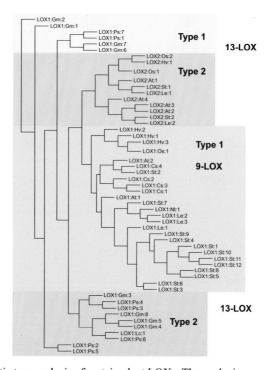

Figure 2 Phylogenetic tree analysis of certain plant LOXs. The analysis was performed with phylip 3.5, and the proteins mentioned in the tree refer to the corresponding accession numbers in the gene bank [For clarification, within the tree only sequences from distinct plant species have been included and have been partially renamed according the nomenclature of (102).]: *Arabidopsis thaliana*: LOX1:At:1 (AtLOX1, Q06327), LOX2:At:1 (AtLOX2, P38418), LOX2:At:2 (AtLOX3, AAF79461), LOX2:At:3 (AtLOX4, AAF21176), LOX1:At:2 (AtLOX5, CAC19365), LOX2:At:4 (AtLOX6, AAG52309); *Cucumis sativa*: LOX1:Cs:1 (AAC61785), LOX1:Cs:2 (CsULOX, AAA79186), LOX1:Cs:3 (CsLBLOX, CAA63483), LOX1:Cs:4 (CAB83038); *Glycine max*: LOX1:Gm:1 (Soybean LOX1, AAA33986), LOX1:Gm:2 (Soybean LOX2, AAA33987), LOX1:Gm:3 (Soybean LOX3, CAA31664), LOX1:Gm:4 (Soybean vlxa, BAA03101), LOX1:Gm:5 (Soybean vlxb, AAB67732), LOX1:Gm:6 (Soybean vlxc, AAA96817), LOX1:Gm:7 (Soybean vlxd, S13381), LOX1:Gm:8 (Soybean vlxe, AAC49159); *Hordeum vulgare*: LOX1:Hv:1 (HvLOXA, Barley LOXA, AAA64893), LOX1:Hv:2 (HvLOXB, Barley LOXB, AAB60715), LOX1:Hv:3 (HvLOXC, Barley LOXC, AAB70865), LOX2:Hv:1 (HvLOX1, AAC12951); *Lycopersicum esculentum*: LOX1:Le:1 (tomLOXA, P38415), LOX1:Le:2 (tomLOXB, P38416), LOX1:Le:3 (tomLOXtox, AAG21691), LOX2:Le:1 (tomLOXC, AAB65766), LOX2:Le:2 (tomLOXD, AAB65767); *Nicotiana tabacum*: LOX1:Nt:1 (NtLOX, S57964); *Oryza sativa*: LOX1:Os:1 (CAA45738), LOX2:Os:1 (ORYSALOXC, BAA03102), LOX2:Os:2 (OsLOX, CAC01439); *Pisum sativum*: LOX1:Ps:1 (AAB71759), LOX1:Ps:2 (CAA55318), LOX1:Ps:3 (CAA55319), LOX1:Ps:4 (Pea LOX, CAA30666), LOX1:Ps:5 (Pea LOX2, CAA34906), LOX1:Ps:6 (Pea LOXG, CAA53730), LOX1:Ps:7 (CAC04380); *Solanum tuberosum*: LOX1:St:1 (SOLTULOX1, S44940), LOX1:St:2 (STLOX, AAD09202), LOX1:St:3 (StLOX1, S73865), LOX1:St:4 (CAA64766), LOX1:St:5 (CAA64765), LOX1:St:6 (POTLX-2, AAB67860) LOX1:St:7 (POTLX-3, AAB67865), LOX1:St:8 (POTLX-1, AAB67858), LOX1:St:9 (AAD04258), LOX1:St:10 (pLOX2, AAB81595), LOX1:St:11 (pLOX1, AAB81594), LOX1:St:12 (CAB65460), LOX2:St:1 (StLOXH1, CAA65268), LOX2:St:2 (StLOXH3, CAA65269).

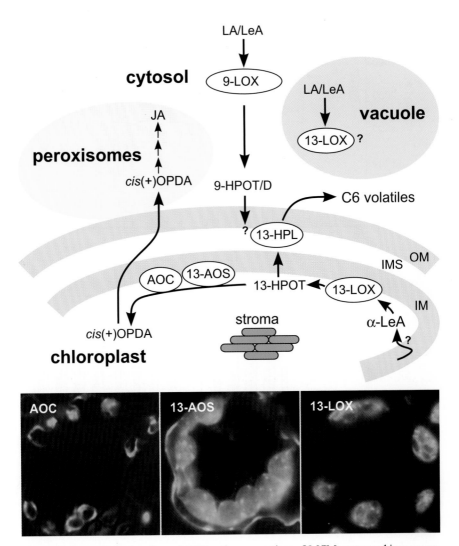

Figure 5 Intracellular location of LOX pathway reactions. OM/IM: outer and inner membrane of the chloroplast envelope; IMS, inter membrane space. Scheme modified according to (30). Immunocytological pictures show mesophyll cells of JA-treated barley leaves (13-LOX, AOS) and JA-treated tomato leaves (AOC). For further details see (29, 80, 136).

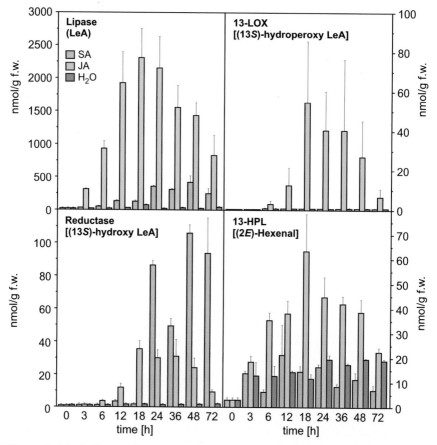

Figure 6 Metabolic profiling. LeA, (13*S*)-hydroperoxy LeA, (13*S*)-hydroxy LeA, and (2*E*)-hexenal were measured in extracts of barley leaf segments floated on water, JA, or SA. For further details on oxylipin analysis see (132). Data on SA treatment are drawn from (132), and data on JA treatment are unpublished.

HPL genes have been isolated. Currently, enzymes with HPL activity are divided into two subfamilies based on their substrate preference. The CYP74B subfamily contains HPLs acting specifically on 13-hydroperoxides (13-HPLs). HPLs accepting either 9- or 13-hydroperoxides as substrates (9/13-HPLs), including the enzymes from cucumber, melon, and *M. truncatula*, are grouped into the subfamily CYP74C. To date, an HPL, which metabolizes specificity 9-hydroperoxides, has not been isolated from potato.

The DESs have recently been recognized as members of the CYP74-family. A DES cDNA has been isolated from tomato (59), *P. patens* (AJ316567), and potato (113). Whereas the tomato and potato enzymes are highly specific for 9-hydroperoxides, the enzyme from the moss *P. patens* has no substrate preference (9/13-DES). The 9-DESs have been grouped into the CYP74D subfamily (59).

PHYLOGENETIC ANALYSIS OF THE CYP74 FAMILY   The increasing availability of sequence information permits a phylogenetic tree analysis of the CYP74 family enzymes (Figure 4*b*), which can contribute to the prediction of biochemical features. As assumed from the proposed catalytic mechanism in Figure 4*a*, DESs and HPLs are more closely related to each other, whereas 9/13-HPLs are more closely related to 13-AOSs on a phylogenetic basis. The 9/13-AOSs, as well as 9/13-DESs, are more closely related to 13-HPLs. Even CYP74 cDNA sequences from the ancient moss *P. patens* fit nicely into this phylogenetic tree.

INTRACELLULAR LOCALIZATION OF CYP74s   Data on the intracellular localization of CYP74s are scarce, but generally all members appear to be particulate enzymes. Several reports described the localization of AOS and HPL in chloroplasts of leaves, first detected at the activity level in spinach (122). Subsequently, activity data revealed copurification with LOX in the chloroplast envelope (9). A chloroplastic location is implied by the fact that many CYP74-cDNAs encode chloroplastic transit peptides (56). This was supported by immunocytochemical analysis in barley leaves (80) and by in vitro import experiments (30). In these import studies, 13-HPL was localized within the outer envelope, and 13-AOS was found within the inner envelope of tomato mesophyll chloroplasts (30). However, the sequence diversity among CYP74s discussed above suggests that compartmentation of these enzymes within the envelope should be more complex. A model summarizing the current knowledge, including the deduced substrate availability for specific enzymes of the LOX pathway, is shown in Figure 5 (see color insert). Because 13-LOXs are not found ubiquitously in the cytosol and, until now, 9-LOXs were only found within this compartment, the cytosol might be primarily the location of (9*S*)-hydroperoxide-derived compounds. In contrast, in the chloroplast, only 13-LOXs have been detected. Therefore, (13*S*)-hydroperoxide metabolizing CYP74s should be predominantly localized in the inner envelope, whereas the (9*S*)-hydroperoxide metabolizing CYP74s should reside in the outer envelope. For 9-LOX, the substrate may derive from the PUFA-CoA ester pool of the cytosol (68). Because the envelope harbors a substantial amount of phospholipids (60), the α-LeA, consumed in the 13-LOX reaction, may be generated by a chloroplast phospholipase A1, which was shown recently to initiate JA biosynthesis at least in the stamen (58a).

FUNCTIONAL ANALYSIS OF CYP74s IN TRANSGENIC PLANTS   In the seven branches of the LOX pathway in which hydroperoxy fatty acids are converted (Figure 3), the CYP74-catalyzed reactions (AOS, HPL, and DES) have been functionally analyzed. This was initially accomplished by transgenic approaches. As described above, these reactions initiate formation of jasmonates (AOS), leaf aldehydes (HPL), and divinyl ethers (DES). Consequently, each group of compounds could be taken as an indicator of transgenic alteration following overexpression or downregulation of the corresponding CYP74 enzyme. In the case of AOS, three different effects were found: (*a*) Constitutive overexpression of a 13-AOS in tobacco or *A. thaliana* (98) led to elevated jasmonate levels only upon wounding, which suggests a lack of AOS substrate in nonwounded tissues (also see below). (*b*) Constitutive overexpression of the 13-AOS from flax in potato revealed 6- to 12-fold higher levels of JA without constitutive expression of JA-responsive genes, which suggests sequestration of elevated JA (47). Upon wounding, the expected JA-/wound-responsive gene expression occurred, but in addition, a rise in the levels of jasmonates was observed (47). (*c*) Inducible expression of flax-13-AOS, lacking the chloroplast transit sequences, in tobacco led to an increase in JA level upon wounding (125). These data indicate that in a plant leaf the generation and activity of the stress signal JA may depend on spatial and temporal expression of its biosynthetic genes, on intracellular compartmentation, and on basal and induced levels of the respective substrates.

The 13-HPL is constitutively expressed in leaves and throughout the floral development of potato plants, preferentially forming (3Z)-hexenal and (3Z)-hexenol (117). The role of these volatiles, and that of the 13-HPL, in defense against sucking insect pests was demonstrated by antisense-mediated depletion of this specific 13-HPL in transgenic potato plants: The decrease in the toxic HPL-derived volatiles increased the performance of aphids (118).

For DES, first hints for a role in defense against plant pathogens such as *Phytophthora infestans* were provided by the detection of increased levels of CnA and CA upon infection of potato (129). Transgenic lines overexpressing DES have not been analyzed thus far.

The Oxylipin Signature—Oxylipin Profiling

In functional genomics approaches, reactions of the LOX pathway have been recorded and linked to environmental cues such as wounding, dehydration, or insect attack (91). As an essential piece of evidence, metabolic profiling should substantiate gene expression analyses. In the lipid metabolism discussed here, it is worthwhile to focus on lipid-derived compounds. Analysis of more than 150 fatty acids and oxylipins can now be performed in parallel and is a valuable tool for demonstrating in vivo activity of the different branches of the LOX pathway. Aldehydes and ketols were initially detected as constituents of leaves (48, 121). Recently, an α-ketol derivative of α-LeA, the (12Z,15Z)-9-hydroxy-10-oxo-12,15-octadecadienoic acid, was isolated and shown to have strong flower-inducing activity (135). In the presence of AOC, the particular enantiomeric form $cis(+)$OPDA is generated from the allene oxide. OPDA accumulates to substantial amounts upon wounding or other stresses (65, 91, 111) and is thought to exhibit signaling properties separate from JA. Its abundant appearance as a covalently linked lipid derivative is a newly found facet of oxylipin diversity (110), and it will be interesting to see whether this membrane-bound OPDA is a storage and/or functionally active compound.

A novel AOS pathway intermediate was identified in *Arabidopsis* and potato (130). Here, a 16-carbon cyclopentenoic acid (dinor OPDA) is a frequent constituent in OPDA- and JA-accumulating tissues. The different relative levels of JA, OPDA, and dinor OPDA observed in various plant species led to the term oxylipin signature (130). Recent microarray expression analysis in *A. thaliana* revealed that this oxylipin signature attributes differentially to JA-dependent gene expression upon wounding, dehydration, and insect feeding (91). Even distinct organs of the tomato flower contain their oxylipin signature, with preferential occurrence of JA in flower stalks and sepals and of OPDA in pistils (49). As revealed after different stress conditions or treatments with JA or salicylate (SA) and as exemplified for barley leaf segments in Figure 6 (see color insert), metabolite profiling can also be indicative of the activity of different branches of the LOX pathway. The 13-LOX reaction (discussed in previous section) of the chloroplast is initiated by liberation of α-LeA, which is presumably derived from chloroplast envelope membranes after JA and SA treatment. JA treatment led to an increase in (13*S*)-hydroperoxy LeA, which is shifted into the HPL branch. The reductase branch, indicated by (13*S*)-hydroxy LeA accumulation, was preferentially activated after SA treatment (132). Interestingly, following an endogenous rise in JA after sorbitol treatment (71), no accumulation of other metabolites of the LOX pathway other than those of the AOS branch was observed (131). Because JA and SA accumulate with different kinetics (17), such a shift in LOX branch activities might vary over time. In addition, metabolites of the DES pathway specifically accumulate. In elicitor-treated potato cells (38), CA is the dominant oxylipin, indicating preferential activity of the DES branch in this type of artificial plant defense system.

The preferential accumulation of compounds generated in the LOX pathway in response to an environmental cue is indicative of distinct physiological functions. As discussed above, many of these compounds are believed to function in plant defense reactions. This may occur via their proteins containing antifungal and/or antimicrobial properties (14). PR gene expression usually occurs in a SA-dependent and JA-independent manner. However, increased resistance against pathogens by establishing induced systemic acquired resistance (ISR) or through expression of defensins and thionins is a common property of jasmonates and octadecanoids (37). JA and related compounds are essential signals in the wound response of leaves, leading to synthesis of proteinase inhibitors (direct defense against herbivores), or may induce phytoalexin synthesis and plant-specific blends of volatiles (19). Apart from terpenoid and sesquiterpene compounds, the HPL-derived (3Z)-hexene-1-ol is a permanent constituent of such a herbivore-induced volatile emission, which acts in direct and indirect defense mechanisms (1, 124). The HPL-derived (3Z)-hexenal is active directly against sucking insects (117).

These data obtained through metabolic profiling suggest that an individual ratio of metabolites may be an additional element, together with altered gene expression and tissue specificity of corresponding enzymes, that allows plants to respond specifically to environmental cues.

Oxylipin metabolite profiling is a measure of enzymatic and autoxidative generation of metabolites. The preferential accumulation of racemic hydro(pero)xy PUFAs, shown for lipid peroxidation products of senescent *A. thaliana* leaves (5) and for HR in tobacco leaves (95), is indicative of autoxidation. In addition, compounds originating from a nonenzymatic hydrogen abstraction at C-14 of α-LeA and leading to hydro(pero)xy derivatives of α-LeA at C-12 and C-16 indicate autooxidation as well (5, 95). Finally, (2E)-4-hydroxy-2-alkenals are specific markers for autooxidation (61, 84, 99).

Regulatory Aspects of the LOX Pathway

A multitude of oxylipins can be generated in the LOX pathway from just a few substrates, mainly α-LeA and LA. The central branch point is presented by the LOX-generated hydroperoxy derivatives, which can be metabolized in at least seven individual reaction sequences (Figure 3). As discussed above for LOXs and the CYP74 family, different substrate specificity, intracellular location, pH optimum, transcriptional activation, and tissue specificity may lead to preferential activation of individual branches. However, we are far away from a detailed understanding of the regulation within the LOX pathway.

Recent data for JA biosynthesis allow a more detailed discussion of this branch. Most genes coding for enzymes of JA biosynthesis are transcriptionally upregulated upon wounding or treatment with jasmonates. Leaves of tomato and potato transiently accumulate mRNAs of 13-LOXs (51), 13-AOSs (56, 105), and AOC. In *Arabidopsis* leaves, mRNAs of AOS (70) and AOC are also wound induced. These observations were discussed in terms of feed-forward regulation in JA biosynthesis (70, 105). However, some data do not support this type of control: (*a*) Despite a transcriptional upregulation of 13-*AOS* by JA or SA, its substrate can be shifted into other branches (132). (*b*) Despite transcriptional upregulation of genes in JA biosynthesis, isotopic dilution analysis revealed no elevated JA formation (81). (*c*) (9Z,12R,13S)-12,13-epoxy-9-octadecanoic acid, which can be formed in the POX branch of the LOX pathway (8), inhibits *AOC* (137). (*d*) Constitutive overexpression of 13-AOS is not necessarily accompanied by elevated levels of JA.

Further data that support a control of substrate availability have accumulated: For example, upon wounding or treatment with (13S)-hydroperoxy LeA, JA levels were elevated in 13-AOS overexpressing lines of tobacco and AOC overexpressing lines of tomato (70, 125; C. Wasternack, unpublished results). Moreover, nontransgenic leaves of *A. thaliana* contain constitutively high levels of 13-LOX, 13-AOS, and AOC, but as mentioned previously, JA is formed only upon wounding (C. Wasternack, unpublished results).

JA biosynthesis and JA action seem to be regulated by the following factors: (*a*) substrate generation as a result of external stimuli (C. Wasternack, unpublished results); (*b*) intracellular sequestration (47); (*c*) tissue-specific generation and accumulation of JA, e.g., in vascular bundles (49; C. Wasternack, unpublished results); (*d*) temporally different patterns of JA accumulation (17, 65); and finally, (*e*) metabolic transformation of JA into its methyl ester (100) or into the volatile degradation product *cis*-jasmone (6), leading to spatially and temporally variable removal of the JA signal. These and other aspects of jasmonates in plant stress responses and development are comprehensively discussed in a recent review (127).

The HPL branch also appears to be regulated by substrate availability (117) on which organ-specific and wound-induced transcriptional activation is superimposed (2, 79). This may contribute to differential activation of the 13-AOS and 13-HPL branch (Figure 6) (56).

Much less is known about the regulation of other branches of the LOX pathway. However, as with JA biosynthesis and JA signaling, we must consider that further regulatory mechanisms exist; these include protein phosphorylation/dephosphorylation (72) and/or protein degradation by ubiquitinylation (134), which may influence the output of a regulatory network such as the LOX pathway.

CONCLUDING REMARKS

PUFA oxidation is implicated in plant development and in permanent adjustment to diverse and variable environmental conditions. The past decade has seen a remarkable increase in our understanding of the complexity of the LOX pathway and its physiological significance. The number and structural as well as functional diversity of LOX isoforms enables the plant to appropriately respond to environmental challenges. The two classes of LOX products, (9S)-hydroperoxy- and (13S)-hydroperoxy derivatives of PUFAs, are considered to be of central importance for the production of a plethora of oxylipins found in plants. The seven branches originating from the hydroperoxide precursors contribute to the plasticity of plant responses, as documented by the stimulus-dependent shift of activity between the branches.

Jasmonates and their octadecanoid precursors were the first oxylipins with an assigned messenger function. A steadily increasing number of studies supports a role of these compounds as a "master switch" in plant development and stress adaptation. More recently, other oxylipins have gained attention as putative signaling molecules. This is exemplified by leaf alcohols, which, as constituents of a plant-specific blend of volatiles, are thought to mediate plant-plant as well as plant-insect interactions.

Reverse genetics approaches have provided initial insight into lipoxygenase function. In addition, generation of oxylipin biosynthesis mutants, as well as ongoing overexpression and inactivation of genes encoding enzymes of the LOX pathway, is expected to broaden our understanding of the physiological role of the corresponding metabolites in plant development and stress adaptation.

ACKNOWLEDGMENTS

We apologize to scientists whose work we overlooked or were not able to include because of length limitations. We thank S. Rosahl and J. Rudd for critical reading of the manuscript and B. Hause for supplying the immunocytological pictures in Figure 5. We are grateful to C. Kaufmann, S. Kunze, and M. Stumpe for their help in preparing the figures and C. Dietel for typing the manuscript.

Visit the Annual Reviews home page at www.annualreviews.org

LITERATURE CITED

1. Baldwin IT, Halitschke R, Kessler A, Schittko U. 2001. Merging molecular and ecological approaches in plant-insect interactions. *Curr. Opin. Plant Biol.* 4:351–58
2. Bate NJ, Sivasankar S, Moxon C, Riley JMC, Thompson JE, Rothstein SJ. 1998. Molecular characterization of an *Arabidopsis* gene encoding hydroperoxide lyase, a cytochrome P-450 that is wound inducible. *Plant Physiol.* 117:1393–400
3. Began G, Sudharshan E, Rao AGA. 1999. Change in the positional specificity of lipoxygenase 1 due to insertion of fatty acids into phosphatidylcholine deoxycholate mixed micelles. *Biochemistry* 38:13920–27
4. Bell E, Creelman RA, Mullet JE. 1995. A chloroplast lipoxygenase is required for wound-induced jasmonic acid accumulation in *Arabidopsis*. *Proc. Natl. Acad. Sci. USA* 92:8675–79
5. Berger S, Weichert H, Wasternack C, Feussner I. 2001. Enzymatic and non-enzymatic lipid peroxidation in leaf development. *Biochem. Biophys. Acta* 1533:266–76
6. Birkett MA, Campbell CA, Chamberlain

K, Guerrieri E, Hick AJ, et al. 2000. New roles for *cis*-jasmone as an insect semiochemical and in plant defense. *Proc. Natl. Acad. Sci. USA* 97:9329–34
7. Blée E. 1998. Biosynthesis of phytooxylipins: the Peroxygenase pathway. *Fett-Lipid* 100:121–27
8. Blée E. 1998. Phytooxylipins and plant defense reactions. *Prog. Lipid Res.* 37:33–72
9. Blée E, Joyard J. 1996. Envelope membranes from spinach chloroplasts are a site of metabolism of fatty acid hydroperoxides. *Plant Physiol.* 110:445–54
10. Brash AR. 1999. Lipoxygenases: occurrence, functions, catalysis, and acquisition of substrate. *J. Biol. Chem.* 274:23679–82
11. Brash AR, Ingram CD, Harris TM. 1987. Analysis of a specific oxygenation reaction of soybean lipoxygenase-1 with fatty acids esterified in phospholipids. *Biochemistry* 26:5465–71
12. Chapman KD. 1998. Phospholipase activity during plant growth and development and in response to environmental stress. *Trends Plant Sci.* 3:419–26
13. Creelman RA, Mullet JE. 1997. Biosynthesis and action of jasmonates in plants.

Annu. Rev. Plant Physiol. Plant Mol. Biol.
48:355–81

14. Croft KPC, Juttner F, Slusarenko AJ. 1993. Volatile products of the lipoxygenase pathway evolved from *Phaseolus vulgaris* (L.) leaves inoculated with *Pseudomonas syringae* pv phaseolicola. *Plant Physiol.* 101:13–24

15. Doderer A, Kokkelink I, Van der Veen S, Valk BE, Schram AW, Douma AC. 1992. Purification and characterization of two lipoxygenase isoenzymes from germinating barley. *Biochim. Biophys. Acta* 1120:97–104

16. Dubbs WE, Grimes HD. 2000. The mid-pericarp cell layer in soybean pod walls is a multicellular compartment enriched in specific lipoxygenase isoforms. *Plant Physiol.* 123:1281–88

17. Engelberth J, Koch T, Schüler G, Bachmann N, Rechtenbach J, Boland W. 2001. Ion channel-forming alamethicin is a potent elicitor of volatile biosynthesis and tendril coiling. Cross talk between jasmonate and salicylate signaling in lima bean. *Plant Physiol.* 125:369–77

18. Esterbauer H, Zollner H, Schaur RJ. 1990. Aldehydes formed by lipid peroxidation: mechanisms of formation, occurrence, and determination. In *Membrane Lipid Oxidation*, ed. C Vigo-Pelfrey, pp. 239–68. Boca Raton: CRC Press

19. Farmer EE. 2001. Surface-to-air signals. *Nature* 411:854–56

20. Feussner I, Bachmann A, Höhne M, Kindl H. 1998. All three acyl moieties of trilinolein are efficiently oxygenated by recombinant His-tagged lipid body lipoxygenase *in vitro*. *FEBS Lett.* 431:433–36

21. Feussner I, Balkenhohl TJ, Porzel A, Kühn H, Wasternack C. 1997. Structural elucidation of oxygenated storage lipids in cucumber cotyledons—implication of lipid body lipoxygenase in lipid mobilization during germination. *J. Biol. Chem.* 272:21635–41

22. Feussner I, Blée E, Weichert H, Rousset C, Wasternack C. 1998. Fatty acid

catabolism at the lipid body membrane of germinating cucumber cotyledons. In *Advances in Plant Lipid Research*, ed. J Sánchez, E Cerdá-Olmedo, E Martínez-Force, pp. 311–33. Sevilla, Spain: Secr. Publ. Univ. Sevilla

23. Feussner I, Hause B, Vörös K, Parthier B, Wasternack C. 1995. Jasmonate-induced lipoxygenase forms are localized in chloroplasts of barley leaves (*Hordeum vulgare* cv *Salome*). *Plant J.* 7:949–57

24. Feussner I, Kindl H. 1992. A lipoxygenase is the main lipid body protein in cucumber and soybean cotyledons during the stage of triglyceride mobilization. *FEBS Lett.* 298:223–25

25. Feussner I, Kindl H. 1994. Particulate and soluble lipoxygenase isoenzymes—comparison of molecular and enzymatic properties. *Planta* 194:22–28

26. Feussner I, Kühn H. 2000. Application of lipoxygenases and related enzymes for the preparation of oxygenated lipids. In *Enzymes in Lipid Modification*, ed. UT Bornscheuer, pp. 309–36. Weinheim, Ger.: Wiley

27. Feussner I, Kühn H, Wasternack C. 2001. The lipoxygenase-dependent degradation of storage lipids. *Trends Plant Sci.* 6:262–67

28. Feussner I, Wasternack C. 1998. Lipoxygenase catalyzed oxygenation of lipids. *Fett-Lipid* 100:146–52

29. Feussner I, Ziegler J, Miersch O, Wasternack C. 1995. Jasmonate- and stress-induced lipoxygenase forms in barley leaf segments (*Hordeum vulgare* cv. Salome). In *Plant Lipid Metabolism*, ed. J-C Kader, P Mazliak, pp. 292–94. Dordrecht: Kluwer Acad.

30. Froehlich JE, Itoh A, Howe GA. 2001. Tomato allene oxide synthase and fatty acid hydroperoxide lyase, two cytochrome P450s involved in oxylipin metabolism, are targeted to different membranes of chloroplast envelope. *Plant Physiol.* 125:306–17

31. Fuller MA, Weichert H, Fischer AM,

Feussner I, Grimes HD. 2001. Activity of soybean lipoxygenase isoforms against esterified fatty acids indicates functional specificity. *Arch. Biochem. Biophys.* 388:146–54

32. Gardner HW. 1989. Soybean lipoxygenase-1 enzymatically forms both 9(S)- and 13(S)-hydroperoxides from linoleic acid by a pH-dependent mechanism. *Biochim. Biophys. Acta* 1001:274–81

33. Gardner HW. 1991. Recent investigations into the lipoxygenase pathway of plants. *Biochim. Biophys. Acta* 1084:221–39

34. Georgalaki MD, Bachmann A, Sotiroudis TG, Xenakis A, Porzel A, Feussner I. 1998. Characterization of a 13-lipoxygenase from virgin olive oil and oil bodies of olive endosperms. *Fett-Lipid* 100:554–60

35. Gerwick WH. 1996. Epoxy allylic carbocations as conceptual intermediates in the biogenesis of diverse marine oxylipins. *Lipids* 31:1215–31

36. Gillmor SA, Villasenor A, Fletterick R, Sigal E, Browner MF. 1998. The structure of mammalian 15-lipoxygenase reveals similarity to the lipases and the determinants of substrate specificity. *Nat. Struct. Biol.* 4:1003–9

37. Glazebrook J. 2001. Genes controlling expression of defense responses in *Arabidopsis*—2001 status. *Curr. Opin. Plant Biol.* 4:301–8

38. Göbel C, Feussner I, Schmidt A, Scheel D, Sanchez-Serrano J, et al. 2001. Oxylipin profiling reveals the preferential stimulation of the 9-lipoxygenase pathway in elicitor-treated potato cells. *J. Biol. Chem.* 276:6267–73

39. Grayburn WS, Schneider GR, Hamilton-Kemp TR, Bookjans G, Ali K, Hildebrand DF. 1991. Soybean leaves contain multiple lipoxygenases. *Plant Physiol.* 95:1214–18

40. Grechkin A. 1998. Recent developments in biochemistry of the plant lipoxygenase pathway. *Prog. Lipid Res.* 37:317–52

41. Grimes HD, Tranbarger TJ, Franceschi VR. 1993. Expression and accumulation patterns of nitrogen-responsive lipoxygenase in soybeans. *Plant Physiol.* 103:457–66

42. Hamberg M. 1988. Biosynthesis of 12-oxo-10,15(Z)-phytodienoic acid: identification of an allene oxide cyclase. *Biochem. Biophys. Res. Commun.* 156:543–50

43. Hamberg M. 1995. Hydroperoxide isomerases. *J. Lipid Mediat. Cell Signal.* 12:283–92

44. Hamberg M. 1999. An epoxy alcohol synthase pathway in higher plants: biosynthesis of antifungal trihydroxy oxylipins in leaves of potato. *Lipids* 34:1131–42

45. Hamberg M, Samuelsson B. 1967. On the specificitiy of the oxygenation of unsaturated fatty acids catalyzed by soybean lipoxidase. *J. Biol. Chem.* 242:5329–35

46. Hamberg M, Sanz A, Castresana C. 1999. α-oxidation of fatty acids in higher plants—identification of a pathogen-inducible oxygenase (PIOX) as an α-dioxygenase and biosynthesis of 2-hydroperoxylinolenic acid. *J. Biol. Chem.* 274:24503–13

47. Harms K, Atzorn R, Brash A, Kühn H, Wasternack C, et al. 1995. Expression of a flax allene oxide synthase cDNA leads to increased endogenous jasmonic acid (JA) levels in transgenic potato plants but not to a corresponding activation of JA-responding genes. *Plant Cell* 7:1645–54

48. Hatanaka A, Harada T. 1973. Formation of cis-3-hexenal, trans-2-hexenal and cis-3-hexenol in macerated *thea sinensis* leaves. *Phytochemistry* 12:2341–46

49. Hause B, Stenzel I, Miersch O, Maucher H, Kramell R, et al. 2000. Tissue-specific oxylipin signature of tomato flowers: allene oxide cyclase is highly expressed in distinct flower organs and vascular bundles. *Plant J.* 24:113–26

50. Hecker M, Ullrich V. 1989. On the mechanism of prostacyclin and thromboxane A2 biosynthesis. *J. Biol. Chem.* 264:141–50

51. Heitz T, Bergey DR, Ryan CA. 1997. A gene encoding a chloroplast-targeted

lipoxygenase in tomato leaves is transiently induced by wounding, systemin, and methyl jasmonate. *Plant Physiol.* 114:1085–93

52. Hildebrand DF. 1989. Lipoxygenases. *Physiol. Plant* 76:249–53

53. Holtman WL, Vredenbregt-Heistek JC, Schmitt NF, Feussner I. 1997. Lipoxygenase-2 oxygenates storage lipids in embryos of germinating barley. *Eur. J. Biochem.* 248:452–58

54. Hornung E, Rosahl S, Kühn H, Feussner I. 2000. Creating lipoxygenases with new positional specificities by site-directed mutagenesis. *Biochem. Soc. Trans.* 28:825–26

55. Hornung E, Walther M, Kühn H, Feussner I. 1999. Conversion of cucumber linoleate 13-lipoxygenase to a 9-lipoxygenating species by site-directed mutagenesis. *Proc. Natl. Acad. Sci. USA* 96:4192–97

56. Howe GA, Lee GI, Itoh A, Li L, DeRocher AE. 2000. Cytochrome P450-dependent metabolism of oxylipins in tomato. Cloning and expression of allene oxide synthase and fatty acid hydroperoxide lyase. *Plant Physiol.* 123:711–24

57. Hughes RK, Lawson DM, Hornostaj AR, Fairhurst SA, Casey R. 2001. Mutagenesis and modelling of linoleate-binding to pea seed lipoxygenase. *Eur. J. Biochem.* 268:1030–40

58. Hughes RK, West SI, Hornostaj AR, Lawson DM, Fairhurst SA, et al. 2001. Probing a novel potato lipoxygenase with dual positional specificity reveals primary determinants of substrate binding and requirements for a surface hydrophobic loop and has implications for the role of lipoxygenases in tubers. *Biochem. J.* 353:345–55

58a. Ishiguro S, Kawai-Oda A, Ueda J, Nishida I, Okada K. 2001. The *DEFECTIVE IN ANTHER DEHISCENCE1* gene encodes a novel phospholipase A1 catalyzing the initial step of jasmonic acid biosynthesis, which synchronizes pollen maturation, anther dehiscence, and flower opening in *Arabidopsis. Plant Cell* 13:2191–209

59. Itoh A, Howe GA. 2001. Molecular cloning of a divinyl ether synthase: identification as a CYP74 cytochrome P450. *J. Biol. Chem.* 276:3620–27

60. Joyard J, Teyssier E, Miege C, Bernyseigneurin D, Marechal E, et al. 1998. The biochemical machinery of plastid envelope membranes. *Plant Physiol.* 118:715–23

61. Kohlmann M, Bachmann A, Weichert H, Kolbe A, Balkenhohl T, et al. 1999. Formation of lipoxygenase-pathway-derived aldehydes in barley leaves upon methyl jasmonate treatment. *Eur. J. Biochem.* 260:885–95

62. Kolomiets MV, Chen H, Gladon RJ, Braun EJ, Hannapel DJ. 2000. A leaf lipoxygenase of potato induced specifically by pathogen infection. *Plant Physiol.* 124:1121–30

63. Kolomiets MV, Hannapel DJ, Chen H, Tymeson M, Gladon RJ. 2001. Lipoxygenase is involved in the control of potato tuber development. *Plant Cell* 13:613–26

64. Kolomiets MV, Hannapel DJ, Gladon RJ. 1996. Potato lipoxygenase genes expressed during the early stages of tuberization. *Plant Physiol.* 112:446

65. Kramell R, Miersch O, Atzorn R, Parthier B, Wasternack C. 2000. Octadecanoid-derived alteration of gene expression and the "Oxylipin Signature" in stressed barley leaves. Implications for different signaling pathways. *Plant Physiol.* 123:177–88

66. Kühn H, Sprecher H, Brash AR. 1990. On singular or dual positional specificity of lipoxygenases. The number of chiral products varies with alignment of methylene groups at the active site of the enzyme. *J. Biol. Chem.* 265:16300–5

67. Kühn H, Wiesner R, Rathmann J, Schewe T. 1991. Formation of ketodienoic fatty acids by the pure pea lipoxygenase-1. *Eicosanoids* 4:9–14

68. Larson TR, Graham IA. 2001. Technical advance: a novel technique for the

sensitive quantification of acyl CoA esters from plant tissues. *Plant J.* 25:115–25

69. Laudert D, Pfannschmidt U, Lottspeich F, Holländer-Czytko H, Weiler EW. 1996. Cloning, molecular and functional characterization of *Arabidopsis thaliana* allene oxide synthase (CYP 74), the first enzyme of the octadecanoid pathway to jasmonates. *Plant Mol. Biol.* 31:323–35

70. Laudert D, Weiler EW. 1998. Allene oxide synthase: a major control point in *Arabidopsis thaliana* octadecanoid signalling. *Plant J.* 15:675–84

71. Lehmann J, Atzorn R, Brückner C, Reinbothe S, Leopold J, et al. 1995. Accumulation of jasmonate, abscisic acid, specific transcripts and proteins in osmotically stressed barley leaf segments. *Planta* 197:156–62

72. Leon J, Rojo E, Sanchez-Serrano JJ. 2001. Wound signalling in plants. *J. Exp. Bot.* 52:1–9

73. Matsui K. 1998. Properties and structures of fatty acid hydroperoxide lyase. *Belg. J. Bot.* 131:50–62

74. Matsui K, Irie M, Kajiwara T, Hatanaka A. 1992. Developmental changes in lipoxygenase activity in cotyledons of cucumber seedlings. *Plant Sci.* 85:23–32

75. Matsui K, Miyahara C, Wilkinson J, Hiatt B, Knauf V, Kajiwara T. 2000. Fatty acid hydroperoxide lyase in tomato fruits: cloning and properties of a recombinant enzyme expressed in *Escherichia coli*. *Biosci. Biotechnol. Biochem.* 64:1189–96

76. Matsui K, Nishioka M, Ikeyoshi M, Matsumura Y, Mori T, Kajiwara T. 1998. Cucumber root lipoxygenase can act on acyl groups in phosphatidylcholine. *Biochim. Biophys. Acta* 1390:8–20

77. Matsui K, Shibutani M, Hase T, Kajiwara T. 1996. Bell pepper fruit fatty acid hydroperoxide lyase is a cytochrome P450 (CYP74B). *FEBS Lett.* 394:21–24

78. Matsui K, Ujita C, Fujimoto S, Wilkinson J, Hiatt B, et al. 2000. Fatty acid 9- and 13-hydroperoxide lyases from cucumber. *FEBS Lett.* 481:183–88

79. Matsui K, Wilkinson J, Hiatt B, Knauf V, Kajiwara T. 1999. Molecular cloning and expression of Arabidopsis fatty acid hydroperoxide lyase. *Plant Cell Physiol.* 40:477–81

80. Maucher H, Hause B, Feussner I, Ziegler J, Wasternack C. 2000. Allene oxide synthases of barley (*Hordeum vulgare* cv. *Salomé*): tissue-specific regulation in seedling development. *Plant J.* 21:199–213

81. Miersch O, Wasternack C. 2000. Octadecanoid and jasmonate signaling in tomato (*Lycopersicon esculentum* Mill.) leaves: endogenous jasmonates do not induce jasmonate biosynthesis. *Biol. Chem.* 381:715–22

82. Nellen A, Rojahn B, Kindl H. 1995. Lipoxygenase forms located at the plant plasma membrane. *Z. Naturforsch. Teil C* 50c:29–36

83. Nelson DR. 1999. Cytochrome P450 and the individuality of species. *Arch. Biochem. Biophys.* 369:1–10

84. Noordermeer MA, Feussner I, Kolbe A, Veldink GA, Vliegenthart JFG. 2000. Oxygenation of (3Z)-alkenals to 4-hydroxy-(2E)-alkenals in plant extracts: a nonenzymatic process. *Biochem. Biophys. Res. Commun.* 277:112–16

85. Noordermeer MA, van Dijken AJH, Smeekens SCM, Veldink GA, Vliegenthart JFG. 2000. Characterization of three cloned and expressed 13-hydroperoxide lyase isoenzymes from alfalfa with unusual N-terminal sequences and different enzyme kinetics. *Eur. J. Biochem.* 267:2473–82

86. Noordermeer MA, Veldink GA, Vliegenthart JFG. 2001. Fatty acid hydroperoxide lyase: a plant cytochrome P450 enzyme involved in wound healing and pest resistance. *Chem. Bio. Chem.* 2:494–504

87. Noordermeer MA, Veldink GA, Vliegenthart JF. 2001. Spectroscopic studies on the active site of hydroperoxide lyase; the influence of detergents on its conformation. *FEBS Lett.* 489:229–32

88. Pan ZQ, Durst F, Werck-Reichhart D, Gardner HW, Camara B, et al. 1995. The major protein of guayule rubber particles is a cytochrome P450. *J. Biol. Chem.* 270:8487–94

89. Parchmann S, Müller MJ. 1998. Evidence for the formation of dinor isoprostanes E_1 from α-linolenic acid in plants. *J. Biol. Chem.* 273:32650–55

90. Rance I, Fournier J, Esquerre-Tugaye MT. 1998. The incompatible interaction between *Phytophthora parasitica* var. *nicotianae* race 0 and tobacco is suppressed in transgenic plants expressing antisense lipoxygenase sequences. *Proc. Natl. Acad. Sci. USA* 95:6554–59

91. Reymond P, Weber H, Damond M, Farmer EE. 2000. Differential gene expression in response to mechanical wounding and insect feeding in *Arabidopsis. Plant Cell* 12:707–20

92. Rosahl S. 1996. Lipoxygenases in plants—their role in development and stress response. *Z. Naturforsch. Teil C* 51:123–38

93. Royo J, Leon J, Vancanneyt G, Albar JP, Rosahl S, et al. 1999. Antisense-mediated depletion of a potato lipoxygenase reduces wound induction of proteinase inhibitors and increases weight gain of insect pests. *Proc. Natl. Acad. Sci. USA* 96: 1146–51

94. Royo J, Vancanneyt G, Perez AG, Sanz C, Störmann K, et al. 1996. Characterization of three potato lipoxygenases with distinct enzymatic activities and different organ-specific and wound-regulated expression patterns. *J. Biol. Chem.* 271:21012–19

95. Rusterucci C, Montillet JL, Agnel JP, Battesti C, Alonso B, et al. 1999. Involvement of lipoxygenase-dependent production of fatty acid hydroperoxides in the development of the hypersensitive cell death induced by cryptogein on tobacco leaves. *J. Biol. Chem.* 274:36446–55

96. Saffert A, Hartmann-Schreier J, Schön A, Schreier P. 2000. A dual function α-dioxygenase-peroxidase and NAD^+ ox-idoreductase active enzyme from germinating pea rationalizing α-oxidation of fatty acids in plants. *Plant Physiol.* 123:1545–52

97. Sanz A, Moreno JI, Castresana C. 1998. PIOX, a new pathogen-induced oxygenase with homology to animal cyclooxygenase. *Plant Cell* 10:1523–37

98. Schaller F. 2001. Enzymes of the biosynthesis of octadecanoid-derived signalling molecules. *J. Exp. Bot.* 52:11–23

99. Schneider C, Tallman KA, Porter NA, Brash AR. 2001. Two distinct pathways of formation of 4-hydroxy-nonenal: mechanisms of non-enzymatic transformation of the 9- and 13-hydroperoxides of linoleic acid to 4-hydroxyalkenals. *J. Biol. Chem.* 276:20831–38

100. Seo HS, Song JT, Cheong JJ, Lee YH, Lee YW, et al. 2001. Jasmonic acid carboxyl methyltransferase: a key enzyme for jasmonate-regulated plant responses. *Proc. Natl. Acad. Sci. USA* 98:4788–93

101. Serhan C, Anderson P, Goodman E, Dunham P, Weissmann G. 1981. Phosphatidate and oxidized fatty acids are calcium ionophores. *J. Biol. Chem.* 256:2736–41

102. Shibata D, Slusarenko A, Casey R, Hildebrand D, Bell E. 1994. Lipoxygenases. *Plant Mol. Biol. Report.* 12:S41–42

103. Shimizu T, Radmark O, Samuelsson B. 1984. Enzyme with dual lipoxygenase activities catalyses leukotriene A4 synthesis from arachidonic acid. *Proc. Natl. Acad. Sci. USA* 81:689–93

104. Siedow JN. 1991. Plant lipoxygenase—structure and function. *Annu. Rev. Plant Physiol. Plant Mol. Biol.* 42:145–88

105. Sivasankar S, Sheldrick B, Rothstein SJ. 2000. Expression of allene oxide synthase determines defense gene activation in tomato. *Plant Physiol.* 122:1335–42

106. Sloane DL, Leung R, Barnett J, Craik CS, Sigal E. 1995. Conversion of human 15-lipoxygenase to an efficient 12-lipoxygenase: the side-chain geometry of

amino acids 417 and 418 determine positional specificity. *Protein Eng.* 8:275–82

107. Slusarenko A. 1996. The role of lipoxygenase in plant resistance to infection. In *Lipoxygenase and Lipoxygenase Pathway Enzymes*, ed. GJ Piazza, pp. 176–97. Champaign, IL: AOCS Press

108. Song WC, Funk CD, Brash AR. 1993. Molecular cloning of an allene oxide synthase—a cytochrome-P450 specialized for the metabolism of fatty acid hydroperoxides. *Proc. Natl. Acad. Sci. USA* 90:8519–23

109. Spiteller P, Kern W, Reiner J, Spiteller G. 2001. Aldehydic lipid peroxidation products derived from linoleic acid. *Biochim. Biophys. Acta* 1531:188–208

110. Stelmach BA, Müller A, Hennig P, Gebhardt S, Schubert-Zsilavecz M, Weiler EW. 2001. A novel class of oxylipins, *sn*1-O-(12-Oxophytodienoyl)-*sn*2-O-(hexadecatrienoyl)-monogalactosyl diglyceride, from *Arabidopsis thaliana*. *J. Biol. Chem.* 276:12832–38

111. Stelmach BA, Müller A, Weiler EW. 1999. 12-oxo-phytodienoic acid and indole-3-acetic acid in jasmonic acid-treated tendrils of *Bryonia dioica*. *Phytochemistry* 51:187–92

112. Deleted in proof

113. Stumpe M, Kandzia R, Göbel C, Rosahl S, Feussner I. 2001. A pathogen-inducible divinyl ether synthase (*CYP74D*) from elicitor-treated potato suspension cells. *FEBS Lett.* 507:371–76

114. Tijet N, Schneider C, Muller BL, Brash AR. 2001. Biogenesis of volatile aldehydes from fatty acid hydroperoxides: molecular cloning of a hydroperoxide lyase (CYP74C) with specificity for both the 9- and 13-hydroperoxides of linoleic and linolenic acids. *Arch. Biochem. Biophys.* 386:281–89

115. Tijet N, Waspi U, Gaskin DJ, Hunziker P, Muller BL, et al. 2000. Purification, molecular cloning, and expression of the gene encoding fatty acid 13-hydro-

peroxide lyase from guava fruit (*Psidium guajava*). *Lipids* 35:709–20

116. Tranbarger TJ, Franceschi VR, Hildebrand DF, Grimes HD. 1991. The soybean 94-kilodalton vegetative storage protein is a lipoxygenase that is localized in paraveinal mesophyll cell vacuoles. *Plant Cell* 3:973–87

117. Vancanneyt G, Sanz C, Farmaki T, Paneque M, Ortego F, et al. 2001. Hydroperoxide lyase depletion in transgenic potato plants leads to an increase in aphid performance. *Proc. Natl. Acad. Sci. USA* 98:8139–44

118. van der Stelt M, Noordermeer MA, Kiss T, Van Zadelhoff G, Merghart B, et al. 2000. Formation of a new class of oxylipins from *N*-acyl(ethanol)amines by the lipoxygenase pathway. *Eur. J. Biochem.* 267:2000–7

119. Veldink GA, Vliegenthart JFG. 1991. Substrates and products of lipoxygenase catalysis. In *Studies in Natural Products Chemistry*, ed. Atta-ur-Rahman, pp. 559–89. Amsterdam: Elsevier

120. Vernooy-Gerritsen M, Leunissen JLM, Veldink GA, Vliegenthart JFG. 1984. Intracellular localization of lipoxygenases-1 and -2 in germinating soybean seeds by indirect labeling with protein A-colloidal gold complexes. *Plant Physiol.* 76:1070–78

121. Vick BA, Zimmerman DC. 1982. Levels of oxygenated fatty acids in young corn and sunflower plants. *Plant Physiol.* 69:1103–8

122. Vick BA, Zimmerman DC. 1987. Pathways of fatty acid hydroperoxide metabolism in spinach leaf chloroplasts. *Plant Physiol.* 85:1073–78

123. Vollenweider S, Weber H, Stolz S, Chetelat A, Farmer EE. 2000. Fatty acid ketodienes and fatty acid ketotrienes: michael addition acceptors that accumulate in wounded and diseased *Arabidopsis* leaves. *Plant J.* 24:467–76

124. Walling LL. 2000. The myriad plant responses to herbivores. *J. Plant Growth Regul.* 19:195–216

125. Wang CX, Avdiushko S, Hildebrand DF. 1999. Overexpression of a cytoplasm-localized allene oxide synthase promotes the wound-induced accumulation of jasmonic acid in transgenic tobacco. *Plant Mol. Biol.* 40:783–93

126. Wardale DA, Lambert EA. 1980. Lipoxygenase from cucumber fruit: localization and properties. *Phytochemistry* 19:1013–16

127. Wasternack C, Hause B. 2002. Jasmonates and octadecanoids—signals in plant stress responses and development. In *Progress in Nucleic Acid Research*, ed. K Moldave, Vol. 72. New York: Academic. In press

128. Wasternack C, Parthier B. 1997. Jasmonate signalled plant gene expression. *Trends Plant Sci.* 2:302–7

129. Weber H, Chetelat A, Caldelari D, Farmer EE. 1999. Divinyl ether fatty acid synthesis in late blight-diseased potato leaves. *Plant Cell* 11:485–93

130. Weber H, Vick BA, Farmer EE. 1997. Dinor-oxo-phytodienoic acid: a new hexadecanoid signal in the jasmonate family. *Proc. Natl. Acad. Sci. USA* 94:10473–78

131. Weichert H, Kohlmann M, Wasternack C, Feussner I. 2000. Metabolic profiling of oxylipins upon sorbitol treatment in barley leaves. *Biochem. Soc. Trans.* 28:861–62

132. Weichert H, Stenzel I, Berndt E, Wasternack C, Feussner I. 1999. Metabolic profiling of oxylipins upon salicylate treatment in barley leaves—preferential induction of the reductase pathway by salicylate. *FEBS Lett.* 464:133–37

133. Wilson RA, Gardner HW, Keller NP. 2001. Cultivar-dependent expression of a maize lipoxygenase responsive to seed infesting fungi. *Mol. Plant-Microbe Interact.* 14:980–87

134. Xie DX, Feys BF, James S, Nietorostro M, Turner JG. 1998. *COI1*: an Arabidopsis gene required for jasmonate-regulated defense and fertility. *Science* 280:1091–94

135. Yokoyama M, Yamaguchi S, Inomata S, Komatsu K, Yoshida S, et al. 2000. Stress-induced factor involved in flower formation of *Lemna* is an α-ketol derivative of linolenic acid. *Plant Cell Physiol.* 41:110–13

136. Ziegler J, Stenzel I, Hause B, Maucher H, Hamberg M, et al. 2000. Molecular cloning of allene oxide cyclase. *J. Biol. Chem.* 275:19132–38

137. Ziegler J, Wasternack C, Hamberg M. 1999. On the specificity of allene oxide cyclase. *Lipids* 34:1005–15

138. Zimmerman DC, Vick BA. 1970. Hydroperoxide isomerase. *Plant Physiol.* 46:445–53

Annu. Rev. Plant Biol. 2002. 53:299–328
DOI: 10.1146/annurev.arplant.53.100301.135207

PLANT RESPONSES TO INSECT HERBIVORY:
The Emerging Molecular Analysis

André Kessler and Ian T. Baldwin

*Department of Molecular Ecology, Max-Planck-Institute for Chemical Ecology, Jena
07745, Germany; e-mail: baldwin@ice.mpg.de, kessler@ice.mpg.de*

Key Words plant-arthropod interaction, insect elicitors, direct defense, indirect defense

■ **Abstract** Plants respond to herbivore attack with a bewildering array of responses, broadly categorized as direct and indirect defenses, and tolerance. Plant-herbivore interactions are played out on spatial scales that include the cellular responses, well-studied in plant-pathogen interactions, as well as responses that function at whole-plant and community levels. The plant's wound response plays a central role but is frequently altered by insect-specific elicitors, giving plants the potential to optimize their defenses. In this review, we emphasize studies that advance the molecular understanding of elicited direct and indirect defenses and include verifications with insect bioassays. Large-scale transcriptional changes accompany insect-induced resistance, which is organized into specific temporal and spatial patterns and points to the existence of herbivore-specific *trans*-activating elements orchestrating the responses. Such organizational elements could help elucidate the molecular control over the diversity of responses elicited by herbivore attack.

CONTENTS

WHY PLANT RESPONSES TO PATHOGENS
AND HERBIVORES FREQUENTLY DIFFER

Autotrophs require sophisticated defenses if they are to survive in a world full of heterotrophs. Attacks from heterotrophs occur on spatial scales ranging from microbes to moose, and plants require defenses that are effective at all levels. Defenses against microbes can be highly effective on small spatial scales; the hypersensitive response (HR), in which cells immediately surrounding the infection site rapidly die and fill with antimicrobial compounds to prevent the spread of the pathogen, is the best-studied example (58). Although the HR is an extremely effective defense, this cellular suicide cannot be used without large costs and requires a sophisticated recognition system to avoid inappropriate deployment. Hence the HR, and the burst of reactive oxygen species (ROS) (114) that frequently precedes the HR and may contribute to apoptosis, are activated when a plant's surveillance system (R-genes, encoding receptor proteins) binds various elicitors from the attacking pathogen (proteins, peptides, lipids, polysaccharides) (58). Although the HR can be effective against sedentary herbivores that attack particular tissues [such as phloem-feeding aphids (73)], it is not effective against most free-living herbivores, which avoid an HR by simply moving to another feeding site. This autonomy and the resulting physiological independence of herbivores from their host plants profoundly expand the spatial scale of the plant-herbivore interaction to include not only whole-plant responses but also the community in which the plant lives (Figure 1).

Most insect herbivores arrive at a plant after their devoted mothers have carefully selected and, in some cases, manipulated [by microbial inoculations: e.g., bark beetles (88); or altering source-sink relationships: e.g., aphids (60)] the host plant and endowed the young herbivore nutritionally and developmentally, enabling it to launch its first attack. The young herbivore is fully equipped with mandibles and other feeding apparatus to force its way through the plant's protective covering, an efficient digestive tract in which plant parts can be digested and assimilated in a milieu controlled by the herbivore (112), and mobility and sensory systems that allow it to move in response to heterogeneity in plant suitability (90). This physiological and behavioral autonomy can account for overarching differences between how plants respond to herbivores and pathogens and in how an attack is perceived.

In contrast to pathogen attack, herbivore attack is frequently associated with wounding, and the "recognition" of herbivore attack frequently involves modifications of a plant's wound response (6, 56, 118). Moreover, the physiological and behavioral autonomy of herbivores also allows plants to use defenses that would

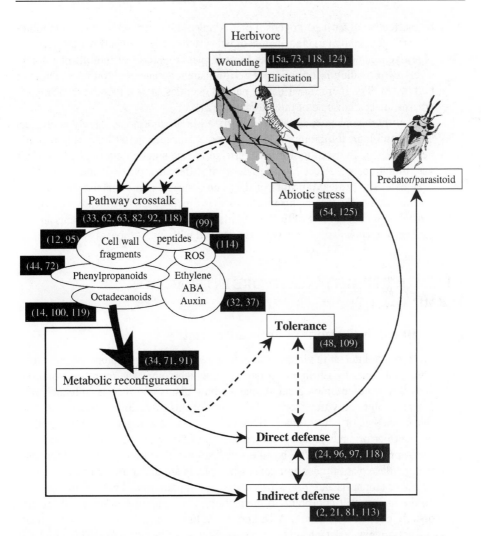

Figure 1   The arena of plant-induced resistance to arthropods. An attacking herbivore wounds the plant and applies or injects elicitors. Wound- and herbivore-specific elicitors in combination with abiotic stresses differentially activate various signaling pathways. These signal cascades interact (pathway crosstalk) to either directly produce volatile signals that function as indirect defenses or effect a fine-tuned metabolic reconfiguration and the expression of defense-related genes. As a consequence of these changes, resources are allocated to regrowth (tolerance) or the production of compounds that directly affect the attacking herbivore (direct defenses: toxic, antinutritive, and antidigestive compounds) or indirectly (indirect defenses) by attracting natural enemies. Recent reviews summarizing the knowledge of particular parts of the arena are listed in black boxes. Evidence is emerging that arthropod-induced resistance results from a coordinated production of specific direct and indirect defenses that complement the existing constitutive defenses of the plant.

be ineffective against pathogens. For example, plants use secondary metabolites that are specifically targeted against organ systems unique to herbivores [nervous, digestive, endocrine, etc. (96)] and use higher trophic-level interactions defensively by providing information or nutritional encouragement to the predators of herbivores. The recruitment of the natural enemies of herbivores drastically increases the spatial scale of the interaction (21).

Plant-herbivore interactions are described in an enormous, largely ecological and entomological literature into which molecular techniques have only recently been injected. To assist molecularly oriented readers in coming to grips with this vast literature, we select recent reviews of different aspects of the interactions (Figure 1) and present a primer of the main concepts and terms used to describe the interaction. We follow this with a review of recent literature on advances in the molecular understanding of the "recognition" of herbivore attack by plants and the resistance mechanisms that have been verified with insect bioassays.

A PRIMER IN PLANT-HERBIVORE INTERACTION TERMINOLOGY

Fitness-Based Evaluations of Resistance Traits

Much of the interest in plant-herbivore interactions among ecologists stems from a seminal paper by Ehrlich & Raven (25), which coined the term "coevolution" and stimulated entomological studies of how plants and insects influence each other's evolutionary trajectories. As a result of this evolutionary focus, the functional analysis of plant traits is frequently evaluated at higher-level integrations of plant performance, namely the correlates of Darwinian fitness (production of seeds, pollen, tubers, etc.). The adjective "defensive" is usually reserved for traits that increase plant fitness correlates when plants are under attack (50). This evolutionary emphasis and fitness-based analysis of plant traits have focused interest on the factors that contribute to maintaining the variability in resistance so frequently observed in nature. A central thrust has been to understand why plants are not always resistant, when it has such clear fitness benefits. This emphasis on understanding variability has focused attention on the fitness costs of different types of resistance: their modes of expression (constitutive or induced) and the environmental, evolutionary, and developmental constraints on them (42, 96, 97). Resistance traits can be broadly categorized into three defense strategies: direct and indirect defenses and tolerance (50). Plants are either constitutively resistant as a result of preformed resistance traits, or they become resistant after an attack as a result of herbivore-induced changes that are either localized to the tissues adjacent to an attack site or systemically expressed throughout the plant.

Costs of Defense

Just as the indiscriminate deployment of the HR would likely severely compromise a plant's performance, the production of resistance traits when they are not needed is likely to be costly for a number of reasons. First, resistance traits can be costly to produce if fitness-limiting resources (such as nitrogen) are invested (5) or if the traits are also toxic to the plant. However, resistance costs can also arise from higher-level ecological processes. For example, specialized herbivores may sequester a plant's defenses and use them for their own defense against predators, or compounds that provide defense against generalist herbivores may attract specialist herbivores, which use them as host-location signals (113). Moreover, the defenses may disrupt important mutualistic interactions, such as pollination, which are also mediated by insects (1). These fitness costs probably provide the selection pressure behind the evolution of inducible resistance, if inducible expression allows plants to forego these fitness costs when the defense is not needed.

Direct Defenses

Direct defenses are any plant traits (e.g., thorns, silica, trichomes, primary and secondary metabolites) that by themselves affect the susceptibility to and/or the performance of attacking arthropods and thus increase plant fitness in environments with herbivores. Defensive secondary metabolites are categorized by their mode of action (24). Proteinase inhibitors (PI) (antidigestive proteins) are inducible by wounding and herbivory and influence herbivore performance by inhibiting insect digestive enzymes (55, 110). Polyphenol oxidases are antinutritive enzymes that decrease the nutritive value of the wounded plant by cross-linking proteins or catalyzing the oxidation of phenolic secondary metabolites to reactive and polymerizing quinones. Toxic compounds (e.g., alkaloids, terpenoids, phenolics) poison generalist herbivores, forcing specialists to invest resources in detoxification mechanisms that in turn incur growth and development costs.

Indirect Defenses

Indirect defenses are plant traits that attract predators and parasitoids of herbivores and increase the carnivore's foraging success and thereby facilitate top-down control of herbivore populations (50). Volatile organic compounds (VOCs) released by herbivore-attacked plants are known to be attractive to arthropod predators and parasitoids in laboratory experiments on agricultural plants (21) and have recently been shown to function defensively under natural conditions (52). The VOC response can be highly specific; parasitic wasps often use this specificity to locate particular hosts or even the particular instars of their hosts (113). However, generalist predators are also attracted by single components of the VOC bouquet, which are commonly emitted after attack from a diverse set of herbivore species (52). In addition to attracting natural enemies of the herbivores, the VOC release can function as a direct defense by repelling the ovipositing herbivores (19, 52). Finally, it may be involved in plant-plant interactions (27). In addition to supplying information to natural enemies about the location and activity of foraging herbivores, plants also provide food and shelter to the enemies of herbivores (3). Extrafloral nectaries increase their rate of nectar secretion after herbivore attack, and these carbohydrates and proteins (from various food bodies or even pollen) provide nutritional encouragement for predators to increase their foraging rate in certain areas of a plant (40). Shelter is provided as specialized structures (leaf domatia) or modifications of existing structures (hollow thorns, stems).

Tolerance

Tolerance decreases the fitness consequences of herbivore attack for a plant. A plant genotype is termed tolerant if it can sustain tissue loss with little or no decrease in fitness relative to that in the undamaged state (109). Although genotypes clearly vary in their tolerance, the mechanisms underlying such variation are not understood. Certain morphological traits, such as meristem sequestration and reactivation, as well as photoassimilate storage in below-ground and stem structures, in addition to physiological responses, such as herbivore-induced increases in photosynthetic capacities, and nutrient uptake, are correlated with compensatory growth following herbivore attack (109). However, the functionally mysterious transcriptional reconfiguration that follows herbivore attack is most likely to hold the key to a more detailed mechanistic understanding of tolerance responses.

ELICITORS FROM HERBIVORES

Any compound that comes from herbivores and interacts with the plant on a cellular level is a potential elicitor. So far, herbivore-specific elicitors have been isolated from oral secretions of lepidopteran species and the oviposition fluid of weevil beetles—the two insect fluids that regularly come in contact with plant wounds (Figure 2). Additionally, some evidence suggests that microbes present in the digestive organs of herbivores are involved in the production of elicitors found in oral secretions (107).

Two classes of elicitors have been isolated from the oral secretions of lepidopteran larvae; both elicit indirect defense responses. The first class includes lytic enzymes, such as β-glucosidase that was isolated from *Pieris brassicae* and elicits the release of terpenoid volatiles from cabbage leaves (66). Other lytic enzymes have been found in the saliva of other lepidopteran species, such as glucose oxidases in *Helicoverpa zea* saliva (29); in piercing insects, such as alkaline phosphatase in whitefly (*Bemisia tabaci*) saliva (31); and in a wide array of watery digestive enzymes from aphid saliva (73); but their roles as elicitors of defense responses have not yet been established.

The second class of elicitors comprises fatty-acid–amino-acid conjugates (FACs), which have been found in the regurgitant of larval Sphingidae (36), Noctuidae, and Geometridae (4, 86). The FAC volicitin [N-(17-hydroxylinolenoyl)-L-glutamine], from *Spodoptera exigua*, induces excised maize seedlings to release the same odor blend of volatile terpenoids and indole that is released when they are damaged by caterpillar feeding (4). Unfortunately a recent study with intact maize seedlings found volicitin to only elicit VOC release when applied to plants at midnight and that the release from excised seedlings was much greater than that from

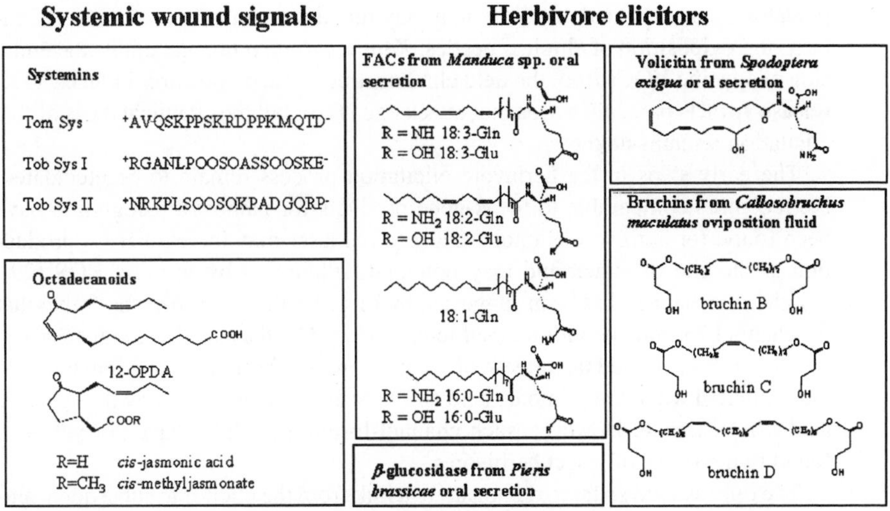

Systemic wound signals   Herbivore elicitors

Figure 2 Examples of systemic wound- and herbivore-specific signals demonstrated to elicit either direct or indirect defenses in plants. Systemins are polypeptide systemic wound signals from solanaceous plants, such as tomato (Tom Sys) and tobacco (Tob Sys I and II), which activate the octadecanoid pathway but are also inducible by various oxylipins (83, 99). Oxylipins, as illustrated by the octadecanoids, 12-oxophytodienoic acid (12-OPDA), jasmonic acid, and methyl jasmonate, elicit defense gene expression, numerous secondary metabolites, and insect resistance (14). Herbivore-specific elicitors have been identified in insect oral secretions and oviposition fluids, the two fluids that commonly come into contact with the wounded plant tissue. *Manduca* and *Spodoptera exigua* oral secretions contain the class of fatty-acid-amino-acid conjugates (FACs) found to actively elicit volatile organic compounds (VOCs), which function as indirect defenses (36, 113). The FACs from *Manduca* also elicit other herbivore-specific changes (Figure 3). *Pieris brassicae* oral secretion contains the enzyme β-glucosidase that also elicits VOC emission (66). The novel class of elicitors from cowpea weevil oviposition fluid (bruchins) elicits neoplasmic tissue formation in peas, which expels the oviposited egg from the pea leaf tissue (23).

intact plants. Moreover, volicitin was less effective than JA in eliciting VOC release (103a). Finally it is interesting to note that none of the separated enantiomers of volicitin were active in lima bean, which is also a host plant of the beet armyworm (108), suggesting that volicitin as an elicitor deserves additional research.

Recently, another class of herbivore-specific elicitors has been determined to induce a novel type of defense response in peas. These elicitors, long chain diols that are mono- and diesterified with 3-hydroxypropanoic acid, are called bruchins because they are excreted with the oviposition fluid of pea and cowpea weevils (Bruchidae) (Figure 2). In certain genotypes of peas, bruchins elicit neoplastic growth on pods, which lifts the egg out of the oviposition site and impedes larval entry into the pod (23). Because the neoplastic growth exposes the larvae to predators, parasites, and desiccation, it may function as an indirect defense. Similarly, the oviposition of elm leaf beetles, *Xanthogaleruca luteola*, elicits the emission of specific VOCs from the field elm host trees. The oviposition-induced VOC release attracts parasitic wasps that attack eggs (70), but the chemical basis of this elicitation remains unknown.

The early steps in the herbivore elicitation process remain to be elucidated. No receptors comparable to the *R*-gene products for pathogen recognition have been found for herbivore elicitors. Glucose oxidases may increase H_2O_2 production at the site of attack and may potentiate elicitation by forming ROS (29). Another mechanism has been suggested by Engelberth et al. (26), who found that the channel-forming peptide alamethicin, produced by the parasitic fungus *Trichoderma viridae*, elicited the release of volatiles in lima bean that are comparable to those elicited when the jasmonic acid (JA) cascade is antagonized by the salicylic acid (SA) cascade. However, such channel-forming peptides have not yet been found in the saliva of insect herbivores.

The only system so far to demonstrate a link from the elicitor to subsequent signaling steps is the *Nicotiana attenuata-Manduca* system (Figure 3) (see Crosstalk below). FACs in the oral secretions of two *Manduca* species elicit JA and ethylene

---

→

Figure 3   Alteration of the wound response of wild tobacco plants (*Nicotiana attenuata*) by *Manduca* caterpillar feeding. (*a*) Wounding of the leaf tissue results in a JA burst, which is amplified by caterpillar feeding and the application of FACs from larval oral secretions to the wound (36, 103). (*b*) Caterpillar attack, or the application of FACs to wounds, but not wounding alone induces the production of volatiles that function as predator attractants in the plant's indirect defense (35, 36). (*c*) Caterpillar feeding and the application of their oral secretions to wounds cause an ethylene burst (49), which (*d*) attenuates the wound- and JA-induced accumulation of nicotine by suppressing the accumulation of transcripts for a key regulatory step in nicotine biosynthesis (pmt: putresine N-methyl transferase) (121). The attenuation of the direct defense, nicotine, may be an adaptation to the feeding of a specialized herbivore, which is able to tolerate high alkaloid concentrations and can potentially use them for its own defense (49). (*e*) Caterpillar attack and the addition of FACs to plant wounds also result in a transcriptional reconfiguration of the plant's wound response (43). This reconfiguration of the wound response consists of three temporal and spatial alterations. Addition of FACs antagonizes the wound-induced increase (W) of transcripts encoding threonine deaminase (TD), representing a type-I expression pattern, which spreads systemically throughout the plant from the wound site. The wound-induced increase in transcripts in a type-IIa expression pattern [exemplified by PIOX (pathogen-induced oxygenase)] is further amplified after application of FACs to wounds. In contrast, the genes with a type-IIb expression pattern are suppressed after wounding and further suppressed with the addition of FACs, as exemplified by the gene encoding the light-harvesting complex subunit LHB C1. Both type-IIa and -IIb patterns are found only in the leaves directly suffering the herbivore attack (102).

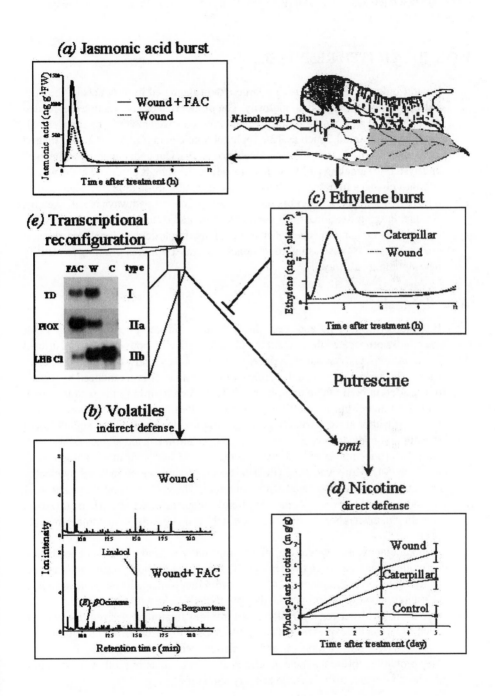

(a) Jasmonic acid burst

— Wound + FAC
---- Wound

Jasmonic acid (ng g⁻¹FW)

Time after treatment (h)

N-linolenoyl-L-Glu

(c) Ethylene burst

— Caterpillar
---- Wound

Ethylene (ng h⁻¹ plant⁻¹)

Time after treatment (h)

(e) Transcriptional reconfiguration

	FAC	W	C	type
TD				I
HOX				IIa
LHB C1				IIb

Putrescine

pmt

(b) Volatiles
indirect defense

Wound

Ion intensity

Linalool

(*E*)-*β*-Ocimene   *cis*-α-Bergamotene

Wound+ FAC

Retention time (min)

(d) Nicotine
direct defense

Wound

Caterpillar

Control

Whole-plant nicotine (m g/g)

Time after treatment (day)

bursts, which are involved in the alterations of the *N. attenuata* wound responses (36, 103, 121).

WOUND-ELICITED RESPONSES

Wound-induced resistance is to a large extent mediated by products of the "octadecanoid" (C_{18}-fatty acids) pathway. The production of various defense-related compounds, e.g., toxins, antinutritive and antidigestive enzymes, requires signaling by octadecaniods, such as 12-oxophytodienoic acid (OPDA), JA, and methyl jasmonate (MJ), all derived from linolenic acid (14). Our understanding of the local and systemic signaling pathways that transduce the signals produced at the wound site into changes in defense-related gene expression throughout the plant and of how herbivore-specific elicitors modify these signaling pathways is still sketchy. An encouraging trend has been the increase in studies using herbivore bioassays to evaluate the consequences of manipulations, in particular signal transduction chains, as well as those using insect attack and transcriptional profiling to identify new signaling pathways.

Wound Signals

Both herbivore feeding and mechanical damage induce systemic responses that are rapidly propagated throughout the plant as well as responses that are restricted to the wound site. Systemic responses require mobile signals; these could be electrical (41), hydraulic (65), and chemical (99). In tomato, PIs are induced by signals transported in both xylem and phloem tissues. Steam girdling the petiole, which kills the phloem but leaves the xylem intact, does not prevent systemic induction of PIs after severe wounding, suggesting a xylem-transported signal. Small crushing wounds do not, however, induce a xylem-mediated signal; rather, they elicit PI production in organs distal to the wound and with intact phloem connections (94). Moreover, *Pin2* (PI II) gene expression can be induced by electrical signals that are associated with wounding. The response and propagation of wound-induced electrical signals apparently requires intact abscisic acid (ABA) signaling pathways because they both are lacking in ABA-deficient tomato mutants (41).

Immediately after wounding, plants transiently produce ROS, such as the superoxide anion, locally in the damaged tissue and H_2O_2 both locally and systemically throughout the plant (114). Because wound-induced oligogalacturonides transiently elicit ROS production, they are thought to be a primary signal of tissue damage (62). Moreover, both oligogalacturonides and fungal-derived chitosan can elicit ROS and the expression of wound-inducible PI genes. However, oligogalacturonides have limited mobility and are also induced by systemins (Figure 2), and they probably represent a local, intermediate step in signaling following systemin production, rather than a mobile primary signal (62).

Systemin, the first described oligopeptide with phytohormonal function, is thought to represent the primary wound signal in some solanaceaous plants (tomato, pepper, black nightshade, tobacco). Cleaved from a 200–amino acid precursor called prosystemin, and probably transported in the phloem, it is active at femtomolar levels and is the best-verified mobile signal (99). Transformed tomato plants expressing prosystemin cDNA in an antisense orientation do not express PI I and II after wounding and are more susceptible to attacking *Manduca sexta* larvae (80). Plants transformed to overexpress the prosystemin gene, in turn, exhibit constitutively activated wound responses and PI transcript expression (69). Systemin-binding proteins in the plasma membrane of tomatoes and on the surface of *Lycopersicon peruvianum* suspension cultured cells initiate a complex wound cascade after binding systemin. The wound cascade concludes with the activation of a phospholipase A_2, which releases linolenic acid from the plasma membrane, supplying the substrate for the initial step in the octadecanoid pathway (99). The systemin-mediated wound cascade is likely to be the main cascade that is subsequently directly modified by herbivore-specific elicitors or indirectly through the recruitment of other signaling cascades. With the recognition that C_{16}-fatty acids and other lipids can be used in the production of JA and other potential signal molecules, the term "octadecanoids" (C_{18}-fatty acids) should be broadened to "oxylipins" (28). However, because the major advances in the endogenous manipulation of wound- and herbivore-induced oxylipins are derived from C_{18}-fatty acid substrates, we retain the term octadecanoids in the following discussion.

Octadecanoids

The importance of the signaling function of octadecanoids for plant-insect interactions has recently been demonstrated by manipulating three enzymes in the pathway: lipoxygenase (LOX), hydroperoxide lyase (HPL), and JA carboxyl methyltransferase (JMT).

LOX, the nonheme iron-containing dioxygenase, catalyzes the oxygenation of linolenic acid to the 9- and 13-hydroperoxides, which are in turn converted to aldehydes and oxoacids. Products from 13-hydroperoxy linolenic acid can be further elaborated by enzymatic cyclization, reduction, and β-oxidation to produce JA. The LOX proteins play roles in plant growth and development, in maturation and senescence, and in the metabolic responses to pathogen attack and wounding that are thought to be mediated by their role in the biosynthesis of oxylipin signals such as JA and OPDA (14, 100).

A number of studies suggest that different LOX isoforms have different functions. In *A. thaliana*, cosuppression-mediated depletion of a LOX isoform led to a decrease in the wound-induced JA levels but did not affect basal JA levels (9). Different LOX isoforms are encoded by multigene families, of which at least three have been found in potato. Transgenic potato plants devoid of one 13-LOX isoform (LOX-H3) through antisense-mediated depletion of the LOX mRNA do not accumulate PIs in response to wounding. Moreover, when Colorado potato beetle larvae and beet armyworm larvae fed on antisense plants, they grew significantly larger than those fed on wild-type plants. More puzzling, however, was the observation that the effect of LOX-H3 on resistance was apparently not through its involvement in the wound-induced increase of JA. JA levels after wounding were similar in both antisense and wild-type plants, and the exogenous application of JA was not able to recover wild-type PI levels in LOX-H3 antisense plants (98). These results suggest that LOX-H3 is producing a JA-independent signal by one of several possible mechanisms: LOX-H3 may directly affect a *pin* gene response; it may be involved in a regulatory network that influences the expression of other LOX genes; or it may affect the wound-responsive production of ethylene that is required, together with JA, for maximal *pin2* expression in tomato (79). It is interesting to note that LOX-H3 antisense potato plants produced more flowers and on average 20% more tuber mass than did wild-type plants, suggesting that the resistance mechanisms mediated by this isoform of LOX incur fitness costs when they are not needed (98).

One of the LOX-catalyzed products, 13(S)-HPOT, is a substrate for several other enzymes that are thought to be important in plant-insect interactions, including HPL, which cleaves the 13-hydroperoxide into C_6-aldehydes and C_{12}-oxoacids (100). The C_6-aldehydes are thought to protect the wound site from microbial infection and to function as a direct defense against some herbivores, whereas the C_{12}-products, which include traumatin and traumatic acid, may be involved in wound healing (78). More recently, the C_6-aldehydes have been suggested to play a role in directly eliciting defense-related gene expression (8), systemin-based signaling (106), and signaling between plants (27).

Recent experiments with HPL-depleted potato plants showed that HPL gene expression is developmentally controlled and the activity levels are posttranscriptionally regulated (115), complementing results from tomato studies (45). Moreover, normal wound-induced gene (*AOS, LOX-H1, LOX-H3, prosystemin, Pin2, LAP*) expression was found in these plants, despite diminished hexenal/3-hexenal production, suggesting that these metabolites do not play a signaling role. It is interesting to note that *Myzus persicae* aphids performed better on HPL-depleted plants than they did on wild-type plants, suggesting that the C_6-aldehydes produced by HPL may play an important role in direct defenses (115).

If the 13(S)-HPOT produced by LOX is not metabolized by HPL, it can become a substrate for allene oxide synthase (AOS), which produces the unstable allene oxide; this, in turn, is cyclized by allene oxide cyclase (AOC) to form enatiomerically pure 9S,13S-OPDA (100). A subsequent reduction and β-oxidation of OPDA produces JA. Both OPDA and JA actively regulate defense gene expression and elicit resistance (14). Most of the genes that encode enzymes in the JA biosynthetic pathway are activated by wounding, and some (e.g., *LOX, AOS, 12-OPDA reductase*) are also upregulated by exogenous JA application, suggesting positive feedback control (62). Overexpression of flax AOS in potato increased constitutive JA concentrations but did not influence the expression of *pin2* genes (38), which suggests a JA-independent signal for *pin*-gene expression. However, the resistance of the AOS overexpressing potato lines against insect attack was not reported.

Although most of the octadecanoids are involved in systemic responses, their function as a mobile signal has yet to be conclusively demonstrated. Applications of ^{14}C labeled JA to leaves of *Nicotiana sylvestris* plants provided evidence for shoot-to-root transport with a kinetic that was identical to the appearance of a transient increase in endogenous JA concentrations in the roots of plants after leaf wounding, suggesting that JA was transported from wounded leaves to roots in these plants (123). Glucuronidase (GUS) reporter gene fusion experiments with the AOS promotor from *A. thaliana* and *N. tabacum* found AOS activation locally and systemically upon wounding. In contrast to the findings from *N. sylvestris*, this induction remained restricted to the application site of JA or OPDA, suggesting that these two octadecanoids were not systemically transported in the transformed plants (57). However, recently the gene for *S*-adenosyl-L-methionine–JMT from *A. thaliana*, which catalyzes the formation of MJ from JA, was cloned (104). JMT transcripts were induced both locally and systemically after wounding and MJ treatments. Transgenic plants overexpressing *JMT* had elevated MJ levels and constitutively expressed JA responsive genes, such as *VSP* and *PDF1.2*, but they exhibited wild-type-JA contents. The expression of defense genes may have contributed to enhanced levels of resistance against the pathogen *Botrytis cinerea* (104). These results suggest that MJ may regulate defensive gene expression and have the potential for mobile intra-and interplant signal function. Earlier work had proposed that MJ was an airborne signal that functioned in interplant communication for defense responses (27, 51).

Crosstalk

Although it is becoming increasingly clear that single signal cascades, as illustrated by the oxylipins, can produce a bewildering array of potential secondary signal molecules with potentially different functions (14, 28, 119), it has also become apparent that herbivore attack frequently involves the recruitment of several signal cascades, the interaction ("crosstalk") among which may explain the specificity of the responses. Reymond & Farmer (92) proposed a tunable dial as a model for the regulation of defensive gene expression, based on the crosstalk of the three signal pathways for JA, ethylene, and SA. According to this model, a plant tailors its defensive responses to a specific attacker by eliciting signal molecules from the three pathways to different degrees. The elicitation of multiple pathways after attack is likely common and may be necessary if plants are to tailor their responses adaptively to the diverse herbivore species that attack them (118). Herbivores frequently attack in guilds, specifically or opportunistically vectoring microbes into the resulting wounds. These interactions of herbivore and pathogen attack and the resulting signal crosstalk may compromise a plant's responses. Some attackers may induce a pathway (e.g., pathogens activating primary SA responsive genes) that influences the activity of another (e.g., chewing herbivores activating JA responsive genes), thereby compromising the plant's defensive reaction against one or both of the enemies (63, 82).

How these responses are fine-tuned to optimize the defense against a particular herbivore species or guild is the subject of recent investigations and reviews (11, 118). One of the clearest differences in defense responses to herbivores exists between chewing caterpillars and phloem-sap-sucking whiteflies or aphids. That attack from *Manduca sexta* caterpillars is "recognized" by the plant as evidenced by a JA burst far greater than that produced by the wounding that herbivores' feeding behavior causes (68, 103) (Figure 3). This JA burst is associated with expression of both wound-responsive genes and a set of novel JA-independent genes. The introduction of oral secretions from the feeding caterpillar into the wound site can account for the differences (36, 56, 102). Aphid feeding, in contrast, induces the expression of pathogen-responsive, SA-, and wound-responsive, JA-regulated genes (118). Green peach aphids, *Myzus persicae*, induced SA-dependent transcription of *PR-1* and *BGL2* in wild-type plants but not in *npr1* mutant plants, which are deficient in SA signaling (76). In addition, plants attacked by aphids had higher mRNA levels of *PDF1.2* (encoding defensin) and *LOX2* (encoding lipoxygenase), both of which are wound inducible and involved in the JA signaling cascade. SA- and JA-dependent genes have also been induced in plants attacked by phloem-feeding whiteflies. These whiteflies induce the expression of genes involved in lignin production, SA biosynthesis, oxidative burst, as well as in pathogenesis-related and JA-responsive PR proteins (118). Moreover, silverleaf whitefly feeding on squash induces accumulation of transcripts encoding *SLW3*, a gene that is not responsive to any known wound or defense signal. This suggests that there are other defense signal cascades waiting to be discovered (116).

The observation that whitefly and aphid feeding elicit both JA- and SA-induced genes runs counter to well-described observations that SA and other cyclooxidase inhibitors can effectively inhibit wound-induced JA production and JA-elicited gene expression (7, 84). Moreover, pathogen-inducible SA, such as is elicited by TMV infection, inhibits wound-inducible JA accumulation and secondary metabolite accumulation (87). The stimulation of SAR (systemic acquired resistance) with the SA mimic, BTH (benzothiadiazole), attenuated the JA-induced expression of polyphenol oxidase in tomato plants and increased the performance of *Spodoptera exigua* caterpillars, suggesting compromised defense responses. Other examples further undermine the belief that SA exclusively mediates pathogen responses and JA exclusively mediates herbivore responses. *Pseudomonas syringae*, a tomato pathogen, induces responses typically associated with SAR and SA signaling, such as PR-protein expression, but also expression of PI genes, which are normally considered to be JA induced. This activation of different pathways may underlie the observed crossresistance of *P. syringae*–attacked tomato plants against both the pathogen and subsequently feeding noctuid larvae (11).

Crosstalk between ethylene and octadecanoid pathways can be either synergistic or antagonistic. Synergistic effects of JA and ethylene have been reported from *A. thaliana*, in the expression of defensive genes [e.g., *PDF1.2* (85)]; from tomato, for maximal induction of *PI* gene expression (79); and from cultivated tobacco, *Nicotiana tobacum*, for the expresssion of two *PR* genes encoding PR1b and osmotin (122). In contrast, in wild tobacco plants (*N. attenuata*), ethylene antagonizes JA-induced transcript accumulation after herbivore damage (Figure 3). In this species, wounding and mammalian herbivore attack increase the production of a potent defense metabolite, nicotine, which in turn is activated by proportional changes in endogenous JA production (7) as well as exogeneous JA applications (5). Attack by the tobacco hornworm, *M. sexta*, a solanaceous specialist, or application of its regurgitant to wounds results in a JA burst (36, 103) and reduces induced nicotine production (68). The attenuation of nicotine accumulation results from an ethylene burst after hornworm feeding (49), which antagonizes the wound-induced transcriptional increase in the nicotine biosynthetic genes *NaPMT1* and *NaPMT2* (121).

Genoud & Metraux (33) summarized examples of the crosstalk between different signal cascades and modeled them as Boolean networks with logical gates and circuits. This model complements that of Reymond & Farmer (92) and makes concrete predictions regarding the outcome of the crosstalk between pathways. The utility of this approach is limited by our incomplete understanding of all the cascades that are involved. Also lacking from the model is how crosstalk translates to ecological interactions among players on the second and third trophic levels and how compromised plant defense responses translate into plant fitness. An understanding of the functional consequences of crosstalk requires a sophisticated understanding of whole-plant function, which the *Manduca-Nicotiana* interaction illustrates (Figure 3).

In this natural interaction, JA and ethylene bursts are both induced when larval-specific elicitors (FACs) (Figure 2) are introduced to the feeding sites and result in reduced nicotine induction. As such, the crosstalk appears to benefit the herbivore and suggests that *Manduca* is feeding in a "stealthy" fashion, reducing its dietary intake of nicotine by suppressing the nicotine (68). However, parallel to the nicotine attenuation, *Manduca* feeding and application of oral secretions or FACs to the wound induce the emission of a suite of VOCs (35) that function as an indirect defense by attracting predators to the feeding herbivore (52). With the downregulation of a direct defense and the parallel upregulation of an indirect defense, *N. attenuata* may be optimizing the defensive function of its volatile release by suppressing nicotine production, which could be sequestered by the herbivore and used against predators attracted by the volatile release. Plant defense compounds are commonly sequestered by adapted herbivores for their own defense, and thus induced nicotine production may wreak havoc with the plant's ability to use "top-down" processes as a defense (113).

MOLECULAR ADVANCES IN DIRECT DEFENSES

Almost any plant trait can be manipulated to function as a direct defense; however, most of the research has focused on the veritable arsenal of secondary metabolites that function as poisons, digestibility reducers, and repellants. Although many potential direct defenses have been identified, definitive proof of the defensive function of a particular metabolite is in large part limited to the few examples in which the molecular basis of production is sufficiently understood to allow its expression to be manipulated. We limit our review to those studies that demonstrably influence a plant-insect interaction or contribute to an understanding of how variability in direct-defense profiles is generated. In the evolutionary arms race between plants and insects, plants are expected to be under strong selection to evolve new defenses as insects evolve resistance to the initial suite of defenses (25). The mechanisms responsible for the generation of variability in direct defenses are therefore of particular interest. Because a majority of the molecular efforts have focused on model solanaceous and brassicaceous plant systems, these are the systems with the most advances.

The defensive function of PIs was first described in solanaceous plants and is now one of the best-verified groups of direct defenses. PIs are expressed in seeds and tubers and also in vegetative tissue after wounding. Wound-induced PIs have been shown to enhance plants' resistance to insects by inhibiting the proteolytic enzymes of the attacking insect. A majority of the described plant PIs are inhibitors of trypsin and chymotrypsin (55), but recent work with maize (110) describes an herbivore-induced PI that inhibits both elastase and chymotrypsin in the midgut of *Spodoptera littoralis* larvae. The defensive effectiveness of PIs depends on their affinity and specificity for the midgut proteinases of the attacking insect and the ability of the insect to alter its proteinase profile and overexpress proteinases, which are PI insensitive after ingestion of PI-laced food (55). Hence a PI that is able to inhibit two types of insect proteinases may be particularly difficult for the herbivore to counter and may be useful in engineering durable resistance in crops.

Because PIs and other defensive proteins are direct gene products, their defensive effects have been tested by genetic transformation in a number of plant species. A recent example is the transformation of white poplar with an *A. thaliana* cystein proteinase inhibitor gene (*Atcys*), which conferred resistance to a major insect pest, *Chrysomela populi*, by inhibiting most of the digestive proteinase activity of this chrysomelid beetle (18). However, when novel PIs are expressed in a host plant, resistance to all of a plant's herbivores is rarely achieved, and the degree of resistance is in part determined by the insect's counter responses. A study that compared the performance of three generalist lepidopteran herbivores on each of three different host plants (tobacco, *Arabidopsis*, and oilseed rape), each transformed to express the mustard trypsin inhibitor MTI-2, demonstrated that the chemical milieu in which the PI is expressed influenced its defensive function (17). To keep one step ahead of rapidly adapting herbivores, plants have evolved mechanisms to produce many new active PIs. Some of these may be able to retain their defensive function in the constantly changing chemical environments of a plant as it matures and senesces. The multidomain structure of some PIs may allow a plant to produce inhibitors against a broad spectrum of proteases that retain their defensive function in different chemical environments.

The serine proteinase inhibitors of the potato type-II inhibitor family consist mostly of two repeated domains (124), whereas in the stigmas of *Nicotiana alata* flowers, the PIs are produced from a precursor protein with six repeat domains (61) that after proteolytic processing eventually produces six single-domain PIs (four trypsin and two chymotrypsin PIs) and a novel two-domain PI. A four-domain PI from the stigmas of the same plant species has recently been discovered (74). These multidomain structures probably allow plants to target a large number of different proteases within a relatively short period of time (74). Over evolutionary time, the reactive sites in the PI genes have accumulated a larger number of mutations, which we would expect to be the signature of an evolutionary arms race to diversify PI properties (39).

Arabidopsis produces many different glucosinolates that can protect plants against generalist herbivores and pathogens but also function as feeding and/or oviposition attractants for Brassicaceae specialists (13). Heterogeneous selection pressures may therefore maintain the variation in metabolite profiles within populations. A model for how the qualitative and quantitative diversity of glucosinolate profiles is maintained was recently proposed by Kliebenstein et al. (53). They examined 39 *Arabidopsis* ecotypes and found that polymorphisms at only five loci, each coding for different branch points in the glucosinolate biosynthetic pathway, were sufficient to generate 14 qualitatively different leaf glucosinolate profiles, including 34 different structures, most of which are derived from methionine by chain elongation. Moreover, a single locus appeared to control a majority (nearly 75%) of the observed quantitative variation.

Transformation has been used to examine the importance of both qualitative and quantitative variation in glucosinolate profiles of *A. thaliana*. For example, the *CYP79A1* gene from *Sorghum bicolor*, which encodes an enzyme that normally converts L-tyrosine to *p*-hydroxyphenylacetaldoxime in the biosynthesis of cyanogenic glycosides, produces *p*-hydroxybenzylgucosinolate, sinalbin (a glucosinolate not normally found in *A. thaliana* plants) when transferred to *A. thaliana* plants (77). The expression of this gene caused a fourfold increase in total glucosinolate levels (largely owing to increased sinalbin production) but did not alter the acceptance of the plants by two brassicaceous specialist flea beetles (*Phyllotreta nemorum* and *P. cruciferae*) in choice tests. This demonstrates that the plant is unlikely to realize lower herbivore loads from these specialists even with an enormous increase in investment in glucosinolate production.

For a plant to rid itself of its specialist herbivore community might require the evolution of an entirely new defense system. Advances in the ability to transform plants with all of the enzymes required for an entire secondary metabolite pathway have allowed researchers to recreate exactly such an evolutionary event. Cyanogenic glucosides, such as dhurrin, are not normally found in any brassicaceous plants, but recently the complete biosynthetic pathway for dhurrin was transferred to *A. thaliana* by expressing two multifunctional microsomal P450 enzymes (CYP79A1 and CYP71E1) and a soluble UDPG-glucosyltransferase (sbHMNGT) (111). It is remarkable that these three enzymes self-organized into a functional complex that efficiently transferred reaction products so that transformed plants were able to produce and store large amounts of dhurrin. Transformed plants were completely resistant to the Brassicaceae-specialist *P. nemorum*. The ability to transfer entire secondary metabolite pathways between species will allow researchers to test a fundamental tenet of the theory by Ehrlich & Raven (25) of plant-herbivore coevolution: that the diversity of secondary metabolites represents constraints on the evolution of herbivore-host selection.

Resistance to herbivores in *Arabidopsis* is correlated with glucosinolate production and breakdown (myrosinase) and the presence of trichomes (67). However, a recent QTL analysis revealed a locus that did not map to any locus of the previously known resistance traits. Jander et al. (47) crossed two commonly studied *Arabidopsis* ecotypes that differed in their susceptibility to the larvae of *Trichoplysia ni* (a generalist noctuid), the Landsberg *erecta* ecotype, and the Columbia ecotype (which is considerably more resistant). Susceptibility mapped to the *TASTY* locus on chromosome 1, which was distinct from genes that affect trichome density, disease resistance, glucosinolate content, and flowering time (47), but close to a recently discovered locus, *esp*, which causes the formation of epithionitriles during the hydrolysis of glucosinolates instead of isothiocyanates (59). Glucosinolates are hydrolyzed during wounding and herbivore attack, and this finding underscores the importance of studying direct defenses not only in planta, but also in the insect digestive system.

Because herbivores take plant material into an environment that they chemically control (112) and adapt their digestive and detoxification systems to neutralize the effects of direct defenses (90), studying these mechanisms and the chemical dynamics that occur in insect guts might provide important insights into the plant traits that provide direct defense. However, it is also clear that the spatial scale needs to be broadened beyond the insect gut to include insect behavior and the natural enemies of insects. For example, when *Brassica napus* plants were transformed with the gene coding for a potato PI, diamondback moth larvae compensated for their decreased digestive efficiency by eating more leaf material (120). Such compensatory responses demonstrate that many direct defenses, particularly those that slow herbivore growth but do not kill them, may not function as defenses if they are not expressed in concert with indirect defenses, namely plant traits that increase the foraging efficiency of the natural enemies of herbivores (75). A combination of defenses that slow the growth of herbivores and increase the probability of their mortality before they become reproductively mature is likely to strongly suppress the growth of herbivore populations and represent a particularly effective defense.

MOLECULAR ADVANCES IN INDIRECT DEFENSES

Analysis of the costs of resistance suggests that certain direct defenses can incur substantial metabolic loads and decrease plant fitness when other demands are made on a plant, such as when they are grown with competitors (5). Perhaps as a consequence of these costs and other constraints on the use and effectiveness of direct defenses, plants have evolved defensive mutualisms in which, for a small investment in information-containing VOC releases or nutritional rewards, insects from higher trophic levels are recruited for a plant's defensive needs. Natural selection should favor both plant genotypes that use traits that enhance effectiveness of natural enemies on one hand and predator genotypes that are able to use such plant traits on the other hand (21). Many plant species express these indirect defense traits when they are attacked by herbivores, and many predators clearly use these traits to increase their foraging efficiency (Figure 1). Although evidence for the effectiveness of these defenses in nature is mounting, much remains to be discovered about the mechanisms responsible for expression of these traits. In the following section, we review the molecular advances in two indirect defenses: One is widespread among plants, herbivore-induced VOC emissions, and one is limited to a few taxa, the induction of extrafloral nectar production.

Volatile Organic Compounds

The VOC emission of more than 15 plant species involved in plant–spider mite–predatory mite, plant-caterpillar-parasitoid (21), plant–leaf beetle–egg parasitoid (70), and plant–caterpillar–predatory bug interactions (52) has been examined. In all systems, the host plant releases wound- and herbivory-inducible volatiles that function as signals in tritrophic interactions. Detailed analysis of the released volatile bouquets has identified many signals that are common to many different plant species, but there are also many compounds that are species specific and are elicited by herbivore-specific cues (21, 36).

The volatiles originate from at least three biosynthetic pathways. First, the so-called green-leaf volatiles, C_6-alcohols and -aldehydes, are produced from α-linolenic acid and linoleic acid via their respective hydroperoxides (78). Some of the green-leaf volatiles may function as direct defenses, as was elegantly demonstrated in a study of transgenic potato plants with depleted HPL (see Octadecanoids), which exhibit lower resistance to aphids (115). Green-leaf volatiles also play a role as kairomones (signals with information content). Cis-3-hexen-1-ol is commonly found in the headspace of plants after herbivore attack (35, 113), and enhancing its release from plants in a field study attracted a generalist predator (52). Trans-2-hexenal, another green-leaf volatile with biocide effects (15) and commonly emitted after herbivore wounding (19), elicits the accumulation of sesquiterpenoid phytoalexins in wounded cotton and Arabidopsis, suggesting a potential role in intra- and interplant signaling. The electrophile α,β-unsaturated carbonyl group of many green-leaf volatiles may confer the ability to induce stress and defense responses in plants (8, 27), but these potential functions need to be verified.

Second are the terpenes derived from the two (mevalonate and nonmevalonate) isoprenoid pathways. Both mono- and sesquiterpenes play a major role as kairomones in attracting predators and parasitoids to attacked plants (113) as well as functioning as phytoalexins (46). Elicitors present in the insect regurgitant induce the release of mono- and sesquiterpenes after herbivore attack (4, 21, 36). Some of the terpenoids induced in lima beans after herbivore damage [homoterpenes 3E-4,8-dimethyl-1,3,7-nonatriene (DMNT) and 3E,7E-4,8,12-trimethyl-1,3,7,11-tridecatetraene] are synthesized de novo in response to herbivore attack (22), and DMNT emission has recently been shown to depend on the herbivore-specific expression of (E)-neridol synthase, which catalyzes the synthesis of the sesquiterpene precursor (3S)-(E)-neridol (16). Many terpenoids are emitted transiently and systemically after arthropod damage (21, 35, 113), and initial evidence suggests that the release of some compounds may be under transcriptional regulation (105).

Two regulatory enzymes in terpenoid biosynthesis are hydroxymethyl glutaryl-CoA reductase (HMGR), which catalyzes the first committed step to the mevalonate terpenoid pathway, and the family of terpenoid synthases, which catalyzes isomerizations and cyclizations of prenyl diphosphates into mono- and sesquiterpenes. The plastidial monoterpene synthases and the cytosolic sesquiterpene synthases share a high degree of sequence similarity and reaction mechanisms (10), which has made functional knockouts difficult to generate. Although the accumulation of HMGR transcripts is induced rapidly by wounding and amplified by herbivore regurgitants (56) and C_6 aldehydes (8), no mutants are currently available for this gene. A first cyclase mutant identified from maize (105) carried an *Ac* (transposition mutation) insertion in the sesquiterpene cyclase, *stc1*, which is normally induced 15- to 30-fold by insect damage, insect oral secretion, and purified volicitin (Figure 2) in wild-type plants. In contrast, *stc1* was not induced in mutant plants, and an analysis of volatiles revealed that *stc1* encodes a synthase for a naphthalene-based sesquiterpene (105). Naphthalene is unfortunately not a major component of the VOC mixture of maize plants, and its function in indirect defense is still unknown.

The third pathway involves a group of volatile compounds emitted by herbivore-damaged plants that are derived from shikimate. This pathway links metabolism of carbohydrates to the biosynthesis of aromatic compounds in microorganisms and plants (44). Methyl salicylate, derived from this pathway, is emitted after herbivore damage but not after mechanical wounding by lima beans and wild tobacco. When applied to lima beans, methyl salicylate was attractive to foraging predatory mites (20) but was not attractive to predatory bugs foraging on wild tobacco in a field experiment (52). Another shikimate-derived metabolite is indole, which is released from maize seedlings after damage by beet armyworm caterpillars but not after mechanical damage. The blend of terpenoids and indole released from maize is attractive to the endoparasitic wasp *Cotesia marginiventris*, which attacks larvae of several Lepidoptera species (113). The enzyme indole-3-glycerol phosphate lyase catalyzes the formation of free indole and is, like naphthalene-containing terpenoids selectively activated by volicitin (Figure 2), an elicitor derived from caterpillar regurgitant (30).

The ecological function of VOCs—namely attracting predators and parasitoids and reducing the herbivore load of the plant—has been demonstrated in nature (52), but direct proof that plants that rely on this indirect defense experience fitness benefits is still lacking. Understanding the mechanisms responsible for the herbivore-induced VOC release and manipulating these mechanisms under field conditions will provide such proof and, additionally, evaluate the agricultural utility of this defense mechanism. Crops that release volatile signals in response to herbivore attack could provide the basis for a new era in sustainable biological control of agricultural pests.

Extrafloral Nectar

In addition to VOC releases, plants use bait to attract the natural enemies of herbivores. These have been particularly elaborated in ant plants, which produce extrafloral nectar as well as Pearl and Mullerian food bodies to attract ants (3). These ant mutualists provide defense, which is so effective that, through evolutionary time, ant-housing acacias have apparently lost their chemical defenses [e.g., cyanogenic glucosides (89)]. Extrafloral nectaries are found in at least 66 families, and several studies have documented the defensive role they play for the plants by attracting wasps and ants that attack herbivorous insects and hence reduce damage from herbivores (3). Several studies have shown that the density of extrafloral nectaries and amino acid concentration of the nectar increases after herbivore attack, but until recently, nothing was known about the underlying mechanisms of elicitation. In *Macaranga tanarius* trees, extrafloral nectar secretion increased after herbivory, mechanical leaf damage, and exogenous JA application (40). In addition, phenidone, an antagonist of endogenous JA biosynthesis, inhibited wound-induced extrafloral nectar secretion, and both the transient increase in JA as well as the wound-induced nectar flow were strongly correlated with the amount of damage. Higher nectar secretion resulted in higher numbers of visitors and defenders, which significantly reduced herbivory. The JA cascade is involved in this indirect defense, as it is in herbivore-induced VOC release.

Ecological research into indirect defenses that are mediated by plant traits has demonstrated that the defensive value of the VOC release and nectar production lies principally in the spatial information these traits provide to the predators. Specifically, plants are helping small predators locate actively foraging herbivores in the vastly larger spatial dimensions of host plants. As a result, if indirect defenses are to be successfully applied in agricultural crops, the spatial information content of the signal must be preserved. If crops were engineered to constitutively release VOCs by, for example, a terpene synthase under control of a constitutive promotor, predators and parasites would rapidly learn to ignore these signals or, worse, to associate them with hunger and thus avoid emitting plants. Therefore the identification of herbivory-responsive regulatory elements will likely be important for the use of these defenses in biotechnology applications.

COMPLEXITY AND COORDINATION OF
INSECT-INDUCED RESPONSES

The changes in plant metabolism in response to herbivore or pathogen attack are probably orchestrated by complex transcriptional changes that include genes coding for both primary and secondary metabolism. Procedures for the analysis of differential expression (microarrays, subtractive libraries, AFLP-cDNA display, and DDRT-PCR) allow researchers to study changes in the "transcriptome," which are elicited in response to herbivore attack or to identify differences in expression between genotypes that differ in resistance. The first results of these techniques, which are just beginning to be applied to plant-insect interactions, suggest that the changes elicited by herbivore attack are comparable in scope and magnitude to those elicited by pathogen attack (64, 101).

The first microarray study of plant-insect interactions analyzed the timing, dynamics, and regulation of the expression of 150 wound-induced genes in *A. thaliana* (93). A time-course analysis of responses elicited by wounding identified groups of genes with similar behaviors, one of which was correlated with the appearance of signals from the jasmonate cascade: OPDA, dnOPDA (dinoroxophytodienoic acid), and JA. But not all genes in this group depended on the JA signaling, as revealed by the comparison of responses from the coronatine-insensitive *coi1-1* mutant (which is JA insensitive) with wild-type plants. Moreover, a comparison of expression patterns in mechanically wounded and *Pieris*-caterpillar-wounded plants revealed very different transcript profiles, particularly in the expression of the water-stress-induced genes, which were reduced in insect-attacked plants. However, the timing and the magnitude of damage caused by insect feeding were not mimicked in the mechanical wound treatment, so it is difficult to know whether the lack of drought-associated expression was a specific response to *Pieris* feeding (93). This microarray study examined a plant-insect interaction with a "boutique" chip consisting of a preselected group of genes. With the availability of microarrays that cover the complete genome of *A. thaliana*, we look forward to the first truly unbiased analyses of the transcriptional changes induced by herbivore attack in this model plant.

For studies with other plant systems that are not supported by genome sequencing projects, other less-expensive approaches provide unbiased analyses of insect-induced transcriptional changes. In one such analysis, differential display (DDRT-PCR) was used to analyze the transcriptional changes in *N. attenuata* after damage by the specialist herbivore *Manduca sexta*. The putative functions of induced and repressed transcripts could be crudely categorized as being involved in photosynthesis, electron transport, cytoskeleton, carbon and nitrogen metabolism, and pathogen response (43). Transcripts involved in photosynthesis were strongly downregulated, whereas transcripts responding to stress, wounding, and invasion of pathogens or involved in shifting carbon and nitrogen were upregulated. From this study, it was estimated that more than 500 genes responded to the attack of this specialized herbivore. To separate the wound-induced changes from the changes elicited by the *M. sexta* oral secretion and regurgitant, a subset of the differentially expressed transcripts was analyzed, and three discrete patterns of expression were identified (102). Regurgitant modified the wound-induced responses by suppressing wound-induced transcripts systemically in the plant (type I) or amplifying the wound response in the attacked leaves (type II). This amplification was either a downregulation of wound-suppressed transcripts (type IIb) or an upregulation of wound-increased transcripts (type IIa) (Figure 3). It is interesting to note that all three patterns of *Manduca*-induced transcriptional changes of the wound response of *N. attenuata* could be fully mimicked by adding minute amounts of FACs to wounds (36) (Figure 3). The amounts of FACs required are so small that they may be transferred to the plant during normal feeding (102).

These two studies demonstrate that herbivore attack causes a coordinated transcriptional reorganization of the plant, which, in turn, points to the existence of herbivore- and wound-specific *trans*-acting factors that mediate the coordinated changes. Although nothing is known about the identity of such transcription factors, the recently discovered ORCA3, a JA-responsive APETALA2 (AP2)-domain transcription factor from *Catharanthus roseus* that regulates the expression of genes from both primary and secondary metabolism required for the production of terpenoid indole alkaloids (117), provides support for the concept. Identification of such regulatory factors will represent a major advance in understanding the bewilderingly complex transcriptional and phenotypic changes that are elicited after herbivore attack.

CONCLUSION

Plant-herbivore interactions, in contrast to plant-pathogen interactions, are characterized by greater physiological independence of the actors, which has two important consequences for future work. First, the physiological independence of insect herbivores means that for many plant-insect interactions, the wound response will play a prominent role. The seminal work of Ryan and colleagues (99) in understanding the wound response provides an important foundation from which to understand how elicitors from herbivores modify these responses. Second, the arena in which the interaction is played out is clearly very large and includes not only the whole plant but its surrounding biotic community. These larger-scale interactions have been extensively studied by ecologists, and whole-organism entomologists and molecular biologists interested in understanding the function of the transcriptional changes observed after insect attack will benefit from establishing collaborations with these research communities.

ACKNOWLEDGMENTS

We apologize to all of the authors whose contribution to this field of research we were not able to cite owing to space restrictions. We thank Emily Wheeler for editorial assistance.

Visit the Annual Reviews home page at www.annualreviews.org

LITERATURE CITED

1. Adler LS, Karban R, Strauss SY. 2001. Direct and indirect effects of alkaloids on plant fitness via herbivory and pollination. *Ecology* 82:2032–44

2. Agrawal AA. 2000. Mechanisms, ecological consequences and agricultural implications of tri-trophic interactions. *Curr. Opin. Plant Biol.* 3:329–35

3. Agrawal AA, Rutter MT. 1998. Dynamic anti-herbivore defense in ant-plants—the role of induced responses. *Oikos* 83:227–36

4. Alborn T, Turlings TCJ, Jones TH, Stenhagen G, Loughrin JH, Tumlinson JH. 1997. An elicitor of plant volatiles from beet armyworm oral secretion. *Science* 276:945–49

5. Baldwin IT. 2001. An ecologically motivated analysis of plant-herbivore interactions in native tobacco. *Plant Physiol.* 127:1449–58

6. Baldwin IT, Halitschke R, Kessler A, Schittko U. 2001. Merging molecular and ecological approaches in plant-insect interactions. *Curr. Opin. Plant Biol.* 4:351–58

7. Baldwin IT, Zhang Z-P, Diab N, Ohnmeiss TE, McCloud ES, et al. 1997. Quantification, correlations and manipulations of wound-induced changes in jasmonic acid and nicotine in *Nicotiana sylvestris*. *Planta* 201:397–404

8. Bate NJ, Rothstein SJ. 1998. C-6-volatiles derived from the lipoxygenase pathway induce a subset of defense-related genes. *Plant J.* 16:561–69

9. Bell E, Creelman RA, Mullet JE. 1995. A chloroplast lipoxygenase is required for wound-induced jasmonic acid accumulation in *Arabidopsis*. *Proc. Natl. Acad. Sci. USA* 92:8675–79

10. Bohlmann J, Meyer-Gauen G, Croteau R. 1998. Plant terpenoid synthases: molecular biology and phylogenetic analysis. *Proc. Natl. Acad. Sci. USA* 95:4126–33

11. Bostock RM, Karban R, Thaler JS, Weyman PD, Gilchrist D. 2001. Signal interactions in induced resistance to pathogens and insect herbivores. *Eur. J. Plant Pathol.* 107:103–11

12. Bowles D. 1998. Signal transduction in the wound response of tomato plants. *Philos. Trans. R. Soc. London Ser. B* 353:1495–510

13. Chew FS. 1988. Biological effects of glucosinolates. *ACS Symp. Ser.* 380:155–81

14. Creelman RA, Mullet JE. 1997. Biosynthesis and action of jasmonates in plants. *Annu. Rev. Plant Physiol. Plant Mol. Biol.* 48:355–81

15. Croft KPC, Juttner F, Slusarenko AJ. 1993. Volatile products of the lipoxygenase pathway evolved from *Phaseolus vulgaris* (L.) leaves inoculated with *Pseudomonas syringae* pv. *phaseolicola*. *Plant Physiol.* 101:13–24

15a. de Bruxelles GL, Roberts MR. 2001. Signals regulating multiple responses to wounding and herbivores. *Crit. Rev. Plant Sci.* 20:487–521

16. Degenhardt J, Gershenzon J. 2000. Demonstration and characterization of (*E*)-nerolidol synthase from maize: a herbivore-inducible terpene synthase participating in (3*E*)-4,8,-dimethyl-1,3, 7-nonatriene biosynthesis. *Planta* 210:815–22

17. De Leo F, Bonade-Bottino M, Ceci LR, Gallerani R, Jouanin L. 2001. Effects of a mustard trypsin inhibitor expressed in different plants on three lepidopteran pests. *Insect Biochem. Mol.* 31:593–602

18. Delledonne M, Allegro G, Belenghi B, Balestrazzi A, Picco F, et al. 2001. Transformation of white poplar (*Populus alba* L.) with a novel *Arabidopsis thaliana*

cysteine proteinase inhibitor and analysis of insect pest resistance. *Mol. Breed.* 7:35–42

19. De Moraes CM, Mescher MC, Tumlinson JH. 2001. Caterpillar-induced nocturnal plant volatiles repel nonspecific females. *Nature* 410:577–80

20. Dicke M, Gols R, Ludeking D, Posthumus Maarten A. 1999. Jasmonic acid and herbivory differentially induce carnivore-attracting plant volatiles in lima bean plants. *J. Chem. Ecol.* 25:1907–22

21. Dicke M, van Loon JJA. 2000. Multitrophic effects of herbivore-induced plant volatiles in an evolutionary context. *Entomol. Exp. Appl.* 97:237–49

22. Donath J, Boland W. 1994. Biosynthesis of acyclic homoterpenes in higher plants parallels steroid hormone metabolism. *J. Plant Physiol.* 143:473–78

23. Doss RP, Oliver JE, Proebsting WM, Potter SW, Kuy SR, et al. 2000. Bruchins: insect-derived plant regulators that stimulate neoplasm formation. *Proc. Natl. Acad. Sci. USA* 97:6218–23

24. Duffey SS, Stout MJ. 1996. Antinutritive and toxic components of plant defense against insects. *Arch. Insect Biochem.* 32:3–37

25. Ehrlich PR, Raven PH. 1964. Butterflies and plants: a study in coevolution. *Evolution* 18:586–608

26. Engelberth J, Koch T, Schueler G, Bachmann N, Rechtenbach J, Boland W. 2000. Ion channel-forming alamethicin is a potent elicitor of volatile biosynthesis and tendril coiling. Crosstalk between jasmonate and salicylate signaling in lima bean. *Plant Physiol.* 125:369–77

27. Farmer EE. 2001. Surface-to-air signals. *Nature* 411:854–56

28. Farmer EE, Weber H, Vollenweider S. 1998. Fatty acid signaling in *Arabidopsis. Planta* 206:167–74

29. Felton GW, Eichenseer H. 1999. Herbivore saliva and its effects on plant defense against herbivores and pathogens.

In *Induced Plant Defenses Against Pathogens and Herbivores: Ecology and Agriculture*, pp. 19–36. St. Paul, MN: Am. Phytopathol. Soc. Press

30. Frey M, Stettner C, Pare PW, Schmelz EA, Tumlinson JH, Gierl A. 2000. An herbivore elicitor activates the gene for indole emission in maize. *Proc. Natl. Acad. Sci. USA* 97:14801–6

31. Funk CJ. 2001. Alkaline phosphatase activity in whitefly salivary glands and saliva. *Arch. Insect Biochem.* 46:165–74

32. Gazzarrini S, McCourt P. 2001. Genetic interactions between ABA, ethylene and sugar signaling pathways. *Curr. Opin. Plant Biol.* 4:387–91

33. Genoud T, Metraux JP. 1999. Crosstalk in plant cell signaling: structure and function of the genetic network. *Trends Plant Sci.* 4:503–7

34. Glazebrook J. 2001. Genes controlling expression of defense responses in *Arabidopsis*—2001 status. *Curr. Opin. Plant Biol.* 4:301–8

35. Halitschke R, Kessler A, Kahl J, Lorenz A, Baldwin IT. 2000. Ecophysiological comparison of direct and indirect defenses in *Nicotiana attenuata. Oecologia* 124:408–17

36. Halitschke R, Schittko U, Pohnert G, Boland W, Baldwin IT. 2001. Molecular interactions between the specialist herbivore *Manduca sexta* (Lepidoptera, Sphingidae) and its natural host *Nicotiana attenuata*. III. Fatty acid–amino acid conjugates in herbivore oral secretions are necessary and sufficient for herbivore-specific plant responses. *Plant Physiol.* 125:711–17

37. Hall MA, Moshkov IE, Novikova GV, Mur LAJ, Smith AR. 2001. Ethylene signal perception and transduction: multiple paradigms? *Biol. Rev.* 76:103–28

38. Harms K, Atzorn R, Brash A, Kuhn H, Wasternack C, et al. 1995. Expression of a flax allene oxide synthase cDNA leads to increased endogenous jasmonic acid (JA) levels in transgenic potato plants but

not to a corresponding activation of JA-responding genes. *Plant Cell* 7:1645–54

39. Haruta M, Major IT, Christopher ME, Patton JJ, Constabel CP. 2001. A Kunitz trypsin inhibitor gene family from trembling aspen (*Populus tremuloides* Michx.): cloning, functional expression, and induction by wounding and herbivory. *Plant Mol. Biol.* 46:347–59

40. Heil M, Koch T, Hilpert A, Fiala B, Boland W, Linsenmair KE. 2001. Extrafloral nectar production of the ant-associated plant, *Macaranga tanarius*, is an induced, indirect, defensive response elicited by jasmonic acid. *Proc. Natl. Acad. Sci. USA* 98:1083–88

41. Herde O, Cortes HP, Wasternack C, Willmitzer L, Fisahn J. 1999. Electric signaling and *Pin2* gene expression on different abiotic stimuli depend on a distinct threshold level of endogenous abscisic acid in several abscisic acid–deficient tomato mutants. *Plant Physiol.* 119:213–18

42. Herms DA, Mattson WJ. 1992. The dilemma of plants: to grow or defend. *Q. Rev. Biol.* 67:283–335

43. Hermsmeier D, Schittko U, Baldwin IT. 2001. Molecular interactions between the specialist herbivore *Manduca sexta* (Lepidoptera, Sphingidae) and its natural host *Nicotiana attenuata*. I. Large-scale changes in the accumulation of growth and defense-related plant mRNAs. *Plant Physiol.* 125:683–700

44. Herrmann KM, Weaver LM. 1999. The shikimate pathway. *Annu. Rev. Plant Physiol. Plant Mol. Biol.* 50:473–503

45. Howe GA, Lee GI, Itoh A, Li L, DeRocher AE. 2000. Cytochrome P450–dependent metabolism of oxylipins in tomato. Cloning and expression of allene oxide synthase and fatty acid hydroperoxide lyase. *Plant Physiol.* 123:711–24

46. Howell CR, Hanson LE, Stipanovic RD, Puckhaber LS. 2000. Induction of terpenoid synthesis in cotton roots and control of *Rhizoctonia solani* by seed treatment with *Trichoderma virens*. *Phytopathology* 90:248–52

47. Jander G, Cui JP, Nhan B, Pierce NE, Ausubel FM. 2001. The TASTY locus on chromosome 1 of *Arabidopsis* affects feeding of the insect herbivore *Trichoplusia ni*. *Plant Physiol.* 126:890–98

48. Juenger T, Lennartsson T. 2000. Tolerance in plant ecology and evolution: toward a more unified theory of plant-herbivore interaction. Preface. *Evol. Ecol.* 14:283–87

49. Kahl J, Siemens DH, Aerts RJ, Gäbler R, Kühnemann F, et al. 2000. Herbivore-induced ethylene suppresses a direct defense but not a putative indirect defense against an adapted herbivore. *Planta* 210:336–42

50. Karban R, Baldwin IT. 1997. *Induced Responses to Herbivory*. Chicago, IL: Chicago Univ. Press. 319 pp.

51. Karban R, Baxter KJ. 2001. Induced resistance in wild tobacco with clipped sagebrush neighbors: the role of herbivore behavior. *J. Insect Behav.* 14:147–56

52. Kessler A, Baldwin IT. 2001. Defensive function of herbivore-induced plant volatile emissions in nature. *Science* 291:2141–44

53. Kliebenstein DJ, Kroymann J, Brown P, Figuth A, Pedersen D, et al. 2001. Genetic control of natural variation in *Arabidopsis* glucosinolate accumulation. *Plant Physiol.* 126:811–25

54. Knight H, Knight MR. 2001. Abiotic stress signaling pathways: specificity and crosstalk. *Trends Plant Sci.* 6:262–67

55. Koiwa H, Bressan RA, Hasegawa PM. 1997. Regulation of protease inhibitors and plant defense. *Trends Plant Sci.* 2:379–84

56. Korth KL, Dixon RA. 1997. Evidence for chewing insect-specific molecular events distinct from a general wound response in leaves. *Plant Physiol.* 115:1299–305

57. Kubigsteltig I, Laudert D, Weiler EW. 1999. Structure and regulation of the *Arabidopsis thaliana* allene oxide synthase gene. *Planta* 208:463–71

58. Lam E, Kato N, Lawton M. 2001. Programmed cell death, mitochondria and the plant hypersensitive response. *Nature* 411:848–53

59. Lambrix VM, Reichelt M, Mitchell-Olds T, Kliebenstein DJ, Gershenzon J. 2001. The *Arabidopsis* epithiospecifier protein promotes the hydrolysis of glucosinolates to nitriles and influences *Trichoplysia ni* herbivory. *Plant Cell.* In press

60. Larson KC, Whitham TG. 1997. Competition between gall aphids and natural plant sinks—plant architecture affects resistance to galling. *Oecologia* 109:575–82

61. Lee MCS, Scanlon MJ, Craik DJ, Anderson MA. 1999. A novel two-chain proteinase inhibitor generated by circularization of a multidomain precursor protein. *Nat. Struct. Biol.* 6:526–30

62. Leon J, Rojo E, Sanchez-Serrano JJ. 2001. Wound signaling in plants. *J. Exp. Bot.* 52:1–9

63. Maleck K, Dietrich RA. 1999. Defense on multiple fronts: How do plants cope with diverse enemies? *Trends Plant Sci.* 4:215–19

64. Maleck K, Levine A, Eulgem T, Morgan A, Schmid J, et al. 2000. The transcriptome of *Arabidopsis thaliana* during systemic acquired resistance. *Nat. Genet.* 26:403–10

65. Malone M, Alarcon JJ. 1995. Only xylem-borne factors can account for systemic wound signalling in the tomato plant. *Planta* 196:740–46

66. Mattiacci L, Dicke M, Posthumus MA. 1995. Beta-glucosidase—an elicitor of herbivore-induced plant odor that attracts host-searching parasitic wasps. *Proc. Natl. Acad. Sci. USA* 92:2036–40

67. Mauricio R. 1998. Costs of resistance to natural enemies in field populations of the annual plant *Arabidopsis thaliana*. *Am. Nat.* 151:20–28

68. McCloud ES, Baldwin IT. 1997. Herbivory and caterpillar regurgitants amplify the wound-induced increases in jasmonic acid but not nicotine in *Nicotiana sylvestris*. *Planta* 203:430–35

69. McGurl B, Orozco-Cardenas M, Pearce G, Ryan CA. 1994. Overexpression of the prosystemin gene in transgenic tomato plants generates a systemic signal that constitutively induces proteinase inhibitor synthesis. *Proc. Natl. Acad. Sci. USA* 91:9799–802

70. Meiners T, Hilker M. 2000. Induction of plant synomones by oviposition of a phytophagous insect. *J. Chem. Ecol.* 26:221–32

71. Memelink J, Verpoorte R, Kijne JW. 2001. ORC anization of jasmonate—responsive gene expression in alkaloid metabolism. *Trends Plant Sci.* 6:212–19

72. Metraux JP. 2001. Systemic acquired resistance and salicylic acid: current state of knowledge. *Eur. J. Plant Pathol.* 107:13–18

73. Miles PW. 1999. Aphid saliva. *Biol. Rev.* 74:41–85

74. Miller EA, Lee MCS, Atkinson AHO, Anderson MA. 2000. Identification of a novel four-domain member of the proteinase inhibitor II family from the stigmas of *Nicotiana alata*. *Plant Mol. Biol.* 42:329–33

75. Moran N, Hamilton WD. 1980. Low nutritive quality as defense against herbivores. *J. Theor. Biol.* 86:247–54

76. Moran PJ, Thompson GA. 2001. Molecular responses to aphid feeding in *Arabidopsis* in relation to plant defense pathways. *Plant Physiol.* 125:1074–85

77. Nielsen JK, Hansen ML, Agerbirk N, Petersen BL, Halkier BA. 2001. Responses of the flea beetles *Phyllotreta nemorum* and *P. cruciferae* to metabolically engineered *Arabidopsis thaliana* with an altered glucosinolate profile. *Chemoecology* 11:75–83

78. Noordermeer MA, Veldink GA, Vliegenthart JFG. 2001. Fatty acid hydroperoxide lyase: a plant cytochrome P450 enzyme involved in wound healing and pest resistance. *Chembiochemisty* 2:494–504

79. O'Donnell PJ, Calvert C, Atzorn R, Wasternack C, Leyser HMO, Bowles DJ. 1996. Ethylene as a signal mediating the wound response of tomato plants. *Science* 274:1914–17

80. Orozco-Cardenas M, McGurl B, Ryan CA. 1993. Expression of an antisense prosystemin gene in tomato plants reduces resistance toward *Manduca sexta* larvae. *Proc. Natl. Acad. Sci. USA* 90:8273–76

81. Pare PW, Tumlinson JH. 1999. Plant volatiles as a defense against insect herbivores. *Plant Physiol.* 121:325–31

82. Paul ND, Hatcher PE, Taylor JE. 2000. Coping with multiple enemies: an integration of molecular and ecological perspectives. *Trends Plant Sci.* 5:220–25

83. Pearce G, Moura DS, Stratmann J, Ryan CA. 2001. Production of multiple plant hormones from a single polyprotein precursor. *Nature* 411:817–20

84. Pena-Cortes H, Albrecht T, Prat S, Weiler Elmar W, Willmitzer L. 1993. Aspirin prevents wound-induced gene expression in tomato leaves by blocking jasmonic acid biosynthesis. *Planta* 191:123–28

85. Penninckx I, Thomma B, Buchala A, Metraux JP, Broekaert WF. 1998. Concomitant activation of jasmonate and ethylene response pathways is required for induction of a plant defensin gene in *Arabidopsis. Plant Cell* 10:2103–13

86. Pohnert G, Jung V, Haukioja E, Lempa K, Boland W. 1999. New fatty acid amides from regurgitant of lepidopteran (Noctuidae, Geometridae) caterpillars. *Tetrahedron Lett.* 55:11275–80

87. Preston CA, Lewandowski C, Enyedi AJ, Baldwin IT. 1999. Tobacco mosaic virus inoculation inhibits wound-induced jasmonic acid–mediated responses within but not between plants. *Planta* 209:87–95

88. Raffa KF, Berryman AA. 1987. Interacting selective pressures in conifer-bark beetle systems as a basis for reciprocal adaptations. *Am. Nat.* 129:234–62

89. Rehr SS, Feeny PP, Janzen DH. 1973. Chemical defenses in Central American non-ant acacias. *J. Anim. Ecol.* 42:405–16

90. Renwick JAA. 2001. Variable diets and changing taste in plant-insect relationships. *J. Chem. Ecol.* 27:1063–76

91. Reymond P. 2001. DNA microarrays and plant defense. *Plant Physiol. Biochem.* 39:313–21

92. Reymond P, Farmer EE. 1998. Jasmonate and salicylate as global signals for defense gene expression. *Curr. Opin. Plant Biol.* 1:404–11

93. Reymond P, Weber H, Damond M, Farmer EE. 2000. Differential gene expression in response to mechanical wounding and insect feeding in *Arabidopsis. Plant Cell* 12:707–19

94. Rhodes JD, Thain JF, Wildon DC. 1999. Evidence for physically distinct systemic signalling pathways in the wounded tomato plant. *Ann. Bot.* 84:109–16

95. Ridley BL, O'Neill MA, Mohnen D. 2001. Pectins: structure, biosynthesis, and oligogalacturonide-related signaling. *Phytochemistry* 57:929–67

96. Rosenthal GA, Berenbaum MR. 1992. *Herbivores: Their Interactions with Secondary Plant Metabolites. Ecological and Evolutionary Processes.* San Diego, CA: Academic

97. Rosenthal GA, Janzen D. 1979. *Herbivores: Their Interactions with Secondary Plant Metabolites.* San Diego, CA: Academic

98. Royo J, Leon J, Vancanneyt G, Albar JP, Rosahl S, et al. 1999. Antisense-mediated depletion of a potato lipoxygenase reduces wound induction of proteinase inhibitors and increases weight

gain of insect pests. *Proc. Natl. Acad. Sci. USA* 96:1146–51

99. Ryan CA. 2000. The systemin signaling pathway: differential activation of plant defensive genes. *Biochim. Biophys. Acta* 1477:112–21

100. Schaller F. 2001. Enzymes of the biosynthesis of octadecanoid-derived signaling molecules. *J. Exp. Bot.* 52:11–23

101. Schenk PM, Kazan K, Wilson I, Anderson JP, Richmond T, et al. 2000. Coordinated plant defense responses in *Arabidopsis* revealed by microarray analysis. *Proc. Natl. Acad. Sci. USA* 97:11655–60

102. Schittko U, Hermsmeier D, Baldwin IT. 2001. Molecular interactions between the specialist herbivore *Manduca sexta* (Lepidoptera, Sphingidae) and its natural host *Nicotiana attenuata*. II. Accumulation of plant mRNA in response to insect-derived cues. *Plant Physiol.* 125:701–10

103. Schittko U, Preston CA, Baldwin IT. 2000. Eating the evidence? *Manduca sexta* cannot disrupt specific jasmonate induction in *Nicotiana attenuata* by rapid consumption. *Planta* 210:343–46

103a. Schmelz EA, Alborn HT, Tumlinson JH. 2001. The influence of intact-plant and excised-leaf bioassay designs on volicitin- and jasmonic acid-induced sesquiterpene volatile release in *Zea mays*. *Planta* 214:171–79

104. Seo HS, Song JT, Cheong JJ, Lee YH, Lee YW, et al. 2001. Jasmonic acid carboxyl methyltransferase: a key enzyme for jasmonate-regulated plant responses. *Proc. Natl. Acad. Sci. USA* 98:4788–93

105. Shen BZ, Zheng ZW, Dooner HK. 2000. A maize sesquiterpene cyclase gene induced by insect herbivory and volicitin: characterization of wild-type and mutant alleles. *Proc. Natl. Acad. Sci. USA* 97:14807–12

106. Sivasankar S, Sheldrick B, Rothstein SJ. 2000. Expression of allene oxide synthase determines defense gene activation in tomato. *Plant Physiol.* 122:1335–42

107. Spiteller D, Dettner K, Boland W. 2000. Gut bacteria may be involved in interactions between plants, herbivores and their predators: microbial biosynthesis of N-acylglutamine surfactants as elicitors of plant volatiles. *Biol. Chem.* 381:755–62

108. Spiteller D, Pohnert G, Boland W. 2001. Absolute configuration of volicitin, an elicitor of plant volatile biosynthesis from lepidopteran larvae. *Tetrahedron Lett.* 42:1483–85

109. Stowe KA, Marquis RJ, Hochwender CG, Simms EL. 2000. The evolutionary ecology of tolerance to consumer damage. *Annu. Rev. Ecol. Syst.* 31:565–95

110. Tamayo MC, Rufat M, Bravo JM, San Segundo B. 2000. Accumulation of a maize proteinase inhibitor in response to wounding and insect feeding, and characterization of its activity toward digestive proteinases of *Spodoptera littoralis* larvae. *Planta* 211:62–71

111. Tattersall DB, Bak S, Jones PR, Olsen CE, Nielsen JK, et al. 2001. Resistance to an herbivore through engineered cyanogenic glucoside synthesis. *Science* 293:1826–28

112. Terra WR. 2001. Special topics issue—biological, biochemical and molecular properties of the insect peritrophic membrane. Preface. *Arch. Insect Biochem.* 47:46

113. Turlings TCJ, Benrey B. 1998. Effects of plant metabolites on the behavior and development of parasitic wasps. *Eco-Science* 5:321–33

114. Van Breusegem F, Vranova E, Dat JF, Inze D. 2001. The role of active oxygen species in plant signal transduction. *Plant Sci.* 161:405–14

115. Vancanneyt G, Sanz C, Farmaki T, Paneque M, Ortego F, et al. 2001. Hydroperoxide lyase depletion in transgenic potato plants leads to an increase in aphid performance. *Proc. Natl. Acad. Sci. USA* 98:8139–44

116. van de Ven WTG, LeVesque CS, Perring

TM, Walling LL. 2000. Local and systemic changes in squash gene expression in response to silverleaf whitefly feeding. *Plant Cell* 12:1409–23

117. van der Fits L, Memelink J. 2000. ORCA3, a jasmonate-responsive transcriptional regulator of plant primary and secondary metabolism. *Science* 289:295–97

118. Walling LL. 2000. The myriad plant responses to herbivores. *J. Plant Growth Regul.* 19:195–216

119. Wasternack C, Parthier B. 1997. Jasmonate signaled plant gene expression. *Trends Plant Sci.* 2:302–7

120. Winterer J, Bergelson J. 2001. Diamondback moth compensatory consumption of protease inhibitor-transformed plants. *Mol. Ecol.* 10:1069–74

121. Winz RA, Baldwin IT. 2001. Molecular interactions between the specialist herbivore *Manduca sexta* (Lepidoptera, Sphingidae) and its natural host *Nicotiana attenuata*. IV. Insect-induced ethylene reduces jasmonate-induced nicotine accumulation by regulating putrescine N-methyltransferase transcripts. *Plant Physiol.* 125:2189–202

122. Xu Y, Chang Pi-Fang L, Liu D, Narasimhan Meena L, Raghothama Kashchandra G, et al. 1994. Plant defense genes are synergistically induced by ethylene and methyl jasmonate. *Plant Cell* 6:1077–85

123. Zhang ZP, Baldwin IT. 1997. Transport of 2-C-14 jasmonic acid from leaves to roots mimics wound-induced changes in endogenous jasmonic acid pools in *Nicotiana sylvestris*. *Planta* 203:436–41

124. Zhou L, Thornburg RW. 1999. Wound-inducible genes in plants. In *Inducible Gene Expression*, ed. PHS Reynolds, pp. 127–58. Portland, OR: Book News

125. Zhu JK. 2001. Cell signaling under salt, water and cold stresses. *Curr. Opin. Plant Biol.* 4:401–6

Annu. Rev. Plant Biol. 2002. 53:329–55
DOI: 10.1146/annurev.arplant.53.100301.135302

PHYTOCHROMES CONTROL PHOTOMORPHOGENESIS BY DIFFERENTIALLY REGULATED, INTERACTING SIGNALING PATHWAYS IN HIGHER PLANTS

Ferenc Nagy[1] and Eberhard Schäfer[2]

[1]*Institute of Plant Biology, Biological Research Center, H-6701 Szeged, Hungary;*
e-mail: nagyf@nucleus.szbk.u-szeged.hu
[2]*Universität Freiburg, Institut für Biologie II/Botanik, D-79104 Freiburg, Germany;*
e-mail: schaegen@sun2.ruf.uni-freiburg.de

Key Words kinetics, mutants, nuclear transport, phototransduction, transcription

■ **Abstract** In this review the kinetic properties of both phytochrome A and B measured by in vivo spectroscopy in *Arabidopsis* are described. Inactivation of phyA is mediated by destruction and that of phyB by fast dark reversion. Recent observations, describing a complex interaction network of various phytochromes and cryptochromes, are also discussed. The review describes recent analysis of light-dependent nuclear translocation of phytochromes and genetic and molecular dissection of phyA- and phyB-mediated signal transduction. After nuclear transport, both phyA- and phyB-mediated signal transduction probably include the formation of light-dependent transcriptional complexes. Although this hypothesis is quite attractive and probably true for some responses, it cannot account for the complex network of phyA-mediated signaling and the interaction with the circadian clock. In addition, the biological function of phytochromes localized in the cytosol remains to be elucidated.

CONTENTS

1040-2519/02/0601-0329$14.00

INTRODUCTION

Perception of changes in the natural environment is of the utmost importance for all living organisms, from prokaryotes to higher eukaryotes. This is especially valid in the case of plants, sessile organisms unable to migrate to more favorable locations in order to adapt to changes in the natural environment. To monitor changes in the ambient light conditions, different photoreceptor classes have evolved: UVB photoreceptors, as yet unidentified at the molecular level (6, 138), the two blue UVA photoreceptor classes cry1 and cry2 (1, 88), the phototropines PHOT1 and PHOT2 (64, 74), and the red/far-red reversible photoreceptors, phytochromes A–E (22, 122). Although it is interesting to note that both phytochrome and cryptochrome genes exist already in prokaryotes (66, 144, 151, 152), their physiological roles in these have not been elucidated.

The description of the role of red/far-red and blue-light photoreceptors in controlling photomorphogenesis and photoperiodism goes back more than seven decades (49). Owing to the red/far-red reversibility of phytochromes, spectroscopical methods could be developed to detect them in vivo, and a protocol could also be devised to purify the pigment (8, 15, 16). An overwhelming amount of physiological data has been collected over all these years, and the modes of phytochrome function have been classified into four groups: VLFR, LFR, red HIR, and far-red HIR. Very low fluence responses (VLFR) are saturated at very low levels of active phytochrome (P_{fr}) either after light pulses (93) or under continuous irradiation leading to responsiveness to near infrared light (116). Low fluence rate responses (LFR) are the classical red/far-red reversible induction responses, i.e., responses that can be induced by a red-light pulse [5 min at moderate fluence rates ($1 \ Wm^{-2}$)] and reverted by a subsequent far-red light pulse (39). Besides these two functional modes, two high irradiance response (HIR) types, one to continuous red, the other to continuous far-red-light, have been described (45, 47). The main characteristics are the dependence on fluence rate and the fact that continuous irradiation can only be substituted by very frequent light pulses (47, 92, 117).

The interpretation of all these physiological data and their correlation with spectroscopical data as well as with properties derived from work on purified phytochromes became problematic when it was shown by genetic studies that five genes encoding phytochrome apoproteins exist in the model plant *Arabidopsis thaliana.* Although the power of genetics has contributed much to unraveling some of these problems, the complexity of five phytochromes with different but overlapping properties makes these analyses quite cumbersome, especially if all the interactions with the other photoreceptors are taken into account. In this review we therefore focus exclusively on the mode of phytochrome action and concentrate primarily on the role of the two dominating phytochromes, phyA and phyB. Briefly, the kinetic properties of phytochromes measured spectroscopically in vivo and in vitro and their intracellular distribution are summarized. In combination with the results of genetic and molecular approaches, this information is used to describe our present limited knowledge about phyA- and phyB-mediated signal transduction.

KINETIC PROPERTIES OF PHYTOCHROMES

In the period between the 1970s and 1990s a large body of information was collected describing the properties of phytochromes using in vivo spectroscopy, in vitro analysis, and immunocytochemistry. Although, in most cases, it is obvious that these data probably describe functions of phyA, we concentrate on more recent data where the properties of the different phytochromes are individually addressed.

Spectroscopical Properties of Phytochromes In Vivo and In Vitro

Even before it became clear that there are several genes encoding different phytochrome apoproteins, in vivo spectroscopical studies (70) and immunochemical studies (23, 123, 142) led to the concept of etiolated and green tissue phytochromes, light-labile and light-stable phytochromes, or type I and type II phytochromes (38). All these terms—except for type I and type II—are quite misleading because, according to analyses of promoter activity (43, 131) and functional studies in photoreceptor mutants, all phytochromes seem to be present in almost all cells at all developmental stages. Furthermore, in cauliflower heads—clearly a light-grown tissue—phytochrome is not stable but shows a rapid dark reversion (15, 73).

In etiolated *Arabidopsis* seedlings the protein levels of phyB-E are so low that by in vivo spectroscopy no photoreversible signal could be detected in *phyA* mutants (50). Thus, in vivo spectroscopical methods allowed investigators to describe the kinetic properties of phyA in WT and *phyB* mutants of *Arabidopsis*. The main characteristics are the following: phyA is synthesized rapidly between 48 h and 96 h after sowing [$\sim$1.5 $\times$ 10^{-6} $\Delta/\Delta A)/\mu$ mg seeds, (50)]; P_{fr} formed by a saturating red-light pulse undergoes rapid proteolytic degradation during a subsequent dark period (half-life: 30 min; $^{1}kd = 0,023$ mn^{-1}). Interestingly, in most ecotypes tested no $P_{fr} \rightarrow P_r$ dark reversion could be detected. So far the only exception is ecotype RLD with 20% dark reversion and a half-life of 5–10 min (33). This is astonishing because PHYA sequences of Columbia and RLD are identical; still only in the latter is dark reversion detected. Thus, in contrast to former assumptions, dark reversion is not a property of the molecule but is a regulated process.

Under continuous irradiation the rate of destruction is dependent on wavelength and photoequilibrium (P_{fr}/P_{tot}) but is saturated at approximately $P_{fr}/P_{tot} = 0.3$ (50). Furthermore, although P_r is probably quite stable—so far only tested in *Cucurbita* (109, 110)—a rapid destruction of P_r can be observed after being cycled through P_{fr} (50). This P_{fr}-induced P_r destruction has previously been described in both monocotyledonous (137) and dicotyledonous seedlings (12, 46). The kinetic model describing all these different spectroscopic data (51) is quite similar to a reaction model proposed many years ago (113). Quantitative analysis showed that the flux through this reaction cycle exhibits the major characteristics of the far-red HIR (51). Clearly the biochemical/cell biological nature of the postulated modification reactions leading to proteolytically degradable P_{fr} and P_r remains obscure.

Because of the very low expression levels of all phytochromes other than phyA, in vivo spectroscopy has only been possible in some special cases (13, 46, 70). These data show that the "second pool" of phytochrome is destroyed slowly, with a half-life of approximately 8–20 h; in some cases, however, a fast dark reversion was reported (73).

To obtain information about the other phytochromes, the chromophore ligase activity of the apoproteins can be used. This activity was first shown by Lagarias & Lagarias (87) in recombinant oat PHYA but could later be demonstrated in PHYB (95, 96) as well as in PHYA, B, C, and E from *Arabidopsis* (32). When expressed in yeast, all different *Arabidopsis* phytochromes were capable of chromophore assembly, formation of photoreversible pigments, and after formation of P_{fr} by a light pulse, a partial but rapid dark reversion ($P_{fr} \rightarrow P_r$) of 20–50% (32). The difference spectra of recombinant phytochrome A and B as well as those expressed and assembled in yeast are undistinguishable; phyC and phyE, however, show a shifted and sharpened far-red peak, indicating that these spectral differences may be useful for the detection of different environmental conditions (32).

In vivo spectroscopical analysis of phyB has recently been described by Eichenberg (31) using the phyB overexpressor line ABO (146) in a phyA⁻ background. Surprisingly, dark reversion in this case is fast (half-life 2–10 min), and almost all P_{fr} molecules (80–95%) undergo dark reversion, leaving at most 15% P_{fr} after a prolonged dark period. The observation that only up to 25% of the phytochrome molecules undergoes dark reversion (120) led to the hypothesis that of the phytochrome dimers only the P_rP_{fr} heterodimers undergo dark reversion. Thus, the almost complete dark reversion in ABO lines indicates that, in this case, $P_{fr}P_{fr}$ homodimers must be destabilized. The hypothesis of different stabilities of heterodimers (P_rP_{fr}) and homodimers ($P_{fr}P_{fr}$) was recently confirmed by Hennig & Schäfer (55).

Photoreceptor Interacting Network

Many years before it became known that phytochromes are encoded by a small gene family (22), physiologists characterized three modes of phytochrome function, i.e., the very low fluence responses (93), the low fluence responses (8), and the high irradiance responses (45, 92, 113). Analysis of photoreceptor mutants showed that the low fluence response curves are mediated by phyB, C, D, or E and the VLFR and HIR by phyA. Very recent data (56) showed that the far-red HIR-induced germination in *Arabidopsis* is absent not only in *phyA* but also in *phyE* mutants. This indicates a requirement of phyE for the phyA-mediated far-red HIR. Interestingly, this requirement is only observed for the germination control and not for other VLFR or HIRs tested, i.e., hook opening, gravitropic sensitivity, inhibition of hypocotyl growth, and anthocyanin accumulation (56). Whereas low fluence response curves have been characterized by red/far-red reversibility and fluence ranges between 10^{-6}–10^{-3} mol m^{-2} (39, 127), for very low fluence response a failure of reversibility and a fluence range between 10^{-12}–10^{-7} mol m^{-2} is assumed (9, 20). Unfortunately, these characterizations are not precise enough and strongly misleading. Based on this interpretation it is generally believed that phyA-mediated responses are not photoreversible. These misinterpretations are based on the simple fact that most, if not all, phyB-E-mediated responses are not very light sensitive, leading to the observation that far-red light pulses cannot induce any responses (124). Therefore, a reversibility from 100 (red light pulse) to 0 (far-red light pulse) response can be observed. In cases of pulse irradiations for phyA-mediated responses a far-red light pulse establishing approximately 2.5% P_{fr} can already lead to 50–100% of the response induced by red-light pulse (29, 125). Thus, to obtain reversibility in phyA-mediated responses long wavelength ($\sim$750 nm) far-red light establishing approximately 0,1% P_{fr} (115) should be used, although in extreme cases even this low amount of P_{fr} can lead to saturation of the responses (116).

Interactions between phytochrome-mediated responses and blue/UVA and UVB photoreceptor signals have already been described by physiological tests. A classical system is the anthocyanin formation in *Sorghum* seedlings (28). This response needs essentially blue, UVA, or UVB light, yet it is also dependent on the presence of activated phytochrome in its P_{fr} form (30). This observation led to the concept of responsiveness amplification and to several contradictory interpretations once photoreceptor mutants became available (3, 19, 99, 106). Unfortunately, owing to the fact that phytochromes also have blue-light absorbance (16, 31, 86), the question whether the activation of a single phytochrome species is essential can only be answered when loss-of-function mutations in all five phytochromes are available. At the moment, as discussed previously (114), we still can only demonstrate that blue/UVA and/or UVB photoreceptors are essential (5). Responsiveness amplification has been observed not only for blue/UVA or UVB and phytochrome signaling but also between VLFR/LFR and HIR (119, 135, 136). Again, the use of photoreceptor mutants clearly demonstrated the positive effect of phyA on phyB, phyB on phyB, cry1 on phyB, and phyA on phyD (17–20, 52–54). Interestingly, negative interactions have also been described, namely of phyA on phyB and vice versa and cry2 on phyB (18, 20, 52, 99). The interaction network is far from being complete because *phyE* and *phyC* mutants have only recently become available and have therefore not been included in these studies.

Mechanistically these processes are not understood at all. Observations of physical interactions between cry1 and phyA (4) and cry2 and phyB (95), the description of the Ca^{2+}-binding protein SUB1, which is involved in cryptochrome and phytochrome coaction (44), and the partial characterization of responsiveness amplification of nuclear transport of phyB(42) are the first hints to the interpretation of this network. Additional interactions in the signal transduction network at the integrating steps at the COP1 and the COP9 signalosome are extremely probable. It is doubtful that additional physiological studies including more mutant combinations will be sufficient for the unraveling of this complex interaction network (for a recent review see 18). Additional studies of the kinetics and concentration dependence of protein/protein interactions in vitro and in vivo probably will be essential for building a model of the interaction network, which then can be analytically described and tested.

INTRACELLULAR LOCALIZATION OF PHYTOCHROMES

Despite the growing wealth of data about the physical characteristics of signaling pathways controlled by these photoreceptors, the primary molecular function of phytochromes remained largely obscure until recently. The competing views as to whether phytochromes are membrane-integrated/associated receptor proteins (48, 49), light-regulated enzymes (151, 152), or function as direct regulators of gene transcription (59) coexisted during this era. Very recent data indicate that all of these competing views were/are partially correct. Notably, there is an increasing body of evidence showing that phyA and phyB have at least two different modes of function: On one hand, they directly interact with transcription factors and act as transcriptional modulators in the nucleus (94), and on the other, they function as light-regulated kinases in the cytoplasm (151). It should be noted, however, that despite the recent progress discussed here several aspects of this problem are still not understood.

It has been evident from the beginning that the subcellular distribution of phytochromes is the key to this problem, and not too surprisingly, the results describing the intracellular localization of phytochromes have been as heterogeneous as the methods used. Early biochemical and immunocytochemical studies demonstrated that the majority of the P_r and P_{fr} forms of phyA were localized in the cytoplasm both in etiolated and light-grown plants. Moreover, these studies established that light treatment induces formation of sequestered areas of phytochrome (SAPs) and that SAPs are colocalized with ubiquitin (96, 107, 133, 134). The interpretation of these findings was that SAPs represent cytosolic degradation of the entire population of the P_{fr} form of phyA (71), and a similar mechanism for the degradation of the lesser-known phy species was also assumed. It is surprising to note that these findings ignored the fact that MacKenzie et al. (91) had reported localization of PHYA into the nuclei of oat seedlings after prolonged irradiation by immunocytochemical methods.

The view that phytochromes are localized in the cytoplasm and that the dominating part of phytochrome-controlled signaling occurs in this cellular compartment was further strengthened, although indirectly, by two types of studies conducted during the early 1990s. First, a series of studies employed microinjection of purified phytochromes and putative signaling intermediates into tomato cells deficient in chromophore biosynthesis. These experiments demonstrated that signaling cascades controlled by phyA and phyB are cell autonomous and mediated via activation of heterotrimeric GTP-binding protein(s). This step triggers induction of a bifurcated signal transduction pathway that modulates the cellular levels of the second messengers cGMP and Ca and activates, ultimately, transcription of light-regulated genes (10, 84, 101). Second, physiological studies in algae, mosses, and ferns showed an action dichroism for chloroplast orientation, polarotropism, and phototropism. These observations suggested an ordered localization of photoreceptors within the cytosol and were, consequently, interpreted as indications that these photoreceptors are localized in or at least associated with the plasma membrane in lower plants (82).

The assumption that phytochromes are localized to the cytosol and their undoubted role in controlling the transcription of specific plant genes invited a search for a molecular mechanism that would regulate, via a phytochrome-controlled signaling cascade, the nucleo/cytoplasmic partitioning of transcription factors mediating expression of light-regulated genes. There is fragmented circumstantial evidence that this type of regulation indeed plays a role in the light-regulated transcription of plant genes (for a review see 98). It appears, however, that this type of regulation does not play a major role and is not responsible for fine tuning the expression of thousands of genes whose transcription is modulated by the concerted action of the phyA-phyE photoreceptors. In contrast, accumulating evidence exists that light-induced transcription is mediated at least partly by a direct interaction of phyA and phyB with transcription factors, an action that is made possible by the regulated import of these phytochrome molecules into the nucleus.

Nucleo/Cytoplasmic Partitioning and Import of phyB into the Nucleus

Sakamoto & Nagatani (112) reported that the subcellular localization of *Arabidopsis* phyB and phyB::GUS fusion proteins is affected by light in protoplasts and in planta. This pioneering study demonstrated that nuclei isolated from light-treated plants contain significantly higher amounts of native phyB or phyB::GUS fusion proteins than those from dark-adapted plants. This finding suggests that, in contrast to phyA, which is supposed to be localized exclusively in the cytosol, the nucleo/cytoplasmic partitioning of phyB is a regulated process, and the import of phyB into the nucleus is part of this regulatory circuit. This novel hypothesis was later substantiated by the same laboratory when an AtphyB::GFP fusion protein was expressed to complement an *Arabidopsis phyB* mutant. These latter studies demonstrated that the phyB::GFP fusion protein is a functional photoreceptor and that the nuclear transport of this molecule is induced by light (150). Parallel with these studies, Kircher et al. (78) demonstrated that the nucleo/cytoplasmic partitioning of a tobacco phyB::GFP fusion protein, like that of AtphyB::GFP, is affected by light. Moreover, these authors provided evidence that the import of phyB into the nucleus is dependent on light quality and quantity, i.e., it is induced by red light and can be reverted by far-red light. These observations were further extended by complementing the *N. plumbaginifolia hlg2* mutant lacking functional phyB with the expression of the AtphyB::GFP and NtphyB::GFP fusion proteins. Additional support for these findings was also provided by analyzing the kinetic and wavelength dependence of the nuclear transport of these fusion proteins in tobacco (42) and of the mutated versions of AtphyB::GFP in transgenic *Arabidopsis* (S. Kircher, P. Gil, L. Kozma-Bognár, E. Fejes, V. Speth, E. Adam, E. Schäfer & F. Nagy, submitted). These studies firmly established that the transport of the endogenous tobacco and *Arabidopsis* phyB as well as that of the phyB::GFP fusion proteins into the nuclei is (*a*) light dependent, (*b*) saturable, (*c*) a relatively slow process with a half saturation time of approximately 1–2 h, (*d*) fluence rate dependent, and (*e*) wavelength dependent. These features show a surprisingly good correlation with those established for PHYB-mediated physiological responses and indicate that the import of the photoreceptor into the nucleus is a major regulatory step in phyB-mediated signaling (41, 42).

As far the molecular aspects of the light-regulated nucleo/cytoplasmic partitioning of the phyB photoreceptor are concerned, however, several issues need to be addressed. First of which is the finding that the import of phyB into the nucleus is always accompanied by a characteristic speckle formation (78, 150) that stops immediately when the light is turned off; the number of speckles then declines gradually in the dark (42).

What Are These Nuclear Speckles?

The low level of endogenous phyB makes immunocytochemical detection and exact localization of the molecule problematic. However, by using the phyB overexpressor line ABO (146), investigators demonstrated both light-induced accumulation of phyB in the nuclear fraction and formation of phyB-containing speckles in defined areas of the nucleus by western blotting and electronmicroscopy, respectively (S. Kircher, P. Gil, L. Kozma-Bognár, E. Fejes, V. Speth, E. Adam, E. Schäfer & F. Nagy, submitted). Moreover, the same authors found that overexpressed, physiologically inactive phyB::GFP fusion proteins accumulate in the nucleus in a light-induced fashion but never form speckles. Thus, it appears that the formation of phyB::GFP-containing speckles is a characteristic feature of the physiologically active photoreceptor rather than an artifact of the GFP fusion protein. When taken together, the size of the speckles and their dynamics make it unreasonable to assume that the speckles simply represent aggregates containing phyB or phyB::GFP. Note, however, that a conformation-dependent interaction of phyB with the transcription factor PIF3 in vitro and in yeast cells has been described (102, 103). Mas et al. (95) reported the formation of speckles containing phyB::RFP and cry2::GFP upon light treatment in tobacco protoplasts and verified the intimate interaction of these photoreceptors by FRET analysis. Several other proteins, among them some putative transcription factors, interact with PHYB. Based on these findings, it is tempting to postulate that the PHYB-containing speckles represent either active transcriptional complexes and/or degradation complexes possibly containing a high number of physically and functionally interacting molecules. To solve this vexing problem, however, the molecular composition of these structures needs to be elucidated.

Is All phyB Transported in the Nucleus?

It should be noted that the kinetic data so far published support the hypothesis that the nucleo/cytoplasmic partitioning of phyB is also regulated by a retention mechanism. In other words, these data indicate that phyB is actively retained in the cytosol in its Pr form in the dark and that the light-mediated $P_r \rightarrow P_{fr}$ photoconversion induces the loss of retention that is followed by the nuclear import of the P_{fr} conformer. If this is the case, why can saturation of phyB import be achieved? Surprisingly, after cell fractionation only approximately 30% of phyB or phyB::GFP was recovered in purified nuclei from tobacco (K. Panigrahi, S. Kircher, F. Nagy & E. Schäfer, unpublished). Thus, even under saturating light conditions the major part of phyB remains cytosolic and must therefore be actively prevented from translocating into the nuclei even in the light. The physiological function of cytosolic phyB as well as the mechanism responsible for retaining phyB in the cytosol in light remains obscure. It could be tentatively suggested that this cytosolic pool of phyB Pfr either may have a role in regulating rapid phytochrome effects such as ion fluxes at the plasma membrane similar to those described in mosses, ferns, and some algae (82) or may affect the orientation of chloroplast and polaro- and phototropism in an as yet unidentified fashion.

How is phyB-Mediated Signaling Inactivated?

Independent of the postulated retention mechanism, the kinetics of the import inactivation of the phyB-dependent signaling cascade is very poorly understood. The rapid dark reversion of phyB P_{fr} to P_r, as measured spectroscopically can lead to a fast stop of nuclear import and, at the same time, to inactivation of signaling in the nucleus. If this scenario applies, how is the prolonged reversibility of phyB-controlled physiological responses achieved? Whether it is mediated by differential dark reversion rates, by differential degradation of phyB localized in the cytosol and the nucleus, by an active export mechanism, or, perhaps, by all of these potential regulatory mechanisms remains to be elucidated.

Nucleo/Cytoplasmic Partitioning and Import of phyA Into the Nucleus

Kircher et al. (78) expressed rice phyA::GFP in transgenic tobacco seedlings and demonstrated for the first time that this fusion protein is also imported into the nucleus and forms speckles in a light-induced manner similar to phyB::GFP. In contrast to phyB, however, the import of phyA into the nucleus is an order of magnitude faster, and it is preceeded by an even more rapid light-dependent cytosolic speckle formation. These observations have been confirmed by analyzing

Arabidopsis phyA::GFP in tobacco (76) and *Arabidopsis* (76; S. Kircher, P. Gil, L. Kozma-Bognár, E. Fejes, V. Speth, E. Adam, E. Schäfer & F. Nagy, submitted) and by immunocytochemical methods in pea seedlings (58). The kinetic characteristics of the import process were quite similar in all these studies, although some differences were found depending on whether the measurements were performed in a heterologous or homologous background. The functionality of the AtphyA::GFP fusion protein was also demonstrated by complementation of an Arabidopsis *phyA* mutant lacking detectable amounts of phyA (76). These studies firmly established that the import of phyA::GFP into the nucleus can be induced by a single light pulse. Even a far-red-light pulse, i.e., a VLF treatment, is sufficient to promote the nuclear import of phyA::GFP. Extended studies of wavelength and fluence rate dependence showed that maximal nuclear import of phyA::GFP could be obtained at 720 nm and high fluence rates (L. Kim, F. Nagy & E. Schäfer, unpublished). These data demonstrate that the import of phyA::GFP into the nucleus shows the characteristics of both VLFR and far-red HIR, as reported for most of phyA-mediated responses. In contrast, the formation of cytosolic speckles can be induced by red light pulses but not by far-red light pulses. We have shown that a 5-min red light pulse leads to strong induction of these phyA::GFP-containing speckles in the cytosol (76). The kinetics and wavelength dependence of the appearance of cytosolic speckles in *Arabidopsis* are quite similar to those obtained by immunocytochemical methods in pea and oat and resemble the formation of SAPs that was described many years ago.

As stated above, the import of phyA into the nucleus shows the characteristics of far-red HIR. This agrees with the reaction scheme recently developed for the interpretation of the kinetics of phyA destruction (51) and implies that the P_{fr} form of phyA, like that of phyB, is the transport-competent conformer. If this theory holds, then continuous far-red irradiation will induce reversion of the imported P_{fr} back to P_r in the nucleus. Indeed, recent data confirmed that the bulk of phyA localized in the nucleus is in its P_r form after continuous far-red irradiation (L. Kim, F. Nagy & E. Schäfer, unpublished). However, these data make it difficult to interpret how phyA can interact in a P_{fr}-dependent fashion with PIF3 (153), RSF1 (36), or other, as yet unidentified, signaling components and can induce signaling in the nucleus under these conditions. Under continuous far-red light more than 95% of the nuclear phyA should be in its P_r form. Either this nuclear P_r is different from the cytosolic and can interact with reaction partners like PIF3 or RSF1/REP1, or an unknown mechanism is responsible for the nuclear function of phyA. Results reported by Hennig et al. (50, 51) and Shinomura et al. (126) indicate that both the half-life of and signals generated by the so-called P_{r+} form of phyA (an intermediate conformer of phyA that has been cycled through P_{fr}) are very short lived. Whether phyA is (also) imported into the nucleus in its P_{r+} form and how the P_{r+} form participates in mediating phyA signaling in the nucleus also remains to be elucidated.

Irrespective of how phyA-induced signaling happens in the nucleus, the long list of not-yet-understood cellular processes regulating signaling by phyB also applies to phyA. For example, it is evident that, just like in the case of phyB, most of phyA remains cytosolic even under light conditions favoring nuclear import. Thus, the function of phyA localized in the cytosol, the molecular mechanisms responsible for cytosolic retardation and nuclear transport of this photoreceptor, as well as the significance of the formation of nuclear speckles and its wavelength and fluence-rate dependence remain to be analyzed.

Nucleo/Cytoplasmic Partitioning and Import of phyC-E Into the Nucleus

For all five phytochromes identified in *Arabidopsis*, cytosolic localization in dark-grown seedlings, light-dependent translocation of phy::GFP fusion proteins into the nucleus, and formation of speckles in the nucleus could be demonstrated (S. Kircher, P. Gil, L. Kozma-Bognár, E. Fejes, V. Speth, E. Adam, E. Schäfer & F. Nagy, submitted). Our preliminary data indicate that the intracellular localization of each of the different phytochromes shows a specific developmental and wavelength dependence. Red and white light induce nuclear import and speckle formation of all phytochromes. Nuclear localization of phyA is very transient owing to the fast proteolytic degradation of the photoreceptor. phyA::GFP is the only fusion protein transported into the nucleus under continuous far-red light.

Our results (78) indicate that the competence for nuclear transport is reached very early, 24–48 h after imbibition of seeds, for phyB and phyE. Surprisingly, at this early developmental stage light-induced import of PHYA cannot be detected, even though phyA is involved in inducing seed germination in *Arabidopsis*. These data indicate that it is a cytosolic rather than nuclear function of phyA that is required for the mediation of this cellular process. The competence for nuclear import of phyC and phyD is significantly delayed in young seedlings as compared to phyB, even though phyD is considered to be the closest homologue of phyB.

Contribution of phyD-E to a number of physiological responses has been revealed by analyzing double null phy mutants (25, 26), yet it is evident that our knowledge about phyC-E dependent signaling is fairly limited both at the molecular and physiological levels. We also note that, in contrast to the rapidly growing number of proteins potentially involved in mediating phyA- and phyB-controlled signaling, either in the nucleus and/or in the cytosol, so far no proteins interacting with phyC-E have been reported. Thus, at present it is premature to discuss the potential significance of the characteristically different cellular distribution of these photoreceptors with respect to signaling.

GENETIC AND MOLECULAR APPROACHES TO UNRAVELING PHYTOCHROME SIGNALING

To obtain information about the components mediating phytochrome-dependent signaling cascades, two experimental approaches have been used nearly exclusively. In one, a number of different genetic screens based on the most contemporary information available were designed and performed within the past 10 years with the aim of isolating phyA and phyB signaling-specific mutations. In the other, a systematic search to identify proteins that interact with selected domains of the phyA and phyB photoreceptor molecules was conducted, mostly taking advantage of the yeast two-hybrid assay system. The combination of these approaches turned out to be productive and led to the identification of a surprisingly large number of putative genes/proteins that seem to be involved in mediating phytochrome-dependent signal transduction. In this review, we concentrate on describing recently identified components and the putative mode of action by which they specifically affect phyA- and/or phyB-induced phototransduction. It is generally accepted that members of the COP/DET/FUS family function as negative regulators of photomorphogenesis that is controlled and induced partly by phyA-phyE. Very recently it became evident that the COP9 signalosome plays a central role in mediating degradation of a number of regulatory proteins, thereby affecting light- (104) and hormone-induced signaling (121). The general function of the COP/FUS/DET system has recently been reviewed (148). Therefore, in this review we refrain from discussing this subject in detail and refer to the COP/DET/FUS pathway only in the context of phyA- and phyB-specific signal transduction.

phyB Phototransduction Pathway

Several screens have been performed using red-light treatments to find mutants that specifically affect phyB signal transduction. A revertant screen using mutagenized ABO seeds resulted in the characterization of RED1 and SLR1, specific to the phyB pathway (67, 145). Unfortunately, to date the genes have not been identified. Another revertant screen using mutagenized phyB-deficient seeds resulted in the isolation of genes affecting auxin-induced signaling (141). The other phyB-specific mutants so far described emerged from screens designed to identify genes responsible for the early flowering phenotype (2). pef2 and pef3 are phyB specific, and pef1 and psi2 also show an altered responsiveness of hypocotyl growth both in red and far-red light (2, 40). However, neither the genes nor the mode of action by which the products of these genes would modulate phyB signaling are yet known. Kretsch et al. (83) used a screening design in which hourly applied red light pulses were followed by subsequent far-red light pulses to isolate mutations that affect the reversibility of the phyB photoreceptor signal transduction. In a large screen (more than one million seedlings were screened) only one stable mutant, designated ohr1, was recovered. The *ohr1* mutant no longer showed red/far-red reversibility of inhibition of hypocotyl growth or of *CAB* gene expression and was characterized as a point mutation within the *PHYB* coding sequence (83). Thus, the numerous screens performed thus far have led to the identification of only a few mutants modulating phyB signal transduction. This indicates that either the phyB-controlled signaling cascade is very short or its components are encoded by genes whose functions significantly overlap or most of the mutations of these genes are lethal.

As an alternative approach to identify signal transduction components specific for phyB, the yeast two-hybrid screening method has been employed. Routinely using yeast strains, investigators found that the full-length PHYB is trans-active in most of the vectors. This limitation explains why various fragments of the PHYB molecule were used as baits to perform most of these screens. For example, by using the very N-terminal domain of PHYB as bait, investigators identified a type-A response regulator, ARR4, as an interaction partner (139). The yeast data were confirmed by demonstrating the interaction of the full-length ARR4 and PHYB proteins by coimmunoprecipitation in vitro and, more importantly, also in extracts derived from transgenic plants. Transgenic plants overexpressing ARR4 show enhanced sensitivity to red light, and the expression of ARR4 is regulated by phyB. It is interesting to note that ARR4 coexpressed with phyB in yeast cells and overexpression of ARR4 in ABO seedlings (146) led to a strong stabilization of phyB in its P_{fr} form. Thus, ARR4 might be a candidate molecule that is required for regulation of $P_{fr} \rightarrow P_r$ dark reversion. Expression of ARR4 is regulated not only by phyB (90) but also by cytokinin (11). The function of ARR4 is phosphorylation dependent, and it is tempting to postulate that the cognate kinase for ARR4 is the recently identified CRE1 histidine kinase, a molecule that also specifically binds cytokinin (69). Taken together, these findings provide the first molecular evidence for an interaction between light- and hormone-dependent signaling cascades with ARR4 acting as a novel molecular switch to mediate this cross-talk (139).

In the revertant screen using mutagenized ABO seeds, a series of intragenic revertants in addition to RED1 were also isolated. Characterization of these mutants showed that the majority of these mutations are single point mutations, clustered and residing within the C-terminal half of the protein. By analyzing the biological function of chimeric photoreceptors, researchers showed that the C-terminal domains of phyA and phyB play an essential role in mediating phytochrome signal transduction (108). Using this C-terminal domain of PHYB as a bait, researchers also have isolated several interacting proteins. From these the function of PIF3 has been characterized in detail (94, 102, 103). PIF3 is a basic helix-loop-helix protein that binds to the C-terminal half of PHYB but also interacts with the intact phyB molecule (102). Mutant derivatives of phyB deficient in signaling in vivo fail to bind PIF3 in vitro. PIF3 is a member of a small gene family and has the characteristics of a transcription factor. The PIF3 protein is localized constitutively in the nucleus. Binding of phyB to PIF3 is conformation dependent and reversibly induced by light, i.e., PIF3 binds, in a highly specific fashion, to the P_{fr} form of phyB (103). Besides binding to the Pfr form of phyB, PIF3 also binds in vitro and in vivo to G-box-containing promoter elements of light-regulated genes. Thus the simultaneous binding of PIF3 to the Pfr form of phyB and to promoters of some of the light-induced genes (LHY, CCA1) points to a straightforward mechanism for a phyB-dependent signaling cascade (94).

This model directly implies that at least some of the signal transduction pathways controlled by phyB are very short, which in turn may explain the low number of phyB-specific mutants characterized so far. We point out, however, that (*a*) overexpression of PIF3 in sense or antisense orientation only results in a very mild modification of the phenotype of wild-type seedlings and (*b*) the phenotype of a PIF3 null mutant is still unknown. One would expect that manipulation of the expression level of a protein that plays a key role in regulating signaling would result in more dramatic phenotypical changes. Alas, the exact role of PIF3 and its homologs in mediating phytochrome signaling in different cell types and at different developmental stages remains to be elucidated.

It is interesting to note that some of the genes whose transcription seems to be mediated at least partly by the PIF3-phyB complex encode proteins that are components of the central circadian clockwork in *Arabidopsis* (118, 147). The photoreceptor that mediates the entrainment of the circadian clock in red light is phyB (130). Additional genes such as ELF3 affecting phototransduction to the central clock (57) and genes belonging to the ZTL/ADO/LKP family were also identified (72, 132). Besides the period length of the circadian rhythm, flowering time and hypocotyl elongation are also affected by altered expression levels of these genes, and there is direct evidence that at least ELF3 (89, 111) and probably GIGANTEA are involved in phyB-controlled signaling (68). Characterization of these proteins showed that ELF3 and ADO2 are localized in the nucleus and bind to the C-terminal half of phyB in yeast. ELF3 could function as a transcription factor, whereas the ZTL/ADO/LKP group of genes may encode novel photoreceptors. These data and the fact that interaction of phyB and the blue light-absorbing photoreceptor cry2 has also been reported (95) indicate that phyB-mediated phototransduction to the circadian clock is controlled by a complex regulatory circuit. The exact way this regulatory circuit functions (24) and simultaneously affects "classical" physiological responses controlled by phyB, however, remains to be elucidated.

phyA Phototransduction Pathway

In recent years many screens employing different designs have been performed to identify phyA-signaling mutants. This was quite easy owing to the special characteristics of phyA-mediated far-red HIR. The strategies included searching for hypersensitive mutants under far-red light (fhy1 and fhy3) (23a, 149, 152a) or after repeated far-red pulses (eid 1, 3, 4, 6) (14, 27), looking for revertants of a weak phyA point mutant (spa1) (61, 62), and screening for mutants resistant to the far-red killing response (pat1) (7). Additional phyA-specific mutants such as fin2 (128), rep1 (129), far1 (65), hfr1/rsf1 (34, 36), and fin219 (63) as well as some defective in both phyA and phyB signaling, including pef1 and psi2 (2, 40), have also been described. The already-high number of mutants is growing, and the screens seem to be unsaturated, although it should be mentioned that in a simple screen looking for hypersensitive mutants only a few alleles of eid1 and spa1 were obtained (Y. Zhou, C. Büche, E. Schäfer & Kretsch, personal communication). Several of the far-red light–specific mutants have also been characterized at the molecular level. Some of these, such as SPA1 (61, 62) FAR1 (65), HFR1/RSF1/REP1 (34, 36, 129), and EID1 (27), are nuclear proteins, whereas others, such as FIN 219 (63) and PAT1 (7), are cytosolic.

The molecular approach using the yeast two-hybrid system led to the identification of three proteins interacting with phyA: PKS1, NPDK2, and PIF3. Originally,

PIF3 was identified as a basic HLH protein functioning in phyA and phyB signaling in vivo (102). More recently it was shown that PIF3 binds not only to phyB (103) but also to phyA, although with almost two orders of magnitude less efficiency and only to the P_{fr} form (153). The Pfr form of phyA can also bind homodimers of PIF3 but not of HFR1/REP1. Just like PIF3, HFR1/REP1 is also a basic loop-helix-loop protein and a positive regulator of phyA signal transduction. Interestingly, the Pfr form of phyA efficiently binds heterodimers of PIF3/HFR1 (34). The abundance of HFR1 mRNA is low in plants irradiated by red light and is significantly increased upon far-red treatment. It is plausible to assume that after red light treatment the Pfr form of phyB interacts in the nuclei with homodimers of PIF3, whereas after far-red light illumination the Pfr form of phyA interacts with either homodimers of PIF3 or heterodimers of PIF3 and HFR1. Thus the differential induction of the transcription of phyA-controlled genes can be determined simply by the different binding specificities of these homo- and heterodimers. In support of this hypothesis we note that in mutant plants lacking detectable amounts of REP1/HFR1/RSF1, only a subset of phyA-controlled responses are affected (129).

PKS1, a novel phosphoprotein localized in the cytoplasm that also has a close homologue termed PKS2, binds to the C-terminal part of PHYA. PKS1 can be phosphorylated by PHYA in vitro on serine residues (151), and its phosphorylated form is also detected in plants (37). The degree of phosphorylation can be enhanced by red light irradiation. Overexpression of PKS1 causes a slightly elongated hypocotyl phenotype in red but not far-red light. Therefore, its role in phyA signaling remains unclear. phyB-mediated phosphorylation of PKS1 has not yet been shown, and it also remains to be proven whether phyB, like phyA, indeed has kinase activity. The phenotype of a PKS1-deficient null mutant is unknown. We note that deletion of the kinase domain of phyB had only a minor effect on phyB-mediated signaling; moreover, the overexpression of this truncated phyB molecule was sufficient to complement a mutant lacking detectable amounts of functional phyB (81). Therefore, the mode in which PKS1 contributes to phytochrome signaling as a negative regulator of phyB-controlled hypocotyl elongation remains to be elucidated. NPDK2 should be a positive regulator in both phyA- and phyB-mediated signal transduction. Loss-of-function alleles have a small but significant effect on the early de-etiolation response (21).

PHYTOCHROME SIGNALING

Based on recent observation it is plausible to assume that light quality- and quantity-dependent import of phyA and phyB into the nucleus is an essential step in signal transduction controlled by these photoreceptors. Our data indicate that light-induced conversion of P_r to P_{fr} is required for translocation; in other words, only the P_{fr} forms of phyA and phyB are transport competent. The kinetics and the wavelength and fluence dependence of the import of phyA and phyB are characteristically different but show good correlation with those established for physiological responses controlled by these photoreceptors. Therefore, the specificity of signaling cascades controlled by phyA and phyB seems to be determined in the cytoplasm.

However, how the rapid translocation of phyA, in contrast to the much slower import of phyB, is achieved is still unclear. One possible explanation is to assume that the translocations of phyB and phyA into the nucleus are modulated by different molecular mechanisms. Alternatively, it may be that the same molecular machinery mediates their import into the nucleus, but the import competence of phyB is regulated by an active retention mechanism. However, both models assume that NLS signal(s) are localized in the C-terminal part of the phyA and phyB proteins, and the NLS signals are masked in the P_r form. These NLS motifs become accessible for interaction with the transport machinery only after conversion to P_{fr}. In phyB it is possible that a second rate-limiting modification of the receptor or a protein(s) interacting with it is required to reach transport competence.

Irrespective of the mechanisms regulating differential import of phyA and phyB, the bulk of these phytochromes remain cytosolic. Even under optimal light conditions to induce nuclear transport, i.e., high-fluence rate of far-red or red light for phyA and phyB, respectively, only a few percents of phyA and approximately 30% of phyB accumulate in the nuclei.

Neither the function of these cytosolic phytochromes nor the mechanism(s) preventing their transport into the nucleus after transformation to P_{fr} are known. In this respect we note that ever since the discovery of the kinase activity of phyA (151) and the identification of PKS1 (37) it has been assumed that the kinase activity of phytochromes might be involved in regulating the nucleo/cytoplasmic distribution of the phytochrome photoreceptors. Alternatively, the kinase activity of phytochrome was proposed to regulate the cellular localization of other components required for light-regulated transcription of plant genes. Regarding function, there is circumstantial evidence that phytochromes permanently localized in the cytoplasm could be involved in regulating the nuclear translocation of a group of transcription factors required for light-induced transcription of a set of genes. The involvement of both phyA and phyB in the induction of the nuclear transport of the CPRF2 transcription factor has been described in parsley suspension cultures (79). Alternatively, as suggested by Nagy et al. (97) and Fankhauser (35), phyA and phyB may regulate, probably indirectly, the retention of phytochromes (mainly phyB) in the cytosol. However, to date kinase activity has been attributed to no phytochrome other than oat phyA, and the deletion of the domain of the phyB molecule responsible for its putative kinase activity has minor effect on phyB-controlled signaling (81). It follows that the mode of action, i.e., how phyA and phyB (and possibly other phytochromes) exert their regulatory function on these cellular processes, remains entirely hypothetical.

It is also not clear how to incorporate the described roles of trimeric G-proteins, Ca^{2+}, calmoduline, and cGMP into this picture. Using microinjection and pharmacological studies, researchers have drawn a bifurcating signal transduction pathway leading from P_{fr} to trimeric G-protein(s), on to anthocyanin accumulation via cGMP, or to CAB-genes expression via Ca^{2+}/calmoduline, and finally to chloroplast development via both branches (10, 84, 101). In addition, Okamoto et al. (103a) reported that overexpression of either the wild-type or constitutively active version of *Arabidopsis* $G\alpha$ subunit protein induced a hypersensitive response to light. This enhanced light sensitivity was more exaggerated in lower light intensity and was observed in white light as well as far-red, red, and blue light. The enhanced responses in far-red and red light required functional phyA and phyB, respectively. Furthermore, the response to far-red light depended on functional FHY1. Thus, these results support the involvement of a heterotrimeric G-protein in the light regulation of *Arabidopsis* seedling development.

Independent of the still-hypothetical regulatory events in the cytoplasm, it seems to be fairly certain that the interaction of phyA and phyB in nuclear localization with a variety of transcription factors is a major regulatory step in phytochrome-mediated signaling. The low number of mutants affecting phyB signaling as well as biochemical data indicate that at least a branch of phyB phototransduction is mediated by a short signaling cascade. phyA-mediated signaling, in good correlation with the specialized role of this photoreceptor, seems to be more complex: The interaction of phyA localized in the nucleus with transcription factors is likely to constitute a major mode of signaling. There also is accumulating evidence that nuclear or cytoplasmic events controlling degradation of phyA and/or other components will severely affect the kinetics and the cell and developmental specificity of the phyA-dependent signaling cascade.

It is generally accepted that COP1 (and maybe other proteins of the COP/DET/FUS group) is a convergence point for upstream signaling pathways dedicated to individual photoreceptors. Nucleo/cytoplasmic partitioning of COP1, as opposed to that of phythochromes, is regulated conversely by light, i.e., it is localized in the nucleus in the dark and is cytoplasmic in the light. There is evidence that COP1 acting as an E3 ligase induces degradation of HY5 (105) in the nucleus in the dark (104). Based on this finding it is suggested that COP1, as a negative regulator of photomorphogenesis, acts by targeting certain proteins to the degradation machinery. Therefore, it is likely that the number of proteins known to interact with COP1 will rapidly increase. Indeed, Hoecker & Quail (60) reported very recently that SPA1, a negative regulator of phyA signaling, specifically interacts with COP1. The model outlined above offers an attractive explanation to the question of how signaling mediated by phytochromes and possibly by other photoreceptors can be regulated by COP1 in dark-grown seedlings. The role of COP1 in affecting phytochrome-mediated signaling in light or at later developmental stages, however, is much less clear.

Besides COP1 and SPA1, EID1 will also probably play a major role in phyA signaling. Whereas SPA1 may be a protein kinase (61), EID1 is a novel F-box protein interacting with both ASK1 and ASK2 (27). Because mutation in EID1 leads to a complete shift of the action spectrum from far-red to red light and extreme hypersensitivity to red light, this nuclear protein must be essential for the nuclear HIR. Again, the most probable function of EID1 is to target the primary nuclear response signal to the degradation machinery. Thus, phyA signaling in the nucleus, in contrast to that mediated by phyB and maybe also by phyC-phyE, will be dependent on a rapid turnover of proteins and thus could turn out to be very similar to the nuclear pathway mediating IAA signaling.

CONCLUSIONS AND PERSPECTIVES

The recent identification of several signal transduction mutants and the observations of light quality- and quantity-dependent nuclear import of phytochromes has dramatically changed and broadened our view about phytochrome signaling. Although the hypothesis of phytochrome-regulated direct transcriptional control and its possible similarity to steroid hormone signaling in animals was proposed more than 30 years ago by H. Mohr, almost nobody followed this line of thinking. The light-dependent nuclear transport of phyB, its P_{fr}-dependent interaction with PIF3, and activation of early target genes like CCA and LHY (both MYB factors) indicate a very short signal transduction chain linking light absorption and target gene activation. Although this pathway is very elegant, it is clearly not the only one, and it seems not to be the dominating pathway because overexpression of PIF3 both in sense and antisense orientation causes a relatively mild phenotype.

Nevertheless, we expect that direct, conformation-dependent interaction of phyA and other phytochromes with transcription factors will eventually be observed. With respect to phyA signaling, it has been reported recently that phyA modulates the transcription of approximately 10% of the genes present in the *Arabidopsis* genome (140). The model presented by the author postulates that phyA in its P_{fr} form interacts in the nucleus with a set of transcription factors (master regulators such as PIF3). These complexes in turn promote light-induced transcription of a diverse set of other transcription factors (CCA1, LHY1, HY5, TOC1, etc.) that are responsible for orchestrating the expression of multiple downstream target genes in various branches of a phyA-regulated transcriptional network. However, identification of other master regulatory genes, besides PIF3, and biochemical characterization of these postulated transcriptional complexes will be essential to understand how specificity is achieved.

A second, complementary pathway, probably controlled by several phytochromes and cryptochrome, involves the induction of the degradation or export of COP1 out of the nucleus. COP1, arguably, has a role in regulating the nuclear degradation of proteins, such as HY5, SPA1, etc., that in turn play key roles in phytochrome-induced signaling. Therefore the regulation of the nucleo/cytoplasmic partitioning of COP1 is expected to be a major regulatory step in phytochrome-dependent signaling. It is not yet clear, however, how the described new mutants can be positioned in this pathway and what the cellular mechanism affecting the partitioning of this key regulator is.

Moreover, we should accept the fact that at present almost nothing is known about the cytosolic pathways controlled by these photoreceptors, even though the majority of phytochromes is located, under all light conditions, in the cytosol.

phyA possesses kinase activity, but its in vivo function remains to be substantiated. There is growing evidence that other signaling cascades such as those induced by cytokinin (139), brassinosteroids (100), and plant defense (140) modulate phytochrome signaling at various levels. However, it has also been proven that light interferes with hormone signaling by regulating the expression/activity of key elements involved in hormone biosynthesis (75). Search for new mutants and novel interaction partners in combination with cell biological and highly specific photophysiological techniques would be useful in untangling a complex, but hopefully not too complex, cytosolic signal transduction network.

The circadian clock in higher plants also controls the expression of nearly all genes regulated by phytochromes. phyA-E plays a role in entraining the circadian clock (130), and recent reports described the first proteins involved in phototransduction from phytochrome to the central clockwork. It has also become evident that the circadian clock regulates the transcription of all phytochrome and cryptochrome genes (80, 143). Our data indicate that speckle formation of the phyB::GFP fusion protein localized in the nuclei exhibits a diurnal rhythm under light/dark conditions (S. Kircher, P. Gil, L. Kozma-Bognár, E. Fejes, V. Speth, E. Adam, E. Schäfer & F. Nagy, submitted). Surprisingly, in all cases, the appearance or disappearance of nuclear spots after dark/light or light/dark transitions, respectively, are much faster than after the first light treatments given to etiolated seedlings. These observations indicate an adaptation of the phytochrome signaling system to light/dark cycles that is due to the circadian clock. This view has been confirmed by analyzing nuclear spot formation in constant darkness or constant light after a diurnal entrainment (41). In this case a circadian oscillation of nuclear speckles of phyB::GFP could be detected with a free running frequency of 24 h. Therefore the physiological consequences of the modification of phytochrome-mediated signaling by the circadian clock under natural conditions should always be kept in mind. Untangling the increasingly complex interaction of the central clockwork with phytochromes, i.e., defining the exact role of these photoreceptors in the functional circadian system, will be an interesting but challenging subject.

ACKNOWLEDGMENTS

Work was supported by the Howard Hughes International Scholarship (55000325) (F.N.) and by the Human Frontiers Science Program (E.S. and F.N.).

Visit the Annual Reviews home page at www.annualreviews.org

LITERATURE CITED

1. Ahmad M, Cashmore AR. 1993. HY4 gene of *A. thaliana* encodes a protein with characteristics of a blue-light photoreceptor. *Nature* 366:162–66
2. Ahmad M, Cashmore AR. 1996. The pef mutants of *Arabidopsis thaliana* define lesions early in the phytochrome signaling pathway. *Plant J.* 10:1103–10
3. Ahmad M, Cashmore AR. 1997. The blue-light receptor cryptochrome 1

shows functional dependence on phytochrome A or phytochrome B in *Arabidopsis thaliana*. *Plant J.* 11:421–27

4. Ahmad M, Jarillo JA, Smirnova O, Cashmore AR. 1998. The CRY1 blue light receptor of Arabidopsis interacts with phytochrome A in vitro. *Mol. Cell.* 1:939–48

5. Batschauer A, Rocholl M, Kaiser T, Nagatani A, Furuya M, Schäfer E. 1996. Blue and UV-A light-regulated CHS expression in Arabidopsis independent of phytochrome A and phytochrome B. *Plant J.* 9:63–69

6. Beggs CJ, Wellmann E. 1994. Photocontrol of flavonoid biosynthesis. See Ref. 75a, pp. 733–51

7. Bolle C, Koncz C, Chua N-H. 2000. PAT1, a new member of the GRAS family, is involved in phytochrome A signal transduction. *Genes Dev.* 14:1269–78

8. Borthwick HA, Hendricks SB, Parker MW, Toole EH, Toole VK. 1952. A reversible photoreaction controlling seed germination. *Proc. Natl. Acad. Sci. USA* 38:662–66

9. Botto JF, Sanchez RA, Whitelam GC, Casal JJ. 1996. Phytochrome A mediates the promotion of seed germination by very low fluences of light and canopy shade light in Arabidopsis. *Plant Physiol.* 110:439–44

10. Bowler C, Neuhaus G, Yamagata H, Chua N-H. 1994. Cyclic GMP and calcium mediate phytochrome phototransduction. *Cell* 77:73–81

11. Brandstatter I, Kieber JJ. 1998. Two genes with similarity to bacterial response regulators are rapidly and specifically induced by cytokinin in *Arabidopsis*. *Plant Cell* 10:1009–19

12. Brockmann J, Rieble S, Kazarinova-Fukshansky N, Seyfried M, Schäfer E. 1987. Phytochrome behaves as a dimer in vivo. *Plant Cell Environ.* 10:105–11

13. Brockmann J, Schäfer E. 1982 Analysis of P_{fr} destruction in *Amaranthus caudatus* L. Evidence for two pools of phytochrome. *Photochem. Photobiol.* 35:555–58

14. Büche C, Poppe C, Schäfer E, Kretsch T. 2000. *Eid1*: a new *Arabidopsis* mutant hypersensitive in phyrochrome A-dependent high-irradiance responses. *Plant Cell* 12:547–58

15. Butler WL, Lane HC, Siegelman HW. 1963. Nonphotochemical transformation of phytochrome in vivo. *Plant Physiol.* 38:514–19

16. Butler WL, Norris KH, Siegelman HW, Hendricks SB. 1959. Detection, assay, and preliminary purification of the pigment controlling photoresponsive development of plants. *Proc. Natl. Acad. Sci. USA* 45:1703–8

17. Casal JJ. 1995. Coupling of phytochrome B to the control of hypocotyl growth in *Arabidopsis*. *Planta* 196:23–29

18. Casal JJ. 2001 Phytochromes, cryptochromes, phototropin: photoreceptor interaction in plants. *Photochem. Photobiol.* 71:1–11

19. Casal JJ, Boccalandro H. 1995. Coaction between phytochrome B and HY4 in *Arabidopsis thaliana*. *Planta* 197:213–18

20. Casal JJ, Sánchez RA, Botto JF. 1998. Modes of action of phytochromes. *J. Exp. Bot.* 49:127–38

21. Choi G, Yi H, Lee J, Kwon YK, Soh MS, et al. 2000. Phytochrome signalling is mediated through nucleoside diphosphate kinase 2. *Nature* 401:610–13

22. Clack T, Matthews S, Sharrock RA. 1994. The phytochrome apoprotein family in Arabidopsis is encoded by five genes: the sequence and expression of PHYD and PHYE. *Plant Mol. Biol.* 25:413–17

23. Cordonnier MM, Greppin H, Pratt LH. 1986. Phytochrome from green *Avena* shoots characterized with a monoclonal antibody to phytochrome from etiolated *Pisum* shoots. *Biochemistry* 25:7657–66

23a. Desnos T, Puente P, Whitelam GC, Harberd NP. 2001. FHY1: a phytochrome

A-specific signal transducer. *Gene Dev.* 15:2980–90

24. Devlin PF, Kay SA. 2000. Cryptochromes are required for phytochrome signaling to the circadian clock but not for rhythmicity. *Plant Cell* 12:2499–509

25. Devlin PF, Patel SR, Whitelam GC. 1998. Phytochrome E influences internode elongation and flowering time in Arabidopsis. *Plant Cell* 10:1479–87

26. Devlin PF, Robson PR, Patel SR, Goosey L, Sharrock RA, Whitelam GC. 1999. Phytochrome D acts in the shade-avoidance syndrome in arabidopsis by contolling elongation growth and flowering time. *Plant Physiol.* 119:909–15

27. Dieterle M, Zhou YC, Schäfer E, Funk M, Kretsch T. 2001. EID1, an F-box protein involved in phytochrome A-specific light signaling. *Genes Dev.* 15:939–44

28. Downs RJ, Siegelman HW. 1963. Photocontrol of anthocyanin synthesis in milo seedlings. *Plant Physiol.* 38:25–30

29. Drumm H, Mohr H. 1974. The dose response curve in phytochrome-mediated anthocyanin synthesis in the mustard seedling. *Photochem. Photobiol.* 20: 151–57

30. Drumm H, Mohr H. 1978. The mode of interaction between blue (UV) light photoreceptors and phytochrome in anthocyanin formation of the *Sorghum* seedling. *Photochem. Photobiol.* 27: 241–48

31. Eichenberg K. 1999. *Die Bedeutung der dynamischen Moleküleigenschaften von Phytochrom für die Lichtsignaltransduktion.* PhD thesis. Albert-Ludwigs-Universität, Freiburg, Ger. 155 pp.

32. Eichenberg K, Bäurle I, Paulo N, Sharrock RA, Rüdiger W, Schäfer E. 2000. *Arabidopsis* phytochromes C and E have different spectral characteristics from those of phytochromes A and B. *FEBS Lett.* 470:107–12

33. Eichenberg K, Hennig L, Martin A, Schäfer E. 2000 Variation in dynamics of phytochrome A in Arabidopsis ecotypes and mutants. *Plant Cell Environ.* 23:311–19

34. Fairchild CD, Schumaker MA, Quail PH. 2000. HFR1 encodes an atypical bHLH protein that acts in phytochrome A signal transduction. *Genes Dev.* 14: 2377–91

35. Fankhauser C. 2001. The phytochromes, a family of red/far-red absorbing photoreceptors. *J. Biol. Chem.* 276:11453–56

36. Fankhauser C, Chory J. 2000. RSF1, an Arabidopsis locus implicated in phytochrome A signal transduction. *Plant Physiol.* 124:39–46

37. Fankhauser C, Yeh KC, Lagarias JC, Zhang H, Elich TD, Chory J. 1999. PKS1, a substrate phosphorylated by phytochrome that modulates light signaling in Arabidopsis. *Science* 284:1539–41

38. Furuya M. 1993. Phytochromes: their molecular species, gene family and functions. *Annu. Rev. Plant Physiol. Plant Mol. Biol.* 44:617–45

39. Furuya M, Schäfer E. 1996. Photoperception and signalling of induction reactions by different phytochromes. *Trends Plant Sci.* 1:301–7

40. Genoud T, Millar AJ, Nishizawa N, Kay SA, Schäfer E, et al. 1998. An *Arabidopsis* mutant hypersensitive to red and far-red light signals. *Plant Cell* 10:889–904

41. Gil P. 2001. *Analysis of the nucleo-cytoplasmatic partitioning of phytochrome B and its differential regulation by light and the circadian clock.* PhD thesis. Albert-Ludwigs-Universität, Freiburg, Ger. 89 pp.

42. Gil P, Kircher S, Adam E, Bury E, Kozma-Bognar L, et al. 2000. Photocontrol of subcellular partitioning of phytochrome-B:GFP fusion protein in tobacco seedlings. *Plant J.* 22:135–45

43. Goosey L, Palecanda L, Sharrock RA. 1997. Differential pattern of expression of the *Arabidopsis PHYB, PHYD,* and *PHYE* phytochrome genes. *Plant Physiol.* 115:959–69

44. Guo H, Mockler T, Duong H, Lin C. 2001. SUB1, an Arabidopsis Ca^{2+}-binding protein involved in cryptochrome and phytochrome coaction. *Science* 284: 487–90

45. Hartmann KM. 1966. A general hypothesis to interpret "high energy phenomena" of photomorphogenesis on the basis of phytochrome. *Photochem. Photobiol.* 5:349–66

46. Heim B, Jabben M, Schäfer E. 1981. Phytochrome destruction in dark- and light-grown *Amaranthus caudatus* seedlings. *Photochem. Photobiol.* 34:89–93

47. Heim B, Schäfer E. 1984 The effect of red and far-red light in the high irradiance reaction of phytochrome (hypocotyl growth in dark-grown *Sinapis alba* L.). *Plant Cell Environ.* 7:39–44

48. Hendricks SB, Borthwick HA. 1967. The function of phytochrome in regulation of plant growth. *Proc. Natl. Acad. Sci. USA* 58:2125–30

49. Hendricks SB, van der Woude WJ. 1983. How phytochrome acts—perspectives on the continuing quest. See Ref. 126a, pp. 4–23

50. Hennig L, Büche C, Eichenberg K, Schäfer E. 1999. Dynamic properties of endogenous phytochrome A in *Arabidopsis* seedlings. *Plant Physiol.* 121: 571–77

51. Hennig L, Büche C, Schäfer E. 2000. Degradation of phytochrome A and the high irradiance response in *Arabidopsis*: a kinetic analysis. *Plant Cell Environ.* 23: 727–34

52. Hennig L, Funk M, Whitelam GC, Schäfer E. 1999. Functional interaction of cryptochrome 1 and phytochrome D. *Plant J.* 20:289–94

53. Hennig L, Poppe C, Sweere U, Martin A, Schäfer E. 2001. Negative interference of endogenous Phytochrome B with Phytochrome A function in *Arabidopsis*. *Plant Physiol.* 125:1036–44

54. Hennig L, Poppe C, Unger S, Schäfer E. 1999. Control of hypocotyl elongation in *Arabidopsis thaliana* by photoreceptor interaction. *Planta* 208:257–63

55. Hennig L, Schäfer E. 2001. Both subunits of the dimeric plant photoreceptor phytochrome require chormophore for stability of the far-red light-absorbing form. *J. Biol. Chem.* 276:7913–18

56. Hennig L, Stoddart WM, Dieterle M, Whitelam GC, Schäfer E. 2001. Phytochrome E controls light-induced germination of *Arabidopsis*. *Plant Physiol.* In press

57. Hicks KA, Millar AJ, Carre IA, Somers DE, Straume M, et al. Conditional circadian dysfunction of the *Arabidopsis* early flowering 3 mutant. *Science* 274:790–92

58. Hisada A, Hanzawa H, Weller JL, Nagatani A, Reid JB, Furuya M. 2000. Light-induced nuclear translocation of endogenous pea phytochrome A visulaized by immunocytochemical procedures. *Plant Cell* 12:1063–78

59. Hock B, Mohr H. 1964. Die Regulation der O_2-Aufnahme von Senfkeimlingen (*Sinapis alba* L.) durch Licht. *Planta* 61:209–27

60. Hoecker U, Quail PH. 2001. The phytochrome A-specific signaling intermediate SPA1 interacts directly with COP1, a constitutive repressor of light signaling in *Arabidopsis*. *J. Biol. Chem.* In press

61. Hoecker U, Tepperman JM, Quail PH. 1999. SPA1, a WD-repeat protein specific to phytochrome A signal transduction. *Science* 284:496–99

62. Hoecker U, Xu Y, Quail PH. 1998. SPA1: a new genetic locus involved in phytochrome A-specific transduction. *Plant Cell* 10:19–33

63. Hsieh HL, Okamoto H, Wang LH, Ang M, Matsui H, et al. 2000. FIN219, an auxin-regulated gene, defines a link between phytochrome A and downstream regulator COP1 in light control of *Arabidopsis* development. *Genes Dev.* 14: 1958–70

64. Huala E, Oeller PW, Liscum E, Han IS,

Larsen E, Briggs WR. 1997. *Arabidopsis* NPH1: a protein kinase with a putative redox-sensing domain. *Science* 278:2120–23

65. Hudson M, Ringli C, Boylan MT, Quail PH. 1999. The FAR1 locus encodes a novel protein specific to phytochrome A signal transduction. *Genes Dev.* 13: 2017–27

66. Hughes J, Lamparter T, Mittman F, Hartmann E, Gärtner W, et al. 1997. A prokaryotic phytochrome. *Nature* 386:663–67

67. Huq E, Kang Y, Halliday KJ, Qin M, Quail PH. 2000. SRL1: a new locus specific to the phyB-signaling pathway in Arabidopsis. *Plant J.* 23:461–70

68. Huq E, Tepperman JM, Quail PH. 2000. GIGANTEA is a nuclear protein involved in phytochrome signaling in *Arabidopsis. Proc. Natl. Acad. Sci. USA* 97: 9789–94

69. Inoue T, Higuchi M, Hashimoto Y, Seki M, Kobayashi M, et al. 2001. Identification of CRE1 as a cytokinin receptor from Arabidopsis. *Nature* 409:1060–63

70. Jabben M, Deitzer GF. 1978. A method for measuring phytochrome in plants grown in white light. *Photochem. Photobiol.* 27:799–802

71. Jabben M, Shanklin J, Vierstra RD. 1989. Red-light-induced accumulation of ubiquitin-phytochrome conjugates in both monocots and dicots. *Plant Physiol.* 90:380–84

72. Jarillo JA, Capel J, Tang RH, Yang HQ, Alonso JM, et al. 2001. An Arabidopsis circadian clock component interacts with both CRY1 and phyB. *Nature* 410:487–90

73. Johnson CB, Hilton J. 1978. Effect of light on phythochrome in cauliflower curd. *Planta* 144:13–17

74. Kagawa T, Sakai T, Suetsugu N, Oikawa K, Ishiguro S, Wada M. 2001. *Arabidopsis* NPL1: a phototropin homolog controlling the chloroplast high-light avoidance response. *Science* 291:2138–41

75. Kang JG, Yan J, Kim DH, Chung KS,

Fuijoka S, et al. 2001. Light and brassinosteroid signals are integrated via a dark induced small G protein in etiolated seedling growth. *Cell* 105:625–36

75a. Kendrick RE, Kronenberg GHM, eds. 1994. *Photomorphogenesis in Plants.* Dordrecht: Kluwer Acad. Publ. 2nd ed.

76. Kim L, Kircher S, Toth R, Adam E, Schäfer E, Nagy F. 2000. Light-induced nuclear import of phytochrome-A:GFP fusion proteins is differentially regulated in transgenic tobacco and *Arabidopsis. Plant J.* 22:125–34

77. Deleted in proof

78. Kircher S, Kozma-Bognar L, Kim L, Adam E, Harter K, et al. 1999. Light quality-dependent nuclear import of the plant photoreceptors phytochrome A and B. *Plant Cell* 11:1445–56

79. Kircher S, Wellmer F, Nick P, Rügner A, Schäfer E, et al. 1999. Nuclear import of the Parsley bZIP transcription factor CPRF2 is regulated by phytochrome photoreceptors. *J. Cell Biol.* 144:201–11

80. Kozma-Bognar L, Hall A, Adam E, Thain SC, Nagy F, Millar AJ. 1999. The circadian clock controls the expression pattern of the circadian input photoreceptor, phytochrome B. *Proc. Natl. Acad. Sci. USA* 96:14652–57

81. Krall L, Reed JW. 2000. The histidine kinase-related domain participates in phytochrome B function but is dispensable. *Proc. Natl. Acad. Sci. USA* 97: 8169–74

82. Kraml M. 1994. Light direction and polarisation. See Ref. 75a, pp. 417–43

83. Kretsch T, Poppe C, Schäfer E. 2000. A new type of mutation in the plant photoreceptor phytochrome B causes loss of photoreversibility and an extremely enhanced light sensitivity. *Plant J.* 22:177–86

84. Kunkel T, Neuhaus A, Batschauer A, Chua NH, Schäfer E. 1996. Functional analysis of yeast-derived phytochrome A and B phycocyanobilin adducts. *Plant J.* 10:625–36

85. Deleted in proof
86. Kunkel T, Tomizawa KI, Kern R, Furuya M, Chua NH, Schäfer E. 1993. In vitro formation of a photoreversible adduct of phycocyanobilin and tobacco apophytochrome B. *Eur. J. Biochem.* 215:587–94
87. Lagarias JC, Lagarias DM. 1989. Self-assembly of synthetic phytochrome holoprotein in vitro. *Proc. Natl. Acad. Sci. USA* 86:5778–80
88. Lin CT, Yang HY, Guo HW, Mockler T, Chen J, Cashmore AR. 1998. Enhancement of blue light sensitivity of Arabidopsis seedlings by a blue light receptor cryptochrome 2. *Proc. Natl. Acad. Sci. USA* 95:2686–90
89. Liu XL, Covington MF, Fankhauser C, Chory J, Wagner DR. 2001. ELF3 encodes a circadian clock-regulated nuclear protein that functions in an Arabidopsis PHYB signal transduction pathway. *Plant Cell* 13:1293–304
90. Lohrmann J, Buchholz G, Keitel C, Sweere U, Kircher S, et al. 1999. Differential expression and nuclear localization of response regulator-like proteins from *Arabidopsis thaliana. Plant Biol.* 1:495–505
91. MacKenzie JM Jr, Coleman RA, Briggs WR, Pratt LH. 1975. Reversible redistribution of phytochrome within the cell upon conversion to its physiologically active form. *Proc. Natl. Acad. Sci. USA* 72:799–803
92. Mancinelli AL, Rabino I. 1978. The "high irradiance responses" of plant photomorphogenesis. *Bot. Rev.* 44:129–80
93. Mandoli DF, Briggs WR. 1981. Phytochrome control of two low-irradiance responses in etiolated oat seedlings. *Plant Physiol.* 67:733–39
94. Martinez-Garcia JF, Huq E, Quail PH. 2000. Direct targeting of light signals to a promoter element-bound transcription factor. *Science* 288:859–63
95. Mas P, Devlin PF, Panda S, Kay SA. 2000. Functional interaction of phyto-chrome B and cryptochrome 2. *Nature* 408:207–11
96. McCurdy D, Pratt LH. 1986. Immunogold electron microscopy of phytochrome in *Avena*: identification of intracellular sites responsible for phytochrome sequestering and enhanced pelletability. *J. Cell Biol.* 103:2541–50
97. Nagy F, Kircher S, Schäfer E. 2001. Intracellular trafficking of photoreceptors during light induced signal transduction. *J. Cell Sci.* 114:475–80
98. Nagy F, Schäfer E. 2000. Nuclear and cytosolic events of light-induced, phyto-chrome-regulated signaling in higher plants. *EMBO J.* 19:157–63
99. Neff MM, Chory J. 1998. Genetic interactions between phytochrome A, phytochrome B and cryptochrome 1 during *Arabidopsis* development. *Plant Physiol.* 118:27–36
100. Neff MM, Nguyen SM, Malancharuvil EJ, Fujioka S, Noguchi T, et al. 1999. BAS1: a gene regulating brassinosteroid levels and light responsiveness in *Arabidopsis. Proc. Natl. Acad. Sci. USA* 211:5316–23
101. Neuhaus G, Bowler C, Kern R, Chua N-H. 1993. Calcium/calmodulin-dependent and -independent phytochrome signal transduction pathways. *Cell* 73:937–52
102. Ni M, Tepperman JM, Quail PH. 1998. PIF3, a phytochrome-interacting factor necessary for normal photoinduced signal transduction, is a basic helix-loop-helix protein. *Cell* 95:657–67
103. Ni M, Tepperman JM, Quail PH. 1999. Binding of phytochrome B to its nuclear signalling partner PIF3 is reversibly induced by light. *Nature* 400:781–84
103a. Okamoto H, Matsui M, Deng XW. 2001. Overexpression of the heterotrimeric G-α protein subunit enhances phytochrome-mediated inhibition of hypocotyl elongation in Arabidopsis. *Plant Cell* 13:1639–51
104. Osterlund MT, Hardtke CS, Wei N, Deng

XW. 2000. Targeted destabilization of HY5 during light-regulated development of Arabidopsis. *Nature* 405:462–66

105. Oyama T, Shimura Y, Okada K. 1997. The Arabidopsis HY5 gene encodes a bZIP protein that regulátes stimulus-induced development of root and hypocotyl. *Genes Dev.* 11:2983–95

106. Poppe C, Sweere U, Drumm-Herrel H, Schäfer E. 1998. The blue light receptor cryptochrome 1 can act independently of phytochrome A and B in *Arabidopsis thaliana*. *Plant J.* 16:465–71

107. Pratt LH. 1994. Distribution and localization of phytochrome within the plant. See Ref. 75a, pp. 163–85

108. Quail PH, Boylan MT, Parks BM, Short TM, Xu Y, Wagner D. 1995. Phytochromes: photosensory perception and signal transduction. *Science* 268:675–80

109. Quail PH, Schäfer E, Marmé D. 1973. *De novo* synthesis of phytochrome in pumpkin hools. *Plant Physiol.* 52:124–27

110. Quail PH, Schäfer E, Marmé D. 1973. Turnover of phytochrome in pumpkin cotyledons. *Plant Physiol.* 52:128–31

111. Reed JW, Nagpal P, Bastow RM, Solomon KS, Dowson-Day MJ, et al. 2000. Independent action of ELF3 and phyB to control hypocotyl elongation and flowering time. *Plant Physiol.* 122:1149–60

112. Sakamoto K, Nagatani A. 1996. Nuclear localization activity of phytochrome B. *Plant J.* 10:859–68

113. Schäfer E. 1975. A new approach to explain the "high irradiance responses" of photomorphogenesis on the basis of phytochrome. *J. Math. Biol.* 2:41–56

114. Schäfer E, Haupt W. 1983. Blue-light effects in phytochrome-mediated responses. See Ref. 126a, pp. 723–44

115. Schäfer E, Lassig TU, Schopfer P. 1975. Photocontrol of phytochrome destruction in grass seedlings. The influence of wavelength and irradiance. *Photochem. Photobiol.* 22:193–202

116. Schäfer E, Lassig TU, Schopfer P. 1982. Phytochrome-controlled exten-

sion growth of *Avena sativa* L. seedlings II. Fluence-rate response relationships and action spectra of mesocotyl and coleoptile responses. *Planta* 154:231–40

117. Schäfer E, Löser G, Heim B. 1983. Formalphysiologische Analyse der Signaltransduktion in der Photomorphogenese. *Berl. Dtsch. Bot. Ges.* 96:497–509

118. Schaffer R, Ramsay N, Samach A, Corden S, Putterill J, et al. 1998. The late elongated hypocotyl mutation of Arabidopsis disrupts circadian rhythms and the photoperiodic control of flowering. *Cell* 93:1219–29

119. Schmidt R, Mohr H. 1981. Time-dependent changes in the responsiveness to light of phytochrome-mediated anthocyanin synthesis. *Plant Cell Environ.* 4: 433–37

120. Schmidt W, Schäfer E. 1974. Dependence of phytochrome dark reactions on the initial photostationary state. *Planta* 116:267–72

121. Schwechheimer C, Serino G, Callis J, Crosby WL, Lyapina S, et al. 2001. Interacting of the COP9 signalosome with the E3 ubiquitin ligase SCF^{TIR1} in mediating auxin response. *Science* 292:1379–82

122. Sharrock RA, Quail PH. 1989. Novel phytochrome sequences in *Arabidopsis thaliana*: structure, evolution and differential expression of a plant regulatory photoreceptor family. *Genes Dev.* 3: 1745–57

123. Shimazaki Y, Pratt LH. 1985. Immunochemical detection with rabbit polyclonal and mouse monoclonal antibodies different pools of phytochrome from etiolated and green Avena shoots. *Planta* 164:333–44

124. Shinomura T, Hanzawa H, Schäfer E, Furuya M. 1998. Mode of phytochrome B action in the photoregulation of seed germination in *Arabidopsis thaliana*. *Plant J.* 13:583–90

125. Shinomura T, Nagatani A, Hanzawa H, Kubota M, Watanabe M, Furuya M. 1996. Action spectra for phytochrome

A- and B-specific photoinduction of seed germination in *Arabidopsis thaliana.* *Proc. Natl. Acad. Sci. USA* 93:8129–33

126. Shinomura T, Uchida K, Furuya M. 2000. Elementary responses of photoperception by phytochrome A for high irradiance response of hypocotyl elongation in Arabidopsis. *Plant Physiol.* 122:147–56

126a. Shropshire WJR, Mohr H, eds. 1983. *Photomorphogenesis.* Berlin/Heidelberg/New York/Tokyo: Springer-Verlag. 2nd ed.

127. Smith H, Whitelam GC. 1990. Phytochrome, a family of photoreceptors with multiple physiological roles. *Plant Cell Environ.* 13:695–707

128. Soh MS, Hong SH, Hanzawa H, Furuya M, Nam HG. 1998. Genetic identification of FIN2, a far-red light specific signaling component of *Arabidopsis thaliana. Plant J.* 16:411–19

129. Soh MS, Kim YM, Han SJ, Song PS. 2000. REP1, a basic helix-loop-helix protein, is required for a branch pathway of phytochrome A signaling in *Arabidopsis. Plant Cell* 12:2061–73

130. Somers DE, Devlin PF, Kay SA. 1998. Phytochromes and cryptochromes in the entrainment of the *Arabidopsis* circadian clock. *Science* 282:1488–90

131. Somers DE, Quail PH. 1995. Temporal and spatial expression patterns of *PHYA* and *PHYB* genes in *Arabidopsis. Plant J.* 7:413–27

132. Somers DE, Schultz TF, Milnamow M, Kay SA. 2000. *ZEITLUPE* encodes a novel clock-associated PAS protein from *Arabidopsis. Cell* 101:319–29

133. Speth V, Otto V, Schäfer E. 1986. Intracellular localization of phytochrome in oat coleoptiles by electron microscopy. *Planta* 168:299–304

134. Speth V, Otto V, Schäfer E. 1987. Intracellular localization of phytochrome and ubiquitin in red-light-irradiated oat coleoptiles by electron microscopy. *Planta* 171:332–38

135. Steinitz B, Bergfeld R. 1977. Pattern formation underlying phytochrome-mediated anthocyanin synthesis in the cotyledons of *Sinapis alba* L. *Planta* 133:229–35

136. Steinitz B, Schäfer E, Drumm H, Mohr H. 1979. Correlation between far-red absorbing phytochrome and response in phytochrome-mediated anthocyanin synthesis. *Plant Cell Environ.* 2:159–63

137. Stone HJ, Pratt PH. 1979. Characterization of the destruction of phytochrome in the red-absorbing form. *Plant Physiol.* 63:680–82

138. Süßlin C. 2001. *Analyse UV-B hyposensitiver Mutanten aus Arabidopsis thaliana.* PhD thesis. Albert-Ludwigs-Universität, Freiburg, Ger. 109 pp.

139. Sweere U, Eichenberg K, Lohrmann J, Mira-Rodado V, Kudla J, et al. 2001. Interaction of the response regulator ARR4 with phytochrome B in modulating red-light signaling. *Science* 294:1108–11

140. Tepperman JM, Zhu T, Chang HS, Quail PH. 2001. Multiple transcription-factor genes are early targets of phytochrome A signaling. *Proc. Natl. Acad. Sci. USA* 98:9437–42

141. Tian Q, Reed JW. 1999. Control of auxin-regulated root development by the *Arabidopsis thaliana* SHY2/IAA3 gene. *Development* 126:711–21

142. Tokuhisa JG, Daniels SM, Quail PH. 1985. Phytochrome in green tissue: spectral and immunochemical evidence for two distinct molecular species of phytochrome in light-grown *Avena sativa. Planta* 164:321–32

143. Toth R, Kevei E, Hall A, Millar AJ, Nagy F, Kozma-Bognar L. 2001. Circadian clock-regulated expression of phytochrome and cryptochrome genes in Arabidopsis. *Plant Physiol.* 127:1607–16

144. Vierstra RD, Davis SJ. 2000. Bacteriophytochromes: new tools for understanding phytochrome signal transduction. *Cell* 11:511–21

145. Wagner D, Hoecker U, Quail PH. 1997. RED1 is necessary for phytochrome B-mediated red light-specific signal transduction in Arabidopsis. *Plant Cell* 9: 731–43

146. Wagner D, Teppermann JM, Quail PH. 1991. Overexpression of phytochrome B induces a short hypocotyl phenotype in transgenic Arabidopsis plants. *Plant Cell* 3:1275–88

147. Wang ZY, Tobin EM. 1998. Constitutive expression of the *Circadian Clock associated 1* (CCA1) gene disrupts circadian rhythms and supresses its own expression. *Cell* 93:1207–17

148. Wei N, Deng XW. 1999. Making sense of the COP9 signalosome: a regulatory complex conserved from *Arabidopsis* to human. *Trends Genet.* 15:98–103

149. Whitelam GC, Johnson E, Peng J, Carol P, Anderson ML, et al. 1993. Phytochrome A null mutants of Arabidopsis display a wild-type phenotype in white light. *Plant Cell* 5:757–68

150. Yamaguchi R, Nakamura M, Mochizuki N, Kay SA, Nagatani A. 1999. Light-dependent translocation of a phytochrome B-GFP fusion protein to the nucleus in transgenic *Arabidopsis. J. Cell Biol.* 145:437–45

151. Yeh KC, Lagarias JC. 1998 Eukaryotic phytochromes: light-regulated serine/threonine protein kinases with histidine kinase ancestry. *Proc. Natl. Acad. Sci. USA* 95:13976–81

152. Yeh KC, Wu SH, Murphy JT, Lagarias JC. 1997. A cyanobacterial phytochrome two-component light sensory system. *Science* 277:1505–8

152a. Zeidler M, Bolle C, Chua NH. 2001. The phytochrome A specific signaling component PAT3 is a positive regulator of Arabidopsis photomorphogenesis. *Plant Cell Physiol.* 42:1193–200

153. Zhu Y, Tepperman JM, Fairchild CD, Quail PH. 2000. Phytochrome B binds with greater apparent affinity than phytochrome A to the basic helix-loop-helix factor PIF3 in a reaction requiring the PAS domain of PIF3. *Proc. Natl. Acad. Sci. USA* 97:13419–24

Annu. Rev. Plant Biol. 2002. 53:357–75
DOI: 10.1146/annurev.arplant.53.100301.135251

THE COMPLEX FATE OF α-KETOACIDS

Brian P. Mooney,[1] Jan A. Miernyk,[1,2] and
Douglas D. Randall[1]
[1]Department of Biochemistry, University of Missouri, Columbia, Missouri 65211;
[2]Plant Genetics Research Unit, USDA, ARS, Columbia, Missouri 65211;
e-mail: mooneyb@missouri.edu, miernykj@missouri.edu, randalld@missouri.edu

Key Words intermediary metabolism, mitochondria, protein structure, pyruvate
dehydrogenase, regulation

■ **Abstract** Plant cells are unique in that they contain four species of α-ketoacid
dehydrogenase complex: plastidial pyruvate dehydrogenase, mitochondrial pyruvate
dehydrogenase, α-ketoglutarate (2-oxoglutarate) dehydrogenase, and branched-chain
α-ketoacid dehydrogenase. All complexes include multiple copies of three components: an α-ketoacid dehydrogenase/decarboxylase, a dihydrolipoyl acyltransferase,
and a dihydrolipoyl dehydrogenase. The mitochondrial pyruvate dehydrogenase complex additionally includes intrinsic regulatory protein-kinase and -phosphatase
enzymes. The acyltransferases form the intricate geometric core structures of the complexes. Substrate channeling plus active-site coupling combine to greatly enhance the
catalytic efficiency of these complexes. These α-ketoacid dehydrogenase complexes
occupy key positions in intermediary metabolism, and a basic understanding of their
properties is critical to genetic and metabolic engineering. The current status of knowledge of the biochemical, regulatory, structural, genomic, and evolutionary aspects of
these fascinating multienzyme complexes are reviewed.

CONTENTS

1040-2519/02/0601-0357$14.00

357

INTRODUCTION

The pyruvate dehydrogenase complexes (PDCs) have strategic roles in plant metabolism. The mitochondrial PDC located in the matrix links cytoplasmic glycolysis with the Krebs cycle and ATP generation by the electron transport chain. The plastidial PDC provides two essential substrates for de novo fatty acid biosynthesis, acetyl-CoA and NADH. PDCs are members of a unique family of large multienzyme structures, the α-ketoacid dehydrogenase complexes. The family includes the PDCs, the α-ketoglutarate dehydrogenase complex (KGDC), and the branched-chain α-ketoacid dehydrogenase complex (BCKDC). These complexes range in molecular mass from 4 to 9 MDa and are among the largest nonviral protein assemblies known. All members include multiple copies of three central enzymes. The E1 component is a specific α-ketoacid dehydrogenase (decarboxylase), E2 is a dihydrolipoyl acyltransferase, and E3 is a dihydrolipoyl dehydrogenase. A generalized reaction scheme is shown in Figure 1.

The overall reaction converts the α-ketoacid, NAD^+, and reduced coenzyme A (CoASH) to CO_2, NADH, and an acyl-CoA. The E1 component catalyzes decarboxylation of the α-ketoacid and employs thiamine pyrophosphate (TPP) as a cofactor producing an acyl-TPP conjugate. The E2, which contains lipoic acid prosthetic groups, transfers the acyl group to CoASH. The E3 component contains FAD and reoxidizes the lipoyl groups of E2 with concomitant formation of NADH.

Multiple copies of E2 form the core of the complexes, to which the other components are noncovalently attached. The close proximity of the three components promotes substrate channeling. Additionally, E2 has a multidomain structure joined by flexible linker regions. This feature allows active-site coupling using a swinging-arm movement of the lipoyl-domain between E1 and E3. These features improve the overall efficiency well beyond that which would be attained with spatially separated enzymes. Trimers of E2 assemble to form the core structures, either a cube (8 trimers) or a pentagonal dodecahedron (20 trimers).

These complexes are defined by their α-ketoacid specificity: pyruvate, α-ketoglutarate, or the branched-chain α-ketoacids. Pyruvate enters mitochondria via a specific membrane carrier where its utilization by PDC links glycolysis with the Krebs cycle. α-ketoglutarate is an intermediate of the Krebs cycle, and α-ketoisocaproate, α-ketomethyvalerate, and α-ketoisovalerate are derived from transamination of leucine, isoleucine, and valine, respectively. These complexes clearly occupy critical positions in intermediary metabolism.

$$R\text{-}CO\text{-}COOH + NAD^+ + CoASH \longrightarrow R\text{-}CO\text{-}S\text{-}CoA + NADH + H^+ + CO_2$$

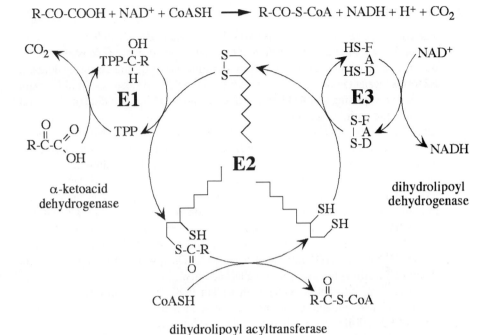

Figure 1 A generalized reaction scheme for the α-ketoacid dehydrogenase complexes. The overall reaction is presented above the contributions of each of the individual component enzymes. The R group defines the specificity of each of the complexes. The saw-tooth pattern represents the lipoyl-lysine side-group of E2. TPP, thiamine pyrophosphate.

This review focuses primarily on the pyruvate dehydrogenase complexes. The reader's attention is directed to recent reviews on nonplant PDC (63, 65) and an excellent review of the glycine decarboxylase complex (GDC) (15).

Previous reviews of plant PDCs (37, 58, 59) have covered the biochemical and regulatory properties of these enzymes. This review provides an update of these properties plus an overview of the molecular and structural features. Additionally, genomic aspects and evolutionary comparisons are considered.

MITOCHONDRIAL PYRUVATE DEHYDROGENASE COMPLEX

The mitochondrial PDC (mtPDC) links glycolysis to the Krebs cycle. Regulation of flux through the PDC controls Krebs cycle activity and ATP synthesis via oxidative phosphorylation. Additionally, carbon entry at other points of the cycle is dependent on acetyl-CoA produced by PDC.

Organization

The E1 component of mtPDC has two subunits (α and β), which form a heterotetramer. E2 is a multidomain protein composed of one or two lipoic acid–binding domains, an E1-binding domain and a C-terminal catalytic and assembly domain. E2 forms the pentagonal dodecahedron core of the PDC. E3 also has a multidomain structure comprising an FAD-binding domain, an NAD-binding domain, a central domain, and an interface domain responsible for dimerization (17). Mammalian (25, 27), yeast (4), and nematode (14, 32) PDCs also contain an E3-binding protein (E3BP, previously designated protein X). E3BP is a multidomain protein typically composed of (from N- to C-termini) a lipoyl domain, an E3-binding domain, and an inner-core domain (25). *Ascaris suum* E3BP is unusual in that it lacks the N-terminal lipoyl domain, suggesting that the prosthetic group is superfluous for E3 binding. Under aerobic conditions, the nematode PDC functions normally without an E3BP. However, during the switch from aerobic to anaerobic respiration, *A. suum* PDC apparently uses the E3BP to recruit additional E3 under conditions where the NADH/NAD$^+$ ratio is elevated (32). Thus, while providing metabolic flexibility, E3BP is not essential for PDC activity.

There is no substantial evidence that plant PDCs contain an E3BP, and none has been cloned or identified during genome sequencing. Many plants contain two types of E2, a di-lipoyl domain type and a mono-lipoyl domain type. Pea (37) and potato (47) mtPDCs contain both types of E2. *Arabidopsis thaliana* contains a di-lipoyl domain E2 (23) plus a mono-lipoyl domain E2 identified in the completed genome sequence (75).

When the two forms of E2 are present, one could act as an E3BP. Alignment of the amino acid sequences of the two types of E2 from *A. thaliana* with that of *Zea mays* E2 (75) reveals that the mono-lipoyl domain–type E2s have conserved subunit-binding domains, whereas that of the di-lipoyl domain type is markedly divergent. It is possible that the differences in the subunit-binding domains reflect different functions of the two E2 proteins: One binds E1, whereas the other functions as an E3BP. We are currently testing this hypothesis and additionally determining if both types of E2 can function as acetyltransferases.

Regulation

Regulation of mtPDC activity, by compartmentalization, product inhibition, and metabolite effectors, has been covered in previous reviews (37, 58, 59). Pea mtPDC activity is also regulated via control of developmental expression (36). MtPDC activity is highest in etiolated plants and in the youngest leaves of light-grown plants. The activity then declines as the leaves mature and is almost absent in senescing leaves. Changes in activity correlate with abundance of E1 and E2 mRNA and protein (36). The expression pattern of the E3 subunit does not follow that of E1 or E2, though this is perhaps not surprising as a single E3 in pea is shared among all the α-ketoacid dehydrogenase complexes (15). A number of plant species show highest mtPDC activity early in development (35, 36, 73, 77), reflecting the biochemical and structural changes during membrane expansion and remodeling.

The most striking level of regulation of mtPDC activity is achieved by reversible phosphorylation (60). It is by this mechanism that mtPDC is inhibited in a light-dependent manner. PDC-kinase (PDK) activity is stimulated by 20 to 80 μM NH_4^+. Stimulation by NH_4^+ is part of the mechanism of light-dependent inactivation of mtPDC (67). Photorespiratory glycine metabolism produces NH_4^+ in the mitochondria, which in turn stimulates the kinase, causing mtPDC inactivation in the light. In support of this conclusion are the results of Budde & Randall (6) and Gemel & Randall (19): Pea plants when illuminated in a low O_2/high CO_2 atmosphere showed reduced light-dependent inactivation of PDC, and inhibitors of photorespiration also prevented the light-dependent inactivation. The NADH from photorespiration can drive oxidative phosphorylation, conserving carbon and making photorespiration less wasteful.

Previous work on reversible phosphorylation used partially purified mtPDC to study the intrinsic kinase and phosphatase. More recently, three cDNAs for plant PDK have been cloned. Two PDKs, with 77% amino acid identity, were cloned from maize (76) and one from *A. thaliana* (72, 87). *Z. mays* PDK1 is constitutively expressed, whereas *Z. mays* PDK2 is upregulated in leaves, possibly allowing acute response to high mitochondrial ATP concentrations during photosynthesis. Recombinant *Z. mays* PDK2 inactivated maize PDC with concomitant phosphorylation of the E1α subunit (76). Mammalian and plant PDKs specifically phosphorylate Ser residues but lack the signature domains of Ser/Thr kinases. On the other hand, all PDKs do contain the five canonical domains of prokaryotic two-component histidine kinases. The reaction mechanism of the two-component His-kinases involves autophosphorylation of a specific His residue followed by phospho-transfer to an Asp residue (reviewed in Ref. 69). Owing to their similarity to His-kinases, it has been proposed that plant PDKs may also use a His residue in their catalytic mechanism (48, 72). When a conserved His in the H-box (H121 of the *A. thaliana* PDK) was mutated, autophosphorylation of the recombinant kinase and transphosphorylation of the E1α subunit of PDC were slower but not abolished (74). This suggested that although H121 might be involved in catalytic activity it was not an essential phosphotransfer His residue. Additionally, when all His residues in *A. thaliana* PDK were chemically modified, both auto- and transphosphorylation were abolished (48). However, more recent data prove that neither of the two absolutely conserved His residues in PDK (H121 and H168 in the *A. thaliana* sequence) are directly involved in the catalytic mechanism (A.T. Tovar-Mendez, unpublished observations). These results reveal that although PDK has sequence similarity with prokaryotic His-kinases it is in fact not a His-kinase but a unique type of Ser kinase (A.T. Tovar-Mendez, unpublished observations). The controversy surrounding interpretation of these results serves to emphasize potential problems in assigning protein function based exclusively on in silico analyses.

Dephosphorylation/reactivation of mtPDC is catalyzed by an E1-specific phospho-pyruvate dehydrogenase phosphatase (PDP). Mammalian PDPs are members of the PP2C class. PDP1 is an $\alpha\beta$ heterodimer consisting of catalytic and regulatory subunits, whereas PDP2 is monomeric (reviewed in Ref. 65). To date, none has been cloned from plants. Similar to early work on PDK, most of the data on plant PDP comes from experiments with partially purified mtPDC. PDP requires divalent cations for activity, in particular Mg^{2+}. PDP activity was inhibited 40% by 10 mM Pi, whereas Krebs cycle intermediates, nucleotides, NAD^+, NADH, acetyl-CoA, CoASH, and polyamines had no effect in mitochondria from light-grown pea seedlings. The ratio of PDK to PDP activity is approximately 5:1 (44). Ca^{2+} activates mammalian PDP (reviewed in Ref. 65), whereas Ca^{2+} is a potential inhibitor of pea PDP, antagonizing the Mg^{2+}-dependent dephosphorylation of mtPDC (44). Our current knowledge of the two regulatory enzymes suggests that control of the phosphorylation state of PDC must be through regulation of PDK. How this is achieved in vivo has not been established. The multilevel regulation of mtPDC allows fine control of its activity and emphasizes the importance of this enzyme to metabolism.

Structure

Broccoli mtPDC has an S coefficient of 59.3, corresponding to a M_r of 5–6 × 10^6 (reviewed in Ref. 37). The subunit stoichiometry of plant mtPDC has not been defined, necessitating use of the mammalian paradigm. Mammalian PDC consists of 20–30 E1 heterotetramers, 1 E2 60mer (20 trimers), and 12 E3 dimers. There are also 12 E3BPs per complex (reviewed in Ref. 63). Mammalian mtPDC is larger than plant mtPDC, with a M_r of 7–9 × 10^6 (reviewed in Ref. 37). The apparent absence of an E3BP in plants could account for some of the size difference. The deduced molecular mass of plant mtPDC, assuming the same subunit stoichiometry as the mammalian complex (minus E3BP), is 7.4 × 10^6.

E2 forms the core of PDC, BCKDC, and KGDC to which E1 and E3 associate noncovalently. The regulatory kinase and phosphatase of mtPDC also bind to E2. Plant mtPDC has a similar structure to those reported from other eukaryotes (37). A minimum mtPDC is schematically presented in Figure 2 (see color insert). Thelen et al. (75) first demonstrated that plant mtPDC E2 forms a 60mer core structure. Purified recombinant maize E2 self-assembled into a high-molecular mass structure consistent with formation of a 60mer. When examined by negative-staining transmission electron microscopy (TEM), maize E2 appeared as a pentagonal dodecahedron structure identical to that previously reported for mammalian E2. The maize E2 core has a diameter of 29 nm and three axes of symmetry characteristic of a pentagonal dodecahedron (75).

Expression of recombinant plant mtPDC E1 has been problematic. A *Bacillus subtilis* secretion system produced very limited amounts of pea E1 (54), and expression by the yeast *Pichia pastoris* was hindered by an unfavorable codon usage bias. Pea E1 was isolated from the yeast and shown to form a catalytically active tetramer with a M_r of approximately 160,000 (53). Other expression systems are being evaluated for production of heterotetrameric *A. thaliana* E1.

Native pea mitochondrial E3 is a homodimer (5). The crystal structure of pea E3 has recently been solved at a resolution of 3.15 Å (17), confirming a twofold axis connecting the subunits through a C-terminal interface domain. The catalytic site comprises residues within the interface domain, explaining the dimerization requirement for full activity (17). *A. thaliana* mitochondrial E3 is encoded by two genes; their deduced amino acid sequences are 92% identical. Both E3s include the conserved interface domains, indicating they likely also form dimers (40). It has been proposed that isoform 1 of E3 (mtLPD1) is most often associated with the GDC, whereas the second E3 isoform (mtLPD2) is associated with PDC, KGDC, and BCKDC. A T-DNA knockout mutant of mtLPD2 had no apparent phenotype, indicating that mtLPD1 is sufficient for all of the mitochondrial enzyme complexes (40).

Nematode PDKs have a native M_r consistent with a dimer (9, 10). Similarly, PDK from mammalian sources is a dimer (65). When *A. thaliana* PDK was expressed in *E. coli* as a fusion with the maltose-binding protein (MBP) and then treated with Factor Xa to remove the MBP moiety, the resulting PDK behaved as a dimer during gel filtration chromatography (74).

PLASTIDIAL PYRUVATE DEHYDROGENASE COMPLEX

The plastidial PDC (ptPDC) catalyzes the same reaction as its mitochondrial counterpart and has a similar enzyme organization. PtPDC provides acetyl-CoA for de novo fatty acid biosynthesis (8, 31) in the stroma. PtPDC was first identified by Reid et al. (64) in leucoplasts of developing castor seeds. Subsequently, ptPDC was found in all plant sources examined (61). In addition to supplying acetyl-CoA, ptPDC is also the primary, if not only, source of stromal NADH necessary for fatty acid biosynthesis (8).

Organization

Cloning of the E1 α and β subunits revealed that they are more closely related to the red algal (*Porphyra purpurea*) sequences than to the mitochondrial subunits (28). The intracellular location of proteins encoded by these cDNAs was confirmed by in vitro import assays using isolated pea chloroplasts (29). PtPDC E2 has a multidomain structure consisting of a transit peptide, a single lipoyl domain, an E1-binding domain, and an inner catalytic/assembly domain (52). Similar to the situation with plant mtPDC, no E3BP has been identified with the plastidial complex, although a second E2 gene was revealed in the *A. thaliana* genome (GenBank accession, AAG12782). It is 60% identical to the previous clone and also has a single lipoyl-domain. Most of the amino acid differences are conservative substitutions distributed throughout the primary structure. The most striking differences are confined to the interdomain linker regions and the transit peptides. The function of this second form of E2 has not been determined. Unlike the two mitochondrial types of E2, the subunit-binding domains of the two plastidial E2 sequences are very similar (B.P. Mooney, unpublished observations). Despite this, an E3BP function for the second type remains a possibility. *A. thaliana* plastidial E3 is encoded by two genes with similar intron/exon structure, suggesting that they are paralogs resulting from a relatively recent gene duplication. The two forms of E3 are approximately 88% identical at the amino acid level, most of the differences being confined to the transit peptide sequences (39). Overall, the plastidial subunits are approximately 30% identical at the amino acid level with their mtPDC counterparts (28, 39, 52).

Regulation

The catalytic properties of the ptPDC are similar to those of mtPDC, although the Km for NAD^+ is approximately half of that for the mitochondrial complex. PtPDC activity is controlled by product inhibition and is more sensitive to the $NADH:NAD^+$ ratio than to the ratio of acetyl-CoA to CoASH (7). PtPDC is more active at alkaline pH (pH optimum of 8.0 vs. 7.2 for mtPDC) and requires higher Mg^{2+} levels, thus reflecting the higher stromal pH and Mg^{2+} concentration that occur during photosynthesis. The activity of ptPDC is therefore higher under photosynthetic conditions, contrasting with the mitochondrial complex, which has decreased activity during photosynthesis. It appears that ptPDC is regulated primarily by changes in the biochemical environment of the stroma during dark to light transitions, i.e., the increased pH and Mg^{2+} levels (8, 43, 82).

Whereas E1α has Ser phosphorylation site 1 conserved (28), ptPDC is not regulated by reversible phosphorylation (59). Plastids do not contain a PDK. Reversible phosphorylation as a means of control of ptPDC was probably selected against as this PDC must be most active during fatty acid biosynthesis, which occurs during the light. Concurrently, there are high levels of ATP, which would obviously tend to keep the ptPDC inactive if a phosphorylation mechanism were used as a means of control (45).

Structure

The ptPDC from pea is similar in size to broccoli mtPDC, with a M_r between that of the *Escherichia coli* and mammalian complexes, 4–7 MDa (8). However, estimating the mass of the plastid complex is difficult because of subunit dissociation during purification (7, 79). Plastidial E1 formed a heterotetramer when the two subunits were coimported by isolated pea chloroplasts (29). When E1β alone was imported, all of the protein remained as a monomer, suggesting there is no pool of E1α subunits with which the newly imported E1β can assemble. However when coimported, E1α and E1β monomer, heterodimer, and heterotetramer forms were found, suggesting the most likely pathway for E1 assembly proceeds through an α/β heterodimer (29). No plastidial E2 structure has yet been determined, however given the size of ptPDC, E2 is probably a pentagonal dodecahedron. Both E3s have conserved interface domains (39), and it is likely that they form active dimers.

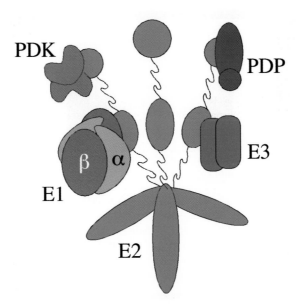

Figure 2 A minimum mitochondrial pyruvate dehydrogenase complex. Subunit stoichiometry is based on plant data, with the exception of PDP, which is assumed to be a heterodimer based on mammalian PDP1. E1 is an α (*orange*) β (*red*) heterotetramer. E2 is a trimer of monolipoyl domain-type subunits (*blue*). E3 is a homodimer (*green*). The intrinsic protein-kinase (PDK) and -phosphatase (PDP) regulatory enzymes are depicted in shades of *purple*.

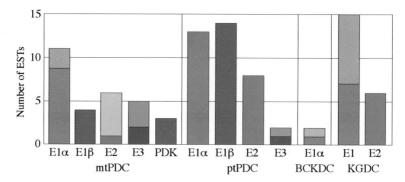

Figure 3 Digital northern analysis of α-ketoacid dehydrogenase complex subunits from *A. thaliana*. The coding sequences of each subunit were used as the search parameter in MEGABLAST (limited to EST and *A. thaliana*) analysis of the GenBank database. Only EST clones sequenced from the 3 end of the cDNAs were scored, and no sequences from the Versailles collection or of less than 100 bp are included. Stacked histograms indicate contributions of the different isoforms of E1, E1α, E2, and E3.

BRANCHED-CHAIN α-KETOACID DEHYDROGENASE COMPLEX

The BCKDC catalyzes the irreversible oxidative decarboxylation of branched-chain α-ketoacids derived from valine, leucine, and isoleucine (30, 56, 57, 83, 85). Mammalian BCKDC is mitochondrial, and although BCKDC activity has not been detected directly in isolated mitochondria from pea leaves, there is substantial evidence that BCKDC is also mitochondrial in plant cells. Metabolic labeling studies using [^{14}C] leucine and NaH[^{14}C]O$_3$ have shown that a branched-chain amino acid degradation pathway (including BCKDC) exists in plant mitochondria (1). Analyses of the deduced amino acid sequences encoded by the isolated BCKDC cDNA clones revealed presumptive N-terminal mitochondrial targeting peptides (38, 50, 51). Isolated pea seedling mitochondria were able to import and process the E1α, E1β, and E2 subunits in an ATP-dependent, valinomycin-inhibited reaction (B.P. Mooney & A.K. Broz, unpublished observations).

Regulation

Like mtPDC, mammalian BCKDC is phosphorylated by an intrinsic E1 kinase (BCDK) (55). Verifying this regulatory mechanism in plants has proven to be difficult. The E1α sequence does include a conserved Ser at phosphorylation site 1 (51), but examination of the recently completed *A. thaliana* genome sequence has failed to identify any candidates for the BCDK. Fujiki et al. (18) reported that E1α and E2 mRNAs were absent in light-grown plants, but expression was induced within 3 h of dark treatment and peaked between 24 h to 48 h in the dark. Additionally, BCKDC activity, using α-keto-β-methylvalerate as substrate, was only detectable in leaves from dark-adapted plants. E2 protein was undetectable in light-grown plants but accumulated to readily detectable levels (by western blotting) following a 72-h dark treatment. Additionally, both E1β and E2 mRNAs were induced by addition of mannitol or the photosynthesis inhibitor DCMU to excised leaves. Transcripts also accumulated during senescence when protein turnover is enhanced (18). The rapid accumulation (and presumably degradation) of BCKDC mRNAs and protein seems like a wasteful regulatory mechanism. Currently we are investigating the possibility that PDK has a dual role in phosphorylating the E1α of BCKDC in addition to mtPDC E1α and whether BCKDC is regulated by reversible phosphorylation.

Structure

The *A. thaliana* BCKDC E1α sequence is 42% identical to its mammalian counterpart (51), whereas E1β is 67% identical (38). Although not experimentally determined, it is likely that plant BCKDC E1 is an $\alpha_2\beta_2$ heterotetramer, similar to the mammalian complex (11). *A. thaliana* BCKDC E2, expressed in *E. coli*, forms a soluble 24mer core structure with octahedral symmetry (49). When examined by negative-staining TEM, the core was an 11-nm square identical to the cubic core of mammalian E2. Analytical ultracentrifugation of *A. thaliana* E2 revealed a single species with a molecular mass of 0.95 MDa to 1.1 MDa, consistent with a 24mer of 45.6-kDa subunits (49).

α-KETOGLUTARATE DEHYDROGENASE COMPLEX

The KGDC is an integral component of the Krebs cycle, catalyzing the oxidative decarboxylation of α-ketoglutarate (2-oxoglutarate) to yield CO_2, succinyl-CoA, and NADH. Organization of the KGDC is similar to PDC and BCKDC, with α-ketoglutarate dehydrogenase (E1), dihydrolipoamide succinyltransferase (E2), and E3 components. Unlike the PDCs and BCKDC, the E1 of KGDC is not composed of α and β subunits. Rather it is a single subunit containing regions of homology with both PDC E1 subunits. The succinyltransferase has a similar domain structure to other E2s. Millar et al. (46) provide evidence suggesting that KGDC is associated with the mitochondrial inner membrane, in contrast to the matrix location of mtPDC.

Regulation

The KGDC isolated from cauliflower mitochondria is activated by 1 mM AMP (46, 80). The binding of AMP to KGDC E1 lowers the *Km* for α-ketoglutarate and increases the *Vmax* (12, 13). The KGDC reaction would also be regulated by the activities of the mtPDC and citrate synthase through competition for the mitochondrial pool of CoASH (16). Like all the α-ketoacid dehydrogenase complexes, the KGDC is substrate specific. The rate of NAD^+ reduction by potato KGDC using pyruvate and isovalerate is less than 2% of the rate with α-ketoglutarate (46).

Structure

The mammalian KGDC has a molecular mass of approximately 2 MDa. E1 and E3 homodimers are arranged around the 24mer E2 core (34). KGDC from potato has an S coefficient of 25 $\pm$ 2, corresponding to a molecular mass of approximately 1.7 MDa. KGDC is significantly smaller than mtPDC as assessed by gel-filtration chromatography (46). Though the structure of plant KGDC E2 has not been studied to any significant extent and the KGDC is the least well understood of all the plant complexes, the similarity in M_r of the mammalian and plant complexes suggests conservation of E2 24mer structure.

GENOMICS

Most subunits of the animal cell α-ketoacid dehydrogenase complexes are encoded by small multigene families. In some instances, this is a function of tissue- or organ-specific gene expression (e.g., 86). In other instances, complex regulatory mechanisms have been postulated based on expression of specific subunit genes in response to physiological changes (e.g., 26). Virtually nothing, however, has been reported about the number of genes encoding subunits of the α-ketoacid dehydrogenase complexes in plant cells. Availability of the complete genomic sequence of the model flowering plant *A. thaliana* (70) has provided a unique opportunity for analysis of subunit gene families. With few exceptions, all subunits of each of the α-ketoacid dehydrogenase complexes from *A. thaliana* are encoded by two genes. The exceptions are ptPDC E1α [1], mtPDC E1β [1], mtPDC E2 [3], PDK [1], BCKDC E1β [1], BCKDC E2 [1], and KGDC E2 [1]. The fragmentary information available for other plants is consistent with the same relatively simple pattern: Each subunit is encoded by one to three gene-family members. With the exception of *A. thaliana* mitochondrial E3s, where one form is associated with the PDC/KGDC and the other with GDC (40), nothing is yet known about specific expression or functional roles of the plant gene-family members.

ANALYSIS OF GENE EXPRESSION

The abundance of a specific cDNA in a random EST collection is proportional, within statistical limits, to the expression of the corresponding gene (3). This is the basis of in silico analysis of gene expression, so-called digital northern analysis.

Figure 3 (see color insert) shows digital northern data for the subunits of the *A. thaliana* α-ketoacid dehydrogenase complexes. MtPDC E1β, E3, and PDK have similar expression profiles, whereas expression of E2 is nearly twofold higher. There is a distinct difference between expression levels of the two subunits of E1; E1α expression is almost threefold that of E1β. Higher expression of the E1α gene might reflect a more rapid turnover rate for E1α, which is common for regulatory proteins. Analog northern data show a similar difference in total expression of pea seedling E1α mRNA relative to E1β (36).

Although expression of the ptPDC subunits in developing seeds (81) is similar to the pattern for mtPDC, the expression levels of ptPDC E1α and E1β are almost identical when all EST collections are considered. PtPDC is not regulated by reversible phosphorylation, which possibly explains the coordinated expression pattern of the two plastidial E1 subunits, as E1α does not play a critical role in ptPDC regulation. PtPDC E2 (52) is expressed at a similar level to mitochondrial E2, and no EST clones have been reported for the second form of ptPDC E2. There is a single EST for each of the two plastidial E3 isoforms, corroborating the data of Lutziger & Oliver (39) that show an almost equal expression for the two E3s.

No EST clones have been reported for BCKDC E1β or E2, and there is only one for each for the two E1α isoforms. This lack of representation in EST collections correlates with the unusual expression pattern observed by Fujiki et al. (18) where the E1β and E2 subunits were absent in the light and only induced after a number of hours in the dark. The KGDC E1 isoforms are the most highly expressed subunits of all the complexes.

It is noteworthy that the mtPDC E1α subunit and E1 of KGDC are the most abundant. Luethy et al. (36) suggested that, as mtPDC E1α is particularly prone to aggregation in the absence of E1β, it might be synthesized at a high level to ensure enough is present to allow formation of the heterotetramer. Furthermore, the abundance of E1α ESTs might reflect higher turnover rates for these subunits.

EVOLUTIONARY CONSIDERATIONS

The α-ketoacid dehydrogenase complexes seem to have appeared early in evolution. They are found in members of the Archaea, Bacteria, and Eukaryota that live under aerobic conditions. A series of unrelated oxidoreductases perform functions similar to those of the α-ketoacid dehydrogenase complexes in anaerobically grown bacteria and amitochondrial members of the Eukaryota (e.g., 66).

There is a single type of molecular architecture for the KGDC. The E2 core is a cube with eightfold symmetry, and E1 is a homodimer (33, 42). Similarly, there is a single molecular architecture for the BCKDC. Again, the E2 core is a cube with eightfold symmetry; however in this instance, E1 is an $\alpha_2\beta_2$-heterotetramer (11, 84). In contrast, there are two types of PDC. Type I, which is limited to the Eubacteria, is similar to KGDC: The E2 core is a cube, and E1 is a homodimer (62). In Type II PDC, the E2 core forms a pentagonal dodecahedron with icosahedral symmetry, and E1 is an $\alpha_2\beta_2$ heterotetramer (63). There are examples of E2 subunits with one (75), two (23, 71), or three lipoyl domains (68), although it is not clear that these differences have a substantial effect upon structure or catalytic activity (24, 41).

---→

Figure 4   An unrooted neighbor-joining distance tree based on the amino acid sequences of the E1 subunits of the α-ketoacid dehydrogenase complexes. Sequences compared are 1, *Zea mays* (AF069911); 2, *Arabidopsis thaliana* (U21214); 3, *Pisum sativum* (U51918); 4, *Homo sapiens* (AAA60232); 5, *Drosophila melanogaster* (AAF45977); 6, *A. suum* (AAA29376); 7, *Saccharomyces cerevisiae* (CAA50657); 8, *Bacillus stearothermophilus* (CAA37628); 9, *Rickettsia prowazekii* (A71681); 10, *Z. mays* (AF069910); 11, *A. thaliana* (U09137); 12, *P. sativum* (U56697); 13, *H. sapiens* (CAA40924); 14, *D. melanogaster* (AAF56855); 15, *A. suum* (AAA29379); 16, *S. cerevisiae* (AAA34583); 17, *B. stearothermophilus* (S14230); 18, *R. prowazekii* (B71681); 19, *Azotobacter vinelandii* (Y15124); 20, *Escherichia coli* (CAA24739); 21, *Pseudomonas aeruginosa* (AAC45353.1); 22, *Pasteurella multocida* (AAK02979); 23, *Neisseria meningitidis* (E81094); 24, *A. thaliana* (U80185); 25, *Phorphyra purpurea* (S73188); 26, *Synechocystis* sp (BAA18592); 27, *Cyanidium caldarium* (AAF12897); 28, *A. thaliana* (U80186); 29, *P. purpurea* (S73187); 30, *C. caldarium* (AAF12898); 31, *Mesostigma viride* (NP_038396); 32, *A. thaliana* (AF077955); 33, *H. sapiens* (AAB59549.1); 34, *D. melanogaster* (AAF54398.1); 35, *Lycopersicon esculentum* (T06589); 36, *P. putida* (5822330); 37, *A. thaliana* (AF061638); 38, *H. sapiens* (NP_000047.1); 39, *D. melanogaster* (AAF45391.1); 40, *Caenorhabditis elegans* (T21454); 41, *P. putida* (AAA65615.1); 42, *S. cerevisiae* (AAA34721); 43, *A. thaliana* (AJ223802.1); 44, *D. melanogaster* (AAF49388); 45, *H. sapiens* (XP_004889); 46, *E. coli* (CAA25280); 47, *R. prowazekii* (CAA14647); 48, *A. vinelandii* (S07776); 49, *C. elegans* (T15098). The sequences were aligned using PHYLIP then analyzed with SEQBOOT (500 rounds of bootstrapping). The scale bar indicates substitutions per site. The tree was constructed using TreeView.

An intriguing feature is the existence of two types of E1: the homodimer and the heterotetramer. The sequences of homodimeric E1 subunits contain features found in both α- and β-subunits of the heterotetrameric E1 sequences. The possibilities that a single ancestral gene was disrupted resulting in two subunits or that two separate genes fused into a single subunit seem equally likely. It cannot be considered that homodimeric E1 is more primitive. Although this form is limited to a few Bacterial PDCs, it is found in all KGDCs.

A phylogenetic analysis of E1 sequences is presented in Figure 4. It is noteworthy that all of the sequences for homodimeric-type E1 subunits are grouped in adjoining clades. Rather than being distributed according to catalytic species, all of the E1α sequences form a clade of clades, as do the E1β sequences. In agreement with the endosymbiotic bases for the origins of organelles (21, 22), all of the mitochondrial E1 sequences are grouped with the *Rickettsia prowazekii* sequence. *R. prowazekii* is thought to be the nearest relative to the mitochondrial ancestor (2). Similarly, the plastidial E1 sequences are grouped with the cyanobacterial

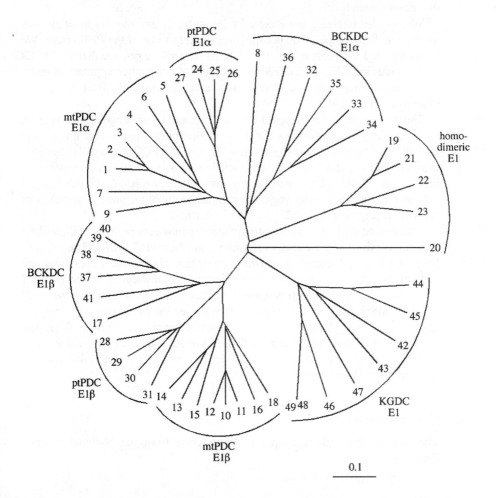

sequences (20). Although there are too few α-ketoacid dehydrogenase complex sequences for an extensive comparison, those extant are distributed in agreement with our current understanding of the tree of life.

PROSPECTUS

Whither the plant α-ketoacid dehydrogenase complexes? Although our laboratory has recently concentrated on the structural aspects of these complexes, the discovery of two types of mtPDC E2s in *A. thaliana* raises a number of important questions about the roles of these similar subunits. Do both types combine in a single hetero-60mer, or are there separate 60mers? In either case, what are the functions of the two types of E2? If there are distinct mono-lipoyl domain and di-lipoyl domain 60mers, are they present within the same mitochondrion, or is there tissue/organ/organelle specific expression? Each of these questions is a focus of current research.

We are also targeting the elusive PDP regulatory enzyme. By analogy with the mammalian complex, the phosphatase is likely to be of the PP2C class. We have identified a number of candidates in the *A. thaliana* genome that are PP2Cs with obvious mitochondrial targeting sequences. A brute-force approach is being employed where all will be cloned, expressed, and their specificity determined in order to identify PDP.

The critical positions occupied by the α-ketoacid dehydrogenase complexes in intermediary metabolism place them in an ideal position for controlling carbon allocation through metabolic engineering. The role of ptPDC, and possibly mtPDC, in providing substrates for de novo fatty acid biosynthesis points to their importance in attempts to alter carbon partitioning into starch or oils. Additionally, novel uses of the α-ketoacid dehydrogenase complexes might include the synthesis of nonnatural polymers such as biodegradable plastics.

Research on plant α-ketoacid dehydrogenase complexes, perhaps owing to their very complexity, has been somewhat limited, and the ptPDC, BCKDC and KGDC have been less well studied. It remains unclear how plants control the expression and activity of the BCKDC. Is reversible phosphoryation a regulatory mechanism? If BCKDC is absent in light-grown plants, do they synthesize and degrade the entire complex during the dark-light transition? What are the functions of the multiple gene-products in *A. thaliana* of the mitochondrial and plastidial PDCs? When and where are they expressed? Do plant PDCs have an E3BP? Clearly, despite over 30 years of research on plant α-ketoacid dehydrogenase complexes, there remain many questions to be answered.

ACKNOWLEDGMENT

The authors' research is supported by a grant from the National Science Foundation.

Visit the Annual Reviews home page at www.annualreviews.org

LITERATURE CITED

1. Anderson MD, Che P, Song J, Nikolau BJ, Wurtele ES. 1998. 3-methylcrotonyl-coenzyme A carboxylase is a component of the mitochondrial leucine catabolic pathway in plants. *Plant Physiol.* 118: 1127–38

2. Andersson SGE, Zomorodipour A, Andersson JO, Sicheritz-Pontén T, Alsmark UCM, et al. 1998. The genome sequence of *Rickettsia prowazekii* and the origin of mitochondria. *Nature* 396:133–43

3. Audic S, Claverie JM. 1997. The significance of digital gene expression profiles. *Genome Res.* 7:986–95

4. Behal RH, Browning KS, Hall TB, Reed LJ. 1989. Cloning and nucleotide sequence of the gene for protein X from *Saccharomyces cerevisiae*. *Proc. Natl. Acad. Sci. USA* 86:8732–36

5. Bourguignon J, Macherel D, Neuburger M, Douce R. 1992. Isolation, characterization, and sequence analysis of a cDNA clone encoding L-protein, the dihydrolipoamide dehydrogenase component of the glycine cleavage system from pea-leaf mitochondria. *Eur. J. Biochem.* 204:865–73

6. Budde RJA, Randall DD. 1990. Pea leaf mitochondrial pyruvate dehydrogenase complex is inactivated in vivo in a light-dependent manner. *Proc. Natl. Acad. Sci. USA* 87:673–76

7. Camp PJ, Miernyk JA, Randall DD. 1988. Some kinetic and regulatory properties of the pea chloroplast pyruvate dehydrogenase complex. *Biochim. Biophys. Acta* 993:269–75

8. Camp PJ, Randall DD. 1985. Purification and characterization of the pea chloroplast pyruvate dehydrogenase complex. A source of acetyl-CoA and NADH for fatty acid biosynthesis. *Plant Physiol.* 77:571–77

9. Chen W, Huang X, Komuniecki PR, Komuniecki R. 1998. Molecular cloning,

functional expression, and characterization of pyruvate dehydrogenase kinase from anaerobic muscle of the parasitic nematode *Ascaris suum*. *Arch. Biochem. Biophys.* 353:181–89

10. Chen W, Komuniecki PR, Komuniecki R. 1999. Nematode pyruvate dehydrogenase kinases: role of the C-terminus in binding of the dihydrolipoyl transacetylase core of the pyruvate dehydrogenase complex. *Biochem. J.* 339:103–9

11. Chuang JL, Wynn RM, Song JL, Chuang DT. 1999. GroEL/GroES-dependent reconstitution of $\alpha_2\beta_2$ tetramers of human mitochondrial branched chain α-ketoacid decarboxylase. *J. Biol. Chem.* 274:10395–404

12. Craig DW, Wedding RT. 1980. Regulation of the 2-oxoglutarate dehydrogenase lipoate succinyltransferase complex from cauliflower by nucleotide. Pre-steady state kinetics and physical studies. *J. Biol. Chem.* 255:5769–75

13. Craig DW, Wedding RT. 1980. Regulation of the 2-oxoglutarate dehydrogenase lipoate succinyltransferase complex from cauliflower by nucleotide. Steady state kinetic studies. *J. Biol. Chem.* 255:5763–68

14. Diaz F, Komuniecki RW. 1994. Pyruvate dehydrogenase complexes from the equine nematode, *Parascaris equorum*, and the canine cestode, *Dipylidium caninum*, helminths exhibiting anaerobic mitochondrial metabolism. *Mol. Biochem. Parasitol.* 67:289–99

15. Douce R, Bourguignon J, Neuburger M, Rébeillé F. 2001. The glycine decarboxylase system: a fascinating complex. *Trends Plant Sci.* 6:167–76

16. Dry IB, Wiskich JT. 1987. 2-oxoglutarate dehydrogenase and pyruvate dehydrogenase activities in plant mitochondria: interaction via a common coenzyme A pool. *Arch. Biochem. Biophys.* 257:92–99

17. Faure M, Bourguignon J, Neuburger M, Macherel D, Sieker L, et al. 2000. Interaction between the lipoamide-containing H-protein and the lipoamide dehydrogenase (L-protein) of the glycine decarboxylase multienzyme system. *Eur. J. Biochem.* 267:2890–98

18. Fujiki Y, Sato T, Ito M, Watanabe A. 2000. Isolation and characterization of cDNA clones for the E1β and E2 subunits of the branched-chain α-ketoacid dehydrogenase complex in Arabidopsis. *J. Biol. Chem.* 275:6007–13

19. Gemel J, Randall DD. 1992. Light regulation of leaf mitochondrial pyruvate dehydrogenase complex. *Plant Physiol.* 100:908–14

20. Giovannoni SJ, Turner S, Olsen GJ, Barns S, Lane DJ, Pace NR. 1988. Evolutionary relationships among cyanobacteria and green chloroplasts. *J. Bacteriol.* 170: 3584–92

21. Gray MW, Burger G, Lang BF. 2001. The origin and early evolution of mitochondria. *Genome Biol.* 2:1018.1–8.5

22. Gray MW, Doolittle WF. 1982. Has the endosymbiont hypothesis been proven? *Microbiol. Rev.* 46:1–42

23. Guan Y, Rawsthorne S, Scofield G, Shaw P, Doonan J. 1995. Cloning and characterization of a dihydrolipoamide acetyltransferase (E2) subunit of the pyruvate dehydrogenase complex from *Arabidopsis thaliana. J. Biol. Chem.* 270:5412–17

24. Guest JR, Lewis HM, Graham LD, Packman LC, Perham RN. 1985. Genetic reconstruction and functional analysis of the repeating lipoyl domains in the pyruvate dehydrogenase multienzyme complex of *Escherichia coli. J. Mol. Biol.* 185:743–54

25. Harris RA, Bowker-Kinley MM, Wu P, Jeng J, Popov KM. 1997. Dihydrolipoamide dehydrogenase-binding protein of the human pyruvate dehydrogenase complex. *J. Biol. Chem.* 272:19746–51

26. Huang YJ, Walker D, Chen W, Klingbeil M, Komuniecki R. 1998. Expression of pyruvate dehydrogenase isoforms during the aerobic/anaerobic transition during the development of the parasitic nematode *Ascaris suum*: altered stoichiometry of phosphorylation/inactivation. *Arch. Biochem. Biophys.* 352:263–70

27. Jilka JM, Rahmatullah M, Kazemi M, Roche TE. 1986. Properties of a newly characterized protein of the bovine kidney pyruvate dehydrogenase complex. *J. Biol. Chem.* 261:1858–67

28. Johnston ML, Luethy MH, Miernyk JA, Randall DD. 1997. Cloning and molecular analyses of the *Arabidopsis thaliana* plastid pyruvate dehydrogenase subunits. *Biochim. Biophys. Acta* 1321:200–6

29. Johnston ML, Miernyk JA, Randall DD. 2000. Import, processing, and assembly of the α- and β-subunits of chloroplast pyruvate dehydrogenase. *Planta* 211:72–76

30. Jones SMA, Yeaman SJ. 1986. Oxidative decarboxylation of 4-methylthio-2-oxobutyrate by branched-chain 2-oxoacid dehydrogenase complex. *Biochem. J.* 237: 621–23

31. Ke J, Behal RH, Back SL, Nikolau BJ, Wurtele ES, Oliver DJ. 2000. The role of pyruvate dehydrogenase and acetyl-coenzyme A synthetase in fatty acid synthesis in developing Arabidopsis seeds. *Plant Physiol.* 123:497–508

32. Klingbeil MM, Walker DJ, Arnette R, Sidawy E, Hayton K, et al. 1996. Identification of a novel dihydrolipoyl dehydrogenase-binding protein in the pyruvate dehydrogenase complex of the anaerobic parasitic nematode, *Ascaris suum. J. Biol. Chem.* 271:5451–57

33. Knapp JE, Mitchell DT, Yazdi MA, Ernst SR, Reed LJ, Hackert ML. 1998. Crystal structure of the truncated cubic core component of the *Escherichia coli* 2-oxoglutarate dehydrogenase multienzyme complex. *J. Mol. Biol.* 280:655–68

34. Koike M, Koike K. 1975. Structure, assembly and function of mammalian alpha-keto acid dehydrogenase complexes. *Adv. Biophys.* 9:187–227

35. Lernmark U, Gardestrom P. 1994. Distribution of pyruvate dehydrogenase complex activities between chloroplast and mitochondria from leaves of different species. *Plant Physiol.* 106:1633–38

36. Luethy MH, Gemel J, Johnston ML, Mooney BP, Miernyk JA, Randall DD. 2001. Developmental expression of the mitochondrial pyruvate dehydrogenase complex in pea (*Pisum sativum*) seedlings. *Physiol. Plant.* 112:559–66

37. Luethy MH, Miernyk JA, David NR, Randall DD. 1996. Plant pyruvate dehydrogenase complexes. See Ref. 55a, pp. 71–92

38. Luethy MH, Miernyk JA, Randall DD. 1998. The nucleotide sequence of a cDNA encoding the E1-beta subunit of the branched-chain alpha-keto acid dehydrogenase from *Arabidopsis thaliana* (Accession No. AF061638) (PGR98-133). *Plant Physiol.* 118:329

39. Lutziger I, Oliver D. 2000. Molecular evidence of a unique lipoamide dehydrogenase in plastids: analysis of plastidic lipoamide dehydrogenase from *Arabidopsis thaliana.* *FEBS Lett.* 484:12–16

40. Lutziger I, Oliver D. 2001. Characterization of two cDNAs encoding mitochondrial lipoamide dehydrogenase from *Arabidopsis thaliana.* *Plant Physiol.* 127:615–23

41. Machado RS, Clark DP, Guest JR. 1992. Construction and properties of pyruvate dehydrogenase complexes with up to nine lipoyl domains per lipoate acetyltransferase chain. *FEMS Microbiol. Lett.* 79:243–48

42. McCartney RG, Rice JE, Sanderson SJ, Bunik V, Lindsay H, Lindsay JG. 1998. Subunit interactions in the mammalian α-ketoglutarate dehydrogenase complex. *J. Biol. Chem.* 273:24158–64

43. Miernyk JA, Camp PJ, Randall DD. 1985. Regulation of plant pyruvate dehydrogenase complexes. *Curr. Top. Plant Biochem. Physiol.* 4:175–90

44. Miernyk JA, Randall DD. 1987. Some properties of pea mitochondrial phosphopyruvate dehydrogenase-phosphatase. *Plant Physiol.* 83:306–10

45. Miernyk JA, Thelen JJ, Randall DD. 1998. Reversible phosphorylation as a mechanism for regulating activity of the mitochondrial pyruvate dehydrogenase complex. In *Plant Mitochondria: From Gene to Function,* ed. IM Møller, P Gardestrom, K Glimelius, E Glaser, pp. 321–27. Leiden, The Neth.: Backhuys

46. Millar AH, Hill SA, Leaver CJ. 1999. Plant mitochondrial 2-oxoglutarate dehydrogenase complex: purification and characterization in potato. *Biochem. J.* 343:327–34

47. Millar AH, Leaver CJ, Hill SA. 1999. Characterization of the dihydrolipoamide acetyltransferase of the mitochondrial pyruvate dehydrogenase complex from potato and comparisons with similar enzymes in diverse plant species. *Eur. J. Biochem.* 264:1–10

48. Mooney BP, David NR, Thelen JJ, Miernyk JA, Randall DD. 2000. Histidine modifying agents abolish pyruvate dehydrogenase complex kinase activity. *Biochem. Biophys. Res. Commun.* 267:500–3

49. Mooney BP, Henzl MT, Miernyk JA, Randall DD. 2000. The dihydrolipoyl acyltransferase (BCE2) subunit of the plant branched-chain α-ketoacid dehydrogenase complex forms a 24-mer core with octagonal symmetry. *Protein Sci.* 9:1334–39

50. Mooney BP, Miernyk JA, Randall DD. 1998. Nucleotide sequence of a cDNA encoding the dihydrolipoylacyltransferase (E2) subunit of the branched-chain α-keto acid dehydrogenase complex from *Arabidopsis thaliana* (Accession No. AF038505) (PGR 98-071). *Plant Physiol.* 117:331

51. Mooney BP, Miernyk JA, Randall DD. 1998. Nucleotide sequence of a cDNA encoding the E1α subunit of the branched-chain α-keto acid dehydrogenase complex from *Arabidopsis thaliana*

(Accession No. AF077955) (PGR98-168). *Plant Physiol.* 118:711

52. Mooney BP, Miernyk JA, Randall DD. 1999. Cloning and characterization of the dihydrolipoamide S-acetyltransferase subunit of the pyruvate dehydrogenase complex (E2) from Arabidopsis. *Plant Physiol.* 120:443–51

53. Moreno JI, David NR, Miernyk JA, Randall DD. 2000. *Pisum sativum* mitochondrial pyruvate dehydrogenase can be assembled as a functional $\alpha_2\beta_2$ heterotetramer in the cytoplasm of *Pichia pastoris*. *Protein Expr. Purif.* 19:276–83

54. Moreno JI, Miernyk JA, Randall DD. 2000. Staphylococcal protein A as a fusion partner directs secretion of the E1α and E1β subunits of pea mitochondrial pyruvate dehydrogenase by *Bacillus subtilis*. *Protein Expr. Purif.* 18:242–48

55. Odessey R. 1982. Purification of rat kidney branched-chain oxo acid dehydrogenase complex with endogenous kinase activity. *Biochem. J.* 204:353–56

55a. Patel MS, Roche TE, Harris RA, eds. 1996. *Alpha-Keto Acid Dehydrogenase Complexes*. Basel: Birkhauser

56. Paxton R, Scislowski PWD, Davis J, Harris RA. 1986. Role of branched-chain 2-oxoacid dehydrogenase and pyruvate dehydrogenase in 2-oxobutyrate metabolism. *Biochem. J.* 234:295–303

57. Pettit FH, Yeaman SJ, Reed LJ. 1978. Purification and characterization of branched-chain α-ketoacid dehydrogenase complex of bovine kidney. *Proc. Natl. Acad. Sci. USA* 75:1881–85

58. Randall DD, Miernyk JA, David NR, Gemel J, Luethy MH. 1996. Regulation of leaf mitochondrial pyruvate dehydrogenase complex activity by reversible phosphorylation. In *Protein Phosphorylation in Plants*, eds. PR Shewry, NG Halford, R Hooley, pp. 87–103. Oxford, UK: Clarendon

59. Randall DD, Miernyk JA, Fang TK, Budde RJA, Schuller KA. 1989. Regulation of the pyruvate dehydrogenase complex in plants. *Ann. NY Acad. Sci.* 573: 192–205

60. Randall DD, Rubin PM, Fenko M. 1977. Plant pyruvate dehydrogenase complex purification, characterization and regulation by metabolites and phosphorylation. *Biochim. Biophys. Acta* 485:336–49

61. Rapp BJ, Randall DD. 1980. Pyruvate dehydrogenase complex from germinating castor bean endosperm. *Plant Physiol.* 65:314–18

62. Reed LJ, Pettit FH, Eley MH, Hamilton L, Collins JH, Oliver RM. 1975. Reconstitution of the Escherichia coli pyruvate dehydrogenase complex. *Proc. Natl. Acad. Sci. USA* 72:3068–72

63. Reed LJ. 2001. A trail of research from lipoic acid to α-keto acid dehydrogenase complexes. *J. Biol. Chem.* 276:38329–36

64. Reid EE, Thompson P, Lyttle CR, Dennis DT. 1977. Pyruvate dehydrogenase complex from higher plant mitochondria and proplastids. *Plant Physiol.* 59:842–48

65. Roche TE, Baker J, Yan X, Hiromasa Y, Gong X, et al. 2001. Distinct regulatory properties of pyruvate dehydrogenase kinase and phosphatase isoforms. *Prog. Nucleic Acid Res. Mol. Biol.* 70:33–75

66. Rotte C, Stejskal F, Zhu G, Keithly JS, Keithly Martin. 2001. Pyruvate:NADP+ oxidoreductase from the mitochondrion of *Euglena gracilis* and from the Apicomplexan *Crytosproidium parvum*: a biochemical relic linking pyruvate metabolism in mitochondriate and amitochondriate protists. *Mol. Biol. Evol.* 18:710–20

67. Schuller KA, Randall DD. 1989. Regulation of pea mitochondrial pyruvate dehydrogenase complex: Does photorespiratory ammonium influence mitochondrial carbon metabolism? *Plant Physiol.* 89:1207–12

68. Stephens PE, Darlison MG, Lewis HM, Guest JR. 1983. The pyruvate dehydrogenase complex of *Escherichia coli* K12. Nucleotide sequence encoding the dihydrolipoamide acetyltransferase component. *Eur. J. Biochem.* 133:481–89

69. Stock AM, Robinson VL, Goudreau PN. 2000. Two-component signal transduction. *Annu. Rev. Biochem.* 69:183–215

70. The Arabidopsis Genome Initiative. 2000. Analysis of the genome sequence of the flowering plant *Arabidopsis thaliana. Nature* 408:796–815

71. Thekkumkara TJ, Ho L, Wexler ID, Pons G, Liu TC, Patel MS. 1988. Nucleotide sequence of a cDNA for the dihydrolipoamide acetyltransferase component of human pyruvate dehydrogenase complex. *FEBS Lett.* 240:45–48

72. Thelen JJ, Miernyk JA, Randall DD. 1998. Nucleotide and deduced amino acid sequences of the pyruvate dehydrogenase kinase from *Arabidopsis thaliana* (Accession No. AF039406) (PGR98-192). *Plant Physiol.* 118:1533

73. Thelen JJ, Miernyk JA, Randall DD. 1999. Molecular cloning and expression analysis of the mitochondrial pyruvate dehydrogenase from maize. *Plant Physiol.* 119:635–43

74. Thelen JJ, Miernyk JA, Randall DD. 2000. Pyruvate dehydrogenase kinase from *Arabidopsis thaliana*: a protein histidine kinase that phosphorylates serine residues. *Biochem. J.* 349:195–201

75. Thelen JJ, Muszynski MG, David NR, Luethy MH, Elthon TE, et al. 1999. The dihydrolipoamide S-acetyltransferase subunit of the mitochondrial pyruvate dehydrogenase complex from maize contains a single lipoyl domain. *J. Biol. Chem.* 274:21769–75

76. Thelen JJ, Muszynski MG, Miernyk JA, Randall DD. 1998. Molecular analysis of two pyruvate dehydrogenase kinases from maize. *J. Biol. Chem.* 273:26618–23

77. Thompson P, Bowsher CG, Tobin AK. 1998. Heterogeneity of mitochondrial protein biogenesis during primary leaf development in barley. *Plant Physiol.* 118:1089–99

78. Deleted in proof

79. Treede H-J, Heise K-P. 1986. Purification of the chloroplast pyruvate dehydrogenase from spinach and maize mesophyll. *Z. Naturforsch. Teil C* 41:149–55

80. Wedding RT, Black MK. 1971. Nucleotide activation of cauliflower alpha-ketoglutarate dehydrogenase. *J. Biol. Chem.* 246:1638–43

81. White JA, Todd J, Newman T, Focks N, Girke T, et al. 2000. A new set of Arabidopsis expressed sequence tags from developing seeds. The metabolic pathway from carbohydrates to seed oil. *Plant Physiol.* 124:1582–94

82. Williams M, Randall DD. 1979. Pyruvate dehydrogenase complex from chloroplasts of *Pisum sativum. Plant Physiol.* 64:1099–103

83. Wynn RM, Davie JR, Meng M, Chuang DT. 1996. Structure, function and assembly of mammalian branched-chain α-ketoacid dehydrogenase complex. See Ref. 55a, pp. 101–17

84. Wynn RM, Davie JR, Zhi W, Cox RP, Chuang DT. 1994. In vitro reconstitution of the 24-meric inner core of bovine mitochondrial branched-chain α-keto acid dehydrogenase complex: requirement for chaperonins GroEL and GroES. *Biochemistry* 33:8962–68

85. Yeaman SJ. 1989. The 2-oxoacid dehydrogenase complexes: recent advances. *Biochem. J.* 257:625–32

86. Young JC, Gould JA, Kola I, Iannello RC. 1998. Review: Pdha-2, past and present. *J. Exp. Zool.* 282:231–38

87. Zou J, Qi Q, Katavic V, Marillia EF, Taylor DC. 1999. Effects of antisense repression of an *Arabidopsis thaliana* pyruvate dehydrogenase kinase cDNA on plant development. *Plant Mol. Biol.* 41:837–49

Annu. Rev. Plant Biol. 2002. 53:377–98
DOI: 10.1146/annurev.arplant.53.100301.135227

MOLECULAR GENETICS OF AUXIN SIGNALING

Ottoline Leyser

Department of Biology, University of York, York YO10 5YW, United Kingdom;
e-mail: hmol1@york.ac.uk

Key Words plant hormones, SCF, Aux/IAA, ARF, ABP

■ **Abstract** The plant hormone auxin is a simple molecule similar to tryptophan, yet it elicits a diverse array of responses and is involved in the regulation of growth and development throughout the plant life cycle. The ability of auxin to bring about such diverse responses appears to result partly from the existence of several independent mechanisms for auxin perception. Furthermore, one prominent mechanism for auxin signal transduction involves the targeted degradation of members of a large family of transcriptional regulators that appear to participate in complex and competing dimerization networks to modulate the expression of a wide range of genes. These models for auxin signaling now offer a framework in which to test how each specific response to auxin is brought about.

CONTENTS

INTRODUCTION

Reviews about auxin traditionally start with a sentence about how very important auxin is to plant growth and development, followed by a despairing comment about how, despite more than a century of research, we know very little about how it works (28, 38). Certainly there has been plenty of time to establish a very complex phenomenology for auxin biology. Exogenous addition of auxin to plants, plant tissues, and plant cells elicits a multitude of responses (7, 16). These include changes in the transcription of various gene families, a range of electrophysiological responses, changes in the rates of cell division and cell elongation, and changes in tissue patterning and differentiation. Which response is triggered depends on a wide variety of factors including cell type, developmental stage, environmental conditions, and the concentration of auxin added. In untreated plants, a similarly wide range of events can be correlated with changes in auxin levels, sensitivity, or transport. This striking range of responses has been central to the auxin mystery. Is it that auxin does just one thing that is linked in an unknown way to all these different responses? Or is it that this simple molecule has astonishing biochemical multifunctionality? Recent progress in understanding the molecular mechanism of auxin signaling is beginning to answer these questions, and the answer seems to be that both explanations are to some extent correct.

AUXIN PERCEPTION

Auxin signaling is assumed to start with the perception of auxin by its interaction with some kind of receptor. Evidence suggests that there are multiple sites for auxin perception, and in this sense, auxin can be considered to be multifunctional in that the auxin signal appears to be transduced through various signaling pathways.

Auxin Binding Protein 1

The search for auxin receptors has naturally focused on the isolation and characterization of proteins that bind auxin (reviewed in 101). Although a variety of such proteins has been identified, conclusive evidence linking them to auxin responses has proved difficult to obtain, and the biochemistry of the proteins has been, depending on your perspective, intriguing, perplexing, or frustrating.

The best-characterized auxin binding protein is ABP1 (reviewed in 73), which was first described in maize (35). Excitement about the role of ABP1 in auxin perception is driven by the high specificity and affinity of its auxin binding, with a K_D for the synthetic auxin NAA of 5×10^{-8} M (73). However, almost none of its other properties are characteristic of a typical receptor. The protein has no homology to any other known receptor family, and the vast majority of it is retained in the endoplasmic reticulum, where the pH is too high for strong auxin binding (34, 92). Some ABP1 apparently escapes to the plasma membrane, where it mediates several cellular responses to applied auxin, including tobacco mesophyll protoplast hyperpolarization (57, 58), the expansion of tobacco and maize cells in culture (10, 44), tobacco mesophyll protoplast division (24), and stomatal closure (26). It is clear that ABP1 acts at the cell surface to mediate these responses because the exogenous addition of anti-ABP1 antibodies, which are unable to enter the cell, can interfere with the ability of auxin to induce the responses. In whole plants, transgenic approaches to change ABP1 levels have resulted in relatively modest phenotypic effects (5, 44). Phenotypes are in general limited to effects on the balance between cell division and cell expansion. For example, overexpression of

ABP1 in tobacco plants results in increases in leaf mesophyll cell size, without affecting final leaf size (44).

Many aspects of ABP1 biology remain mysterious, but recently two extremely important tools have been added to the collection available for the investigation of ABP1 function. First an insertional mutant in the *Arabidopsis ABP1* gene has been recovered, allowing the phenotype of complete loss of ABP1 function to be assessed for the first time (10). The phenotype of plants homozygous for the mutation is embryo lethality early in the globular stage. This demonstrates the essential role that ABP1 plays in plant growth, but it makes analysis of the postembryonic role of ABP1 difficult, requiring conditional mutations.

The second major advance is the solving of the crystalline structure of ABP1 to a resolution of 1.9 Angstroms (107). The combination of these new genetic and biochemical tools will allow better analysis of the events immediately up- and downstream of ABP1 (103) so that its role in auxin signaling can be clarified.

Intracellular Sites for Auxin Perception

The existing evidence suggests that there are multiple auxin receptors, and hence the work on ABP1 is expected to answer only part of the question of how the auxin signal is perceived. For example, although ABP1 appears to act at the cell surface, there is good evidence for intracellular perception of auxin, much of which is derived from comparing the effects of auxins that differ in their transport properties into and out of cells (e.g., 12). This approach has been strengthened by the isolation of mutants in *Arabidopsis* that differ in their ability to transport auxins. For example, the *AUX1* gene of *Arabidopsis* encodes a protein with homology to amino acid permeases and is thought to act as an auxin uptake carrier (6, 65). Loss of *AUX1* function results in a variety of phenotypes including auxin-resistant root elongation and reduced root gravitropism (6, 65, 81). The roots of *aux1* mutants are resistant to the effects of membrane-impermeable auxins such as 2,4-D. However, *aux1* mutants respond normally to the membrane-permeable auxin NAA, and addition of NAA to *aux1* mutant roots can restore graviresponse (65, 111). This suggests that intracellular auxin is important for root growth inhibition.

Potential intracellular auxin binding proteins have been identified. For example, a 57-kDa soluble auxin binding protein has been identified in rice (47). This protein appears to interact directly with the plasma membrane proton pumping ATPase, suggesting a very short signal transduction pathway from auxin to increased proton pumping, cell wall acidification, and hence cell elongation (48, 49).

Other Routes for Auxin Perception

There is considerable speculation about the possible role of auxin transporters in auxin signaling (for a review of auxin transport see 78). These proteins most certainly interact with auxin, and it is possible that auxin levels in the cell are monitored by measuring the flux of auxin through either influx or efflux carriers, or both, like counting sheep through a gate. It is also possible that specific transporter family members do not in fact act as transporters but rather have specialized receptor function. An interesting precedent is seen in the budding yeast *Saccharomyces cerevisiae*, where recent evidence suggests that amino acid levels are detected by a permease-like receptor, Ssy1 (22). That auxin signaling evolved from an amino acid signaling pathway is an attractive hypothesis, and hence it is interesting to note that the budding yeast amino acid signal is transduced from Ssy1, via a pathway dependent on regulated protein stability, to bring about changes in transcription (8, 40). Strong evidence to support a role for a very similar signal transduction pathway acting in the auxin response is now available (see below).

AUXIN SIGNAL TRANSDUCTION

Although there is currently little to link auxin perception with downstream events, rapid progress in the area of auxin signal transduction has been made recently through the use of genetic approaches in *Arabidopsis* (reviewed in 59). Large screens for mutants with altered auxin sensitivity were used to define genes whose normal function is required for wild-type auxin response. Among the loci defined by these screens are *AXR1*, *AXR2*, *AXR3*, *AXR4*, and *AXR6*. A sixth locus, *TIR1*, was originally identified because mutations in it result in resistance to inhibitors of polar auxin transport, but these mutations were subsequently found also to confer auxin resistance (85). The molecular basis for the *axr4* and *axr6* phenotypes (36, 37) has not yet been determined, but the remaining loci have been cloned and they appear to function in a single pathway involved in auxin-regulated ubiquitin-dependent protein turnover (Figure 1).

The Role of Ubiquitin-Mediated Protein Degradation

Ubiquitin-mediated proteolysis occurs through the conjugation of a multiubiquitin chain to target proteins, which are subsequently degraded by the 26S proteasome (reviewed in 43, 80, 102). Conjugation of ubiquitin is a three-step process. Ubiquitin first is activated by ubiqutin activating enzyme, E1. This requires ATP and results in the formation of a high-energy thiol ester linkage between ubiquitin and a conserved cysteine in the E1. Ubiquitin is then passed to a ubiquitin conjugating enzyme, E2, which acts in concert with a ubiquitin protein ligase, E3, to link ubiquitin to a lysine residue of the target protein. A fourth step in which a multiubiquitin chain is extended from this first ubiquitin may require a multiubiquitin chain assembly factor, E4 (4, 54).

Much of the substrate specificity in the selection of proteins for degradation lies in the E3 enzyme, and this is reflected in the fact that hundreds of E3s of various classes are encoded by the eukarytotic genomes so far sequenced (21, 80, 102, 110). Prominent among these are the so-called SCF-type E3s (reviewed in 21, 110). SCFs take their name from three of their subunits; they are multimeric enzymes consisting of at least four subunits. Two of these subunits, members of the Cullin (also known as CDC53) and RBX1 (also known as ROC1 or HRT1) families,

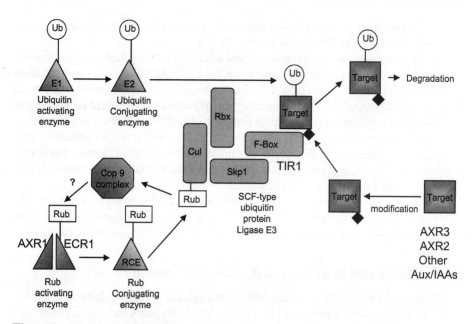

Figure 1 Auxin regulates the ubiquitination of target proteins, marking them for degradation by the 26S proteasome. The figure shows key components of this pathway. Ubiquitin (Ub) must be activated before conjugation to specific targets (*top left*). Target selection is mediated by the F-box–containing subunit of an SCF-type ubiquitin protein ligase (*center*). Auxin-regulated modification of the targets, which include the Aux/IAA proteins, is likely to be required for recognition by the F-box protein (*right-hand side*). Efficient activity of the SCF requires conjugation and deconjugation of a ubiquitin-related protein of the Rub family to the Cullin (Cul) subunit of the SCF (*bottom left*). Like ubiquitin, Rub must be activated before conjugation by a dimeric enzyme with homology to ubiquitin-activating enzyme. Deconjugation of Rub requires the Cop 9 complex. It is not clear if Rub protein is recycled during this process. In *Arabidopsis*, mutations, including *axr1*, *ecr1*, *axr2*, *axr3*, and *tir1*, in components of this pathway result in defective auxin response.

form a dimer able to catalyse multiubiquitin chain formation (88). Consistent with this function, RBX1 family members contain a RING-H2 finger domain common to many ubiquitin ligases (21). The Cullin/RBX1 dimer is linked to an F-box–containing protein through a member of the SKP1 protein family (15, 50, 79). The amino-terminal F-box motif, characteristic of F-box proteins, is required for the interaction between the F-box protein and the SKP1. The carboxyl-terminus of the F-box protein consists of any one of a variety of protein-protein interaction domains, and this domain interacts with the ubiquitination target and hence selects the substrate for degradation (50, 79). SCF-dependent signaling pathways are found throughout the plant, fungal, and animal kingdoms (15, 19, 21, 43, 79). For example, as mentioned above, in budding yeast, amino acid signaling is mediated through targeted degradation of transcription factors by an SCF complex including the F-box protein Grr1 (8, 40).

Crucial evidence for the involvement of this type of regulated protein turnover pathway in auxin signaling came when the *TIR1* gene was cloned and found to encode an F-box protein (85). TIR1 has subsequently been shown to participate in an SCF complex, SCF^{TIR1}, supporting the hypothesis that targeted protein degradation is required for normal auxin signaling (27). The auxin resistance phenotype conferred by complete loss of TIR1 function is relatively weak, but sequence comparisons with other *Arabidopsis* F-box proteins have identified several close homologues of TIR1, suggesting that TIR1 is likely to be functionally redundant in part (85, 110). The carboxyl-terminal domain of TIR1 and its close homologues consist of leucine-rich repeats, which are presumably involved in target selection (39, 50). So what then are the targets for SCF^{TIR1}? Recent evidence suggests that the products of the *AXR2* and *AXR3* genes interact with SCF^{TIR1} and are destabilized as a result.

SCF^{TIR1}-Mediated Instability of the Aux/IAA Protein Family

The *AXR2* and *AXR3* loci were defined by dominant mutations that confer a range of auxin-related phenotypes (61, 95, 96, 106). Both genes have been cloned and found to encode members of the Aux/IAA gene family (72, 84). Aux/IAA proteins are found throughout higher plants and are characterized by four highly conserved domains (reviewed in 3, 31) (Figure 2). The dominant mutations in the *axr2* and *axr3* alleles map to domain II (72, 84). Similar dominant mutations in domain II of other Aux/IAA family members have been recovered from a range of screens for auxin-related phenotypes such as tropism defects, photomorphogenesis defects, and altered root branching patterns (83, 93). The generality of these results indicates that domain II is of great importance for wild-type Aux/IAA function.

Several lines of evidence demonstrate that domain II functions in protein destabilization. Those Aux/IAA proteins that have been examined in detail are localized in the nucleus and have very short half-lives, ranging from a few minutes to a few hours (2, 29, 77). The fusion of Aux/IAA proteins to entirely unrelated reporter proteins, such as luciferase or beta glucuronase (GUS), results in the destabilization of the reporter protein (29, 108). This indicates that the Aux/IAAs contain a transferable destabilization sequence, a so-called degron. This degron has now been further defined as a 13–amino acid region from the core of domain II because fusion of just these 13 amino acids to luciferase is sufficient to confer a short half-life on the luciferase protein, providing that it is localized to the nucleus (82). A second line of evidence for the importance of domain II in regulating Aux/IAA stability comes from the analysis of the molecular basis for the phenotypes conferred by *axr3-1*, *axr2-1*, and similar Aux/IAA mutants. The dominant domain II mutations found in such alleles increase protein half-life dramatically, in a range of direct and indirect assays, without affecting nuclear localization (29, 77, 108). These data confirm that domain II and nuclear localization are necessary and sufficient to give the Aux/IAA proteins their short half-life and that this rapid degradation is required for a wild-type auxin response.

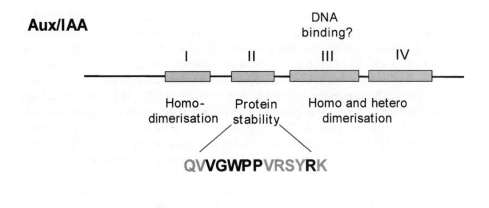

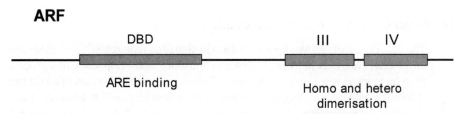

Figure 2 The linear structure of typical family members of the Aux/IAAs (*top*) and auxin response factors (ARFs) (*bottom*). The proteins share homology at their carboxyl termini in two domains (III and IV), which mediate homo- and heterodimerization. At the amino terminus, Aux/IAA proteins have two domains, one of which (II) is necessary and sufficient to mediate auxin-regulated destabilization of the protein. The consensus sequence of this domain is shown, with the invariant amino acids in black. ARFs have an amino-terminal DNA binding domain that binds to the auxin response elements (AREs) of auxin-regulated genes.

The requirement for both functional SCF^{TIR1} and rapid turnover of Aux/IAA proteins for normal auxin signaling has led to the hypothesis that Aux/IAAs are targeted for degradation by SCF^{TIR1}. Consistent with this idea, inhibitors of the 26S proteasome stabilize Aux/IAA-reporter fusion proteins, indicating that Aux/IAA destruction occurs via the 26S proteasome (29, 82). Evidence for the involvement of SCF^{TIR1} comes from two complementary experimental approaches (29). First, so-called pull-down assays suggest a physical interaction between SCF^{TIR1} and AXR3 and AXR2. Glutathione S-transferase (GST)–tagged AXR3 or AXR2 proteins were produced in bacteria and added to extracts from plants expressing c-*myc* epitope–tagged TIR1, and glutathione-agarose was used to purify any GST-associated proteins from these extracts. The c-*myc*-tagged TIR1 protein was among the proteins collected in this way, along with other generic SCF components. These data support a direct or indirect physical interaction between Aux/IAAs and SCF^{TIR1}. Second, genetic evidence suggests that this interaction targets the Aux/IAAs for degradation. Support for this idea comes from experiments involving the amino-terminal half of the AXR3 protein, including domain II and a nuclear localization sequence, fused to the GUS reporter protein and introduced into transgenic plants under the control of a heat-shock inducible promoter. After heat shock, a limited amount of the fusion protein accumulates, but this pool was rapidly depleted in comparison to GUS alone, confirming that fusion to the amino-terminal half of AXR3 can confer reduced protein stability on GUS. However, the stability of the fusion protein was greatly increased when the construct was crossed into a *tir1* mutant background. This suggests that the instability of Aux/IAA proteins requires TIR1 in vivo. Consistent with this idea, the dominant domain II mutations such as *axr3-1* or *axr2-1*, which confer auxin response phenotypes, simultaneously increase the stability of the mutant proteins and reduce or abolish their interaction with SCF^{TIR1}. Hence there is a tight correlation between Aux/IAA protein stability in planta and their ability to interact with SCF^{TIR1} in the pull-down assay (29).

The Effect of Auxin on Aux/IAA Stability

The data described above suggest that auxin signaling requires SCF^{TIR1}-mediated turnover of Aux/IAA proteins. In order to transduce auxin responses, this process must be in some way regulated by auxin. Indeed, exogenous addition of auxin can reduce the half-lives of Aux/IAA proteins below their already low level (29, 112). However, the mechanism by which auxin influences SCF^{TIR1} activity and Aux/IAA turnover is not clear. In theory, auxin could act at a variety of levels by influencing the recognition of the Aux/IAAs by SCF^{TIR1} or the overall activity of SCF^{TIR1} in ubiquitinating the Aux/IAAs. Alternatively, auxin could play a role in transferring the multiubiquitinated Aux/IAAs to the proteasome. Evidence to date is limited but in general supports a role for auxin in regulating the SCF^{TIR1}-Aux/IAA interaction because the auxin-induced destabilization of Aux/IAAs correlates with an increase in the abundance of Aux/IAA-SCF^{TIR1} complexes in the pull-down assay, without affecting the amount of TIR1 present (29). Indeed, the auxin dose response curve for the reduction in Aux/IAA stability and the dose response curve for Aux/IAA-SCF^{TIR1} interaction are remarkably similar (29). These observations suggest that auxin increases the affinity of Aux/IAAs for SCF^{TIR1}. This hypothesis is attractive because it mirrors the established mechanism of action of several similar targeted protein degradation signaling pathways in other species.

In most of the best-understood systems, the interaction of the F-box protein with the degradation target is dependent on the modification of the target, usually by phosphorylation (90). For example, phosphorylation of the NF-Kappa B transcription factors is essential for their recognition by an SCF-type ubiquitin-protein ligase and their subsequent degradation (reviewed in 45). One model then is that auxin regulates a kinase cascade that results in phosphorylation of the Aux/IAA proteins, increasing their affinity for SCF^{TIR1}. There is certainly an increasing body of biochemical and genetic evidence supporting a role for various kinases in auxin signaling. For example, auxin induces a MAP kinase activity in *Arabidopsis* roots (67), Aux/IAA proteins can be phosphorylated in vitro by the photoreceptor phytochrome A (14), and mutations in the PINOID serine-threonine protein kinase of *Arabidopsis* confer a range of auxin signaling defects (11). However, it is unlikely that direct phosphorylation of Aux/IAAs plays a role in their desta-bilization because the 13–amino acid region from domain II that is sufficient to confer auxin-inducible destabilization on a reporter protein does not contain any conserved phosphorylation sites (82, 112). Therefore, if the Aux/IAAs conform to the phosphorylation paradigm established in other species, the interaction between the Aux/IAA proteins and TIR1 must be indirect, being mediated by some kind of adaptor protein, with the interaction between the adaptor protein and the Aux/IAA being phosphorylation-dependent.

Although most well-characterized target-SCF interactions are phosphorylation dependent, counter examples are now beginning to emerge. One such example comes from the degradation of the human transcription factor, hypoxia-inducible factor (HIF), by an SCF-related E3 where substrate selection is mediated by the von Hippel–Lindau tumor suppressor protein (VHL) (reviewed in 43). HIF is required for the activation of a variety of responses to hypoxia, which are kept switched off when the oxygen supply is adequate by the rapid destruction of HIF by VHL-mediated ubiquitination. The interaction of HIF with VHL is dependent on the hydroxylation of a HIF proline residue by an oxygen-dependent proline hydroxylase (41, 42). It is tempting to speculate that a similar modification could regulate the interaction of the Aux/IAA degron and SCF^{TIR1} because the most striking feature of the consensus domain II degron is two consecutive, completely conserved proline residues. These prolines are clearly required for Aux/IAA in-stability because mutations in the prolines are the most common cause of the dominant Aux/IAA stabilizing mutations and are found in alleles such as *axr3-1* and *axr2-1* (72, 83, 84, 93).

Mechanisms of Regulation of SCF^{TIR1} Activity

SCF^{TIR1} ABUNDANCE   Apart from regulation of the interaction between the sub-strate and the SCF complex, the abundance or activity of the SCF complex pro-vides additional possible sites for regulation of the pathway, by either auxin or other interacting signals. The abundance of SCF^{TIR1} can certainly influence auxin sensitivity. Overexpression of *TIR1* in transgenic plants results in increased auxin responses (27). These auxin-response phenotypes are dependent on the additional *TIR1* participating in an SCF complex because overexpression of a *TIR1* variant in which the F-box had been mutated did not result in auxin response phenotypes (27). Furthermore, loss-of-function *tir1* alleles appear to be partially dominant, suggesting haploinsufficiency (85). Therefore auxin responses could in theory be brought about by changes in *TIR1* expression levels. This explanation could ac-count for some of the difference in auxin responsiveness observed between tissues because *TIR1* is not expressed uniformly throughout the plant (27).

RUB1 CONJUGATION TO SCF^{TIR1} Apart from regulation of SCF abundance, SCF activity can apparently be regulated in a variety of ways. New SCF-interacting components have recently been identified and are likely to have regulatory roles.

These components include the SKP1-interacting proteins SGT1 (51) and SnRK protein kinase (23). Another factor that clearly has a dramatic impact on SCF function is the modification of the cullin subunit by the conjugation of a ubiquitin-like protein called RUB1 (homologous to human NEDD8). The conjugation of RUB1 to cullin appears to increase the activity of the SCF (69, 109), and there is increasing evidence that cycles of RUB1 addition and removal are required for efficient SCF function (64, 87). RUB1 is added to cullin by a chain of events very similar to ubiquitination (62). RUB1 is activated by a dimeric RUB1 activating enzyme, the subunits of which are homologous to the amino- and carboxyl-terminal parts of ubiquitin activating enzyme (17, 18, 20, 56, 62). The activated RUB1 is passed to a RUB1 conjugating enzyme and then onto the cullin. It is unclear whether this process requires an E3-like enzyme, however the purified *Arabidopsis* RUB1 activating enzyme and an E2-like enzyme named RCE1 can conjugate RUB1 to AtCul1 in vitro, indicating that an E3-like activity is not essential (18). To date, cullins are the only proteins known to be modified in this way.

Although the precise function of RUB1 modification of cullin is unknown, genetic and biochemical evidence indicates that such modification is essential for wild-type auxin responses. The evidence originates from the analysis of one of the best-characterized auxin response mutants: those in the *AXR1* gene of *Arabidopsis*. These mutants are auxin resistant in every response so far analyzed and show morphological phenotypes consistent with a globally reduced sensitivity to auxin (30, 63, 91, 94). Importantly, Aux/IAA-reporter protein fusions are stabilized in an *axr1* mutant background (29), and *axr1*, *tir1* double mutants have a synergistic phenotype, indicating that AXR1 and TIR1 act in overlapping pathways to destabilize Aux/IAA proteins (85). This hypothesis is supported by the biochemical characterization of *AXR1*. *AXR1* encodes a subunit of a RUB1 activating enzyme, homologous to the amino-terminal half of ubiquitin activating enzyme (60) (Figure 1). In *axr1* loss-of-function alleles, RUB1 conjugation to AtCul1, and hence SCF^{TIR1}, is reduced (17, 20). Some conjugation still occurs, and this is likely to be through the activity of a gene closely homologous to *AXR1* that is found in the *Arabidopsis* genome (17). AXR1 acts in concert with a protein that is related to the carboxyl-terminal half of ubiquitin activating enzyme, named ECR1 (17, 18). The AXR1-ECR1 dimer activates RUB1 by the formation of a thiol ester linkage between RUB1 and a cysteine in ECR1 (20). Transgenic plants overexpressing a mutant form of ECR1, in which this cysteine is replaced by an alanine, show *axr1*-like auxin response phenotypes (17). Presumably the ECR1 cysteine to alanine substitution acts as a dominant negative mutation by titrating out AXR1 into inactive heterodimers. Consistent with this idea, the transgenic lines show reduced conjugation of RUB1 to AtCul1 (17). Taken together, these data indicate that RUB1 modification of SCF^{TIR1} is essential for normal auxin response.

RUB1 REMOVAL FROM SCF^{TIR1} Recent results suggest that removal of RUB1 from SCF^{TIR1} is also important for auxin signaling (87). RUB1 deconjugation is apparently mediated by the COP9 signalosome. The COP9 signalosome is a multiprotein complex found throughout the plant, fungal, and animal kingdoms, with the notable exception of budding yeast, which has only one of the subunits (reviewed in 86, 104). The COP9 signalosome was originally identified in *Arabidopsis* in screens for light signaling mutants (9). Mutation in any one of the COP9 subunits results in photomorphogenesis even in dark-grown seedlings. In addition to this phenotype, null mutations in the COP9 subunits also result in seedling lethality, but partial loss of function of the CSN5 subunit (also called JAB1) was recently achieved through an antisense approach, and plants carrying the transgene survived to adulthood despite reduced COP9 levels (87). Unexpectedly, the adult phenotype of these plants is reminiscent of the *axr1* mutant phenotype and includes auxin-resistant root elongation, reduced root branching, and increased shoot branching (87). When the antisense CSN5 construct was crossed into the *axr1-3* mutant background, a synergistic effect on the phenotype was observed, suggesting that the COP9 complex and AXR1 act in overlapping pathways (87). Consistent with this idea, the expression of an Aux/IAA-luciferase fusion protein in the CSN5 antisense lines revealed increased stability of the fusion protein compared to its stability in a wild-type background (87). Furthermore, the COP9 signalosome coimmunoprecipitates with SCF^{TIR1}, and AtCul1 interacts with the CSN2 subunit in the yeast two-hybrid system (87). Despite the similarities between the effects of AXR1/ECR1 and COP9 loss of function, paradoxically, COP9 signalosome mutants accumulate AtCul1-RUB1 conjugates, opposite to the effect observed in *axr1* and *ecr1* mutants. This suggests that not only RUB1 addition but also RUB1 removal from AtCul1 is required for full SCF^{TIR1} activity.

POSSIBLE BIOCHEMICAL FUNCTIONS FOR RUB1 CONJUGATION As mentioned above, the precise biochemical function of these proposed cycles of RUB1 conjugation and deconjugation to AtCul1 is not clear. Evidence to date suggests that the primary effect is on SCF activity and not on SCF-target recruitment (64). One possible role could be in subcellular localization. This is suggested by the observation that the cullin found localized to the centrosome of cultured mammalian cells is disproportionately NEDD8 modified (25). Furthermore, conjugation of the ubiquitin-like protein SUMO-1 of humans targets cytosolic RanGAP1 to the nuclear pore complex (70), although other functions for SUMO conjugation have also been identified (reviewed in 66). SCF activity and ubiquitination of Aux/IAAs requires nuclear localization, but it is possible that the degradtion occurs in the cytoplasm or in a nuclear subcompartment. Therefore, RUB1 modification could be required for subcellular targeting of the SCF. However, so far the data do not support this hypothesis because AtCul1 appears to be normally localized in the nucleus in *axr1* mutant plants where RUB1 conjugation is severely compromised (17).

An alternative hypothesis is that RUB1 conjugation is involved in mediating SCF-proteasome interactions. In this context it is interesting to note that the subunits of the COP9 signalosome, with which the SCF interacts, are homologous to the subunits of the lid subcomplex of the 26S proteasome (105). Evidence that the COP9 signalosome interacts with the proteasome also exists (55).

The activity of the RUB1 conjugation pathway in *Arabidopsis* is apparently unaffected by auxin addition, indicating that it is not a direct route for auxin signal transduction (17). Indeed, the fact that most, if not all, of the *axr1* phenotypes can be explained in terms of reduced auxin response is something of a mystery because it appears that AXR1 is involved in conjugation of RUB1 to all SCF complexes. Why then do *axr1* mutants show such specific auxin-response phenotypes? It seems likely that this reflects the particular sensitivity of auxin-regulated protein degradation to RUB1 modification of cullins. There is certainly evidence from yeast that different pathways are varyingly sensitive to loss of RUB1 modification activities. In fission yeast, the RUB1 conjugation-deconjugation pathway is essential for viability (71, 76). In contrast, budding yeast lacks most of the COP9 signalosome subunits (71), and the RUB1 conjugation pathway is not required for viability (56, 62). However, RUB1 conjugation is necessary for efficient functioning of SCF^{CDC4}, with other SCFs being apparently unaffected (56).

Aux/IAA Protein Function

Whatever the mechanisms for auxin input into the system, the auxin-regulated degradation of Aux/IAA proteins is an essential part of the auxin response. An important question then is what are the effects of the auxin-induced changes in Aux/IAA protein abundance?

DIMERIZATION WITHIN THE AUX/IAA FAMILY   Aux/IAA proteins appear to act as transcriptional regulators through the formation of a variety of dimers. First, they can dimerize with other members of the Aux/IAA family (46). These interactions require domains III and IV, and possibly domain I (Figure 2) (46, 77). In *Arabidopsis*, there are at least 24 different Aux/IAAs (83). Although only a subset of Aux/IAAs are expressed in any one tissue (1) and not all Aux/IAA family members may dimerize with high affinity, the possibility of an enormous number of different Aux/IAA dimers still exists. The function of Aux/IAA dimers is not clear. Domain III and the surrounding region have homology to bacterial transcriptional repressor proteins of the beta alpha alpha class (68). Dimerization of such proteins results in the formation of an unusual sequence-specific DNA binding domain based on beta-sheet. It is therefore possible that Aux/IAA proteins can bind DNA directly and regulate transcription. Their nuclear location is certainly consistent with this idea, although to date there is no published evidence to support it directly.

DIMERIZATION BETWEEN AUX/IAAs AND AUXIN RESPONSE FACTORS (ARFs)   A second possibility is that dimerization of Aux/IAAs within the Aux/IAA family prevents dimerization of Aux/IAAs with other partners. The only other protein family known to be able to heterodimerize with Aux/IAAs is the auxin response factor (ARF) family of transcription factors (46, 97, 99) (Figure 2). ARFs interact with Aux/IAAs through carboxyl-terminal domains with homology to Aux/IAA domains III and IV. ARF proteins can certainly bind DNA directly through an amino-terminal B3-type DNA binding domain, similar to that found in the maize VP1 protein (97, 99). The ARFs tested to date bind specifically to the TGTCTC-containing auxin response element (ARE) of auxin-regulated genes. Synthetic palindromic or direct repeats of these six nucleotides are sufficient to bind ARFs and confer auxin regulation on the transcription of a reporter gene (97–99). Hence ARFs appear to mediate auxin-regulated gene expression through binding to AREs and Aux/IAAs have the potential to alter the transcription of auxin-regulated genes through interacting with ARFs.

ARFs also form a large gene family, consisting of at least 10 members in *Arabidopsis* (99). Much of the work to characterize the effects of ARFs and Aux/IAAs on transcription from auxin-regulated promoters has been carried out using a carrot cell suspension culture protoplast system in which ARF or Aux/IAA-derived genes are coexpressed with a synthetic or natural auxin-inducible promoter-reporter fusion (98, 100). Expression of ARFs in this system has produced an interesting range of results (98). ARFs can be grouped into subfamilies depending on their effect on gene expression in this system, which correlates with their predicted protein sequence between the amino-terminal DNA binding domain and the carboxyl-terminal dimerization domains. Expression of ARF1, which is P/S/T-rich in this region, suppresses both basal and auxin-inducible expression from ARE-containing promoters. In contrast, expression of ARF5, ARF6, ARF7, or ARF8, which are Q-rich in their middle regions, increases both basal and auxin-inducible expression from the ARE-containing promoters, whereas expression of ARF2, ARF3, ARF4, or ARF9 has no effect. These regulatory characteristics are independent of the ARE binding because essentially identical results were obtained when the ARF DNA binding domain was replaced with the DNA binding domain from the budding yeast GAL4 protein and the ARE-reporter gene fusion was replaced simultaneously with a GAL4 binding element-reporter gene fusion (98). These data suggest that auxin inducibility of ARE-containing promoters depends on the middle and C-terminal regions of the ARFs because a Q-rich middle region with a C-terminal Aux/IAA-like dimerization domain is sufficient to confer auxin inducibility on a GAL4 DNA binding domain-promoter element system.

When the DNA binding domain was removed from the various ARFs, and these truncated versions were introduced into a carrot protoplast with an ARE-reporter construct, the same effects on transcription as with the full-length proteins were observed (98). These effects were abolished if the dimerization domains were also removed (98). One explanation for these results is that the truncated ARFs can dimerize with endogenous ARFs that occupy the AREs through their DNA binding domains, and their middle regions can subsequently regulate transcription. Certainly, domain-swapping experiments suggest that the middle region determines whether an ARF will activate or repress transcription, and all these effects are dependent on the dimerization domains because expression of the middle regions alone did not affect transcription from the ARE-regulated reporter (98).

In a similar result using this assay, expression of several Aux/IAA family members has been shown to inhibit auxin-inducible transcription from ARE-containing promoters (100). This result is consistent with the idea that Aux/IAA dimerization with ARE-bound ARFs can prevent the ability of ARFs to active transcription in response to auxin. An attractive hypothesis is that this occurs by Aux/IAAs competing with ARFs for dimerization through domains III and IV. However, the presence of endogenous ARFs on the AREs in this system has not been proven, and there is currently no direct evidence to show ARF-Aux/IAA association on a promoter.

A MODEL FOR AUXIN-REGULATED GENE EXPRESSION   Despite the caveats, these results have led to a model to explain auxin-regulated transcription from ARE-containing promoters (Figure 3). In this model, it is proposed that ARFs permanently occupy the AREs of auxin-regulated genes, regardless of auxin levels. When auxin levels are low, Aux/IAA proteins are stable, and they dimerize with ARF proteins on the AREs, blocking ARF function. When auxin levels rise, the

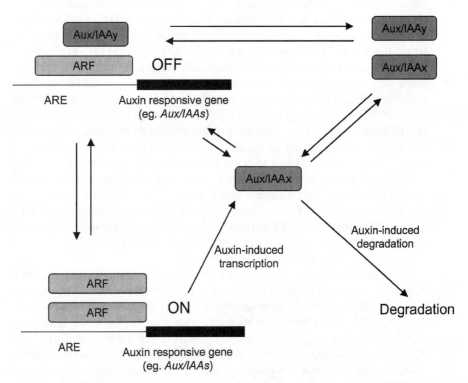

Figure 3   A model for Aux/IAA protein action. Aux/IAA proteins are able to form a variety of dimers both within the Aux/IAA family and with members of the ARF family (*top*). The equilibrium between these dimers has the potential to regulate transcription from ARE-containing promoters. The abundance of any one member of the Aux/IAA family (e.g., Aux/IAAx; *center*) is regulated by transcription and degradation, both of which are regulated by auxin (*bottom*). The dynamics of this regulatory network are clearly very complex.

Aux/IAAs are destabilized, and the resulting drop in Aux/IAA abundance allows ARF-ARF dimerization at the AREs and hence auxin-regulated transcription of the associated genes.

Although there is little direct evidence to support this model, a large body of compelling circumstantial evidence is accumulating. In addition to the work described above, mutations in two of the Q-rich ARFs have been recovered, and these confer a range of auxin-related phenotypes including reduced expression from ARE-containing auxin-inducible promoters (32, 33). This is consistent with the idea that Q-rich ARFs are required to activate expression of auxin-inducible genes in vivo. There is also good evidence that transcription from ARE-regulated genes is usually kept inactive by very unstable auxin-inactivated repressor proteins because transcription from ARE-containing promoters can be activated by either auxin or inhibitors of protein synthesis such as cycloheximide (53). The possibility that these unstable transcriptional inhibitors are the Aux/IAA proteins is supported by the observation that the dominant, stabilizing mutations in domain II of the Aux/IAA proteins often result in the constitutive repression of transcription from auxin-inducible ARE-containing promoters (1, 72, 75, 83).

Although this model is compelling, it is very much complicated by the fact that most *Aux/IAA* genes contain ARE elements in their promoters and are consequently auxin inducible at the level of transcription (reviewed in 3, 31). In fact, the *Aux/IAA* gene family was originally defined because of this auxin inducibility, with some members being induced within a few minutes of auxin addition (reviewed in 3, 31). Hence at the same time as their protein levels are being depleted by auxin-induced SCF^{TIR1}-mediated destabilization, they are being replenished by increased transcript accumulation. The auxin-induced increases in transcript levels often persist for many hours and outlast more modest increases in Aux/IAA protein levels (1, 74). This is not the pattern of message and protein accumulation that would be expected if Aux/IAA proteins repress their own transcription. On the contrary if this were the case, then Aux/IAA mRNA levels should decrease as Aux/IAA protein levels rise.

However, these predictions do not take into account the fact that all Aux/IAAs are different, with each Aux/IAA protein potentially regulating the transcription of a specific subset of *Aux/IAA* and other auxin-responsive genes. For example, the transcription of a very rapidly induced *Aux/IAA* gene may be inactivated by the product of an *Aux/IAA* family member that is induced much later after the increase in auxin levels. Such specificity of Aux/IAA gene function is demonstrated by several lines of evidence. First, very similar stabilizing mutations in different Aux/IAA genes confer quite different and even opposite phenotypes. For example, the *axr3-1* mutant has increased adventitious rooting, whereas the *axr2-1* mutant has fewer adventitious roots than wild type (72). Furthermore, the transcriptional inductions of different Aux/IAA family members show different temporal patterns, different auxin-dose response kinetics, and different tissue specificities (1). The transcription of some members requires new protein synthesis, and some are not even auxin induced (1, 83). In addition, if there is competition for dimer formation both within and between the Aux/IAAs and the ARFs, then depending on the relative affinities of particular protein combinations, increased levels of some Aux/IAAs could increase Aux/IAA-Aux/IAA dimerization at the expense of Aux/IAA-ARF dimerization. Hence ARF-ARF dimerization might be increased (Figure 3). In support of this idea, stabilizing mutations in some Aux/IAAs appear to result in increased auxin responses, including ectopic activation of auxin-inducible genes (13, 52, 61).

In *Arabidopsis*, there are at least 24 Aux/IAAs, which can have different expression patterns, half-lives, auxin-induced destabilization dynamics, and dimerization affinities, and at least 10 ARFs with similar variations in patterns of expression, effects on transcription, and dimerization affinities. Hence when auxin levels change, the resulting changes in Aux/IAA stability can trigger a mind-boggling array of possible effects on transcription. In this way, this model for auxin response perhaps can go some way toward explaining how such diverse responses can be induced by a simple signal, acting through a single signaling pathway. The diversity of possible responses is expanded still further by considering the possibility of additional degradation targets for SCF^{TIR1} and its homologues. For example, the stability of the auxin efflux carrier PIN2/EIR1 appears to be regulated by auxin in an AXR1-dependent manner (89).

CONCLUSIONS

These recent advances provide a framework in which to understand auxin signaling. They highlight four key challenges remaining. First, a better characterization of the events that link auxin to changes in Aux/IAA stability is required. Second, although the current model encompasses sufficient complexity to allow for diverse responses to auxin, the exact manner in which this complexity is encoded to produce each specific auxin response is still unknown. Cracking this code will require an understanding of which Aux/IAAs and ARFs are involved in each auxin response and of the changes in interactions that occur between them in response to changing auxin levels and hence changing Aux/IAA levels. Third, it will be necessary to determine the mechanisms by which the resulting changes in gene expression mediate specific auxin responses. Fourth, some auxin responses are likely to be independent of the SCF system, and these alternative pathways also need to be investigated.

Visit the Annual Reviews home page at www.annualreviews.org

LITERATURE CITED

1. Abel S, Nguyen MD, Theologis A. 1995. The Ps-IAA4/5-like family of early auxin-inducible messenger-RNAs in *Arabidopsis thaliana*. *J. Mol. Biol.* 251:533–49
2. Abel S, Oeller PW, Theologis A. 1994.

Early auxin-induced genes encode short-lived nuclear proteins. *Proc. Natl. Acad. Sci. USA* 91:326–30
3. Abel S, Theologis A. 1996. Early genes and auxin action. *Plant Physiol.* 111:9–7

4. Azevedo C, Santos-Rosa MJ, Shirasu K. 2001. The U-box protein family in plants. *Trends Plant Sci.* 6:354–38

5. Bauly JM, Sealy IM, Macdonald H, Brearley J, Droge S, et al. 2000. Overexpression of auxin-binding protein enhances the sensitivity of guard cells to auxin. *Plant Physiol.* 124:1229–38

6. Bennett MJ, Marchant A, Green HG, May ST, Ward SP, et al. 1996. *Arabidopsis AUX1* gene: a permease-like regulator of root gravitropism. *Science* 273:948–50

7. Berleth T, Sachs T. 2001. Plant morphogenesis: long-distance coordination and local patterning. *Curr. Opin. Plant Biol.* 4: 57–62

8. Bernard F, Andre B. 2001. Ubiquitin and the SCF^{Grr1} ubiquitin ligase complex are involved in the signaling pathway activated by external amino acids in *Saccharomyces cerevisiae. FEBS Lett.* 496:81–85

9. Chamovitz DA, Wei N, Osterlund MT, vonArnim AG, Staub JM, et al. 1996. The COP9 complex, a novel multisubunit nuclear regulator involved in light control of a plant developmental switch. *Cell* 86:115–21

10. Chen JG, Ullah H, Young JC, Sussman MR, Jones AM. 2001. ABP1 is required for organized cell elongation and division in *Arabidopsis* embryogenesis. *Genes Dev.* 15:902–11

11. Christensen SK, Dagenais N, Chory J, Weigel D. 2000. Regulation of auxin response by the protein kinase PINOID. *Cell* 100:469–78

12. Claussen M, Luthen H, Bottger M. 1996. Inside or outside? Localization of the receptor relevant to auxin-induced growth. *Physiol. Plant* 98:861–67

13. Collett CE, Harberd NP, Leyser O. 2000. Hormonal interactions in the control of hypocotyl elongation. *Plant Physiol.* 124: 553–62

14. Colon-Carmona A, Chen DL, Yeh KC, Abel S. 2000. Aux/IAA proteins are phosphorylated by phytochrome in vitro. *Plant Physiol.* 124:1728–38

15. Craig KL, Tyers M. 1999. The F-box: a new motif for ubiquitin dependent proteolysis in cell cycle regulation and signal transduction. *Prog. Biophys. Mol. Biol.* 72:299–328

16. Davies PJ, ed. 1995. *Plant Hormone Physiology, Biochemistry and Molecular Biology*. Dordrecht: Kluwer Acad. Publ.

17. del Pozo JC, Dharmasiri S, Hellmann H, Walker L, Gray WM, et al. 2002. AXR1-ECR1-dependent conjugation of RUB1 to the *Arabidopsis* cullin AtCul1 is required for auxin response. *Plant Cell.* In press

18. del Pozo JC, Estelle M. 1999. The *Arabidopsis* cullin AtCUL1 is modified by the ubiquitin-related protein RUB1. *Proc. Natl. Acad. Sci. USA* 96:15342–47

19. del Pozo JC, Estelle M. 2000. F-box proteins and protein degradation: an emerging theme in cellular regulation. *Plant Mol. Biol.* 44:123–28

20. del Pozo JC, Timpte C, Tan S, Callis J, Estelle M. 1998. The ubiquitin-related protein RUB1 and auxin response in *Arabidopsis. Science* 280:1760–63

21. Deshaies RJ. 1999. SCF and cullin/RING-H2-based ubiquitin ligases. *Annu. Rev. Cell Dev. Biol.* 15:435–67

22. Didion T, Regenberg B, Jorgensen MU, Kielland-Brandt MC, Andersen HA. 1998. The permease homologue Ssy1p controls the expression of amino acid and peptide transporter genes in *Saccharomyces cerevisiae. Mol. Microbiol.* 27: 643–50

23. Farras R, Ferrando A, Jasik J, Kleinow T, Okresz L, et al. 2001. SKP1-SnRK protein kinase interactions mediate proteasomal binding of a plant SCF ubiquitin ligase. *EMBO J.* 20:2742–56

24. Fellner M, Ephritikhine G, Barbier-Brygoo H, Guern J. 1996. An antibody raised to a maize auxin-binding protein has inhibitory effects on cell division of tobacco mesophyll protoplasts. *Plant Physiol. Biochem.* 34:133–38

25. Freed E, Lacey KR, Huie P, Lyapina SA, Deshaies RJ, et al. 1999. Components of

an SCF ubiquitin ligase localize to the centrosome and regulate the centrosome duplication cycle. *Genes Dev.* 13:2242–57

26. Gehring CA, McConchie RM, Venis MA, Parish RW. 1998. Auxin-binding protein antibodies and peptides influence stomatal opening and alter cytoplasmic pH. *Planta* 205:581–86

27. Gray WM, del Pozo JC, Walker L, Hobbie L, Risseeuw E, et al. 1999. Identification of an SCF ubiquitin-ligase complex required for auxin response in *Arabidopsis thaliana*. *Genes Dev.* 13:1678–91

28. Gray WM, Estelle M. 2000. Function of the ubiquitin-proteasome pathway in auxin response. *Trends Biochem. Sci.* 25:133–38

29. Gray WM, Kepinski S, Rouse D, Leyser O, Estelle M. 2001. Auxin regulates SCF^{TIR1}-dependent degradation of Aux/IAA proteins. *Nature* 414:271–76

30. Gray WM, Ostin A, Sandberg G, Romano CP, Estelle M. 1998. High temperature promotes auxin-mediated hypocotyl elongation in Arabidopsis. *Proc. Natl. Acad. Sci. USA* 95:7197–202

31. Guilfoyle T, Hagen G, Ulmasov T, Murfett J. 1998. How does auxin turn on genes? *Plant Physiol.* 118:341–47

32. Hardtke CS, Berleth T. 1998. The *Arabidopsis* gene *MONOPTEROS* encodes a transcription factor mediating embryo axis formation and vascular development. *EMBO J.* 17:1405–11

33. Harper RM, Stowe-Evans EL, Luesse DR, Muto H, Tatematsu K, et al. 2000. The *NPH4* locus encodes the auxin response factor ARF7, a conditional regulator of differential growth in aerial *Arabidopsis* tissue. *Plant Cell* 12:757–70

34. Henderson J, Bauly JM, Ashford DA, Oliver SC, Hawes CR, et al. 1997. Retention of maize auxin-binding protein in the endoplasmic reticulum: quantifying escape and the role of auxin. *Planta* 202:313–23

35. Hertel R, Thomson K-St, Russo VEA.

1972. In vitro auxin binding to particulate cell fractions from corn coleoptiles. *Planta* 107:325–40

36. Hobbie L, Estelle M. 1995. The *axr4* auxin-resistant mutants of *Arabidopsis thaliana* define a gene important for root gravitropism and lateral root initiation. *Plant J.* 7:211–20

37. Hobbie L, McGovern M, Hurwitz LR, Pierro A, Liu NY, et al. 2000. The *axr6* mutants of *Arabidopsis thaliana* define a gene involved in auxin response and early development. *Development* 127:23–32

38. Hooley R. 1998. Auxin signaling: homing in with targeted genetics. *Plant Cell* 10:1581–83

39. Hsiung WG, Chang HC, Pellequer JL, La Valle R, Lanker S, et al. 2001. F-box protein Grr1 interacts with phosphorylated targets via the cationic surface of its leucine-rich repeat. *Mol. Cell. Biol.* 21:2506–20

40. Iraqui I, Vissers S, Bernard F, De Craene JO, Boles E, et al. 1999. Amino acid signaling in *Saccharomyces cerevisiae*: a permease-like sensor of external amino acids and F-box protein Grr1p are required for transcriptional induction of the *AGP1* gene, which encodes a broad-specificity amino acid permease. *Mol. Cell. Biol.* 19:989–1001

41. Ivan M, Kondo K, Yang HF, Kim W, Valiando J, et al. 2001. HIF alpha targeted for VHL-mediated destruction by proline hydroxylation: implications for O_2 sensing. *Science* 292:464–68

42. Jaakkola P, Mole DR, Tian YM, Wilson MI, Gielbert J, et al. 2001. Targeting of HIF-alpha to the von Hippel–Lindau ubiquitylation complex by O_2-regulated prolyl hydroxylation. *Science* 292:468–72

43. Jackson PK, Eldridge AG, Freed E, Furstenthal L, Hsu JY, et al. 2000. The lore of the RINGs: substrate recognition and catalysis by ubiquitin ligases. *Trends Cell Biol.* 10:429–39

44. Jones AM, Im KH, Savka MA, Wu MJ, DeWitt NG, et al. 1998. Auxin-dependent

cell expansion mediated by overexpressed auxin-binding protein 1. *Science* 282: 1114–17

45. Karin M, Ben-Neriah Y. 2000. Phosphorylation meets ubiquitination: the control of NF-kappa B activity. *Annu. Rev. Immunol.* 18:621–63

46. Kim J, Harter K, Theologis A. 1997. Protein-protein interactions among the Aux/IAA proteins. *Proc. Natl. Acad. Sci. USA* 94:11786–91

47. Kim YS, Kim DH, Jung J. 1998. Isolation of a novel auxin receptor from soluble fractions of rice (*Oryza sativa* L.) shoots. *FEBS Lett.* 438:241–44

48. Kim YS, Kim D, Jung J. 2000. Two isoforms of soluble auxin receptor in rice (*Oryza sativa* L.) plants: binding property for auxin and interaction with plasma membrane H+-ATPase. *Plant Growth Regul.* 32:143–50

49. Kim YS, Min JK, Kim D, Jung J. 2001. A soluble auxin-binding protein, ABP (57)—purification with anti-bovine serum albumin antibody and characterization of its mechanistic role in the auxin effect on plant plasma membrane H+-ATPase. *J. Biol. Chem.* 276:10730–36

50. Kishi T, Yamao F. 1998. An essential function of Grr1 for the degradation of Cln2 is to act as a binding core that links Cln2 to Skp1. *J. Cell Sci.* 111:3655–61

51. Kitagawa K, Skowyra D, Elledge SJ, Harper JW, Hieter P. 1999. SGT1 encodes an essential component of the yeast kinetochore assembly pathway and a novel subunit of the SCF ubiquitin ligase complex. *Mol. Cell* 4:21–33

52. Knee EM, Hangarter RP. 1996. Differential IAA dose response relations of the *axr1* and *axr2* mutants of *Arabidopsis*. *Physiol. Plant* 98:320–24

53. Koahiba T, Ballas N, Wong L-M, Theologis A. 1995. Transcriptional regulation of *PS-IAA4/5* and *PS-IAA6* early gene expression by indoleacetic acid and protein synthesis inhibitors in pea (*Pisum sativum*). *J. Mol. Biol.* 253:396–413

54. Koegl M, Hoppe T, Schlenker S, Ulrich HD, Mayer TU, et al. 1999. A novel ubiquitination factor, E4, is involved in multiubiquitin chain assembly. *Cell* 96:635–44

55. Kwok SF, Staub JM, Deng XW. 1999. Characterization of two subunits of *Arabidopsis* 19S proteasome regulatory complex and its possible interaction with the COP9. *J. Mol. Biol.* 285:85–95

56. Lammer D, Mathias N, Laplaza JM, Jiang W, Liu Y, et al. 1998. Modification of yeast Cdc53p by the ubiquitin-related protein Rub1p affects function of the SCF^{Cdc4} complex. *Genes Dev.* 12:914–26

57. Leblanc N, David K, Grosclaude J, Pradier JM, Barbier-Brygoo H, et al. 1999. A novel immunological approach establishes that the auxin-binding protein, Nt-abp1, is an element involved in auxin signaling at the plasma membrane. *J. Biol. Chem.* 274:28314–20

58. Leblanc N, Perrot-Rechenmann C, Barbier-Brygoo H. 1999. The auxin-binding protein Nt-ERabp1 alone activates an auxin-like transduction pathway. *FEBS Lett.* 449:57–60

59. Leyser HMO. 1997. Auxin: lessons from a mutant weed. *Physiol. Plant* 100:407–14

60. Leyser HMO, Lincoln CA, Timpte CS, Lammer D, Turner JC, et al. 1993. *Arabidopsis* auxin-resistance gene *AXR1* encodes a protein related to ubiquitin-activating enzyme E1. *Nature* 304:161–64

61. Leyser HMO, Pickett FB, Dharmasiri S, Estelle M. 1996. Mutations in the *AXR3* gene of *Arabidopsis* result in altered auxin response including ectopic expression from the *SAUR-AC1* promoter. *Plant J.* 11:403–13

62. Liakopoulos D, Doenges G, Matuschewski K, Jentsch S. 1998. A novel protein modification pathway related to the ubiquitin system. *EMBO J.* 17:2208–14

63. Lincln C, Britton JH, Estelle M. 1990. Growth and development of the *axr1* mutants of *Arabidopsis*. *Plant Cell* 2:1071–80

64. Lyapina S, Cope G, Shevchenko A, Serino

G, Tsuge T, et al. 2001. Promotion of NEDD8-CUL1 conjugate cleavage by COP9 signalosome. *Science* 292:1382–85

65. Marchant A, Kargul J, May ST, Muller P, Delbarre A, et al. 1999. AUX1 regulates root gravitropism in *Arabidopsis* by facilitating auxin uptake within root apical tissues. *EMBO J.* 18:2066–73

66. Melchior F. 2000. SUMO—nonclassical ubiquitin. *Annu. Rev. Cell Dev. Biol.* 16: 591–626

67. Mockaitis K, Howell SH. 2000. Auxin induces mitogenic activated protein kinase (MAPK) activation in roots of *Arabidopsis* seedlings. *Plant J.* 24:785–96

68. Morgan KE, Zarembinski TI, Theologis A, Abel S. 1999. Biochemical characterization of recombinant polypeptides corresponding to the predicted beta alpha alpha fold in Aux/IAA proteins. *FEBS Lett.* 454:283–87

69. Morimoto M, Nishida T, Honda R, Yasuda H. 2000. Modification of cullin-1 by ubiquitin-like protein Nedd8 enhances the activity of SCF^{skp2} toward p27 (kip1). *Biochem. Biophys. Res. Commun.* 270:1093–96

70. Muller S, Matunis MJ, Dejean A. 1998. Conjugation with the ubiquitin-related modifier SUMO-1 regulates the partitioning of PML within the nucleus. *EMBO J.* 17:61–70

71. Mundt KE, Porte J, Murray JM, Brikos C, Christensen PU, et al. 1999. The COP9/signalosome complex is conserved in fission yeast and has a role in S phase. *Curr. Biol.* 9:1427–30

72. Nagpal P, Walker LM, Young JC, Sonawala A, Timpte C, et al. 2000. *AXR2* encodes a member of the Aux/IAA protein family. *Plant Physiol.* 123:563–73

73. Napier RM. 1995. Towards an understanding of ABP1. *J. Exp. Bot.* 46:1787–95

74. Oeller PW, Theologis A. 1995. Induction kinetics of the nuclear proteins encoded by the early indoleacetic acid-inducible genes *Ps-IAA4/5* and *Ps-IAA6* in pea (*Pisum sativum* L.). *Plant J.* 7:37–48

75. Oono Y, Chen QG, Overvoorde PJ, Kohler C, Theologis A. 1998. *age* mutants of Arabidopsis exhibit altered auxin-regulated gene expression. *Plant Cell* 10:1649–62

76. Osaka F, Saeki M, Katayama S, Aida N, Toh-e A, et al. 2000. Covalent modifier NEDD8 is essential for SCF ubiquitin-ligase in fission yeast. *EMBO J.* 19:3475–84

77. Ouellet F, Overvoorde PJ, Theologis A. 2001. IAA17/AXR3: biochemical insight into an auxin mutant phenotype. *Plant Cell* 13:829–41

78. Deleted in proof

79. Patton EE, Willems AR, Tyers M. 1998. Combinatorial control in ubiquitin-dependent proteolysis: don't Skp the F-box hypothesis. *Trends Genet.* 14:236–43

80. Pickart CM. 2001. Mechanisms underlying ubiquitination. *Annu. Rev. Biochem.* 70:503–33

81. Pickett FB, Wilson AK, Estelle M. 1990. The *aux1* mutation of *Arabidopsis* confers both auxin and ethylene resistance. *Plant Physiol.* 94:1462–66

82. Ramos JA, Zenser N, Leyser O, Callis J. 2001. Rapid degradation of Aux/IAA proteins requires conserved amino acids of domain II and is proteasome-dependent. *Plant Cell* 13:2349–60

83. Rogg LE, Lasswell J, Bartel B. 2001. A gain-of-function mutation in IAA28 suppresses lateral root development. *Plant Cell* 13:465–80

84. Rouse D, Mackay P, Stirnberg P, Estelle M, Leyser O. 1998. Changes in auxin response from mutations in an *AUX/IAA* gene. *Science* 279:1371–73

85. Ruegger M, Dewey E, Gray WM, Hobbie L, Turner J, et al. 1998. The TIR1 protein of *Arabidopsis* functions in auxin response and is related to human SKP2 and yeast Grr1p. *Genes Dev.* 12:198–207

86. Schwechheimer C, Deng XW. 2000. The COP/DET/FUS proteins—regulators of eukaryotic growth and development. *Semin. Cell Dev. Biol.* 11:495–500

87. Schwechheimer C, Serino G, Callis J,

Crosby WL, Lyapina S, et al. 2001. Interactions of the COP9 signalosome with the E3 ubiquitin ligase SCF^{TIR1} in mediating auxin response. *Science* 292:1379–82

88. Seol JH, Feldman RMR, Zachariae W, Shevchenko A, Correll CC, et al. 1999. Cdc53/cullin and the essential Hrt1 RING-H2 subunit of SCF define a ubiquitin ligase module that activates the E2 enzyme Cdc34. *Genes Dev.* 13:1614–26

89. Sieberer T, Seifert GJ, Hauser MT, Grisafi P, Fink GR, et al. 2000. Post-transcriptional control of the *Arabidopsis* auxin efflux carrier EIR1 requires AXR1. *Curr. Biol.* 10:1595–98

90. Skowyra D, Craig KL, Tyers M, Elledge SJ, Harper JW. 1997. F-box proteins are receptors that recruit phosphorylated substrates to the SCF ubiquitin-ligase complex. *Cell* 91:209–19

91. Stirnberg P, Chatfield SP, Leyser HMO. 1999. *AXR1* acts after lateral bud formation to inhibit lateral bud growth in *Arabidopsis*. *Plant Physiol.* 121:839–47

92. Tian HC, Klambt D, Jones AM. 1995. Auxin-binding protein-1 does not bind auxin within the endoplasmic reticulum despite this being the predominant subcellular location for this hormone receptor. *J. Biol. Chem.* 270:26962–69

93. Tian Q, Reed JW. 1999. Control of auxin-regulated root development by the *Arabidopsis thaliana SHY2/IAA3* gene. *Development* 126:711–21

94. Timpte C, Linclon C, Pickett FB, Turner J, Estelle M. The *AXR1* and *AUX1* genes of *Arabidopsis* function in separate auxin-response pathways. *Plant J.* 8:561–69

95. Timpte C, Wilson AK, Estelle M. 1994. The *axr2-1* mutation of *Arabidopsis thaliana* is a gain-of-function mutation that disrupts an early step in auxin response. *Genetics* 138:1239–49

96. Timpte CS, Wilson AK, Estelle M. 1992. Effects of the *axr2* mutation of *Arabidopsis* on cell shape in hypocotyl and inflorescence. *Planta* 188:271–78

97. Ulmasov T, Hagen G, Guilfoyle TJ. 1997.

ARF1, a transcription factor that binds to auxin response elements. *Science* 276: 1865–68

98. Ulmasov T, Hagen G, Guilfoyle TJ. 1999. Activation and repression of transcription by auxin-response factors. *Proc. Natl. Acad. Sci. USA* 96:5844–49

99. Ulmasov T, Hagen G, Guilfoyle TJ. 1999. Dimerization and DNA binding of auxin response factors. *Plant J.* 19:309–19

100. Ulmasov T, Murfett J, Hagen G, Guilfoyle TJ. 1997. Aux/IAA proteins repress expression of reporter genes containing natural and highly active synthetic auxin response elements. *Plant Cell* 9:1963–71

101. Venis MA, Napier RM. 1995. Auxin receptors and auxin binding proteins. *Crit. Rev. Plant Sci.* 14:27–47

102. Voges D, Zwickl P, Baumeister W. 1999. The 26S proteasome: a molecular machine designed for controlled proteolysis. *Annu. Rev. Biochem.* 68:1015–68

103. Warwicker J. 2001. Modelling of auxin-binding protein 1 suggests that its C-terminus and auxin could compete for a binding site that incorporates a metal ion and tryptophan residue 44. *Planta* 212:343–47

104. Wei N, Deng XW. 1999. Making sense of the COP9 signalosome—a regulatory protein complex conserved from *Arabidopsis* to human. *Trends Genet.* 15:98–103

105. Wei N, Tsuge T, Serino G, Dohmae N, Takio K, et al. 1998. The COP9 complex is conserved between plants and mammals and is related to the 26S proteasome regulatory complex. *Curr. Biol.* 8:919–22

106. Wilson AK, Pickett FB, Turner JC, Estelle M. 1990. A dominant mutation in *Arabidopsis* confers resistance to auxin, ethylene and abscisic acid. *Mol. Gen. Genet.* 222:377–83

107. Woo EJ, Bauly J, Chen JG, Marshall J, Macdonald H, et al. 2000. Crystallization and preliminary X-ray analysis of the auxin receptor ABP1. *Acta Crystallogr. D* 56:1476–78

108. Worley CK, Zenser N, Ramos J, Rouse D,

Leyser O, et al. 2000. Degradation of Aux/IAA proteins is essential for normal auxin signaling. *Plant J.* 21:553–62

109. Wu K, Chen A, Pan ZQ. 2000. Conjugation of Nedd8 to CUL1 enhances the ability of the ROC1-CUL1 complex to promote ubiquitin polymerization. *J. Biol. Chem.* 275:32317–24

110. Xiao WY, Jang JC. 2000. F-box proteins in *Arabidopsis*. *Trends Plant Sci.* 5:454–57

111. Yamamoto M, Yamamoto KT. 1998. Differential effects of 1-naphthaleneacetic acid, indole-3-acetic acid and 2,4-dichlorophenoxyacetic acid on the gravitropic response of roots in an auxin-resistant mutant of *Arabidopsis*, *aux1*. *Plant Cell Physiol.* 39:660–64

112. Zenser N, Ellsmore A, Leasure C, Callis J. 2001. Auxin modulates the degradation rate of Aux/IAA proteins. *Proc. Natl. Acad. Sci. USA.* 98:11795–800

Annu. Rev. Plant Biol. 2002. 53:399–419
DOI: 10.1146/annurev.arplant.53.092401.134447

RICE AS A MODEL FOR COMPARATIVE GENOMICS OF PLANTS

Ko Shimamoto and Junko Kyozuka

*Laboratory of Plant Molecular Genetics, Nara Institute of Science and Technology,
630-0101 Japan; e-mail: simamoto@bs.aist-nara.ac.jp, junko@bs.aist-nara.ac.jp*

Key Words gene function, evolution, reproductive development, defense

■ **Abstract** Rapid progress in rice genomics is making it possible to undertake detailed structural and functional comparisons of genes involved in various biological processes among rice and other plant species, such as *Arabidopsis*. In this review, we summarize the current status of rice genomics. We then select two important areas of research, reproductive development and defense signaling, and compare the functions of rice and orthologous genes in other species involved in these processes. The analysis revealed that apparently orthologous genes can also display divergent functions. Changes in functions and regulation of orthologous genes may represent a basis for diversity among plant species. Such comparative genomics in other plant species will provide important information for future work on the evolution of higher plants.

CONTENTS

1040-2519/02/0601-0399$14.00

INTRODUCTION

Rice is emerging as a model cereal for molecular biological studies (27, 43). The main reasons for this are as follows: (*a*) The complete genome has been sequenced. Although the complete rice genome sequence is the proprietary information of a private company, an international group is expected to generate complete genome sequences that will be accessible to anyone in the near future. In addition, a large expressed sequence tag (EST) database is available to researchers, facilitating a quick identification of genes of interest. (*b*) Tools for functional genomics are available. Transposon- and T-DNA-tagged rice populations exist, and the microarray technology for studying mRNA expression profiles is available. (*c*) Production of transgenic plants is relatively easy compared to that of other major cereals. Efficient use of *Agrobacterium*-mediated transformation (36) made routine use of transgenic rice possible for a variety of research purposes.

Because each topic listed above has been recently reviewed, this article does not review in detail the current progress in rice genomics. Because of the progress in rice genomics, a number of novel studies are becoming possible. One exciting area of research is the comparative analysis of gene functions among rice and other cereals and between rice and *Arabidopsis*. One important outcome of such comparative gene analysis is the insight that will be gained into the evolutionary aspects of gene functions in higher plants. Moreover, because rice genes are structurally and functionally homologous to a high degree with corresponding genes in other major cereals, knowledge obtained from rice genes can be used to study them in other cereals.

For this review, we chose two important areas of plant biology, reproductive development and defense signaling, and compared the structures and functions of rice genes with corresponding genes in other species, mainly those in *Arabidopsis*. Similar analyses with many other plant species will become possible in the future, and we hope that this review will provide one of the first attempts using available information on rice and *Arabidopsis* genomes.

RICE GENOMICS

Genome Sequencing

The complete sequencing of the genome of a japonica variety, Nipponbare, was reported by a private company in early 2001 although it has not been released to the public. The sequence information, which covers approximately 80% of the rice genome, was generated by another private company, and it is accessible to researchers worldwide. The International Rice Genome Sequencing Project (IRGSP) is under way, and it will completely sequence the genome of the same japonica variety in two years. Therefore, at the time of this writing (August 2001), a large body of rice sequence information is available worldwide to any researcher. Current information on the progress of the public effort is available on the World Wide

Web (http://rgp.dna.affrc.go.jp/). The amount of information on the rice genome that is currently publicly available is enormous, and it is of great help in the identification of genes for various research purposes (65). At the moment, comparable amounts of genome information are not available for any other monocots.

EST Database

With respect to the EST database of rice genes, approximately 50,000 ESTs are currently registered. Furthermore, sequencing of full-length cDNAs is under way by the Rice Genome Research Project (RGP) in Tsukuba, Japan, and these data will be available in the near future. The information on the full-length cDNA will be extremely important for the functional genomics and proteomics of rice, and because no complete full-length cDNA data are available for any other plant genomes, it will have great impact on future studies of plant genomics in general.

Gene-Tagged Lines

Several experimental approaches have been undertaken to develop rice lines in which genes are randomly tagged by insertion elements (31, 39, 41, 45). Since the first introduction of maize *Ac/Ds* into rice (40, 91), variously modified constructs have been introduced into rice (14, 32, 75). Their usefulness for the functional genomics of rice has been shown (22), but more information about their transposition in rice plants over several generations is needed to develop an efficient gene-tagging system. One major concern is that transpositions in rice may not be particularly stable, which would make it possible for *Ac/Ds* to become silent over several generations in the majority of the plant population (41). As a more systematic application of *Ac/Ds* is being tried on a larger scale, a more efficient use of the materials is anticipated for the functional identification of rice genes.

Tos17 is a retrotransposon of rice, and its transposition is activated through tissue culture (38). Because Tos17 has this property, it has been developed as a tag for the functional genomics of rice. Currently, a large population of plants derived from tissue cultures, in which *Tos 17* is distributed over the entire genome, is being developed (39). The *Tos 17* system is useful for tagging new genes based on mutant phenotypes as well as for screening knockouts of genes of known sequences. Because the *Tos 17* system is endogenous, an advantage of its use is that there is no special regulation for mutant screening when conducting a large-scale cultivation of plants.

Because *Agrobacterium*-mediated transformation is efficiently used for rice (36), T-DNA can also be used as a tag for rice genes. An analysis of T-DNA-tagged lines of rice was performed (47), and the current major question concerns the determination of the maximum background mutation rate caused by tissue culture. It would be ideal to have different gene-tagging systems for rice because each tagging system has advantages and disadvantages and each element may have preferences for sequences of insertion sites.

Microarray Technology

A microarray technology for rice gene-expression studies has been developed by RGP and is now being applied in a number of studies (115). Oligonucleotide-based probe array (GeneChip) technology, which has been commercially applied for *Arabidopsis* genes, is also being developed by a private company (119). We hope that these technologies will immediately be made available to researchers worldwide. Availability of these technologies for various gene expression studies will be essential for future studies on the functions of rice genes.

Positional Cloning

In addition to the use of various tagged lines for gene cloning, it is very useful to have methods to clone genes based on their map positions. There are rich sources of mutations in rice generated by various physical and chemical methods, and many of those have been mapped to specific locations of chromosomes. Recently, it has become feasible to isolate rice genes based on their map positions owing to the development of numerous numbers of molecular markers and the accumulation of practical knowledge required for cloning (4, 9, 94, 98, 104, 112, 118). The availability of a large amount of information on genomic sequences will have a significant impact on the speed with which gene cloning develops. Furthermore, less space for growing rice plants and less cost for mapping mutations are now required for positional cloning in rice. So far, fewer than 10 genes have been isolated by this method, but the number will increase rapidly with progress in the genome sequencing of rice. Progress in positional cloning of rice genes is very important because it is one of a few methods to isolate genes with novel functions.

One technical difficulty that had been associated with the positional cloning of rice genes was that rice transformation was used to confirm the functions of isolated candidate genes. Until recently, it had been difficult to introduce many gene constructs in rice for the functional identification of isolated genes. Improvement of *Agrobacterium*-mediated transformation methods (55) made such experiments possible within a reasonable length of time.

Rice genomics has progressed greatly in recent years. All the necessary tools for structural and functional studies of rice genes are rapidly being developed. These changes are expected to spur dramatic and prompt progress in the study of rice genes, and, without doubt, they will contribute toward the better general understanding of plant biology.

RICE REPRODUCTIVE DEVELOPMENT

Rice reproductive development exhibits several features that are not observed in *Arabidopsis*. One big difference between rice and *Arabidopsis* is found in their photoperiodic responses. Transition to the reproductive phase is induced under short-day (SD) conditions in rice, whereas it is enhanced under long day (LD) in *Arabidopsis*. Differences in their inflorescence morphologies are also of great interest. Furthermore, grass species show a highly derived flower morphology

(30, 66, 87). Molecular genetic approaches that mainly use *Arabidopsis* have revealed a number of key genes involved in the control of reproductive development (3, 62, 80, 107, 113). Knowledge obtained from the studies on dicot species is useful for the understanding of mechanisms controlling reproductive development of rice and other grass species. Furthermore, comparative analysis of key genes in diverse plant species, such as rice and *Arabidopsis*, should produce important insights into how morphologies and developmental patterns have evolved (18, 59). Here, we overview recent progress in molecular and genetic approaches for the elucidation of the mechanisms controlling reproductive development of rice and other grass species with an emphasis on the comparative aspects of genes and genetic systems.

Genes Involved in Floral Transition

Recently, several genes involved in the determination of flowering time have been isolated from rice. *Hd1* and *Hd6* correspond to qualitative trait loci (QTLs) controlling the heading date of rice, and both genes were isolated by the use of fine-scale high-resolution mapping (98, 112). As the whole genome sequencing project progresses in rice, cloning of genes by map-based techniques is becoming feasible for rice genes, as well as for QTLs. *Hd1* encodes a homolog of *CONSTANS (CO)*, which functions in the photoperiodic control of flowering in *Arabidopsis* (82). A study using natural allelic variants and transformation analysis indicates that *Hd1* might function in the promotion of flowering under SD conditions and in inhibition under LD conditions. This is in contrast to *CO*, which functions only under the LD condition, the promotive condition for flowering in *Arabidopsis*. *Hd6* is a weaker QTL than *Hd1*. It encodes an α subunit of casein kinase2 (CK2). Casein kinase2 interacts with the CCA1 protein, a putative circadian oscillator of *Arabidopsis* (105). Previous studies have suggested the possible involvement of CK2 in the photoperiodic control of flowering; however, so far, supporting evidence has been limited (96). Thus, the isolation of *CK2α* as *Hd6* provided direct evidence that CK2 was an important gene for the photoperiod control locus of rice. These studies indicated the possibility that similar genes (putative orthologs) are involved in the control of the day-length responses of flowering in short-day (rice) and long-day plants (*Arabidopsis*). Apparently, the next question is how different responses to photoperiods are generated using a conserved set of genes. In the *photoperiodic sensitivity 5 (se5)* mutant of rice, photoperiod sensitivity is completely lost, which results in very early flowering under both SD and LD (117). Cloning of the *SE5* gene by a candidate-cloning method revealed that it encodes a putative heme functioning in phytochrome chromophore biosynthesis (42). Because the photoperiod response is not greatly altered in the loss-of-function mutants of *HY1*, a putative *SE5* ortholog in *Arabidopsis* (74), the roles of phytochromes in the photoperiod control may be different in rice and *Arabidopsis*.

Rice Inflorescence Development

The grass inflorescence, called either a spike or a panicle, according to its branching pattern, is formed from the original apical meristem at the top of a plant.

Rice inflorescence is a panicle because it is highly branched (Figure 1, see color insert). Generally, in most angiosperm species, the arrangement of flowers and the branching pattern are used to describe the inflorescence form (106). In grass species, flowers, called florets, are produced into a group that is enclosed by a pair of small leaf-like structures called glumes (5). The group of florets enclosed by glumes is a spikelet, and an arrangement of spikelets, rather than florets, is used to describe grass inflorescence (20). Nevertheless, because the structure of spikelets is also a useful trait to classify the grass species, the number of florets produced in a spikelet and their arrangement are also genetically defined and are considered to be a species-specific characteristic. For example, in maize, short branches consisting of a pair of spikelets, each containing two florets, are produced on the main spike and lateral branches. Only a single floret is formed in a rice or a barley spikelet. On the other hand, a wheat spikelet contains several undefined numbers of florets.

During the inflorescence development in grass species, a series of axillary meristems with distinct identities is produced. The development of a rice panicle has four sequential steps. The first axillary structure produced by the primary shoot apical meristem after the phase transition results in several panicle branches (primary branches). The primary shoot apical meristem then degenerates prematurely after it forms several primary branches and leaves a scar called a degenerated point. This degeneration of the primary apical meristem is unique to rice and not observed in other grass species. The newly formed shoot apical meristems of primary branches initiate the next order of axillary meristems. The first few are secondary branch shoots, and the meristems produced afterward are lateral spikelet meristems. The transition from a spikelet meristem to a floret meristem is unclear in rice because a spikelet contains only a single floret. Finally, the shoot apical meristem of each branch is transformed to a terminal flower. Because the shoot apical meristem of the branch panicle is consumed to form a terminal flower, the growth of rice coinflorescences is determinate (Figure 1).

Meristem Identity Genes in Grasses

Recent molecular genetic approaches using mainly *Arabidopsis* have successfully revealed a number of key genes involved in the establishment of plant architecture. In particular, extensive progress has been made in the understanding of the genetic network controlling floral meristem initiation and development (62, 80, 107, 113). In *Arabidopsis*, the cooperative activity of the floral meristem identity genes, *LEAFY (LFY)* (108), *APETALA1 (AP1)* (64), and *CAULIFLOWER (CAL)* (53), assigns a floral fate to lateral meristems (23). On the other hand, *TFL1* is required for the inflorescence meristem to maintain the indeterminate state of the shoot apex (2, 7, 90).

From a comparative genomics point of view, the function of the homologs of the floral meristem identity genes in rice and other grass species is of great interest. It is possible that the changes of a few key genes account for the evolutionary variations found among diverse plant species, such as rice and *Arabidopsis*. Putative orthologs of dicot floral meristem identity genes have been isolated from several grass species (26, 44, 57, 58, 61, 68). We isolated two MADS box genes, *RAP1A* and *RAP1B*, showing extensive sequence similarities to *AP1* (57). The *RAP1A* was specifically expressed in very young floral meristems and outer whorls of the young rice floret, as *AP1* is in *Arabidopsis* (57, 64). On the other hand, the expression pattern of *RFL*, rice *LFY*, was distinct from that of *LFY* (58, 108). *RFL* expression started in the inflorescence meristem from a very early stage of rice panicle development, whereas *LFY* is expressed in floral meristems. In contrast to the situation in rice, a later onset of an *LFY* ortholog expression was observed in lolium, a grass species (26). This suggested the possibility that the genetic hierarchy controlling the expression of floral meristem identity genes may not be conserved in grasses. Ectopic expression of *LFY* in *Arabidopsis* confers a striking change in the inflorescence form from indeterminate to determinate with the production of a terminal flower in addition to an extreme early flowering phenotype (109). In contrast, ectopic expressions of the same gene in rice did not cause a dramatic change in panicle morphology but only conferred weak early-flowering phenotypes (35), suggesting a divergence in the genetic network controlling floral meristem initiation, especially in downstream targets of the LFY, a transcription factor, between the two species. In addition to these differences, the amino acid sequences of grass *LFY* homologs are also divergent. So far, *LFY* homologs have been isolated from two grass species, rice and ryegrass. Phylogenetic analysis using 42 *LFY* homologs indicated that the evolutionary rates of monocot *LFY* homologs are significantly higher than those of other angiosperms (37) (Figure 2), suggesting that the *LFY* homolog proteins in monocot species may have divergent functions. Taken together, the three aspects described here, expression patterns, phenotypes caused by the ectopic expression of *LFY*, and a function as a protein, grass *LFY* homologs diverge from *LFY*. *LFY* encodes a transcription factor unique to plants and a single-copy gene in the *Arabidopsis* genome (108). Results obtained so far suggest that *RFL* is also a single *LFY* homolog in the rice genome (J. Kyozuka, unpublished results). This makes it unlikely that another duplicated copy of the *LFY* homolog plays an orthologous role in the development of rice reproduction. To solve this puzzling situation concerning *RFL* function in rice, analysis of the loss-of-function phenotype of *RFL* will be required.

In rice *fzp* [described as *fzp2* (37)] mutant plants, spikelets are led to the indeterminate generation of meristems. Thus, the *fzp* phenotype can be interpreted as the transformation of the floral meristems to inflorescence shoots, as shown in the *lfy* mutant as well as mutants of *LFY* orthologs in other dicot species (19, 58, 71, 83, 95, 108). However, our analysis showed that it is very unlikely that *RFL* is the causal gene of the *fzp* mutant (56). Isolation of the *FZP* gene, which is now in progress, may provide a clue to understand the mechanisms controlling floral meristem initiation and the function of floral meristem identity genes in rice.

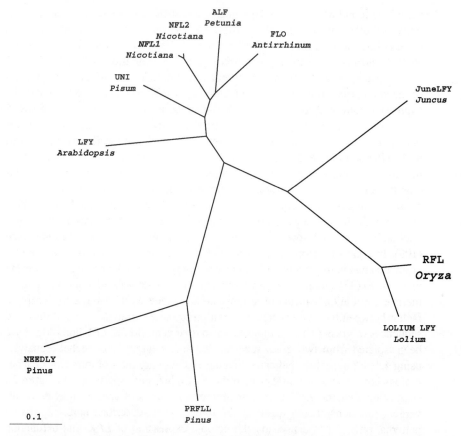

Figure 2 Phylogenetic tree of the *FLO/LFY* genes. Genus name is indicated beneath the name of each *FLO/LFY* homolog.

Determinacy of Meristems in Grass Inflorescences

The determinacy of the meristem is an important feature for the establishment of the inflorescence structure. When the growth of the main axis ends in a flower, the inflorescence is classified as determinate, whereas in an indeterminate inflorescence, the main axis continues to produce lateral structures without turning into a terminal flower. The primary and secondary inflorescences of *Arabidopsis* do not produce a terminal flower; thus, *Arabidopsis* inflorescences are indeterminate (Figure 1). This is also the case in maize inflorescences (66).

The fate of the branch meristem apices in rice inflorescences has been controversial. Some suggest that the panicle branch meristem remains indeterminate and that all the spikelets, including the one formed at the top of each branch, are thus produced laterally. However, in a generally accepted view, the terminal spikelet meristem is transformed from the panicle branch meristem. In the weak allele of the *lax* mutant of rice, *lax-1*, the production of lateral spikelets is severely suppressed, although terminal spikelets are normally produced and set seeds (56). The phenotype of the lax mutant indicates that the initiation of the terminal and lateral spikelets is controlled by different genetic programs, supporting the notion that the shoot apical meristem of rice panicle branches, which is the equivalent of the secondary inflorescences of *Arabidopsis*, turns into a terminal flower meristem.

The genetic mechanisms controlling the determinacy of the inflorescence have been most intensively studied in *Arabidopsis* and *Antirrhinum*. In these two species, *TFL1* and its *Antirrhinum* ortholog *CENTRORADIALIS (CEN)*, respectively, are required for the inflorescence meristem to maintain its indeterminate growth fate. Mutations in *TFL1* and *CEN* result in the inflorescence meristem becoming a terminal flower (2, 6, 7, 90). The expression patterns of the meristem identity genes established by their mutual regulation explain the molecular basis for the indeterminate nature of the *Arabidopsis* inflorescence. In *Arabidopsis*, *TFL1* inhibits the expression of *LFY* and *AP1* at the center of the shoot apex to prevent the inflorescence meristem from becoming a floral meristem (60); in turn, *LFY* and *AP1* inhibit *TFL1* expression in the lateral meristems committed to a floral fate (23, 108).

By analogy with the function of the *TFL1* gene in *Arabidopsis*, there are several possibilities to explain the mechanism that controls the production of the terminal flower in the rice panicle. First, the *TFL1* functions may be lacking in rice. Second, *TFL1* functions are present in rice; however, owing to the divergence in their expression patterns, they do not prevent the expression of floral meristem identity genes at the center of the shoot apex, allowing the formation of a terminal flower. An alternative possibility is that a different mechanism operates to control the determinacy of the inflorescence in rice. At least two *TFL1* homologs, which can rescue *Arabidopsis tfl1* mutation, exist in the rice genome (J. Kyozuka, unpublished results). Therefore, the first possibility is very unlikely. In addition, *LpTFL1*, a *TFL1* homolog of ryegrass, conferred similar effects to *Arabidopsis* development when overexpressed by the constitutive 35S promoter, indicating that the LtpTFL1 protein possesses a conserved biochemical function as TFL1 (44). Taking all the above findings into consideration, the function of *TFL1/CEN* homologs is probably conserved in grass species. To test the second possibility and reveal how *TFL1* functions are involved in the control of inflorescence architecture of grass species, the precise determination of their expression patterns is required.

If an insufficiency of the TFL1-like function in the shoot apex is a cause of the formation of a terminal flower in the panicle branches in rice, a supply of TFL1-like functions should avoid the terminal flower formation and change the fate of a panicle branch meristem from determinate to indeterminate. Overexpression of rice *TFL1* homologs in transgenic rice plants causes delays in the transition from the vegetative to the reproductive phase and from branch shoot to floral meristems (J. Kyozuka, unpublished results). These phenotypes are similar to those observed in transgenic dicot plants overexpressing *TFL1* homologs (84), suggesting that rice has a pathway that can respond to the overexpressed *TFL1* homologs. However, a terminal flower was still produced on the top of each panicle branch. This indicates that the supply of the overexpressed *TFL1*-like functions is not sufficient to change the nature of the branch meristem from determinate to indeterminate. Thus, the formation of the terminal flower in the rice panicle might be governed by molecular mechanisms that are independent of *TFL1*-like functions.

As a grass primary shoot apical meristem undergoes a few distinct identities, determinacy can be considered at each phase. The determinacy of a spikelet meristem is controlled by the *IDS* gene in maize. In the *indeterminate spikelet* (*ids*) mutant, the spikelet meristem acquires a partial indeterminacy, leading to the production of additional florets in a single spikelet instead of the two florets observed in a normal maize spikelet (15). The *IDS1* gene was cloned and shown to have a strong homology to *APETALA2* (49), although it is still unclear whether it is the closest homolog of AP2 in maize and what the function of the IDS ortholog is in other grasses.

The determinacy of a floral meristem of *Arabidopsis* is controlled by *AGA-MOUS* (*AG*) (114). Rice and maize *AG* orthologs have been isolated and their loss-of-function phenotypes reported (50, 67, 88). Reduction of floral meristem determinacy was observed in maize and rice plants, in which the function of the *AG* orthologs, *ZAG1* and *OsMADS3*, respectively, was decreased (50, 67). Therefore, the determinacy of the floral meristem in grasses is probably controlled by a conserved system in *Arabidopsis*.

Another MADS-box gene of rice, *OsMADS1*, is also involved in the determinacy of a floret (46). In a strong allele of the rice *lhs1* mutant, an additional floret is produced in a spikelet. Detailed observation revealed that the extra florets arose from a whorl of stamens or a carpel. This observation suggested that the extra florets produced are axillary. In this respect, *lhs1* resembles the *ap1* mutant of *Arabidopsis*, in which extra flowers often arise in the axils of the first whorl organs. In addition to the production of extra florets, defects in the floret organ number and development were also clear in the *lhs1* mutant. Overexpression of *OsMADS1* results in an extremely early flowering phenotype (46). Although *RAP1A* and *RAP1B* are closer to *AP1* than *OsMADS1* on the basis of their sequences, *OsMADS1* seems to have a closer function than *AP1* with respect to its function.

MADS-box genes have a wide range of functions in plant development (76). Among their multiple functions, the functions of MADS-box genes are best understood during reproductive development. An increasing number of MADS-box genes have been isolated from rice and other grass species (16, 17, 51, 57, 72, 73, 92). Further comparative analysis of those genes will shed light on their roles in rice reproductive development.

Genes Involved in Floral Organ Development

Grass species have flowers with highly derived structures (Figure 3, see color insert) (30, 66, 87). Recently, questions regarding the evolution of floral organs in grass species are being addressed with the aid of the ABC model (113). Emerging analysis suggests that the basic mechanisms of flower development are probably conserved between grasses and dicots; therefore, the ABC model can be extended to grass species (30, 66, 87). As the evolutionary aspects of grass flower development have been discussed in a number of recent reviews, we only present a summary of the current status of the applicability of the ABC model to grass species. The expression patterns of putative rice and maize class B and C genes strongly suggest the applicability of the ABC model to grass species. Furthermore, the functions of class B and C gene orthologs have been studied in rice and maize, and those studies have provided useful information regarding the identity of lodicules, which had been unclear for a long time. In the mutant of the maize *AP3* ortholog, *silky1* (*sil1*), stamens and lodicules are homeotically transformed into carpels and palea/lemma-like structures, respectively (1). Reduction of rice *PI* ortholog (*OsMADS4*) function by an antisense method also resulted in a similar phenotype (50). These findings imply that there is a homologous relationship between lodicules and petals and that the B function is conserved in grass flowers. *OsMADS16* was isolated by a yeast two-hybrid screening using OsMADS4 as a bait (72). Although the function of *OsMADS16* remains to be reported, based on its high sequence similarity with *AP3*, a physical interaction with OsMADS4, which is a putative *PI* ortholog, and its spatial expression pattern, *OsMADS16* is most likely to be a rice *AP3*.

Although the ABC model was established based on phenotypes of loss-of-function mutants, results obtained from gain-of-function analysis also fit the model well. Ectopic expression of *AG* by a constitutive promoter causes homeotic transformation from sepals to carpels and petals to stamens in dicot flowers (63, 70). An advantage of using rice for the analysis is that transformation is exceptionally easy as compared with other grass species. In rice, homeotic conversion from lodicules to stamens is caused by the ectopic expression of *OsMADS3*, a rice *AG* ortholog, by a strong *Actin1* promoter (J. Kyozuka, unpublished data). This added further strong evidence for the interpretation of the lodicule identity. However, in contrast to the progress made toward understanding the identity of lodicules, the nature of the lemma and palea is still unclear. The expression pattern of *RAP1A* and the phenotype of the *sil1* mutant suggest that the palea/lemma is probably the equivalent of the sepals. To confirm this assumption, analysis of the loss-of-function mutants of the grass class A genes is necessary.

DEFENSE SIGNALING IN RICE

Disease Resistance in Rice

Disease resistance has been extensively studied in rice because it is one of the most important traits for rice breeding (85, 102). Among pathogens, interactions with rice blast (*Magnaporthe grisea*) and bacterial blight (*Xanthomonas oryzae* pv. *oryzae* or *Xoo*) have been most extensively studied, and their genetic studies have indicated that the interactions conform to the gene-for-gene hypothesis. Key defense responses such as the hypersensitive response, oxidative burst, phytoalexin synthesis, and defense gene expression during infection of pathogens have been observed during infection of rice pathogens (54, 69). Moreover, a number of lesion-mimic mutants have been recently characterized (97, 116). Although molecular biological studies of disease resistance in rice have only recently begun, rapid progress in the near future is expected because of the progress in rice genomics and the ease in the production of transgenic plants.

Disease Resistance Genes

Although a large number of disease resistance genes (*R* genes) have been isolated from dicotyledonous species (20), only recently have the *R* genes of rice been isolated. The first rice *R* gene isolated is *Xa21*, which confers resistance to bacterial blight. *Xa21* encodes a receptor kinase protein carrying leucine-rich repeats (LRR) (94). Until recently, this had been the only *R* gene encoding this type of protein. However, the FLS2 of *Arabidopsis*, which is thought to be a receptor for the flagellin molecule, has been shown to have a similar structure (29). In both proteins, LRR functions as a receptor and the kinase domain as a signal transducer (28, 34). A swapping experiment between Xa21 and BRI1, a putative receptor for brassinosteroids (BR), demonstrated that, when the extracellular motif of Xa21 was exchanged with that of BRI1, rice cells expressing this chimeric protein exhibited a defense response upon BR addition, indicating that the LRR motif functions as a receptor (34). These two genes are a unique class of R genes in plants; however, it remains to be determined which pathogens are specifically recognized by the *Arabidopsis* FLS2 protein.

Another cloned rice *R* gene for bacterial blight is *Xa1*, which encodes a protein carrying a nucleotide-binding site (NBS) and LRR (118); the encoded protein is structurally similar to the most typical group of *R* genes: *RPS2*, *RPM1*, and *RPP5* of *Arabidopsis*, *N* of tobacco, and L^6 and *M* of flax. One unique feature of *Xa1* compared with other *R* genes having the same structure is that its expression is induced by infection of bacterial blight, although whether induction is required for resistance remains to be determined. Therefore, two rice *R* genes conferring resistance to different races of bacterial blight encode proteins of distinct structures.

Two blast resistance genes of rice, *Pi-b* (104) and *Pi-ta* (9), have been isolated, and both of them were shown to encode proteins of the NBS-LRR class. One significant observation with the *Pi-ta* gene was that the Pi-ta protein was able to physically interact with the AVR-Pita protein in the yeast two-hybrid system and in an in vitro binding assay (48). Despite extensive studies with other *R* genes in dicots, no such direct interaction has been demonstrated with R proteins and corresponding AVR proteins. Whether this is unique to *Pi-ta* remains to be seen. Thus, the four rice *R* genes isolated to date are of two types, a receptor kinase with an LRR motif and the NBS-LRR protein. Characterization of these rice *R* genes and comparison of these genes with the *R* genes of *Arabidopsis* clearly indicate that no structural divergence between rice and *Arabidopsis R* genes occurred. How differences between the interactions of rice and *Arabidopsis* with their respective pathogens generate downstream signaling will be of great interest in the future. To perform comparative genomics of *R* genes, more genes need to be cloned and characterized in rice.

Rac Small G-Protein Genes

The Rac family of small GTPases (also termed Rop) is involved in defense signaling in various plant species (52, 78, 86). In rice, the constitutively active form of a rice Rac gene induces production of reactive oxygen species (ROS), cell death, and increased disease resistance in transgenic rice plants (52, 78). ROS production by Rac genes has been also demonstrated in *Arabidopsis* (81), tobacco (86), and soybean cells (79). Rac consists of a multigene family, and 11 members have been identified in *Arabidopsis* (110). Analysis of the gene family allowed classification of the members into two groups mainly based on the sequences in the insert region that is thought to be involved in signaling and the C-terminal sequences that are important for lipid modification of the Rac proteins. The first group (Group I), which consists of eight members, AtRAC1–6, AtRAC9, and AtRAC11 (110), has a conserved insert region and CXXL at the C-terminal, a typical motif for prenylation. This group of Rac genes seemed to undergo duplications after diversification into two groups during evolution. The second group (Group II), which consists of only three members, AtRAC7, AtRAC8, and AtRAC10, has a distinct insert region and an extension at the C-terminal; in addition, it does not carry the typical CXXL for prenylation. This seemed to be caused by insertion of an additional intron in the C-terminal region (110). It was also suggested that Group II Rac genes have evolved only in vascular plants (110).

In rice, 10 members of the Rac gene family have been identified so far based on EST and the genome sequence database of RGP (K. Shimamoto & J. Kyozuka, unpublished results). In contrast to the situation in *Arabidopsis*, 7 of 10 members in rice are classified into Group II. Extensive duplication of certain members of Group II, such as OsRac2, seems to have occurred. OsRac1, which has been most extensively studied among rice Rac genes, belongs to Group II. The other three members can be classified into Group I. Therefore, the majority of the Rac genes in rice belong to Group II, which is specific to vascular plants. This is in striking contrast to the situation in *Arabidopsis*. Thus, it will be extremely interesting to carry out a comparative analysis of the functions of Rac genes in rice and *Arabidopsis*.

Recent studies have indicated that plant Rac proteins show diverse cellular activities such as defense, polar growth in root tips and pollen tubes, ABA-regulated closure of guard cells, and hormone signaling, suggesting that they are involved in many of the important cellular activities of higher plants (101). Therefore, striking differences between rice and *Arabidopsis* with respect to the number and sequences of Rac genes may represent evolutionary differences in growth, development, and adaptation to environments between the two species.

Heterotrimeric G Protein

Heterotrimeric G proteins are a major group of signaling molecules in mammals and are mainly responsible for various cellular responses to external signals (33). Heterotrimeric G proteins consist of α, β, and γ subunits, and each subunit consists of a multigene family in mammals (33). In contrast to mammals, only one or two genes for each subunit exist in plants (24). Nevertheless, a large number of studies using inhibitors and agonists suggests that the heterotrimeric G protein is involved in a variety of signaling in plants (8, 24).

Recent studies using knockout mutants of the single-copy Gα subunit gene in *Arabidopsis* indicate that the heterotrimeric G protein is involved in auxin signaling (100) and the regulation of ion channels and ABA signaling in guard cells (103). In another study, in which the same gene was overexpressed, the heterotrimeric G protein was shown to be involved in the phytochrome-mediated inhibition of hypocotyl elongation in *Arabidopsis*, although the effect appeared to be limited to certain branches of the phytochrome signaling pathways (77). In rice, mutations in the corresponding single-copy Gα gene result in dwarfism, dark-green leaf, and small round seeds (4, 25). It has been shown that this rice mutation causes partial loss of gibberellin response (99). Apparent involvement of the heterotrimeric G protein in several distinct signaling pathways raises an interesting question. How does a single Gα protein mediate multiple signaling functions? Furthermore, an intriguing question is whether the heterotrimeric G protein is involved in distinct signaling pathways in *Arabidopsis* and rice.

We have recently found that the heterotrimeric G protein is required for disease resistance of rice against fungal and bacterial pathogens (U. Suharsono, T. Kawasaki, H. Satoh & K. Shimamoto, unpublished results). Furthermore, all the key events required for resistance, such as ROS production, induction of defense-related genes, and cell death, are regulated by the heterotrimeric G protein. This information adds another signaling role to the heterotrimeric G protein in rice. One possible mechanism to explain its role in multiple signaling pathways is that the heterotrimeric G protein may function as an enhancer of certain signaling intermediates commonly used for each signaling pathway. In this case, the heterotrimeric G protein interacts with structurally similar but functionally distinct signaling intermediates. This system works nicely for the crosstalk between different signaling pathways; however, it is not easy to maintain the specificity of signaling. Detailed analysis of how the heterotrimeric G protein is involved in various signaling pathways in rice and *Arabidopsis* should reveal the diversity in its regulation and function in plants. This will give important insights into comparative genomics of plant genes involved in signal transduction. With this information, a comparative analysis of the heterotrimeric G proteins can be eventually extended even to animals.

NPR1/NIM1 Gene

The *NPR1/NIM1* gene of *Arabidopsis* (10) is a key regulator of salicylic acid (SA)–mediated systemic acquired resistance (SAR) (21). *NPR1/NIM1* expression is increased by treatment of *Arabidopsis* with inducers of SAR including SA, 2,6-dichloroisonicotinic acid, and benzothiadiazole (BTH) (21). Overexpression of the *NPR1/NIM1* gene causes enhanced disease resistance to bacterial and oomycete pathogens (11). The NPR1/NIM1 protein interacts with the TGA family of bZIP transcription factors (21).

Recently, the *NPR1/NIM1* gene of *Arabidopsis* was introduced into rice, and resultant transgenic rice plants showed increased resistance to bacterial blight

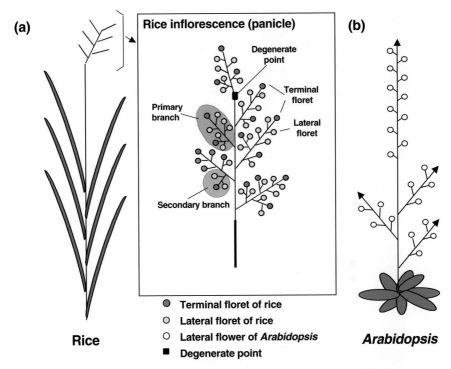

Figure 1 Comparison of the inflorescence architecture between rice and A*arabidopsis*. (*a*) Rice, (*b*) *Arabidopsis*. One big difference between the rice and *Arabidopsis* inflorescences is in the determinacy of the inflorescence meristems. The primary shoot apical meristem of rice loses its activity and leaves a scar called a degenerate point. Then, the inflorescence meristem on each panicle branch is transformed into a terminal floret. In contrast, the development of the primary and secondary inflorescences is indeterminate in *Arabidopsis*.

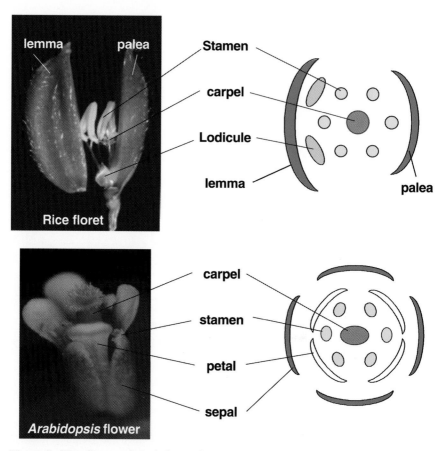

Figure 3 Rice floret and *Arabidopsis* flower. The rice floret is composed of a lemma, palea, two lodicules, six stamens, and a carpel. An *Arabidopsis* flower contains four sepals, four petals, six stamens, and two carpels.

(12). Furthermore, it was shown that the *Arabidopsis* NPR1/NIM1 protein could interact with the rice TGA family of transcription factors (12). More recently, a rice ortholog of the *NPR1/NIM1* gene was isolated, and its overexpression in transgenic rice led to enhanced resistance to bacterial blight (13). These studies suggest that the *NPR1/NIM1*-mediated disease-resistance pathway is conserved in rice and that its molecular mechanism may be similar in rice and *Arabidopsis*.

These findings are very interesting from a comparative genomics point of view. Because SAR has not been well established in cereals, including rice, the role of the NPR1/NIM1 protein in the defense signaling of rice was unexpected. SAR in dicots such as *Arabidopsis*, tobacco, and cucumber is mediated by increased SA upon pathogen infection. In contrast, rice has constitutively very high levels of SA (93). Furthermore, SA-deficient rice generated by introduction of the *nahG* gene does not show alterations in defense gene expression (111). These findings seem to suggest that SAR, which is induced by increased SA in dicots, does not exist in rice. One contrasting observation is that one SAR-inducing compound, BTH, induces disease resistance to rice blast and expression of certain defense genes (89). Therefore, one possible hypothesis explaining all these observations is that the SAR-like defense pathway is operating in rice and that the NPR1/NIM1 protein is a regulator of this pathway. However, it is independent of SA, and perhaps BTH enhances this resistance pathway. It will be of extreme interest to find out whether such a signaling pathway exists in rice and, if it does, what the key genes in this pathway are. Moreover, it will be interesting to determine to what extent this pathway is conserved in dicots.

CONCLUDING REMARKS

It is clear from our limited comparative analysis of the gene functions between rice and other plant species such as *Arabidopsis* that structural similarities do not always indicate functional similarities. For a particular multigene family, we often find more copies in rice than in *Arabidopsis*. Moreover, the number and kinds of members of a gene family are different between the two species. Most importantly, homologous genes are often differently regulated in rice and other species. A detailed elucidation of the structures and functions of homologous genes in monocots and dicots should provide valuable insights into the diversity of higher plants with respect to development, signal transduction, and responses to environmental signals at the molecular level in the future. With the completion of rice genome sequencing, the comparative genomics of rice and *Arabidopsis* will provide the foundation for such studies in plants. When the genome analysis of other plants is accomplished, comparative genomics will be extended to more diverse plant species.

ACKNOWLEDGMENTS

We thank former and current members of the Laboratory of Plant Molecular Genetics at the Nara Institute of Science and Technology for various contributions that made it possible for us to write this review.

Visit the Annual Reviews home page at www.annualreviews.org

LITERATURE CITED

1. Ambrose BA, Lerner DR, Ciceri P, Padilla CM, Yanofsky M, Schmidt RJ. 2000. Molecular and genetic analysis of the *Silky1* gene reveal conservation in floral organ specification between eudicots and monocots. *Mol. Cell* 5:569–79

2. Alvarez J, Guli CL, Yu X-H, Smyth DR. 1992. *Terminal flower*: a gene affecting inflorescence development in *Arabidopsis thaliana*. *Plant J.* 2:103–16

3. Araki T. 2000. Transition from vegetative to reproductive phase. *Curr. Opin. Plant Biol.* 4:63–68

4. Ashikari M, Wu J, Yano M, Sasaki T, Yoshimura A. 1999. Rice gibberellin-insensitive dwarf mutant gene *Dwarf 1* encodes the alpha-subunit of GTP-binding protein. *Proc. Natl. Acad. Sci. USA* 96:10284–89

5. Bell AD. 1991. *Plant Form*. New York: Oxford Univ. Press

6. Bradley D, Carpenter R, Copsey L, Vincent C, Rothstein R, Coen E. 1996. Control of inflorescence architecture in *Antirrhinum*. *Nature* 379:791–97

7. Bradley D, Ratcliffe O, Vincent C, Carpenter R, Coen E. 1997. Inflorescence commitment and architecture in *Arabidopsis*. *Science* 275:80–83

8. Blumwald E, Aharon GS, Lamb C-H. 1998. Early signal transduction pathways in plant-pathogen interactions. *Trends Plant Sci.* 3:342–46

9. Bryan GT, Wu K-S, Farrall L, Jia Y, Hershey HP, et al. 2000. A single amino acid difference distinguishes resistant and susceptible alleles of the rice blast resistance gene *Pi-ta*. *Plant Cell* 12:2033–45

10. Cao H, Glazebrook J, Clarke J, Volko S, Dong X. 1997. The *Arabidopsis npr1* gene that controls systemic aquired resistance encodes a novel protein containing ankyrin repeats. *Cell* 88:57–63

11. Cao H, Li X, Dong X. 1998. Generation of broad-spectrum disease resistance by overexpression of an essential regulatory gene in systemic acquired resistance. *Proc. Natl. Acad. Sci. USA* 95:6531–36

12. Chern M-S, Fitzgerald HA, Yadav RC, Canlas PE, Dong X, et al. 2001. Evidence for a disease-resistance pathway in rice similar to the *NPR1*-mediated signaling pathway in *Arabidopsis*. *Plant J.* 27:101–13

13. Chern M-S, Canlas PE, Fitzgerald HA, Yadav RC, Dong X, et al. 2001. Components of a rice disease resistance pathway similar to the NPR1-mediated pathway in Arabidopsis. *Int. Congr. Mol. Plant-Microbe Interact., 10th*, Abstr. 299

14. Chin HG, Choe MS, Lee SH, Park SH, Koo JC, et al. 1999. Molecular analysis of rice plants harboring an *Ac/Ds* transposable element-mediated gene trapping system. *Plant J.* 19:615–23

15. Chuc G, Meeley RB, Hake S. 1998. The control of maize spikelet meristem fate by the *Apetala2*-like gene indeterminate spikelet1. *Genes Dev.* 12:1145–54

16. Chung Y-Y, Kim S-R, Finkel D, Yanofsky M, An G. 1994. Early flowering and reduced apical dominance result from ectopic expression of a rice MADS box gene. *Plant Mol. Biol.* 26:657–65

17. Chung YY, Kim SR, Kang HG, Noh YS, Park MC, et al. 1995. Characterization of two rice MADS box genes homologous to *GLOBOSA*. *Plant Sci.* 109:45–56

18. Coen ES, Nugent JM. 1994. Evolution of flowers and inflorescences. *Development* Suppl.:107–16

19. Coen ES, Romero JM, Doyle S, Elliott R, Murphy G, Carpenter R. 1990. *Floricaula*: a homeotic gene required for flower development in *Antirrhinum majus*. *Cell* 63:1311–22

20. Dangl JL, Jones JDG. 2001. Plant pathogens and integrated defence responses to infection. *Nature* 411:826–33

21. Dong X. 2001. Genetic dissection of systemic acquired resistance. *Curr. Opin. Plant Biol.* 4:309–14

22. Enoki H, Izawa T, Kawahara M, Komatsu M, Koh S, et al. 1999. *Ac* as a tool for functional genomics of rice. *Plant J.* 19:605–13

23. Ferrandiz C, Gu Q, Martinssen R, Yanofsky MF. 2000. Redundant regulation of meristem identity and plant architecture by *FRUITFULL, APETALA1*, and *CAULIFLOWER. Development* 127:725–34

24. Fujisawa Y, Kato T, Iwasaki Y. 2001. Structure and function of heterotrimeric G proteins in plants. *Plant Cell Physiol.* 42:789–94

25. Fujisawa Y, Kato T, Ohki S, Ishikawa A, Kitano H, et al. 1999. Suppression of the heterotrimeric G protein causes abnormal morphology, including dwarfism in rice. *Proc. Natl. Acad. Sci. USA* 96:7575–80

26. Gocal GFW, King RW, Blundell CA, Schwartz OM, Andersen CH, Weigel D. 2001. Evolution of floral meristem identity genes. Analysis of *Lolium temulentum* genes related to *APETALA1* and *LEAFY* of *Arabidopsis. Plant Physiol.* 125:1788–801

27. Goff SA. 1999. Rice as a model for cereal genomics. *Curr. Opin. Plant Biol.* 2:86–89

28. Gomez-Gomez L, Bauer Z, Boller T. 2001. Both the extracellular leucine-rich repeat domain and the kinase activity of FSL2 are required for flagellin binding and signaling in *Arabidopsis. Plant Cell* 13:1155–63

29. Gomez-Gomez L, Boller T. 2000. FLS2: an LRR receptor-like kinase involved in the perception of the bacterial elicitor flagellin in *Arabidopsis. Mol. Cell* 5:1003–11

30. Goto K, Kyozuka J, Bowman JL. 2001. Turning floral organs into leaves, leaves into flowers. *Curr. Opin. Genet. Dev.* 11:449–56

31. Greco R, Ouwerkerk PBF, Sallaud C, Kohli A, Colombo L, et al. 2001. Transposon insertional mutagenesis in rice. *Plant Physiol.* 125:1175–77

32. Greco R, Ouwerkerk PBF, Taal AJ, Favalli C, Beguiristain T, et al. 2001. Early and multiple *Ac* transposition in rice suitable for efficient insertional mutagenesis. *Plant Mol. Biol.* 46:215–27

33. Hamm HE. 1998. The many faces of G protein signaling. *J. Biol. Chem.* 273:669–72

34. He Z, Wang ZY, Li J, Zhu O, Lamb C, et al. 2000. Perception of brassinosteroids by the extracellular domain of the receptor kinase BRI1. *Science* 288:2360–63

35. He Z, Zhu Q, Dabi T, Li D, Weigel D, Lamb C. 2000. Transformation of rice with the *Arabidopsis* floral regulator *LEAFY* causes early heading. *Transgenic Res.* 9:223–27

36. Hiei Y, Ohta S, Komari T, Kumashiro T. 1994. Efficient transformation of rice (*Oryza sativa* L.) mediated by *Agrobacterium* and sequence analysis of the boundaries of the T-DNA. *Plant J.* 6:271–82

37. Himi S, Sano R, Nishiyama T, Tanahashi T, Kato M, et al. 2001. Evolution of MADS-box gene induction by *FLO/LFY* genes. *J. Mol. Evol.* 53:387–93

38. Hirochika H, Sugimoto K, Ohtsuki Y, Tsugawa H, Kanda M. 1996. Retrotransposon of rice involved in mutation induced by tissue culture. *Proc. Natl. Acad. Sci. USA* 93:7783–88

39. Hirochika H. 2001. Contribution of the *Tos17* retrotransposon to rice functional genomics. *Curr. Opin. Plant Biol.* 4:118–22

40. Izawa T, Miyazaki C, Yamamoto M, Terada R, Iida S, et al. 1991. Introduction and transposition of the maize transposable element *Ac* in rice (*Oryza sativa* L.). *Mol. Gen. Genet.* 227:391–96

41. Izawa T, Ohnishi T, Nakano T, Ishida N, Enoki H, et al. 1997. Transposon tagging in rice. *Plant Mol. Biol.* 35:219–29

42. Izawa T, Oikawa T, Tokutomi S, Okuno K,

Shimamoto K. 2000. Phytochromes confer the photoperiodic control of flowering in rice (a short-day plant). *Plant J.* 22:391–99

43. Izawa T, Shimamoto K. 1996. Becoming a model plant: the importance of rice to plant science. *Trends Plant Sci.* 1:95–99

44. Jensen CS, Salchert K, Nielsen KK. 2001. A *TERMINAL FLOWER1*–like gene from perennial ryegrass involved in floral transition and axillary meristem identity. *Plant Physiol.* 125:1517–28

45. Jeon J, An G. 2001. Gene tagging in rice: a high throughput system for functional genomics. *Plant Sci.* 161:211–19

46. Jeon JS, Jang S, Nam J, Kim C, Lee S-H, et al. 2000. *Leafy hull sterile 1* is a homeotic mutation in a rice MADS box gene affecting rice flower development. *Plant Cell* 12:871–84

47. Jeon JS, Lee S, Jung KH, Jun SH, Jeong DH, et al. 2000. T-DNA insertional mutagenesis for functional genomics of rice. *Plant J.* 22:561–70

48. Jia Y, McAdams SA, Bryan GT, Hershey HP, Valent B. 2000. Direct interaction of resistance gene and avirulence gene products confers rice blast resistance. *EMBO J.* 19:4004–14

49. Jofuku KD, den Boer BG, Van Montagu M, Van Montagu Okamuro. 1994. Control of Arabidopsis flower and seed development by the homeotic gene *APETALA2*. *Plant Cell* 6:1211–25

50. Kang HG, Jeon J-S, Lee S, An G. 1998. Identification of class B and class C organ identity genes from rice plants. *Plant Mol. Biol.* 38:1021–29

51. Kang HG, Noh YS, Chung YY, Costa M, An K, An G. 1995. Phenotypic alterations of petals and sepals by ectopic expression of a rice MADS box gene in tobacco. *Plant Mol. Biol.* 29:1–10

52. Kawasaki T, Henmi K, Ono E, Hatakeyama S, Iwano M, et al. The small GTP-binding protein Rac is a regulator of cell death in plants. *Proc. Natl. Acad. Sci. USA* 96:10922–26

53. Kempin SA, Savidge B, Yanofsky MF. 1995. Molecular basis of the cauliflower phenotype in *Arabidopsis*. *Science* 267: 522–25

54. Koga H. 1994. Hypersensitive death, autofluorescence and ultrastructural changes in cells of leaf sheaths of susceptible and resistant near-isogenic lines of rice (*Pi-z^t*) in relation to penetration and growth of *Pyricularia oryzae*. *Can. J. Bot.* 72:1463–77

55. Komari T, Hiei Y, Ishida Y, Kumashiro T, Kubo T. 1998. Advances in cereal gene transfer. *Curr. Opin. Plant Biol.* 1:161–65

56. Komatsu M, Maekawa M, Shimamoto K, Kyozuka J. 2001. The *LAX1* and *FRIZZY PANICLE2* genes determine the inflorescence architecture of rice by controlling rachis-branch and spikelet development. *Dev. Biol.* 231:364–73

57. Kyozuka J, Kobayashi T, Morita M, Shimamoto K. 2000. Spatially and temporally regulated expression of rice MADS box genes with similarity to *Arabidopsis* class A, B, C genes. *Plant Cell Physiol.* 41:710–18

58. Kyozuka J, Konishi S, Nemoto K, Izawa T, Shimamoto K. 1998. Down-regulation of *RFL*, the *FLO/LFY* homolog of rice, accompanied with panicle branch initiation. *Proc. Natl. Acad. Sci. USA* 95:1979–82

59. Lawton-Rauh AL, Alvarez-Buylla ER, Purugganan MD. 2000. Molecular evolution of flower development. *Trends Ecol. Evol.* 15:144–49

60. Liljegren S, Gustafson-Brown C, Pinyopich A, Ditta GS, Yanofsky MF. 1999. Interactions among *APETALA1*, *LEAFY*, and *TERMINAL FLOWER1* specify meristem fate. *Plant Cell* 11:1007–18

61. Lim J, Moon YH, An G, Jang SK. 2000. Two rice MADS domain proteins interact with *OsMADS1*. *Plant Mol. Biol.* 44:513–27

62. Ma H. 1998. To be, or not to be, a flower—control of floral meristem identity. *Trends Genet.* 14:26–32

63. Mandel MA, Bowman JL, Kempin SA,

Ma H, Meyerowitz EM, Yanofsky MF. 1992. Manipulation of flower structure in transgenic tobacco. *Cell* 71:133–43

64. Mandel MA, Gustafson-Brown C, Savidge B, Yanofsky MF. 1992. Molecular characterization of the *Arabidopsis* floral hopmeotic gene *APETALA1*. *Nature* 360:273–77

65. Matsumoto T, Wu J, Baba T, Katayose Y, Yamamoto K, et al. 2001. Rice genomics: current status of genome sequencing. *Novartis Found. Symp.* 236:28–36

66. McSteen P, Laudencia-Chingcuanco D, Colasanti J. 2000. A floret by any other name: control of meristem identity in maize. *Trends Plant Sci.* 5:61–66

67. Mena M, Ambrose BA, Meeley RB, Briggs SP, Yanofsky M, Schmidt RJ. 1996. Diversification of C-function activity in maize flower development. *Science* 274:1537–40

68. Mena M, Mandel MA, Lerner DR, Yanofsky MF, Schmidt RJ. 1995. A characterization of the MADS-box gene family in maize. *Plant J.* 8:845–54

69. Midoh N, Iwata M. 1997. Cloning and characterization of a probenazole-inducible gene for an intracellular pathogenesis-related protein in rice. *Plant Cell Physiol.* 37:9–18

70. Mizukami Y, Ma H. 1992. Ectopic expression of the floral homeotic gene *AGAMOUS* in transgenic *Arabidopsis* plants alters floral organ identity. *Cell* 71:119–31

71. Molinero-Rosales N, Jamilena M, Zurita S, Gomez P, Capel J, Lozano R. 1999. *FALSIFLORA*, the tomato orthologue of FLORICAULA and LEAFY, controls flowering time and floral meristem identity. *Plant J.* 20:685–93

72. Moon YH, Jung JY, Kang HG, An G. 1999. Identification of a rice *APETALA3* homologue by yeast two-hybrid screening. *Plant Mol. Biol.* 40:167–77

73. Moon YH, Kang HG, Jung JY, Jeon JS, Sung SK, An G. 1999. Determination of the motif responsible for interaction between the rice APETALA1/AGAMOUS-like nine family proteins using yeast two-hybrid system. *Plant Physiol.* 120:1193–204

74. Muramoto T, Kohchi T, Yokota A, Hwang I, Goodman HM. 1999. The *Arabidopsis* photomorphogenic mutant *hy1* is deficient in phytochrome chromophore biosynthesis as a result of a mutation in a plastid heme oxygenase. *Plant Cell* 11:335–47

75. Nakagawa Y, Machida C, Machida Y, Toriyama K. 2000. Frequency and pattern of transposition of the maize transposable element *Ds* in transgenic rice plants. *Plant Cell Physiol.* 41:733–42

76. Ng M, Yanofsky MF. 2001. Function and evolution of the plant MADS-box gene family. *Nat. Rev.* 2:186–95

77. Okamoto H, Matsui M, Deng XW. 2001. Overexpression of the heterotrimeric G-protein α-subunit enhances phytochrome-mediated inhibition of hypocotyls elongation in Arabidopsis. *Plant Cell* 13:1639–51

78. Ono E, Wong H-L, Kawasaki T, Hasegawa M, Kodama O, et al. 2001. Essential role of the small GTPase Rac in disease resistance of rice. *Proc. Natl. Acad. Sci. USA* 98:759–64

79. Park J, Choi HJ, Lee S, Lee T, Yang Z, et al. 2000. Rac-related GTP-binding protein in elicitor-induced reactive oxygen generation by suspension-cultured soybean cells. *Plant Physiol.* 124:725–32

80. Pidkowich MS, Klenz JE, Haughn GW. 1999. The making of a flower: control of floral meristem identity in *Arabidopsis*. *Trends Plant Sci.* 4:64–70

81. Potikha TS, Collins CC, Johnson DI, Delmer DP, Levine A. 1999. The involvement of hydrogen peroxide in the differentiation of secondary walls in cotton fibers. *Plant Physiol.* 119:849–58

82. Putterill J, Robson F, Lee K, Simon R, Coupland G. 1995. The *CONSTANS* gene of *Arabidopsis* promotes flowering and encodes a protein showing similarities

to zincfinger transcription factors. *Cell* 80:847–57

83. Ratcliff F, Martin-Hernandez AM, Baulcombe DC. 2001. Tobacco rattle virus as a vector for analysis of gene function by silencing. *Plant J.* 25:237–45

84. Ratcliffe OJ, Amaya I, Vincent CA, Rothstein S, Carpenter R, et al. 1998. A common mechanism controls the life cycle and architecture of plants. *Development* 124:1609–15

85. Ronald PC. 1997. The molecular basis of disease resistance in rice. *Plant Mol. Biol.* 35:179–86

86. Schiene K, Puhler A, Niehaus K. 2000. Transgenic tobacco plants that express an antisense construct derived from a *Medicago sativa* cDNA encoding a Rac-related small GTP-binding protein fail to develop necrotic lesions upon elicitor infiltration. *Mol. Gen. Genet.* 263:761–70

87. Schmidt RJ, Ambrose BA. 1998. The blooming of glass flower development. *Curr. Opin. Plant Biol.* 1:60–67

88. Schmidt RJ, Veit B, Mandel MA, Mena M, Hake S, Yanofsky M. 1993. Identification and molecular characterization of *ZAG1*, the maize homolog of the *Arabidopsis* floral homeotic gene *AGAMOUS*. *Plant Cell* 5:729–37

89. Schweizer P, Schlagenhauf E, Schaffrath U, Dudler R. 1999. Different patterns of host genes are induced in rice by *Pseudomonas syringae*, a biological inducer of resistance, and the chemical inducer benzothiadiazole (BTH). *Eur. J. Plant Pathol.* 105:659–65

90. Shannon S, Meeks-Wagner R. 1991. A mutation in the *Arabidopsis TFL1* gene affects inflorescence meristem development. *Plant Cell* 3:877–92

91. Shimamoto K, Miyazaki C, Hashimoto H, Izawa T, Itoh K, et al. 1993. Transactivation and stable integration of the maize transposable element *Ds* cotransfected with the *Ac* transposase gene in transgenic rice plants. *Mol. Gen. Genet.* 239:354–60

92. Shinozuka Y, Kojima S, Shomura A, Ichimura H, Yano M, et al. 1999. Isolation and characterization of rice MADS box gene homologues and their RFLP mapping. *DNA Res.* 6:123–29

93. Silverman P, Seskar M, Kanter D, Schweizer P, Metraux J-P. 1995. Salicylic acid in rice: biosynthesis, conjugation, and possible role. *Plant Physiol.* 108:633–39

94. Song WY, Wang GL, Chen LL, Kim HS, Pi LY, et al. 1995. A receptor-kinase–like protein encoded by the rice disease resistance gene *Xa21*. *Science* 270:1804–6

95. Souer E, van der Krol A, Kloos A, Bliek M, Mol J, Koes R. 1998. Genetic control of branching pattern and floral identity during *Petunia* inflorescence development. *Development* 125:733–42

96. Sugano S, Andronis C, Ong MS, Green RM, Tobin M. 1999. The protein kinase CK2 is involved in regulation of circadian rhythms in *Arabidopsis*. *Proc. Natl. Acad. Sci. USA* 96:12362–66

97. Takahashi A, Kawasaki T, Henmi K, Shii K, Kodama O, et al. 1999. Lesion-mimic mutants of rice with alterations in early signaling events of defense. *Plant J.* 17:535–45

98. Takahashi Y, Shomura A, Sasaki T, Yano M. 2001. *Hd6*, a rice quantitative trait locus involved in photoperiod sensitivity, encodes the subunit of protein kinase CK2. *Proc. Natl. Acad. Sci. USA* 98:7922–27

99. Ueguchi-Tanaka M, Fujisawa Y, Kobayahi M, Ashikari M, Iwasaki Y, et al. 2000. Rice dwarf mutant d1, which is defective in the alpha subunit of the heterotrimeric G protein, affects gibberellin signal transduction. *Proc. Natl. Acad. Sci. USA* 97:11638–43

100. Ullah H, Chen J-G, Young JC, Im K-H, Sussman MR, et al. 2001. Modulation of cell proliferation by heterotrimeric G protein in *Arabidopsis*. *Science* 292:2066–69

101. Valster AH, Hepler PK, Chernoff J. 2000. Plant GTPases: the Rhos in bloom. *Trends Cell Biol.* 10:141–46

102. Wang G-L, Leung H. 2000. Molecular biology of host-pathogen interactions in rice disease. In *Molecular Biology of Rice*, ed. K Shimamoto, pp. 201–32. Tokyo, Japan: Springer-Verlag

103. Wang X-Q, Ullah H, Jones AM, Assmann SM. 2001. G protein regulation of ion channels and abscisic acid signaling in Arabidopsis guard cells. *Science* 292:2070–72

104. Wang Z-X, Yano M, Yamanouchi U, Iwamoto M, Monna L, et al. 1999. The Pib gene for rice blast resistance belongs to the nucleotide binding and leucine-rich repeat class of plant disease resistance genes. *Plant J.* 19:55–64

105. Wang ZY, Tobin EM. 1998. Constitutive expression of the *CIRCADIAN CLOCK ASSOCIATED 1 (CCA1)* gene disrupts circadian rhythms and suppresses its own expression. *Cell* 93:1207–17

106. Webering F. 1989. *Morphology of Flowers and Inflorescences*. New York: Cambridge Univ. Press. 405 pp.

107. Weigel D. 1995. The genetics of flower development: from floral induction to ovule morphogenesis. *Annu. Rev. Plant Physiol. Plant Mol. Biol.* 29:19–39

108. Weigel D, Alvarez J, Smyth DR, Yanofsky MF, Meyerowitz EM. 1992. *LEAFY* controls floral meristem identity in *Arabidopsis*. *Cell* 69:843–59

109. Weigel D, Nilsson O. 1995. A developmental switch sufficient for flower initiaiton in diverse plants. *Nature* 377:495–500

110. Winge P, Brembu T, Kristensen R, Bones AM. 2000. Genetic structure and evolution of RAC-GTPases in *Arabidopsis thaliana*. *Genetics* 156:1959–71

111. Yang Y, Qi M. 2001. Anti-oxidative role of salicylic acid in rice plants. *Int. Congr.*

Mol. Plant-Microbe Interact., 10th, Abstr. 195

112. Yano M, Katayose Y, Ashikari M, Yamanouchi U, Monna L, et al. 2000. *Hd1*, a major photoperiod sensitivity quantitative trait locus in rice, is closely related to the *Arabidopsis* flowering time gene *CONSTANS*. *Plant Cell* 12:2473–83

113. Yanofsky MF. 1995. Floral meristems to floral organs: genes controlling early events in *Arabidopsis* flower development. *Annu. Rev. Plant Physiol. Plant Mol. Biol.* 46:167–88

114. Yanofsky MF, Bowman JL, Drews G, Feldmann K, Meyerowitz EM. 1990. The protein encoded by the *Arabidopsis* homeotic gene *AGAMOUS* resembles transcription factors. *Nature* 346:35–39

115. Yazaki J, Kishimoto N, Nakamura K, Fujii F, Shimbo K, et al. 2000. Embarking on rice functional genomics via cDNA microarray: use of 3′ UTR probes for specific gene expression analysis. *DNA Res.* 7:367–70

116. Yin Z, Chen J, Zeng L, Goh L, Leung H, et al. 2000. Characterizing rice lesion mimic mutants and identifying a mutant with a broad-spectrum resistance to rice blast and bacterial blight. *Mol. Plant-Microbe Interact.* 13:869–76

117. Yokoo M, Okuno K. 1993. Genetic analysis of earliness mutations induced in rice cultivar Norin 8. *Jpn. J. Breed.* 43:1–11

118. Yoshimura S, Yamanouchi U, Katayose Y, Toki S, Wang ZX, et al. 1998. Expression of *Xa1*, a bacterial blight-resistance gene in rice, is induced by bacterial inoculation. *Proc. Natl. Acad. Sci. USA* 95:1663–68

119. Zhu T, Wang X. 2000. Large-scale profiling of the Arabidopsis transcriptome. *Plant Physiol.* 124:1472–76

Annu. Rev. Plant Biol. 2002. 53:421–47
DOI: 10.1146/annurev.arplant.53.100301.135158
Copyright © 2002 by Annual Reviews. All rights reserved

ROOT GRAVITROPISM: An Experimental Tool to Investigate Basic Cellular and Molecular Processes Underlying Mechanosensing and Signal Transmission in Plants

K. Boonsirichai, C. Guan, R. Chen, and P. H. Masson

Laboratory of Genetics, University of Wisconsin-Madison, 445 Henry Mall, Madison, Wisconsin 53706; e-mail: phmasson@facstaff.wisc.edu

Key Words gravity signal transduction, auxin transport, auxin response

■ **Abstract** The ability of plant organs to use gravity as a guide for growth, named gravitropism, has been recognized for over two centuries. This growth response to the environment contributes significantly to the upward growth of shoots and the downward growth of roots commonly observed throughout the plant kingdom. Root gravitropism has received a great deal of attention because there is a physical separation between the primary site for gravity sensing, located in the root cap, and the site of differential growth response, located in the elongation zones (EZs). Hence, this system allows identification and characterization of different phases of gravitropism, including gravity perception, signal transduction, signal transmission, and curvature response. Recent studies support some aspects of an old model for gravity sensing, which postulates that root-cap columellar amyloplasts constitute the susceptors for gravity perception. Such studies have also allowed the identification of several molecules that appear to function as second messengers in gravity signal transduction and of potential signal transducers. Auxin has been implicated as a probable component of the signal that carries the gravitropic information between the gravity-sensing cap and the gravity-responding EZs. This has allowed the identification and characterization of important molecular processes underlying auxin transport and response in plants. New molecular models can be elaborated to explain how the gravity signal transduction pathway might regulate the polarity of auxin transport in roots. Further studies are required to test these models, as well as to study the molecular mechanisms underlying a poorly characterized phase of gravitropism that is independent of an auxin gradient.

CONTENTS

INTRODUCTION

Most animals have the ability to move from location to location, giving them great flexibility for survival. Humans are no exception to this rule, often coping with detrimental environmental, political, or socioeconomical conditions by seeking new locations that can sustain a "better and happier life." Most plants, on the other hand, are entirely unable to move; they are tightly anchored to the soil at the site of their germination. Thus, they must spend their entire life cycle at the same site. Such a mode of existence is made possible in part by the fact that most plants have rather limited demands on the type of nutrients they need in order to survive (i.e., light, CO_2, H_2O, and mineral ions).

Even so, plants have to cope with the availability and heterogeneity of light and nutrients in their immediate environment. For instance, light is very limited underground, whereas water and mineral ions are mostly available in the soil. Consequently, different plant organs have evolved in varying ways to efficiently perform the essential functions needed to sustain life. For instance, photosynthesis occurs mainly in shoots, whereas roots are specialized to anchor the plant to its substratum as well as to take up water and mineral ions.

Gravity is a ubiquitous force applied to all objects. It is directed toward the center of the planet and, as such, can be used to define up and down. This directional environmental cue has been acting upon organisms, organs, cells, and intracellular structures since their appearance on earth, and most organisms have acquired ways to use it as a guide for a variety of biological processes. Plants are no exception, with their ability to guide the growth of their organs at a specific angle from the gravity vector (39). This "gravitational set point angle" (GSPA) varies from species to species, organ to organ, and between environmental conditions (39). In general, the corresponding process, named gravitropism, dictates upward shoot growth and downward root growth.

The ability of plant organs to use gravity as a guide for growth was first recognized by Knight in 1806 (62). It has since received much attention. Some of the molecular processes that underlie gravitropism in higher plants have recently been uncovered through careful studies of root, hypocotyl, inflorescence-stem, and coleoptile gravitropism. In this review, we discuss the current status of our knowledge of the molecular mechanisms governing root gravitropism, provide a model that attempts to explain some features of this response, and compare the general features of root gravitropism with those of shoots and coleoptiles.

ROOT GRAVITROPISM

Root Growth and Morphogenesis

Typically, the root of a higher plant is radially symmetrical and composed of a limited number of concentric cell types, making up different tissue types: epidermis, cortex, endodermis, pericycle, and vasculature. Its tip is covered by a cellular structure that protects the apical meristem from damage during growth in soil and serves multiple sensory functions for the root. This structure, the root cap, is composed of central columella, peripheral, and tip cells (illustrated in Figure 1A, see color insert).

The columella cells of the root cap are highly polarized. Their center is deprived of organelles but filled with a network of small and dynamic microfilaments that may connect to the cell membrane. Most of the endoplasmic reticulum (ER) is located at the periphery and essentially constitutes a tubular network. However, highly specialized nodal ER structures are also found in columella cells and appear more abundant along the outer membrane of the most peripheral columella cells. The peripheral region of columella cells also contains other organelles, transverse microfilaments, and microtubules. The nucleus is located at the top of these cells. Amyloplasts, on the other hand, sediment to the physical bottom, on top of the distal ER (25, 34, 133, 136).

Roots grow using a combination of cell division within the meristem and cell elongation in the EZ. The root meristem is located between the root cap and the main part of the EZ. Meristematic cells are derived from the division of a defined number of initials in the root promeristem (Figure 1A).

Cells within and immediately proximal to the meristematic area of a root expand in all three dimensions as their distance from the promeristem increases owing to new divisions at the meristem. The relative contribution of elongation (expansion parallel to the root axis) to the overall expansion process also increases as the cells distance themselves from the cap. Hence, cell expansion progressively converts from a quasi-isotropic to a distinctly anisotropic process. The EZ is a small region of the root tip (less than 700 μm in *Arabidopsis thaliana*) where cell expansion occurs. It can be subdivided into two regions, the distal and central elongation zones (DEZ and CEZ, respectively) (4, 55). The DEZ, which partly overlaps with the meristem, is composed of distal cells whose rate of elongation is below 30% of its maximal value. The CEZ, on the other hand, is located basally to the DEZ and is made up of cells whose rate of elongation is centered on the region of maximal elongation rate (54).

This distinction between DEZ and CEZ appears biologically relevant, as DEZ cells also display specific cytological and physiological properties [reviewed in (4, 55)]. For instance, DEZ cells are the first ones involved in root responses to thigmostimulation (53), mechanical impedance, and electrotropic stimulation (55), as well as low water potential (6, 70, 104) and auxin treatments (54). The first cells involved in the differential growth that accompanies a gravitropic curvature are also located within the DEZ (discussed below).

Root Gravitropism Involves a Complex
Tip-Curvature Response

When roots are accidentally or experimentally reoriented within the gravity field, they undergo a tip-curvature response that ultimately results in a resumption of tip growth along its GSPA (Figure 1*B*). The cytological properties of this curvature response have been studied in detail in several plant species, most notably mung bean, corn, and *A. thaliana*. In all cases, it appears that gravitropic curvature is initiated by differential regulation of cellular elongation at opposite flanks of the DEZ. Specifically, a small group of cells at the upper flank of the DEZ increases their rates of elongation, whereas cells at the lower flank of the DEZ and CEZ, and also at the topside of the CEZ, stop elongating. This allows for the development of a curvature at the tip of the root. As the root grows further, the curvature becomes fully expressed in the CEZ and eventually becomes part of the mature zone. By that time, the tip has resumed growth along its GSPA (54, 56) (Figure 1). The existence of such a root-tip-curvature response to gravistimulation implies that roots can (*a*) perceive gravity, (*b*) interpret this mechanical information and transduce it into a physiological signal, and (*c*) respond to the physiological signal by differential cellular elongation at the DEZ and CEZ (Figure 1*B*).

Gravisensing in Roots

THE STARCH-STATOLITH HYPOTHESIS OF GRAVITY SENSING   The earliest experiments aimed at identifying the gravity-sensing regions of plant organs involved the use of centrifugation. The principle behind these experiments is that the portion of a plant organ located at the axis of a centrifuge is subjected only to the force of gravity (g). Those located at some distance away from the centrifugal axis are subjected to a force resulting from both gravitational and centrifugal accelerations. Hence, if the gravity-sensing region of a plant organ is lined up with the centrifugal axis, that organ should grow as if it were not subjected to centrifugation and should follow its GSPA relative to gravity. If, on the other hand, the gravity-sensing region is located at some distance from the centrifuge axis, the organ should curve and grow along a GSPA relative to the resulting vector dictated by both gravity and centrifugal acceleration.

For roots, this approach implicated the root tip as the principle site of gravity sensing. Unfortunately, the approach lacked the precision necessary to pinpoint specific locales within the tip that serve as hosts for the gavity-sensing machinery. Hence, the gravity-sensing region of roots was delimited to a rather large region of the tip, which included the cap, meristem, and part of the DEZ [reviewed in (87)]. Similar conclusions were drawn in experiments where roots were allowed to grow into a curved glass tube. In these experiments, a gravitropic response was observed only when the tip of the roots was maintained at some angle from the gravity vector; stimulation of the EZs did not elicit a gravitropic response [discussed in (100)].

The gravity-sensing region of roots defined in these early experiments included the root cap. Recognizing that the columella cells at the center of the root cap contain starch-filled plastids (amyloplasts), which sediment to the physical bottom of the cells, Haberlandt (45) proposed the starch-statolith hypothesis. According to this hypothesis, the perception of gravity is mediated by the sedimentation of amyloplasts within these specialized cells (statocytes). Because amyloplasts in other cell types do not sediment, those present in the columella cells appear quite specialized and were named statoliths. In roots, statocytes are found in the columella cells of the cap, whereas in shoots they constitute the endodermis that surrounds the vasculature (45, 82). The starch-statolith hypothesis has received a great deal of attention since its inception. Yet, it has been highly controversial even though recent data support, at least in part, an involvement of amyloplasts in gravity sensing.

Early ablation experiments had demonstrated that the root cap is essential for root gravitropism. When the root cap was removed, the root tip did not develop a gravitropic curvature even though in most cases the ablation had no or little effect on overall root growth (23, 26, 57, 64, 135). Hence, the results of these experiments were compatible with an involvement of the root cap in gravity sensing and/or signal transduction even though its requirement for other phases of the gravitropic response (87) remains a possibility.

Unfortunately, these early surgical ablation experiments did not allow an analysis of the involvement of specific root-cap cells in gravitropism. The advent of *A. thaliana* (with its rather simple and well-characterized root) as a model system in the genetic analysis of plant growth and development, and the construction of high-precision laser ablation devices allowed researchers to ablate specific cell types in otherwise intact root caps. This strategy allowed the identification of specific cap cells as important gravity sensors (14). In these experiments, specific groups of root-cap cells were laser ablated, and the corresponding roots were analyzed for their ability to respond to different periods of gravistimulation. Ablated roots were gravistimulated by horizontal positioning over specific periods of time. At the end of stimulation, plates carrying the stimulated seedlings were rotated for several hours on a rotary platform that randomizes their orientation within the gravity field to allow the development of a tip curvature in response to the previously given stimulus. The presentation time, defined as the minimal period of gravistimulation needed for a plant organ to commit to a curvature response, was then derived and used to characterize the gravitropic sensitivity of ablated roots. This analysis demonstrated that the first two layers of columella cells provide the largest contribution to gravity sensing, whereas the contribution of peripheral and tip cells was minimal. Furthermore, ablating the entire root cap resulted in an almost complete elimination of gravitropism, even though it did not affect the actual rate of root growth. Because the first two layers of columella cells in Arabidopsis root caps contain the largest amyloplasts, whose sedimentation is maximal, the results appear consistent with the starch-statolith hypothesis (14).

The laser-ablation experiments described above were rather short term, as ablated cells are rapidly replaced by new cells generated by the division of initials in the apical meristem. A novel genetic ablation strategy was devised to analyze the effect of long-term ablations of root-cap cells on root growth, development, and gravitropism. In this case, *A. thaliana* was transformed with a construct carrying the diphteria toxin A gene under the control of the root-cap specific *RCP1* promoter. Transgenic plants were recovered, shown to have both short roots that stopped growing after a few days of growth and a disorganized root tip, and displayed no evidence of a gravitropic response (121). The deficiency in root gravitropism displayed by these plants was in agreement with the results of the laser-ablation studies described above. However, the results of these two ablation experiments also differed in that long-term ablation resulted in a cessation of root growth, whereas the short-term ablation did not. This difference suggests that the root cap might also contribute to the regulation of overall root growth (121). In any case, these two experiments converged in identifying the root cap as an important modulator of root gravitropism.

Several independent lines of evidence point to the root cap as the possible primary site of gravity sensing and to the columella amyloplasts as possible gravity susceptors. To test the possibility that amyloplasts might be involved as susceptors in gravity perception, the Hasenstein lab (65) tested whether artificial lateral displacement of root-cap amyloplasts might promote a curvature response in the direction of displacement. They took advantage of the fact that amyloplasts contain starch, a dense material with diamagnetic properties. Like any diamagnetic particle, starch can be displaced by the application of ponderomotive forces derived from high-gradient magnetic fields (HGMF). This property was confirmed when the application of HGMF near the root cap resulted in a lateral displacement of amyloplasts within the columella cells. Importantly, the root tip developed a curvature in response to this magnetophoretic displacement of amyloplasts. Furthermore, amyloplast displacement and tip curvature were not observed when starchless *Arabidopsis pgm* mutants were used. The results strongly suggest that amyloplast displacement is sufficient to promote a curvature at the tip of the root, in full agreement with the starch-statolith hypothesis (65).

THE HYDROSTATIC MODEL OF GRAVISENSING   Unfortunately, the story is not as clear cut as it appears at first glance. First, the laser-ablation experiments described above showed that capless roots still display some evidence of gravitropism. Second, similar conclusions were made when starchless *A. thaliana* mutants were analyzed. In this case, some gravitropism was evident, even though amyloplasts did not sediment (19, 61). Third, changes in electrical potential of DEZ cortical cells occurred much faster (within 30 sec) than anticipated if they were to derive from a chemical signal originating at the root cap upon gravistimulation (52). Similarly, root-cap-dependent changes in proton secretion associated with changes in wall pH were observed on the topside of gravistimulated DEZ and CEZ. These electrophysiological changes at the CEZ were altered in starch-deficient mutants, whereas wild-type and mutant DEZs responded equally well to gravistimulation (8, 37). Fourth, rice roots exposed to high-density solutions during gravistimulation did not develop a graviresponse, arguing against the involvement of intracellular gravisusceptors (112).

Hence, a secondary mechanism of gravity sensing may exist in roots, either in the root cap or in another region of the root tip, such as the DEZ. Even though its identity remains unknown, this mechanism could comply with the hydrostatic model of gravisensing, which was elaborated in an attempt to explain the results of elegant studies on the influence of gravity on cytoplasmic streaming polarity in *Characean* internodal cells (111, 127).

The orientation of a *Chara* internodal cell within the gravity field appears to be sensed by machinery that allows subtraction of the total force exerted by the protoplast on two opposite sides of the cell. This subtraction effectively eliminates the isotropic osmotic pressure from the overall pressure exerted by the protoplast on its cell wall. Because plant protoplasts are connected to their cell walls by transmembrane cytoskeletal linkages, changes in cell orientation within the gravity field promote changes in the distribution of plasma-membrane tensions at these focal attachment points. Consequently, mechanosensitive ion channels may be activated, allowing for increases in cytosolic Ca^{2+} levels. The latter increases may activate a gravity signal transduction pathway responsible for the cellular response, which, in this case, results in a change in the polarity of cytoplasmic streaming (111).

THE GRAVIRECEPTOR(S) IN ROOTS REMAINS UNCHARACTERIZED   For gravisensing to occur, physical information derived from amyloplast or protoplast sedimentation must be converted into a physiological signal, which must be transmitted to the EZs where it can regulate the differential cellular growth responsible for gravitropic curvature. In order for this signal transduction pathway to proceed to the final response, some form of receptor must be activated.

Although the identity of this gravity receptor remains unknown, several indirect lines of evidence point to the possible involvement of mechanosensitive ion channels. Sievers and collaborators (106, 108) have proposed that sedimenting amyloplasts may exert tension at the plasma membrane by pulling on membrane-associated actin filaments. Under their model, this action would promote the opening of stretch-activated channels, which would result in cytosolic Ca^{2+} pulses that trigger the gravity signal transduction pathway (106, 108).

The "tensegrity" model of gravisensing (136), formulated by Staehelin and collaborators, also emphasizes the importance of the actin-based cytoskeletal network in integrating the signal generated by the sedimentation of multiple statocytes in the columella cells (136). However, this model postulates that sedimenting amyloplasts disrupt postulated links between the central actin-based cytoskeletal network and stretch receptors at the plasma membrane. Nodal ER structures, which are located at the interface between the ER-free central region and the peripheral ER network of the columella cells and accumulate along the outer tangential walls of the flanking columella cells, might locally shield such connections from disruption by sedimenting amyloplasts, thereby modulating gravisensing (136).

Several laboratories have studied the distribution of microfilaments in root-cap statocytes. The existence of a discrete and dynamic network of actin microfilaments in the central region and at the periphery of columella cells was reported. This network was denser around the amyloplasts and nucleus, supporting the model described above (25, 34, 49, 133, 136). Treating roots with cytochalasins B and D, which interfere with actin microfilament integrity, resulted in an altered distribution of plastids and nuclei, increased sedimentation of amyloplasts within the statocytes, and a decrease in the drop of membrane potential within root statocytes upon gravistimulation (48, 72, 108, 109). Coincidentally, amyloplasts migrated toward a more proximal position within the statocytes after roots were exposed to microgravity for a short period of time during parabolic rocket flights, suggesting that they might be connected to actin microfilaments (15, 126).

A quantitative analysis of the influence of the actin cytoskeletal network on the sedimentation of amyloplasts within the statocytes would allow investigators to test and refine these models. Recent advances in the utilization of optical tweezers to manipulate the position of amyloplasts within the statocytes and to quantitate the forces acting upon them (105a) should provide the tool needed for such an analysis.

Gravity Signal Transduction and Signal Transmission

GRAVITY SIGNAL TRANSDUCTION IN ROOTS MAY UTILIZE Ca^{2+} AND/OR CYTOSOLIC pH AS SECOND MESSENGERS   There is indirect evidence to support a role for Ca^{2+} in gravity signal transduction. Pharmacological agents believed to interfere with the activity of Ca^{2+} channels, calmodulin, calmodulin-related proteins, and/or Ca^{2+}-ATPases affect gravitropism (10, 12, 16, 88, 107, 110, 114). Additionally, root-cap statocytes have been observed to contain elevated levels of Ca^{2+} and calmodulin relative to other cells in the root (3, 16, 20, 27, 110). Similarly, the irradiation of *Zea mays* (Merit cultivar) roots with white or red light, which is necessary to render them orthogravitropic, led to an elevated level of calmodulins in the root-cap statocytes (38).

Increased levels of calmodulin in root-cap statocytes allow for small changes in cytosolic Ca^{2+} levels to effectively activate signal transduction pathways (110). However, gravity-induced changes in cytosolic Ca^{2+} levels have not been identified yet despite several attempts to do so (67). Such changes may be too small to be detected by the technology currently available. Alternatively, the changes in Ca^{2+} levels may be highly localized within the statocytes and not detectable with the current Ca^{2+} imaging technology. It is also possible that gravistimulation simply does not promote Ca^{2+} fluctuations in statocytes. More work is needed to answer these important questions.

Alterations in root-cap cytoplasmic and apoplastic pH are also likely to be involved in early phases of gravity signal transduction. Fast changes in cytosolic pH have been observed within the columella cells of the root cap, changing from a resting value of 7.2 to a value of 7.6 within 2 min of gravistimulation. These changes preceded the initiation of a gravitropic response at the DEZ by at least 10 min (37, 101). They may be related to the rapid changes in membrane potential observed in *Lepidium* root-cap columella cells and to the changes in ion fluxes observed around the root cap that occur in response to gravistimulation (7, 13).

One research team (101) reported that the gravity-induced changes in cytosolic pH occurred mainly in the L2 layer of the root-cap columella, with faster changes on the lower cells within that tier than in the upper ones. Cells in the L3 layer were shown by that same group to undergo an acidification within the same period (101). Another report (37), however, showed that cells in all layers of the columella responded similarly to gravistimulation by an alkalinization of their cytoplasm (37). Nevertheless, both reports agreed that apoplastic pH within the root cap acidified early after gravistimulation. Also, the changes in cytosolic pH were found to be important for the gravitropic response because disruption of cytoplasmic pH with modifiers at concentrations that did not affect the rate of root growth did affect root gravitropism (37, 101).

Even though changes in cytosolic pH appear important for early phases of gravity signal transduction, their mode of action remains unclear. Because lateral differences across the root cap are minimal, it is likely that these changes do not contribute to establishing a lateral polarity across the tip, such as the auxin transport polarity that appears to be needed to establish a lateral auxin gradient across the gravistimulated root cap (see below). It appears more likely that these changes contribute to conditioning the root-cap columella cells for establishment of such a polarity or for an enhanced ability to utilize that polarity when established. Again, more work is needed to determine the precise role of these pH changes in gravitropism.

THE GRAVITY SIGNAL TRANSDUCTION PATHWAY ALSO INVOLVES THE *ARG1/RHG* AND *ARL2* GENE PRODUCTS   The genetic analysis of gravitropism has identified multiple genes involved in the process. Thus far, however, only a few of them have turned out to encode proteins involved in early phases of gravity signal transduction. The *ARG1/RHG* gene of *A. thaliana* appears to be one of them. Mutations in this gene affected root and hypocotyl gravitropism, without affecting either the rate of organ growth or root-growth resistance to phytohormones or polar auxin transport inhibitors (40, 102). Furthermore, *arg1* mutant seedlings contain starch in their columella and shoot endodermal cells, suggesting that the mutation does not affect starch biosynthesis or accumulation. Mutant hypocotyls also displayed wild-type kinetics of phototropism, indicating that the mutation did not affect the ability of plant organs to curve by differential cell expansion in response to environmental factors other than gravity (40, 102). Based on the phenotype of *arg1* mutant seedlings, it has been proposed that the *ARG1/RHG* gene encodes a protein involved in the early phases of gravity signal transduction (40, 102). This hypothesis is supported by the observation that targeted expression of *ARG1* in the root cap of *arg1-2* mutant plants rescues gravitropism in roots but not in hypocotyls, whereas apparent expression in the endodermal cells rescues the gravitropic defect of hypocotyls but not roots (K. Boonsirichai & P. H. Masson, in preparation).

The *ARG1/RHG* gene has been cloned and found to encode a dnaJ-like protein that carries a J domain at its amino terminus (102). Such proteins have been implicated as molecular chaperones mediating the folding of other proteins, targeting proteins either to specific cellular compartments or for degradation, mediating macromolecular complex assembly, or facilitating specific signal transduction pathways (17, 60). They carry out these functions either separately or in association with HSP70 and, sometimes, HSP90 (60). In some instances, the corresponding macromolecular complexes have been found to interact with the actin cytoskeleton (83). The ARG1/RHG protein contains a predicted coiled-coil domain that shares homology with a number of coiled coils found in proteins known to interact with microtubules or microfilaments in other systems (102). Therefore, ARG1/RHG may function in gravity signal transduction by interacting directly or indirectly with the actin cytoskeleton within the statocytes, possibly contributing to the formation of a macromolecular signal-transducing complex in proximity of that network (102).

The *A. thaliana* genome contains a large number of genes encoding dnaJ-like proteins. However, only two of these genes encode proteins that share a high level of homology with ARG1 throughout their lengths. These genes, named *ARG1-Like1* (*ARL1*) and *ARL2*, have been analyzed recently. ARL2 was also found to function in gravitropism. Mutations in this gene resulted in gravitropic defects in both roots and hypocotyls that were very similar to those of *arg1/rhg*, although somewhat weaker. Interestingly, double mutants *arg1*; *arl2* displayed an altered gravitropic phenotype similar to that of *arg1*, suggesting that both proteins interact in the same gravity signal transduction pathway in roots and hypocotyls (E. S. Rosen, K. Boonsirichai, C. Guan, K. Poff & P. H. Masson, in preparation). It is interesting to note that *arg1* and *arl2* mutant roots and hypocotyls are still capable of gravitropism, even though the properties of the response are altered.

We have performed a genetic screen for modifiers of the gravitropic phenotypes associated with *arg1-2*, as a means of identifying new components of the gravity signal transduction pathway (K. Boonsirichai, J. C. Sedbrook & P. H. Masson, in preparation). This screen has yielded several enhancers of the gravitropic phenotype associated with *arg1-2*. Two of these enhancers, named *modifier of arg1-1* (*mar1*) and *mar2*, appear to almost completely eliminate the gravitropic response of *arg1-2* roots. Amazingly, the remaining response, found in a small fraction of the seedlings, was opposite compared to wild type (negative response instead of positive for roots). These important results suggest that *ARG1*, *MAR1*, and *MAR2* are implicated in gravity signal transduction, contributing to interpretation of the gravitropic signal (K. Boonsirichai, J. C. Sedbrook & P. H. Masson, in preparation). Work is in progress to clone these two novel signal transducers.

THE GRAVITY SIGNAL TRANSDUCTION PATHWAY PROMOTES AUXIN REDISTRIBUTION ACROSS THE ROOT TIP   More than 70 years ago, Cholodny (22) and Went (129) proposed that tropic stimuli, including gravity and lateral light, promoted the lateral transport of auxin toward the bottom of plant organs at the EZs. Because auxin promotes cell elongation in shoot and inhibits it in root, the corresponding auxin gradient was hypothesized to promote upward and downward curvatures in shoots and roots, respectively (22, 129). Ever since its inception, this model has influenced research on gravitropism, and an interesting discussion on its current relevance can be found (120). In this review, we focus on summarizing the results of more recent molecular and genetic studies, which suggest that auxin redistribution may occur across the root tip upon gravistimulation and support a role for auxin in the gravitropic response. Recent data supporting the existence of an auxin-gradient-independent phase in gravitropism are also discussed.

The Cholodny-Went theory does not appear to fully explain gravitropism in roots. As noted earlier, the main site for gravity sensing is located in the root cap, whereas the curvature response develops in the EZ. One attempt to resolve this discrepancy is the "fountain" model of auxin transport (63, 64) (Figure 2, see color insert). According to this model, indole-3-acetic acid (IAA), an auxin, is mainly synthesized in young shoot tissues and transported through the vasculature into the root tip (acropetal transport), where it adds to another pool of de novo synthesized auxin. Auxin is then redistributed laterally to more peripheral tissues and transported back toward the EZs (basipetal transport) where it regulates cellular elongation. Upon gravistimulation, lateral auxin redistribution across the root tip is altered, favoring preferential downward movement across the cap. This net downward auxin movement results in the establishment of a lateral auxin gradient, which would then be transported basipetally to the responding EZs where it would promote downward tip curvature (63, 64).

The "fountain" model of auxin transport is supported by several studies of auxin transport polarity in roots. For instance, several studies indicate that auxin is transported basipetally through peripheral tissues in the root tip (28, 29). In these experiments, the overall polarity of auxin transport was reverted from basipetal to acropetal when root tips were decapped, supporting a role for the root cap in the lateral redistribution of auxin to peripheral tissues (47). Upon gravistimulation, a downward redistribution of auxin across the root cap was observed. This lateral auxin redistribution preceded the curvature response, in agreement with the "fountain" model (134). Both active metabolism and the presence of Ca^{2+} appear to be prerequisites for the redistribution (47).

If the "fountain" model of auxin transport and gravitropism is correct, then polar auxin transport would be essential for root gravitropism. Consistent with this, inhibitors of polar auxin transport have been shown to affect root gravitropism at concentrations that were insufficient to affect root growth rates (78). Similar studies have also suggested that the basipetal transport of auxin along the root tip is necessary for the gravitropic response (90). Unfortunately, recent observations suggest that some auxin transport inhibitors may affect a broader range of cellular processes, such as vesicular trafficking, casting some doubts on the value of these pharmacological experiments (42). It should be noted however that mutations in genes directly implicated in polar auxin transport also affect root gravitropism (discussed below), which directly supports a role for auxin transport in gravitropism.

The "fountain" model implies that one should be able to find in vivo evidence for auxin redistribution across the root tip in response to gravistimulation. New molecular studies appear to support this conclusion, showing that auxin-responsive genes are differentially expressed on opposite sides of gravistimulated root tips with enhanced expression at the bottom side (73, 99). It is now believed that the acropetal auxin transport stream in roots regulates lateral root development (95), whereas the basipetal transport controls gravitropism and, possibly, the initial cell divisions that initiate lateral root formation (18, 90).

AUXIN INFLUX AND EFFLUX CARRIER COMPLEXES ARE ESSENTIAL FOR ROOT GRAVITROPISM   Early biochemical studies indicated that polar auxin transport involves both auxin influx and efflux carrier complexes within transporting cells (32, 33, 92). The chemiosmotic model of polar auxin transport in plants attempts to explain this process (92). This model accounts for the fact that a substantial fraction of IAA found in the apoplast was in the protonated form, owing to low pH. Protonated lipophilic IAA can penetrate a plant cell by passive diffusion through the plasma membrane. IAA can also enter a cell through auxin influx carriers, which could act as H^+-IAA symporters. Once within the basic environment of a cell's cytoplasm, IAA deprotonates and accumulates owing to the low membrane permeability of the deprotonated form. To be transported out of the cell, deprotonated IAA would require an auxin efflux carrier complex.

The polarity of auxin transport within cell files may be determined by the polar distribution of auxin efflux carriers within transporting cells. Accordingly, basipetal auxin transport would occur when the efflux carrier machinery is preferentially located in the basal membrane of transporting cells. Under this model, the driving force for auxin transport along cell files would correspond to the proton-motive force across the plasma membrane (71, 92, 96, 98).

Recent molecular evidence supports several aspects of this model. Molecular genetic analysis of root gravitropism in *A. thaliana* allowed the identification of a potential component of the auxin influx carrier machinery in roots (9, 74). Loss-of-function mutations in *AUX1* result in altered root gravitropism associated with increased root-growth resistance to IAA, 2,4-dichlorophenoxy acetic acid (2,4-D), and ethylene (74). The *AUX1* gene has been cloned and shown to encode a putative transmembrane protein with sequence similarity to bacterial amino-acid permeases (9). The gravitropic phenotype of *aux1* mutant roots can be rescued by adding small amounts of 1-naphthalene acetic acid (1-NAA) to the medium, whereas 2,4-D and IAA have little or no effect (75, 132). Unlike 2,4-D and IAA, which require an auxin influx carrier, 1-NAA diffuses efficiently through the plasma membrane (33). Hence, *aux1* affects a gene that appears to encode a component of the auxin influx carrier machinery (75, 132). This is consistent with the observation that the saturable uptake of 2,4-D is reduced in *aux1* mutant root cultures compared to wild type (75). The *AUX1* gene appears to be expressed in most tissues of the root tip, with the notable exception of the upper three layers of the columella (75).

Other molecules, such as chromosaponin I, a γ-pyronyl-tripterpenoid saponin isolated from *Pisum sativum* L., may interact with AUX1 to regulate its function in auxin transport and gravitropism (89). However, the regulation of its production and mechanism of action on AUX1 are still unknown. It is also not clear whether AUX1 functions in auxin import on its own or as part of a multimeric complex with other proteins. Biochemical studies have allowed the identification of 40- and 42-kD membrane proteins that bind to ^{14}C-IAA (50, 51). These proteins appear to interact with other proteins to form a multimeric complex that is hypothesized to function as a channel or transporter (50, 51). Unfortunately, the putative channel or transporter activity for this complex remains untested, and AUX1 has not been shown to belong to such a complex yet.

Based on early biochemical studies, it appears that the auxin efflux carrier complex may include a transmembrane component, a regulatory protein that binds naphthylphthalamic acid (NPA, a polar auxin transport inhibitor), and possibly a third linker component (77). A root-specific transmembrane component of this complex has recently been characterized and shown to be important for root gravitropism, as discussed below.

Mutations in the *AGR1/EIR1/PIN2/WAV6* gene (named *AGR1* in this review) of *A. thaliana* confer increased root-growth sensitivity to high concentrations of 1-NAA and increased root-growth resistance to ethylene and to polar auxin transport inhibitors (21, 73, 80, 125). Mutant roots also display altered gravitropism and defects in auxin transport. *AGR1* encodes a transmembrane protein that shares homology with bacterial transporters. When expressed in yeast, this protein confers increased resistance to toxic IAA derivatives and also increases the ability of yeast cells to export preloaded radiolabeled IAA (21, 73, 80, 125). The AGR1 protein was immunolocalized to the basal side of epidermal and cortical cells in the root DEZ (80). Taken together, these data strongly suggest that the AGR1 protein is a transmembrane component of the auxin efflux carrier complex in roots. It should be noted that *AGR1* belongs to a gene family consisting of eight genes (116a). One of these genes, *PIN1*, encodes a protein that is believed to function as a transmembrane component of the auxin efflux carrier complex in vascular tissues. *pin1* mutations result in alterations in the morphogenesis of stems, which often terminate in pin-like structures. Auxin transport is highly defective in these tissues (41, 84). The corresponding protein has been localized to the vasculature of plant organs, to the basal side of the cells in stems, and to the distal side in root (41).

As discussed above, probable transmembrane components of both auxin influx and auxin efflux carrier complexes in the root tip have been characterized and shown to be essential for root gravitropism. The corresponding proteins localize to regions of the tip where basipetal auxin transport occurs. However, these proteins have not been observed in the root cap, where lateral auxin distribution in response to gravistimulation is believed to occur. Taken together with the fact that auxin response mutants also display altered root gravitropism (see below), the data strongly suggest that auxin is needed for the gravitropic response of roots.

Unfortunately, these experiments do not address whether the AGR1-carrying auxin efflux carrier complex controls the flow of auxin between the root cap and the EZs in a gravity-dependent manner. This could happen if the activity of the complex is sensitive to a gravity-induced signal generated in the root cap and transmitted to the EZ cells. Apoplastic Ca^{2+} and electrical signals have been identified in this region of the root and could serve such a purpose (4, 11, 66). Alternatively, it is also possible that the DEZ cells are capable of sensing gravity (37, 54). In either case, gravistimulation could directly or indirectly activate a signal transduction pathway within *AGR1*-expressing cells, allowing for regulation of auxin efflux activity within the DEZ. Differential activation of a signal transduction pathway on the top and bottom sides of gravistimulated root-tip DEZs could result in differential basipetal flows of auxin on opposite sides of the root. This pathway could also allow for lateral transport of auxin from top to bottom of stimulated tips. Indeed, the AGR1 protein is localized not only at the basal side of the epidermal and cortical DEZ cells, but also in the outer periclinal membranes of cortical cells (80). In either case, this pathway would contribute to either the formation or the generation of a lateral auxin gradient across gravistimulated root tips, thereby allowing for the differential growth responsible for gravitropic curvature.

In this regard, it is interesting to note that posttranscriptional processes may be involved in the regulation of auxin transport by affecting the activity or abundance of efflux carriers. For instance, protein phosphorylation may modulate auxin efflux activity in plants (32, 91). Furthermore, pharmacological experiments on suspension-cultured tobacco cells and experiments involving the analysis of expression of *AGR1-GUS* transcriptional and translational fusions suggest the existence of a feedback regulatory loop affecting the stability of auxin efflux carriers in response to auxin (32, 105). Coincidentally, some components of the auxin efflux carrier complex turn over very quickly at the plasma membrane, exhibiting a half-life of less than 10 min (32).

AGR1-LIKE PROTEINS MAY CONTRIBUTE TO THE ESTABLISHMENT OF A LATERAL AUXIN GRADIENT ACROSS GRAVISTIMULATED ROOT CAPS   As discussed above, a lateral auxin gradient appears to be generated across the root cap in response to gravistimulation (134). Although neither *PIN1* nor *AGR1* are expressed in the root cap, other *AGR1*-like genes may contribute to lateral auxin redistribution to more peripheral root tissues in the cap (21, 73, 80, 105, 125). Some members of the *AGR1* gene family are expressed in the root-cap columella cells (R. Chen & P. H. Masson, unpublished data). At least one of these root-cap-expressed AGR1-like proteins appears to be localized symmetrically at the periphery of columella cells and quiescent center in vertically grown roots. However, protein localization becomes polarized upon gravistimulation, with accumulation of higher protein levels at the new physical bottom of the statocytes (R. Chen & P. H. Masson, in preparation). Hence, the preliminary data suggest that the gravity signal transduction pathway may play an important role in the regulation of targeted AGR1-like protein trafficking, stability, and/or phosphorylation in the root tip. These posttranscriptional regulatory processes could contribute to the control of auxin transport polarity across the root tip, allowing for formation of the lateral auxin gradient previously discussed in this review (Figure 3, see color insert) (R. Chen & P. H. Masson, in preparation). Interestingly, recent studies indicate that the basal localization of the AGR1-like PIN1 protein results from rapid actin-dependent cycling between the plasma membrane and endosomal compartments and that inhibitors of auxin transport, known to affect gravitropism, function by blocking PIN1 cycling (42, 113). The localization of AGR1 and other AGR1-like proteins might be regulated by a similar process.

Root-Curvature Response to Gravistimulation Involves Auxin

AUXIN IS NEEDED FOR PROPER GRAVITROPIC RESPONSE IN ROOTS   Based on the above data, we can conclude that auxin transport is essential for root gravitropism. This conclusion is reinforced by the fact that many auxin-response mutants display defects in root gravitropism. Typical auxin-response mutants show defects in root-growth response to auxin as well as various morphological defects, including alterations in apical dominance, vasculature development, and organ growth. Such mutants can be organized in two categories based on the mode of action of the corresponding genes. One group of auxin-response mutations affects genes implicated in protein degradation, whereas the other group affects genes encoding nuclear proteins that have been implicated in the regulation of gene expression.

SOME AUXIN-RESPONSE MUTANTS AFFECT A PROTEIN-DEGRADATION PATHWAY
Molecular genetic analysis of auxin response in *A. thaliana* has revealed the involvement of a protein-degradation pathway. In brief, upon activation, an E3 ubiquitin-ligase complex called SCF^{TIR1}, made up of a cullin (AtCUL1), ASK1 (a homolog of the yeast Skp1p protein), RBX1, and the F-box-containing protein TIR1, appears to ubiquitinate specific proteins that act as repressors of auxin action, targeting them for degradation (43).

Activation of this ubiquitin-ligase SCF^{TIR1} complex occurs through the attachment of a ubiquitin-like RUB1 protein to AtCUL1 (30). AtCUL1 rubination requires RUB1 activation by another protein complex that includes AXR1, a protein that shares similarity with the N-terminus of E1 ubiquitin-activating enzymes and ECR1, which shares homology with the C-terminal end of E1s (30, 31, 43). Interestingly, mutations in several of the genes involved in this pathway, including AXR1, TIR1, and ASK1, affect root gravitropism and auxin response (43, 68, 97) (M. Estelle, personal communication). Furthermore, overexpression of TIR1 promotes auxin response, suggesting that SCF^{TIR1} is limiting in the process (43).

OTHER AUXIN-RESPONSE MUTANTS AFFECT GENES INVOLVED IN TRANSCRIPTIONAL REGULATION   Several of the mutations that affect gravitropism and auxin response correspond to genes that encode proteins that appear to control the expression of auxin-response genes in an auxin-dependent fashion. The *AUX/IAA* genes were originally identified as a group of genes whose expression was rapidly upregulated by auxin (2, 117). Subsequent database homology searches and yeast two-hybrid screenings allowed for the identification of other Aux/IAA proteins (1, 59, 124). Structurally, the Aux/IAA proteins share four conserved domains, I through IV. Domain I is the least conserved of the four domains, and no functions have been attributed to it yet. Domain IV contains a nuclear localization domain that appears functional (2). Conserved domain II appears to serve as a destabilizer for the protein (131). Genetic analysis in *A. thaliana* has demonstrated that the functional integrity of this domain is necessary for both protein destabilization and auxin responsiveness (131). Hence, the Aux/IAA proteins are short-lived nuclear proteins that appear to mediate some aspects of auxin responsiveness.

Domains III and IV appear to function in Aux/IAA protein dimerization. Interestingly, Aux/IAA proteins interact not only with a specific subset of other Aux/IAA proteins, but also with a group of transcription factors named auxin-response factors (ARFs) (59, 122, 124). These are transcription factors that carry domains III and IV at their C-terminus, along with a DNA-binding domain at the N-terminus and, most often, a transcription activation domain at the middle (44). These factors interact specifically with auxin-response elements located within the promoter region of auxin-responsive genes, which confer auxin responsiveness when inserted in the promoter of reporter genes (122–124). Some Aux/IAA proteins repress expression of several auxin-responsive genes (124), whereas others appear to function as activators, at least in some tissues (69). Therefore, Aux/IAA and ARF proteins may function together in regulating the auxin-dependent expression of auxin-responsive genes [reviewed in (93)]. Auxin responsiveness of this regulatory network appears to require the degradation of Aux/IAA proteins. Hence, these Aux/IAA regulators may be the targets of the SCF^{TIR1} protein-degradation system described above [reviewed in (93)].

Several mutations in *A. thaliana AUX/IAA* genes have been identified and shown to affect both auxin responsiveness and gravitropism of various plant organs. For instance, the dominant mutation *auxin-resistant 2-1* (*axr2-1*), shown to alter both gravitropism in all plant organs and auxin responsiveness (119, 130), affects domain II of *IAA7*, a gene that belongs to the Aux/IAA family (81). Similarly, the *suppressor of hy2* (*shy2*), *axr3*, and *solitary root* (*slr*) mutations, which result in defective root and hypocotyl gravitropism, also affect members of the Aux/IAA family: *IAA3*, *IAA17*, and *IAA14*, respectively (58, 69, 94, 115, 118). Furthermore, the auxin-resistant, hypocotyl-gravitropism mutant *massugu2* (*msg2*) affects the *IAA19* gene (116). All of these mutations are semidominant, specifically missense mutations in domain II, which suggests that they affect the turnover of the corresponding Aux/IAA proteins. However, loss-of-function mutations in *SHY2/IAA3* also affect root gravitropism, root-growth response to auxin, and lateral root formation, supporting its direct involvement in these processes (118).

Mutations in some *ARF* genes also appear to affect gravitropism, further supporting a role for this regulatory network in the regulation of gravitropism and auxin

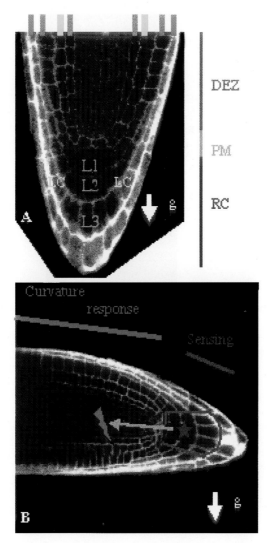

Figure 1 Zones of the *Arabidopsis thaliana* root tip involved in early phases of gravitro-pism. (*A*) Confocal image of a propidium iodide-stained root tip showing the root cap (RC), the promeristem (PM), and the distal elongation zone (DEZ). The root cap is made of three layers of columella cells at the center (L1, L2, and L3), lateral cells (LC), and tip cells (TC). The different tissues that make up the root are represented by color-coded bars: lateral cap (*red*), epidermis (*brown*), cortex (*yellow*), endodermis (*green*), and vasculature (*horizontal blue line*). (*B*) Confocal image of a propidium iodide-stained gravistimulated root tip showing the site of gravity sensing (columella of the root cap) (*outlined in blue and marked by a blue star*) and the site of graviresponse (DEZ) (*lightning sign*). The *green arrow* illustrates the need for signal transmission between the root cap and the elongation zones. The gravity vector (g) is represented by a *white arrow* in both panels.

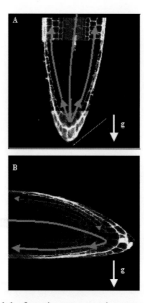

Figure 2 The fountain model of auxin transport in roots. Auxin, mainly synthesized in young shoot tissues, is transported through the vasculature into the root tip, where it is redistributed to more peripheral tissues. It is then transported back basipetally to the elongation zones where it regulates cell expansion. (*A*) When a root grows vertically downward, auxin redistribution to lateral tissues is symmetrical. (*B*) Upon gravistimulation, auxin is preferentially transported laterally toward the bottom side of the tip, resulting in the formation of a lateral auxin gradient. The resulting gradient is then transmitted to the elongation zones, where it is partly responsible for the differential growth that underlies the gravitropic curvature. *Green arrows* represent the direction of auxin transport. Their width is a relative representation of auxin fluxes. *White arrows* represent the gravity vector (g). These pictures show confocal images of the propidium iodide stained *Arabidopsis thaliana* root tips represented in Figure 1.

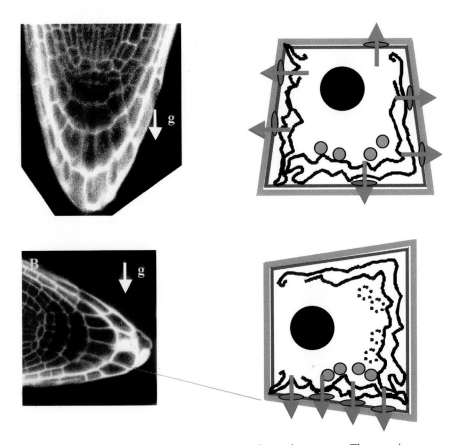

Figure 3 Proposed model to explain gravity sensing at the root cap. The two pictures on the left are close-ups of the confocal images of (*A*) vertically grown and (*B*) gravistimulated *Arabidopsis thaliana* root tips shown in Figure 1. One columella cell is schematized on the right side of this figure, (*A*) before and (*B*) after gravistimulation. This columella cell is surrounded by a plasma membrane (*blue*) and by a cell wall (*brown*). It is highly polarized, containing a nucleus (*black circle*) at the top, peripheral endoplasmic reticulum (*black lines* at the periphery of the cytoplasm), and dense amyloplasts (*brown circles*) that sediment to the bottom. In vertically grown roots, auxin efflux carrier complexes (*red ovals*) are believed to localize symmetrically at the plasma membrane, allowing for auxin efflux on all sides of the cell. Upon gravistimulation, amyloplasts sediment from their original position (*dotted white circle*) to the new physical bottom of the cell. This triggers a gravity signal transduction pathway, which is proposed to promote the accumulation of efflux carrier complexes at the bottom side of the cell. This allows for a preferential downward redistribution of auxin across the root cap, as illustrated in Figure 2. *Green arrows* illustrate auxin efflux, and *white arrows* represent the gravity vector (g).

response. For example, mutations in *NONPHOTOTROPIC 4 (NPH4)/MASSUGU 1 (MSG1)*, which encodes ARF7, affect both hypocotyl gravitropism and auxin response (46).

From the preceding discussion, we can conclude that several *AUX/IAA* and *ARF* genes mediate both auxin response and gravitropism. Considering the fact that mutations in the SCF^{TIR1} protein-degradation pathway also result in altered root gravitropism (43, 68, 97), the data support a role for auxin in the gravitropic response of roots.

The *A. thaliana* genome contains 29 genes encoding Aux/IAA proteins and 23 genes encoding ARF variants [reviewed in (93)]. Each gene is characterized by a specific temporal and spatial expression pattern, and the corresponding protein interacts with only a subgroup of Aux/IAA and/or ARF proteins. Furthermore, the transcription of many of these genes is also regulated by auxin, although with different regulation kinetics, and the function of several of these genes is needed for gravitropism. Therefore, a thorough analysis of this complex auxin-dependent regulatory network is required if one wants to fully understand the mechanisms underlying gravitropism (93).

An Auxin-Gradient-Independent Phase May Also Exist

Because mutations in the auxin transport and response pathways affect root gravitropism, as discussed above, we conclude that auxin plays an important role in this root-growth response to its environment. However, it could do so either as a polar signal carrying the vectorial information derived from graviperception in the root cap to the EZs or as a growth regulator that is simply needed for the graviresponse but is deprived of vectorial information. Of course, the role of auxin in gravitropism could also lie somewhere in between these two models.

Information outlined in this review argues that auxin may carry some vectorial information to the EZs. Indeed, a lateral gradient of auxin appears to be generated across the root cap in response to gravistimulation, and its transmission to the EZs is consistent with it carrying vectorial information to the graviresponding regions of the root. However, several experimental data suggest that the graviresponse might be more complex than one would anticipate if it were a simple response to a gravity-induced auxin gradient across the tip. For instance, early phases of the gravitropic response involve differential growth at the DEZ that results from activation of cell elongation at the top side and inhibition at the bottom side (5, 54, 56, 103, 138). At the same time, there appears to be an inhibition of cell elongation at both the top and bottom sides of the CEZ (5, 54, 56, 103, 138). Such complex changes are not compatible with the Cholodny-Went theory. Also, Ishikawa & Evans (52) observed rapid changes (30 sec) in intracellular potentials of the outer cortical cells at the DEZ of mung bean roots upon gravistimulation. Such changes were too fast to reflect a response to the gravity-induced auxin gradient transmitted from the root cap (52).

Another phenomenon that cannot easily be explained by the Cholodny-Went theory is a reversal of the differential growth accompanying the gravitropic curvature when a root reaches the vertical. This observation suggests that auxin sensitivity of cells in the EZs might be modulated, as these cells are subjected to the gravity-induced auxin gradient (35).

Even more puzzling is the observation of a vigorous graviresponse when roots were exposed to high auxin levels, sufficient to completely eliminate root growth (54, 78). In this case, the entire graviresponse occurred at the DEZ, suggesting that this region of the root is prone to respond to gravistimulation in an auxin-gradient-independent manner (54, 78).

If the initial phases of a graviresponse, involving differential growth at the DEZ, is independent of an auxin gradient generated at the cap, one has to consider other mechanisms as potential mediators of the response. Several models have been postulated to explain this interesting phase of gravitropism, even though most of them suffer from a lack of experimental confirmation. One such model postulates that Ca^{2+} gradients, generated across gravistimulated root tips, may mediate early phases of gravitropism. This model is supported by the observation that gravistimulation promotes lateral movement of apoplastic Ca^{2+} toward the bottom side of the root apex (11, 66). Furthermore, asymmetric application of Ca^{2+} on one side of the DEZ promotes a curvature toward the site of application, and application of the Ca^{2+}-chelator BAPTA to maize root tips prevents the curvature (11). It is interesting to note that Ca^{2+} has previously been shown to be essential for lateral auxin redistribution across the root cap (47). Hence, one could hypothesize that a lateral Ca^{2+} gradient may also differentially regulate the basipetal transport of auxin at the top and bottom sides of the root, along the DEZ, if it is transmitted to that region. This could contribute to the establishment of a lateral auxin gradient along the DEZ that could be responsible for the second auxin-gradient-dependent phase of the response. If that process could somehow be physiologically shielded from the contribution of exogenously supplied auxin, this mechanism would also have the potential of contributing to the first phase of the response. Unfortunately, mechanisms underlying the establishment of Ca^{2+} gradients across the cap, their transport from the root cap to the EZs without leakage in the bathing medium, and their effect on cellular elongation at the DEZ, are still unknown.

Another mechanism that might regulate early phases of the graviresponse involves differential proton influx and/or efflux, leading to pH asymmetry across gravistimulated root tips. In vertical roots, surface pH is most acidic at the fastest elongating region, including the CEZ, and less acidic at the DEZ and meristematic areas (13, 37, 79, 86). Upon gravistimulation, surface pH becomes less acidic on the bottom side and more acidic on the upper side. Furthermore, the region of acidic wall pH extends into the DEZ on the top side (79), suggesting that the rapid increase in cell elongation observed at that region of the root upon gravistimulation may be mediated by proton efflux (24, 36). This conclusion appears to be reinforced by measurements of proton fluxes along the tip of gravistimulated roots, which indicate increases in proton efflux along the top side of gravistimulated roots compared to vertical roots (137).

Similar observations have also been made when wall pH was analyzed at various sites of gravistimulated roots (37). In these experiments, wall acidification was found on the top side of the DEZ and preceded initiation of the curvature response by 5 min. Furthermore, the extent of wall acidification of top-side CEZ cells was lower in starchless *pgm* mutants than in wild type, whereas the *pgm* wall-acidification response of the DEZ was wild type in appearance. Hence, wall acidification at the DEZ appears independent of the presence or absence of starch in the columella cells of the root cap, further supporting the possibility that these cells may respond to different gravitropic signals (37).

Even though it is likely that the surface and wall acidification at the CEZ induced by gravistimulation may be part of a response to the lateral auxin gradient generated in response to gravistimulation, the changes observed at the DEZ may have nothing to do with auxin responses (12, 37). Indeed, polar auxin transport inhibitors do not interfere with gravity-induced changes in electrical currents near the tip, even though it remains possible that these currents may involve ions other than protons (12).

Current flows may also contribute to regulating the DEZ response to gravistimulation (36). It has been demonstrated that the patterns of electrical currents around the root tip are modified upon gravistimulation. Whereas vertical roots witness currents flowing out of the main EZ into the meristematic area, gravistimulated roots bore two opposing currents: one flowing acropetally along the top side of the DEZ and one flowing basipetally along the bottom side (8, 24, 128). These changes in electrical currents precede the initiation of gravitropic curvature and are preceded by asymmetrical changes in membrane potentials in root-cap statocytes (7). The possibility that such currents may be based on proton fluxes at least in some regions of the root is under debate (13, 24).

CONCLUDING REMARKS

Even though the involvement of gravity as a guide for the growth of plant organs has been recognized for almost two centuries, little information is available on the molecular mechanisms controlling it. This changed recently when several advances in the study of root gravitropism revealed the importance of cell biological and molecular processes in regulating different phases of gravitropism, as discussed in this review. Such advances have been made possible partly because of the physical separation between the principle sites of gravity sensing and growth response in roots, which permits a better dissection of each phase of the response. Substantial evidence discussed in this review supports a role for amyloplasts in gravity perception in roots, even though an alternative mechanism may coexist. The resulting gravity signal transduction pathway, from which several components have now been identified, appears to regulate the polarity of auxin transport within the columella of the cap and to regulate the activity of plasma-membrane molecules that affect proton secretion in that region. Auxin appears to be transported from the root cap to the EZs using a complex set of transmembrane proteins that belong to auxin influx and efflux carrier machinery. The localization of the efflux carriers appears to determine the polarity of transport. Furthermore, analyses of gravitropism and auxin response mutants revealed the existence of a protein-degradation pathway as a necessary component of these responses, as discussed above. Also, they allowed the identification of a group of transcriptional regulators as important targets for the degradation pathway. In this review, we propose a role for the gravity signal transduction pathway in controlling the localization of auxin efflux carrier complexes within the columella cells, thereby permitting the lateral transport of auxin during gravistimulation.

Additional evidence coming from analyses of shoot, coleoptile, and pulvinus gravitropism in higher plants also suggests involvement of other molecules, such as inositol 1,4,5-trisphosphates (85), and organelles, such as vacuoles (115), in various phases of gravitropism [reviewed in (76)], even though root, shoot, and hypocotyl gravitropism are genetically separable [reviewed in (76, 115)]. We can anticipate that novel research tools brought about by the completion of major plant genome sequencing projects (genetics, reverse genetics, genomics, proteomics) and by recent technological breakthroughs in the disciplines of cell biology, biochemistry, and physiology will accelerate the pace of discoveries in this exciting field. Such anticipated progress will likely have profound implications for our understanding of plant biology in general because many of the molecular processes that underlie gravitropism, including mechanotransduction and auxin transport and response, also regulate a number of other aspects of plant growth and development.

ACKNOWLEDGMENTS

We thank Christen Yuen, Billy Hung, Li-Sen Young, and Jessica Will for their critical comments on this manuscript. Work in our laboratory is supported by grants from the National Aeronautic and Space Administration (NASA grants NAG2-1189, NAG2-1336, and NAG2-1492), the National Science Foundation (NSF grant MCB-9905675), and HATCH funds (WIS04310). This is manuscript #3592 of the Laboratory of Genetics.

Visit the Annual Reviews home page at www.annualreviews.org

LITERATURE CITED

1. Abel S, Nguyen MD, Theologis A. 1995. The PS-IAA4/5-like family of early auxin-inducible mRNAs in *Arabidopsis thaliana*. *J. Mol. Biol.* 251:533–49

2. Abel S, Oeller P, Theologis A. 1994. Early auxin-induced genes encode short-lived nuclear proteins. *Proc. Natl. Acad. Sci. USA* 91:326–30

3. Allan E, Trewavas A. 1985. Quantitative changes in calmodulin and NAD kinase

during early cell development in the root apex of *Pisum sativum. Planta* 165:493–501

4. Baluska F, Volkmann D, Barlow P. 1996. Specialized zones of development in roots: view from the cellular level. *Plant Physiol.* 112:3–4

5. Barlow P, Rathfelder E. 1985. Distribution and redistribution of extension growth along vertical and horizontal gravireacting maize roots. *Planta* 165:134–41

6. Baskin T, Meekes H, Liang B, Sharp R. 1999. Regulation of growth anisotropy in well-watered and water-stressed maize roots. II. Role of cortical microtubules and cellulose microfibrils. *Plant Physiol.* 119:681–92

7. Behrens H, Gradmann D, Sievers A. 1985. Membrane-potential responses following gravistimulation in roots of *Lepidium sativum* L. *Planta* 163:463–72

8. Behrens H, Weisenseel M, Sievers A. 1982. Rapid changes in the pattern of electric current around the root tip of *Lepidium sativum* L. following gravistimulation. *Plant Physiol.* 70:1079–1083

9. Bennett MJ, Marchant A, Green HG, May ST, Ward SP, et al. 1996. Arabidopsis *AUX1* gene: a permease-like regulator of root gravitropism. *Science* 273:948–50

10. Biro R, Hale C, Wiegang O, Roux S. 1982. Effect of chlorpromazine on gravitropism in Avena coleoptiles. *Ann. Bot.* 50:737–42

11. Björkman T, Cleland R. 1991. The role of extracellular free Ca^{2+} gradients in gravitropic signalling in maize roots. *Planta* 185:379–84

12. Björkman T, Leopold A. 1987. Effect of inhibitors of auxin transport and of calmodulin on a gravisensing-dependent current in maize roots. *Plant Physiol.* 84:847–50

13. Björkman T, Leopold A. 1987. An electric-current associated with gravity sensing in maize roots. *Plant Physiol.* 84:841–46

14. Blancaflor E, Fasano J, Gilroy S. 1998. Mapping the functional roles of cap cells in the response of Arabidopsis primary roots to gravity. *Plant Physiol.* 116:213–22

15. Buchen B, Braun M, Hejnowicz Z, Sievers A. 1993. Statoliths pull on microfilaments. Experiments under microgravity. *Protoplasma* 172:38–42

16. Busch M, Sievers A. 1993. Membrane traffic from the endoplasmic reticulum to the Golgi apparatus is disturbed by an inhibitor of the Ca^{2+} ATPase in the ER. *Protoplasma* 177:23–31

17. Caplan AJ, Cyr DM, Douglas MG. 1993. Eukaryotic homologues of *Escherichia coli* dnaJ: a diverse protein family that functions with HSP70 stress proteins. *Mol. Biol. Cell* 4:555–63

18. Casimiro I, Marchant A, Bhalero R, Beeckman T, Dhooge S, et al. 2001. Auxin transport promotes Arabidopsis lateral root initiation. *Plant Cell* 13:843–52

19. Caspar T, Pickard B. 1989. Gravitropism in a starchless mutant of Arabidopsis: implications for the starch-statolith theory of gravity sensing. *Planta* 177:185–97

20. Chandra S, Chabot J, Morrison G, Leopold A. 1982. Localization of Ca^{2+} in amyloplasts of root cap cells using ion microscopy. *Science* 216:1221–23

21. Chen RJ, Hilson P, Sedbrook J, Rosen E, Caspar T, Masson P. 1998. The *Arabidopsis thaliana AGRAVITROPIC 1* gene encodes a component of the polar-auxin-transport efflux carrier. *Proc. Natl. Acad. Sci. USA* 95:15112–17

22. Cholodny H. 1928. Beiträge zur hormonalen theorie von tropismen. *Planta* 6:118–34

23. Ciesielski T. 1872. Untersuchungen über die Abwartskrummung der Wurzel. *Beitr. Biol. Pflanz.* 1:1–30

24. Collings D, White R, Overall R. 1992.

Ionic current changes associated with the gravity-induced bending response in roots of *Zea mays* L. *Plant Physiol.* 100:1417–26

25. Collings D, Zsuppan G, Allen N, Blancaflor E. 2001. Demonstration of prominent actin filaments in the root columella. *Planta* 212:392–403

26. Darwin C. 1880. *The Power of Movement in Plants.* London: Murray

27. Dauwalder M, Roux S, Hardison L. 1986. Distribution of calmodulin in pea seedlings: immunochemical localization in plumules and root apices. *Planta* 168:461–70

28. Davies P, Doro J, Tarbox A. 1976. The movement and physiological effect of indoleacetic acid following point applications to root tips of *Zea mays*. *Physiol. Plant.* 36:333–37

29. Davies P, Mitchell E. 1972. Transport of indoleacetic acid in intact roots of *Phaseolus coccineus*. *Planta* 105:139–54

30. del Pozo J, Estelle M. 1999. The Arabidopsis cullin AtCUL1 is modified by the ubiquitin-related protein RUB1. *Proc. Natl. Acad. Sci. USA* 96:15342–47

31. del Pozo J, Timpte C, Tan S, Callis J, Estelle M. 1998. The ubiquitin-related protein RUB1 and auxin response in Arabidopsis. *Science* 280:1760–63

32. Delbarre A, Muller P, Guern J. 1998. Short-lived and phosphorylated proteins contribute to carrier-mediated efflux, but not to influx, of auxin in suspension-cultured tobacco cells. *Plant Physiol.* 116:833–44

33. Delbarre A, Muller P, Imhoff V, Guern J. 1996. Comparison of mechanisms controlling uptake and accumulation of 2, 4-dichlorophenoxy acetic acid, naphthalene-1-acetic acid, and indole-3-acetic acid in suspension-cultured tobacco cells. *Planta* 198:532–41

34. Driss-Ecole D, Vassy J, Rembur J, Guivarc'h A, Prouteau M, et al. 2000. Immunolocalization of actin in root stato-cytes of *Lens culinaris* L. *J. Exp. Bot.* 51:521–28

35. Evans M. 1991. Gravitropism: interaction of sensitivity modulation and effector redistribution. *Plant Physiol.* 95:1–5

36. Evans M, Ishikawa H. 1997. Cellular specificity of the gravitropic motor response in roots. *Planta* 203:S115–22

37. Fasano J, Swanson S, Blancaflor E, Dowd P, Kao T, Gilroy S. 2001. Changes in root cap pH are required for the gravity response of the Arabidopsis root. *Plant Cell* 13:907–21

38. Feldman L, Gildow V. 1984. Effect of light on protein patterns in gravitropically stimulated root caps of corn. *Plant Physiol.* 74:208–12

39. Firn R, Digby J. 1997. Solving the puzzle of gravitropism—has a lost piece been found? *Planta* 203:S159–63

40. Fukaki H, Fujisawa H, Tasaka M. 1997. The *RHG* gene is involved in root and hypocotyl gravitropism in *Arabidopsis thaliana*. *Plant Cell Physiol.* 38:804–10

41. Gälweiler L, Guan C, Müller A, Wisman E, Mendgen K, et al. 1998. Regulation of polar auxin transport by AtPIN1 in Arabidopsis vascular tissue. *Science* 282:2226–30

42. Geldner N, Friml J, Stierhof Y-D, Jürgens G, Palme K. 2001. Auxin transport inhibitors block PIN1 cycling and vesicle trafficking. *Nature* 413:425–28

43. Gray WM, del Pozo JC, Walker L, Hobbie L, Risseeuw E, et al. 1999. Identification of an SCF ubiquitin-ligase complex required for auxin response in *Arabidopsis thaliana*. *Genes Dev.* 13:1678–91

44. Guilfoyle T, Ulmasov T, Hagen G. 1998. The ARF family of transcription factors and their role in plant hormone-responsive transcription. *Cell Mol. Life Sci.* 54:619–27

45. Haberlandt G. 1900. Ueber die perzeption des geotropischen reizes. *Ber. Dtsch. Bot. Ges.* 18:261–72

46. Harper R, Stowe-Evans E, Lusse D, Muto H, Tatematsu K, et al. 2000. The

NPH4 locus encodes the auxin response factor ARF7, a conditional regulator of differential growth in aerial Arabidopsis tissue. *Plant Cell* 12:757–70

47. Hasenstein K, Evans M. 1988. Effects of cations on hormone transport in primary roots of *Zea mays*. *Plant Physiol.* 86:890–94

48. Hensel W. 1985. Cytochalasin B affects the structural polarity of statocytes from cress roots (*Lepidium sativum* L.). *Protoplasma* 129:178–87

49. Hensel W. 1989. Tissue slices from living root caps as a model system in which to study cytodifferentiation of polar cells. *Planta* 177:296–303

50. Hicks G, Rayle D, Jones A, Lomax T. 1989. Specific photoaffinity labeling of two plasma membrane polypeptides with an azido auxin. *Proc. Natl. Acad. Sci. USA* 86:4948–52

51. Hicks G, Rice M, Lomax T. 1993. Characterization of auxin binding proteins from zucchini plasma membrane. *Planta* 189:83–90

52. Ishikawa H, Evans M. 1990. Gravity-induced changes in intracellular potentials of elongating cortical cells of mung bean roots. *Plant Cell Physiol.* 31:457–62

53. Ishikawa H, Evans M. 1992. Induction of curvature in maize roots by calcium or by thigmostimulation: role of the postmitotic isodiametric growth zone. *Plant Physiol.* 100:762–68

54. Ishikawa H, Evans M. 1993. The role of the distal elongation zone in the response of maize roots to auxin and gravity. *Plant Physiol.* 102:1203–10

55. Ishikawa H, Evans M. 1995. Specialized zones of development in roots. *Plant Physiol.* 109:725–27

56. Ishikawa H, Hasenstein K, Evans M. 1991. Computer-based video digitizer analysis of surface extension in maize roots: Kinetics of growth rate changes during gravitropism. *Planta* 183:381–90

57. Juniper B, Groves S, Landau-Schachar

B, Audus L. 1966. Root cap and the perception of gravity. *Nature* 209:93–94

58. Kim B, Sho M, Hong S, Furuya M, Nam H. 1998. Photomorphogenetic development of the Arabidopsis *shy2-1D* mutation and its interaction with phytohormones in darkness. *Plant J.* 15:61–68

59. Kim J, Harter K, Theologis A. 1997. Protein-protein interactions among the Aux/IAA proteins. *Proc. Natl. Acad. Sci. USA* 94:11786–91

60. Kimura Y, Yahara I, Lindquist S. 1995. Role of the protein chaperone YDJ1 in establishing Hsp90-mediated signal transduction pathways. *Science* 268:1362–65

61. Kiss J, Hertel R, Sack F. 1989. Amyloplasts are necessary for full gravitropic sensitivity in roots of *Arabidopsis thaliana*. *Planta* 177:198–206

62. Knight T. 1806. On the direction of the radicle and germen during the vegetation of seeds. *Philos. Trans. R. Soc.* 99:108–20

63. Konings H. 1967. On the mechanism of transverse distribution of auxin in geotropically exposed pea roots. *Acta Bot. Neerl.* 16:161–76

64. Konings H. 1968. The significance of the root cap for geotropism. *Acta Bot. Neerl.* 17:203–11

65. Kuznetsov O, Hasenstein K. 1996. Intracellular magnetophoresis of amyloplasts and induction of root curvature. *Planta* 198:87–94

66. Lee J, Mulkey T, Evans M. 1983. Gravity induced polar transport of calcium across root tips of maize. *Plant Physiol.* 73:874–76

67. Legue V, Blancaflor E, Wymer C, Perbal G, Fantin D, Gilroy S. 1997. Cytoplasmic free Ca^{2+} in Arabidopsis roots changes in response to touch but not gravity. *Plant Physiol.* 114:789–800

68. Leyser H, Lincoln C, Timpte C, Lammer D, Turner J, Estelle M. 1993. Arabidopsis auxin-resistance gene *AXR1* encodes

a protein related to ubiquitin activating enzyme E1. *Nature* 364:161–64

69. Leyser H, Pickett F, Dharmasiri S, Estelle M. 1996. Mutation in the *AXR3* gene of Arabidopsis results in altered auxin response including ectopic expression from the SAUR-AC1 promoter. *Plant J.* 10:403–13

70. Liang B, Sharp R, Baskin T. 1997. Regulation of growth anisotropy in well-watered and water-stressed maize roots. I. Spatial distribution of longitudinal, radial and tangential expansion rates. *Plant Physiol.* 115:101–11

71. Lomax T, Mehlhorn R, Briggs W. 1985. Active auxin uptake by zucchini membrane vesicles: quantitation using ESR volume and pH determinations. *Proc. Natl. Acad. Sci. USA* 82:6541–45

72. Lorenzi G, Perbal G. 1990. Actin filaments responsible for the location of the nucleus in the lentil statocyte are sensitive to gravity. *Biol. Cell* 68:259–63

73. Luschnig C, Gaxiola R, Grisafi P, Fink G. 1998. EIR1, a root-specific protein involved in auxin transport, is required for gravitropism in *Arabidopsis thaliana*. *Genes Dev.* 12:2175–87

74. Maher E, Martindale S. 1980. Mutants of *Arabidopsis thaliana* with altered responses to auxin and gravity. *Biochem. Genet.* 18:1041–53

75. Marchant A, Kargul J, May S, Muller P, Delbarre A, et al. 1999. AUX1 regulates root gravitropism in Arabidopsis by facilitating auxin uptake within root apical tissues. *EMBO J.* 18:2066–73

76. Masson P, Tasaka M, Morita M, Guan C, Chen R, Boonsirichai K. 2002. *Arabidopsis thaliana*: a model for the study of root and shoot gravitropism. In *Arabidopsis*, ed. E Meyerowitz, C Somerville. Rockville, MD: Am. Soc. Plant Biol. In press

77. Morris D, Rubery P, Jarman J, Sabater M. 1991. Effect of inhibitors of protein synthesis on transmembrane auxin transport in *Cucurbita pepo* L. hypocotyl segments. *J. Exp. Bot.* 42:773–83

78. Muday G, Haworth P. 1994. Tomato root growth, gravitropism, and lateral root development: correlation with auxin transport. *Plant Physiol. Biochem.* 32:193–203

79. Mulkey T, Evans M. 1981. Geotropism in corn roots: evidence for its mediation by differential acid efflux. *Science* 212:70–71

80. Müller A, Guan C, Gälweiler L, Tanzler P, Huijser P, et al. 1998. AtPIN2 defines a locus of Arabidopsis for root gravitropism control. *EMBO J.* 17:6903–11

81. Nagpal P, Walker L, Young J, Sonawala A, Timpte C, et al. 2000. AXR2 encodes a member of the Aux/IAA protein family. *Plant Physiol.* 123:563–74

82. Nemec B. 1900. Ueber die art der wahrnehmung des schwekraftreizes bei den pflanzen. *Ber. Dtsch. Bot. Ges.* 18:241–45

83. Nishida E, Koyasu S, Sakai H, Yahara I. 1986. Calmodulin-regulated binding of the 90-kDa heat shock protein to actin filaments. *J. Biol. Chem.* 261:16033–36

84. Okada K, Ueda J, Komaki M, Bell C, Shimura Y. 1991. Requirement of the polar auxin transport system in early stages of Arabidopsis floral bud formation. *Plant Cell* 3:677–84

85. Perera I, Heilmann I, Chang S, Boss W, Kaufman P. 2001. A role for inositol 1,4,5-trisphosphate in gravitropic signaling and the retention of cold-perceived gravistimulation of oat shoot pulvini. *Plant Physiol.* 125:1499–507

86. Pilet P, Versel J, Mayor R. 1983. Growth distribution and surface pH patterns along maize roots. *Planta* 158:398–402

87. Poff K, Martin H. 1989. Site of graviperception in roots: a re-examination. *Physiol. Plant.* 76:451–55

88. Pont-Lezica R, McNally J, Pickard B. 1993. Wall-to-membrane linkers in onion epidermis: some hypotheses. *Plant Cell Environ.* 16:111–23

89. Rahman A, Ahamed A, Amakawa T, Goto N, Tsurumi S. 2001. Chromosaponin I specifically interacts with AUX1 protein in regulating the gravitropic response of Arabidopsis roots. *Plant Physiol.* 125:990–1000

90. Rashotte A, Brady S, Reed R, Ante S, Muday G. 2000. Basipetal auxin transport is required for gravitropism in roots of Arabidopsis. *Plant Physiol.* 122:481–90

91. Rashotte A, DeLong A, Muday G. 2001. Genetic and chemical reductions in protein phosphatase activity alter auxin transport, gravity response and lateral root growth. *Plant Cell* 13:1683–96

92. Raven J. 1975. Transport of indoleacetic acid in plant cells in relation to pH and electrical potential gradients, and its significance for polar IAA transport. *New Phytol.* 74:163–72

93. Reed J. 2001. Roles and activities of Aux/IAA proteins in Arabidopsis. *Trends Plant Sci.* 6:420–25

94. Reed J, Elumalai R, Chory J. 1998. Suppressors of an *Arabidopsis thaliana phyB* mutation identify genes that control light signaling and hypocotyl elongation. *Genetics* 148:1295–310

95. Reed R, Brady S, Muday G. 1998. Inhibition of auxin movement from the shoot into the root inhibits lateral root development in Arabidopsis. *Plant Physiol.* 118:1369–78

96. Rubery P, Sheldrake A. 1974. Carrier-mediated auxin transport. *Planta* 188:101–21

97. Ruegger M, Dewey E, Gray WM, Hobbie L, Turner J, Estelle M. 1998. The TIR1 protein of Arabidopsis functions in auxin response and is related to human SKP2 and yeast Grr1p. *Genes Dev.* 12:198–207

98. Sabater M, Rubery P. 1987. Auxin carriers in Cucurbita vesicles. *Planta* 171:507–13

99. Sabatini S, Beis D, Wolkenfelt H, Murfett J, Guilfoyle T, et al. 1999. An auxin-dependent distal organizer of pattern and polarity in the Arabidopsis root. *Cell* 99:463–72

100. Sack F. 1991. Plant gravity sensing. *Int. Rev. Cytol.* 127:193–252

101. Scott A, Allen N. 1999. Changes in cytosolic pH within Arabidposis root columella cells play a key role in the early signaling pathway for root gravitropism. *Plant Physiol.* 121:1291–98

102. Sedbrook J, Chen R, Masson P. 1999. *ARG1* (*Altered Response to Gravity*) encodes a novel DnaJ-like protein which potentially interacts with the cytoskeleton. *Proc. Natl. Acad. Sci. USA* 96:1140–45

103. Selker J, Sievers A. 1987. Analysis of extension and curvature during the graviresponse in Lepidium roots. *Am. J. Bot.* 74:1863–1971

104. Sharp R, Silk W, Hsiao T. 1988. Growth of the maize primary root at low water potentials. I. Spatial distribution of expansive growth. *Plant Physiol.* 87:50–57

105. Sieberer T, Seifert G, Hauser M, Grisafi P, Fink G, Luschnig G. 2000. Posttranscriptional control of the Arabidopsis auxin efflux carrier EIR1 requires AXR1. *Curr. Biol.* 10:1595–98

105a. Sievers A, Buchen B, Hodick D. 1996. Gravity sensing in tip-growing cells. *Trends Plant Sci.* 1:273–79

106. Sievers A, Buchen B, Volkmann D, Hejnowicz Z. 1991. Role of the cytoskeleton in gravity perception, In *The Cytoskeletal Basis of Plant Growth and Form*, ed. C Lloyd, pp. 169–82. London: Academic

107. Sievers A, Busch M. 1992. An inhibitor of the Ca^{2+}-ATPases in the sarcoplasmic reticula inhibits transduction of the gravity stimulus in cress roots. *Planta* 188:619–22

108. Sievers A, Kruse S, Kuo-Huang L, Wendt M. 1989. Statoliths and microfilaments in plant cells. *Planta* 179:275–78

109. Sievers A, Sondag C, Trebacz K, Hejnowicz Z. 1995. Gravity induced changes in intracellular potentials in

statocytes of cress roots. *Planta* 197:392–98

110. Sinclair W, Oliver I, Maher P, Trewavas A. 1996. The role of calmodulin in the gravitropic response of *Arabidopsis thaliana agr-3* mutant. *Planta* 199:343–51

111. Staves M. 1997. Cytoplasmic streaming and gravity sensing in Chara internodal cells. *Planta* 203:S79–84

112. Staves M, Wayne R, Leopold A. 1997. The effect of external medium on the gravitropic curvature of rice (*Oryza sativa*, Poaceae) roots. *Am. J. Bot.* 84:1522–29

113. Steinmann T, Geldner N, Grebe M, Mangold S, Jackson CL, et al. 1999. Coordinated polar localization of auxin efflux carrier PIN1 by GNOM ARF GEF. *Science* 286:316–18

114. Stinemetz C, Kuzmanoff K, Evans M, Jarret H. 1987. Correlations between calmodulin activity and gravitropic sensitivity in primary roots of maize. *Plant Physiol.* 84:1337–42

115. Tasaka M, Kato T, Fukaki H. 2001. Genetic regulation of gravitropism in higher plants. *Int. Rev. Cytol.* 206:135–54

116. Tatematsu K, Watahiki M, Yamamoto K. 1999. Evidences for a dominant mutation of *IAA19* that disrupts hypocotyl growth curvature responses and alters auxin sensitivity. *Abstr. 8-39. 10th Int. Conf. Arabidopsis Res.*, The Univ. Melbourne, Aust.

116a. The Arabidopsis Genome Initiative. 2001. Analysis of the genome sequence of the flowering plant *Arabidopsis thaliana. Nature* 408:796–815

117. Theologis A, Huynh T, Davis R. 1985. Rapid induction of specific mRNAs by auxin in pea epicotyl tissue. *J. Mol. Biol.* 183:53–68

118. Tian Q, Reed JW. 1999. Control of auxin-regulated root development by the *Arabidopsis thaliana SHY2/IAA3* gene. *Development* 126:711–21

119. Timpte CS, Wilson AK, Estelle M. 1992. Effects of the *axr2* mutation of Arabidopsis on cell shape in hypocotyl and inflorescence. *Planta* 188:271–78

120. Trewavas A. 1992. What remains of the Cholodny-Went theory? *Plant Cell Environ.* 15:759–94

121. Tsugeki R, Fedoroff NV. 1999. Genetic ablation of root cap cells in Arabidopsis. *Proc. Natl. Acad. Sci. USA* 96:12941–46

122. Ulmasov T, Hagen G, Guilfoyle T. 1997. ARF1, a transcription factor that binds to auxin response elements. *Science* 276:1865–68

123. Ulmasov T, Liu Z, Hagen G, Guilfoyle T. 1995. Composite structure of auxin response elements. *Plant Cell* 7:1611–23

124. Ulmasov T, Murfett J, Hagen G, Guilfoyle T. 1997. Aux/IAA proteins repress expression of reporter genes containing natural and highly active synthetic auxin response elements. *Plant Cell* 9:1963–71

125. Utsuno K, Shikanai T, Yamada Y, Hashimoto T. 1998. *AGR*, an agravitropic locus of *Arabidopsis thaliana*, encodes a novel membrane-protein family member. *Plant Cell Physiol.* 39:1111–18

126. Volkmann D, Buchen B, Hejnowicz Z, Tewinkel M, Sievers A. 1991. Oriented movement of statoliths studied in a reduced gravitational field during parabolic flights or rockets. *Planta* 185:153–61

127. Wayne R, Staves M, Leopold A. 1992. The contribution of the extracellular matrix to gravisensing in characean cells. *J. Cell Sci.* 101:611–23

128. Weisenseel M, Becker H, Ehlgötz J. 1992. Growth, gravitropism, and endogenous ion currents of cress roots (*Lepidium sativum* L.). *Plant Physiol.* 100:16–25

129. Went F. 1928. Wuchstoff und Wachstum. *Rec. Trav. Bot. Neerl.* 25:1–116

130. Wilson AK, Pickett FB, Turner JC, Estelle M. 1990. A dominant mutation in Arabidopsis confers resistance to auxin, ethylene and abscisic acid. *Mol. Gen. Genet.* 222:377–83

131. Worley CK, Zenser N, Ramos J, Rouse D, Leyser O, et al. 2000. Degradation of Aux/IAA proteins is essential for normal auxin signalling. *Plant J.* 21:553–62

132. Yamamoto M, Yamamoto K. 1998. Differential effects of 1-naphtaleneacetic acid, indole-3-acetic acid and 2,4-dichlorophenoxyacetic acid on the gravitropic response of roots in an auxin-resistant mutant of Arabidopsis, *aux1*. *Plant Cell Physiol.* 39:660–64

133. Yoder T, Zheng H, Todd P, Staehelin L. 2001. Amyloplast sedimentation dynamics in maize columella cells support a new model for the gravity-sensing apparatus of roots. *Plant Physiol.* 125:1045–60

134. Young L, Evans M, Hertel R. 1990. Correlations between gravitropic curvature and auxin movement across gravistimulated roots of *Zea mays*. *Plant Physiol.* 92:792–96

135. Younis A. 1954. Experiments on the growth and geotropism of roots. *J. Exp. Bot.* 5:357–72

136. Zheng H, Staehelin L. 2001. Nodal endoplasmic reticulum, a specialized form of endoplasmic reticulum found in gravity-sensing root tip columella cells. *Plant Physiol.* 125:252–65

137. Zieschang H, Köhler K, Sievers A. 1993. Changing proton concentrations at the surfaces of gravistimulated Phleum roots. *Planta* 190:546–54

138. Zieschang H, Sievers A. 1991. Graviresponse and the localization of its initiating cells in roots of *Phleum pratense* L. *Planta* 184:468–77

Annu. Rev. Plant Biol. 2002. 53:449–75
DOI: 10.1146/annurev.arplant.53.100301.135233

RUBISCO: Structure, Regulatory Interactions, and Possibilities for a Better Enzyme

Robert J. Spreitzer
Department of Biochemistry, Institute of Agriculture and Natural Resources,
University of Nebraska, Lincoln, Nebraska 68588-0664; e-mail: rspreitzer1@unl.edu

Michael E. Salvucci
Western Cotton Research Laboratory, United States Department of Agriculture,
Agricultural Research Service, Phoenix, Arizona 85040-8830;
e-mail: msalvucci@wcrl.ars.usda.gov

Key Words   carbon dioxide, catalysis, chloroplast, photosynthesis, protein structure

■ **Abstract**   Ribulose-1,5-bisphosphate (RuBP) carboxylase/oxygenase (Rubisco) catalyzes the first step in net photosynthetic CO_2 assimilation and photorespiratory carbon oxidation. The enzyme is notoriously inefficient as a catalyst for the carboxylation of RuBP and is subject to competitive inhibition by O_2, inactivation by loss of carbamylation, and dead-end inhibition by RuBP. These inadequacies make Rubisco rate limiting for photosynthesis and an obvious target for increasing agricultural productivity. Resolution of X-ray crystal structures and detailed analysis of divergent, mutant, and hybrid enzymes have increased our insight into the structure/function relationships of Rubisco. The interactions and associations relatively far from the Rubisco active site, including regulatory interactions with Rubisco activase, may present new approaches and strategies for understanding and ultimately improving this complex enzyme.

CONTENTS

INTRODUCTION

As the entry point of CO_2 into the biosphere, ribulose-1,5-bisphosphate carboxylase/oxygenase (Rubisco) is central to life on earth. Its very slow catalytic rate of a few per second, the low affinity for atmospheric CO_2, and the use of O_2 as an alternative substrate for the competing process of photorespiration together make Rubisco notoriously inefficient as the initial CO_2-fixing enzyme of photosynthesis. Consequently, land plants must allocate as much as 50% of their leaf nitrogen to Rubisco, making this single enzyme the most abundant protein in the world (35).

As the rate-limiting step of photosynthesis in both C_3 (55) and C_4 plants (144), Rubisco is often viewed as a potential target for genetic manipulation to improve plant yield (83, 126, 132). The food, fiber, and fuel needs of an ever-increasing human population and shortages in the availability of water for agriculture are challenges of the twenty-first century that would be impacted positively by successful manipulation of Rubisco in crop plants. During the past 10 years, resolution of a variety of Rubisco atomic structures has increased our understanding of the reaction mechanism of the enzyme. Based on this information, the new conventional wisdom is that improving Rubisco will not be simple but will require multiple mutations that subtly change the positioning of critical residues within the active site. Considering the explosion of new knowledge that has occurred in just the past few years regarding the interactions and associations that impact the structure, function, and regulation of Rubisco, there is reason to believe that successful manipulation of Rubisco may yet be achieved.

GENERAL STRUCTURAL AND FUNCTIONAL CONSIDERATIONS

Wealth of Rubisco Structures and Sequences

Besides being one of the slowest, Rubisco is also one of the largest enzymes in nature, with a molecular mass of 560 kDa. In land plants and green algae, the chloroplast *rbcL* gene encodes the 55-kDa large subunit, whereas a family of *rbcS* nuclear genes encodes nearly identical 15-kD small subunits (reviewed in 28, 124). Following posttranslational processing of both subunits, small subunits are added to a core of chaperone-assembled large subunits in the chloroplast (reviewed in 108, 126). The resulting Form I Rubisco holoenzyme is composed of eight large and eight small subunits (Figure 1A, see color insert). Variations on this theme include the Form II Rubisco of some prokaryotes and dinoflagellates consisting of a dimer of only large subunits (89, 132, 151) and the Form I Rubisco of nongreen algae produced from *rbcL* and *rbcS* genes that are both chloroplast encoded (132). In still another variation, the Rubisco from archaebacteria is neither Form I or II, but a decamer comprising five large-subunit dimers (82).

More than 20 Rubisco X-ray crystal structures now exist within the Protein Data Bank (10). These range from the homodimeric holoenzyme of *Rhodospirillum rubrum* (4) that provided the first glimpse of the active site to the high-resolution structures of spinach Rubisco with bound substrate, product, and transition-state analogs (3, 68, 133–136). The C-terminal domain of the large subunit of every Rubisco enzyme forms a classic α/β-barrel. Residues predominately in the loops between β strands and α helices interact with the transition-state analog 2-carboxy-arabinitol 1,5-bisphosphate (CABP) (Figure 1*B*). Because several N-terminal-domain residues of a neighboring large subunit also participate as designated "active-site" residues (Figure 1*C*), the functional unit structure of Rubisco is a large-subunit dimer (Figure 1*A*). Four pairs of associated large subunits are capped on each end by four small subunits, each of which interacts with three large subunits (Figure 1*A*). Because large subunits of Form II enzymes contain all the structural elements required for catalysis, the origin and role of the small subunit in Form I enzymes remain enigmatic.

More than 2000 *rbcL* and 300 *rbcS* sequences now reside within GenBank, generated primarily for phylogenetic reconstructions (21, 22, 63). The deduced large-subunit sequences are fairly conserved, and any differences in length occur primarily at the N and C termini. (Throughout this review, numbering of large-subunit residues will be based on the sequence of the spinach large subunit.) *rbcL*-like sequences have also been found in prokaryotes that do not possess photo-autotrophy via the Calvin cycle (reviewed in 132). However, the deduced products of these genes appear to lack certain residues essential for carboxylation, indicating that they may have other functions in the cell (48).

Small subunits are more divergent than large subunits. (Throughout this review, numbering of small-subunit residues will be based on species-specific sequences.) Whereas land-plant and green-algal small subunits generally have larger loops between β strands A and B, some prokaryotes and all nongreen algae have longer C-terminal extensions. Small but significant regions of sequence identity have been observed between the *Synechococcus* small subunit and a protein associated with carboxysome assembly (98), indicating that there may be a shared interaction with the Rubisco large subunit (reviewed in 62) or divergence from a common ancestral protein of unknown function.

Catalytic Mechanisms and Kinetic Insights

FIRST PARTIAL REACTION   Much is known about the Rubisco catalytic mechanisms from chemical modification, directed mutagenesis, and structural studies (reviewed in 23, 53, 124). Carbamylated Rubisco with bound Mg^{2+} (see below) binds RuBP and converts it to the 2,3-enediol(ate) form (enol-RuBP). This first partial reaction requires that a proton be abstracted from the C-3 of RuBP (66). Whereas some controversy surrounded the nature of the requisite base (68, 79), it is now thought that the free oxygen atom of the activator carbamate on Lys-201 is in a suitable environment to accept this proton (3, 23, 91). The basicity of the free carboxylate oxygen appears to be influenced by Asp-203 and Glu-204, two residues that together with carbamylated Lys-201 ligate the Mg^{2+} (Figure 1C). These residues are essential for catalysis (45), but they have not yet been altered in ways to directly test their involvement in proton abstraction. Once abstracted from RuBP, the proton is then shuttled in turn to the oxygen atom at C-2 of enol-RuBP, ε-amino group of Lys-175, and then to 3-phosphoglycerate (50, 51).

CONFORMATIONAL CHANGES   Carbamylation of Lys-201 occurs spontaneously at slightly alkaline pH (80) and is stabilized by Mg^{2+} and various active-site residues (reviewed in 23). Carbamylation is required to "activate" the enzyme, converting it from a catalytically incompetent to a catalytically competent form. The participation of the carbamate and Mg^{2+} in the catalytic mechanism constrains the ways in which the active site can be modified to impact specificity or turnover (16, 49, 53).

Carbamylation causes only minor changes in the conformation of the Rubisco large subunit, primarily affecting residues in the loop between β strands B and C of the N-terminal domain (118, 133). In contrast, binding of RuBP and other phosphorylated ligands induces several major structural changes in and around the active site (91, 118, 133, 136). Most obvious is a 12-Å shift of α/β-barrel loop 6 from a retracted (open) to an extended (closed) position (118, 133) (Figure 1B). In the extended position, Lys-334 of loop 6 interacts with the C-1 phosphate (P1) of the bound ligand, as well as with Glu-60 at the C-terminal end of β strand B of a neighboring large subunit (Figure 1C). Other changes include rotation of the N-terminal domain, which positions the βB-βC loop over the bound substrate, and extension of the C-terminus across the face of the subunit, which may stabilize loop 6 in the extended conformation via ionic interactions (3, 68) (Figure 2, see color insert). These interactions involve Asp-473, a residue that has been proposed to be a latch site for holding the loops in the extended positions (32). The extended loops in the closed conformation shield the active site from solvent.

SECOND PARTIAL REACTION AND CO_2/O_2 SPECIFICITY   It is the second, irreversible partial reaction, the addition of gaseous CO_2 or O_2 to the enol-RuBP, that determines the specificity and rate of the overall reaction (15, 23, 95). The closure of loop 6 affects the position of the ε-amino group of Lys-334 (Figure 1C) that, along with Mg^{2+}, stabilizes quite similar transition states arising from the reaction of either CO_2 or O_2 with enol-RuBP (16, 23). The resulting products, 2-carboxy-3-ketoarabinitol 1,5-bisphosphate or 2-peroxy-3-ketoarabinitol 1,5-bisphosphate (12, 52, 95), are then protonated and hydrated to produce two molecules of 3-phosphoglycerate or one molecule of 3-phosphoglycerate and one molecule of 2-phosphoglycolate, respectively.

The preference of Rubisco for CO_2 versus O_2 is represented by the specificity factor (Ω), the ratio of the catalytic efficiency (V_{max}/K_m) of carboxylation (V_c/K_c) to that of oxygenation (V_o/K_o). Ω, which determines the ratio of the velocity of carboxylation (v_c) to that of oxygenation (v_o) at any specified concentrations of CO_2 and O_2 (57, 70), can be expressed as

$$\Omega = V_c K_o / V_o K_c = v_c / v_o \times [O_2]/[CO_2]. \tag{1}$$

Because Ω is determined by the second, rate-limiting partial reaction (95), one can further define this relationship based on transition-state theory as

$$\ln \Omega = \ln k_c / k_o = \left(\Delta G_o^{\ddagger} - \Delta G_c^{\ddagger} \right)/RT, \tag{2}$$

where k_c and k_o are the rate constants for the competing partial reactions, $\Delta G_c^{\ddagger}$ and $\Delta G_o^{\ddagger}$ are the carboxylation and oxygenation free energies of activation, R is the gas constant, and T is the absolute temperature (15, 16). Because $\Delta G_o^{\ddagger}$ is greater than $\Delta G_c^{\ddagger}$ (and k_o is smaller than k_c), k_o increases faster with temperature than does k_c (16). This accounts for observed decreases in Ω as temperature increases (15, 58, 139). For Rubisco enzymes from organisms as divergent as *R. rubrum* ($\Omega = 15$) and spinach ($\Omega = 80$), $\Delta G_o^{\ddagger} - \Delta G_c^{\ddagger}$ differs by less than 6 kJ mol^{-1} (16). Because this is less energy than that of a single hydrogen bond, rather subtle changes in protein structure are likely responsible for substantial changes in the gaseous substrate specificity of Rubisco.

Defining a Better Enzyme

Although Ω receives much attention when discussions turn toward the quest for a better enzyme (83), Ω does not provide a direct measure of the rates, rate constants, or catalytic efficiencies of either carboxylation or oxygenation. It represents the difference between the free energies of activation for the oxygenation and carboxylation transition states (Equation 2). Ω is a constant for each Rubisco enzyme, equal to the ratio of carboxylation to oxygenation catalytic efficiencies at a given temperature (Equation 1). As a ratio, Ω does not measure how fast the enzyme fixes CO_2 at any given concentrations of CO_2 and O_2. Consequently, to determine if a Rubisco enzyme is truly "better," one needs to measure the amount of CO_2 fixed minus the amount potentially lost via photorespiration. Laing et al. (70) used Rubisco kinetics to define net photosynthesis (Pn) as

$$Pn = v_c - tv_o = (V_c K_o([CO_2] - [O_2]/\Omega))/(K_o[CO_2] + K_c K_o + K_c[O_2]), \tag{3}$$

where t equals 0.5, i.e., the moles of CO_2 lost per mole of O_2 fixed (92). Thus, it is useless to replace a low-Ω enzyme with a high-Ω enzyme (83) if the individual kinetic constants indicate that Pn will not increase at the CO_2 and O_2 concentrations that occur in vivo.

DIVERGENCE OF KINETIC PROPERTIES Ω values differ between Rubisco enzymes from divergent species (57). Form II enzymes of certain prokaryotes and dinoflagellates have the lowest values ($\Omega = 15$) (57, 151), whereas Form I enzymes from nongreen, eukaryotic algae have the highest ($\Omega = 100$–240) (101, 102, 139). The Ω values of the less-divergent Form I enzymes of eubacteria, e.g., cyanobacteria ($\Omega = 40$), green algae ($\Omega = 60$), and land plants ($\Omega = 80$–100), fall between these two extremes (57, 126). Because there is an inverse correlation of Ω with V_c (57, 102, 139), it is difficult to tell if one enzyme is better than another in vivo. For example, the Form II Rubisco of *R. rubrum*, an obligate anaerobe for photosynthetic CO_2 fixation, has a very low Ω value, but a high V_c (57). The low Ω value is inconsequential in this microorganism because anaerobic growth coupled with the high V_c ensures a high v_c in vivo. Cyanobacteria, green algae, and C_4 plants have CO_2-concentrating mechanisms (reviewed in 62, 85, 123), reducing the importance of Ω and K_c in vivo. Although the thermophilic red alga *Galdieria partita* has the highest Ω value yet reported for a Rubisco enzyme assayed at 25°C ($\Omega = 240$) (139), this value falls below 80 (and its k_c and k_o both increase) when the enzyme is assayed at 45°C, the temperature at which *Galdieria* is normally cultured. Because the Rubisco enzymes of C_3 plants have evolved under conditions of ambient temperature and high O_2 concentration, one is tempted to conclude that these must be the "best" Rubisco enzymes. However, if transferred to a cyanobacterium, which has a CO_2-concentrating mechanism (reviewed in 62), the relatively low V_c values of the land-plant enzymes would result in a decrease in Pn, even though the Ω value has doubled.

SELECTION FOR A BETTER ENZYME   Attempts have been made to directly select for "better" Rubisco enzymes in vivo under CO_2-limited conditions (29, 96, 131; reviewed in 124, 125). However, one cannot culture enough single-celled organisms to select for more than a few substitutions simultaneously (124). Clearly, if a single amino-acid substitution could increase Pn, it would have already been selected during evolution. Interactions between residues close in the tertiary structure may be important, but such interacting residues may be far apart in primary structure, limiting in vitro combinatorial mutagenesis. Furthermore, deletions, insertions, or associations between subunits are difficult to manipulate by random genetic-selection strategies, even though such "irreversible" changes during evolution may account for major limitations on Rubisco catalytic efficiency.

Schemes for genetic selection are further confounded by the diversity of cellular environments. For example, Rubisco enzymes from photosynthetic microorganisms that contain CO_2-concentrating mechanisms generally have higher K_c values than those of Rubisco enzymes from C_3 plants that lack such concentrating mechanisms (57). One might succeed in selecting a better enzyme by growing these organisms under CO_2-limited conditions (131) only to find that a lower K_c or higher K_o/K_c, which is often obtained at the expense of a decrease in V_c (reviewed in 125), provides no benefit because the CO_2-concentrating mechanism provides sufficient CO_2 to saturate the enzyme. One can, of course, argue indefinitely about the utility of selection strategies, but no Rubisco enzyme has yet been selected that improves Pn in vivo except for normal-Ω revertants of low-Ω Rubisco mutants of *Chlamydomonas* (reviewed in 125).

DISTANT EFFECTS AND RATIONALE FOR IMPROVEMENT   The most obvious targets for improving Rubisco are the 20 residues in van der Waals contact with CABP in the Rubisco active site (3). Directed mutagenesis of these "active-site" residues in *R. rubrum*, *Synechococcus*, *Chlamydomonas*, or tobacco Rubisco has shown that all are required for maximal rates of catalysis (19, 153; reviewed in 53, 124, 125). Furthermore, these active-site residues cannot account for the variation in kinetic parameters observed between Rubisco enzymes from different species (57, 102, 139, 151) because they are nearly 100% conserved among thousands of large-subunit sequences. Thus, analysis of active-site residues is important for understanding the details of the Rubisco catalytic mechanisms, but these highly conserved and essential residues might not be the best targets for engineering a better enzyme. A more promising strategy for the eventual design of an improved enzyme is to examine variable residues and regions farther from the active site.

INTERACTIONS IN THE RUBISCO LARGE SUBUNIT

Whereas all Rubisco X-ray crystal structures show very similar $C\alpha$ backbone structures, there is substantial divergence in amino-acid side chains. Thus, it is difficult to correlate differences in structure with differences in kinetic properties. It has also been a challenge to apply genetic methods for investigating divergent Rubisco enzymes. Because eukaryotic Rubisco holoenzymes fail to assemble when their subunits are expressed in *Escherichia coli* (24, 42), most directed mutations have been made in the *R. rubrum* or *Synechococcus* Rubisco enzymes expressed in *E. coli* (reviewed in 53, 124). Only in tobacco and the green alga *Chlamydomonas reinhardtii* has it become possible to transform the chloroplast genome as a means for analyzing the effects of mutant large subunits on the function of eukaryotic Rubisco (153, 158, 159). Although progress has been made in eliminating or substituting *rbcL* genes in photosynthetic prokaryotes (38, 96, 99), these systems have yet to be exploited for the genetic dissection of Rubisco structure/function relationships (reviewed in 124, 132).

Mutational Approaches

The first attempt to step outside the sphere of the active-site residues was made using classical genetics with *Chlamydomonas* (reviewed in 125). Because this organism can survive in the absence of photosynthesis when supplied with acetate as a source of carbon and energy and continues to synthesize a complete photosynthetic apparatus even when grown in darkness, a number of *rbcL* missense mutants were recovered by screening acetate-requiring strains (G54D, G171D, T173I, R217S, G237S, L290F, V331A) (reviewed in 124, 125). Four of the missense mutants (G54D, R217S, L290F, V331A) and their suppressors have defined regions relatively far from the active site that can influence Ω. These regions include the secondary structural elements close to the loops that contain Lys-201 and Lys-334 (14, 17, 137), as well as regions buried within the N-terminal domain and at the interface between large and small subunits (13, 31, 54, 128). Despite the fact that Rubisco is required for the survival of land plants, two *rbcL* missense mutations (S112F and G322S) that disrupt holoenzyme assembly have been identified in variegated mutants of tobacco (8, 121). Missense mutations recovered by screening are not distributed randomly within the gene because only those that affect essential structural or functional properties will be observed (124). Thus, most of the resulting amino-acid substitutions are likely to provide useful information about catalysis or assembly. For reasons that are not readily apparent, no *rbcL* structural-gene mutation has yet been described from the screening of photosynthesis-deficient prokaryotes (2, 132).

Scanning mutagenesis, in which all residues of a certain type are replaced by directed mutagenesis, has been used to examine those Gly residues that are conserved among all Rubisco large subunits (20, 71). When each of the 22 Gly residues in the *Synechococcus* enzyme was replaced with either Ala or Pro, only the G47A, G47P, G122A, G171A, G179A, G403A, G405A, and G416A enzymes retained some carboxylase activity. However, in the absence of detailed kinetic analysis, it is not known whether further study of the regions affected by these substitutions would be informative. Other conserved large-subunit residues distant from the active site have been replaced by directed mutagenesis with effects on Rubisco function or stability (9, 31, 88; reviewed in 124). For example, when Cys-172 was replaced with Ser in the *Chlamydomonas* large subunit, an increase in holoenzyme stability in vivo was observed under oxidative stress conditions that triggered holoenzyme degradation (88). However, as noted above, replacement of conserved residues is not expected to account for differences in catalysis among divergent enzymes.

Most directed mutagenesis studies aimed at examining distant interactions have relied on a phylogenetic approach in which the identities of residues are changed from those of one species to those of another species of Rubisco. The few conserved differences in sequences between C_3 and C_4 plant Rubisco and between Rubisco enzymes that display different specificities for interaction with Rubisco activase have been identified. However, because a variety of land-plant Rubisco enzymes cannot be genetically engineered, conclusions must be based primarily on alterations of the *Synechococcus* or *Chlamydomonas* enzyme (86, 93). The major limitation to the phylogenetic approach is, once again, deciding which of the many divergent regions are worth analyzing (e.g., 86, 93, 140). Nonetheless, by combining this approach with others, two large-subunit regions have been investigated intensely. One of these regions is close to the active site, and the other is distant.

Loop-6 Amino-Acid Substitutions

Prior to solving the X-ray crystal structure of the dynamic loop 6 (68), mutant screening and selection in *Chlamydomonas* revealed that a V331A substitution in this loop caused decreases in V_c and Ω (14). The second-site suppressor substitutions T342I and G344S in α-helix 6 complemented the V331A substitution and increased V_c and Ω to levels sufficient for restoring photoautotrophic growth to the mutant cells (14, 17). When the T342I substitution was created alone in *Synechococcus* Rubisco, little or no change occurred in Ω, but V_c was decreased and K_m (RuBP) was increased (46, 107). Whereas T342I appears to complement V331A by replacing bulk in the hydrophobic core of the loop (14, 68), the recently solved crystal structure of *Chlamydomonas* Rubisco (135a) confirmed that the G344S substitution may complement V331A by a different mechanism (Figure 2A). Nonetheless, changes in these residues likely affect the discrimination between CO_2 and O_2 by altering the placement of active-site Lys-334 (15, 49) (Figure 2). Deletion of loop-6 residues eliminates function (27, 72), and every conserved loop-6 residue that has been substituted causes a dramatic decline in V_c or Ω (where measured) (46, 75, 107, 153; reviewed in 124, 125). However, engineered substitutions at the conserved, N-terminal-domain Lys-128, which may hydrogen bond with the carboxyl group of Val-331 (Figure 2), also causes decreases in V_c and Ω (9), indicating that changes in loop 6 might influence catalysis by altering the orientation of other active-site residues like Asn-123 (18, 122, 158).

Taking a phylogenetic approach, several investigators have changed the *Synechococcus* α-helix 6 sequence DKAS (residues 338–341) to the EREI or ERDI sequence characteristic of land plants (Figure 2, compare panels *B* and *C*) (46, 61, 94). In all of these studies, V_c was decreased by 8–40%. The lower V_c is characteristic of land-plant Rubisco, but the Ω value did not increase to that of a land-plant enzyme. Similarly, single substitutions in loop 6 of the *Synechococcus* enzyme had either minimal or negative influence on the catalytic properties of Rubisco (81, 94).

Following the discovery that Rubisco enzymes of nongreen, eukaryotic algae generally have higher Ω values than those of land-plant enzymes (101, 102, 139) and are quite divergent in loop-6 sequences (Figure 2), the *Synechococcus* sequence KASTL (residues 339–343) was changed to either the PLMIK or the PLMVK sequence characteristic of *Cylindrotheca* or related nongreen algae (100, 107). However, these mutations reduced both V_c and Ω, and a variety of single residue substitutions produced similar results (100, 107). The *Synechococcus* sequences VDL (residues 346–348) was also changed to YNT or YHT, but these substitutions blocked the assembly of a functional holoenzyme in *E. coli* (100, 107). Loop-6 residues that differ between two species are likely to be complemented by additional residues outside of this loop that also differ (Figure 2). However, to examine only eight residues that differ between two enzymes (in all single, pairwise, and multiple combinations) would require the daunting task of creating and analyzing 256 mutant enzymes!

Only two studies have examined divergent residues involving loop 6 that are far apart in primary structure but close together in tertiary structure (46, 159). Because the C-terminal end of the large subunit interacts with loop 6 (Figure 2), a *Synechococcus* mutant enzyme was created in which both the ETMDKL C-terminal end and the DKAS loop-6 residues (nos. 338–341) were changed to those of spinach Rubisco (PAMDTV and ERDI, respectively). However, the properties of this mutant enzyme were not different from the properties of the mutant enzymes in which each of the two regions was changed separately (i.e., 30% decrease in V_c but no change in Ω) (46).

In contrast to *Synechococcus*, there are only three residues in loop 6 that differ between *Chlamydomonas* (Leu-326, Val-341, Met-349) and land-plant (Ile-326, Ile-341, Leu-349) Rubisco. Whereas a V341I substitution in the *Chlamydomonas* enzyme had little effect, an L326I substitution decreased holoenzyme stability in vitro and in vivo (159). This decrease in stability could be partially complemented by an M349L substitution on the opposite side of the loop (Figure 2, compare panels *A* and *B*). Thus, the divergent pair of residues in van der Waals contact at the base of loop 6 may be retained during evolution owing to its contribution to holoenzyme stability. However, the L326I/M349L double-mutant enzyme had substantial decreases in V_c and Ω (159). The catalytic efficiency and specificity of land-plant Rubisco must be maintained by other residues that complement Ile-326 and Leu-349, and these residues must be different from those that complement the analogous loop-6 residues in *Chlamydomonas*. Because only four of the residues that interact with this region of loop 6 differ between *Chlamydomonas* and spinach (Figure 2, compare panels *A* and *B*), it should be possible to create and analyze a complete set of mutant enzymes.

Bottom of the Barrel

By mutant screening and selection in *Chlamydomonas*, an L290F substitution at the bottom of β-strand 5 was found to decrease V_c and Ω (13), and A222T (in α-helix 2) and V262L (below β-strand 4) suppressor substitutions were recovered that restored Ω to the wild-type level (54). The L290F mutant is a temperature-conditional, acetate-requiring strain. Its mutant Rubisco supports photoautotrophic growth at 25°C, but the enzyme is unstable and degraded at 35°C in vivo (13). The A222T and V262L substitutions not only improve the catalytic efficiency of L290F, but also its thermal stability in vivo and in vitro (31, 54). When present alone, the A222T and V262L suppressor substitutions each improved the thermal stability of otherwise wild-type Rubisco in vitro with only a slight decrease in V_c (31). Although A222T and V262L are improved enzymes with respect to thermal stability, *Chlamydomonas* does not live at temperatures where this enhanced thermal stability can be manifested (60°C). Nonetheless, these results indicate that it may be possible to select for improved Rubisco under conditions not previously encountered by the organism during evolution.

The "long distance" interactions between residue 290 and residues 222 and 262 are particularly interesting because all three residues are in contact with the βA-βB loop of the small subunit (Figure 3*A*, see color insert). The importance of this small-large-subunit interface, which is approximately 20 Å away from the active site, was recently confirmed when small-subunit N54S and A57V suppressor substitutions were found that restored Ω and thermal stability of the L290F enzyme back to wild-type values (Figure 3*A*) (30). All of these substitutions may affect a hydrogen-bond network (68) that extends from the region of Leu-290 and culminates at active-site residue His-327 (30, 31, 54). Nonetheless, subtle changes quite far from the active site can influence catalytic efficiency, indicating that small-subunit residues may also be appropriate targets for genetic engineering of an improved Rubisco.

Only three large-subunit residues (nos. 256, 258, 265) differ between *Chlamydomonas* and spinach Rubisco in the region surrounding the small-subunit βA-βB loop (Figure 3, compare panels *A* and *B*). In a recent study (Y. C. Du & R. J. Spreitzer, unpublished), C256F, K258R, and I265V substitutions were created by directed mutagenesis and chloroplast transformation, thereby converting *Chlamydomonas* residues to the corresponding residues of spinach Rubisco (Figure 3, panels *A* and *B*). Whereas the single- and double-mutant substitutions have only minor effects on catalysis, Rubisco from the C256F/K258R/I265V triple-mutant strain, which can survive photoautotrophically, has a 10% decrease in Ω, largely owing to a substantial decline in V_c. Once again, there must be different residues in spinach Rubisco that complement Phe-256, Arg-258, and Val-265. However, these divergent residues may reside within the closely associated small subunit (Figure 3).

INTERACTIONS INVOLVING THE SMALL SUBUNIT

The existence of Form II Rubisco enzymes, composed of only large subunits, indicates that small subunits are not absolutely essential for carboxylase activity. Furthermore, *Synechococcus* large-subunit octamers (void of small subunits) retain $\sim$1% carboxylase activity and have a normal Ω value (5, 44, 74, 87). However, V_c is drastically reduced in such minimal enzymes, and a variety of side products are produced by misprotonation of RuBP (87). Scanning mutagenesis of cyanobacterial small subunits, expressed with large subunits in *E. coli*, has also shown that substitutions at some of the conserved residues can decrease V_c and holoenzyme assembly (41, 69; reviewed in 124, 126). Although it is apparent that small subunits can influence catalysis indirectly, it has been difficult to determine whether divergent small-subunit residues play a role in the differences in kinetic constants among Rubisco enzymes from different species. *R. rubrum* Rubisco lacks small subunits, small subunits of nongreen algae are encoded by the chloroplast genome (which has yet to be transformed), and land plants have a family of *rbcS* genes in the nucleus that cannot be readily eliminated (43, 65, 152; reviewed in 124, 126).

Hybrid Holoenzymes

Because small-subunit primary structures are more divergent than large-subunit sequences, it is reasonable to consider whether small subunits contribute to the phylogenetic differences in Rubisco catalytic efficiency and Ω. When large subunits from *Synechococcus* ($\Omega = 40$) were assembled with small subunits from spinach ($\Omega = 80$) or *Alcaligenes eutrophus* ($\Omega = 74$), the hybrid enzymes had Ω values comparable to that of the *Synechococcus* holoenzyme despite substantial decreases in V_c (7, 74). In contrast, coexpression of large subunits from *Synechococcus*

($\Omega = 40$) and small subunits from the diatom *Cylindrotheca* ($\Omega = 107$) in *E. coli* produced a hybrid holoenzyme that had an intermediate Ω value of 65 (101). By exploiting chloroplast transformation of tobacco, large subunits from sunflower ($\Omega = 98$) were assembled with the resident small subunits of tobacco ($\Omega = 85$) to produce a hybrid holoenzyme that also had an intermediate Ω value of 89 (60). Because the Ω values of hybrid Rubisco enzymes can, in some cases, be influenced by the contributed small subunits, one might hope to improve Rubisco by providing foreign small subunits. However, the resulting hybrid holoenzymes had greater than 80% decreases in V_c (60, 101). Furthermore, because there are substantial differences in small-subunit sequences (29 residues differ between tobacco and sunflower), it is difficult to determine which part(s) of the small subunit may contribute to differences in catalytic efficiency and Ω.

Small-Subunit βA-βB Loop

The small subunits of prokaryotic and nongreen-algal Rubisco lack 10 residues of a 22-residue loop between β strands A and B that is characteristic of land-plant small subunits (Figure 3). This βA-βB loop contains 27 residues in the small subunits of green algae (Figure 3A). In land plants and green algae, the loop extends between and over the ends of two large subunits from the bottom side of the α/β barrel and interacts with large-subunit α-helices 2 and 8, as well as with the βA-βB loops of two neighboring small subunits (3, 68, 135a). The long C-terminal extensions of some prokaryotic and nongreen-algal small subunits form β loops that place residues into positions similar to the residues in the βA-βB loops of plants and green algae (47, 130) (Figure 3, compare panel *D* with panels *A* and *B*). Because the βA-βB loop region is the most divergent structural feature of Rubisco enzymes, it may account for differences in catalytic efficiency and Ω.

To examine the significance of these residues, the *Synechococcus* βA-βB loop was replaced with the βA-βB loop of the pea small subunit (148). Whereas the wild-type *Synechococcus* small subunit could not assemble with pea large subunits in isolated pea chloroplasts, the chimeric small subunit was now able to assemble owing to the presence of the pea βA-βB loop. A number of amino-acid substitutions were created within the pea βA-βB loop (R53E, E54R, H55A, P59A, D63G, D63L, Y66A), but only R53E blocked holoenzyme assembly (1, 40). Because these studies relied on import of small-subunit precursors into isolated chloroplasts, insufficient amounts of Rubisco could be isolated to determine the influence of the βA-βB-loop substitutions on Rubisco catalysis.

Because N54S and A57V suppressor substitutions in the βA-βB loop of the *Chlamydomonas* small subunit could increase the V_c, Ω, and thermal stability of the large-subunit L290F mutant enzyme (30) (Figure 3A), it seemed likely that the βA-βB loop might also have a direct influence over Rubisco catalysis. Using a mutant of *Chlamydomonas* that lacks the *rbcS* gene family as a host for transformation (65), five βA-βB-loop residues conserved in sequence between *Chlamydomonas* and land plants, but different or absent from the corresponding loops of prokaryotes and nongreen algae, were each replaced with Ala (i.e., *Chlamydomonas*

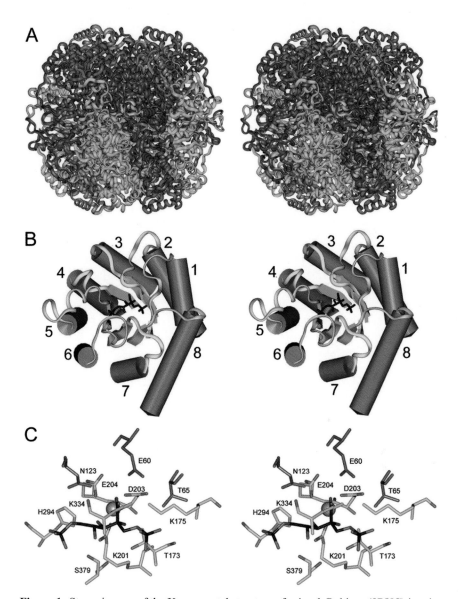

Figure 1 Stereo images of the X-ray crystal structure of spinach Rubisco (8RUC) in relation to catalysis (3). (*A*) The holoenzyme is composed of eight large subunits (*dark blue*, *light blue*) and eight small subunits (*red*, *orange*). Active sites that form between two neighboring large subunits are denoted by loop 6 (*yellow*). (*B*) The C-terminal domain of each large subunit forms an α/β-barrel. Loops (*yellow*) between β strands (*green*) and α helices (*red*) contain residues that interact with the transition-state analog CABP (*black*). (*C*) C-terminal-domain residues (*light blue*) from one large subunit and N-terminal-domain residues (*dark blue*) from a neighboring large subunit interact with CABP (*black*). Mg^{2+} is denoted as a *gray sphere*. Oxygen atoms are colored *red*.

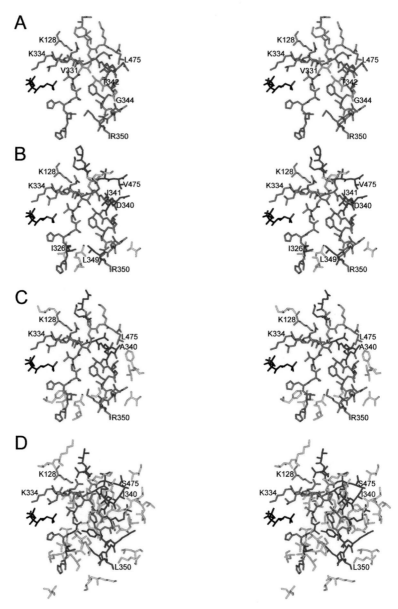

Figure 2 Stereo images of the loop-6 region of Rubisco from (*A*) *Chlamydomonas reinhardtii* at 1.4-Å resolution (135a), (*B*) spinach (8RUC) (3), (*C*) *Synechococcus* (1RBL) (91), and (*D*) *Galdieria* (1BWV) (130). Loop-6 residues (*red*) affected by complementing mutant substitutions in *Chlamydomonas* are colored *green* (17). Loop-6 residues and C-terminal residues (*orange*) that differ relative to those of *Chlamydomonas* are colored *blue*. Other residues that differ from those of *Chlamydomonas* within 5 Å of the loop-6 or C-terminal residues are colored *yellow*. CABP is colored *black*.

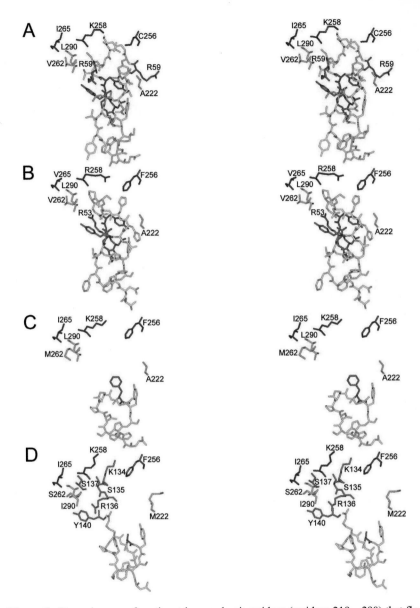

Figure 3 Stereo images of pertinent large-subunit residues (residues 219—290) that flank the small-subunit βA-βB loop (*yellow*) of (*A*) *Chlamydomonas* at 1.4-Å resolution (135a), (*B*) spinach (8RUC) (3), (*C*) *Synechococcus* (1RBL) (91), and (*D*) *Galdieria* (1BWV) (130). Conserved and analogous residues within the small-subunit βA-βB loops are colored *red*. Arg-59 from a neighboring *Chlamydomonas* small subunit is colored *orange*. Residues altered by screening and selection in *Chlamydomonas* are colored *green* (13, 30, 54). *Galdieria* C-terminal small-subunit residues are colored *violet*.

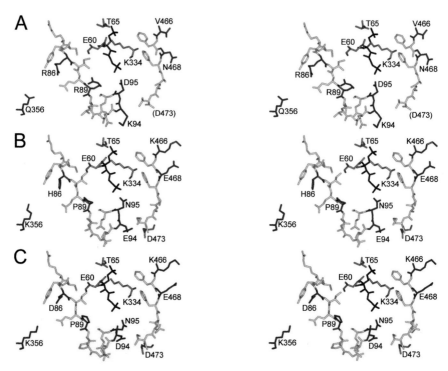

Figure 4 Stereo images of Rubisco large-subunit residues that may account for the species specificity of interaction with Rubisco activase (147). (*A*) tobacco (4RUB) (118), (*B*) spinach (8RUC) (3), and (*C*) *Chlamydomonas* at 1.4-Å resolution (135a). Surface residues that differ in charge between Solanaceae (tobacco) and non-Solanaceae (spinach and *Chlamydomonas*) species are in *blue* (73, 93). Pertinent active-site residues and the Asp-473 latch residue (32) (which is not visible in the tobacco structure) are colored *red*. CABP is colored *black*.

Arg-59, Tyr-67, Tyr-68, Asp-69, Arg-71) (Figure 3*A*) (127). Although none of the substitutions eliminated holoenzyme assembly, most of the mutant enzymes had decreased V_c values, and the R71A enzyme had a reduction in Ω (127). Because Arg-71 can influence Ω and differs relative to the analogous residues of *Synechococcus* (Phe-53) and *Galdieria* (Ala-46) (Figure 3), this residue, and those with which it interacts, may contribute to the differences in catalytic properties between divergent Rubisco enzymes. Thus, the small subunit, and the βA-βB loop in particular, may also be a suitable target for future attempts at engineering an improved Rubisco.

REGULATORY INTERACTIONS OF RUBISCO ACTIVASE

The active site of Rubisco assumes a closed conformation with certain phosphorylated ligands regardless of the carbamylation state of Lys-201 (3, 133–136). This finding is consistent with kinetic evidence indicating that RuBP and its epimer, xylulose bisphosphate, bind very tightly to uncarbamylated sites, and even tighter ($\sim 10^3$ times) than to sites that are carbamylated (56, 157). Duff et al. (32) have proposed that a 9.2-Å distance between the P1 and P2 phosphates of a bisphosphorylated compound is required for the closed conformation. The crystal structure of Rubisco with its carboxylation product, 3-phosphoglyceric acid, is in an open conformation (134), indicating that C—C bond cleavage of the carboxylated C_6 or oxygenated C_5 intermediate during normal catalysis is apparently sufficient to open a closed active site (6, 32). The very tight binding of RuBP to uncarbamylated sites indicates that, once closed, these sites are very slow to open. Thus, the closed conformation represents a potential dead end for uncarbamylated sites because these sites are unable to trigger opening via C—C bond cleavage. That binding of RuBP to the uncarbamylated, low-Ω Rubisco of photosynthetic bacteria is much less tight (56) indicates that tight binding of RuBP to uncarbamylated sites may have developed during evolution as a consequence of increased specificity for CO_2.

Opening the Closed Active Site

In the absence of catalysis, conversion of Rubisco from the closed to the open conformation is extremely slow. To facilitate the process, plants contain Rubisco activase, an ATP-dependent enzyme that releases tight-binding sugar phosphates from the Rubisco active site (reviewed in 97, 113). Activase is an AAA^+ protein, a member of a superset of proteins related to the AAA (ATPases associated with a variety of cellular activities) family that includes a wide variety of proteins with chaperone-like functions (90). Common to each is a core AAA^+ module that contains 11 motifs, some of which, like the P-loop and Walker A and B sequences, are highly conserved among ATPases (90).

Activase interacts with Rubisco, somehow facilitating the release of bound sugar phosphates from the active site (145). The interaction almost certainly involves changing the conformation of Rubisco in a way that promotes opening of the closed configuration. Once activase opens a closed site, the sugar phosphate can dissociate and free the site for activation via spontaneous carbamylation and metal binding (145, 149). Studies with the activase-minus *Arabidopsis rca* mutant (115) and antisense tobacco and *Arabidopsis* plants (33, 84) have shown that photosynthesis at atmospheric levels of CO_2 is severely impaired when plants lack activase because Rubisco becomes sequestered in an inactive form. Thus, activase would be required in all photosynthetic organisms that contain a Rubisco whose active sites are prone to forming a tightly closed conformation with RuBP in the uncarbamylated state.

Activase protein has been detected in all plant species examined, including both C_3 and C_4 plants and green algae (116). GenBank contains entries for activase gene sequences from at least 21 different land-plant species and three species of green algae. An activase-like gene has also been identified in the cyanobacterium *Anabaena* (76). The recombinant product of this activase-like gene catalyzed ATP hydrolysis but was not functional in relieving inhibition of Rubisco by carboxyarabinitol 1-phosphate (77).

Because of the dependence on ATP hydrolysis (105, 129) and, thus, stromal ATP, the controlled switching of Rubisco active sites from the closed to open conformation by activase forms the basis for the regulation of Rubisco by light (97, 113). Activase also facilitates the release of compounds that induce the closure of loop 6 upon binding to carbamylated sites (78, 104, 106, 145). Thus, activase is also required in organisms that contain certain tight-binding inhibitors that sequester carbamylated active sites in the closed conformation. The occurrence, synthesis, and properties of these compounds, which include 3-ketoarabinitol bisphosphate, a compound produced at the active site by misprotonation of RuBP, and carboxyarabinitol 1-phosphate, a naturally occurring inhibitor in some plants, have been discussed in detail by others (6, 34, 64, 119, 157).

Structure/Function Relationships of Activase

ACTIVASE STRUCTURE   For many molecular chaperones, the AAA^+ module is linked covalently to domains that determine the actual cellular function (90). The ATPase activity of the AAA^+ module acts as a motor driving the intermolecular interactions that are required for function. The active molecule is usually multimeric, composed of many AAA^+ subunits often assembled in rings (90 and references therein). In the case of activase, the active form appears to be an oligomer of 14 or perhaps 16 subunits (78, 146).

The activase subunits are highly self-associating, increasing in activity with the extent of oligomerization (111, 146). Oligomerization occurs in response to the binding of ATP or its nonhydrolyzable analog ATPγS. However, ATPγS does not substitute for ATP in Rubisco activation, indicating that oligomerization of activase occurs first, followed by hydrolysis of ATP either before or during the interaction with Rubisco to drive conformational changes in the latter. Mixing experiments with mutant and wild-type activases expressed in *E. coli* have also shown that the subunits function cooperatively (111, 142), consistent with the presence of interactive domains on adjacent subunits, a common feature in AAA^+

proteins (90). The consensus AAA$^+$ module of activase is most similar in sequence to the cell-division-cycle protein 48 and the ATP-dependent regulatory subunit of the eukaryotic 26S protease, a member of the Clp/Hsp100 family. Another member of this family, Hsp104, exhibits ATP- and protein-concentration-dependent association and subunit cooperativity very reminiscent of activase (117). Preliminary analysis of the activase-Rubisco complex by electron microscopy indicates that activase subunits may encircle Rubisco (11), similar to the interaction of the GroEL-type chaperonins with unfolded proteins, including the Form II Rubisco subunits (143).

INTERACTIONS WITH RUBISCO   Several studies have attempted to define the regions of activase necessary for its interaction with Rubisco. Taking advantage of the species specificity of activase (see below), Esau et al. (36) used chimeric activase proteins to show that the C-terminus of activase is important for recognizing Rubisco. Studies with mutants truncated at the N-terminus have shown that the first 50 amino acids of the mature activase protein are not required for ATPase activity but are required for Rubisco activation (141). Within the N-terminus of activase, a conserved Trp residue at position 16 appears to be involved in the interaction between Rubisco and activase (37, 141). Thus, both the C- and N-terminal regions of activase, which lie outside the AAA$^+$ module, appear to be involved in the interaction with Rubisco.

Opening of the closed conformation of Rubisco may involve a physical interaction between activase and Rubisco. Although no stable complex has been isolated, evidence from cross-linking studies (154), immunoprecipitation (154, 156), and electron microscopy (11) favor the possibility of a direct interaction. The most compelling evidence is the species specificity exhibited by activase from members of the Solanaceae like tobacco (147). Activase from tobacco is an inefficient activator of Rubisco from non-Solanaceae land plants and the green-alga *Chlamydomonas* but is active toward tobacco Rubisco and Rubisco from two other Solanaceae plants. The opposite result was obtained with spinach activase. P89R and D94K amino-acid substitutions, introduced into the large subunit of *Chlamydomonas* Rubisco to change specific surface residues to those characteristic of Solanaceae Rubisco (97), altered the species specificity of activase (73, 93). The results suggest the possibility of an activase-recognition region formed, in part, by the loop between β-strands C and D on the surface of the large-subunit N-terminal domain (Figure 4, see color insert).

The mechanism by which activase alters the active site of Rubisco from a closed to an open conformation is unknown. The process is coupled to ATP hydrolysis, either for priming activase for its interaction or during the actual conformation-changing event. Duff et al. (32) have suggested that the activase-Rubisco interaction may involve Asp-473, the latch site that they proposed stabilizes the closed conformation of Rubisco (Figure 4). The C-terminus of the large subunit also contains two residues that differ between Solanaceae and non-Solanaceae enzymes, and these residues reside near the surface of the N-terminal domain of a neighboring large subunit when Rubisco is in the closed conformation (Figure 4, compare panels *A* and *B*).

ACTIVASE ISOFORMS   An interesting but complicating feature of activase is the presence of two subunits of approximately 47 kDa and 42 kDa in many plant species (116). In all cases that have been reported to date, alternative splicing of a pre-mRNA produces these two proteins that are identical except for the presence of an extra 27–36 amino acids at the C-terminus of the longer form (109, 138, 150). The two activase polypeptides are active both in ATP hydrolysis and Rubisco activation (120), but they differ in kinetic properties (26, 120, 155), as well as in their thermal stabilities (26). Zhang & Portis (155) have shown that the longer subunit type is subject to redox regulation via thioredoxin-f-mediated reduction of a pair of Cys residues in the C-terminal extension. Reduction decreases the sensitivity of activase to inhibition by ADP, both the longer form per se and heteromers containing both forms of activase. Redox regulation of activase would serve a regulatory role, adjusting activase activity to light intensity by fine tuning the sensitivity of the enzyme to inhibition by ADP.

Alternative splicing is usually responsible for production of the two forms of activase. However, the two forms are encoded by different activase genes in cotton (M. E. Salvucci, unpublished). Another exception is barley, which has two activase genes, only one of which is alternatively spliced (109). In several plant species, including tobacco, tomato, maize, and *Chlamydomonas*, only a single activase subunit type is produced under nonstress conditions (67, 103, 116). Although these species contain the shorter form of activase that does not harbor the two Cys residues necessary for redox regulation, irradiance levels still affect the activation state of Rubisco in these plants in much the same way as in plants that contain both forms of activase (112). Thus, questions remain concerning activase and Rubisco regulation in plants containing only the shorter form of activase and the occurrence and function of multiple activase genes.

Activase as a Potential Target for Increasing Photosynthesis

Because most strategies for improving Rubisco require a change in the structure of the enzyme, it is necessary to consider how each change will affect the interaction of Rubisco with activase, chaperonins, and other proteins necessary for assembly or function (reviewed in 126). For example, improvements that involve replacement of Rubisco subunits, even if properly assembled, could be ineffective in vivo if activase is unable to recognize the Rubisco and reverse formation of dead-end complexes (60). This problem might even occur with single amino-acid substitutions if they alter recognition of Rubisco by activase (73, 93). Thus, each strategy for improving Rubisco should be mindful of the possible need to co-design activase. Redesigning activase will require a more complete understanding of the mechanism of action and the sites for interaction with Rubisco.

Because the activation state of Rubisco limits photosynthesis under conditions of high CO_2 and temperature, improvements in activase may stimulate photosynthesis under certain conditions (25). For example, the decrease in Rubisco activation that occurs in response to elevated CO_2 appears to involve limitation of activase by [ATP] (25, 110). Engineering activase to be less sensitive to inhibition by ADP, either by directed mutagenesis (59) or by altering the relative expression of the two forms (26, 155), may improve the performance of plants under high CO_2. At elevated temperatures, Rubisco activation decreases to levels that limit photosynthesis because activase is unable to keep pace with the much faster rate of Rubisco deactivation (25, 39). The poor performance of activase at high temperature is caused, in part, by its exceptional thermal lability (114). Thus, changes in activase that improve its thermal stability or increase its amount represent possible approaches for increasing photosynthesis at elevated temperatures. The latter approach may be especially useful in C_4 plants, which have elevated levels of CO_2 at the site of Rubisco but do not exhibit a marked stimulation of photosynthesis by temperature because of lower Rubisco activation (S. J. Crafts-Brandner & M. E. Salvucci, unpublished).

PROSPECTS FOR IMPROVEMENT

Gross changes in the gaseous composition of the earth's atmosphere has selected for land-plant Rubisco with a relatively high Ω. By comparison, the evolutionarily pressure to optimize the land-plant enzyme for performance in controlled agricultural settings has been minimal and indirect. Instead, land plants have evolved under natural conditions where the availability of water and/or nitrogen often limits photosynthesis. As shown by the many studies involving CO_2-enrichment, increasing the rate of carboxylation by Rubisco will increase plant yield, provided that sufficient nitrogen is available for increased protein synthesis. Improving the catalytic efficiency of Rubisco will have the same effect but require less nitrogen to implement.

As the new century opens, the prospects for improving Rubisco are excellent. Elucidation of the X-ray crystal structures of Rubisco in its many forms has provided a structural framework for understanding Rubisco function and evaluating the effects of mutations (3, 32, 91, 135a). Coupled with thousands of Rubisco sequences, these structures may guide phylogenetic and bioinformatic inquiries into the diversity of kinetic parameters. Successful efforts to produce mutations in the chloroplast-encoded large subunit (153, 158) and to replace subunits (60, 65, 127) have laid the groundwork for applying new approaches to understanding Rubisco structure/function relationships. Finally, the realization that Rubisco function is tied to and can be limited by its interaction with activase (25) indicates that improvements in activase may provide a totally new approach for enhancing photosynthesis.

ACKNOWLEDGMENTS

We thank Vijay Chandrasekaran, Patrick D. McLaughlin, and Sriram Satagopan for preparing the figures and acknowledge research support from the U.S. Department of Agriculture (and its National Research Initiative) and Department of Energy.

Visit the Annual Reviews home page at www.annualreviews.org

LITERATURE CITED

1. Adam Z. 1995. A mutation in the small subunit of ribulose-1,5-bisphosphate carboxylase/oxygenase that reduces the rate of its incorporation into holoenzyme. *Photosynth. Res.* 43:143–47

2. Andersen K. 1979. Mutations altering the catalytic activity of a plant-type ribulose bisphosphate carboxylase/oxygenase in *Alcaligenes eutrophus*. *Biochim. Biophys. Acta* 585:1–11

3. Andersson I. 1996. Large structures at high resolution: the 1.6 Å crystal structure of spinach ribulose-1,5-bisphosphate carboxylase/oxygenase complexed with 2-carboxyarabinitol bisphosphate. *J. Mol. Biol.* 259:160–74

4. Andersson I, Knight S, Schneider G, Lindqvist Y, Lundqvist T, et al. 1989. Crystal structure of the active site of ribulose-bisphosphate carboxylase. *Nature* 337:229–34

5. Andrews TJ. 1988. Catalysis by cyanobacterial ribulose-bisphosphate carboxylase large subunits in the complete absence of small subunits. *J. Biol. Chem.* 263:12213–19

6. Andrews TJ. 1996. The bait in the Rubisco mousetrap. *Nat. Struct. Biol.* 3:3–7

7. Andrews TJ, Lorimer GH. 1985. Catalytic properties of a hybrid between cyanobacterial large subunits and higher plant small subunits of ribulose bisphosphate carboxylase-oxygenase. *J. Biol. Chem.* 260:4632–36

8. Avni A, Edelman M, Rachailovich I, Aviv D, Fluhr R. 1989. A point mutation in the gene for the large subunit of ribulose-1,5-bisphosphate carboxylase/oxygenase affects holoenzyme assembly in *Nicotiana tabacum*. *EMBO J.* 8:1915–18

9. Bainbridge G, Anralojc PJ, Madgwick PJ, Pitts JE, Parry MAJ. 1998. Effect of mutation of lysine-128 of the large subunit of ribulose bisphosphate carboxylase/oxygenase from *Anacystis nidulans*. *Biochem. J.* 336:387–93

10. Berman HM, Westbrook J, Feng Z, Gilliland G, Bhat TN, et al. 2000. The Protein Data Bank. *Nucleic Acids Res.* 28: 235–42

11. Buchen-Osmond C, Portis AR Jr, Andrews TJ. 1992. Rubisco activase modifies the appearance of Rubisco in the electron microscope. *Proc. Int. Congr. Photosynth., 9th, Nagoya*, 3:653–56. Dordrecht: Kluwer

12. Chen YR, Hartman FC. 1995. A signature of the oxygenase intermediate in catalysis by ribulose-bisphosphate carboxylase/oxygenase as provided by a site-directed mutant. *J. Biol. Chem.* 270: 11741–44

13. Chen Z, Chastain CJ, Al-Abed SR, Chollet R, Spreitzer RJ. 1988. Reduced CO_2/O_2 specificity of ribulose-1,5-bisphosphate carboxylase/oxygenase in a temperature-sensitive chloroplast mutant of *Chlamydomonas reinhardtii*. *Proc. Natl. Acad. Sci. USA* 85:4696–99

14. Chen Z, Spreitzer RJ. 1989. Chloroplast intragenic suppression enhances the low CO_2/O_2 specificity of mutant ribulose-bisphosphate carboxylase/oxygenase. *J. Biol. Chem.* 264:3051–53

15. Chen Z, Spreitzer RJ. 1991. Proteolysis and transition-state analog binding of mutant forms of ribulose-1,5-bisphosphate carboxylase/oxygenase from *Chlamydomonas reinhardtii*. *Planta* 83: 597–603

16. Chen Z, Spreitzer RJ. 1992. How various factors influence the CO_2/O_2 specificity of ribulose-1,5-bisphosphate carboxylase/oxygenase. *Photosynth. Res.* 31:157–64

17. Chen Z, Yu W, Lee JH, Diao R, Spreitzer RJ. 1991. Complementing amino-acid

substitutions within loop 6 of the α/β-barrel active site influence the CO_2/O_2 specificity of chloroplast ribulose-1, 5-bisphosphate carboxylase/oxygenase. *Biochemistry* 30:8846–50

18. Chene P, Day AG, Fersht AR. 1992. Mutation of asparagine 111 of Rubisco from *Rhodospirillum rubrum* alters the carboxylase/oxygenase specificity. *J. Mol. Biol.* 225:891–96

19. Chene P, Day AG, Fersht AR. 1997. Role of isoleucine-164 at the active site of Rubisco from *Rhodospirillum rubrum*. *Biochem. Biophys. Res. Commun.* 232:482–86

20. Cheng ZQ, McFadden BA. 1998. A study of conserved in-loop and out-of-loop glycine residues in the large subunit of ribulose bisphosphate carboxylase/oxygenase by directed mutagenesis. *Protein Eng.* 11:457–65

21. Clegg MT. 1993. Chloroplast gene sequences and the study of plant evolution. *Proc. Natl. Acad. Sci. USA* 90:363–67

22. Clegg MT, Cummings MP, Durbin ML. 1997. The evolution of plant nuclear genes. *Proc. Natl. Acad. Sci. USA* 94: 7791–98

23. Cleland WW, Andrews TJ, Gutteridge S, Hartman FC, Lorimer GH. 1998. Mechanism of Rubisco: the carbamate as general base. *Chem. Rev.* 98:549–61

24. Cloney LP, Bekkaoui DR, Hemmingsen SM. 1993. Co-expression of plastid chaperonin genes and a synthetic plant Rubisco operon in *Escherichia coli*. *Plant Mol. Biol.* 23:1285–90

25. Crafts-Brandner SJ, Salvucci ME. 2000. Rubisco activase constrains the photosynthetic potential of leaves at high temperature and CO_2. *Proc. Natl. Acad. Sci. USA* 97:13430–37

26. Crafts-Brandner SJ, van de Loo FJ, Salvucci ME. 1997. The two forms of ribulose-1,5-bisphosphate carboxylase/oxygenase activase differ in sensitivity to elevated temperature. *Plant Physiol.* 114:439–44

27. Day AG, Chene P, Fersht AR. 1993. Role of phenylalanine-327 in the closure of loop 6 of ribulosebisphosphate carboxylase/oxygenase from *Rhodospirillum rubrum*. *Biochemistry* 32:1940–44

28. Dean C, Pichersky E, Dunsmuir P. 1989. Structure, evolution, and regulation of *RbcS* genes in higher plants. *Annu. Rev. Plant Physiol. Plant Mol. Biol.* 40:415–39

29. Delgado E, Parry MAJ, Lawlor DW, Keys AJ, Medrano H. 1993. Photosynthesis, ribulose-1,5-bisphosphate carboxylase and leaf characteristics of *Nicotiana tabacum* L. genotypes selected by survival at low CO_2 concentrations. *J. Exp. Bot.* 44:1–7

30. Du YC, Hong S, Spreitzer RJ. 2000. *RbcS* suppressors enhance the CO_2/O_2 specificity and thermal stability of *rbcL*-mutant ribulose-1,5-bisphosphate carboxylase/oxygenase. *Proc. Natl. Acad. Sci. USA* 97:14206–11

31. Du YC, Spreitzer RJ. 2000. Suppressor mutations in the chloroplast-encoded large subunit improve the thermal stability of wild-type ribulose-1,5-bisphosphate carboxylase/oxygenase. *J. Biol. Chem.* 275:19844–47

32. Duff AP, Andrews TJ, Curmi PM. 2000. The transition between the open and closed states of rubisco is triggered by the inter-phosphate distance of the bound bisphosphate. *J. Mol. Biol.* 298:903–16

33. Eckardt NA, Snyder GW, Portis AR Jr, Ogren WL. 1997. Growth and photosynthesis under high and low irradiance of *Arabidopsis thaliana* antisense mutants with reduced ribulose-1,5-bisphosphate carboxylase/oxygenase activase content. *Plant Physiol.* 113:575–86

34. Edmondson DL, Kane HJ, Andrews TJ. 1990. Substrate isomerization inhibits ribulosebisphosphate carboxylase-oxygenase during catalysis. *FEBS Lett.* 260: 62–66

35. Ellis RJ. 1979. The most abundant

protein in the world. *Trends Biochem. Sci.* 4:241–44

36. Esau BD, Snyder GW, Portis AR Jr. 1996. Differential effects of N and C terminal deletions on the two activities of Rubisco activase. *Arch. Biochem. Biophys.* 326:100–5

37. Esau B, Snyder GW, Portis AR Jr. 1998. Activation of ribulose bisphosphate carboxylase/oxygenase (Rubisco) with chimeric activase proteins. *Photosynth. Res.* 58:175–81

38. Falcone DL, Tabita FR. 1991. Expression of endogenous and foreign ribulose-1,5-bisphosphate carboxylase-oxygenase (RubisCO) genes in a RubisCO deletion mutant of *Rhodobacter sphaeroides. J. Bacteriol.* 173:2099–108

39. Feller U, Crafts-Brandner SJ, Salvucci ME. 1998. Moderately high temperatures inhibit ribulose-1,5-bisphosphate carboxylase/oxygenase (Rubisco) activase-mediated activation of Rubisco. *Plant Physiol.* 116:539–46

40. Flachmann R, Bohnert HJ. 1992. Replacement of a conserved arginine in the assembly domain of ribulose-1,5-bisphosphate carboxylase/oxygenase small subunit interferes with holoenzyme formation. *J. Biol. Chem.* 267:10576–82

41. Flachmann R, Zhu G, Jensen RG, Bohnert HJ. 1997. Mutations in the small subunit of ribulose-1,5-bisphosphate carboxylase/oxygenase increase the formation of the misfire product xylulose-1, 5-bisphosphate. *Plant Physiol.* 114:131–36

42. Gatenby AA, van der Vies SM, Rothstein SJ. 1987. Coexpression of both the maize large and wheat small subunit genes of ribulosebisphosphate carboxylase in *Escherichia coli. Eur. J. Biochem.* 168:227–31

43. Getzoff TP, Zhu G, Bohnert HJ, Jensen RG. 1998. Chimeric *Arabidopsis thaliana* ribulose-1,5-bisphosphate carboxylase/oxygenase containing a pea small subunit protein is compromised in

carbamylation. *Plant Physiol.* 116:695–702

44. Gutteridge S. 1991. The relative catalytic specificities of the large subunit core of *Synechococcus* ribulosebisphosphate carboxylase/oxygenase. *J. Biol. Chem.* 266:7359–62

45. Gutteridge S, Lorimer G, Pierce J. 1988. Details of the reactions catalyzed by mutant forms of Rubisco. *Plant Physiol. Biochem.* 26:675–82

46. Gutteridge S, Rhoades DF, Herrmann C. 1993. Site-specific mutations in a loop region of the C-terminal domain of the large subunit of ribulosebisphosphate carboxylase/oxygenase that influence substrate partitioning. *J. Biol. Chem.* 268:7818–24

47. Hansen S, Vollan VB, Hough E, Andersen K. 1999. The crystal structure of Rubisco from *Alcaligenes eutrophus* reveals a novel central eight-stranded β-barrel formed by β-strands from four subunits. *J. Mol. Biol.* 288:609–21

48. Hanson TE, Tabita FR. 2001. A ribulose-1,5-bisphosphate carboxylase/oxygenase (RubisCO)-like protein from *Chlorobium tepidum* that is involved with sulfur metabolism and the response to oxidative stress. *Proc. Natl. Acad. Sci. USA* 98:4397–402

49. Harpel MR, Hartman FC. 1994. Chemical rescue by exogenous amines of a site-directed mutant of ribulose-1,5-bisphosphate carboxylase/oxygenase that lacks a key lysyl residue. *Biochemistry* 33: 5553–61

50. Harpel MR, Hartman FC. 1996. Facilitation of the terminal proton transfer reaction of ribulose 1,5-bisphosphate carboxylase/oxygenase by active-site Lys 166. *Biochemistry* 35:13865–70

51. Harpel MR, Larimer FW, Hartman FC. 1998. Multiple catalytic roles of His 287 of *Rhodospirillum rubrum* ribulose-1, 5-bisphosphate carboxylase/oxygenase. *Protein Sci.* 7:730–38

52. Harpel MR, Serpersu EH, Lamerdin

JA, Huang ZH, Gage DA, Hartman FC. 1995. Oxygenation mechanism of ribulose-bisphosphate carboxylase/oxygenase. Structure and origin of 2-carboxytetritol 1,4-bisphosphate, a novel O_2-dependent side product generated by a site-directed mutant. *Biochemistry* 34: 11296–306

53. Hartman FC, Harpel MR. 1994. Structure, function, regulation, and assembly of D-ribulose-1,5-bisphosphate carboxylase/oxygenase. *Annu. Rev. Biochem.* 63:197–234

54. Hong S, Spreitzer RJ. 1997. Complementing substitutions at the bottom of the barrel influence catalysis and stability of ribulose-bisphosphate carboxylase/oxygenase. *J. Biol. Chem.* 272: 11114–17

55. Hudson GS, Evans JR, von Caemmerer S, Arvidsson YBC, Andrews TJ. 1992. Reduction of ribulose-1,5-bisphosphate carboxylase/oxygenase content by antisense RNA reduces photosynthesis in transgenic tobacco plants. *Plant Physiol.* 98:294–302

56. Jordan DB, Chollet R. 1983. Inhibition of ribulose bisphosphate carboxylase by substrate ribulose 1,5-bisphosphate. *J. Biol. Chem.* 258:13752–58

57. Jordan DB, Ogren WL. 1981. Species variation in the specificity of ribulosebisphosphate carboxylase/oxygenase. *Nature* 291:513–15

58. Jordan DB, Ogren WL. 1984. The CO_2/O_2 specificity of ribulose-1,5-bisphosphate carboxylase/oxygenase. Dependence on ribulosebisphosphate concentration and temperature. *Planta* 161: 308–13

59. Kallis RP, Ewy RG, Portis AR Jr. 2000. Alteration of the adenine nucleotide response and increased Rubisco activation activity of *Arabidopsis* Rubisco activase by site-directed mutagenesis. *Plant Physiol.* 123:1077–86

60. Kanevski I, Maliga P, Rhoades DF, Gutteridge S. 1998. Plastome engineering of ribulose-1,5-bisphosphate carboxylase/oxygenase in tobacco to form a sunflower large subunit and tobacco small subunit hybrid. *Plant Physiol.* 119:133–41

61. Kane HJ, Viil J, Entsch B, Paul K, Morell MK, Andrews TJ. 1994. An improved method for measuring the CO_2/O_2 specificity of ribulosebisphosphate carboxylase-oxygenase. *Aust. J. Plant Physiol.* 21:449–61

62. Kaplan A, Reinhold L. 1999. CO_2 concentrating mechanisms in photosynthetic microorganisms. *Annu. Rev. Plant Physiol. Plant Mol. Biol.* 50:539–70

63. Kellogg EA, Juliano ND. 1997. The structure and function of Rubisco and their implications for systematic studies. *Am. J. Bot.* 84:413–28

64. Keys AJ, Major I, Parry MAJ. 1995. Is there another player in the game of Rubisco regulation? *J. Exp. Bot.* 46:1245–51

65. Khrebtukova I, Spreitzer RJ. 1996. Elimination of the *Chlamydomonas* gene family that encodes the small subunit of ribulose-1,5-bisphosphate carboxylase/oxygenase. *Proc. Natl. Acad. Sci. USA* 93:13689–93

66. King WA, Gready JE, Andrews TJ. 1998. Quantum chemical analysis of the enolization of ribulose bisphosphate: the first hurdle in the fixation of CO_2 by Rubisco. *Biochemistry* 37:15414–22

67. Klein RR, Salvucci ME. 1995. Rubisco, Rubisco activase and ribulose-5-phosphate kinase gene expression and polypeptide accumulation in a tobacco mutant defective in chloroplast protein synthesis. *Photosynth. Res.* 43:213–23

68. Knight S, Andersson I, Branden CI. 1990. Crystallographic analysis of ribulose 1,5-bisphosphate carboxylase from spinach at 2.4 Å resolution. *J. Mol. Biol.* 215:113–60

69. Kostov RV, Small CL, McFadden BA. 1997. Mutations in a sequence near the N-terminus of the small subunit

alter the CO_2/O_2 specificity factor for ribulose bisphosphate carboxylase/oxygenase. *Photosynth. Res.* 54:127–34

70. Laing WA, Ogren WL, Hageman RH. 1974. Regulation of soybean net photosynthetic CO_2 fixation by the interaction of CO_2, O_2 and ribulose 1,5-diphosphate carboxylase. *Plant Physiol.* 54:678–85

71. Larimer FW, Harpel MR, Hartman FC. 1994. β-elimination of phosphate from reaction intermediates by site-directed mutants of ribulose-bisphosphate carboxylase/oxygenase. *J. Biol. Chem.* 269: 11114–20

72. Larson EM, Larimer FW, Hartman FC. 1995. Mechanistic insights provided by deletion of a flexible loop at the active site of ribulose-1,5-bisphosphate carboxylase/oxygenase. *Biochemistry* 34: 4531–37

73. Larson EM, O'Brien CM, Zhu G, Spreitzer RJ, Portis AR Jr. 1997. Specificity for activase is changed by a Pro-89 to Arg substitution in the large subunit of ribulose-1,5-bisphosphate carboxylase/oxygenase. *J. Biol. Chem.* 272: 17033–37

74. Lee B, Read BA, Tabita FR. 1991. Catalytic properties of recombinant octameric, hexadecameric, and heterologous cyanobacterial/bacterial ribulose-1, 5-bisphosphate carboxylase/oxygenase. *Arch. Biochem. Biophys.* 291:263–69

75. Lee GJ, McDonald KA, McFadden BA. 1993. Leucine 332 influences the CO_2/O_2 specificity factor of ribulose-1,5-bisphosphate carboxylase/oxygenase from *Anacystis nidulans*. *Protein Sci.* 2:1147–54

76. Li L-A, Gibson JL, Tabita FR. 1997. The Rubisco activase (*rca*) gene is located downstream from rbcS in *Anabaena* sp. strain CA and is detected in other *Anabaena/Nostoc* strains. *Plant Mol. Biol.* 21:753–64

77. Li L-A, Zianni MR, Tabita FR. 1999. Inactivation of the monocistronic *rca* gene in *Anabaena variabilis* suggests a physi-

ological ribulose bisphosphate carboxylase/oxygenase activase-like function in heterocystous cyanobacteria. *Plant Mol. Biol.* 40:467–78

78. Lilley RM, Portis AR Jr. 1997. ATP hydrolysis activity and polymerization state of ribulose-1,5-bisphosphate carboxylase oxygenase activase: Do the effects of Mg^{2+}, K^+, and activase concentrations indicate a functional similarity to actin? *Plant Physiol.* 114:605–13

79. Lorimer GH, Hartman FC. 1988. Evidence supporting lysine 166 of *Rhodospirillum rubrum* ribulosebisphosphate carboxylase as the essential base which initiates catalysis. *J. Biol. Chem.* 263:6468–71

80. Lorimer GH, Miziorko HM. 1980. Carbamate formation on the ε-amino group of a lysyl residue as the basis for the activation of ribulosebisphosphate carboxylase by CO_2 and Mg^{2+}. *Biochemistry* 19:5321–28

81. Madgwick PJ, Parmar S, Parry MAJ. 1998. Effect of mutations of residue 340 in the large subunit polypeptide of Rubisco from *Anacystis nidulans*. *Eur. J. Biochem.* 253:476–79

82. Maeda N, Kitano K, Fukui T, Ezaki S, Atomi H, et al. 1999. Ribulose bisphosphate carboxylase/oxygenase from the hyperthermophilic archaeon *Pyrococcus kodakaraensis* KOD1 is composed solely of large subunits and forms a pentagonal structure. *J. Mol. Biol.* 293: 57–66

83. Mann CC. 1999. Genetic engineers aim to soup up crop photosynthesis. *Science* 283:314–16

84. Mate CJ, Hudson GS, von Caemmerer S, Evans JR, Andrews TJ. 1993. Reduction of ribulose bisphosphate carboxylase activase levels in tobacco (*Nicotiana tabacum*) by antisense RNA reduces ribulose bisphosphate carboxylase carbamylation and impairs photosynthesis. *Plant Physiol.* 102:1119–28

85. Matsuoka M, Furbank RT, Fukayama

H, Miyao M. 2001. Molecular engineering of C$_4$ photosynthesis. *Annu. Rev. Plant Physiol. Plant Mol. Biol.* 52:297–314

86. Morell MK, Kane HJ, Hudson GS, Andrews TJ. 1992. Effects of mutations at residue 309 of the large subunit of ribulosebisphosphate carboxylase from *Synechococcus* PCC 6301. *Arch. Biochem. Biophys.* 299:295–301

87. Morell MK, Wilkin JM, Kane HJ, Andrews TJ. 1997. Side reactions catalyzed by ribulose-bisphosphate carboxylase in the presence and absence of small subunits. *J. Biol. Chem.* 272:5445–51

88. Moreno J, Spreitzer RJ. 1999. Cys-172 to Ser substitution in the chloroplast-encoded large subunit affects stability and stress-induced turnover of ribulose-bisphosphate carboxylase/oxygenase. *J. Biol. Chem.* 274:26789–93

89. Morse D, Salois P, Markovic P, Hastings JW. 1995. A nuclear-encoded form II RuBisCO in dinoflagellates. *Science* 268:1622–24

90. Neuwald AF. 1999. AAA$^+$: a class of chaperone-like ATPases associated with the assembly, operation and disassembly of protein complexes. *Genome Res.* 9:27–43

91. Newman J, Gutteridge S. 1993. The X-ray structure of *Synechococcus* ribulose-bisphosphate carboxylase/oxygenase-activated quaternary complex at 2.2-Å resolution. *J. Biol. Chem.* 268:25876–86

92. Ogren WL. 1984. Photorespiration: pathways, regulation, and modification. *Annu. Rev. Plant Physiol.* 35:415–42

93. Ott CM, Smith BD, Portis AR Jr, Spreitzer RJ. 2000. Activase region on chloroplast ribulose-1,5-bisphosphate carboxylase/oxygenase: nonconservative substitution in the large subunit alters species specificity of protein interaction. *J. Biol. Chem.* 275:26241–44

94. Parry MAJ, Madgwick P, Parmar S, Cornelius MJ, Keys AJ. 1992. Mutations in loop six of the large subunit of ribulose-1,5-bisphosphate carboxylase affect substrate specificity. *Planta* 187:109–12

95. Pierce J, Andrews TJ, Lorimer GH. 1986. Reaction intermediate partitioning by ribulose-bisphosphate carboxylase with different substrate specificities. *J. Biol. Chem.* 261:10248–56

96. Pierce J, Carlson TJ, Williams JGK. 1989. A cyanobacterial mutant requiring the expression of ribulose bisphosphate carboxylase from a photosynthetic anaerobe. *Proc. Natl. Acad. Sci. USA* 86:5753–57

97. Portis AR Jr. 1995. The regulation of Rubisco by Rubisco activase. *J. Exp. Bot.* 46:1285–91

98. Price GD, Howitt SM, Harrison K, Badger MR. 1993. Analysis of a genomic DNA region from the cyanobacterium *Synechococcus* sp. strain PCC7942 involved in carboxysome assembly and function. *J. Bacteriol.* 175:2871–79

99. Qian Y, Tabita FR. 1998. Expression of *glnB* and a *glnB*-like gene (*glnK*) in a ribulose bisphosphate carboxylase/oxygenase-deficient mutant of *Rhodobacter sphaeroides*. *J. Bacteriol.* 180: 4644–49

100. Ramage RT, Read BA, Tabita FR. 1998. Alteration of the α helix region of cyanobacterial ribulose 1,5-bisphosphate carboxylase/oxygenase to reflect sequences found in high substrate specificity enzymes. *Arch. Biochem. Biophys.* 349:81–88

101. Read BA, Tabita FR. 1992. A hybrid ribulosebisphosphate carboxylase/oxygenase enzyme exhibiting a substantial increase in substrate specificity factor. *Biochemistry* 31:5553–59

102. Read BA, Tabita FR. 1994. High substrate specificity factor ribulose bisphosphate carboxylase/oxygenase from eukaryotic marine algae and properties of recombinant cyanobacterial Rubisco containing "algal" residue modifications. *Arch. Biochem. Biophys.* 312:210–18

103. Roesler KR, Ogren WL. 1990. Primary structure of *Chlamydomonas reinhardtii* ribulose-1,5-bisphosphate carboxylase/oxygenase activase and evidence for a single polypeptide. *Plant Physiol.* 94: 1837–41

104. Robinson SP, Portis AR Jr. 1988. Release of the nocturnal inhibitor, carboxyarabinitol-1-phosphate, from ribulose bisphosphate carboxylase/oxygenase by rubisco activase. *FEBS Lett.* 233:413–16

105. Robinson SP, Portis AR Jr. 1989. Adenosine triphosphate hydrolysis by purified Rubisco activase. *Arch. Biochem. Biophys.* 268:93–99

106. Robinson SP, Portis AR Jr. 1989. Ribulose-1,5-bisphosphate carboxylase/oxygenase activase protein prevents the in vitro decline in activity of ribulose-1, 5-bisphosphate carboxylase/oxygenase. *Plant Physiol.* 90:968–71

106a. Rochaix JD, Goldschmidt-Clermont M, Merchant S, eds. 1998. *The Molecular Biology of Chloroplasts and Mitochondria in Chlamydomonas.* Dordrecht: Kluwer

107. Romanova AK, Zhen-Qi-Cheng Z, McFadden BA. 1997. Activity and carboxylation specificity factor of mutant ribulose 1,5-bisphosphate carboxylase/oxygenase from *Anacystis nidulans. Biochem. Mol. Biol. Int.* 42:299–307

108. Roy H, Andrews TJ. 2000. Rubisco: assembly and mechanism. In *Photosynthesis: Physiology and Metabolism*, ed. RC Leegood, TD Sharkey, S von Caemmerer, pp. 53–83. Dordrecht: Kluwer

109. Rundle SJ, Zielinski RE. 1991. Organization and expression of two tandomly oriented genes encoding ribulosebisphosphate carboxylase/oxygenase activase in barley. *J. Biol. Chem.* 266:4677–85

110. Ruuska SA, Andrews TJ, Badger MR, Price GD, von Caemmerer S. 2000. The role of chloroplast electron transport and metabolites in modulating Rubisco activity in tobacco. Insights from transgenic plants with reduced amounts of cytochrome b/f complex or glyceraldehydes 3-phosphate dehydrogenase. *Plant Physiol.* 122:491–504

111. Salvucci ME. 1992. Subunit interactions of Rubisco activase: polyethylene glycol promotes self-association, stimulates ATPase and activation activities, and enhances interactions with Rubisco. *Arch. Biochem. Biophys.* 298:688–96

112. Salvucci ME, Anderson JC. 1987. Factors affecting the activation state and the level of total activity of ribulose bisphosphate carboxylase in tobacco protoplasts. *Plant Physiol.* 85:66–71

113. Salvucci ME, Ogren WL. 1996. The mechanism of Rubisco activase: insights from studies of the properties and structure of the enzyme. *Photosynth. Res.* 47:1–11

114. Salvucci ME, Osteryoung KW, Crafts-Brandner SJ, Vierling E. 2001. Exceptional sensitivity of Rubisco activase to thermal denaturation in vitro and in vivo. *Plant Physiol.* 127:1053–64

115. Salvucci ME, Portis AR Jr, Orgen WL. 1986. Light and CO_2 response of ribulose-1,5-bisphosphate carboxylase/oxygenase activation in *Arabidopsis* leaves. *Plant Physiol.* 80:655–59

116. Salvucci ME, Werneke JM, Ogren WL, Portis AR Jr. 1987. Purification and species distribution of Rubisco activase. *Plant Physiol.* 84:930–36

117. Schirmer EC, Ware DM, Quetsch C, Kowal AS, Lindquist SL. 2001. Subunit interactions influence the biochemical and biological properties of Hsp104. *Proc. Natl. Acad. Sci. USA* 98:914–19

118. Schreuder HA, Knight S, Curmi PMG, Andersson I, Cascio D, et al. 1993. Crystal structure of activated tobacco rubisco complexed with the reaction-intermediate analogue 2-carboxyarabinitol 1,5-bisphosphate. *Protein Sci.* 2: 1136–46

119. Seemann JR, Kobza J, Moore BD. 1990. Metabolism of carboxyarabinitol 1-phosphate and regulation of ribulose-1,

5-bisphosphate carboxylase. *Photo-synth. Res.* 23:119–30

120. Shen JB, Orozco EM Jr, Ogren WL. 1991. Expression of the two isoforms of spinach ribulose 1,5-bisphosphate carboxylase activase and essentiality of the conserved lysine in the consensus nucleotide-binding domain. *J. Biol. Chem.* 266:8963–68

121. Shikanai T, Foyer CH, Dulieu H, Parry MAJ, Yokota A. 1996. A point mutation in the gene encoding the Rubisco large subunit interferes with holoenzyme assembly. *Plant Mol. Biol.* 31:399–403

122. Soper TS, Larimer FW, Mural RJ, Lee EH, Hartman FC. 1992. Role of asparagine-111 at the active site of ribulose-1,5-bisphosphate carboxylase/oxygenase from *Rhodospirillum rubrum* as explored by site-directed mutagenesis. *J. Biol. Chem.* 267:8452–57

123. Spalding MH. 1998. CO_2 concentrating mechanism. See Ref. 106a, pp. 529–47

124. Spreitzer RJ. 1993. Genetic dissection of Rubisco structure and function. *Annu. Rev. Plant Physiol. Plant Mol. Biol.* 44:411–34

125. Spreitzer RJ. 1998. Genetic engineering of Rubisco. See Ref. 106a, pp. 515–27

126. Spreitzer RJ. 1999. Questions about the complexity of chloroplast ribulose-1,5-bisphosphate carboxylase/oxygenase. *Photosynth. Res.* 60:29–42

127. Spreitzer RJ, Esquivel MG, Du YC, McLaughlin PD. 2001. Alanine-scanning mutagenesis of the small-subunit βA-βB loop of chloroplast ribulose-1,5-bisphosphate carboxylase/oxygenase: substitution at Arg-71 affects thermal stability and CO_2/O_2 specificity. *Biochemistry* 40:5615–21

128. Spreitzer RJ, Thow G, Zhu G. 1995. Pseudoreversion substitution at large-subunit residue 54 influences the CO_2/O_2 specificity of chloroplast ribulose-bisphosphate carboxylase/oxygenase. *Plant Physiol.* 109:681–86

129. Streusand VJ, Portis AR Jr. 1987. Ru-bisco activase mediates ATP-dependent activation of ribulosebisphosphate carboxylase. *Plant Physiol.* 85:152–54

130. Sugawara H, Yamamoto H, Shibata N, Inoue T, Okada S, et al. 1999. Crystal structure of carboxylase reaction-oriented ribulose-1,5-bisphosphate carboxylase/oxygenase from a thermophilic red alga, *Galdieria partita. J. Biol. Chem.* 274:15655–61

131. Suzuki K. 1995. Phosphoglycolate phosphatase-deficient mutants of *Chlamydomonas reinhardtii* capable of growth under air. *Plant Cell Physiol.* 36:95–100

132. Tabita FR. 1999. Microbial ribulose-1,5-bisphosphate carboxylase/oxygenase: a different perspective. *Photosynth. Res.* 60:1–28

133. Taylor TC, Andersson I. 1996. Structural transitions during activation and ligand binding in hexadecameric Rubisco inferred from the crystal structure of the activated unliganded spinach enzyme. *Nat. Struct. Biol.* 3:95–101

134. Taylor TC, Andersson I. 1997. Structure of a product complex of spinach ribulose-1,5-bisphosphate carboxylase/oxygenase. *Biochemistry* 36:4041–46

135. Taylor TC, Andersson I. 1997. The structure of the complex between rubisco and its natural substrate ribulose 1,5-bisphosphate. *J. Mol. Biol.* 265:432–44

135a. Taylor TC, Backlund A, Bjorhall K, Spreitzer RJ, Andersson I. 2001. First crystal structure of Rubisco from a green alga, *Chlamydomonas reinhardtii. J. Biol. Chem.* 276:48159–64

136. Taylor TC, Fothergill MD, Andersson I. 1996. A common structural basis for the inhibition of ribulose 1,5-bisphosphate carboxylase by 4-carboxyarabinitol 1,5-bisphosphate and xylulose 1,5-bisphosphate. *J. Biol. Chem.* 271:32894–99

137. Thow G, Zhu G, Spreitzer RJ. 1994. Complementing substitutions within loop regions 2 and 3 of the α/β-barrel active site influence the CO_2/O_2 specificity

of chloroplast ribulose-1,5-bisphosphate carboxylase/oxygenase. *Biochemistry* 33:5109–14

138. To KY, Suen DF, Chen SCG. 1999. Molecular characterization of ribulose-1,5-bisphosphate carboxylase/oxygenase activase in rice leaves. *Planta* 209: 66–76

139. Uemura K, Anwaruzzaman M, Miyachi S, Yokota A. 1997. Ribulose-1,5-bisphosphate carboxylase/oxygenase from thermophilic red algae with a strong specificity for CO_2 fixation. *Biochem. Biophys. Res. Commun.* 233:568–71

140. Uemura K, Shibata N, Anwaruzzaman M, Fujiwara M, Higuchi T, et al. 2000. The role of structural intersubunit microheterogeneity in the regulation of the activity in hysteresis of ribulose-1,5-bisphosphate carboxylase/oxygenase. *J. Biochem.* 128:591–99

141. van de Loo FJ, Salvucci ME. 1996. Activation of ribulose-1,5-bisphosphate carboxylase/oxygenase (Rubisco) involves Rubisco activase Trp16. *Biochemistry* 35:8143–48

142. van de Loo FJ, Salvucci ME. 1998. Involvement in two aspartate residues of Rubisco activase in coordination of the ATP γ-phosphate and subunit cooperativity. *Biochemistry* 37:4621–25

143. Viitanen PV, Todd MJ, Lorimer GH. 1994. Dynamics of the chaperonin ATPase cycle: implications for facilitating protein folding. *Science* 265:659–66

144. von Caemmerer S, Millgate A, Farquhar GD, Furbank RT. 1997. Reduction of ribulose-1,5-bisphosphate carboxylase/oxygenase by antisense RNA in the C4 plant *Flaveria bidentis* leads to reduced assimilation rates and increased carbon isotope discrimination. *Plant Physiol.* 113:469–77

145. Wang ZY, Portis AR Jr. 1992. Dissociation of ribulose-1,5-bisphosphate bound to ribulose-1,5-bisphosphate carboxylase/oxygenase and its enhancement by ribulose-1,5-bisphosphate carboxyl-

ase/oxygenase activase-mediated hydrolysis of ATP. *Plant Physiol.* 99:1348–53

146. Wang ZY, Ramage RT, Portis AR Jr. 1993. Mg^{2+} and ATP or adenosine 5'-[γ-thio]-triphosphate (ATPγS) enhances intrinsic fluorescence and induces aggregation which increases the activity of spinach Rubisco activase. *Biochim. Biophys. Acta* 1202:47–55

147. Wang ZY, Snyder GW, Esau BD, Portis AR Jr, Ogren WL. 1992. Species-dependent variation in the interaction of substrate-bound ribulose-1,5-bisphosphate carboxylase/oxygenase (Rubisco) and Rubisco activase. *Plant Physiol.* 100: 1858–62

148. Wasmann CC, Ramage RT, Bohnert HJ, Ostrem JA. 1989. Identification of an assembly domain in the small subunit of ribulose-1,5-bisphosphate carboxylase. *Proc. Natl. Acad. Sci. USA* 86:1198–202

149. Werneke JM, Chatfield JM, Ogren WL. 1988. Catalysis of ribulosebisphosphate carboxylase/oxygenase activation by the product of a Rubisco activase cDNA clone expressed in *Escherichia coli*. *Plant Physiol.* 87:917–20

150. Werneke JM, Chatfield JM, Ogren WL 1989. Alternative mRNA splicing generates the two ribulosebisphosphate carboxylase/oxygenase activase polypeptides in spinach and *Arabidopsis*. *Plant Cell* 1:815–25

151. Whitney SM, Andrews TJ. 1998. The CO_2/O_2 specificity of single-subunit ribulose-bisphosphate carboxylase from the dinoflagellate, *Amphidinium carterae*. *Aust. J. Plant Physiol.* 25:131–38

152. Whitney S, Andrews T. 2001. The gene for the ribulose-1,5-bisphosphate carboxylase/oxygenase (Rubisco) small subunit relocated to the plastid genome of tobacco directs the synthesis of small subunits that assemble into Rubisco. *Plant Cell* 13:193–205

153. Whitney SM, von Caemmerer S, Hudson GS, Andrews TJ. 1999. Directed

mutation of the Rubisco large subunit of tobacco influences photorespiration and growth. *Plant Physiol.* 121:579–88

154. Yokota A, Tsujimoto N. 1992. Characterization of ribulose 1,5-bisphosphate carboxylase/oxygenase carrying ribulose-1,5-bisphosphate at its regulatory sites and the mechanism of interaction of this form of the enzyme with ribulose 1,5-bisphosphate carboxylase/oxygenase activase. *Eur. J. Biochem.* 204:901–9

155. Zhang N, Portis AR Jr. 1999. Mechanism of light regulation of Rubisco: a specific role for the larger Rubisco activase isoform involving reductive activation by thioredoxin-f. *Proc. Natl. Acad. Sci. USA* 96:9438–43

156. Zhang Z, Komatsu S. 2000. Molecular cloning and characterization of cDNAs encoding two isoforms of ribulose-1,5-bisphosphate carboxylase/oxygenase activase in rice (*Oryza sativa* L.). *J. Biochem.* 128:383–89

157. Zhu G, Jensen RG. 1991. Xylulose bisphosphate synthesized by ribulose 1,5-bisphosphate carboxylase/oxygenase during catalysis binds to decarbamylated enzyme. *Plant Physiol.* 97:1348–53

158. Zhu G, Spreitzer RJ. 1994. Directed mutagenesis of chloroplast ribulose-bisphosphate carboxylase/oxygenase: substitutions at large subunit asparagine 123 and serine 379 decrease CO_2/O_2 specificity. *J. Biol. Chem.* 269:3952–56

159. Zhu G, Spreitzer RJ. 1996. Directed mutagenesis of chloroplast ribulose-1,5-bisphosphate carboxylase/oxygenase: loop-6 substitutions complement for structural stability but decrease catalytic efficiency. *J. Biol. Chem.* 271:18494–98

Annu. Rev. Plant Biol. 2002. 53:477–501
DOI: 10.1146/annurev.arplant.53.100301.135202

A NEW MOSS GENETICS: Targeted Mutagenesis in *Physcomitrella patens*

Didier G. Schaefer

*Institut d'Écologie, Laboratoire de Phytogénétique Cellulaire, Bâtiment de Biologie,
Université de Lausanne, CH-1015 Lausanne, Switzerland;
e-mail: Didier.schaefer@ie-pc.unil.ch*

Key Words   model system, gene targeting, homologous recombination, functional genomics

■ **Abstract**   The potential of moss as a model system to study plant biology is associated with their relatively simple developmental pattern that nevertheless resembles the basic organization of the body plan of land plants, the direct access to cell-lineage analysis, their similar responses to plant growth factors and environmental stimuli as those observed in other land plants, and the dominance of the gametophyte in the life cycle that facilitates genetic approaches. Transformation studies in the moss *Physcomitrella patens* have revealed a totally unique feature for plants, i.e., that foreign DNA sequences integrate in the genome preferentially at targeted locations by homologous recombination, enabling for the first time in plants the application of the powerful molecular genetic approaches used routinely in bacteria, yeast, and since 1989, the mouse embryonic stem cells. This article reviews our current knowledge of *Physcomitrella patens* transformation and its unique suitability for functional genomic studies.

CONTENTS

INTRODUCTION

A major goal of modern biology is to understand the relationship between biological systems and the presence and activity of genes. Model systems of biology have provided both the experimental material and the biological context for framing questions that explore this relationship. The investigator's ability to use several different model systems allows the function of related genes to be tested in a variety of biological and/or methodological contexts. Genomics, along with the conservation of gene structure and function found among diverse organisms, represents a "golden thread" that links model systems and allows the comparison between them.

During the past 15 years, an enormous effort by the biological community has led to the completion of nucleotide sequences of genomes from both prokaryotic and eukaryotic model systems such as *Escherichia coli* (9), *Bacillus subtilis* (62), *Saccharomyces cerevisiae* (20), and more recently *Caenorhabditis elegans* (15) and *Drosophila melanogaster* (2). The goal of these efforts has been to establish gene catalogues for particular organisms and to provide insight into the function of the corresponding proteins. In plants, the recently completed genome sequence of *Arabidopsis thaliana* (3a) will soon be followed by the completion of a genome sequence for rice (*Oriza sativa*) (6), and we can expect that further genomic sequences will be produced for other plants in the future. The next challenge for the biological community is to assess the precise function of the sequenced genes, a general approach that is referred to as functional genomics.

Several strategies have been applied to identify new genes and decipher protein function in plants. The most basic approach is sequence comparison of a gene or a protein with existing databases. Homology to another protein whose function has been previously established can provide important clues as to cellular function and biochemical properties. However, beyond this, plant functional genomics is essentially performed by the two strategies described below. The first approach, like homology comparisons, provides only an indirect indication of gene function. It makes use of global gene expression profiles and is based either on the study of populations of cDNAs or expressed sequence tags (ESTs) deposited on DNA microarrays (34, 98) or on the study of populations of proteins separated by two-dimensional gel electrophoresis (proteomics) (122). Transcriptomic and proteomic approaches identify known or new genes or proteins that are specifically up- or downregulated and give valuable information on the expression profile of gene networks in response to experimental conditions. Analysis of such expression data by high-throughput computational methods is useful for assigning genes to particular regulatory networks and for identifying new networks and interactions. However, this is nonetheless a descriptive approach and alone does not tell the investigator what function a gene performs in a cell or in an organism. Similar considerations apply to the emerging areas of metabolomics and phenomics.

The most widely used approach for performing true, genome-wide functional genomics has been insertional mutagenesis. Insertional mutagenesis by transgenesis and the phenotypic and molecular characterization of tagged mutated lines has been used extensively to study plant gene function (128). Several collections of T-DNA and transposon-tagged transformants of *Arabidopsis* (86) and rice (51) have been generated and successfully used to identify and characterize genes involved in different biological processes. Tagged mutagenesis usually generates loss-of-function mutations and rarely identifies genes that act redundantly.

However, activation tagging, based on the random insertion of enhancers in the genome, allows the identification of gain-of-function mutations (129), whereas gene trap screens (117) allow gene activity to be monitored in planta by generating fusion between endogenous genetic elements and a reporter gene. These approaches significantly complement the tools available to plant functional geneticists. Recent collections generated in *Arabidopsis* have generated more than 100,000 insertional lines, and this saturation of the genome essentially ensures that at least one insertional event can be recovered for any given gene. This is particularly useful where fully sequenced genomes are available because the location of an element within the genome can be determined simply by sequencing genomic sequences that flank each insertional element.

Nonetheless, insertional mutagenesis has several conceptual disadvantages: (*a*) It is based on random integration of tags in the genome and thus can only generate stochastically distributed allelic mutations in the genome; (*b*) it requires the generation of large numbers of transformants to saturate the genome with tags; (*c*) it only identifies detectable viable phenotypes and does not generate conditional mutations, which is a prerequisite for the identification and the characterization of essential genes; (*d*) the identification of the tagged mutant or gene can be very difficult and time consuming because multiple insertion or major chromosomal rearrangements can be associated with the insertion of the tag (78); and (*e*) it does not enable the generation of specific point mutations in a gene, which is a prerequisite for detailed functional analysis and for generating an allelic series of mutations within a specific gene that allow a full range of possible phenotypes to be explored.

Gene targeting (GT) is the generation of specific mutations in a genome by homologous recombination-mediated integration of foreign DNA sequences. GT circumvents the limitations associated with stochastic mutational approaches and represents the ultimate genetic tool for functional genomics. GT provides the methodological core of functional genetic studies in bacteria, yeast, and several filamentous fungi because the integration of foreign DNA into their genome occurs efficiently at targeted locations by homologous recombination (HR). It reaches its maximal efficiency in *S. cerevisiae*, and this accounts for an important part in the extensive use of budding yeast as a model system (100, 120). Yet, GT is not routinely used in animals and plants because the integration of foreign DNA in their genome occurs orders of magnitude more frequently at random locations by illegitimate recombination, thus preventing the identification of the rare targeted mutant in a large population of transformants. In 1989, the observation that the ratio of targeted to random integration events upon transformation of a mouse embryonic stem (ES) cell line ranges from 0.1–10% (16, 123) accounts for its exponential development as a model system in animal biology (77). Since 1988, the feasibility of gene targeting in plants has been demonstrated, following either direct gene transfer to protoplasts (87) or *Agrobacterium*-mediated transformation (65, 80) into either artificial (80, 87) or natural loci (65). Yet, despite numerous studies, the ratio of targeted to random integration events observed so far in plants hardly reaches 10^{-4}, which prevents the general use of gene targeting approaches for plant functional genomics (74, 89, 126).

In contrast to all other plants tested so far, integration of foreign DNA sequences in the genome of the moss *Physcomitrella patens* occurs predominantly at targeted locations by HR (107, 110). This methodological progress represents a true revolution in our strategies to address gene function in plants and places *Physcomitrella patens* in a unique position among model systems in multicellular eukaryotes (108). This ability has been exploited in a number of studies to directly address gene function, and these are discussed here for both the technical lessons that can be learned and for the value of these studies for understanding basic biological processes. I review our current knowledge of genetic transformation and gene targeting in *Physcomitrella patens* and emphasize how this moss can be used as a model system that advantageously complements other tools available for functional genomic approaches.

THE *PHYSCOMITRELLA PATENS* MODEL SYSTEM

The major features of the life cycle of *Physcomitrella* are indicated in Figure 1 (see color insert). *Physcomitrella* has a relatively simple plant architecture that is nevertheless composed of the same elements as those present in other land plants. Its life cycle is dominated by the haploid gametophyte, which presents two distinct developmental phases: the protonema that displays characteristic one-dimensional filamentous growth and the gametophore that develop by three-dimensional caulinary growth into a typical small plant. The former allows the study of basic biological processes with microbiological culture technology, whereas more complex developmental processes such as organogenesis and the establishment of a body plan are adequately addressed in the latter. Most of the biological studies have been performed on the protonema, which is ideal for following biological processes and developmental patterning at the single cell level.

The biology of *Physcomitrella* and its potential as a model genetic system to study plant biological processes have been discussed in several excellent reviews and are not presented here (21–28, 33, 56, 57, 92, 94, 111, 113, 114). These reviews cover most of the work that has been conducted in *Physcomitrella* over the past 20 years and highlight the similarity of its biology with that of higher plants. These studies have shown that this moss provides an adequate system to study a broad range of biological processes including cell division, cell growth, cell polarity, cellular ultrastructure, photomorphogenesis, photo- and gravitropic responses, hormone-mediated responses, signal transduction pathways, chloroplast development, filamentous growth, organogenesis, and plant development.

The wild-type strain of *Physcomitrella patens* grown in different laboratories derives from a single spore isolated in England, thus ensuring genetic homogeneity among studies conducted in different laboratories. The size of the genome has been estimated by flow cytometry to be approximately 480 Mbp, and karyotypic analysis has shown that it is composed of 27 very small chromosomes (94, 95).

To date, no physical or genetic map of the *Physcomitrella* genome has been developed. Recently, the *Physcomitrella* EST Program (90) conducted by the University of Leeds (U.K.) and the Washington University in St. Louis (U.S.A.) has provided the scientific community with approximately 16,000 ESTs in the database, while the German agrochemical company BASF has generated an EST database of *Physcomitrella* that contains approximately 120,000 entries representing more than 20,000 genes. Preliminary analysis of genes and ESTs strongly indicate that these sequences are highly similar to those of the corresponding higher plant genes. These similarities extend to codon usage (68, 96), intron-exon structure (18, 61, 67), and multigene family complexity (18, 61, 95). It is clear that the number of *Physcomitrella* genes and EST sequences accessible in the database will increase rapidly, and each entry directly enables the generation of the corresponding moss mutant by targeted transgenesis.

Increasing interest to study mosses also comes from the field of evolutionary developmental genetics. Experimental model systems in animal developmental biology include organisms from contrasting taxonomic groups, such as worm, fly, and mammal. By contrast, model systems in plant biology have essentially focused on flowering plants. The phylogenetic basal position of Bryophytes among land plants places them in an ideal situation to address fundamental questions such as how have plants evolved from simple to complex forms (29). In this respect, functional analysis of the recently identified *Physcomitrella* gene homolog to key players in plant development such as the MADS (61) and homeobox genes (18, 102) of higher plants may give us valuable information about the ancestral mechanisms that govern the development of land plants (121). One can predict that this is only the beginning.

GENETIC TRANSFORMATION OF THE MOSS *PHYSCOMITRELLA PATENS*

Transformation With DNA Without Homology

METHODOLOGICAL CONSIDERATIONS   Protoplasts of *Physcomitrella* provide the ideal material for developing methods in genetic transformation owing to the ease with which they can be isolated and to their high physiological and regenerative capacities. The first successful genetic transformation of *Physcomitrella* was achieved by polyethylene glycol (PEG)-mediated direct gene transfer into protoplasts (112). Since then, the protocol has been optimized (107, 109) and used routinely for transforming *Physcomitrella* and for transient gene expression assays (4, 41, 42, 46a, 47, 48, 52, 66, 79, 97, 101, 110, 119, 130, 131). Transient gene expression was also reported following biolistic delivery of DNA-coated microbeads into protonemal tissue (58, 106), but this method seems to be less efficient than PEG-mediated transformation as there has been only one report of a transgenic strain produced this way (58). Genetic transformation of *Physcomitrella* using *Agrobacterium*-based methods has also been attempted without success in several laboratories (28, 94, 109). It is nevertheless possible that *Physcomitrella* might be transformed with hyper-virulent *Agrobacterium* strains (63) or upon transformation with T-DNA carrying moss genomic sequences.

To date, the neomycin resistance gene *nptII* (7), the hygromycin resistance gene *aphIV* (45), and the sulfadiazine resistance gene *sul* (52) have been used as positive selectable markers in *Physcomitrella*. The resistance genes are usually driven by the CaMV 35S promoter (4, 41, 42, 47, 48, 58, 79, 107, 110, 112, 119) or the NOS promoter (4, 66, 97, 101, 112), although it seems that the level of resistance achieved by NOS-driven resistance cassettes is lower than that observed with the 35S promoter. Selection conditions for antibiotic resistant clones are stringent and usually use antibiotic concentrations that kill untransformed protoplasts within a week of selection and correspond to 10 times the lethal dose 50 (LD_{50}, i.e., 50 mg/L geneticin sulfate, 25 mg/L hygromycin B, or 150 mg/L sulfadiazine). Meanwhile, the cytological markers encoding β-glucuronidase (GUS) (58, 107, 130) and green fluorescent protein (GFP) (17, 55, 57) were shown to be suitable for transformation experiments in *Physcomitrella*.

Survival rates of *Physcomitrella* protoplasts after PEG-mediated transformation range from 15–30%, and selection for antibiotic resistance is initiated one week after transformation, when protoplasts have regenerated colonies of 5–10 cells. Initial relative transformation frequencies (RTF; the percentage of antibiotic resistant colonies in the regenerating population) monitored 14 days after transfer to selective medium are extremely high, ranging from 3–30% (107). These values correspond to the percentage of protoplasts expressing a cytoplasmic GFP observed 48 h after PEG-mediated transfection with a 35S-GFP plasmid (17). A typical petri dish after 15 days of selection shows two classes of resistant clones (57, 107). The majority of the resistant colonies display poor growth and altered regeneration and are episomal transformants, whereas a small percentage display growth and regeneration comparable to protoplasts plated on nonselective medium and represent integrative transformants.

EPISOMAL REPLICATIVE TRANSFORMATION   Episomal antibiotic resistant clones have been obtained with every transforming DNA tested so far and have been routinely referred to as unstable clones. Their phenotypic and molecular characteristics are as follows: (*a*) Unstable clones are recovered at extremely high frequencies (RTF 10–30% and 1–5% with supercoiled and linearized DNA, respectively) that correspond to up to several hundred clones per μg DNA (107). (*b*) They can be propagated for years as protonemal cultures as long as constant selective pressure is maintained but lose their resistance phenotype and the transforming DNA following a 14-day growth period on nonselective medium (4, 107, 109). (*c*) They display reduced growth and altered regeneration on selective medium, both of which are positively correlated with the antibiotic concentration (107). (*d*) Unstable clones are formed by a mosaic of transformed and untransformed cells, as observed by the presence of resistant and sensitive sectors and of GUS- or GFP-labeled and unlabeled cells in protonema regenerating on selective medium (4, 107).

(*e*) Phenotypic and molecular analyses show that maintenance of the transgene is tissue specific, being restricted to the filamentous protonema, and is not transmitted to the gametophore (4, 107, 109). Consistent with this observation, no resistant spores have been recovered from strains regenerated under nonselective conditions (107). (*f*) At the molecular level, analysis of uncut plant DNA by Southern blotting reveals that the transgene sequences comigrate with high molecular weight DNA and that these high molecular weight arrays are formed by direct head-to-tail repeats of the transforming plasmid (4, 107). (*g*) Genomic reconstruction indicates that for DNA isolated from protonema the number of plasmid copies per haploid genome ranges between 3 and 40 (4, 107). This is probably an underestimate of the real copy number per resistant cell given the strong mosaicism observed in protonema. These values fall to below one copy per haploid genome when DNA is isolated from older cultures that have differentiated into gametophores (107). (*h*) Direct evidence that transforming sequences are effectively replicated in moss cells was provided by Southern blot analyses that showed that the bacterial Dam methylation pattern of the transfected DNA was lost in DNA isolated from these clones (D. G. Schaefer, unpublished data).

These characteristics demonstrate that in episomal transformants the transforming sequences are concatenated to form high molecular weight extrachromosomal elements that are replicated in moss cells but poorly partitioned during mitosis (4, 107, 109). Replicative transformation in *Physcomitrella* provides the first example of the successful episomal maintenance of bacterial plasmids in plant cells. Though rare, a similar pattern of behavior has also been observed for plasmid DNA introduced into *Xenopus* early embryos (70), *Caenorabditis elegans* (73, 118), and several other animal eukaryotes. That bacterial plasmids can be concatenated and replicated in moss cells raises very interesting questions as to the mechanisms underlying this process. It also represents the first step toward the development of episomal vectors and moss artificial chromosomes in *Physcomitrella*. Episomal transformation vectors provide extremely valuable tools for molecular genetic approaches in yeast. Their development has been achieved by the combined addition of functional elements, such as origins of DNA replication (yeast ARS element), centromere (yeast CEN), and telomere from yeast or *Tetrahymena* into bacterial plasmids, to generate a palette of tools ranging from ARS plasmid to yeast artificial chromosome (YAC) vectors (13, 120). A similar approach should be tested for vectors used to transform *Physcomitrella*, and the mitotic stability and the developmental partition of these sequences during its life cycle should be assessed. The observation that replicative transformants in *Physcomitrella* obtained with a YAC-derived vector (11) appear to display improved mitotic stability and occasional transmission of the transgenes to the gametophore supports the potential for such approaches in the near future (107). Furthermore, the development of episomal vectors with improved mitotic stability [moss ARS-like plasmids (MARS)] would directly enable the general use of high frequency replicative transformation as an approach to complement mutations in *Physcomitrella*.

In practice, these replicative transformants can be distinguished from true integrative transformants after transformation by adopting a selection screen that alternates periods of growth on selective and nonselective medium. Integrative transgenic strains display unrestricted growth during all stages of selection, whereas replicative transformants either die or show antibiotic sensitive sectors when transferred back to selective medium (56, 107, 109).

ILLEGITIMATE INTEGRATIVE TRANSFORMATION   Foreign DNA sequences can integrate by illegitimate recombination in the genome of *Physcomitrella*. Efficiencies are low (58, 66, 97, 101, 107, 112), and RTF range around 0.001% (1 in 10^5 regenerating colonies) after transformation with supercoiled plasmid and are increased approximately five times (RTF 0.005%) when linearized DNA is used (4, 97, 107, 109, 112). In comparison with the high replicative transformation frequencies observed during the same experiments, these values most likely represent the true level of illegitimate integrative transformation in *Physcomitrella*. These transgenic strains can complete their life cycle normally on selective medium and maintain antibiotic resistance when grown on nonselective medium. Compiled data from experiments performed with antibiotic resistance and cytological markers demonstrate that the new trait is mitotically stable and is expressed in every cell and at all stages of development (4, 58, 107, 109, 112). At the molecular level, Southern blot analysis shows that between 1 and 50 direct head-to-tail repeats of the transforming plasmid are integrated in the genome, usually at a single locus (4, 97, 107, 112, 130), although integration at two loci has also been observed (97). Finally, genetic analysis shows that the transgene is meiotically stable and segregates in a Mendelian way as a single locus character that is independent from strain to strain (52, 107, 112).

Transformation With DNA Carrying Homologous Sequences

CONCEPTS OF TARGETED MUTAGENESIS   The basic principles and strategies of gene targeting (GT) and allele replacement in eukaryotes have been defined in budding yeast (100, 120) and applied to develop targeted mutagenesis in the mouse ES cell system (77). This methodology has been further improved when associated with the use of site-specific recombination systems (such as bacteriophage P1 Cre/lox), which facilitates the subsequent elimination of undesired sequences integrated in the genome upon transformation. This allows very accurate targeted transgenesis to be performed in the yeast and mouse genome (104, 105). These approaches are based on the use of two main types of targeting vectors: insertion and replacement vectors. Each type generates different types of mutations, and their basic characteristics and the different patterns of integration observed so far in *Physcomitrella* are schematized in Figure 2 (see color insert).

The three main criteria used to assess the efficiency of both gene disruption by targeted insertion and point mutagenesis by targeted allele replacement are (*a*) the ratio of targeted to random integration events observed upon integrative transformation, (*b*) the extent of sequence homology required to efficiently target chromosomal loci, and (*c*) the ratio of insertion versus replacement events monitored upon targeted integration of replacement vectors. Each of these parameters has now been assessed in *Physcomitrella* and is reviewed in detail below.

TRANSFORMATION WITH INSERTION VECTORS   The efficiency of GT with insertion vectors has been assessed in three sets of experiments designed to target six artificial loci (107), three independent single copy genomic sequences (110), and a specific member of the highly conserved multigene family encoding for chlorophyll a/b binding proteins (cab) (47). Taking advantage of the fact that pUC-derived transformation vectors share approximately 3 kb of sequence homology, I have retransformed independent kanamycin- or hygromycin-resistant transformants carrying several repeats of either plasmid pHP23b (35S-neo) (88) or pGL2 (35S-hygro) (8) with the other plasmid to assess GT efficiency to artificial loci (107). The successful targeting of artificial loci with a 90% efficiency was supported by the following experimental evidence: (*a*) Retransformation frequencies were on average one order of magnitude higher in transgenic strains as compared to wild type, with a maximum RTF value of 0.1% (107); (*b*) cosegregation of the hygromycin and kanamycin resistance markers occurred in progeny of 90% of tested clones (107); and (*c*) PCR and Southern blot analyses provided molecular evidence for the integration of direct repeats of the second plasmid by HR within the tandem repeats of the previously integrated plasmid (S. Vlach & D. G. Schaefer, unpublished data). Independently, cosegregation of positive selectable markers used for sequential transformation was also observed in transformants generated to develop transposon-mediated mutagenesis in *Physcomitrella* (25, 52). The successful targeting of three independent single copy genomic loci ranging in size from 2.4 to 3.6 kb definitively demonstrated that targeted integration by HR was the dominant path of integration of foreign DNA sequences in the genome of *Physcomitrella* (110). This conclusion was supported by the following experimental evidence: (*a*) Transformation rates with vectors carrying moss genomic sequences were on average one order of magnitude higher (RTF up to 0.1%) than those observed with nonhomologous control plasmids, (*b*) targeted integration of tandem repeats [2–30] of the vectors by HR was confirmed by Southern blot analyses in 75–100% of the plants analyzed, and (*c*) integration of the targeting plasmids in single loci and meiotic stability of the targeted loci was demonstrated by segregation analyses.

The specificity of GT in *Physcomitrella* was further assessed in experiments designed to disrupt the *ZLAB1* gene (67), one of the 15 members of the cab multigene family (95). Using an insertion vector carrying 1 kb of the ZLAB1 genomic sequence, disruption of only the true homologous gene family member by targeted integration of tandem repeats of the vector occurred in 30% of the transformants analyzed, although nucleotide sequence homology between different members was as high as 87–93% (47).

Taken together, these data demonstrate that the efficiency of targeted insertional mutagenesis with insertion vectors sharing more than 2.0 kb of sequence homology with the moss genome is in the range of 90%, which is unique in the plant kingdom, is much more efficient than it is in mouse ES cells (32), and is comparable with efficiencies observed in budding yeast (108, 110). Targeted insertion of several [1–50] direct repeats of the transforming plasmid in the corresponding target locus occurs by a single HR event, allowing the direct generation of insertional mutations such as loss-of function (see Figure 2*AI*). The specificity of the recombination process enables the disruption of a specific member within a gene family. The data also indicate that integrated transgenes provide accessible targets for the subsequent integration of additional transforming vectors, a feature that must be considered in designing experiments for the sequential transformation of strains.

TRANSFORMATION WITH REPLACEMENT VECTORS   The minimal and optimal amounts of sequence homology required to target genomic sequences and the pattern of targeted integration of replacement vectors have been defined in experiments designed to target the *Physcomitrella* adenine phosphoribosyl transferase gene (*Ppapt*) (D. G. Schaefer, M. Chakhparonian, S. Vlach, K. von Schwartzenberg, N. Houba-Hérin, C. Pethe, J.-P. Zrÿd & M. Laloue, unpublished data). The *apt* gene provides an ideal model locus for GT studies in haploid organisms as its loss of function confers resistance to adenine analogues such as 2,6-diaminopurine (DAP) and can thus be directly selected (76). Using both cDNA- and gDNA-based replacement vectors, we have been able to establish the following parameters: (*a*) Targeted integration by HR is possible using stretches of continuous sequence homology ranging from 50–200 bp (48) and reaches a maximum when targeting sequences extending over 1–2 kb are used; (*b*) under optimal conditions, RTF for the *PPapt* locus are in the range of 0.1%, and targeting efficiencies are above 90% (i.e., more than 9/10 transformation events involve HR); (*c*) targeted replacement of genomic sequences mediated by two HR events (i.e., allele replacement) (Figures 2*BII,III*) and targeted insertion mediated by a single HR coupled with a nonhomologous end-joining reaction (NHEJ) (Figure 2*BV*) occur at similar rates when replacement vectors are used; (*d*) replacement of the genomic sequence by a single copy of the replacement cassette is observed in approximately half of the replacement events (Figure 2*BII*), whereas the structure of the other half of replacement events is characterized by the presence of direct repeats of either the replacement cassette (Figure 2*BIII*) or the entire transforming plasmid (not illustrated); (*e*) targeted insertion of replacement vectors is characterized by the integration of tandem direct repeats [1–30 copies] of either the replacement cassette or the entire plasmid (Figure 2*BIV*); (*f*) NHEJs that occur between direct repeats of the transforming DNA or at the junctions between plasmid sequences and the genome are identical within one set of direct repeats but vary in different targeted transformants. These reactions follow the main features described in other plants for NHEJ events, i.e., deletion of plasmid sequences followed by rejoining within short stretches of microhomology (43).

Taken together, these data demonstrate that targeted mutagenesis with replacement vectors is as efficient in *Physcomitrella* as in *S. cerevisiae*. Sequence homology requirements and the ratio of targeted to random integration events are comparable to those observed for targeted mutagenesis in budding yeast (100) and are more favorable here than for any other multicellular eukaryotic system amenable to GT (108). Yet the integration pattern of replacement vectors in *Physcomitrella* is different from that observed in budding yeast, where replacement vectors integrate essentially by allele replacement, and is instead highly similar to targeted integration events observed in Chinese hamster ovary cells (1). This suggests that the mechanisms accounting for targeted integration in moss are different from the models proposed for budding yeast (85). The synthesis-dependent-strand-annealing model or the recently proposed break-induced replication model (60) account for the observations described above, but clearly further studies on the mechanisms of GT in wild-type *Physcomitrella* and in strains mutated in recombination genes will be needed to clarify our understanding of the mechanisms of HR in this moss.

Analysis of Gene Function by Targeted Disruption

Efficient GT to natural loci prompted several groups to initiate functional studies by targeted disruption of genes involved in four completely unrelated biological processes: the division of chloroplasts, the biosynthesis of unsaturated fatty acids, the ubiquitin-mediated proteolytic pathway, and the purine salvage pathway. Using a cDNA-based replacement vector carrying approximately 1 kb of discontinuous homologous sequence, Reski's group in Freiburg successfully disrupted the *ftsZ1* gene of *Physcomitrella* in 14% of the transgenic plants analyzed (119). In bacteria, the ftsZ protein plays a major role in the formation of the dividing ring during cytokinesis, and nuclear-encoded eukaryotic ftsZ homologues have been found in plants (55, 82). The observation that cells of *ftsZ* knock-out moss strains contain a single giant chloroplast instead of the 50 lens-shaped chloroplasts usually found in wild-type cells provided the first functional evidence that FtsZ proteins are essential for chloroplast division. This observation was subsequently confirmed in *Arabidopsis*, using antisense approaches (81, 82). Subsequently, a second ftsZ gene was isolated from *Physcomitrella* and shown to be functionally nonredundant to the first one (55). More remarkably, Kiessling and coworkers used transient expression of *ftsZ*-GFP fusion cassettes to demonstrate that both FtsZ proteins form a permanent, highly organized filamentous scaffold within moss chloroplasts. This newly identified subcellular structure is reminiscent of cellular cytoskeleton, and the authors proposed that it may function to maintain plastid shape and named it the plastoskeleton (55). Whether similar structures also exist in chloroplasts of vascular plants awaits further studies, but the work conducted in moss provided new insights into our understanding of the mechanisms of plastid division.

Lipids of *Physcomitrella* are composed of up to 30% arachidonic acid (44), a polyunsaturated fatty acid (PUFA) found only in lower plants. Searching for new genes involved in the biosynthesis of this PUFA, the group of E. Heinz in Hamburg isolated a novel Δ6-acyl-group desaturase gene from *Physcomitrella* (*PPDES6*) and disrupted it with a genomic DNA-based replacement vector carrying approximately 2 kb of sequence homology (41). Disruption of *PPDES6* was supported by PCR and by Southern and Northern blot analyses in 5/5 randomly chosen transformants. These knock-out plants displayed no visible developmental phenotype, but analyses of their fatty acid composition revealed a clear biochemical phenotype. Plants displayed reduced levels of arachidonic acid and other PUFA whose biosynthesis involves a Δ6-desaturation step and a concomitant accumulation of putative precursors. That *PPDES6* encodes for a real Δ6-desaturase was further confirmed by feeding experiments in moss and expression studies in yeast. Thus, this work showed that targeted disruption in *Physcomitrella* can be used to identify novel plant genes and to describe new biochemical pathways involved in the biosynthesis of metabolites.

One major route for the selective degradation of proteins in eukaryotic cells is the ubiquitin-proteasome pathway. The 26S proteasome, an ATP-dependent protease complex that degrades ubiquitin-tagged proteins, is composed of a core proteinase, the 20S proteasome, associated with a pair of regulatory complexes known as the 19S complex (127). The RPN10 protein is a subunit of the 19S complex that has affinity for multiubiquitin chains in vitro and may function in recognizing and recruiting ubiquitinated proteins to be degraded by the proteasome. Yet, its function is still a matter of debate because disruption of the *rpn10* gene in *S. cerevisiae* did not cause any obvious phenotype (such as growth defects) except for an increased sensitivity to amino-acid homologues (124). However the same disruption did lead to an embryonic lethal phenotype in mouse (53). In our group, Girod and colleagues isolated and disrupted the *rpn10* gene of *Physcomitrella* (*PPrpn10*) with a cDNA-based replacement vector carrying 1300 bp of discontinuous homologous sequence (42). Successful disruption of the gene in 2 out of 55 transgenic strains was confirmed by PCR, Southern blot analysis, and immunological analyses. The knock-out moss strains were viable but displayed a dramatic developmental phenotype characterized by altered protonemal development and the impairment of bud formation. This developmental phenotype was complemented following expression of *PPrpn10* gene at ectopic locations, but mutated versions of the gene failed to fully complement the phenotype, providing insight into the role of the different domains within the *PPrpn10* gene. Thus this work provides a telling example of the need to use different model systems for the functional characterization of complex regulatory mechanisms such as the ubiquitin/proteasome proteolytic pathway. The experiments performed in *Physcomitrella* revealed an interaction between this pathway and development that was not observed in yeast and that could not be further investigated in mouse owing to the embryonic lethality of the knock-out phenotype.

The adenine phosphoribosyl transferase gene (*apt*) encodes for a maintenance enzyme of the purine salvage pathway that recycles adenine into AMP. In *Arabidopsis*, *apt* genes form a small family that contains at least three expressed members (35), and the *apt* mutant BM3 (75) carries a mutation in the *Ataptl* gene (35). This mutant has approximately 1% APRTase activity, is resistant to DAP, displays reduced growth rate and male sterility due to abortive pollen development (91), and is unable to form callus or regenerate in vitro (64), but whether the observed phenotype could be attributed to impaired purine or cytokinin metabolism was unclear (35, 75). *Physcomitrella* has only one *apt* gene (*Ppapt*) that closely resembles the *Arabidopsis Atapt1* gene at the level of genetic structure and biochemical activity of the encoded protein. *Physcomitrella apt* null alleles display a strong developmental phenotype characterized by abortive gametophore development. In addition, these mutants display increased sensitivity to exogenously supplied adenine, which prompted us to reanalyze the *Arabidopsis* mutant and to observe that BM3 also displayed hypersensitivity to adenine, an observation that had not been made previously. Finally, we could show that expression of the *Arabidopsis Atapt1* or *Atapt2* genes at ectopic locations fully complements the *apt* null phenotype in *Physcomitrella*, giving rise to fertile plants that are sensitive to 2,6-diaminopurine and resistant to adenine (D. G. Schaefer, M. Chakhparonian, S. Vlach, K. von Schwartzenberg, N. Houba-Hérin, C. Pethe, J.-P. Zrÿd & M. Laloue, unpublished data). Thus, disrupting the *apt* gene in *Physcomitrella* provided two novel pieces of information: First, *Arabidopsis* homologues can functionally complement mutations in *Physcomitrella*. Second, the developmental phenotype observed in moss *apt* null strains was associated with the inability of the plant to recycle adenine, an observation that was correlated in *Arabidopsis*.

The studies reviewed above are the first examples of the use of efficient GT in *Physcomitrella* to provide rapid access to valuable new information about genetic networks that control diverse biological processes.

Additional Methodological Development

FUNCTIONAL STUDIES OF PROMOTERS IN *PHYSCOMITRELLA*   The generalized use of *Physcomitrella* for targeted transgenesis will require the use of several well-characterized promoters for the construction of constitutive and inducible expression vectors, as a result of the following features of *Physcomitrella* transformation: (*a*) Promoter sequences integrated in the moss genome provide target sites for the subsequent integration of an expression cassette driven by the same promoter. (*b*) Preliminary evidence indicates that when two transgenes are driven by the same promoter, they may be stochastically silenced by a mechanism that remains to be determined (D. G. Schaefer, unpublished observation; 52). Two types of promoters, endogenous moss promoters and heterologous promoters, are discussed below. Their use in transformation experiments will enable very different types of studies to be performed.

The presence of an endogenous moss promoter in a transformation vector will result in the predominant targeted integration of the vector into the homologous promoter locus. This provides a straightforward tool for monitoring the activity of the corresponding promoter in a native chromosomal regulatory context and for characterizing promoter functional elements after replacement of the native sequence with an in vitro mutated one. Such efficient and accurate promoter-trap strategies will certainly be broadly applied to characterize moss promoters and to manipulate expression levels of endogenous genes.

Yet the use of endogenous moss promoters is clearly not ideal for functional studies in which investigators must be able to alter gene expression levels, and this requires the use of several different heterologous promoters to drive expression cassettes that are to be introduced in the same transformed strain. So far, the rice actin-1 gene promoter (72) and the maize ubiquitin-1 gene promoter (19) have been shown to promote GUS expression levels following transient expression in protoplasts that were comparable to those observed in rice and maize (17, 131; D. G. Schaefer & M. Uzé, unpublished data). In contrast, the 35S-GUS construct pBI221 (50) promoted only weak GUS activity (17, 107, 131). Yet transfection with an improved version of pBI221 [pNcoGUS (12)] carrying an optimized translation initiation context as defined by Kozak (59) resulted in a 10-fold increase in GUS activity (17, 107, 130, 131), whereas the introduction of the 5' untranslated intron of the rice actin-1 gene promoter between the 35S promoter and the *uidA* gene [pBCGA4 (72)] resulted in an additional 5-fold increase in GUS activity as compared to pNcoGUS, activities that were in the same range as those observed with constructs driven by the rice actin-1 gene promoter (17). These observations indicate that optimizing the expression of transfected genes in moss follows the same basic principles as those observed in other plants and that characterized heterologous promoters from higher plants may provide valuable tools for the future construction of constitutive expression cassettes for *Physcomitrella*.

HETEROLOGOUS CONDITIONAL PROMOTERS   The development of a tightly regulated conditional gene expression system for *Physcomitrella* is essential for switching transgenes on and off and for the generation of conditional mutations. This is especially important for the study of essential genes in the haploid background of the moss gametophyte because such experiments can then directly be performed by replacing the natural promoter with a conditional one that can be turned on during the selection period and subsequently switched off for functional studies. Chemically inducible gene expression systems are well suited for such approaches and should ideally meet three objectives: (*a*) The inducible promoter should not be leaky under noninduced conditions; (*b*) the induction factor should reach at least two orders of magnitude, and its modulation should be positively correlated with the concentration of the chemical inducer; and (*c*) the chemical inducer should not interfere with plant biological processes (38).

Knight and colleagues have studied the regulation of the abscissic acid (ABA)– and osmotic stress–induced wheat *Em* gene promoter (69). In transgenic strains, *Em*-driven GUS activity was readily detectable upon noninduced conditions (probably in response to low concentrations of endogenous ABA) and was induced by factors of 5- and 30-fold following treatment with 0.44 M mannitol or 10^{-4} M ABA, respectively (58). The Em-GUS cassette was strongly expressed in transient gene expression assay, most likely in response to the presence of mannitol in the protoplast culture medium, and could be induced fourfold further upon ABA treatment. Detectable GUS activity under noninduced conditions suggested that some of the molecular mechanisms involved in ABA and stress responses might be conserved between mosses and cereals (58). This assumption was supported by gel retardation assay and DNAseI footprint analysis, which showed that factors present in moss nuclear protein extracts bind specifically to the same abscissic acid response element (ABRE) previously identified from studies with wheat nuclear extracts (46). Thus, although the wheat Em promoter shows strong inducible activity in *Physcomitrella*, this promoter is clearly not ideal for modulating gene expression levels in this moss as it will be continually induced by endogenous fluctuations in ABA levels in the plant.

Tetracycline-regulated gene expression in *Physcomitrella* was reported using the tetracycline-repressible system developed by Gatz et al. (130). In the presence of 1–10 mg/L tetracycline in culture medium, GUS activities measured both by transient gene expression assays and in transgenic strains were hardly detectable. In the absence of tetracycline, GUS expression was induced approximately 100-fold within 24 h to reach specific activities in the range of those observed with GUS reporter cassettes driven by the rice actin-1 gene promoter (130). Expression of the GUS gene could subsequently be repressed upon transfer to tetracycline-supplemented medium, but the kinetics of repression could not be determined owing to the relatively long half-life of the GUS protein in vivo. Although continuous growth on tetracycline may be deleterious to moss growth (M. Laloue, personal communication), tetracycline-regulated gene expression provides a very promising system for conditional expression of genes in *Physcomitrella*.

The two-component glucocorticoid-inducible gene expression GVG system developed by Aoyama & Chua (3) is composed of a constitutive cassette expressing the glucocorticoid receptor fused to a transcriptional activator and of a reporter construct that carries the inducible promoter driving a reporter gene. This system has great potential as it should not, a priori, interfere with plant biological processes. Moreover, wild-type *Physcomitrella* grown on medium supplemented with the synthetic glucocorticoid dexamethasone at concentrations of up to 0.5 mM did not display any obvious alteration of growth or development. Using transient gene expression assay in protoplasts, Chakhparonian (17) showed that GUS expression can be induced up to 50-fold in protoplasts grown in the presence of 30 μM dexamethasone. Yet, transfection with the reporter construct alone yielded detectable GUS activities, indicating that the inducible promoter was leaky, an observation that was previously reported in tobacco (3). Nevertheless, glucocorticoid-regulated gene expression represents a potentially useful system for modulating gene expression in *Physcomitrella*.

Conditional expression systems such as those based on the Gal4 promoter in budding yeast (54) or the *nmt1* promoter in fission yeast (5, 71) have provided essential tools for the functional dissection of genetic networks in these model organisms. Preliminary data obtained in *Physcomitrella* indicated that both the tetracycline-regulated and the glucocorticoid-inducible systems could be used to modulate the expression of genes in this moss, a picture that is similar to that observed in higher plants (39). Additional studies are required to further improve conditional gene expression in *Physcomitrella*, and the recently developed TGV system that combines dexamethasone induction and tetracycline repression (10) as well as the oestrogen-inducible XVE system that appears less leaky than GVG (132) remain to be tested.

CRE/LOX-MEDIATED SITE-SPECIFIC RECOMBINATION IN *PHYSCOMITRELLA*   The site-specific recombination Cre/lox system of bacteriophage P1 is used to introduce site-specific mutations in a genome. It is composed of the Cre recombinase and a small DNA-recognition target site, the *loxP* site, that consists of two 13-bp inverted repeats flanking an 8-bp asymmetric spacer that defines the orientation of the site. Recombination events mediated by the Cre recombinase lead to the excision or the inversion of DNA sequences located between two direct or inverted repeats of *loxP* sites, respectively. In plants, the Cre/lox system was successfully used to generate mutations as diverse as translocations, deletions, or inversions and to regulate conditional gene expression or site-specific integration (83, 84, 126). One of its direct applications in plant biotechnology is the excision of selectable markers and plasmid sequences in genetically modified crops (31, 133). Yet, it is in combination with efficient GT that this system realizes its full potential because the latter allows *loxP* sites to be introduced at specific locations within the genome. The utility of this approach is well illustrated by the extensive use of the Cre/lox system to manipulate the genomes of yeast (103) and mouse (105). In this respect, setting up methodologies for accurate targeted mutagenesis in *Physcomitrella* necessitates the development of an efficient site-specific recombination system.

The possibility of using the Cre/lox system to excise plasmid repeats inserted in the genome of *Physcomitrella patens* was recently demonstrated by Chakhparonian (17) in two independent strains carrying one or five copies of a plasmid containing a 35S nptII cassette flanked by two *loxP* sites in direct orientation. Following transient expression of a 35S Cre cassette [pMM23 (30)], protoplasts of these strains were screened for plasmid excision, based on loss of the resistance marker. Loss of plasmid repeats by Cre-mediated site-specific recombination was observed in 3–10% of protoplast-derived colonies and was confirmed by PCR, Southern blot analyses, and sequence analysis. Such an efficient excision is in the range of that observed in other plant systems following transient expression of Cre. This methodology permits the rapid and easy identification of recombined clones in a small population. This finding has one immediate consequence for strategies of targeted mutagenesis in *Physcomitrella*: *loxP* sites should be built into the design of every vector to be used for transformation. Their presence allows the generation of point mutations by gene conversion (as illustrated in Figure 2*B*), the recycling of selectable markers for sequential transformation, or the elimination of integrated plasmid repeats. Ultimately the combination of efficient GT with site-specific recombination will enable the study of plant biological processes in *Physcomitrella* with the type of accurate targeted mutagenesis strategies that have been developed in budding yeast and mouse (77, 103).

See text page C-2

Figure 1 (page C-1) The life cycle of the moss *Physcomitrella patens* (*a–c*). The three developmental stages of its life cycle: (*a*) the juvenile gametophyte or protonema displaying characteristic filamentous growth (bar = 1 cm); (*b*) the adult gametophyte composed of a colony of gametophores developing by caulinary growth (5 mm); (*c*) the parasitic diploid sporophyte that develops at the apex of a gametophore to form a spore capsule where spores are differentiated (1.5 mm). (*d–h*) Closer view of the different stages of protonema development. (*d*) A germinating protoplast 72 h after isolation displays a highly polarized structure; it has already divided once, and the daughter cell has resumed apical growth (25 μm). (*e*) Typical filamentous growth pattern of a branched chloronemal filament with its tip cell undergoing a developmental transition to form a caulonema (100 μm). (*f–g*) Two cell types with distinct ultrastructural and biological characteristics form the protonema: The photosynthetic chloronema (*f*) is filled with chloroplast and has cell walls perpendicular to the axis of the filament. The adventitious caulonema (*g*) contains less chloroplasts and forms cell walls that are oblique with respect to the axis of the filament (40 μm). (*h*) Transition from filamentous growth to caulinary growth can be pinpointed to a single cell, the caulonemal side-branch initial that differentiates into a primitive meristem, the bud (50 μm). Caulonemal side-branch initials can also form chloronemal or caulonemal cells, and the relative ratio of these different developmental fates can be modulated by chemical and environmental factors and accounts for the general morphology of the protonema. (*i–k*) The gametophore displays a body plan that is simpler but analogous to higher plants. (*i*) A fully differentiated gametophore is composed of a non-vascularized photosynthetic stem carrying simple leaves formed by a single layer of photosynthetic cells and displaying phyllotaxis along the stem axis and of filamentous pigmented rhizoids that radiate from the base of the stem (2 mm). Reproductive organs differentiate in the apical part of the gametophore: the male antheridia (*j*), borne in the leaf axils, and the female archegonia (*k*), borne terminally (100 μm). Ciliated antherozoids formed in the antheridia swim and fuse with the egg cell located in the archegonia to give rise to a diploid cell that further differentiates into the sporophyte (*c*).

(A) INSERTION VECTOR

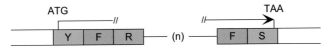

(I) Insertional gene disruption

(B) REPLACEMENT VECTOR

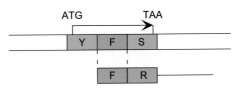

(II) Disruption by allele replacement

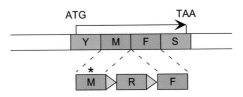

(III) Disruption by multi-copy replacement

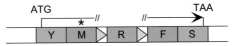

(IV) Point mutation following site-specific recombination

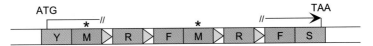

(V) Insertional disruption (homologous recombination in 5')

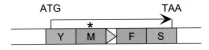

See text page C-4

Figure 2 (page C-3) (*A*) Insertion vectors contain one homologous targeting fragment (*F box*) of your favorite sequence (YFS) cloned next to a marker encoding a resistance gene (*R box*). Targeted integration occurs by a single HR event (*dashed lines*) and results in the insertion of one or several copies of the vector flanked by two repeats of the targeting sequence (*F*). Insertion vectors are suitable for generating insertional gene disruption (*AI*) but not for generating point mutations. (*B*) In replacement vectors, the resistance gene (*R box*) flanked by 2 *loxP* sites in direct orientation (*yellow triangle*) is inserted between two homologous targeting fragments (*M** and *S boxes*, where M* carries a point mutation) of your most favorite sequence (YMFS). Targeted integration occurring by two HR events within M and F leads to the replacement either of the endogenous chromosomal sequence by one (*BII*) or several repeats of the replacement cassette (*BIII*), or of the entire plasmid (*not shown*). When the selectable marker flanked by two *loxP* sites is cloned in an intron, subsequent expression of the Cre recombinase eliminates the selectable marker and the plasmid repeats to generate accurate targeted point mutations (*BIV*) (M → M*). Yet, replacement vectors can also integrate by a single HR event within either the M or F sequence leading to an insertional disruption, as described with insertion vectors (V for HR in M).

TAGGED MUTAGENESIS AND GENE- AND ENHANCER-TRAP SCREENS   Thus far, I have discussed the potential of targeted transgenesis in *Physcomitrella* for the functional analysis of plant genes. The unique opportunity that this offers in the plant kingdom contrasts singularly with the limited genetic knowledge we have of the biology of *Physcomitrella*. This prompted Hasebe's group to use large-scale insertional mutagenesis approaches in combination with efficient GT to generate collections of tagged and trapped lines of *Physcomitrella*. In these collections, new phenotypes and gene expression patterns can be identified and immediately further characterized (46a, 79). The first report from Nishiyama et al. (79) described the construction of two moss genomic libraries mutagenized by shuttle mutagenesis (116) with a mini-transposon designed either to tag moss genes with a *nptII* marker or to trap them with a *uidA* marker. Following transformation of *Physcomitrella* with these libraries, tagged lines displaying a visible developmental phenotype and trapped lines identified by GUS histochemical staining were recovered at frequencies of 4% and 3%, respectively (79). A preliminary analysis of these transformants indicated that such an approach was suitable for generating mutants that displayed a broad range of developmental phenotypes and for identifying gene expression patterns that cover many of the different developmental stages or cell types found in *Physcomitrella*. Using the same overall experimental strategy, Hiwatashi et al. (46a) developed further *uidA*-tagged gene-trap and enhancer-trap elements and used them to transform *Physcomitrella* under both homologous and nonhomologous conditions. Transformation rates were one order of magnitude higher under homologous conditions, suggesting that the use of mutagenized genomic libraries enhanced the yield of tagged and trapped lines. GUS-positive gene-trap and enhancer-trap lines were recovered at frequencies of 4% (235/5637) and 29% (1073/3726), respectively, which represent very high frequencies for the application of such strategies. More importantly, this work demonstrates that the trapped gene can be efficiently identified using RACE PCR with *uidA*-specific primers even though multiple copies of the mini-transposon were inserted in a trap line (46a). To account for the presence of multiple copies of the mini-transposon in tagged and trapped lines, the authors proposed that mini-transposon-tagged sequences concatenate extrachromosomally to generate large tandem arrays of mutagenized genomic sequences that subsequently integrate by HR in a single locus within the terminal genomic sequence of the array (46a), a hypothesis that fits well with the structural feature of replicative and integrative transformation of *Physcomitrella* reviewed here. This enables the study of moss genes or enhancers trapped upon shuttle mutagenesis without the simultaneous disruption of their original locus in the plant, a situation that presents advantages in the haploid gametophyte of *Physcomitrella*.

The two studies reviewed above represent major progress in the development of *Physcomitrella* for plant functional genomics. They provide the first large-scale collection of tagged insertional mutants in *Physcomitrella* available to the public in which a broad range of new phenotypes and new gene expression patterns in vivo can be identified. They also show that subsequent isolation of the mutated gene is facilitated by the presence of the mini-transposon tag and provide the first example of a complete reverse genetic analysis successfully completed in *Physcomitrella*. Finally, they show that, in a manner similar to its application in budding yeast (14), shuttle mutagenesis can be used to undertake large-scale analysis of gene expression, protein localization, and gene disruption in *Physcomitrella*.

CONCLUSION

The data reviewed above demonstrate that GT efficiency in *Physcomitrella* is amazingly high and can only be compared with that observed in budding yeast (108). Under optimal conditions, such efficiency enables within five weeks the isolation of dozens of targeted mutants from a single transformation experiment (41, 107, 110). The efficiency of GT in *Physcomitrella*, combined with the possibility of using the site-specific Cre/lox system, allows the application of the sophisticated strategies of targeted mutagenesis, which so far have been applied only in yeast and mouse ES cells. Thus, at the methodological level, *Physcomitrella* fulfills the ultimate requirement for its development as a model system because the generation of predetermined point mutations at any location of its genome is directly feasible.

Prospects for improving gene targeting frequencies in animals (125) and plants (74) (126) have been reviewed recently. The ideal situation would be to achieve a useful rate of GT in a wild-type background, as this would allow unambiguous interpretation of the data. This requirement is completely fulfilled by the *Physcomitrella* system. The strategy based on the generation of a recombination substrate in vivo as described by Rong & Golic in *Drosophila* (99) has yet to be tested in higher plants. The use of promoter trap positive-negative screens as applied in mammalian cells (115) also looks promising. Although it will not increase the yield of targeted integration events, it will mainly strengthen the screen for such events. Current strategies also include attempts to improve the ratio of targeted to random integration events by modulating the activity either of endogenous genes involved in DNA recombination and repair or following the overexpression of heterologous genes involved in these pathways (74, 126). Yet, as emphasized by Vasquez et al. (125), such strategies should be adopted with great caution and ideally should be performed only by transient expression of advantageous genes rather than by their inactivation or constitutive expression, as such mutations may seriously impair the maintenance of genome integrity. This is dramatically illustrated by the observation that loss of function of the Rad50 gene in *Arabidopsis* stimulates intrachromosomal recombination in planta (40) but simultaneously leads to plant sterility (36) and accelerated senescence due to telomere instability (37). In this respect again, *Physcomitrella patens* provides an ideal model system.

Why is gene targeting so efficient in *Physcomitrella*? We (108, 110, 111) previously proposed that it could be correlated with the dominance of the haploid gametophyte in its life cycle, whereas Reski (93) proposed that efficient GT could be correlated with the fact that protoplasts are highly synchronized and blocked in G2 during transformation. Both hypotheses can be tested experimentally.

In budding yeast, extensive genetic and biochemical studies have shown that efficient GT is tightly associated with the repair of DNA double-strand breaks (85). Yet a mutation that would specifically alter the ratio of targeted to random integration events observed upon integrative transformation has not yet been identified, indicating that the mechanisms that control this ratio are not easy to unravel (85). In this respect, *Physcomitrella* provides a valuable alternate system to study this process. The identification and functional analysis of moss genes involved in DNA repair and recombination may provide insight into these mechanisms and ultimately facilitate the development of strategies to target genes efficiently in higher eukaryotes.

In my opinion, *Physcomitrella* is poised to become an extremely valuable model system in biology. The efficiency of GT in *Physcomitrella* can and should be used immediately, as it fills a major methodological gap and advantageously complements the tools available for the study of plant functional genomics. The suitability of the *Physcomitrella* system to study plant biology is supported by the structural and functional similarities observed between mosses and higher plants both in biological processes and at the genetic level. The plant scientific community could advantageously benefit from a whole genome sequencing program to gain the required information about *Physcomitrella* genes for future functional genomic studies. The field is moving fast, and we have all the tools in our hands to develop the *Physcomitrella* system as a "green yeast" for addressing plant biological questions. Furthermore, the use of GT technology in *Physcomitrella* allows for the study of complex developmental processes such as organogenesis or the establishment of a body plan, processes that cannot be studied in budding yeast. As such, this plant may prove to be in a unique position to unravel the genetic networks controlling complex developmental processes in multicellular Eukaryotes.

ACKNOWLEDGMENTS

I would like to thank Jean-Pierre Zrÿd for his constant support and the valuable discussions we had over the past 15 years. I acknowledge my former collaborators Susan Vlach and Mikhail Chakhparonian for their contribution and enthusiasm and Michel Laloue and his collaborators in Versailles for the fruitful collaboration we have had over the past seven years. I am very grateful to Michael Lawton for his critical review and valuable suggestions on the manuscript.

Visit the Annual Reviews home page at www.annualreviews.org

LITERATURE CITED

1. Adair GM, Scheerer JB, Brotherman A, McConville S, Wilson JH, et al. 1998. Targeted recombination at the Chinese hamster APRT locus using insertion versus replacement vectors. *Somat. Cell Mol. Genet.* 24:91–105

2. Adams MD, Celniker SE, Holt RA, Evans CA, Gocayne JD, et al. 2000. The genome

sequence of *Drosophila melanogaster.* *Science* 287:2185–95

3. Aoyama T, Chua N-H. 1997. A glucocorticoid-mediated transcriptional induction system in transgenic plants. *Plant J.* 11:605–11

3a. Arabidopsis Genome Initiative. 2000. Analysis of the genome sequence of the flowering plant *Arabidopsis thaliana. Nature* 408:796–815

4. Ashton NW, Champagne CEM, Weiler T, Verkoczy LK. 2000. The bryophyte *Physcomitrella patens* replicates extrachromosomal transgenic elements. *New Phytol.* 146:391–402

5. Basi G, Schmidt E, Maundrell K. 1993. TATA box mutations in the *Schizosaccharomyces pombe* nmt1 promoter affect transcription efficiency but not transcription start point or thiamine repressibility. *Gene* 123:131–36

6. Bennetzen JL. 1999. Plant genomics takes root, branches out. *Trends Genet.* 15:85–87

7. Bevan MW, Flavell RB, Chilton M-D. 1983. A chimaeric antibiotic resistance gene as a selectable marker for plant cell transformation. *Nature* 304:184–87

8. Bilang R, Iida S, Peterhans A, Potrykus I, Paszkowski J. 1991. The 3′ terminal region of the hygromycin-B-resistance gene is important for its activity in *Escherichia coli* and *Nicotiana tabacum. Gene* 100:247–50

9. Blattner FR, Plunkett GI, Bloch CA, Perna NT, Burland V, et al. 1997. The complete genome sequence of *Escherichia coli* K-12. *Science* 277:1453–74

10. Böhner S, Lenk I, Rieping M, Herold M, Gatz C. 1999. Transcriptional activator TGV mediates dexamethasone-inducible and tetracycline-inactivable gene expression. *Plant J.* 19:87–95

11. Bonnema AB, Peytavi R, Van Daelen RAJ, Zabel P, Grimsley N. 1993. Development of an in vivo complementation system for identification of plant genes using yeast artificial chromosomes (YACS).

In *Bacterial Wilt*, ed. GL Hartmann, AC Hayward, pp. 177–82. Canberra: Aust. Cent. Int. Agric. Res.

12. Bonneville J-M, Sanfaçon H, Fütterer J, Hohn T. 1989. Posttranscriptional *trans*-activation in cauliflower mosaic virus. *Cell* 59:1135–43

13. Burke DT, Carle GF, Olson MV. 1987. Cloning of large segments of exogenous DNA into yeast by means of artificial chromosome vectors. *Science* 236:806–12

14. Burns N, Grimwade B, Rossmacdonald PB, Choi EY, Finberg K, et al. 1994. Large-scale analysis of gene expression, protein localization and gene disruption in *Saccharomyces cerevisiae. Genes Dev.* 8:1087–105

15. *Caenorhabditis elegans* sequencing consortium. 1998. Genome sequence of the nematode *C. elegans*: a platform for investigating biology. *Science* 282:2012–19

16. Capecchi MR. 1989. Altering the genome by homologous recombination. *Science* 244:1288–92

17. Chakhparonian M. 2001. *Développement d'outils de la mutagenèse ciblée par recombinaison homologue chez Physcomitrella patens.* PhD thesis. Univ. Lausanne, Switz. www.unil.ch/lpc/docs/

18. Champagne C, Ashton N. 2001. Ancestry of KNOX genes revealed by bryophyte (*Physcomitrella patens*) homologs. *New Phytol.* 150:23–36

19. Christensen AH, Quail PH. 1996. Ubiquitin promoter-based vectors for high-level expression of selectable and/or screenable marker genes in monocotyledonous plants. *Trans. Res.* 5:213–18

20. Clayton RA, White O, Ketchum KA, Venter JC. 1997. The first genome from the third domain of life. *Nature* 387:459–62

21. Cove DJ. 1983. Genetics of Bryophyta. In *New Manual of Bryology*, ed. RM Schuster, pp. 222–31. Tokyo, Jpn.: Hattori Bot. Lab.

22. Cove DJ. 1992. Regulation of development in the moss, *Physcomitrella patens.*

In *Developmental Biology. A Molecular Genetic Approach*, ed. S Brody, DJ Cove, S Ottolenghi, VEA Russo, pp. 179–93. Heidelberg: Springer-Verlag

23. Cove DJ. 2000. The moss, *Physcomitrella patens. J. Plant Growth Regul.* 19:275–83

24. Cove DJ, Ashton NW. 1988. Growth regulation and development in *Physcomitrella patens*: an insight into growth regulation and development of bryophytes. *Bot. J. Linn. Soc.* 98:247–52

25. Cove DJ, Kammerer W, Knight CD, Leech MJ, Martin CR, et al. 1991. Developmental genetic studies of the moss *Physcomitrella patens*. In *Molecular Biology of Plant Development*, ed. GI Jenkins, W Schuch, pp. 31–43. Cambridge, UK: Co. Biol.

26. Cove DJ, Knight CD. 1987. Gravitropism and phototropism in the moss *Physcomitrella patens*. In *Developmental Mutants in Higher Plants*, ed. H Thomas, D Grierson, pp. 181–96. Cambridge, UK: Cambridge Univ. Press

27. Cove DJ, Knight CD. 1993. The moss *Physcomitrella patens*, a model system with potential for the study of plant reproduction. *Plant Cell* 5:1483–88

28. Cove DJ, Knight CD, Lamparter T. 1997. Mosses as model systems. *Trends Plant Sci.* 2:99–105

29. Cronk QCB. 2001. Plant evolution and development in a post-genomic context. *Nat. Rev. Genet.* 2:607–19

30. Dale EC, Ow DW. 1990. Intra- and intermolecular site-specific recombination in plants mediated by bacteriophage P1 recombinase. *Gene* 91:79–85

31. Dale EC, Ow DW. 1991. Gene transfer with subsequent removal of the selection gene from the host genome. *Proc. Natl. Acad. Sci. USA* 88:558–62

32. Deng C, Capecchi MR. 1992. Reexamination of gene targeting frequency as a function of the extent of homology between the targeting vector and the target locus. *Mol. Cell. Biol.* 12:3365–71

33. Doonan JH, Duckett JD. 1988. The bryophyte cytoskeleton: experimental and immunofluorescence studies of morphogenesis. *Adv. Bryol.* 3:1–31

34. Eisen MB, Brown PO. 1999. DNA arrays for analysis of gene expression. *Methods Enzymol.* 303:179–205

35. Gaillard C, Moffat BA, Blacker M, Laloue M. 1998. Male sterility associated with APRT deficiency is *Arabidopsis thaliana* results from a mutation in the gene APT1. *Mol. Gen. Genet.* 257:348–53

36. Gallego ME, Jeanneau M, Granier F, Bouchez D, Bechtold N, et al. 2001. Disruption of the Arabidopsis RAD50 gene leads to plant sterility and MMS sensitivity. *Plant J.* 25:31–41

37. Gallego ME, White CI. 2001. RAD50 function is essential for telomere maintenance in Arabidopsis. *Proc. Natl. Acad. Sci. USA* 98:1711–16

38. Gatz C. 1996. Chemically inducible promoters in transgenic plants. *Curr. Opin. Biotech.* 7:168–72

39. Gatz C. 1997. Chemical control of gene expression. *Annu. Rev. Plant Physiol. Plant Mol. Biol.* 48:89–108

40. Gherbi H, Gallego ME, Jalut N, Lucht J, Hohn B, et al. 2000. Homologous recombination in planta is stimulated in the absence of Rad50. *EMBO J.* 2:287–91

41. Girke T, Schmidt H, Zähringer U, Reski R, Heinz E. 1998. Identification of a novel Δ6-acyl-group desaturase by targeted disruption in *Physcomitrella patens. Plant J.* 15:39–48

42. Girod PA, Fu HY, Zryd JP, Vierstra RD. 1999. Multiubiquitin chain-binding subunit MCB1 (RPN10) of the 26S proteasome is essential for developmental progression in *Physcomitrella patens. Plant Cell* 11:1457–71

43. Gorbunova V, Levy AA. 1999. How plants make ends meet: DNA double-strand break repair. *Trends Plant Sci.* 4:263–69

44. Grimsley NH, Grimsley JM, Hartmann E. 1980. Fatty acid composition of mutants of the moss *Physcomitrella patens. Phytochemistry* 20:1519–24

45. Gritz L, Davies J. 1983. Plasmid-encoded hygromycin B resistance: the sequence of hygromycin B phosphotransferase gene and its expression in *Escherichia coli* and *Saccharomyces cerevisiae. Gene* 25:179–88

46. Guiltinan MJ, Marcotte WR, Quatrano RS. 1990. A plant leucin zipper protein that recognizes an abscissic acid response element. *Science* 250:267–71

46a. Hiwatashi Y, Nishiyama T, Fujita T, Hasebe M. 2001. Establishment of gene-trap and enhancer-trap systems in the moss *Physcomitrella patens. Plant J.* 28: 105–16

47. Hofmann AH, Codon AT, Ivascu C, Russo VEA, Knight C, et al. 1999. A specific member of the cab multigene family can be efficiently targeted and disrupted in the moss *Physcomitrella patens. Mol. Gen. Genet.* 261:92–99

48. Houba-Hérin N, Reynolds S, Schaefer D, von Schwartzenberg K, Laloue M. 1997. Molecular characterization of homologous recombination events in the moss *Physcomitrella patens. Plant Sci., Coll. Gén. Soc. Fr. Physiol. Vég.*, ed. JC Petch, A Latché, M Bouzayen, 3:22–23. Toulouse: SFPV-INRA

49. Deleted in proof

50. Jefferson RA, Kavanagh TA, Bevan MW. 1987. GUS fusions: β-glucuronidase as a sensitive and versatile gene fusion marker in higher plants. *EMBO J.* 6:3901–7

51. Jeon JS, Lee S, Jung KH, Jun SH, Jeong DH, et al. 2000. T-DNA insertional mutagenesis for functional genomics in rice. *Plant J.* 22:561–70

52. Kammerer W, Cove DJ. 1996. Genetic analysis of the result of re-transformation of transgenic lines of the moss, *Physcomitrella patens. Mol. Gen. Genet.* 250: 380–82

53. Kawahara H, Kasahara M, Nishiyama A, Ohsumi K, Goto T, et al. 2000. Developmentally regulated, alternative splicing of the *Rpn10* gene generates multiple forms of 26S proteasome. *EMBO J.* 19:4144–53

54. Keegan L, Gill G, Ptashne M. 1986. Separation of DNA binding from the transcription-activating function of eukaryotic regulatory protein. *Science* 231:699–704

55. Kiessling J, Kruse S, Rensing SA, Harter K, Decker EL, et al. 2000. Visualization of a cytoskeleton-like FtsZ network in chloroplasts. *J. Cell Biol.* 151:945–50

56. Knight CD. 1994. Studying plant development in mosses: the transgenic route. *Plant Cell Environ.* 17:669–74

57. Knight CD. 2000. Moss, molecular tools for phenotypic analysis. In *Encyclopedia of Cell Technology*, ed. E Spier, pp. 936–44. New York: Wiley

58. Knight CD, Sehgal A, Atwal K, Wallace JC, Cove DJ, et al. 1995. Molecular responses to abscissic acid and stress are conserved between moss and cereals. *Plant Cell* 7:499–506

59. Kozak M. 1991. Structural features in eukaryotic mRNAs that modulate the initiation of translation. *J. Biol. Chem.* 266: 19867–70

60. Kraus E, Leung W-Y, Haber JE. 2001. Break-induced replication: a review and an example in budding yeast. *Proc. Natl. Acad. Sci. USA* 98:8255–62

61. Krogan N, Ashton N. 2000. Ancestry of plant MADS-box genes revealed by bryophyte (*Physcomitrella patens*) homologues. *New Phytol.* 147:505–17

62. Kunst F, Ogasawara N, Moszer I, Albertini AM, Alloni G, et al. 1997. The complete genome sequence of the gram-positive bacterium *Bacillus subtilis. Nature* 390:249–56

63. Lazo GR, Stein PA, Ludwig RA. 1991. A DNA transformation-competent *Arabidopsis* genomic library in *Agrobacterium. BioTechnology* 9:963–68

64. Lee D, Moffat BA. 1994. Adenine salvage activity during callus induction and plant growth. *Physiol. Plant* 87:483–92

65. Lee KY, Lund P, Lowe K, Dunsmuir P. 1990. Homologous recombination in plant cells after *Agrobacterium*-mediated transformation. *Plant Cell* 2:415–25

66. Leech MJ, Kammerer W, Cove DJ, Martin C, Wang TL. 1993. Expression of *myb*-related genes in the moss *Physcomitrella patens. Plant J.* 3:51–61

67. Long Z, Wang S-H, Nelson N. 1989. Cloning and nucleotide sequence analysis of genes coding for the major chlorophyll-binding protein of the moss *Physcomitrella patens* and of the halotolerant alga *Dunaliella salin. Gene* 76:299–312

68. Machuka J, Bashiardes S, Ruben E, Spooner K, Cuming A, et al. 1999. Sequence analysis of expressed sequence tags from an ABA-treated cDNA library identifies stress response genes in the moss *Physcomitrella patens. Plant Cell Physiol.* 40: 378–87

69. Marcotte WR, Bayley CC, Quatrano RS. 1988. Regulation of a wheat promoter by abscissic acid in rice protoplasts. *Nature* 335:454–57

70. Marini NJ, Etkin LD, Benbow RM. 1988. Persistence and replication of plasmid DNA microinjected into early embryos of *Xenopus laevis. Dev. Biol.* 127:421–34

71. Maundrell K. 1993. Thiamine-repressible expression vectors pREP and pRIP for fission yeast. *Gene* 123:127–30

72. McElroy D, Blowers AD, Jenes B, Wu R. 1991. Construction of expression vectors based on the rice actin 1 (Act1) 5′ region for the use in monocot transformation. *Mol. Gen. Genet.* 231:150–60

73. Mello CC, Kramer JM, Stinchcomb D, Ambros V. 1991. Efficient gene transfer in *C. elegans*: extrachromosomal maintenance and integration of transforming sequences. *EMBO J.* 10:3959–70

74. Mengiste T, Paszkowski J. 1999. Prospects for the precise engineering of plant genomes by homologous recombination. *Biol. Chem.* 380:749–58

75. Moffat B, Pethe C, Laloue M. 1991. Metabolism of benzyladenine is impaired in a mutant of *Arabidopsis thaliana* lacking adenine phosphoribosyl transferase activity. *Plant Physiol.* 95:900–8

76. Moffat BA, Sommerville C. 1988. Posi-tive selection for male-sterile mutants of *Arabidopsis* lacking adenine phosphoribosyl transferase activity. *Plant Physiol.* 86:1150–54

77. Müller U. 1999. Ten years of gene targeting: targeted mouse mutants, from vector design to phenotype analysis. *Mech. Dev.* 82:3–21

78. Nacry P, Camilleri C, Courtial B, Caboche M, Bouchez D. 1998. Major chromosomal rearrangements induced by T-DNA transformation in Arabidopsis. *Genetics* 149:641–50

79. Nishiyama T, Hiwatashi Y, Sakakibara K, Kato M, Hasebe M. 2000. Tagged mutagenesis and gene trap in the moss, *Physcomitrella patens* by shuttle mutagenesis. *DNA Res.* 7:9–17

80. Offringa R, de Groot JA, Haagsman HJ, Does MP, van den Elzen PJM, et al. 1990. Extrachromosomal homologous recombination and gene targeting in plant cells after *Agrobacterium*-mediated transformation. *EMBO J.* 9:3077–84

81. Osteryoung KW, Pyke KA. 1998. Plastid division: evidence for a prokaryotically derived mechanism. *Curr. Opin. Plant Biol.* 1:475–79

82. Osteryoung KW, Stokes KD, Rutherford SM, Percival AL, Lee WY. 1998. Chloroplast division in higher plants requires members of two functionally divergent gene families with homology to bacterial FtsZ. *Plant Cell* 10:1991–2004

83. Ow DW. 1996. Recombinase-directed chromosome engineering in plants. *Curr. Opin. Biotechnol.* 7:181–86

84. Ow DW, Medberry SL. 1995. Genome manipulation through site-specific recombination. *Crit. Rev. Plant Sci.* 14:239–61

85. Paques F, Haber JE. 1999. Multiple pathways of recombination induced by double-strand breaks in *Saccharomyces cerevisiae. Microb. Mol. Biol. Rev.* 63: 349–404

86. Parinov S, Sundaresan V. 2000. Functional genomics in *Arabidopsis*: large-scale insertional mutagenesis complements the

genome sequencing project. *Curr. Opin. Biotechnol.* 11:157–61

87. Paszkowski J, Baur M, Bogucki A, Potrykus I. 1988. Gene targeting in plants. *EMBO J.* 7:4021–26

88. Paszkowski J, Shillito RD, Saul M, Mandak V, Hohn T, et al. 1984. Direct gene transfer to plants. *EMBO J.* 3:2717–22

89. Puchta H, Hohn B. 1996. From centiMorgans to base pairs: homologous recombination in plants. *Trends Plant Sci.* 1:340–48

90. Quatrano R, Bashiardes S, Cove D, Cuming A, Knight C, et al. 1999. Leeds Washington University Moss EST Project. http://www.ncbi.nlm.nih.gov

91. Regan S, Moffat BA. 1990. Cytochemical analysis of pollen development in wildtype Arabidopsis and a male-sterile mutant. *Plant Cell* 2:877–89

92. Reski R. 1998. Development, genetics and molecular biology of mosses. *Bot. Acta* 111:1–15

93. Reski R. 1998. Physcomitrella and Arabidopsis: the David and Goliath of reverse genetics. *Trends Plant Sci.* 3:209–10

94. Reski R. 1999. Molecular genetics of Physcomitrella. *Planta* 208:301–9

95. Reski R, Faust M, Wang X-H, Wehe M, Abel WO. 1994. Genome analysis of a moss, *Physcomitrella patens*. *Mol. Gen. Genet.* 244:352–59

96. Reski R, Reynolds S, Wehe M, Kleberjanke T, Kruse S. 1998. Moss (*Physcomitrella patens*) expressed sequence tags include several sequences which are novel for plants. *Bot. Acta* 111:143–49

97. Reutter K, Atzorn R, Hadeler B, Schmülling T, Reski R. 1998. Expression of the bacterial ipt gene in Physcomitrella rescues mutations in budding and plastid division. *Planta* 206:196–203

98. Richmond T, Somerville S. 2000. Chasing the dream: plant EST microarrays. *Curr. Opin. Plant Biol.* 3:108–16

99. Rong YS, Golic KG. 2000. Gene targeting by homologous recombination in Drosophila. *Science* 288:2013–18

100. Rothstein R. 1991. Targeting, disruption, replacement and allele rescue: integrative DNA transformation in yeast. *Methods Enzymol.* 194:281–301

101. Russell AJ, Knight MR, Cove DJ, Knight CD, Trewavas AJ, et al. 1996. The moss, *Physcomitrella patens*, transformed with apoaequorin cDNA responds to cold shock, mechanical perturbation and pH with transient increases in cytoplasmic calcium. *Trans. Res.* 5:167–70

102. Sakakibara K, Nishiyama T, Kato M, Hasebe M. 2001. Isolation of homeodomain-leucine zipper genes from the moss *Physcomitrella patens* and the evolution of homeodomain-leucine zipper genes in land plants. *Mol. Biol. Evol.* 18:491–502

103. Sauer B. 1993. Manipulation of the transgene by site-specific recombination: use of *cre* recombinase. *Methods Enzymol.* 225:890–900

104. Sauer B. 1994. Recycling selectable markers in yeast. *Biotechnology* 16:1086–88

105. Sauer B. 1998. Inducible gene targeting in mice using the Cre/lox system. *Methods* 14:381–92

106. Sawahel W, Onde S, Knight C, Cove D. 1992. Transfer of foreign DNA into *Physcomitrella patens* protonemal tissue by using the gene gun. *Plant Mol. Biol. Rep.* 10:314–15

107. Schaefer DG. 1994. *Molecular genetic approaches to the biology of the moss Physcomitrella patens*. PhD Thesis. Univ. Lausanne, Switz. www.unil.ch/lpc/docs/DSThesis.htm

108. Schaefer DG. 2001. Gene targeting in *Physcomitrella patens*. *Curr. Opin. Plant Biol.* 4:143–50

109. Schaefer DG, Bisztray G, Zrÿd J-P. 1994. Genetic transformation of the moss *Physcomitrella patens*. In *Plant Protoplasts and Genetic Engineering V*, ed. YPS Bajaj, pp. 349–64. Berlin/Heidelberg/New York: Springer-Verlag

110. Schaefer DG, Zrÿd J-P. 1997. Efficient gene targeting in the moss *Physcomitrella patens*. *Plant J.* 11:1195–206

111. Schaefer DG, Zrÿd J-P. 2001. The moss *Physcomitrella patens* now and then. *Plant Physiol.* 127:1430–38

112. Schaefer DG, Zrÿd J-P, Knight CD, Cove DJ. 1991. Stable transformation of the moss *Physcomitrella patens*. *Mol. Gen. Genet.* 226:418–24

113. Schumaker KS, Dietrich MA. 1997. Programmed changes in form during moss development. *Plant Cell* 9:1099–107

114. Schumaker KS, Dietrich MA. 1998. Hormone-induced signaling during moss development. *Annu. Rev. Plant Physiol. Plant Mol. Biol.* 49:501–23

115. Sedivy JM, Dutriaux A. 1999. Gene targeting and somatic cell genetics—a rebirth or a coming of age? *Trends Genet.* 15:88–90

116. Seifert HS, Chen EY, So M, Heffron F. 1986. Shuttle mutagenesis: a method of transposon mutagenesis for *Saccharomyces cerevisiae*. *Proc. Natl. Acad. Sci. USA* 83:735–39

117. Springer PS. 2000. Gene traps: tools for plant development and genomics. *Plant Cell* 12:1007–20

118. Stinchcomb DT, Shaw JE, Carr SH, Hirsh D. 1985. Extrachromosomal DNA transformation of *Caenorhabditis elegans*. *Mol. Cell. Biol.* 5:3484–96

119. Strepp R, Scholz S, Kruse S, Speth V, Reski R. 1998. Plant nuclear gene knockout reveals a role in plastid division for the homolog of the bacterial cell division protein FtsZ, an ancestral tubulin. *Proc. Natl. Acad. Sci. USA* 95:4368–73

120. Struhl K. 1983. The new yeast genetics. *Nature* 305:391–97

121. Theissen G, Münster T, Henschel K. 2001. Why don't mosses flower? *New Phytol.* 150:1–8

122. Thiellement H, Bahrman N, Damerval C, Plomion C, Rossignol M, et al. 1999. Proteomics for genetic and physiological studies in plants. *Electrophoresis* 20:2013–26

123. Thomas KR, Capecchi MR. 1987. Site-directed mutagenesis by gene targeting in mouse embryo derived stem cells. *Cell* 51:503–12

124. van Nocker S, Sadis S, Rubin DM, Glickman M, Fu H, et al. 1996. The multiubiquitin-chain-binding protein mcb1 is a component of the 26S proteasome is *Saccharomyces cerevisiae* and plays a nonessential, substrate-specific role in protein turnover. *Mol. Cell. Biol.* 16:6020–28

125. Vasquez KM, Marburger K, Intody Z, Wilson JH. 2001. Manipulating the mammalian genome by homologous recombination. *Proc. Natl. Acad. Sci. USA* 98:8403–10

126. Vergunst AC, Hooykaas PJJ. 1999. Recombination in the plant genome and its application in biotechnology. *Crit. Rev. Plant Sci.* 18:1–31

127. Voges D, Zwickl P, Baumeister W. 1999. The 26S proteasome: a molecular machinery designed for controlled proteolysis. *Annu. Rev. Biochem.* 68:1015–68

128. Walbot V. 1992. Strategies for mutagenesis and gene cloning using transposon tagging and T-DNA insertional mutagenesis. *Annu. Rev. Plant Physiol. Plant Mol. Biol.* 43:49–82

129. Weigel D, Ahn JH, Blazquez MA, Borevitz JO, Christensen SK, et al. 2000. Activation tagging in Arabidopsis. *Plant Physiol.* 122:1003–13

130. Zeidler M, Gatz C, Hartmann E, Hughes J. 1996. Tetracycline-regulated reporter gene expression in the moss *Physcomitrella patens*. *Plant Mol. Biol.* 30:199–205

131. Zeidler M, Hartmann E, Hughes J. 1999. Transgene expression in the moss *Ceratodon purpureus*. *J. Plant Physiol.* 154:641–50

132. Zuo J, Niu Q-W, Chua N-H. 2000. An estrogen receptor-based transactivator XVE mediates highly inducible gene expression in transgenic plants. *Plant J.* 24:265–73

133. Zuo JM, Niu Q-W, Møller SG, Chua N-H. 2001. Chemical-induced, site-specific DNA excision in transgenic plants. *Nat. Biotechnol.* 19:157–61

Annu. Rev. Plant Biol. 2002. 53:503–21
DOI: 10.1146/annurev.arplant.53.100301.135212

COMPLEX EVOLUTION OF PHOTOSYNTHESIS

Jin Xiong
*Department of Biology, Texas A&M University, College Station, Texas 77843;
e-mail: jxiong@mail.bio.tamu.edu*

Carl E. Bauer
*Department of Biology, Indiana University, Bloomington, Indiana 47405;
e-mail: cbauer@bio.indiana.edu*

Key Words   phylogeny, bacteriochlorophyll, photosynthetic reaction center

■ **Abstract**   The origin of photosynthesis is a fundamental biological question that has eluded researchers for decades. The complexity of the origin and evolution of photosynthesis is a result of multiple photosynthetic components having independent evolutionary pathways. Indeed, evolutionary scenarios have been established for only a few photosynthetic components. Phylogenetic analysis of Mg-tetrapyrrole biosynthesis genes indicates that most anoxygenic photosynthetic organisms are ancestral to oxygen-evolving cyanobacteria and that the purple bacterial lineage may contain the most ancestral form of this pigment biosynthesis pathway. The evolutionary path of type I and type II reaction center apoproteins is still unresolved owing to the fact that a unified evolutionary tree cannot be generated for these divergent reaction center subunits. However, evidence for a cytochrome *b* origin for the type II reaction center apoproteins is emerging. Based on the combined information for both photopigments and reaction centers, a unified theory for the evolution of reaction center holoproteins is provided. Further insight into the evolution of photosynthesis will have to rely on additional broader sampling of photosynthesis genes from divergent photosynthetic bacteria.

CONTENTS

1040-2519/02/0601-0503$14.00

INTRODUCTION

Photosynthesis is a biological process that harvests solar energy for the formation of chemical bonds. The advent of photosynthesis, especially oxygenic photosynthesis, has fundamentally changed the landscape of Earth. Oxygen-evolving photosynthesis has generated most of the atmospheric oxygen that is used for respiration as well as formation of an ozone layer to shield Earth from UV radiation (1). Thus, the process of photosynthesis is arguably the most important chemical reaction on Earth that has led to the development of advanced life forms.

There are two major speculations regarding the origin of photosynthesis (29, 30). One suggests that photosynthesis originated in a prebiotic environment on Earth and was coupled with the origin of life. The second, which is based on more recent molecular phylogenetic analyses, suggests that photosynthesis evolved after chemolithotrophy-based life was developed in the last common ancestor. However, the evolutionary history of how multiple biochemical components were incorporated in the photosynthetic process has long remained murky. The questions that remain in debate are: What is the earliest ancestor for photosynthetic organisms, what is its closest living descendent, and how did photosynthesis evolve into diverse lineages?

Photosynthesis has been widely accepted as a bacterially derived process given that there are no Archaeal species that synthesize Mg-tetrapyrrole-based photosystems and that the photosynthetic properties of eukaryotes were acquired from cyanobacteria through endosymbiosis. Because photosynthesis evolved from bacteria, an understanding of the origin and evolution of photosynthesis requires detailed phylogenetic analysis of photosynthesis genes from the known bacterial branches that synthesize photosystems. These branches include (*a*) purple bacteria that synthesize a simplified, nonoxygen-evolving type II photosystem; (*b*) green nonsulfur bacteria that also synthesize a nonoxygen-evolving type II photosystem; (*c*) green sulfur bacteria that synthesize a type I photosystem with a homodimeric reaction center; (*d*) the gram-positive heliobacteria that synthesize a similar homodimeric type I reaction center; and (*e*) cyanobacteria that contain both a type I photosystem and a more complex oxygen-evolving type II photosystem, both of which contain a heterodimeric reaction center core.

This review covers recent evolutionary analyses of these five photosynthetic lineages containing either one or two photosystem types. The goal is to reveal the complexity of the evolutionary pathway of photosynthesis, in which no simple linear branching scheme can be drawn to account for distribution of various photosynthetic components in diverse photosynthetic organisms. With a focus on Mg-tetrapyrrole biosynthesis enzymes and photosynthetic reaction center apoproteins, we discuss various scenarios derived from the use of advanced molecular phylogenetic methods for the early evolutionary history of the reaction centers.

GEOCHEMICAL EVIDENCE AND THEORIES FOR THE ORIGIN OF PHOTOSYNTHESIS

Earth is over 4.5 giga-years (Gyrs) old (1 Gyr $= 1 \times 10^9$ years) (26). During the first 0.5 Gyr (Hadean stage), Earth's environment was characterized by frequent and massive meteorite bombardment. Some of the impacts are thought to have been capable of evaporating the entire ocean. The Archean stage is characterized by the end of the bombardment and the beginning of life, probably starting approximately 3.8 to 4.0 Gyrs ago. The earliest Archean fossil stromatolites that contain evidence for biological carbon fixation (38) are dated at 3.8 Gyrs old. The earliest cyanobacteria-like cellular microfossils are 3.5 Gyrs old (39, 40), only 0.3 Gyr after the first record of life. However, Earth was beginning to be dominated by cyanobacteria starting 2.7 Gyrs ago, coinciding with the early rise of oxygen on Earth (9). Within approximately 1 Gyr, cyanobacteria increased the O_2 level to approximately one fourth of the current level. The success of cyanobacteria was suggested to be due to not only the energetic advantage of photosynthesis, but also the inhibition of competitors through the creation of the potentially toxic agent O_2 (27). The first emergence of eukaryotes was 1.8 Gyrs ago. It took almost another 0.6–0.8 Gyr for photosynthetic cyanobacteria to be incorporated into eukaryotes as chloroplasts. Algae subsequently lifted the atmospheric oxygen to the current level. The descendents of algae, land plants, appeared on Earth 0.5 Gyrs ago, creating the landscape that we know today. Some of the major historical landmarks on Earth are depicted in Figure 1, with schematic representation of the rise of oxygen levels in atmosphere due to oxygenic photosynthesis.

Because the origin of oxygenic photosynthesis is key to understanding the early evolution of photosynthesis, it is necessary to mention the current debate regarding the 3.5-Gyr-old cellular fossil record, which was determined to be of filamentous cyanobacterial origin. Based on this finding, it has been suggested that photosynthetic organisms were the earliest cellular organisms on Earth and that oxygenic photosynthesis started in the early Archean (30, 35). However, molecular analyses have suggested that the earliest forms of life may not be phototrophic, despite the energetic advantages of photosynthesis. In addition, the identification of the microfossils, which is based mainly on morphological resemblance to current microorganisms, presents its own challenges. Pierson (34) pointed out that the same morphological characteristics can be easily attributed to other filamentous nonoxygen-evolving phototrophs such as green nonsulfur bacteria, including *Chloroflexus*, that are microscopically indistinguishable from many cyanobacteria. Controversies surrounding the biogenicity of certain specific microfossils and stromatolites also exist (5). Another problem with linking the microfossils to cyanobacteria is that it is impossible to determine whether the organisms in question really contained photosynthesis-related genes, especially in light of the pervasive lateral gene transfers that had occurred during early evolution of prokaryotes. (The mosaic and dynamic nature of bacterial genomes are discussed in detail below.) Because no definite geochemical evidence exists, tracing the origin of the biochemical process of photosynthesis using geological evidence alone will be difficult.

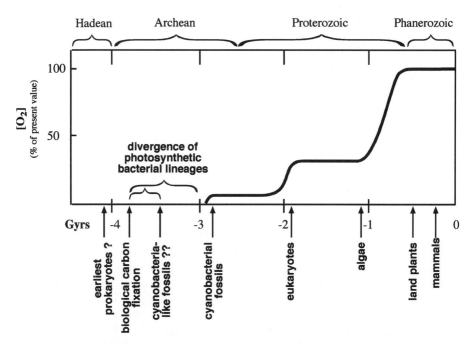

Figure 1 Schematic representation of the rise of oxygen level on Earth during the early history of life [modified from (27)]. Major evolutionary landmarks are indicated by arrows on the lower x-axis, and major geological periods are indicated by brackets on the upper x-axis. Putative stages of early divergence of photosynthetic prokaryotes are indicated by brackets above the lower x-axis: one assumes ~3–3.8 Gyrs and the other 3.5–3.8 Gyrs, depending on what date is accepted as the starting point for oxygenic photosynthesis.

There are many geologically oriented hypotheses regarding the origin of photosynthesis. The overall consensus appears to be that primitive photosynthetic pigments first evolved from chemoautotrophic organisms living in an environment where local chemical disequilibrium can be easily exploited, leading to a capability of cells to use pigments to exploit light as an additional source of energy. Full-fledged photosynthesis would develop later to allow cells to use sunlight as their sole energy source (26). A hypothesis was proposed by Nisbet et al. (25) suggesting that anoxygenic photosynthesis could have evolved from purple bacteria that used infrared phototaxis as an intermediate step. Based on a close match of the emission spectra of geothermal light and the absorption spectra of bacteriochlorophylls *a* and *b*, the authors suggested that photosynthesis arose from bacteriochlorophyll *a*- or *b*-containing organisms near oceanic hydrothermal vents where weak infrared radiation could be detected. They proposed that the organisms initially used the bacteriochlorophyll pigments to sense infrared light, which guided them in their phototaxis behavior. The bacteria capable of detecting geothermal light and phototaxis could preferentially occupy an optimum habitat, which may render them with evolutionary advantages in the competition for nutrient resources. Rudimentary photosynthesis would have subsequently evolved as a supplement to chemotrophy, giving the organisms an added selective advantage. Through further adaptation of a primitive photosystem, the organisms would start making use of the near-infrared part of sunlight when moving to shallow water. Eventually, chlorophylls would be developed to make use of higher energy (visible) light to split water. The hypothesis appears to be supported by the observation that purple bacteria are indeed capable of infrared phototaxis behavior (36). Furthermore, the amount of energy emitted from the geothermal light appears to be sufficient to drive photochemical reactions (45). The hypothesis also received support from our phylogenetic analysis (detailed below), suggesting that purple photosynthetic bacteria may indeed represent the first group of organisms to evolve photosynthesis.

EVOLUTION OF NONPHOTOSYNTHESIS-RELATED MOLECULAR MARKERS

Owing to the difficulty of using geochemical markers to trace the origin of photosynthesis, another approach to reconstruct the evolutionary history of photosynthesis is to use phylogenetic methods to analyze molecular sequence information. Before sequence information for a large number of photosynthesis genes was available, most phylogenetic analysis relied on the characterization of nonphotosynthesis genes. The most-often used phylogenetic marker is small subunit ribosomal rRNA (SS rRNA or 16S rRNA for prokaryotes). Several versions, with minor variations, of SS rRNA trees exist; all of which consistently place green nonsulfur bacteria as the earliest photosynthetic lineage (28, 32, 47), with heliobacteria branching off as the second photosynthetic lineage, followed by green sulfur bacteria, cyanobacteria, and then purple bacteria (proteobacteria). Gruber & Bryant (13) constructed phylogenetic trees using sigma 70 factor sequences within a large number of bacterial groups. They found that heliobacteria (Firmicutes) branch closer to proteobacteria than to cyanobacteria, which was very different from the branching patterns observed with 16S rRNA. Gupta et al. (14) used heat shock proteins (Hsp60 and Hsp70) as markers and concluded that the earliest evolving photosynthetic lineage is the heliobacterial group that subsequently diverged to green nonsulfur bacteria and then to cyanobacteria, green sulfur bacteria, and finally purple bacteria.

The above controversies [see (50)] with the use of these nonphotosynthesis gene markers significantly hindered the development of a clear picture of the evolutionary history of photosynthesis. These controversies also reinforced recent suggestions by various authors that lateral gene transfers played a major role in prokaryotic evolution (10, 17, 20). Thus, attempts to extrapolate the evolution of photosynthesis, which is a very ancient event, from the characterization of nonphotosynthesis genes may be a futile endeavor [for details see (50)].

EVOLUTION OF MG-TETRAPYRROLE BIOSYNTHESIS

Genes and enzymes involved in (bacterio)chlorophyll biosynthesis are intimately involved in the evolutionary process of photosynthesis. Because these pigments are an integral part of the reaction centers, they are ideal molecular indicators for a wide-scale phylogenetic comparison of photosynthetic complexes. The biosynthetic enzymes are relatively conserved in sequence, structure, and function throughout all types of photosynthetic organisms and therefore readily reveal the ancient evolutionary history of the pigment biosynthesis.

Recently, Xiong et al. (50, 51) conducted comprehensive sequence analysis of many pigment biosynthesis genes and enzymes from a variety of diverse photosynthetic bacteria in an effort to understand the molecular evolution of photosynthesis. Nine different bacteriochlorophyll biosynthesis genes and enzymes, one concatenated set of genes and enzymes of all prokaryotic photosynthetic groups, and several algal and plant taxa were analyzed (50). Phylogenetic trees were constructed using three major phylogenetic methods, i.e., distance neighbor joining (NJ), maximum parsimony (MP), and maximum likelihood (ML). The direction of evolution was resolved by rooting the trees with properly selected outgroup sequences. [An outgroup is a sequence that is homologous to the sequences under consideration but is found only in distantly related proteins or organisms and/or has diverged to a different function within the same organism (2).] For example, the nitrogenase complex was used as an outgroup for the light-independent protochlorophyllide reductase complex. In addition to sequence similarities, the two types of protein complexes have been experimentally verified as carrying out different but related enzymatic activities (11). For the Mg-chelatase complex, cobalt chelatase and putative nickel chelatase were used as an outgroup. All of these enzymes have related but modified functions in similar but distinct metabolic pathways.

Using the rooted phylogenetic approach (mentioned above), Xiong et al. (50) observed that most bacteriochlorophyll-synthesizing anoxygenic photosynthetic bacterial groups evolved earlier than chlorophyll-synthesizing oxygenic photosynthetic groups. This implies that anoxygenic photosynthesis evolved prior to oxygenic photosynthesis. This conclusion is also supported by the ferrous iron oxidation experiment that shows that purple nonsulfur bacteria can naturally oxidize colorless Fe^{2+} to brown Fe^{3+} and reduce CO_2 to cell material, implying that the Precambrian banded iron formations may be of anoxygenic origin (46). One surprise finding from the rooted phylogenetic analysis is that the purple photosynthetic group emerges as the earliest branching taxon. Another surprising finding is that green sulfur bacteria and green nonsulfur bacteria are sister groups that together form the second-earliest branching order on the phylogenetic tree. This result is striking in that these different green bacterial lineages contain vastly different photosystems, with green sulfur bacteria containing a type I photosystem and green nonsulfur bacteria containing a type II photosystem. Except for the fact that the two groups of organisms synthesize a common antenna pigment, bacteriochlorophyll

c, and share a similar light-harvesting complex, the chlorosome, they are otherwise unrelated in many aspects of cellular metabolism (2). Xiong et al. (50, 51) also observed that heliobacteria form a sister group with cyanobacteria and the rest of the oxygenic photosynthetic taxa. The conclusion is equally striking in that, according to biochemical analyses, heliobacteria contain a far more primitive photosynthetic apparatus than any other known photosynthetic organisms (2). The heliobacterial reaction center contains a single pigment in a homodimer protein complex without any peripheral antenna complexes. Moreover, this sister relationship between heliobacteria and cyanobacteria is supported by other recent analysis on ribosomal protein genes (15) and cytochrome b proteins (41, 51).

One of the major implications of these results is a reversal of the often-cited Granick hypothesis (12), which states that pigment biosynthesis recapitulates the evolutionary appearance of the pigment itself. As chlorophyll biosynthesis is one of the intermediate steps in bacteriochlorophyll a biosynthesis [see review in (43)], the Granick hypothesis of pigment biosynthesis can be interpreted to mean that chlorophyll-containing oxygenic photosynthetic organisms predate bacteriochlorophyll-containing anoxygenic photosynthetic organisms. However, molecular phylogeny clearly indicates that bacteriochlorophyll a is indeed a more ancient pigment. This implies that the chlorophyll a–containing cyanobacterial reaction centers are a more recent evolutionary product and that the evolution of pigment biosynthesis from bacteriochlorophylls to chlorophylls may have involved gene loss and shortening of the pathway, rather than stepwise additions.

By comparing bacteriochlorophyll biosynthesis gene-based phylogeny with 16S rRNA–based phylogeny (32), it is clear that the evolutionary pathways of the two sets of genes are drastically different. This incongruence may be attributed to lateral gene transfer events for bacteriochlorophyll biosynthesis genes. With more and more prokaryotic genomes being sequenced, the chimeric or mosaic nature of their genomes is being recognized. In fact, lateral gene transfers may have so pervasively occurred among ancient prokaryotes that they may be the driving force of prokaryotic evolution (20). This supports the notion that a gene phylogeny in prokaryotes maximally reflects the evolution of a metabolic process rather than the whole genome (50). As Doolittle (10) recently pointed out, the evolutionary history of life should be best represented by a "mangrove-like network" with many crossovers between branches, rather than by the well-known "universal tree."

Although directions of lateral gene transfers are often difficult to resolve, there are methods to address this issue. As further analysis of the evolution of (bacterio)chlorophyll biosynthesis genes, Xiong & Bauer (48) recently carried out a Baysian analysis (52) to infer ancestral sequences of (bacterio)chlorophyll biosynthesis genes at internal nodes of phylogenetic trees. The Baysian analysis (Figure 2) shows that the ancestral sequence for all in-group sequences indeed belongs to purple bacteria. The ancestral sequence at the node before the divergence of green bacteria and cyanobacteria/heliobacteria groups is the green sulfur bacterial lineage, whereas the nodal sequence before green sulfur and green nonsulfur is also

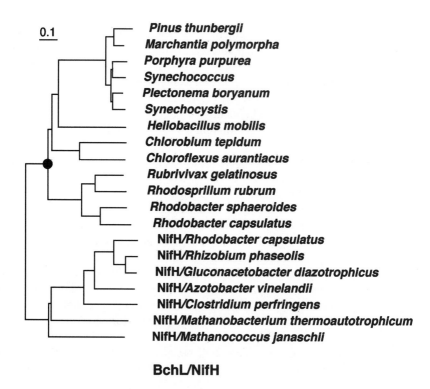

BchL/NifH

```
AKVFAVYGKGGIGKSTTSSNLSAALAKKGKKVLQIGCDPKHDSTFTLTGR
LIPKTVIDVLEEVNFHPEDLRPEDVVFEGFGGVDCVEAGGPPAGTGCGGY
VVGETVKLLKELGVFDEYDVVVFDVLGDVVCGGFAAPLQENYADYALIVT
SNDFDSIFAANRICAAIQEKAKSGYKVRLAGIIANRSDATDGTDLIDKFA
ERIGTRLLGRVPRLDVIRRSRLKKKTLFEMEEDPELSEVAAEYLELAEQL
LEKKGEDAVIPNPLTDREIFELLGGFDLP
```

Figure 2 Phylogenetic tree of BchL (*top*) with ancestral sequence (*middle*) at the common internal node (*solid circle*) inferred by Baysian analysis. Direction of lateral gene transfer is indicated below the sequence (*bottom*).

green sulfur. The sequence before the divergence of cyanobacteria and heliobacteria indicates a closer relationship to cyanobacterial sequences. Thus, a more accurate route for the bacteriochlorophyll biosynthesis gene transfer is obtained (shown in Figure 2). The resolution of the direction of evolution by the Baysian analysis is largely consistent with the phylogenetic interpretation by Xiong et al. (50). The one surprising exception is that bacteriochlorophyll g synthesized by heliobacteria may have evolved from chlorophyll a of cyanobacteria. This is an interesting observation given that the two pigments have highly related structures, with the only difference between bacteriochlorophyll g and chlorophyll a being an isomerization of the double bond at ring II, which occurs spontaneously in bacteriochlorophyll g upon exposure to air.

With the illustration of significant lateral gene transfer events among prokaryotic genomes and the phylogenetic consensus that purple bacteria are the earliest photosynthetic group, a rethinking of the significance of the 3.5-Gyr-old cyanobacteria-like microfossils is warranted. Indeed, linking microfossil morphology with no information about the metabolic nature of the object is problematic. This is because genomes from ancient cyanobacteria-like fossils may not have encoded any photosynthesis genes at the 3.5-Gyr period.

ORIGIN OF PHOTOSYNTHETIC REACTION CENTER APOPROTEINS

In addition to Mg tetrapyrroles, another key part of the photosynthetic apparatus is the photosynthetic reaction center apoproteins. The origin and evolutionary pathway of the photosynthetic reaction center apoproteins has long been the subject of debate. This is mainly owing to the lack of obvious sequence similarity between type I and type II reaction center apoproteins. Based on similarity in photochemical charge separation of both types of reaction centers, it was proposed that both reaction centers may have derived from a common origin (1, 22). More-refined structural analysis of the two types of reaction centers supported the proposal that both types I and II of reaction centers are homologous. Schubert et al. (42) have shown that both reaction centers have a common structural motif with similar topology for transmembrane helices and cofactor arrangement. The most-recent high-resolution crystal structures of both type I and type II reaction centers confirmed significant conservation in the topology of transmembrane helices as well as of the positions of the core Mg-tetrapyrrole photopigments (18, 53). The structural data have often been interpreted as evidence that these two reaction centers most likely evolved from a common origin. However, despite significant structural conservation, the lack of primary sequence conservation between the type I and type II reaction center apopolypeptides still calls into question the possibility of convergent evolution (51). In addition, the nature of the common ancestor for the reaction center apoproteins remains unresolved.

Because a unified reaction center apopolypeptide phylogeny that includes all photosynthetic lineages cannot be generated, there have been several biochemically oriented hypotheses proposed regarding the origin of the reaction centers. Mulkidjanian & Junge (24) suggested that a putative 11-transmembrane helix protein providing a UV protection role may be an ancestor for the reaction centers. Because of the unclear nature of the proposed UV protector, this hypothesis has remained unsupported. Meyer (23) suggested that photosynthetic reaction centers may have originated from the cytochrome b subunit of the cytochrome bc_1 complex. This proposal was based on the observation that both types of proteins are membrane spanning, with bound tetrapyrroles and quinones functioning in electron transfers. However, because this hypothesis lacks supporting evidence based on sequence analysis, it has not been broadly accepted.

In our own analysis (J. Xiong & C. Bauer, submitted), we discovered that type II reaction center core polypeptides share weak sequence similarity with cytochrome b of the bc_1 complex. Further refined analysis revealed a region of high similarity, which includes three contiguous transmembrane helices (B, C, and D) of cytochrome b that bind two hemes and three contiguous transmembrane helices (B, C, and D) of type II reaction centers that bind cofactors such as (bacterio)chlorophylls, (bacterio)pheophytin, quinone, and a nonheme iron. The core regions with $\sim$130 residues show a sequence identity of 20–30% between the two groups. Furthermore, three of the four heme ligands in cytochrome b are conserved with the cofactor ligands in the reaction center polypeptides [the special pair (bacterio)chlorophyll histidine ligand, the nonheme iron histdine ligand, and the chlorophyll$_Z$ ligand (for oxygenic reaction center only)]. In addition, there is relatively strong statistical support from PRSS tests (33) for sequence comparison and DALI search (16) for structural comparison for the two groups of membrane proteins.

Recent studies have shown a ubiquitous existence of cytochrome b and the Rieske iron-sulfur protein, essential components of the cytochrome bc complex, in Archaea, Bacteria, and Eukarya (41). Phylogenetic analyses of several electron transfer proteins related to aerobic respiration (7, 8, 41) also indicate that many respiratory components including the cytochrome bc complex and cytochrome c oxidase may have existed in the last common ancestor of Archaea and Bacteria. In contrast, Mg-tetrapyrrole-based photosystems are found in Bacteria only, inclusive of the chloroplast lineage, and are therefore less likely to have existed in the common ancestor of both Archaea and Bacteria. This finding led to the proposal that photosynthetic metabolism must have appeared after the advent of respiratory metabolism (6). Nitschke et al. (27) further proposed that photochemical reaction centers may have evolved by integrating into an existing respiratory electron transport chain. The evidence for the existence of respiratory components prior to photosynthetic ones is contrary to the common belief that oxygenic photosynthesis must precede respiration because the respiratory process requires oxygen as a substrate. However, the primitive form of cytochrome bc complex may have performed either an anaerobic type of respiration for energy conversion or, as recently suggested, an aerobic type of respiration in the presence of an extremely low level of oxygen in a primitive Earth environment (19, 37) that was resulted from photolysis of water molecules.

EVOLUTIONARY PATHWAY OF REACTION CENTER APOPROTEINS

Although structural similarities of the type I and type II reaction centers provide a convincing argument for a common evolutionary ancestor, it is impossible to infer phylogenetic relationships between the type I and type II bacterial photosynthetic reaction center apoproteins owing to extensive sequence divergence. Previous analysis on the evolutionary pathways of reaction centers has been thus largely based on subunit composition and functionality. There are three competing models to account for the evolutionary paths of the two types of reaction centers in oxygenic cyanobacteria. One is the selective-loss model by Olson & Pierson (31), which suggests that both types of photosystem were developed from a single cyanobacterial origin. Subsequent loss of one of the photosystem types during evolutionary diversification would give rise to the existing single reaction center types in anoxygenic photosynthetic lineages. Another model is the fusion model (1, 22), which suggests that a primordial homodimeric photosystem would give rise to either type I or type II anoxygenic photosynthetic reaction centers. Subsequent "fusion" or arrival of these two types of anoxygenic photosystems in a single organism and the development of water-oxidizing capabilities would give rise to the current oxygenic cyanobacterial photosystems. A third model was put forward by Vermaas (44), suggesting that a homodimeric anoxygenic type I reaction center is the common ancestor to all of the current reaction centers. According to this model, the cyanobacterial reaction centers would have been derived from duplication and fragmentation of an ancestral type I reaction center. The purple bacterial reaction center would be the latest development through a subsequent selective loss of the type I photosystem in the cyanobacterial lineage.

For this review, we carried out new phylogenetic analysis for both types of reaction center core proteins. To help determine what the first reaction center was like at the protein sequence level, we calculated the ancestral sequences using Baysian analysis (52). The reconstructed ancestral sequence at the internal node for the type II reaction centers (Figure 3A) is shown to be a purple bacterial L subunit-like protein. Thus, a simplified scenario for the type II reaction center gene flow appears to be a diversification from purple bacteria to both green nonsulfur and cyanobacterial lineages. Similar analysis for type I reaction centers indicates that the ancestral sequence belongs to the heliobacterial type (Figure 3B), suggesting that both green nonsulfur and the cyanobacterial type I reaction center are results of diversification from a primitive heliobacterial reaction center. The analysis result does not come as a complete surprise as the finding that the heliobacterial reaction center, with

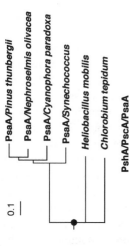

A

0.1

D2/*Spinacia oleracea*
D2/*Synechocystis*
D1/*Spinacia oleracea*
D1/*Synechocystis*
PufM/*Chloroflexus aurantiacus*
PufM/*Rhodobacter capsulatus*
PufL/*Chloroflexus aurantiacus*
PufL/*Rhodobacter capsulatus*
PetB/*Paracoccus denitrificans*
PetB/*Rhodobacter sphaeroides*

DVKGGLWQLITLHATVAFFGWMLRQYIEISRKLGMGSYHIPVAFSGA
IAAYLVYLFVIRPMMMGSAWMGYALPYGGYAYGNFFYNPFFSLHMLG
IAGLFGSALLLAMHGWALVLSANANPYGAEVVRRESKEKGKKPTAPDH
EDTYFIVRDL

Green nonsulfur
bacteria
Purple
bacteria
Cyanobacteria

B

0.1

PsaA/*Pinus thunbergii*
PsaA/*Nephroselmis olivacea*
PsaA/*Cyanophora paradoxa*
PsaA/*Synechococcus*
Heliobacillus mobilis
Chlorobium tepidum

PshA/PscA/PsaA

LTISPPEREAEKVKPAGVKPKGTVPPPKWNKPGPKAKGLPGGPSMAT
ADAAFNPRAQVFEWFKDKVPATRGAVLKAHINHLGMVAGFLSFVLVH
GFILAIYLSWLSDQVLFAPTPQVVWPIVGQGILNGDFGGGWFYLIFY
ARLYQLGLDASARSADALMWARLHLLAAIGFWLIGGWFLHFKTPRED
EWLKNVSNPFGKTLVGQFHFLALVGTLWGMHMAYIGVRGANGGIVPT
GLSFPLLFLGWGDMFGPITGGFFAPIPAPLAGNHVAPGAFLFLGGFF
ALGIYWWHLPPNAAIHLNDDVFHHFAGYLTKRFAFFEKDWEAVLSVS
AQVFAPHFATVLFAMIIWNRPDQPILSFYFMGDYALSNAKEFYAAPE
YRELASQNPGFLIVILGHLVPGVLFVIGGVFHGAHFMLRATNDPKLY
EALKDFKMLKRDAICYDHDFQKKFLGLIMFGAFLPNDTLRALGRPQD
IFVSYGIATHPVFANTIADLHHLAKAGMFANMTYINSLAFGGGDVIG
TPLHDAIFGSHGTVSDFVAAHVIAGGLHFTMVPLWRMVFFSKVSPWT
TKVGMKAKRDYEFPCLGPAYGGTCSISLVDQFYLAIFWSLQVIAPVW
FYLDGVATSSEVYGQAAEDFKANPTWFSLHAVSNFTSEGYFSYFLIS
QTTRMFKYYDGHLVQAGLLLLGAHFIWAFTFSMLFQYRGSRDEGAEV
LKWAHEQVGLKFAGKVYNRALSLKEGKAIGTFLYFKGTVLCMWCFAM
VGFYQVG

Green sulfur
bacteria
Heliobacteria
Cyanobacteria

Figure 3 Phylogenetic trees for type II (*A*) and type I (*B*) reaction centers (*top*). Calculated ancestral sequences (*middle*) at common ancestral nodes (*solid circles*) are shown below the trees. Only partial ancestral sequence for type II reaction center was derived using a refined and reliable alignment region for type II reaction center proteins and cytochrome *b*. Direction of lateral gene transfer for both types of photosynthetic reaction centers are indicated below the sequences (*bottom*).

no known antenna complexes (2), is the most primitive photosynthetic apparatus. However, a paradox exists given that heliobacteria also appear to contain the most recently acquired Mg-tetrapyrrole biosynthesis genes. These paradoxal observations underscore the difficulty in unraveling photosynthesis evolution, which involves components that contain different evolutionary paths.

Our ancestral sequence analysis supports the fusion model as well as a derivative of that model presented by Xiong et al. (51). Our current analysis is clearly in disagreement with the selective-loss model because the two types of cyanobacterial reaction center apoproteins are developed from two different anoxygenic sources. It is also in disagreement with the model by Vermaas because the cyanobacterial type II reaction center is not ancestral to the purple bacterial reaction center. If both types of reaction centers indeed share a common ancestry (18, 42, 53), the type II purple bacterial reaction center may be more ancestral relative to the type I reaction center because the type II reaction center is more closely related to the putative cytochrome b origin. In this scenario, the type I reaction center would be a result of gene fusion for a type II reaction center and ancestral CP43/CP47 proteins. Close sequence similarity of photosystem (PS) II antenna proteins CP43/CP47 with the light-harvesting domain of type I reaction centers has been demonstrated by Vermaas (44) and Xiong et al. (51). Structural similarities between CP43/CP47 and the light-harvesting domain of PSI have been verified by high-resolution crystal structures of both types of photosynthetic reaction centers (18, 53).

UNIFIED MODEL FOR THE ORIGIN AND EVOLUTION
OF REACTION CENTERS

In an attempt to reconcile the differences between the evolutionary pathways for Mg tetrapyrroles and for reaction center apoproteins, we further proposed an "apoprotein-early" hypothesis (J. Xiong & C. Bauer, submitted). The proposal was based on combined analysis results showing that almost all photosynthetic groups independently received pigment biosynthesis genes and reaction center apoprotein genes from at least two different sources. It is almost impossible for both sets of genes to have arrived at a given lineage at the same time. This raises the question of which set of genes, the reaction center genes or Mg-tetrapyrrole genes, would have arrived first. We propose that Mg-tetrapyrrole biosynthesis genes may have evolved in organisms containing preexisting reaction center apoproteins that may have still functioned as cytochromes. This hypothesis is also augmented by the argument that if Mg tetrapyrroles had arrived alone without polypeptides to bind Mg tetrapyrroles that assist in photochemistry, then there would be a waste of cellular resources for biosynthesis and, more importantly, a generation of damaging free radicals by photoexcitation, which would be highly detrimental to the cell's survival—a clearly unfavorable and unlikely evolutionary outcome.

The combined evolutionary information for both reaction centers and Mg tetrapyrroles allowed us to construct a unified theory for the evolution of reaction center holoproteins (Figure 4, see color insert). In this hypothesis, we propose that an initial gene duplication event for cytochrome *b* gave rise to a cytochrome that became the precursor of a type II reaction center apoprotein. Once Mg tetrapyrroles evolved, the cytochrome *b*–like primitive form of the reaction center apoprotein would then combine with Mg tetrapyrroles to become a functional reaction center holoprotein. For Mg-tetrapyrrole biosynthesis, the pigment biosynthesis genes may have first derived from the purple photosynthetic bacterial lineage, making purple bacteria the most ancestral photosynthetic form. The pigment biosynthesis genes would then have been transferred to green sulfur bacteria and then bifurcated to both green nonsulfur bacteria and cyanobacteria and eventually to heliobacteria, making bacteriochlorophyll *g* the latest development of photosynthetic pigment. Although heliobacteria received the photopigment last, they may have developed the earliest type I reaction center through a gene fusion process of genes for the type II reaction center protein and ancestral CP43/CP47 proteins. For cyanobacteria to perform oxygenic photosynthesis with both types of reaction centers, their type II reaction center apoproteins would have been laterally transferred from the purple bacterial lineage; their type I reaction center apoproteins would have been derived from the heliobacterial lineage. In all cases, the aforementioned reaction center as cytochrome *b* scenario would apply before the bacterial lineages received Mg tetrapyrroles.

EVOLUTION OF OXYGENIC PHOTOSYNTHESIS

The development of oxygenic photosynthesis was one of the most important events in Earth's history as it fundamentally changed the redox balance on Earth and permitted development of aerobic metabolism and more-advanced life forms (2). As discussed above, oxygenic photosynthesis is likely a late event during evolution that was preceded by anoxygenic photosynthesis. The key in developing oxygenic photosynthesis is the development of the manganese complex that is capable of water oxidation. Blankenship & Hartman (3) have proposed that hydrogen peroxide (H_2O_2) may be a transitional donor to PSII before water was used as an electron donor. This mechanism may have been developed first as a means to detoxify excess peroxide. The authors (3) also proposed that the proteinaceous portion of PSII involved in binding the manganese cluster may have developed through domain swapping with catalase. This proposal is based on the observation that photosynthetic water oxidation is mechanistically similar to that of catalases involved in breaking down H_2O_2 to water and oxygen. Specifically, they observed that the binuclear Mn center in catalase is similar to one half of the tetranuclear Mn center within the water oxidation complex. Indeed, under certain conditions, the PSII water oxidation complex can act as a catalase. The redox potential for oxidation of H_2O_2 to O_2 is $+0.27$ V, which is lower than the oxidation of purple bacterial reaction center P870 to P870$^+$ at $+0.5$ V. Although this theory is interesting, it lacks geochemical support showing that there was indeed a significant

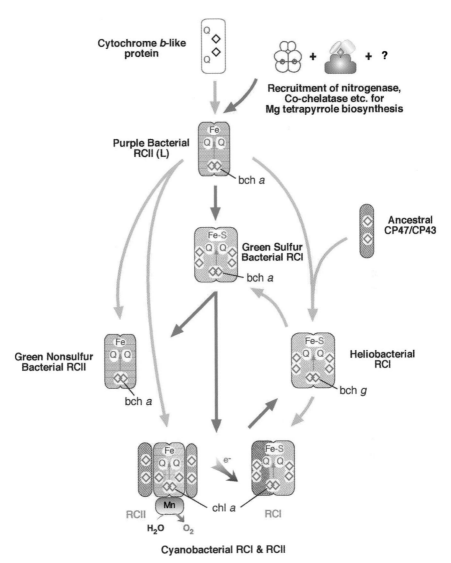

Figure 4 Model of evolutionary pathway of photosynthetic reaction center holoproteins based on combined phylogenetic information for both reaction center apoproteins and Mg-tetrapyrrole biosynthesis enzymes. *Green lines* indicate the direction of evolution for pigment biosynthesis, *yellow lines* for reaction center apoproteins.

amount of H_2O_2 on Earth (30). There is also no phylogenetic or structural evidence for relatedness between catalase and the water oxidation complex.

EVOLUTION OF THE CYTOCHROME *BC* COMPLEX

Many types of cytochromes are involved in photochemistry either directly or indirectly. The cytochrome *bc* complex appears to be conserved among all five photosynthetic groups and is involved intimately in the photosynthetic process. A recent review on the early evolution of cytochrome *bc* complexes was provided by Schütz et al. (41) in which the primary structures, functional characteristics, and phylogenetic positions of subunits of cytochrome *bc* complexes from diverse bacterial and Archaeal groups were analyzed. Similar comparison was also made by Xiong et al. (51). Both groups observed interesting differences in the physical organization of the structural genes in various organisms. For example, in the cytochrome *bc* complex of green sulfur bacteria, cytochrome *b* is a fusion protein of cytochrome b_6 and subunit IV, whereas in heliobacteria the two proteins are encoded by two separate but linked open reading frames. In nonphotosynthetic *Bacillus subtilis*, subunit IV, whose gene is downstream of the cytochrome *b* gene, is instead fused with cytochrome *c*. The phylogenetic tree topology from both studies share many similarities, though the work of Schütz et al. (41) includes more phylogenetic taxa and provides a case of lateral gene transfer of cytochrome *b* from proteobacteria to *Aquifex*.

The conserved nature of subunits of the cytochrome *bc* complex and its ubiquitous presence in all photosynthetic lineages appear to fit the criteria for them to be used for wide-scale phylogenetic comparisons. However, these molecular markers suffer from not being able to truly represent the photosynthetic process as the cytochrome *bc* complex is found in most nonphotosynthetic taxa where it is involved in the respiratory electron transport chain. Cytochrome *bc* genes are thus arguably no more effective than 16S rRNA as a molecular marker for tracing the evolution of photosynthesis.

EVOLUTION OF CAROTENOID BIOSYNTHESIS GENES

Carotenoids are found in reaction centers and light-harvesting complexes where they fulfill the light-harvesting and free-radical scavenging roles. However, carotenoids are not unique to photosynthesis and are synthesized ubiquitously in all three domains, Bacteria, Archaea, and Eukarya, serving as retinoid components, hormones, and membrane components. There are few evolutionary studies involving carotenoid biosynthesis genes. Boucher & Doolittle (4) as well as Lange et al. (21) surveyed enzymes for isoprenoid biosynthesis among a large number of eubacterial species. The reconstructed evolutionary history of the enzymes bears no resemblance with other known phylogenies. All results have clearly indicated that lateral gene transfer events have contributed to the early distribution of the carotenoid biosynthesis genes across the prokaryotic genomes (4, 21).

Although the antiquity and wide distribution of carotenoids may shed light on early cell evolution, owing to their secondary role in photosynthesis, carotenoid biosynthetic enzymes may be of less value in determining the early evolutionary routes in photosynthesis.

CONCLUDING REMARKS

The overall evolution of photosynthesis is a complex process involving distinct origins and pathways of many different components. Consequently, the evolutionary development of photosynthesis cannot be described by a simple, linear branching pathway. Evidence has shown that various photosynthetic components have distinctively different evolutionary histories, indicating that these components may have been recruited or modified from several other preexisting metabolic pathways during different stages of evolutionary development (2). Although photosynthesis as a mosaic process appears to have no single well-defined origin, what is certain is that the emergence of Mg tetrapyrroles and reaction center apoproteins was the key event that led to the advent of photosynthesis (1). The evolution of (bacterio)chlorophyll pigments may be a limiting step in the overall evolution of reaction center holoprotein (48, 50). Thus, the evolutionary pathway of (bacterio)chlorophyll pigments provides a unified view for both oxygenic and anoxygenic photosynthesis. One issue that has been satisfactorily solved with overwhelming support is that anoxygenic photosynthesis evolved prior to oxygenic photosynthesis (48, 50, 51).

Despite recent developments, there are still many unresolved key issues within the evolution of photosynthesis, such as the precise nature of the earliest photosystems and how water-oxidizing capability evolved in cyanobacteria. Because of the inherent difficulty of studying deep evolution and the presence of highly diversified photosynthetic components, we believe that new molecular insight on evolution of photosynthesis can only be obtained through larger sampling of photosynthesis genes in more diverse phototrophic organisms. With the increased pool of molecular data, in particular for photosynthesis genes, and with the use of more sophisticated phylogenetic analysis tools, a clearer picture of evolutionary history of photosynthesis will continue to emerge.

ACKNOWLEDGMENTS

The work is supported by an NIH grant (GM53940) and an Arizona State University NASA astrobiology grant to C.E.B. as well as a faculty startup fund from Texas A&M University to J.X.

LITERATURE CITED

1. Blankenship RE. 1992. Origin and early evolution of photosynthesis. *Photosynth. Res.* 33:91–111

2. Blankenship RE. 2001. Molecular evidence for the evolution of photosynthesis. *Trends Plant Sci.* 6:4–6

3. Blankenship RE, Hartman H. 1998. The origin and evolution of oxygenic photosynthesis. *Trends Biochem. Sci.* 23:94–97

4. Boucher Y, Doolittle WF. 2000. The role of lateral gene transfer in the evolution of isoprenoid biosynthesis pathways. *Mol. Microbiol.* 37:703–16

5. Buick R. 1993. Microfossil recognition in Archean rocks: an appraisal of spheroid and filaments from a 3500 M.Y. old chertbarite unite at North Pole, Western Australia. *Palaios* 5:441–59

6. Castresana J, Saraste M. 1995. Evolution of energetic metabolism: the respiration-early hypothesis. *Trends Biochem. Sci.* 20:443–48

7. Castresana J, Lübben M, Saraste M, Higgins DG. 1994. Evolution of cytochrome oxidase, an enzyme older than atmospheric oxygen. *EMBO J.* 13:2516–25

8. Castresana J, Lübben M, Saraste M. 1995. New archaebacterial genes coding for redox proteins: implications for the evolution of aerobic mechanism. *J. Mol. Biol.* 250:202–10

9. Des Marais DJ. 2000. When did photosynthesis emerge on Earth? *Science* 289:1703–5

10. Doolittle WF. 1999. Phylogenetic classification and the universal tree. *Science* 284:2124–28

11. Fujita Y, Bauer CE. 2000. Reconstitution of light-independent protochlorophyllide reductase from purified Bchl and BchN-BchB subunits. *In vitro* confirmation of nitrogenase-like features of a bacteriochlorophyll biosynthesis enzyme. *J. Biol. Chem.* 275:23583–88

12. Granick S. 1965. Evolution of heme and chlorophyll. In *Evolving Genes and Proteins*, ed. V Bryson, HJ Vogel, pp. 67–88. New York: Academic

13. Gruber TN, Bryant DA. 1997. Molecular systematic studies of eubacteria, using σ^{70-}-type sigma factors of group 1 and group 2. *J. Bacteriol.* 179:1734–47

14. Gupta RS, Mukhtar T, Singh B. 1999. Evolutionary relationships among photosynthetic prokaryotes (*Heliobacterium chloroum, Chloroflexus aurantiacus*, cyanobacteria, *Chlorobium tepidum* and proteobacteria): implications regarding the origin of photosynthesis. *Mol. Microbiol.* 32:893–906

15. Hansmann S, Martin W. 2000. Phylogeny of 33 ribosomal and six other proteins encoded in an ancient gene cluster that is conserved across prokaryotic genomes: influence of excluding poorly alignable sites from analysis. *Int. J. Syst. Evol. Microbiol.* 50:1655–63

16. Holm L, Sander C. 1996. Mapping the protein universe. *Science* 273:595–602

17. Jain R, Rivera MC, Lake JA. 1999. Horizontal gene transfer among genomes: the complexity hypothesis. *Proc. Natl. Acad. Sci. USA* 96:3801–6

18. Jordan P, Fromme P, Witt HT, Klukas O, Saenger W, Krauss N. 2001. Three-dimensional structure of cyanobacterial photosystem I at 2.5 Å resolution. *Nature* 411:909–17

19. Kasting JF. 1993. Earth's early atmosphere. *Science* 259:920–26

20. Koonin EV, Aravind L, Kondrashov AS. 2000. The impact of comparative genomics on our understanding of evolution. *Cell* 101:573–76

21. Lange BM, Rujan T, Martin W, Croteau R. 2000. Isoprenoid biosynthesis: the evolution of two ancient and distinct pathways

across genomes. *Proc. Natl. Acad. Sci. USA* 97:13172–77

22. Mathis P. 1990. Compared structure of plant and bacterial photosynthetic reaction centers. Evolutionary implications. *Biochim. Biophys. Acta* 1018:163–67

23. Meyer TE. 1994. Evolution of photosynthetic reaction center and light harvesting chlorophyll proteins. *Biosystems* 33:167–75

24. Mulkidjanian AY, Junge W. 1997. On the origin of photosynthesis as inferred from sequence analysis. *Photosynth. Res.* 51:27–42

25. Nisbet EG, Cann JR, van Dover CL. 1995. Origins of photosynthesis. *Nature* 373:479–80

26. Nisbet EG, Sleep NH. 2001. The habitat and nature of early life. *Nature* 409:1083–91

27. Nitschke W, Muhlenhoff U, Liebl U. 1998. Evolution. In *Photosynthesis: A Comprehensive Treatise*, ed. A Raghavendra, pp. 286–304. Cambridge, UK: Cambridge Univ. Press

28. Olsen GJ, Woese CR, Overbeek R. 1994. The winds of (evolutionary) change: breathing new life into microbiology. *J. Bacteriol.* 176:1–6

29. Olson JM. 1999. Early evolution of chlorophyll-based photosystems. *Chemtracts* 12:468–82

30. Olson JM. 2001. Evolution of photosynthesis (1970), re-examined thirty years later. *Photosynth. Res.* 68:95–112

31. Olson JM, Pierson BK. 1987. Evolution of reaction centers in photosynthetic prokaryotes. *Int. Rev. Cytol.* 108:209–48

32. Pace NR. 1997. A molecular view of microbial diversity and the biosphere. *Science* 276:734–40

33. Pearson WR, Lipman DJ. 1988. Improved tools for biological sequence comparison. *Proc. Natl. Acad. Sci. USA* 85:2444–48

34. Pierson BK. 1994. The emergence, diversification, and role of photosynthetic eubacteria. In *Early Life on Earth*, Nobel Symp.,

No. 84, ed. S. Bengston, pp. 161–80. New York: Columbia Univ. Press

35. Pierson BK, Olson JM. 1989. Evolution of photosynthesis in anoxygenic photosynthetic prokaryotes. In *Microbial Mat: Physiological Ecology of Benthic Microbial Communities*, ed. Y Cohen, E Rosenberg, pp. 402–27. Washinton, DC: Am. Soc. Microbiol.

36. Ragatz L, Jiang Z-Y, Bauer CE, Gest H. 1994. Phototactic purple bacteria. *Nature* 370:104

37. Saraste M, Castresana J, Higgins D, Lübben M, Wilmanns M. 1996. Evolution of cytochrome oxidase. In *Origin and Evolution of Biological Energy Conversion*, ed. H Baltscheffsky, pp. 255–89. New York: Wiley

38. Schidlowski M. 1988. A 3800-million-year isotopic record of life from carbon in sedimentary rocks. *Nature* 333:313–18

39. Schopf JW. 1993. Microfossils of the early Archean apex chert: new evidence of the antiquity of life. *Science* 260:640–46

40. Schopf JW, Packer BM. 1987. Early Archean (3.3-billion- to 3.5-billion-year-old) microfossils from Warrawoona Group, Australia. *Science* 237:70–73

41. Schütz M, Brugna M, Lebrun E, Baymann F, Huber R, et al. 2000. Early evolution of cytochrome *bc* complexes. *J. Mol. Biol.* 300:663–75

42. Schubert WD, Klukas O, Saenger W, Witt HT, Fromme P, Krauss N. 1998. A common ancestor for oxygenic and anoxygenic photosynthetic systems: a comparison based on the structural model of photosystem I. *J. Mol. Biol.* 280:297–314

43. Suzuki JY, Bollivar DW, Bauer CE. 1997. Genetic analysis of chlorophyll biosynthesis. *Annu. Rev. Genet.* 31:61–89

44. Vermaas WFJ. 1994. Evolution of heliobacteria: implications for photosynthetic reaction center complexes. *Photosynth. Res.* 41:285–94

45. White SN, Chave AD, Reynolds GT, Gaidos EJ, Tyson JA, van Dover CL. 2000. Variations in ambient light emission from

black smokers and flange pools on the Juan de Fuca Ridge. *Geophys. Res. Lett.* 27:1151–54

46. Widdel F, Schnell S, Heising S, Ehrenreich A, Assmus B, Schink B. 1993. Ferrous iron oxidation by anoxygenic photosynthetic bacteria. *Nature* 362:834–36

47. Woese CR. 1997. Bacterial evolution. *Microbiol. Rev.* 51:221–71

48. Xiong J, Bauer CE. 2001. Further evidence and new molecular insight on the origin and early evolution of photosynthesis. *Proc. Int. Photosynth. Congr., 12th.* In press

49. Deleted in proof

50. Xiong J, Fischer WM, Inoue K, Nakahara M, Bauer CE. 2000. Molecular evidence for the early evolution of photosynthesis. *Science* 289:1724–30

51. Xiong J, Inoue K, Bauer CE. 1998. Tracking molecular evolution of photosynthesis by characterization of a major photosynthesis gene cluster from *Heliobacillus mobilis*. *Proc. Natl. Acad. Sci. USA* 95:14851–56

52. Yang Z, Kumar S, Nei M. 1995. A new method of inference of ancestral nucleotide and amino acid sequences. *Genetics* 141:1641–50

53. Zouni A, Witt HT, Kern J, Fromme P, Krauss N, et al. 2001. Crystal structure of photosystem II from *Synechococcus elongatus* at 3.8 Å resolution. *Nature* 409:739–43

Annu. Rev. Plant Biol. 2002. 53:523–50
DOI: 10.1146/annurev.arplant.53.100301.135242

CHLORORESPIRATION

Gilles Peltier and Laurent Cournac

CEA Cadarache, Laboratoire d'Ecophysiologie de la Photosynthèse, Département d'Ecophysiologie Végétale et de Microbiologie, UMR 163 CNRS-CEA, Direction des Sciences du Vivant, Université Mediterranée, CEA 1000, F-13108 Saint-Paul-lez-Durance, France; e-mail: gilles.peltier@cea.fr, laurent.cournac@cea.fr

Key Words chloroplasts, complex I, photosynthesis, alternative oxidase, respiration

■ **Abstract** Chlororespiration has been defined as a respiratory electron transport chain (ETC) in interaction with the photosynthetic ETC in thylakoid membranes of chloroplasts. The existence of chlororespiration has been disputed during the last decade, with the initial evidence mainly obtained with intact algal cells being possibly explained by redox interactions between chloroplasts and mitochondria. The discovery in higher-plant chloroplasts of a plastid-encoded NAD(P)H-dehydrogenase (Ndh) complex, homologous to the bacterial complex I, and of a nuclear-encoded plastid terminal oxidase (PTOX), homologous to the plant mitochondrial alternative oxidase, brought molecular support to the concept of chlororespiration. The functionality of these proteins in nonphotochemical reduction and oxidation of plastoquinones (PQs), respectively, has recently been demonstrated. In thylakoids of mature chloroplasts, chlororespiration appears to be a relatively minor pathway compared to linear photosynthetic electron flow from H_2O to $NADP^+$. However, chlororespiration might play a role in the regulation of photosynthesis by modulating the activity of cyclic electron flow around photosystem I (PS I). In nonphotosynthetic plastids, chlororespiratory electron carriers are more abundant and may play a significant bioenergetic role.

CONTENTS

INTRODUCTION

In photosynthetic cells, two main bioenergetic processes, photosynthesis and respiration, involve electron transport reactions coupled to ATP synthesis. Whereas
photosynthesis converts light energy into chemical energy (NADPH and ATP) and
allows the net fixation of carbon, respiration converts reducing power contained in
carbohydrates into phosphorylating power. In eukaryotic cells, these processes take
place in two separate organelles: Photosynthesis occurs in chloroplasts and respiration in mitochondria. In photosynthetic bacteria and cyanobacteria, photosynthetic
and respiratory electron transport chains (ETCs) operate in close interaction, sharing identical electron carriers such as the plastoquinone (PQ) pool (127). In the
early 1960s, Goedheer (53), by studying photosynthetic luminescence transients in
unicellular green algae, postulated that a dark oxidation of intersystem electron carriers could occur through "some kind of chloroplast respiration." Bennoun (8, 9),
based on the effects of respiratory inhibitors on chlorophyll fluorescence induction
curves in unicellular green algae, proposed the existence in thylakoid membranes
of a respiratory chain connected to the photosynthetic ETC. This respiratory activity, called chlororespiration, was suggested to originate from the bacterial ancestor
of chloroplasts (127). According to the initial model of chlororespiration, reducing
equivalents would be supplied in the dark from a stromal pool of NAD(P)H to PQs
and further to a putative chloroplastic oxidase (Figure 1). As in the case of bacterial
and mitochondrial respiration, chlororespiratory electron transfer was thought to
be accompanied by a vectorial transport of protons generating a transthylakoidal
electrochemical gradient. The existence of chlororespiration has been seriously
questioned during the past decade, with chlorophyll fluorescence measurements
performed on intact cells being alternately explained in terms of redox interactions
between chloroplasts and mitochondria (10, 11, 65, 95). The discovery in higher
plants of electron transport carriers possibly involved in chlororespiration, first
a plastid-encoded NAD(P)H-dehydrogenase (Ndh) complex (17, 22, 56, 66, 76,

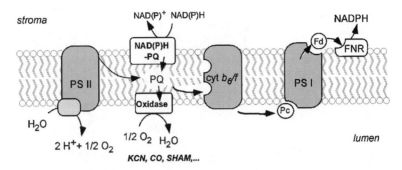

Figure 1   Early model of chlororespiration as proposed by Bennoun (8). Chlororespiration would involve a NAD(P)H-plastoquinone oxidoreductase activity, the thykaloid plastoquinone (PQ) pool, and a putative terminal oxidase and would interact with the photosynthetic electron transport chain at the PQ pool. According to this model, inhibitors of mitochondrial oxidases [e.g., KCN, CO, SHAM (salicylhydroxamic acid)] would affect the chlororespiratory oxidase, thus leading to net reduction of the PQ pool. Cyt, cytochrome; PS, photosystem; Pc, plastocyanin; Fd, ferredoxin; FNR, ferredoxin-NADP$^+$ reductase.

126, 131) and more recently a plastid terminal oxidase (19, 32, 150), have supplied molecular support to the concept of chlororespiration. The purpose of this review is to revisit the initial model of chlororespiration in light of these recent findings.

INTERACTIONS BETWEEN PHOTOSYNTHESIS AND RESPIRATION WITHIN PLANT CELLS

Initial experiments leading to the concept of chlororespiration relied on the effect of respiratory oxidase inhibitors, like cyanide, CO, or salicylhydroxamic acid (SHAM), on the redox state of PQs in intact unicellular algae (8). These experiments were based on the assumption that chloroplasts and mitochondria are not in redox communication. However, such effects could also be explained by an inhibition of mitochondrial respiration and the existence of redox interactions between mitochondria and chloroplasts (Figure 2). Indeed, metabolic interactions occur between chloroplasts and mitochondria [for reviews, see (49, 65, 114)]. They rely on the exchange of reducing equivalents [NAD(P)H] or phosphorylating power (ATP) between chloroplasts and cytosol, owing to the participation of metabolic shuttles, such as the phosphate or dicarboxylate translocators located in the inner chloroplast envelope and the oxaloacetate translocator located in the inner mitochondrial membrane (60, 63, 65, 114). Stromal reducing equivalents can reduce the PQ pool through a nonphotochemical electron pathway (51, 52). Addition of inhibitors of mitochondrial respiration to *Chlamydomonas reinhardtii* cells was reported to

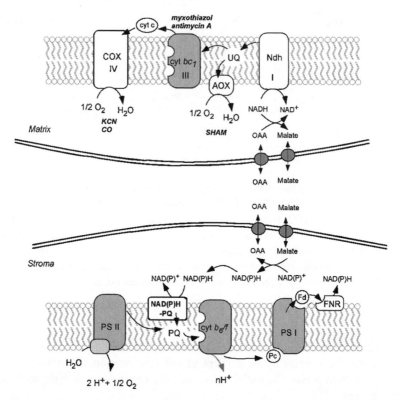

Figure 2 Interactions between the photosynthetic and mitochondrial electron transport chains (ETCs). Reducing equivalents are exchanged between chloroplasts and mitochondria by metabolic shuttles, such as the dicarboxylate and the oxaloacetate (OAA) translocators. Inhibition of the mitochondrial ETC leads to an increase in the internal pools of matrix and stromal NAD(P)H and to a subsequent reduction of the plastoquinone (PQ) pool, owing to a thykaloid NAD(P)H-PQ oxidoreductase activity. AOX, alternative oxidase; COX, cytochrome c oxidase; PC, plastocyanin; PS, photosystem; SHAM, salicylhydroxamic acid; UQ, ubiquinone.

result in a decrease in ATP and a rise in the reduction levels of both pyridine nucleotide and PQ pools (47, 118). The use of a *Chlamydomonas* mitochondrial cyt *b* mutant resistant to myxothiazol clearly demonstrated that PQ reduction observed in *Chlamydomonas* in response to myxothiazol treatment results from the inhibition of mitochondrial activity (10).

Another approach to study interactions between photosynthesis and chlororespiration in algae was based on polarographic or mass spectrometric measurements of O_2 exchange (35, 107). The nonlinearity, induced by anaerobiosis, in the light saturation curve of O_2 evolution was explained by an interaction between PQs, a nonphotochemical reductant (likely NADPH), and O_2 (35). In other experiments using either polarography or mass spectrometry and $^{18}O_2$, perturbations of O_2 exchange induced by a single turnover flash showed that photosynthesis is closely linked to a respiratory process (107). The study of photosynthetic mutants of *Chlamydomonas* deficient in photosystem (PS) II or PS I further established that flash-induced PS I activity generates a transitory inhibition of a respiratory process (107), whereas flash-induced PS II activity generates a stimulation of this process (116). The flash-induced transient observed in intact *Chlamydomonas* cells was insensitive to the inhibition of mitochondrial respiration (106, 107) and was concluded to result from an interaction between photosynthesis and chlororespiration.

Both models of interaction (model 1: chlororespiration in interaction with photosynthesis vs. model 2: mitochondrial respiration in interaction with photosynthesis) may explain changes in the redox status of intersystem electron carriers in response to various respiratory inhibitors. Notably, both models involve an activity catalyzing nonphotochemical reduction of PQs (Figures 1 and 2) but differ with regard to the involvement of a plastid terminal oxidase. In model 1, a plastid terminal oxidase would allow nonphotochemical oxidation of PQs (Figure 1), whereas in model 2, a chloroplast oxidase would not be required (Figure 2). In the following sections, we review experimental evidence showing the existence of nonphotochemical reduction and oxidation pathways of PQs and discuss the molecular nature of the electron carriers involved in these processes.

NONPHOTOCHEMICAL REDUCTION OF PQS

Experimental Evidence for a Nonphotochemical Reduction of PQs

The primary evidence for nonphotochemical reduction of PQs came from a study of H_2 metabolism in microalgae. Under anaerobic conditions, unicellular green algae such as *Scenedesmus obliquus* or *Chlamydomonas* are able to catalyze either the reduction of CO_2, using H_2 as an electron donor, or the production of H_2 (46). These reactions require the activity of PS I and a period of adaptation under anaerobic conditions in order to initiate the biosynthesis and/or activation of a reversible hydrogenase. Both H_2 production and H_2-dependent CO_2 reduction are insensitive to 3-(3,4-dichlorophenyl)-1,1-dimethylurea (DCMU), and therefore electrons can originate from a source other than PS II. These processes are, however, sensitive to 2,5-dibromo-3-methyl-6-isopropyl-*p*-benzoquinone (DBMIB), a quinone analog inhibitor of the cyt b_6/f complex, thus showing involvement of the PQ pool (50, 52, 84).

Chlorophyll fluorescence techniques have been widely used to study nonphotochemical reduction of PQs in different systems, including algae (8, 36), higherplant chloroplasts (28, 37, 90), and leaves (42, 55). The basal fluorescence level F_0, determined under weak (nonactinic) light, is related to the redox state of Q_A, the first quinone acceptor of PS II, which is in redox equilibrium with the PQ pool. Addition of exogenous NADPH or NADH to osmotically lysed spinach or potato chloroplasts induces an increase in the F_0 chlorophyll fluorescence level, which was interpreted as a donation of electrons to PQs (28, 37, 90). In leaves, the PQ pool remains reduced in the dark for several hours following a period of illumination (42, 55). This effect, which was particularly marked in plants showing a NAD-malic enzyme-type of C_4 metabolism, was attributed to the reduction of the PQ pool by stromal donors (42, 55). $P700^+$ re-reduction kinetics following flash illumination (3, 36) or changes in the redox state of intersystem cytochromes (48), measured by absorption changes at specific wavelengths, were also interpreted as a nonphotochemical reduction of PQs.

Thus, nonphotochemical reduction of PQs is a well-established phenomenon that occurs in most photosynthetic organisms. The critical question to be answered is which enzymatic activity or activities account for this phenomenon?

A Plastid-Encoded Ndh Complex Homologous to Complex I

Sequencing of different higher-plant plastid genomes revealed the presence of open reading frames homologous to genes encoding subunits of the mitochondrial NADH dehydrogenase complex (Complex I) (100, 101, 135). These genes are transcribed (88), and some of the encoded polypeptides have been detected in thylakoid membranes (12, 23, 56, 78, 86, 96, 145). The isolation and partial characterization of a plastidial Ndh complex (also called Ndh1) has also been reported (22, 33, 56, 124). Based on its subunit composition, the plastidial Ndh complex resembles the proton-pumping complex of bacteria and cyanobacteria (45). In *Escherichia coli*, the NADH dehydrogenase complex is composed of 14 different subunits that bind several redox factors, including six to eight iron-sulfur clusters and one flavin mononucleotide; three of these subunits are involved in NADH binding (80). It is surprising to note that plastid and cyanobacterial genomes do not contain sequences homologous to genes encoding NADH-binding subunits, which raises the question whether NADH or another compound acts as a stromal electron donor to the plastidial complex (45). It was suggested that ferredoxin-$NADP^+$ reductase (FNR) may be associated with the Ndh complex and may allow the use of NADPH as an electron donor (45, 56, 113). This appears unlikely, however, because the purified Ndh1 complex prefers NADH over NADPH (22, 33, 124), whereas FNR displays strong preference for NADPH (136). Another possibility is that genes encoding NADH-binding subunits are located in the nucleus, as is the case for a NADH-binding subunit of the plant mitochondrial Ndh complex (54). However, no chloroplast-targeted subunit corresponding to the NADH-oxidizing domain of mitochondrial complex I has been identified in the *Arabidopsis thaliana* genome (121), thus indicating that electrons may come from a source other than NADH. A genetic approach, based on mutant screening by chlorophyll fluorescence imaging, may help in the future to identify nuclear-encoded subunits of the Ndh complex involved in substrate binding (130).

Complete chloroplast DNA sequences have been determined in many plant species, including land plants and green algae. *Ndh* genes are present in plastid genomes of *Marchantia polymorpha* (100), *Nicotiana tabacum* (132), *Oryza sativa* (64), *Zea mays* (82), *Arabidopsis* (123), *Lotus japonicus* (73), and *Spinacia oleracea* (128) but are absent or present as pseudogenes in *Pinus thunbergii* (144). The absence of *ndh* genes does not appear to be a general feature of gymnosperms because the *ndh* F gene was detected in all vascular plant divisions, including gymnosperm species (92). In contrast, *ndh* genes are absent from the chloroplast genomes of most algal species, including the green algae *Euglena gracilis* (57), *Chlorella vulgaris* (143), *Chlamydomonas reinhardtii* (D. Stern, personal communication), and the red alga *Porphyra purpurea* (120). However, plastidial *ndh* genes are found in most ancestral algae, e.g., *Nephroselmis olivacea* (140) or *Mesostigma viride* (81). It is notable that *ndh* genes are absent in *Epifagus virginia*, a nonphotosynthetic parasitic plant that has lost most of the photosynthetic genes (34, 148).

In species lacking chloroplast *ndh* genes, like *Chlamydomonas*, a relocation to the nucleus would have probably not occurred during the evolution of species. Indeed, it is generally considered that key integral membrane proteins of chloroplasts and mitochondria cannot be imported across the organellar outer membranes because of their high hydrophobicity (112), which would explain the retention of the corresponding genes in the organelle genome (104). Paradoxically, the existence of a plastid Ndh complex sensitive to rotenone has been reported in *Chlamydomonas* (51, 52, 83). This was taken as evidence supporting the existence of chlororespiration (8, 107). In addition, an 18-kDa polypeptide related to *frx*B (or *ndh*I) (see 87) was assumed to be a subunit of an Ndh complex (93, 151). It should be noted that thylakoid membrane fractions prepared from *Chlamydomonas* cells contain some contaminating mitochondrial material (4), which may explain the presence in thylakoid preparations of mitochondrial proteins like Ndh1 (51) or various cytochromes (4, 108). The absence of a functional Ndh1 complex in *Chlamydomonas* chloroplasts is supported by the insensitivity of H_2 photoproduction to rotenone and piericidin, both potent inhibitors of complex I, in intact *Chlamydomonas* cells treated with DCMU (29). Thus, nonphotochemical reduction of PQs, which sustains H_2 photoproduction in these conditions, is unlikely to rely on the presence of an Ndh1 complex.

The absence of *ndh* genes in some species raises the question of the functional significance of these genes. Are *ndh* gene products and the Ndh complex essential to the functioning of the ETC or to other plastid functions? New insights have been provided by experiments of inactivation of chloroplast *ndh* genes in tobacco, which were performed independently by four laboratories (17, 66, 76, 131) using plastid transformation (102, 137). Although one of these studies proposed that the Ndh complex is essential (76), it was finally concluded that the Ndh complex is dispensable for plant growth under normal conditions (17, 66, 85, 95, 131). However, some effects on electron transport properties were observed, including the disappearance in *ndh* mutants of the transient increase in F_0 chlorophyll fluorescence after a period of actinic illumination. Although nonessential for growth, the Ndh complex is functional and is involved in nonphotochemical reduction of PQs (17, 66, 76, 131).

Involvement of Other PQ-Reductase Activities

Several lines of evidence suggested the existence of different pathways of non-photochemical reduction of PQs, the most conclusive being based on the analysis of tobacco chloroplast *ndh* mutants. In osmotically lysed chloroplasts, NAD(P)H donates electrons to the PQ pool with the same efficiency in both WT and *ndh* mutants, thus revealing a pathway for nonphotochemical reduction of PQs independent of the Ndh1 complex (29, 39). A study of the effect of heat stress on tobacco *ndh* mutants also provided support for the involvement of a PQ reduction pathway different from that of the Ndh complex (125, 152).

The most obvious candidate for an alternative pathway of nonphotochemical reduction of PQs is a putative ferredoxin-PQ reductase (FQR). FQR has been described as a component of cyclic electron transfer reactions around PS I, which would explain the sensitivity of this process to antimycin A (6, 25, 91). The increase in chlorophyll fluorescence observed following addition of NADPH to osmotically lysed chloroplasts supplemented with ferredoxin is inhibited by antimycin A (37, 90) and is therefore attributed to the activity of FQR (37). However, in marked contrast to the Ndh complex, FQR is still uncharacterized at a biochemical or molecular level.

FNR might also be involved in the donation of electrons to the PQ pool (90) and/or in cyclic electron transfer reactions around PS I (67, 129), but the possible role of this flavoprotein in these reactions has been controversial (28, 37). Because FNR cannot reduce PQs directly, participation of this enzyme in nonphotochemical PQ reduction is probably via association with either FQR (6) or the Ndh complex (45, 56, 113).

Other enzymes, such as alternative NADH dehydrogenases, might also be involved in the nonphotochemical reduction of PQs. In potato chloroplasts, the increase in chlorophyll fluorescence induced by NAD(P)H involves neither the Ndh1 complex (homologous to mitochondrial complex I) nor FNR. Based on the effect of various inhibitors and on its substrate specificity, this phenomenon was concluded to result from a flavoenzyme-containing NAD(P)-PQ oxidoreductase activity such as a DT-diaphorase, also called Ndh2 (28). Ndh2 enzymes are monosubunit flavoproteins, which can transfer electrons from NAD(P)H to the ubiquinone pool in a nonelectrogenic manner. It is interesting to note that three open reading frames in *Synechocystis* PCC 6803 code for type 2 dehydrogenases. However, because deletion studies could not provide evidence that Ndh2 activity contributes to respiratory electron flow, these enzymes were proposed instead to act as redox sensors (68). Eight alternative Ndh2 homologues are also found in the *Arabidopsis* genome, three of them being possibly targeted to chloroplasts (Table 1). Whether some of these gene sequences actually encode chloroplast-targeted Ndh2 involved in chlororespiration will require further investigation.

Another possible candidate for the nonphotochemical reduction of PQ is succinate dehydrogenase, which would represent a chloroplast homologue of the mitochondrial complex II. The presence of succinate dehydrogenase activity has been documented in *Chlamydomonas* chloroplasts where succinate supplies electrons

TABLE 1   *Arabidopsis thaliana* and *Chlamydomonas reinhardtii* gene sequences encoding possible components of the chlororespiratory electron transport chain (ETC) other than the plastid Ndh1 complex

Arabidopsis thaliana

Proteins	Genes	Predicted targeting		Similarities
		(*a*)	(*b*)	
Ndh2s	AT3g44190	—	—	Unknown proteins; low similarity to Ndh2s
	AT4g28220	M	M	External rotenone-insensitive Ndh2 (*Solanum tuberosum*)
	AT2g20800	M	M	External rotenone-insensitive Ndh2 (*Solanum tuberosum*)
	AT4g21490	SP	SP	External rotenone-insensitive Ndh2 (*Solanum tuberosum*)
	AT2g29990	C	M	Internal rotenone-insensitive Ndh2 (*Solanum tubersum*)
	AT1g07180	C	M	Internal rotenone-insensitive Ndh2 (*Solanum tubersum*)
	AT4g05020	M	M	External rotenone-insensitive Ndh2 (*Solanum tuberosum*)
	AT5g08740	C	C	Ndh2 (*Synechocystis* sp. PCC6803)
Alternative oxidases	AT5g64210	M	M	Higher plant mitochondrial AOXs
	AT3g22370	M	M	Higher plant mitochondrial AOXs
	AT3g22360	M	M	Higher plant mitochondrial AOXs
	AT3g27620	—	—	Higher plant mitochondrial AOXs
	AT1g32350	M	M	Higher plant mitochondrial AOXs
	AT4g22260	C	C	Low similarity to higher plant AOXs, identified as *IMMUTANS*

Chlamydomonas reinhardtii[*]

Protein	Contig	ESTs	Predicted targeting		Similarity:
			(*a*)	(*b*)	
Alternative oxidase	200111023.6219.2	15	M	C	*IMMUTANS*

Gene sequences were retrieved from the TIGR *A. thaliana* database (www.tigr.org) and from the ChlamyEST database (www.biology.duke.edu/chlamy_genome/crc.html). Blasts were performed at NCBI. Targeting predictions were made using (*a*) iPSORT (www.hypothesiscreator.net/iPSORT) or (*b*) TargetP (www.cbs.dtu.dk/services/TargetP). M, mitochondrial targeting; C, chloroplastic targeting; —, no predicted targeting; SP, signal peptide.

[*]Note that several ESTs showing homologies to Ndh2s are found in the ChlamyEST database.

to PS I (147). The genome of the cyanobacterium *Synechocystis* PCC 6803 contains two open reading frames that show significant similarities to genes coding for succinate:quinol oxidoreductases. These genes encode subunits of a succinate dehydrogenase complex capable of reducing PQs in cyanobacterial thylakoid membranes (26). Succinate and succinate dehydrogenase were recently proposed to be the major pathway of nonphotochemical PQ reduction in cyanobacteria (27). Note, however, that genes encoding a chloroplast-targeted succinate dehydrogenase have not been found in the *Arabidopsis* genome (121).

Nonphotochemical PQ reduction is a widely documented and well-established phenomenon. Inactivation studies concluded that the plastid Ndh1 complex is involved in this process in higher-plant chloroplasts. Alternative activities (Ndh2, FQR, etc.), which have been proposed to account for nonphotochemical PQ reduction in algal chloroplasts and may also be present in higher-plant chloroplasts, remain to be characterized at a molecular level.

NONPHOTOCHEMICAL OXIDATION OF PQS

Experimental Evidence for a Nonphotochemical Oxidation of PQs

During linear electron flow in photosynthesis, PQs are reduced by PS II and reoxidized by the cytochrome b_6/f complex in cooperation with PS I. From chlorophyll fluorescence measurements, it was concluded that, following its reduction during strong illumination, the PQ pool is reoxidized in the dark by a pathway involving O_2 (58, 142). PS I–deficient mutants of *Chlamydomonas* provide an independent way to study alternative PQ oxidation pathways (31, 109). In these mutants, a limited but significant electron flow from PS II to molecular O_2 can be measured using $^{18}O_2$ and mass spectrometry. This pathway involves the PQ pool but not the cyt b_6/f complex (31, 109). We discuss below the various mechanisms that have been proposed to account for nonphotochemical oxidation of PQs.

Reverse Electron Flow Through the Ndh Complex?

In intact PS I–deficient cells, electron flow from PS II to O_2 is inhibited by mitochondrial respiration inhibitors (32, 109). It was initially proposed that NADH could be formed in the absence of PS I by reverse functioning of an Ndh complex (109, 119), similar to the process operating in photosynthetic bacteria (75). In such a scheme, the Ndh complex would catalyze "uphill" electron transfer from plastoquinol to NAD^+ using the electrochemical gradient generated by PS II. The reducing equivalents would subsequently be transferred to mitochondria via metabolic shuttles, such as the dicarboxylate translocator (Figure 2). This mechanism, which was proposed to explain the dark reoxidation of PQs (10), would require an electrogenic Ndh complex, which is likely absent from *Chlamydomonas* chloroplasts (see above). Also, the PS II–dependent electron flow to O_2 was reported to occur in isolated chloroplasts from PS I–deficient mutants, thus localizing the PQ-oxidation activity to chloroplasts (32). Therefore, the observed effect of mitochondrial inhibitors on photosynthetic electron transport in these mutants likely results from chloroplast-mitochondrial redox interactions, the inhibition of mitochondria leading to an overreduction of NAD(P) and, thus, PQs (47) and to the subsequent inhibition of PS II. However, in higher-plant chloroplasts, we cannot exclude a reverse electron flow through the Ndh1 complex, which is likely to be electrogenic, based on its homology with bacterial complex I (5).

The Plastid Terminal Oxidase (PTOX)

The study of an *Arabidopsis* mutant (*immutans*) showing a variegated pheno-type and considerable growth retardation led to the long-awaited identification of a plastid oxidase (19, 150). In this mutant, there is an impairment in phytoene desaturation, an essential step in carotenoid biosynthesis that requires PQ as an electron acceptor. The inactivated gene possessed sequence homologies with mito-chondrial alternative oxidases (AOX; see Table 1) and was thus named *PTOX* for plastid terminal oxidase (30, 72). Expression of PTOX in *E. coli* conferred a cyanide-resistant quinol oxidase activity to this bacterium, which was sensitive to propyl gallate (32, 72) and significantly less sensitive to SHAM (30). Both com-pounds are well-known inhibitors of plant mitochondrial AOXs (13). Similarly, the PS II–dependent electron pathway occurring in *Chlamydomonas* PS I–less mutants was strongly inhibited by propyl gallate and, to a lesser extent, by SHAM (32). Based on these features and the detection of a thylakoid protein that crossreacted with an antibody raised against PTOX, it was proposed that chlororespiratory O_2 uptake is due to a *Chlamydomonas* homologue of PTOX (32). It is notable that a gene sequence homologous to PTOX and possibly targeted to chloroplasts is found in a *Chlamydomonas* contig (Table 1). The overexpression of *Arabidopsis* PTOX in tobacco and the effect of propyl gallate were recently used to show that PTOX facilitates the reoxidation of PQ either during the induction of photosynthesis or after a period of strong illumination. It was concluded that PTOX is connected to both chlororespiratory and photosynthetic ETCs (T. Joët, B. Genty, E-M. Josse, M. Kuntz, L. Cournac & G. Peltier, submitted).

Involvement of Other PQ-Oxidase Activities

The *Arabidopsis* mutant *immutans* that lacks PTOX accumulates carotenoids in the dark but at slower rates than wild type, suggesting the existence of other path-way(s) for PQ reoxidation (18). Apart from PTOX, no other gene sequence encod-ing homologues to AOX (or PTOX) and predicted to be targeted to the chloroplast is present in the *Arabidopsis* genome (Table 1). Several authors concluded that a cyanide-sensitive chloroplast oxidase is involved in chlororespiration (14, 79, 107). In *Pleurochloris meiringensis*, a xanthophytic alga, it was reported that the cyanide concentrations needed to affect the re-reduction of the cytochrome f^+/c_{553}^+ electron pool are much higher than those necessary to completely block mitochon-drial activity (14). Similarly, based on the effect of cyanide on the electrochromic bandshift in tobacco protoplasts, the chlororespiratory oxidase was concluded to be much less sensitive to cyanide than the mitochondrial cytochrome oxidase (79). An effect of high cyanide concentrations (5 mM) on the PS II–dependent elec-tron pathway to O_2 occurring in thylakoids of PS I–deficient *Chlamydomonas* mutants was also reported (32). Note, however, that analysis of the *Arabidopsis* genome does not reveal obvious gene sequences encoding chloroplast-targeted quinol or cytochrome oxidases that could be involved in such alternative pathways (121).

From studies performed in a reconstituted system containing the thylakoid Ndh complex and a thylakoid hydroquinone peroxidase, researchers proposed that H_2O_2 produced by the action of superoxide dismutase on $O_2 \cdot^-$ generated by Mehler reactions may be used as a substrate for chlororespiration (22). The possible involvement of a superoxide dismutase in chlororespiration may explain cyanide effects on the redox state of intersystem electron carriers, whereas the involvement of a peroxidase could account for the effects of CO (22). Other candidates for the oxidation of PQs have been proposed. For instance, based on kinetic comparisons between the reoxidation of PQs measured in spinach chloroplasts and cytochrome b-559 autooxidation in isolated PS II particles, it was suggested that dark reoxidation of the PQ pool could be mediated by the low-potential form of cytochrome b-559 (77). It is clear that more work needs to be done to characterize nonphotochemical PQ oxidative pathways other than PTOX and to determine whether they rely on enzymatic or nonenzymatic processes.

CURRENT MODEL OF CHLORORESPIRATION

In higher-plant species containing plastid *ndh* genes, chlororespiration would involve the activities of an Ndh complex and a terminal oxidase (PTOX) (Figure 3*A*). In species lacking plastid *ndh* genes (i.e., most green algae and some *Pinus* species), nonphotochemical reduction of PQs would be achieved by another activity (for instance, an Ndh2-like enzyme) (Figure 3*B*). Note that, as inferred from the study of tobacco *ndh* mutants, alternate activities are also likely to be present in species containing the plastid-encoded Ndh complex, and these activities could complement that of the Ndh complex (29, 69, 125).

The Ndh complex, like the PS I and ATP synthase complexes, is absent from appressed thylakoids and is mainly located in nonappressed membranes (12, 66, 124). In C_4 plants of the NADP-malic enzyme type, the thylakoid membrane system of bundle-sheath plastids is composed almost entirely of nonappressed lamellae and lacks PS II activity. In such C_4 species, different subunits of the Ndh complex are more abundant in bundle-sheath than in mesophyll plastids (78). Recently, PTOX was also shown to be located in nonappressed lamellae of tobacco chloroplasts (T. Joët, B. Genty, E-M. Josse, M. Kuntz, L. Cournac & G. Peltier, submitted). Based on these data, we propose that chlororespiration would be restricted to stroma lamellae (Figure 3).

Activity of Chlororespiration

Estimates of the electron flow rate of chlororespiration have been presented by a number of laboratories (8, 14, 42, 107). These estimates should be considered with

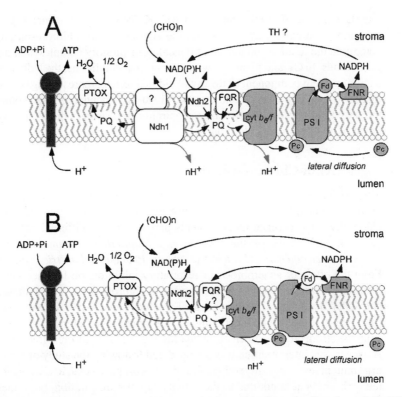

Figure 3 Current models of chlororespiration. Based on the location of the Ndh complex and PTOX, chlororespiration would be restricted to nonappressed lamellae and absent from appressed thylakoids. (*A*) In chloroplasts of higher plants, chlororespiration involves the Ndh1 complex and PTOX. Alternative pathways of nonphotochemical reduction of PQs, including an Ndh2 activity or a putative FQR, are also likely present. TH, putative transhydrogenase. (*B*) In species lacking plastidic *ndh* genes (e.g., most green algae), chlororespiration would involve PTOX and an Ndh2 activity or a putative FQR. In both cases, alternative pathways of PQ oxidation (not represented in the figure) are also likely present.

extreme caution. Indeed, some of these estimates have been based on measurements related to the interaction between PQ, stromal electron donors, and O_2 (8, 42), and this phenomenon is known to be affected by variations in mitochondrial respiration. Other estimates rely on the sensitivity of the chlororespiratory oxidase to cyanide (14, 107), which has not been observed in the PTOX terminal oxidase recently proposed to be involved in chlororespiration (32). Based on the effect of propyl gallate, an inhibitor of the chlororespiratory quinol oxidase (32, 72) on O_2 uptake rates in *Chlamydomonas* cells, the chlororespiratory electron flow rate was estimated at approximately 50 nmol O_2 consumed $\times$ min^{-1} $\times$ mg^{-1} chlorophyll, which represents approximately 2% of the maximal photosynthetic electron flow rate (L. Cournac, G. Latouche, Z. Cerovic, K. Redding, J. Ravenel & G. Peltier, submitted). Another way to evaluate the significance of chlororespiration is to determine the relative abundance of the chlororespiratory complexes in thylakoid membranes. In C_3 land plants grown under normal conditions, the Ndh complex is a minor protein component of the thylakoid membrane, representing only approximately 0.2% of total thylakoid proteins (124). In conclusion, when compared to the activity of photosynthesis, chlororespiration appears as a quantitatively minor process in higher-plant chloroplasts.

Electrogenicity of Chlororespiration

The existence in the dark of a pH gradient across thylakoid membranes has been postulated, based on luminescence measurements in green algae (70) and chlorophyll fluorescence measurements performed with different unicellular organisms, such as the prasinophycean alga *Mantoniella squamata* (146), *Pleurochloris meiringensis* (15), and the diatom *Phaeodactylum tricornutum* (139). From the effect of anaerobiosis or respiratory inhibitors on luminescence kinetics in a *Chlamydomonas* mutant deficient in chloroplastic ATP synthase, it was concluded that the dark electrochemical gradient is due to two different processes: (*a*) the reverse hydrolytic functioning of ATP synthases resulting in proton transfer from the stroma to the lumen of thylakoids and (*b*) the chlororespiratory ETC (8). However, this interpretation was reconsidered following the study of a mitochondrial mutant resistant to myxothiazol, and it was proposed that, in *Chlamydomonas*, the dark gradient is not due to the activity of chlororespiration, but rather to the existence of a new ATP-driven gradient generator (115). The effects of anaerobiosis and respiratory inhibitors were explained by interactions with mitochondrial respiration (10). In the diatom *P. tricornutum*, both the effects of anaerobiosis and antimycin A on the energization of the thylakoid membrane were interpreted to involve chlororespiration, but possible interactions with mitochondrial respiration were not considered (139).

In *Chlamydomonas*, and perhaps in other plant species lacking the plastidial Ndh complex, the chlororespiratory pathway probably involves a nonelectrogenic PQ-reductase activity (which might be an Ndh2-like, FQR, or succinate dehydrogenase) together with a quinol oxidase (see Figure 3*B*). Based on its homology with mitochondrial AOX, the plastid quinol oxidase is also likely to be nonelectrogenic (32). In these species, the chlororespiratory electron pathway may be either nonelectrogenic or poorly electrogenic (32). Under these conditions, the dark electrochemical gradient would be attributed to other mechanisms (10, 115). In higher-plant chloroplasts containing a functional Ndh1 complex showing homologies with the proton-pumping bacterial NADH dehydrogenase (45), chlororespiratory electron flow could be electrogenic (see Figure 3*A*). However, the existence of a dark gradient in higher-plant chloroplasts is not as well documented as in green algae, and experimental data supporting the electrogenicity of either the plastidial Ndh complex or the chlororespiratory chain per se are lacking.

Interactions Between Chlororespiration, Mitochondrial Respiration, and Photosynthesis

The existence within plant cells of different ETCs, including a photosynthetic chain located in chloroplasts and two respiratory chains, one mitochondrial and one plastidic, raises the question of the bioenergetic interactions between these systems. Although direct interactions can occur between ETCs located in thylakoid membranes (photosynthesis and chlororespiration), more complex interactions based on metabolic exchange between organelles can also occur between chloroplastic and mitochondrial ETCs. This results in a complex network of interactions, some of which are favored depending on experimental conditions and/or the physiological state of the cells. This was recently illustrated by a kinetic analysis of flash-induced O_2 exchange transients in intact *Chlamydomonas* cells. Flash illumination triggered a rapid perturbation of chlororespiration (observed in a timescale of 300 ms) when both pyridine nucleotide and PQ pools were highly reduced and a slower perturbation of mitochondrial respiration (timescale of 3 s) when they were more oxidized (L. Cournac, G. Latouche, Z. Cerovic, K. Redding, J. Ravenel & G. Peltier, submitted). In plant mitochondria, the activity of alternative oxidase is modulated by some respiratory substrates (NADPH, pyruvate, etc.) and is stimulated when the ubiquinone pool is highly reduced (141). Based on the homology between PTOX and the mitochondrial alternative oxidase, we propose that PTOX might contribute to regulate the PQ redox poise and might be notably stimulated under conditions where the pyridine nucleotide and PQ pools are highly reduced.

ROLE OF CHLORORESPIRATION AND CHLORORESPIRATORY ELECTRON CARRIERS

When considering a possible role for chlororespiration, one should distinguish between the potential role of the entire chlororespiratory ETC, from NAD(P)H to PQ and O_2, and that of individual electron carriers within the chlororespiratory chain such as the Ndh complex or the terminal oxidase. In the dark, functioning of chlororespiration may allow the oxidation of reduced stromal compounds for various metabolic pathways and/or the establishment of a trans-thylakoidal pH gradient. During illumination, a possible contribution of chlororespiration appears more difficult to evaluate owing to the interconnection with photosynthesis and to the apparently low electron flow occurring in chlororespiration relative to that in photosynthesis.

Plastid Carbon Metabolism in the Dark

A possible function of a chlororespiratory ETC is the reoxidation of reducing equivalents produced by plastid carbon metabolism. Within chloroplasts, starch degradation leads to the formation of hexose and hexose-P (105), which are in turn converted into CO_2, dihydroxyacetone-P, and 3-P-glycerate by the reactions of either glycolysis and/or the pentose-P pathway (24, 134). It was proposed that reduced pyridine nucleotides produced during this process are reoxidized by a process involving PQs (50). From the effects of rotenone, antimycin A, and sodium azide on $^{14}CO_2$ release measured in intact spinach and green-algal chloroplasts supplied with [^{14}C]glucose or [^{14}C]fructose, it was concluded that a chlororespiratory ETC is involved in this process (133). This chloroplastic CO_2 release was estimated to be between 10% and 20% of total algal respiration. Reducing equivalents formed during the oxidation of glyceraldehyde-3-P to 3-P-glycerate could be diverted to the thylakoid ETC at the level of PQs through an NADH dehydrogenase. More recently, it was reported that metabolism of exogenous acetate supplied to heterotrophically grown *Chlamydomonas* cells provides reducing equivalents to chloroplasts and stimulates electron flow toward the intersystem electron carriers (36).

Nonphotosynthetic Plastids and Chloroplast Biogenesis

In the absence of a photosynthetic ETC in proplastids during chloroplast biogenesis, or in differentiated nongreen plastids (such as etioplasts or chromoplasts), chlororespiration has the potential to reoxidize reduced compounds and to generate a transmembrane pH gradient. Based on the observation that large amounts of different subunits of the Ndh complex (Ndh-H, -K, and -F) are found in etioplasts (23, 44), it was proposed that this complex may serve to generate a proton gradient across the prothylakoid membrane (44). This gradient would be necessary for the integration of certain proteins into these membranes when a functional photosynthetic ETC is not yet operative (149). However, the validity of this hypothesis is questioned when one considers that no alteration in the development of chloroplasts has been reported in the absence of the Ndh complex, which is the only potentially electrogenic electron carrier of chlororespiration (17, 66, 85, 131).

The role of the chlororespiratory oxidase PTOX in carotenoid biosynthesis has been clearly demonstrated. In chromoplasts of *Narcissus pseudonarcissus*, phytoene desaturation, a key metabolic step in carotenoid biosynthesis, has been reported to be linked to a plastid respiratory pathway (94). Study of the *immutans* mutant established the involvement of PTOX in this process during chloroplast biogenesis (19, 150). The PTOX polypeptide was also found in an achlorophyllous membrane fraction of ripening pepper chromoplasts (72). Based on the abundance of chlororespiratory components in nongreen plastids, a role for chlororespiration in the bioenergetic metabolism [either in ATP supply or in reoxidation of the NAD(P)H pools] of these nonphotosynthetic organelles can be proposed but needs further work to be established.

Transcriptional and Posttranscriptional Regulatory Mechanisms Controlled by the Redox State of PQs

The redox state of the PQ pool is a key element in the control of different regulatory processes occurring within plant cells. For example, the redox state of this pool controls the transcription of genes encoding both PS I and PS II reaction-center apoproteins (111) as well as *cab* genes (41), which code for light-harvesting protein complex apoproteins. Also, the association of the mobile light-harvesting complex II with PS II or PS I depends on the former's phosphorylation state, which is controlled by a thylakoid protein kinase. This phenomenon, called state transition, is controlled by the redox state of PQs and the cyt b_6/f complex (1, 7). Electron carriers involved in chlororespiration may participate in such regulation by modulating the redox state of the PQ pool. In *Chlamydomonas*, transition from state 1 (mobile light-harvesting complex associated with PS II) to state 2 (mobile light-harvesting complex associated with PS I) was observed in the dark in response to mitochondrial inhibitors (16). This effect was attributed to a higher NAD(P)H/NAD(P) ratio, resulting in nonphotochemical reduction of PQ, kinase activation, and thereby transition to state 2. Aside from their effects on the redox state of the PQ pool, chlororespiratory electron carriers may also directly act as sensors in redox signaling. Such a possible role, already documented for the cbb_3 oxidase of *Rhodobacter sphaeroides* (99) and proposed for the Ndh2 dehydrogenases of *Synechocystis* (68), remains to be established for the chlororespiratory complexes.

Cyclic Electron Transfer Around PS I

In photosynthetic bacteria, photosynthesis is achieved by cyclic electron transport reactions around a single photosystem, thereby generating a proton gradient that promotes ATP synthesis. The existence of cyclic photophosphorylation around PS I during oxygenic photosynthesis was postulated nearly 40 years ago (138). In higher-plant thylakoids, this phenomenon is sensitive to antimycin A and requires the cyt b_6/f complex and the addition of ferredoxin as a soluble cofactor (6, 138). Cyclic photophosphorylation might provide additional ATP for optimal functioning of photosynthetic CO_2 assimilation (61, 62). By performing photoacoustic measurements with *Chlamydomonas* cells, Ravenel et al. (117) observed two cyclic pathways around PS I. One of these pathways was sensitive to antimycin A and was assumed to be ferredoxin dependent. The other pathway may involve synthesis of NADPH and re-entry of electrons into the PQ pool through a NAD(P)H dehydrogenase activity. In C_4 plants, it was proposed that the chloroplastic Ndh complex, which is preferentially expressed in the PS II–deficient bundle-sheath cells, may be involved in cyclic electron flow around PS I (78). In C_3 plants, experimental evidence showing the involvement of the Ndh complex in cyclic electron transfer reactions around PS I was obtained from the study of tobacco *ndh* mutants. The transient postillumination increase in chlorophyll fluorescence, which is likely an aftereffect of cyclic electron flow, was absent in all *ndh* mutants (17, 29, 76, 131). More recently, based on the more-pronounced effect of antimycin A on photosynthesis in a *ndhB*-inactivated mutant compared to the control, it was concluded that two pathways of cyclic electron flow operate around PS I in higher-plant chloroplasts (69). One of these pathways would be sensitive to antimycin A and would involve the putative FQR; the other would involve the Ndh1 complex

(69). Tobacco mutants lacking the Ndh1 complex show growth retardation in response to moderate water deficit, this effect being related to a reduced ability of the mutants to cope with low intracellular CO_2 concentrations (66). It was proposed that the increased ATP demand due to the stimulation of photorespiration (103) could not be fulfilled in such *ndh* mutants (66). From the effect of antimycin A on photosynthetic activity of tobacco *ndh* mutants, investigators recently concluded that both pathways of cyclic electron transport around PS I are required for optimal photosynthesis under conditions of high ATP demand (e.g., high photorespiration) but can independently fulfil this metabolic demand when it is lower (69).

Of interest is the finding that a cyanobacterial mutant inactivated in the *ndh*B gene (98) requires high CO_2 levels for growth. This was explained by a requirement of the inorganic carbon (Ci)-concentrating mechanism for ATP supplied by cyclic photophosphorylation (98). Whether such a Ci-concentrating mechanism, requiring an extra-ATP supply from cyclic electron flow, occurs in algal and higher-plant chloroplasts remains to be established.

In addition to a direct role of the Ndh complex in cyclic electron flow, chlororespiration might be involved in the initiation of cyclic electron transfer reactions around PS I. Indeed, the establishment of cyclic electron-transport pathways requires a fine poising of the redox state of electron carriers (2, 110, 153). Particularly, O_2 and NADPH concentrations appear to play an important role during this process. Initially, O_2 was suggested to poise electron carriers by acting as an electron acceptor on the reducing side of PS I through the Mehler reactions (153), but chlororespiration, by regulating the redox state of the intersystem electron carriers, may fulfil such a role (58). Plastoquinol diffusion between the appressed and nonappressed thylakoids is recognized as a slow process, and the PQ pool in the latter membranes might be functionally isolated from PS II (71). As a consequence, this pool could be mainly involved in cyclic electron flow. In this context, adequate poising of electron carriers, which is required for optimal functioning of cyclic electron flow, can be achieved by chlororespiratory electron carriers, using NAD(P)H as electron donor and O_2 as the terminal acceptor. In such a scheme, the regulation of the redox state of PQ by chlororespiration could also affect the efficiency of NAD(P)H-independent cyclic pathways, such as the antimycin A–sensitive pathway.

Chlororespiration and Abiotic Stress Responses

In response to environmental stress, both the activity of chlororespiration and the expression of the Ndh complex increase. In green algae, nitrogen starvation enhances the donation of electrons from a stromal pool to PQs, thereby triggering transition to state 2 and stimulating the chlororespiratory pathway (108). Similar phenomena likely occur in response to sulfur starvation (89) or hyperosmotic stress (38). In higher plants, heat stress increases the nonphotochemical reduction of intersystem electron carriers (59, 79, 152). Also, photooxidative stress increases the expression of the chloroplast Ndh complex and plastid peroxidase activity (20–22, 86).

Chlororespiration has been suggested to exert a protective role in photosynthesis (97). Several mechanisms can be envisioned. For example, PTOX (or the electrogenic Ndh complex via reverse electron flow) may act as a safety valve to avoid overreduction of PS II acceptors in intense light. Cyclic electron flow around PS I may assist in energy dissipation or supply ATP for protection and repair mechanisms (43). The trans-thylakoidal pH gradient generated by chlororespiration may facilitate the integration of membrane proteins during the recovery from photoinhibition (44). Also, a putative pathway involving a PQH_2-dependent peroxidase activity may help to scavenge active O_2 species generated by Mehler reactions (20, 22). The possible involvement of chlororespiratory electron carriers in photoprotection is supported by the finding that tobacco mutants lacking the Ndh1 complex are more sensitive to excessive light (40). However, considerably more work remains to be done to clearly establish whether chlororespiratory components really have a protective function and to determine the mechanisms involved in this protection.

CONCLUSIONS AND PERSPECTIVES

Although highly controversial in the past, the existence of chlororespiration now appears to be well established both in algal and higher-plant chloroplasts. In higher plants, chlororespiration is likely electrogenic and involves a plastid-encoded Ndh1 complex and a PTOX, both enzymes being preferentially located in nonappressed thylakoid membranes. In unicellular green algae such as *Chlamydomonas*, chlororespiration likely involves a PTOX-like activity but does not involve an Ndh1 complex. In chloroplasts of mature leaves, both the Ndh and PTOX appear as relatively minor protein components of thylakoids membranes, and the rate of electron flow in chlororespiration is very low compared to photosynthesis. However, during chloroplast biogenesis or in nongreen plastids, the amounts of chlororespiratory enzymes are higher than in mature green plastids. The chlororespiratory oxidase PTOX is involved in carotenoid biosynthesis. Chlororespiration could also contribute to the bioenergetic metabolism of nongreen plastids by supplying ATP or reoxidizing metabolic compounds, but such a role remains to be clearly demonstrated.

Inactivation of *ndh* genes by plastid transformation in tobacco has shown that the Ndh complex is functional and participates in cyclic electron flow around PS I, supplying extra ATP for optimal photosynthesis, particularly under conditions where CO_2 is limiting (e.g., water stress). Whether this participation reflects a direct involvement of the Ndh complex in cyclic electron flow or an indirect regulation of the cyclic reactions by a fine tuning of the redox state of intersystem electron carriers needs further investigation.

To date, the only chlororespiratory electron carriers clearly identified are the plastid-encoded Ndh complex and the nuclear-encoded PTOX. Different lines of evidence suggest the existence of alternative pathways of nonphotochemical oxidation and reduction of the PQ pool such as Ndh2 or FQR. Future studies must be tailored to elucidate the molecular components involved in these putative pathways.

ACKNOWLEDGMENTS

We thank T. Joët, P. Rich, D. Rumeau, and D. Stern for providing preprints or unpublished results and B. Genty, M. Havaux, B. Prickril, and K. Redding for helpful discussions and critical reading of the manuscript.

Visit the Annual Reviews home page at www.annualreviews.org

LITERATURE CITED

1. Allen JF, Bennett J, Steinback KE, Arntzen CJ. 1981. Chloroplast protein phosphorylation couples plastoquinone redox state to distribution of excitation energy between photosystems. *Nature* 291:25–29

2. Arnon DI, Chain RK. 1975. Regulation of ferredoxin-catalyzed photosynthetic phosphorylations. *Proc. Natl. Acad. Sci. USA* 72:4961–65

3. Asada K, Heber U, Schreiber U. 1992. Pool size of electrons that can be donated to P700$^+$, as determined in intact leaves: donation to P700$^+$ from stromal components via the intersystem chain. *Plant Cell Physiol.* 33:927–32

4. Atteia A, de Vitry C, Pierre Y, Popot J-L. 1992. Identification of mitochondrial proteins in membrane preparations from *Chlamydomonas reinhardtii. J. Biol. Chem.* 267:226–34

5. Barber J, Peltier G, Niyogi K, Foyer CH, Laisk A, Matthijs HCP. 2000. Flexibility in photosynthetic electron transport: a newly identified chloroplast oxidase involved in chlororespiration. Discussion. *Philos. Trans. R. Soc. London Ser. B* 355:1453–54

6. Bendall DS, Manasse RS. 1995. Cyclic photophosphorylation and electron transport. *Biochim. Biophys. Acta* 1229:23–38

7. Bennett J. 1991. Protein phosphorylation in green plant chloroplasts. *Annu. Rev. Plant Physiol. Plant Mol. Biol.* 42:281–311

8. Bennoun P. 1982. Evidence for a respiratory chain in the chloroplast. *Proc. Natl. Acad. Sci. USA* 79:4352–56

9. Bennoun P. 1983. Effects of mutations and of ionophore on chlororespiration in *Chlamydomonas reinhardtii. FEBS Lett.* 156:363–65

10. Bennoun P. 1994. Chlororespiration revisited: mitochondrial-plastid interactions in *Chlamydomonas. Biochim. Biophys. Acta* 1186:59–66

11. Bennoun P. 1998. Chlororespiration, sixteen years later. In *The Molecular Biology of Chloroplasts and Mitochondria in Chlamydomonas*, ed. J-D Rochaix, M Goldschmidt-Clermont, S Merchant, pp. 675–83. Dordrecht, The Neth.: Kluwer

12. Berger S, Ellersiek U, Westhoff P, Steinmuller K. 1993. Studies on the expression of NDH-H, a subunit of the NAD(P)H-plastoquinone oxidoreductase of higher plant chloroplasts. *Planta* 190:25–31

13. Berthold DA. 1998. Isolation of mutants of the *Arabidopsis thaliana* alternative oxidase (ubiquinol:oxygen oxidoreductase) resistant to salicylhydroxamic acid. *Biochim. Biophys. Acta* 1364:73–83

14. Büchel C, Garab G. 1995. Evidence for the operation of a cyanide-sensitive oxidase in chlororespiration in the thylakoids of the chlorophyll c-containing alga *Pleurochloris meiringensis* (*Xanthophyceae*). *Planta* 197:69–75

15. Büchel C, Wilhelm C. 1990. Wavelength-independent state transitions and light-regulated chlororespiration as mechanisms to control the energy status in the chloroplast of *Pleurochloris meiringensis*. *Plant Physiol. Biochem.* 28:307–14

16. Bulté L, Gans P, Rebeille F, Wollman F-A. 1990. ATP control on state transitions *in vivo* in *Chlamydomonas reinhardtii*. *Biochim. Biophys. Acta* 1020:72–80

17. Burrows PA, Sazanov LA, Svab Z, Maliga P, Nixon PJ. 1998. Identification of a functional respiratory complex in chloroplasts through analysis of tobacco mutants containing disrupted plastid *ndh* genes. *EMBO J.* 17:868–76

18. Carol P, Kuntz M. 2001. A plastid terminal oxidase comes to light: implications for carotenoid biosynthesis and chlororespiration. *Trends Plant Sci.* 6:31–36

19. Carol P, Stevenson D, Bisanz C, Breitenbach J, Sandmann G, et al. 1999. Mutations in the *Arabidopsis* gene *immutans* cause a variegated phenotype by inactivating a chloroplast terminal oxidase associated with phytoene desaturation. *Plant Cell* 11:57–68

20. Casano LM, Martin M, Sabater B. 2001. Hydrogen peroxide mediates the induction of chloroplastic Ndh complex under photooxidative stress in barley. *Plant Physiol.* 125:1450–58

21. Casano LM, Martin M, Zapata JM, Sabater B. 1999. Leaf age– and paraquat concentration–dependent effects on the levels of enzymes protecting against photooxidative stress. *Plant Sci.* 149:13–22

22. Casano LM, Zapata JM, Martin M, Sabater B. 2000. Chlororespiration and poising of cyclic electron transport—plastoquinone as electron transporter between thylakoid NADH dehydrogenase and peroxidase. *J. Biol. Chem.* 275:942–48

23. Catala R, Sabater B, Guera A. 1997. Expression of the plastid *ndhF* gene product in photosynthetic and non-photosynthetic tissues of developing barley seedlings. *Plant Cell Physiol.* 38:1382–88

24. Chen CG, Gibbs M. 1991. Glucose respiration in the intact chloroplast of *Chlamydomonas reinhardtii*. *Plant Physiol.* 95:82–87

25. Cleland RE, Bendall DS. 1992. Photosystem I cyclic electron transport—measurement of ferredoxin-plastoquinone reductase activity. *Photosynth. Res.* 34:409–18

26. Cooley JW, Howitt CA, Vermaas WF. 2000. Succinate: quinol oxidoreductases in the cyanobacterium *Synechocystis* sp. strain PCC 6803: presence and function in metabolism and electron transport. *J. Bacteriol.* 182:714–22

27. Cooley JW, Vermaas WFJ. 2001. Succinate dehydrogenase and other respiratory pathways in thylakoid membranes of *Synechocystis* sp. strain PCC 6803: capacity comparisons and physiological function. *J. Bacteriol.* 183:4251–58

28. Corneille S, Cournac L, Guedeney G, Havaux M, Peltier G. 1998. Reduction of the plastoquinone pool by exogenous NADH and NADPH in higher plant chloroplasts—characterization of a NAD(P)H-plastoquinone oxidoreductase activity. *Biochim. Biophys. Acta* 1363:59–69

29. Cournac L, Guedeney G, Joët T, Rumeau D, Latouche G, et al. 1998. Non-photochemical reduction of intersystem electron carriers in chloroplasts of higher plants and algae. In *Photosynthesis: Mechanism and Effects*, ed. G Garab, pp. 1877–82. Dordrecht, The Neth.: Kluwer

30. Cournac L, Josse EM, Joët T, Rumeau D, Redding K, et al. 2000. Flexibility in photosynthetic electron transport: a newly identified chloroplast oxidase involved in chlororespiration. *Philos. Trans. R. Soc. London Ser. B* 355:1447–53

31. Cournac L, Redding K, Bennoun P, Peltier G. 1997. Limited photosynthetic electron flow but no CO_2 fixation in *Chlamydomonas* mutants lacking photosystem I. *FEBS Lett.* 416:65–68

32. Cournac L, Redding K, Ravenel J, Rumeau D, Josse EM, et al. 2000. Electron

flow between photosystem II and oxygen in chloroplasts of photosystem I–deficient algae is mediated by a quinol oxidase involved in chlororespiration. *J. Biol. Chem.* 275:17256–62

33. Cuello J, Quiles MJ, Albacete ME, Sabater B. 1995. Properties of a large complex with NADH dehydrogenase activity from barley thylakoids. *Plant Cell Physiol.* 36:265–71

34. dePamphilis CW, Palmer JD. 1990. Loss of photosynthetic and chlororespiratory genes from the plastid genome of a parasitic flowering plant. *Nature* 348:337–39

35. Diner B, Mauzerall D. 1973. Feedback controlling oxygen production in a cross-reaction between two photosystems in photosynthesis. *Biochim. Biophys. Acta* 305:329–52

36. Endo T, Asada K. 1996. Dark induction of the non-photochemical quenching of chlorophyll fluorescence by acetate in *Chlamydomonas reinhardtii. Plant Cell Physiol.* 37:551–55

37. Endo T, Mi HL, Shikanai T, Asada K. 1997. Donation of electrons to plastoquinone by NAD(P)H dehydrogenase and by ferredoxin-quinone reductase in spinach chloroplasts. *Plant Cell Physiol.* 38:1272–77

38. Endo T, Schreiber U, Asada K. 1995. Suppression of quantum yield of photosystem II by hyperosmotic stress in *Chlamydomonas reinhardtii. Plant Cell Physiol.* 36:1253–58

39. Endo T, Shikanai T, Sato F, Asada K. 1998. NAD(P)H dehydrogenase–dependent, antimycin A–sensitive electron donation to plastoquinone in tobacco chloroplasts. *Plant Cell Physiol.* 39:1226–31

40. Endo T, Shikanai T, Takabayashi A, Asada K, Sato F. 1999. The role of chloroplastic NAD(P)H dehydrogenase in photoprotection. *FEBS Lett.* 457:5–8

41. Escoubas JM, Lomas M, LaRoche J, Falkowski PG. 1995. Light intensity regulation of *cab* gene transcription is signaled by the redox state of the plasto-

quinone pool. *Proc. Natl. Acad. Sci. USA* 92:10237–41

42. Feild TS, Nedbal L, Ort DR. 1998. Non-photochemical reduction of the plastoquinone pool in sunflower leaves originates from chlororespiration. *Plant Physiol.* 116:1209–18

43. Finazzi G, Barbagallo RP, Bergo E, Barbato R, Forti G. 2001. Photoinhibition of *Chlamydomonas reinhardtii* in state 1 and state 2—damages to the photosynthetic apparatus under linear and cyclic electron flow. *J. Biol. Chem.* 276:22251–57

44. Fischer M, Funk E, Steinmüller K. 1997. The expression of subunits of the mitochondrial complex I–homologous NAD(P)H-plastoquinone-oxidoreductase during plastid development. *Z. Naturforsch. Teil C* 52:481–86

45. Friedrich T, Steinmuller K, Weiss H. 1995. The proton-pumping respiratory complex I of bacteria and mitochondria and its homologue in chloroplasts. *FEBS Lett.* 367:107–11

46. Gaffron H, Rubin J. 1942. Fermentative and photochemical production of hydrogen in algae. *J. Gen. Physiol.* 26:219–40

47. Gans P, Rebeillé F. 1990. Control in the dark of the plastoquinone redox state by mitochondrial activity in *Chlamydomonas reinhardtii. Biochim. Biophys. Acta* 1015:150–55

48. Garab G, Lajko F, Mustardy L, Marton L. 1989. Respiratory control over photosynthetic electron transport in chloroplasts of higher plant cells—evidence for chlororespiration. *Planta* 179:349–58

49. Gardestrom P, Lernmark U. 1995. The contribution of mitochondria to energetic metabolism in photosynthetic cells. *J. Bioenerg. Biomembr.* 27:415–21

50. Gfeller RP, Gibbs M. 1985. Fermentative metabolism of *Chlamydomonas reinhardtii.* 2. Role of plastoquinone. *Plant Physiol.* 77:509–11

51. Godde D. 1982. Evidence for a membrane-bound NADH-plastoquinone oxidoreductase in *Chlamydomonas reinhardtii* CW-15. *Arch. Microbiol.* 131:197–202

52. Godde D, Trebst A. 1980. NADH as electron donor for the photosynthetic membrane of *Chlamydomonas reinhardtii*. *Arch. Microbiol.* 127:245–52

53. Goedheer JC. 1963. A cooperation of two-pigment systems and respiration in photosynthetic luminescence. *Biochim. Biophys. Acta* 66:61–71

54. Grohmann L, Rasmusson AG, Heiser V, Thieck O, Brennicke A. 1996. The NADH-binding subunit of respiratory chain complex I is nuclear-encoded in plants and identified only in mitochondria. *Plant J.* 10:793–803

55. Groom QJ, Kramer DM, Crofts AR, Ort DR. 1993. The non-photochemical reduction of plastoquinone in leaves. *Photosynth. Res.* 36:205–15

56. Guedeney G, Corneille S, Cuine S, Peltier G. 1996. Evidence for an association of *ndhB*, *ndhJ* gene products and ferredoxin-NADP-reductase as components of a chloroplastic NAD(P)H dehydrogenase complex. *FEBS Lett.* 378:277–80

57. Hallick RB, Hong L, Drager RG, Favreau MR, Monfort A, et al. 1993. Complete sequence of *Euglena gracilis* chloroplast DNA. *Nucleic Acids Res.* 21:3537–44

58. Harris GC, Heber U. 1993. Effects of anaerobiosis on chlorophyll fluorescence yield in spinach (*Spinacia oleracea*) leaf discs. *Plant Physiol.* 101:1169–73

59. Havaux M. 1996. Short-term responses of photosystem I to heat stress. Induction of a PSII-independent electron transport through PSI fed by stromal components. *Photosynth. Res.* 47:85–97

60. Heber U. 1974. Metabolite exchange between chloroplasts and cytoplasm. *Annu. Rev. Plant Physiol. Plant Mol. Biol.* 25:393–421

61. Heber U, Egneus H, Hanck U, Jensen M, Köster S. 1978. Regulation of photosynthetic electron transport and phosphorylation in intact chloroplasts and leaves of *Spinacia oleracea*. *Planta* 143:41–49

62. Heber U, Walker D. 1992. Concerning a dual function of coupled cyclic electron transport in leaves. *Plant Physiol.* 100:1621–26

63. Heldt HW, Heineke D, Heupel R, Krömer S, Riens B. 1990. Transfer of redox equivalents between subcellular compartments of a leaf cell. In *Current Research in Photosynthesis*, ed. M Batscheffsky, pp. 15.1–5.7. Dordrecht, The Neth.: Kluwer

64. Hiratsuka J, Shimada H, Whittier R, Ishibashi T, Sakamoto M, et al. 1989. The complete sequence of the rice (*Oryza sativa*) chloroplast genome—intermolecular recombination between distinct transfer-RNA genes accounts for a major plastid DNA inversion during the evolution of the cereals. *Mol. Gen. Genet.* 217:185–94

65. Hoefnagel MHN, Atkin OK, Wiskich JT. 1998. Interdependence between chloroplasts and mitochondria in the light and the dark. *Biochim. Biophys. Acta* 1366:235–55

66. Horvath EM, Peter SO, Joët T, Rumeau D, Cournac L, et al. 2000. Targeted inactivation of the plastid *ndhB* gene in tobacco results in an enhanced sensitivity of photosynthesis to moderate stomatal closure. *Plant Physiol.* 123:1337–49

67. Hosler JP, Yocum CF. 1985. Evidence for two cyclic photophosphorylation reactions concurrent with ferredoxin-catalyzed non-cyclic electron transport. *Biochim. Biophys. Acta* 808:21–31

68. Howitt CA, Udall PK, Vermaas WFJ. 1999. Type 2 NADH dehydrogenases in the cyanobacterium *Synechocystis* sp. strain PCC 6803 are involved in regulation rather than respiration. *J. Bacteriol.* 181:3994–4003

69. Joët T, Cournac L, Horvath EM, Medgyesy P, Peltier G. 2001. Increased sensitivity of photosynthesis to antimycin A

induced by inactivation of the chloroplast *ndhB* gene. Evidence for a participation of the NADH-dehydrogenase complex to cyclic electron flow around photosystem I. *Plant Physiol.* 125:1919–29

70. Joliot P, Joliot A. 1980. Dependence of delayed luminescence upon adenosine-triphosphatase activity in *Chlorella*. *Plant Physiol.* 65:691–96

71. Joliot P, Lavergne J, Beal D. 1992. Plastoquinone compartmentation in chloroplasts. 1. Evidence for domains with different rates of photoreduction. *Biochim. Biophys. Acta* 1101:1–12

72. Josse EM, Simkin AJ, Gaffe J, Laboure AM, Kuntz M, Carol P. 2000. A plastid terminal oxidase associated with carotenoid desaturation during chromoplast differentiation. *Plant Physiol.* 123:1427–36

73. Kato T, Kaneko T, Sato S, Nakamura Y, Tabata S. 2000. Complete structure of the chloroplast genome of a legume, *Lotus japonicus*. *DNA Res.* 7:323–30

74. Deleted in proof

75. Knaff DB. 1978. Reducing potentials and the pathway of NAD$^+$ reduction. In *The Photosynthetic Bacteria*, ed. RK Clayton, WR Sistrom, pp. 629–40. New York: Plenum

76. Kofer W, Koop HU, Wanner G, Steinmüller K. 1998. Mutagenesis of the genes encoding subunits A, C, H, I, J and K of the plastid NAD(P)H-plastoquinone-oxidoreductase in tobacco by polyethylene glycol-mediated plastome transformation. *Mol. Gen. Genet.* 258:166–73

77. Kruk J, Strzalka K. 1999. Dark reoxidation of the plastoquinone pool is mediated by the low-potential form of cytochrome b-559 in spinach thylakoids. *Photosynth. Res.* 62:273–79

78. Kubicki A, Funk E, Westhoff P, Steinmüller K. 1996. Differential expression of plastome-encoded *ndh* genes in mesophyll and bundle-sheath chloroplasts of the C$_4$ plant *Sorghum bicolor* indicates that the complex I–homologous

NAD(P)H-plastoquinone oxidoreductase is involved in cyclic electron transport. *Planta* 199:276–81

79. Lajko F, Kadioglu A, Borbely G, Garab G. 1997. Competition between the photosynthetic and the (chloro)respiratory electron transport chains in cyanobacteria, green algae and higher plants. Effect of heat stress. *Photosynthetica* 33:217–26

80. Leif H, Sled VD, Ohnishi T, Weiss H, Friedrich T. 1995. Isolation and characterization of the proton-translocating NADH: ubiquinone oxidoreductase from *Escherichia coli*. *Eur. J. Biochem.* 230: 538–48

81. Lemieux C, Otis C, Turmel M. 2000. Ancestral chloroplast genome in *Mesostigma viride* reveals an early branch of green plant evolution. *Nature* 403:649–52

82. Maier RM, Neckermann K, Igloi GL, Kossel H. 1995. Complete sequence of the maize chloroplast genome—gene content, hotspots of divergence and fine tuning of genetic information by transcript editing. *J. Mol. Biol.* 251:614–28

83. Maione TE, Gibbs M. 1986. Association of the chloroplastic respiratory and photosynthetic electron transport chains of *Chlamydomonas reinhardtii* with photoreduction and the oxyhydrogen reaction. *Plant Physiol.* 80:364–68

84. Maione TE, Gibbs M. 1986. Hydrogenase-mediated activities in isolated chloroplasts of *Chlamydomonas reinhardtii*. *Plant Physiol.* 80:360–63

85. Maliga P, Nixon PJ. 1998. Judging the homoplastomic state of plastid transformants. *Trends Plant Sci.* 3:376–77

86. Martin M, Casano LM, Sabater B. 1996. Identification of the product of *ndhA* gene as a thylakoid protein synthesized in response to photooxidative treatment. *Plant Cell Physiol.* 37:293–98

87. Matsubara H, Oh OH, Takahashi Y, Fujita Y. 1995. Three iron-sulfur proteins encoded by three ORFs in chloroplasts and cyanobacteria. *Photosynth. Res.* 46:107–15

88. Matsubayashi T, Wakasugi T, Shinozaki K, Yamaguchi-Shinozaki K, Zaita N, et al. 1987. Six chloroplast genes (*ndhA-F*) homologous to human mitochondrial genes encoding components of the respiratory chain NADH dehydrogenase are actively expressed: determination of the splice sites in *ndhA* and *ndhB* premRNAs. *Mol. Gen. Genet.* 210:385–93

89. Melis A, Zhang LP, Forestier M, Ghirardi ML, Seibert M. 2000. Sustained photobiological hydrogen gas production upon reversible inactivation of oxygen evolution in the green alga *Chlamydomonas reinhardtii*. *Plant Physiol.* 122:127–35

90. Mills JD, Crowther D, Slovacek RE, Hind G, McCarty RE. 1979. Electron transport pathways in spinach chloroplasts. Reduction of the primary acceptor of photosystem II by reduced nicotinamide adenine dinucleotide phosphate in the dark. *Biochim. Biophys. Acta* 547:127–37

91. Moss DA, Bendall DS. 1984. Cyclic electron transport in chloroplasts. The Q-cycle and the site of action of antimycin. *Biochim. Biophys. Acta* 767:389–95

92. Neyland R, Urbatsch LE. 1996. The *ndhF* chloroplast gene detected in all vascular plant divisions. *Planta* 200:273–77

93. Nie ZQ, Chang DY, Wu M. 1987. Protein-DNA interaction within one cloned chloroplast DNA replication origin of *Chlamydomonas*. *Mol. Gen. Genet.* 209:265–69

94. Nievelstein V, Vandekerchove J, Tadros MH, Lintig JV, Nitschke W, Beyer P. 1995. Carotene desaturation is linked to a respiratory redox pathway in *Narcissus pseudonarcissus* chromoplast membranes. Involvement of a 23-kDa oxygen-evolving-complex-like protein. *Eur. J. Biochem.* 233:864–72

95. Nixon PJ. 2000. Chlororespiration. *Philos. Trans. R. Soc. London Ser. B* 355:1541–47

96. Nixon PJ, Gounaris K, Coomber SA, Hunter CN, Dyer TA, Barber J. 1989. *psbG* is not a photosystem II gene but may be an *ndh* gene. *J. Biol. Chem.* 264:14129–35

97. Niyogi KK. 2000. Safety valves for photosynthesis. *Curr. Opin. Plant Biol.* 3:455–60

98. Ogawa T. 1991. A gene homologous to the subunit-2 gene of NADH dehydrogenase is essential to inorganic carbon transport of *Synechocystis* PCC6803. *Proc. Natl. Acad. Sci. USA* 88:4275–79

99. Oh JI, Kaplan S. 2000. Redox signaling: globalization of gene expression. *EMBO J.* 19:4237–47

100. Ohyama K, Fukuzawa H, Kohchi T, Shirai H, Sano T, et al. 1986. Chloroplast gene organization deduced from complete sequence of liverwort *Marchantia polymorpha* chloroplast DNA. *Nature* 322:572–74

101. Ohyama K, Kohchi T, Sano T, Yamada Y. 1988. Newly identified groups of genes in chloroplasts. *Trends Biochem. Sci.* 13:19–22

102. O'Neill C, Horvath GV, Horvath E, Dix PJ, Medgyesy P. 1993. Chloroplast transformation in plants: polyethylene glycol (PEG) treatment of protoplasts is an alternative to biolistic delivery systems. *Plant J.* 3:729–38

103. Osmond CB. 1981. Photorespiration and photoinhibition. Some implications for the energetics of photosynthesis. *Biochim. Biophys. Acta* 639:77–98

104. Palmer JD. 1997. Organelle genomes: going, going, gone. *Science* 275:790–91

105. Peavey DG, Steup M, Gibbs M. 1977. Characterization of starch breakdown in intact spinach chloroplast. *Plant Physiol.* 60:305–8

106. Peltier G, Havaux M, Ravenel J. 1995. Chlororespiration in unicellular green algae. Studies with mitochondrial respiration deficient mutants. In *Photosynthesis: From Light to Biosphere*, ed. P Mathis, pp. 887–90. Dordrecht, The Neth.: Kluwer

107. Peltier G, Ravenel J, Verméglio A. 1987. Inhibition of a respiratory activity by short saturating flashes in *Chlamydomonas*:

evidence for a chlororespiration. *Biochim. Biophys. Acta* 893:83–90

108. Peltier G, Schmidt GW. 1991. Chlororespiration: an adaptation to nitrogen deficiency in *Chlamydomonas reinhardtii. Proc. Natl. Acad. Sci. USA* 88:4791–95

109. Peltier G, Thibault P. 1988. Oxygen exchange studies in *Chlamydomonas* mutants deficient in photosynthetic electron transport: evidence for a photosystem II–dependent oxygen uptake *in vivo. Biochim. Biophys. Acta* 936:319–24

110. Peters F, Vanspanning R, Kraayenhof R. 1983. Studies on well-coupled photosystem I–enriched subchloroplast vesicles— optimization of ferredoxin-mediated cyclic photophosphorylation and electric potential generation. *Biochim. Biophys. Acta* 724:159–65

111. Pfannschmidt T, Nilsson A, Allen JF. 1999. Photosynthetic control of chloroplast gene expression. *Nature* 397:625–28

112. Popot J-L, de Vitry C. 1990. On the microassembly of integral membrane proteins. *Annu. Rev. Biophys. Biophys. Chem.* 19:369–403

113. Quiles MJ, Cuello J. 1998. Association of ferredoxin-NADP oxidoreductase with the chloroplastic pyridine nucleotide dehydrogenase complex in barley leaves. *Plant Physiol.* 117:235–44

114. Raghavendra AS, Padmasree K, Saradadevi K. 1994. Interdependence of photosynthesis and respiration in plant cells— interactions between chloroplasts and mitochondria. *Plant Sci.* 97:1–14

115. Rappaport F, Finazzi G, Pierre Y, Bennoun P. 1999. A new electrochemical gradient generator in thylakoid membranes of green algae. *Biochemistry* 38:2040–47

116. Ravenel J, Peltier G. 1992. Stimulation of the chlororespiratory electron flow by photosystem II activity in *Chlamydomonas reinhardtii. Biochim. Biophys. Acta* 1101:57–63

117. Ravenel J, Peltier G, Havaux M. 1994. The cyclic electron pathways around photosystem I in *Chlamydomonas reinhardtii* as determined *in vivo* by photoacoustic measurements of energy storage. *Planta* 193:251–59

118. Rebeillé F, Gans P. 1988. Interaction between chloroplasts and mitochondria in microalgae. *Plant Physiol.* 88:973–75

119. Redding K, Cournac L, Vassiliev IR, Golbeck JH, Peltier G, Rochaix JD. 1999. Photosystem I is indispensable for photoautotrophic growth, CO_2 fixation, and H_2 photoproduction in *Chlamydomonas reinhardtii. J. Biol. Chem.* 274:10466–73

120. Reith M, Munholland J. 1995. Complete nucleotide sequence of the *Porphyra purpurea* chloroplast genome. *Plant Mol. Biol.* 13:327–32

121. Rich PR, Fisher N, Lennon A, Prommeenate P, Purton S, et al. 2001. An assessment of the pathways of dark reduction and oxidation of the plastoquinone pool in thylakoid membranes of higher plants and green algae. *Proc. Int. Congr. Photosynth., 12th, Brisbane, Aust.* Victoria, Aust.: CSIRO. http://www.publish.csiro.au/ps2001/cf/home/index.cfm

122. Deleted in proof

123. Sato S, Nakamura Y, Kaneko T, Asamizu E, Tabata S. 1999. Complete structure of the chloroplast genome of *Arabidopsis thaliana. DNA Res.* 6:283–90

124. Sazanov LA, Burrows P, Nixon PJ. 1996. Detection and characterization of a complex I–like NADH-specific dehydrogenase from pea thylakoids. *Biochem. Soc. Trans.* 24:739–43

125. Sazanov LA, Burrows PA, Nixon PJ. 1998. The chloroplast Ndh complex mediates the dark reduction of the plastoquinone pool in response to heat stress in tobacco leaves. *FEBS Lett.* 429:115–18

126. Sazanov LA, Burrows PA, Nixon PJ. 1998. The plastid *ndh* genes code for an NADH-specific dehydrogenase: isolation of a complex I analogue from pea thylakoid membranes. *Proc. Natl. Acad. Sci. USA* 95:1319–24

127. Scherer S. 1990. Do photosynthetic and respiratory electron transport chains share

redox proteins? *Trends Biochem. Sci.* 15: 458–62

128. Schmitz-Linneweber C, Maier RM, Alcaraz JP, Cottet A, Herrmann RG, Mache R. 2001. The plastid chromosome of spinach (*Spinacia oleracea*): complete nucleotide sequence and gene organization. *Plant Mol. Biol.* 45:307–15

129. Shahak Y, Crowther D, Hind G. 1981. The involvement of ferredoxin-NADP⁺ reductase in cyclic electron transport in chloroplasts. *Biochim. Biophys. Acta* 636:234–43

130. Shikanai T, Endo T. 2000. Physiological function of a respiratory complex, NAD(P)H dehydrogenase in chloroplasts: dissection by chloroplast reverse genetics. *Plant Biotechnol.* 17:79–86

131. Shikanai T, Endo T, Hashimoto T, Yamada Y, Asada K, Yokota A. 1998. Directed disruption of the tobacco *ndhB* gene impairs cyclic electron flow around photosystem I. *Proc. Natl. Acad. Sci. USA* 95:9705–9

132. Shinozaki K, Ohme M, Tanaka M, Wakasugi T, Hayashida N, et al. 1986. The complete nucleotide sequence of the tobacco chloroplast genome: its gene organization and expression. *EMBO J.* 5:2043–49

133. Singh KK, Chen C, Gibbs M. 1992. Characterization of an electron transport pathway associated with glucose and fructose respiration in the intact chloroplasts of *Chlamydomonas reinhardtii* and spinach. *Plant Physiol.* 100:327–33

134. Stitt M, Rees TA. 1980. Carbohydrate breakdown by chloroplasts of *Pisum sativum*. *Biochim. Biophys. Acta* 627: 131–43

135. Sugiura M. 1992. The chloroplast genome. *Plant Mol. Biol.* 19:149–68

136. Süss KH. 1982. Purification of stromal and membrane-bound ferredoxin-NADP⁺ reductase. In *Methods in Chloroplast Molecular Biology*, ed. M Edelman, RB Hallick, N-H Chua, pp. 957–71. Amsterdam, The Neth.: Elsevier Biomed. Press

137. Svab Z, Maliga P. 1993. High-frequency plastid transformation in tobacco by selection for a chimeric *aadA* gene. *Proc. Natl. Acad. Sci. USA* 90:913–17

138. Tagawa K, Arnon DI, Tsujimoto HY. 1963. Role of chloroplast ferredoxin in energy conversion process of photosynthesis. *Proc. Natl. Acad. Sci. USA* 49:567–72

139. Ting CS, Owens TG. 1993. Photochemical and nonphotochemical fluorescence quenching processes in the diatom *Phaeodactylum tricornutum*. *Plant Physiol.* 101: 1323–30

140. Turmel M, Otis C, Lemieux C. 1999. The complete chloroplast DNA sequence of the green alga *Nephroselmis olivacea*: insights into the architecture of ancestral chloroplast genomes. *Proc. Natl. Acad. Sci. USA* 96:10248–53

141. Vanlerberghe GC, McIntosh L. 1997. Alternative oxidase: from gene to function. *Annu. Rev. Plant Physiol. Plant Mol. Biol.* 48:703–34

142. Vernotte C, Etienne A-L, Briantais J-M. 1979. Quenching of the system II chlorophyll fluorescence by the plastoquinone pool. *Biochim. Biophys. Acta* 545:519–27

143. Wakasugi T, Nagai T, Kapoor M, Sugita M, Ito M, et al. 1997. Complete nucleotide sequence of the chloroplast genome from the green alga *Chlorella vulgaris*: the existence of genes possibly involved in chloroplast division. *Proc. Natl. Acad. Sci. USA* 94:5967–72

144. Wakasugi T, Tsudzuki J, Ito S, Nakashima K, Tsudzuki T, Sugiura M. 1994. Loss of all *ndh* genes as determined by sequencing the entire chloroplast genome of the black pine *Pinus thunbergii*. *Proc. Natl. Acad. Sci. USA* 91:9794–98

145. Whelan J, Young S, Day DA. 1992. Cloning of *ndhK* from soybean chloroplasts using antibodies raised to mitochondrial complex I. *Plant Mol. Biol.* 20:887–95

146. Wilhelm C, Duval J-C. 1990. Fluorescence induction kinetics as a tool to detect a chlororespiratory activity in the

prasinophycean alga, *Mantoniella squamata. Biochim. Biophys. Acta* 1016:197–202

147. Willeford KO, Gombos Z, Gibbs M. 1989. Evidence for chloroplastic succinate dehydrogenase participating in the chloroplastic respiratory and photosynthetic electron transport chains of *Chlamydomonas reinhardtii. Plant Physiol.* 90:1084–87

148. Wolfe KH, Morden CW, Palmer JD. 1992. Function and evolution of a minimal plastid genome form a nonphotosynthetic parasitic plant. *Proc. Natl. Acad. Sci. USA* 89:10648–52

149. Wollman FA, Minai L, Nechushtai R. 1999. The biogenesis and assembly of photosynthetic proteins in thylakoid membranes. *Biochim. Biophys. Acta* 1411:21–85

150. Wu DY, Wright DA, Wetzel C, Voytas DF, Rodermel S. 1999. The *immutans* variegation locus of *Arabidopsis* defines a mitochondrial alternative oxidase homolog that functions during early chloroplast biogenesis. *Plant Cell* 11:43–55

151. Wu M, Nie ZQ, Yang J. 1989. The 18-kD protein that binds to the chloroplast DNA replicative origin is an iron-sulfur protein related to a subunit of NADH dehydrogenase. *Plant Cell* 1:551–57

152. Yamane Y, Shikanai T, Kashino Y, Koike H, Satoh K. 2000. Reduction of Q(A) in the dark: another cause of fluorescence Fo increases by high temperatures in higher plants. *Photosynth. Res.* 63:23–34

153. Ziem-Hanck U, Heber U. 1980. Oxygen requirement of photosynthetic CO_2 assimilation. *Biochim. Biophys. Acta* 591:266–74

Annu. Rev. Plant Biol. 2002. 53:551–80
DOI: 10.1146/annurev.arplant.53.100301.135238

Structure, Dynamics, and Energetics of the Primary Photochemistry of Photosystem II of Oxygenic Photosynthesis

Bruce A. Diner[1] and Fabrice Rappaport[2]

[1]CR&D, Experimental Station, E. I. du Pont de Nemours & Co., Wilmington, Delaware 19880-0173; e-mail: Bruce.A.Diner@usa.dupont.com
[2]Service de Photosynthèse/UPR-CNRS 1261, Institut de Biologie Physico-Chimique, 75005 Paris, France; e-mail: rappaport@ibpc.fr

Key Words P680, reaction center, primary charge separation, chlorophyll, pheophytin

■ **Abstract** Recent progress in two-dimensional and three-dimensional electron and X-ray crystallography of Photosystem II (PSII) core complexes has led to major advances in the structural definition of this integral membrane protein complex. Despite the overall structural and kinetic similarity of the PSII reaction centers to their purple nonsulfur photosynthetic bacterial homologues, the different cofactors and subtle differences in their spatial arrangement result in significant differences in the energetics and mechanism of primary charge separation. In this review we discuss some of the recent spectroscopic, structural, and mutagenic work on the primary and secondary electron transfer reactions in PSII, stressing what is experimentally novel, what new insights have appeared, and where questions of interpretation remain.

CONTENTS

1040-2519/02/0601-0551$14.00

INTRODUCTION

In the past few years there have been a considerable number of advances at the spectroscopic, biochemical, and structural level that have improved our understanding of structure/function relationships in the Photosystem II (PSII) reaction center. We attempt to integrate here many of these findings as they relate to primary and secondary electron transfer, stressing what is new and what issues still remain unresolved. We refer the reader to a number of excellent review articles (22, 24, 27, 46, 88, 90, 133) that cover in more detail some of the older data or are more focussed in covering a subset of the topics addressed here.

STRUCTURE

Really spectacular progress has been made very recently in the description of the physical structure of the PSII reaction center and core complexes. The term reaction center has been used to describe a biochemically isolated entity comprised of polypeptides D1 and D2 (PsbA and PsbD, respectively, plus redox cofactors), cytochrome b_{559} (PsbE and F, plus heme) and PsbI (56a, 77a). In this review, unless otherwise indicated, this term refers only to the D1/D2 complex including the associated cofactors. The PSII core complex, also biochemically isolated, contains up to 25 different integral membrane and extrinsic polypeptide subunits (50), including those of the biochemically isolated reaction center and chlorophyll-protein complexes CP43 (PsbC) and CP47 (PsbB). Beginning with an appreciation of the structural and functional homologies between the reaction centers of PSII and those of the purple nonsulfur photosynthetic bacteria (75, 125), increasingly detailed models have been proposed for the arrangement of the prosthetic groups and of the D1 (L) and D2 (M) subunits that coordinate them (97, 112, 148). After many years of trying to do protein crystallography, investigators have had major successes recently that are allowing them to place the structural models on a much firmer footing. First, a spinach subcore complex containing the reaction center and CP47 formed, upon detergent dialysis, two-dimensional crystals (91) that yielded an 8-Å structure (92), following image processing of electron micrographs. This structure has now been refined to 6 Å (90). More recently, three-dimensional crystals of dimeric O_2-evolving PSII core complexes from *Synechococcus elongatus* have, in a remarkable advance, yielded a structure with 3.8-Å resolution (149). Although the resolution is still too low to resolve the side chains of the amino acid residues, most of the chromophores, the α-helices, the β-sheets, and the Mn cluster have been localized with good precision (Figure 1, see color insert). The overall protein complex extends well outside the membrane, 10 Å on the stromal side and 55 Å on the lumenal side. A band that is 40 Å thick is considered to be within the membrane. At least 17 subunits are present, 14 of which are integral membrane polypeptides, contributing 36 transmembrane α-helices. A total of 35 chlorophylls (Chls) have been identified, with 16 assigned to polypeptide CP47, 13 to polypeptide CP43, and 3 each to polypeptides D1 and D2 (137). Two pheophytins (Pheos) are associated with the D1-D2 complex, which forms the heart of the reaction center. These chlorins are arranged in a pseudo-C_2-symmetrical fashion around the nonheme iron, as are the CP47 and D2 polypeptides on one side and CP43 and D1 on the other. The symmetry is broken by the cytochrome b_{559} heme on the stromal side of the complex and coordinated by PsbE and PsbF, which are located near the D2 subunit but 27 Å from Chl_{ZD2} (see below); by the Mn cluster associated with the lumenal side of D1 and located 15 Å from the pseudo-C_2-symmetry axis; and by cytochrome c_{550} and PsbO, which are extrinsically associated with the lumenal side of CP43 and of D1, respectively.

Analysis of this structure as well as that at the primary, amino acid level have indicated significant differences with the reaction centers of *Rhodopseudomonas viridis* and *Rhodobacter sphaeroides*. The central reaction center chlorins (Figure 1) are designated as belonging to the A or B branch of the reaction center, the presumed electron transfer active and inactive branches, respectively. The special pair chlorophylls, P_A (P_{D1}) and P_B (P_{D2}) located on the lumenal side of the complex and implicated in primary charge separation (see below), have parallel ring planes 5 Å apart and are 10 Å apart center to center (91, 149). These are more widely separated (Figure 2) than the special pair bacteriochlorophylls (Bchlorophylls) in the bacterial reaction centers where the respective ring plane separation and center to center distance are 3.5 Å and 7.4–7.6 Å [e.g., Reference (36)]. This wider separation in PSII greatly weakens the excitonic coupling that characterizes the bacterial reaction center homologues. This difference has important consequences for excitation energy localization within the reaction center (see below). Two monomeric chlorophylls, B_A and B_B, located 9.8 Å and 10 Å, respectively, center to center from P_A and P_B, are inclined at a 30° angle with respect to the membrane plane as are their Bchlorophyll homologues in the bacterial reaction centers. The homologous histidines that coordinate these Bchlorophylls in the bacterial reaction centers are missing in PSII. $Pheo_A$ and $Pheo_B$ are located at 10.7 Å and 10.6 Å, respectively, (center to center) toward the stromal side from B_A and B_B, with their head groups perpendicular to the membrane plane. The primary quinone electron acceptor, Q_A, is located an additional 12.0 Å (center to center) toward the stromal side from $Pheo_A$ and 10.5 Å from the nonheme iron. Electron transfer in the bacterial reaction centers originates from the P_A/P_B special pair in its lowest excited singlet state to B_A then to $Pheo_A$ and ultimately to Q_A [for review see Reference (147)]. Despite the presence of homologous redox components in PSII, kinetic deconvolution of energy transfer and primary charge separation in PSII is complicated by the spectral congestion that exists within its reaction center

Center-to-center distances (Å)

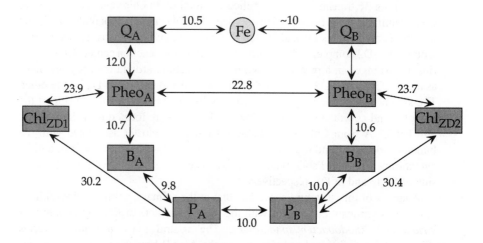

PSII - *Synechococcus elongatus* (149)

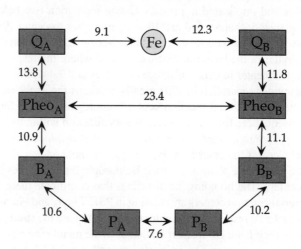

BRC - *Rhodobacter sphaeroides* (36)

Figure 2 The center to center distances in Å between the cofactors of the *Synechococcus elongatus* PSII reaction centers (149) and of the *Rb. sphaeroides* reaction centers (36). Adapted from (36).

(see below). An additional symmetrically localized pair of chlorophylls, for which there is no homologue in the bacterial reaction center, Chl_{ZD1} and Chl_{ZD2}, is located at 30.2 Å and 30.4 Å, respectively, from P_A and P_B. One or both of these can be oxidized in a low quantum yield process (see below) that may regulate PSII charge separation in a nonphotochemical quenching process [e.g., Reference (106)].

The current level of resolution of the PSII crystal structure permits only a description of the $C\alpha$ trace of the most-ordered α-helical and β-sheet regions of the polypeptides that comprise the PSII core complex. The identity and orientation of the amino acid side chains remain unknown. Despite this lack of definition, there are a number of residues, some of which are bacterial reaction center homologues, that have been identified by site-directed mutations coupled with spectroscopic analysis as being ligands to prosthetic groups. These are discussed below as we describe each of the redox active components of the PSII reaction center.

ENERGY TRANSFER

New structural data have permitted the localization of 37 of the chlorin rings (35 chlorophylls and two pheophytins) (137, 149). The position (within 1 Å) and orientation (within 10°) of the ring planes are known with considerable accuracy. Although the orientation of the transition dipoles cannot be deduced from the crystal structure, these are to some extent known from linear dichroism and photoselection experiments. This structural information has promoted renewed interest in modeling of the kinetics of energy transfer from the antenna to the reaction center. There appear to be two camps regarding this issue. In one of these (Figure 3), the Reversible Radical Pair Model (3, 12, 21, 102), there is rapid equilibration of the excitation energy between the antenna and the reaction center [$\tau \leq 15$ ps; e.g.,

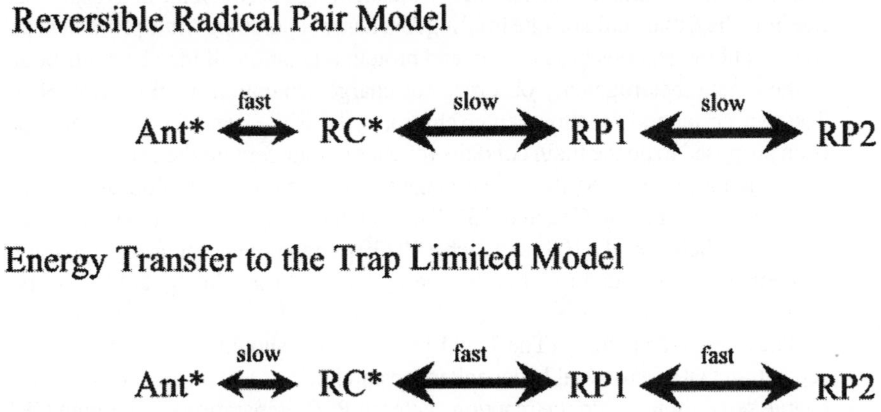

Reversible Radical Pair Model

$$\text{Ant*} \xleftrightarrow{\text{fast}} \text{RC*} \xleftrightarrow{\text{slow}} \text{RP1} \xleftrightarrow{\text{slow}} \text{RP2}$$

Energy Transfer to the Trap Limited Model

$$\text{Ant*} \xleftrightarrow{\text{slow}} \text{RC*} \xleftrightarrow{\text{fast}} \text{RP1} \xleftrightarrow{\text{fast}} \text{RP2}$$

Figure 3 Two contrasting models for the kinetic limitation of primary charge separation in PSII. In the upper model, the rate is limited by electron transfer within the reaction center. In the lower model, the rate is limited by energy transfer to the reaction center.

(21)]. The latter is a shallow trap, and charge separation is trap limited. Barter et al. (3) have recently provided additional support for this model by showing that the effect of the antenna size on the Chl* singlet state lifetime could be well simulated by a three-state model [(AntennaRC)* ↔ RP1 ↔ RP2] in which the excitation energy is rapidly equilibrated between the antenna and reaction center, and two radical pair states (RP1 and RP2) are involved in the slower, two-stage trapping process. The lifetime of the singlet excited state increases with antenna size, ranging from reaction centers (8 chlorin pigments) up to PSII membrane fragments from spinach (BBY, ~250 chlorin pigments) (5).

In a contrasting model (Figure 3), equilibration of light energy within the antenna complexes [$\tau < 5$ ps (21, 137)] and within the reaction center [$\tau \leq 400$ fs, (35, 73)] is considered to be rapid, but energy transfer between the antenna and the reaction center is slow and rate limiting. Based on careful examination of the PSII X-ray crystal structure, Vasil'ev et al. (137) concluded that two chlorophylls, C12 and C30 of CP43 and CP47, respectively, are the core antenna pigments that are responsible for 50% of energy transfer to the reaction center. These authors concluded, however, that the distance and orientation of these pigments with respect to the reaction center pigments would put energy transfer to the reaction center in the 100 ps range, making energy transfer to the center rate limiting for charge separation. This may be a common feature in photosynthesis as rate limitation by energy transfer to the reaction center has also been observed in *Rb. sphaeroides* (4) and in PSI (127) [for review see (133)]. An additional conclusion of Vasil'ev et al. (137) is that a lower limit for the intrinsic rate of charge separation in PSII is $(0.7 \text{ ps})^{-1}$, a rate consistent with a rapid component for charge separation upon direct excitation of P680 as reported by Groot et al. (47) ($\tau = 0.4$ ps at 240 K). This rate is substantially faster than most reports of charge separation, which range from 1–20 ps for PSII (32, 44, 61, 95, 103, 139) and for bacterial reaction centers (38, 133). Groot et al. (47) suggest that even within the PSII reaction center there could be kinetic components for energy transfer (e.g., between the active and inactive branches) that could be rate limiting for charge separation. Multiple radical pair states, inhomogeneous broadening, and protein relaxation all may be contributing to the kinetic heterogeneity observed for charge separation in PSII (see below). The two peripheral reaction center chlorophylls, Chl_{ZD1} and Chl_{ZD2}, which had been proposed to be the main conduits for energy transfer into the center (70, 136), have orientations and positions that make for weaker coupling to the antenna and reaction center than do C12 and C30. This conclusion is consistent with measurements of Schelvis et al. (103) in which a 20–30 ps component for charge separation was attributed to a rate limitation by energy transfer from Chl_{ZD1} and Chl_{ZD2} [see also (136)].

The Energy Transfer To The Trap Limited Model should show a rate of charge separation that is only weakly dependent on antenna size and on the intrinsic rate of charge separation within the reaction center in PSII preparations containing CP43 and CP47, i.e., preserving the integrity of the rate-limiting step. In the D_1D_2 cyt b_{559} PSII reaction centers that lack these two subunits, the observed rate should, in this model, more closely reflect the intrinsic rate of charge separation. At variance with this expectation, Barter et al. (3) found a rather monotonic dependence of the singlet excited state lifetime upon the antenna size. It could be argued, however, that the D_1D_2 cyt b_{559} PSII reaction centers, the smallest photoactive complexes, were missing Q_A, whereas the larger ones had Q_A reduced. This difference could distort the apparent dependence of the excited state lifetime on antenna size. Indeed, the redox state of Q_A could influence, likely via electrostatic effects, the singlet state lifetimes independent of antenna size. It has been observed that the redox state of the quinone has a large influence on the rise (95, 102) kinetics of primary radical pair formation. Gibasiewicz et al. (41), however, showed that the kinetics of radical pair formation as detected by a photovoltaic method were largely independent of the redox state of Q_A, although the decay was greatly accelerated by Q_A^- as opposed to Q_A or Q_AH_2. This issue remains to be resolved. Ways in which the two models of Figure 3 might begin to overlap could include the discovery of additional chlorophylls that have not yet been resolved in the 3.8-Å crystal structure that might speed antenna-center excitation equilibration or, as Groot et al. (47) suggest, there might be energy transfer rate limitations even within the reaction center.

COMPARISON WITH BACTERIAL REACTION CENTERS

Primary charge separation in the reaction centers of the purple photosynthetic bacteria is initiated by the excitation of the special pair Bchlorophylls, P_A and P_B, following energy transfer from the light-harvesting complexes. These P_A and P_B are located close to each other (36) (see above) (Figures 1 and 2) and are excitonically coupled (500–1000 cm^{-1}), thereby constituting a long wavelength trap for the reaction center excitation energy (e.g., 870 nm in *Rb. sphaeroides*) (147). The splitting between the special pair Bchlorophylls (P_A/P_B) and the accessory Bchlorophylls (B_A and B_B) is ~1000 cm^{-1} and between the special pair Bchlorophylls and the bacteriopheophytins (Bpheophytins) ~1500 cm^{-1} (147). $(P_A/P_B)^*$ in bacterial reaction centers decays in a multiexponential process to yield P^+Pheo^- with time constants ranging from 0.9 to 4 ps, which may involve P^+B^- as an intermediate, but there is not unanimity of opinion in this regard [see Reference (147)]. In PSII, the reaction center chlorophylls are much more congested spectrally as their absorbance spectra heavily overlap. In contrast to the bacterial reaction centers, the overall absorption envelope of the PSII reaction centers is only approximately 500 cm^{-1} at half height. The excitonic coupling between the reaction center pigments is weak, with the homologous P_A and P_B chlorophylls suggested to show a coupling of approximately 85–140 cm^{-1} (11, 118, 134). The structural homology between the bacterial reaction centers and PSII and the similar time domains over which charge separation occurs would lead one to assume that primary charge separation in PSII occurs mechanistically as it does in the bacterial reaction center. The kinetics of primary charge separation in PSII, however, have been reported to show multiple phases for the formation of the P^+Pheo^- charge pair (32, 44, 45, 60, 77). In addition, the triplet state of P, formed by P^+Pheo^- recombination at liquid helium temperature, has been shown in PSII (131, 135) to be localized on a monomeric chlorophyll with an orientation (ring plane 30° with respect to the membrane plane) more like that of B_A or B_B rather than P_A and P_B as is seen in the bacterial reaction centers (49, 119). These observations have led a number of authors to suggest that primary charge separation might occur differently in PSII, potentially initiated by excitation of B_A rather than P_A/P_B. A number of groups have shown in bacterial reaction centers that it is possible to observe, under certain conditions, the formation of radical pair states upon direct excitation of B_A (69, 128, 130, 142). Dekker & van Grondelle (24) and Rutherford and coworkers (98, 99) have suggested that in PSII the lack of spectral differentiation and the orientation of ^3P, respectively, might be consistent with a contribution of B_A^* to primary radical formation. Key questions then are (*a*) where is primary charge separation initiated and (*b*) are the oxidized primary donor cation radical and the triplet localized on the same or on different chlorophylls, and if the latter, which one of the two has migrated and on what timescale?

SPECTRA OF PSII REACTION CENTER CHLORINS

P_A, the Primary Site for P^+ Cation Localization

Although it has been difficult to track the reaction center triplet and primary cation radical of PSII because of the spectral congestion mentioned above, there has been some success using Fourier transform infrared spectroscopy (FTIR). Noguchi et al. (80) and Breton and coworkers (101) have been able to show, based on the vibrational frequency of the $C_{13^1}=O$ ($C_9=O$) carbonyl, that the chlorophyll on which the reaction center triplet, ^3P, is localized at low temperature (1669 cm^{-1}) is different from the chlorophyll(s) on which the P^+ cation is localized (1679 or 1704 cm^{-1}).

More recently, Diner and coworkers (30) have introduced site-directed mutations that have permitted optical tagging of some of the reaction center chlorophylls. These tags have permitted the assignment of the Qy and Soret absorbance maxima to P_A, P_B, and B_A. Histidines D1-His198 and D2-His197 are the PSII homologues of histidines L-His173 and M-His200(202) of the bacterial reaction centers responsible for the coordination of chlorophylls P_A and P_B, respectively. Both residues were replaced in PSII by a variety of amino acids that alter the reduction potentials of the coordinated chlorophylls and introduce displacements to the blue of their absorbance spectra. The latter is explained by the loss of the more polarizable His axial ligand, which stabilizes the chlorophyll excited state. Replacement of the P_A ligand, D1-His198, with Gln resulted in the largest displacement (3 nm) to the blue of the P_A absorbance spectrum for the Soret (433 → 430 nm, 298 K) and Qy transitions (672.5 → 669.5 nm, 80 K) (30). That the P^+-P difference spectrum at all temperatures shows this same shift argues that P_A is the primary location of the P^+ cation independent of the temperature. Measurements of absorbance changes polarized parallel and perpendicular to the membrane plane indicate that the orientation of the Qy transition of the chlorophyll responsible for $P^+(P_A^+)$ is parallel to the membrane plane (71) (Figure 1).

Most surprising among the site-directed replacements reported at D1-His198 and D2-His197 were those that did not introduce coordinating side chains (e.g., Ala, Val) (30). Introduction of Ala gave photoautotrophic strains in both cases. However, larger noncoordinating resides such as Leu (30) and Tyr (138) gave no reaction center. It is likely that in the case of the D1-His198Ala and the D2-His197Ala mutations a water molecule has replaced the imidazole ring of histidine as the axial ligand to Mg^{2+}. A residue like Leu would displace the water molecule preventing coordination, and deprotonation of the water ligand or O—H bond stretch would explain the substantial stabilization ($\Delta E'$ $P^+/P \sim -80$ mV) of P^+ by the Ala mutation relative to the wild-type strain. A similar replacement of the His coordination with a water molecule has been reported in the L-His173Gly and the M-His200Gly mutants of *Rb. sphaeroides* (43, 105) and in the B-His656Glu mutant of PSI (66, 144). In contrast to PSII, however, replacement of each His ligand with Leu does allow assembly of the bacterial reaction center but with incorporation of a Bpheophytin in place of Bchlorophyll at the mutated site (14a, 58a). It is possible, in the case of PSII, that the binding energy contributed by the axial coordination is required for chlorin incorporation. In bacterial reaction centers, other residues must contribute sufficiently to allow Bchlorin binding.

The oxidation of redox active tyrosines, Y_Z and Y_D, produces band shifts in the Soret and the Qy regions of the chlorophyll absorption spectra. For Y_Z, the band shift is sensitive to the D1-His198Gln mutation. The band shift is centered at 434 nm in wild type and at 432 nm in the mutant (30), indicating that P_A is a major probe of the oxidation of this tyrosine. This observation and the P^+-P difference spectrum indicate that the absorbance maxima for P_A are located at 433 and 672.5 nm in wild type.

P_B

By symmetry, P_B should be an analogous probe of the oxidation of Y_D. The Soret band shift for $Y_Z\cdot$-Y_Z is centered at 433–434 nm (consistent with P_A), whereas for $Y_D\cdot$-Y_D it is centered at 436 nm (31). The Soret absorption maximum for P_B is therefore located at 436 nm. This 2–3-nm displacement of the absorbance maximum to the red for P_B relative to P_A is likely conserved in the Qy region, placing the probable absorbance maximum of P_B at 675 nm.

B_A

A band shift to the blue of electrochromic origin accompanies the formation of P^+. This band is centered at $\sim$684 nm at 5 K (54) and at 681–682 nm at 80 K (30). This same chlorophyll is also bleached upon formation of ^3P (684 nm at 5 K, 682 nm at 80 K) (see below for the mechanism of triplet formation) (30, 54). Both the P^+-induced band shift and the ^3P-^1P difference spectrum are insensitive to the D1-His198Gln site-directed mutation. This insensitivity, the observation of an electrochromic shift of an accessory Bchlorophyll to the blue upon formation of P^+ in bacterial reaction centers (147), the primary localization of P^+ on P_A, and the reaction center triplet orientation all support the assignment of the 681–682-nm (684 nm at 5 K) feature to the Qy transition of B_A.

Pheo$_A$ and Pheo$_B$

Borohydride reduction of the inactive branch pheophytin, Pheo$_B$ (109) and exchange of Pheo$_B$ (39, 40, 108) and partial exchange of Pheo$_A$ (40, 39) with 13^1-deoxo-13^1-hydroxy-pheophytin a indicate clearly that both pheophytins have their Qy absorption maxima at 676–680 nm (at 6 K). Linear dichroism indicates that the Qy transitions are perpendicular to the membrane plane (39) (Figure 1). The Qx transitions of both pheophytins are located at 543 nm (at 6 K) (40). In bacterial reaction centers, the Qx transition is sensitive to hydrogen bonding to the $C_{13^1}=O$ of the exocyclic ring (15, 18). Moënne-Loccoz et al. (76) have shown using Resonance Raman spectroscopy that Pheo$_A$ in PSII is likely hydrogen bonded (as it is in bacterial reaction centers), with residue D1-130 being the likely homologue of the bacterial L-104. Indeed, Giorgi et al. (42) have shown that *Synechocystis* site-directed mutations D1-Gln130Glu and D1-Gln130Leu displace the Qx band to longer and shorter wavelength, respectively, consistent with a strengthening and a disappearance of the $C_{13^1}=O$ hydrogen bond. This finding was recently confirmed by a high-field electron paramagnetic resonance (EPR) spectroscopy study of PSII core complexes bearing mutations at this position (33). That the Qx transitions of both PSII pheophytins are located at 543 nm implies that they are bound with hydrogen bonds of equal strength. Resonance Raman spectra of PSII core complexes in which each of the pheophytins is replaced by a chlorophyll (D. Force, A. Pascal, B. Robert & B. Diner, in preparation) show, in each case, the loss of a similar feature at 1685 cm^{-1} arising from a hydrogen-bonded carbonyl group (A. Pascal, personal communication). This observation also indicates directly that both pheophytins have similar hydrogen-bond strengths to the $C_{13^1}=O$ carbonyl. It had been proposed earlier that one possible distinction between the active and inactive branches of the reaction center might be the presence of a hydrogen-bonded pheophytin on the active branch and the absence of a hydrogen bond on the inactive branch (76). The above observations rule out this hypothesis.

B$_B$

Replacement of the inactive branch Pheo$_B$ with 13^1-deoxo-13^1-hydroxy-pheophytin a results in the loss of the Pheo Qy transition at 676–680 nm but also causes the blue shift of a transition at 680 nm that is attributed to B$_B$ (39). Replacement of Pheo$_A$ induces a blue-shifted transition at approximately 682 nm that is attributed to B$_A$, in agreement with the above assignment.

Chl$_{ZD1}$ and Chl$_{ZD2}$

A linear dichroism study (39) and characterization of a PSII reaction center complex containing five chlorophylls (126) both attribute to Chl$_{ZD1}$ and to Chl$_{ZD2}$ a 670-nm Qy transition oriented nearly perpendicular to the membrane plane

(Figure 1). The assignments of the absorption maxima of the reaction center chlorins are summarized in Figure 1.

REACTION CENTER TRIPLET, ^{3}P

The energetic consequences of the spectral assignments described above provide new insights into the mechanism of charge separation and the localization of the reaction center triplet state, ^{3}P. Under conditions where P^+Pheo^- cannot relax to form $P^+Q_A^-$ either because Q_A is reduced or absent, there is a decorrelation of the spins on the radical pair resulting in singlet/triplet mixing. Recombination of the triplet state of the radical pair gives rise to a reaction center triplet state that forms with a $\tau \sim 30$ ns at room temperature (20) and slows as a function of temperature (81, 141) to 100–200 ns at 10 K (58, 134). The energy of the ^{3}P is estimated to be 0.54 eV below the singlet state (141). The triplet resides on a monomeric chlorophyll (25, 34, 100, 114, 134) and has the gz component of its anisotropic g-tensor oriented at 60° with respect to the membrane plane (68, 135); both findings are consistent with ^{3}P residing on B_A or B_B. This localization is in contrast to what is found in the bacterial reaction centers where the triplet resides on one or both P_A and P_B (49, 119). The comparison of the optical difference spectrum ^{3}P-^{1}P with that of P^+-P (see above) indicates that the triplet is localized on a chlorophyll absorbing at 684 nm at 5 K (30, 54, 55). This chlorophyll was associated with the electrochromic band shift observed upon formation of P_A^+ and so is attributed primarily to B_A rather than to B_B in PSII (30).

Assuming very weak coupling of the singlet states of the reaction center chlorophylls, the energy of the triplet state should track that of the singlet state (72). As B_A is the longest wavelength chlorophyll of the reaction center, it is now understandable why the triplet is localized on B_A at 5 K rather than on P_A, the major site of cation localization at all temperatures. It is not clear whether the triplet is actually generated at P_A or $Pheo_A$ by recombination of $P_A^+Pheo_A^-$ (or $P_A^+B_A^-$ or $B_A^+Pheo_A^-$), migrating to B_A at a rate faster than triplet formation, or whether it is the cation that migrates to B_A or the anion that migrates to B_A from $P_A^+Pheo^-$. At this point, the triplet would be formed, respectively, by $B_A^+Pheo_A^-$ or $P_A^+ B_A^-$ recombination, thus directly forming the triplet on B_A.

Although the triplet is clearly localized on B_A at ≤ 80 K, the localization of ^{3}P changes as the temperature rises (10, 58, 78, 80). At elevated temperature (≥ 150 K), another chlorophyll(s) begins to contribute to the reaction center triplet population. This chlorophyll has a $C_{13^1}{=}O$ stretch like that of P_A (80, 101) and is oriented perpendicular to the membrane plane (58). Over the same temperature range, the ^{3}P-^{1}P difference spectrum begins to show a short wavelength shoulder, the position of which is sensitive to the D1-His198Gln mutation (30). All of these observations are consistent with an increased sharing of the reaction center triplet with P_A at elevated temperature. The temperature dependence of the triplet localization is consistent with an energy gap ΔE of 8–13 meV that would correspond to the difference in the triplet state energies of B_A and P_A (10, 58, 78, 80).

SHARING OF THE CATION RADICAL

Diner et al. (30) have proposed that the P^+ cation radical is primarily localized on P_A^+. A mid-IR band at 1940 cm^{-1} detected by FTIR (80; J. Breton, personal communication) and methyl hyperfine couplings measured by electron nuclear double resonance (ENDOR) (93, 117) both gave indications of some delocalization of the P^+ cation. The ENDOR measurements were consistent with an approximate 80:20 distribution between two or more chlorophylls with one chlorophyll dominating (now known to be P_A^+). The D2-His197Ala mutation gave ~1-nm displacement of the P^+-P difference spectrum to the red, whereas the same mutation on D1-His198Ala produced a 2-nm shift to the blue (30). This observation is consistent with a displacement of the P^+ cation toward P_B in the D2-His197Ala mutant through stabilization of P_B^+. This ability to control the position of the cation suggests the possibility that its position might be subject to electrostatic control as well (see below).

PRIMARY CHARGE SEPARATION

It would appear from the spectral assignments discussed above that B_A is the longest wavelength chlorin located within the reaction center. Additional support for this conclusion comes from a study by Konermann & Holzwarth (63) of the spectral decomposition of the PSII reaction center at 10 K, which led these authors to conclude that B_A or B_B was responsible for the longest wavelength emission. Furthermore, Peterman et al. (85) have measured at 5 K the vibronic fine structure of a line-narrowed emission spectrum upon excitation of PSII reaction centers at 684–686.1 nm. The emission spectrum, which should arise from the longest wavelength-emitting component of the reaction center, shows a C_{13^1}=O (C_9=O) carbonyl stretch at 1669 cm^{-1}. Such a feature has been assigned by Noguchi and coworkers to B_A (80, 81).

At 5 K, the ratio of finding the excitation energy on B_A (684 nm) versus P_A (672.5 nm) B_A^*/P_A^* (Figure 4) is approximately 10^{31}, based on Boltzmann considerations (singlet state energy difference of 31 meV). This partition, for *Synechocystis*, is somewhat more extreme than for spinach and *Synechococcus* (10^{21}–10^{24}) where

$\longrightarrow$

Figure 4   The relative localization of the excitation energy on the central components of the PSII reaction center at 5, 77, and 298 K. The shaded area corresponds to the percent of the total excitation on that component, calculated from the Boltzmann equation using the absorbance maxima of Figure 1. This distribution is calculated at each temperature based on the absorbance maxima of the chromophores at 5 K. The use of the 5-K spectrum is certainly an over simplification at 298 K, but the energy is broadly distributed anyway at the latter temperature.

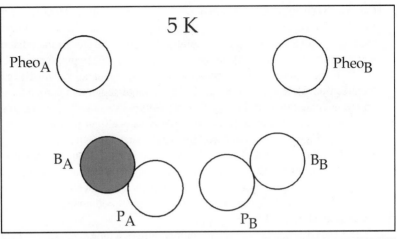

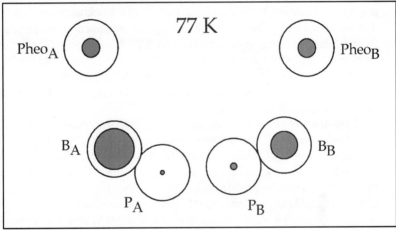

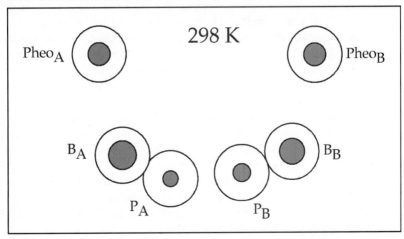

the absorbance maxima of P_A are likely located at 676 nm and 675 nm, respectively (54, 55). For B_A and $Pheo_A$ (680 nm), the $B_A^*/Pheo_A^*$ ratio at 5 K is 5.6×10^{10}. As the rate of charge separation is generally thought to be slower than that of energy equilibration [but see (47)] between the reaction center pigments (100–250 fs) (35, 77), the excited state responsible for charge separation will be exclusively localized on B_A at this temperature. That charge separation can occur at ≤ 20 K on the ps timescale with high quantum yield ($\sim 75\%$) (47, 57, 143) implies that it must be occurring from B_A^*. The two likely radical pair products are $P_A^+B_A^-$ and $B_A^+Pheo_A^-$ (Figure 5). van Broderode et al. (129) have shown that direct excitation of B_A will produce $P_A^+B_A^-$ as the initial radical pair in *Rb. sphaeroides* reaction centers. These authors have also shown that excitation of B_A in reaction centers in which the special pair Bchlorophylls have been replaced by a Bchlorophyll:Bpheophytin heterodimer (M-His202Leu mutant) will produce predominantly $B_A^+Pheo^-$ as the initial radical pair (130).

The observations in PSII at low temperature of a long-lived singlet excited state arising from a chlorophyll absorbing at 683 nm (63, 64) with spectral similarities to the charge-separating state (85) would imply that the radical pair state(s) generated from B_A^* is(are) nearly equipotential with B_A^*. Were there a distribution of the energies of these states, then some fraction of them would require thermal activation and at 5 K might be inaccessible, giving rise to long-lived fluorescence (24, 85). Indeed, Groot et al. (47) have shown that charge separation in PSII reaction centers is to some extent an activated process.

Konermann et al. (64) and Peterman et al. (85) were unable to detect fluorescence emission from Pheo* at 5–10 K. However, at 77 K, Pheo* is apparent with a ratio of Chl*/Pheo* of 4:1 (64) (Figure 4). This observation is consistent with a 4-nm (10.5 meV) difference between the energies of the lowest excited singlet states of B_A (684 nm) and Pheo (680 nm). It is possible then that Pheo* could begin to contribute to radical pair formation as the temperature increases, generating $B_A^+Pheo^-$ as the initial product. At 298 K, the energy is likely to be distributed over all of the central pigments of the reaction center with B_A^* representing no more than 25% of the total (Figures 4 and 5). All of the active branch pigments could contribute to radical pair formation, although the relative contributions to charge separation from the different excited states are likely to be different depending on their rates of charge separation and on the relative energies of the radical pair and excited states. The observed heterogenity in charge separation in reaction centers is then likely to arise from the heterogeneity of excited states that contribute to it, the inhomogeneous broadening of the optical transitions (48), a distribution of energy levels for a single radical pair state (48, 62), a succession of increasingly stabilized radical pair states accessed through electron transfer (24, 62), and relaxations within the protein matrix (Dynamic Solvation Model) (24, 62, 84). Proposals by Dekker & van Grondelle (24) situate primary radical pair formation in ≤ 2 ps, expanded radical pair formation in ~ 8 ps, and protein relaxation on the 50 ps time scale. Where excited, the Chl_{ZD1} and Chl_{ZD2} would contribute 20–30 ps components that are limited by energy transfer to the central pigments of the reaction center.

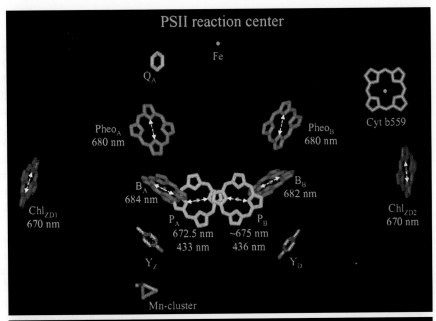

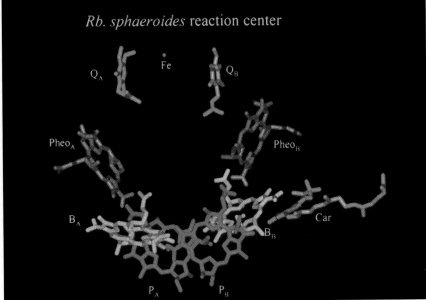

Figure 1 Redox active components of the *Synechococcus elongatus* Photosystem II [adapted from Reference (149)] and the *Rb. sphaeroides* reaction centers [adapted from Reference (17)]. The view in both cases is in the plane of the membrane. A and B refer to the active branch and inactive branch components, respectively. The absorbance maxima for each of the chromophores in the *Synechocystis* PSII reaction center are indicated. The *double-headed dashed arrows* refer to the approximate orientation of the Qy transitions.

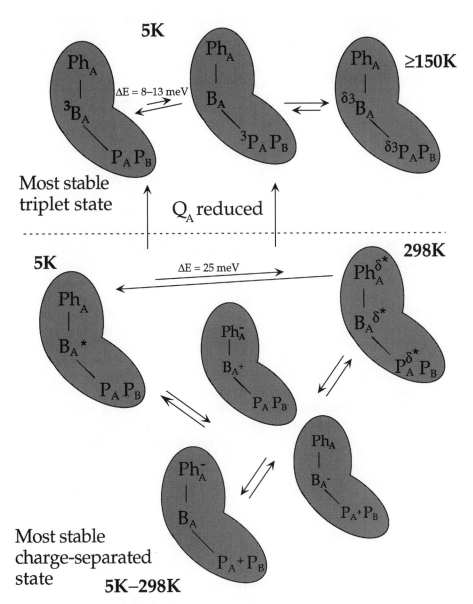

Figure 5 The localization as a function of temperature of the singlet excited states before charge separation and of the triplet and radical pair states after charge separation, in the active branch of the PSII reaction centers.

A number of groups have directly measured the depth of the PSII trap. In plants, the free-energy gap between the equilibrated singlet excited state and P^+Pheo^-, measured at 60 ps following actinic excitation at 298 K, is only -27 meV (74), increasing to -46 to ~-110 meV on the nanosecond timescale (8, 48, 62, 141), consistent with the Dynamic Solvation Model mentioned above. Site-directed mutations constructed in *Synechocystis* 6803 that increase the reduction potential of P^+/P (D1-His198Gln, $\Delta G \sim 16$ meV) or lower the reduction potential of $Pheo/Pheo^-$ by removing a hydrogen bond to the $C_{13^1}{=}O$ carbonyl (D1-Gln130Leu, $\Delta G \sim 53$ meV) produce significant decreases in the concentration of the radical pair detected at 60 ps. In *Synechocystis*, the D1-Gln130 hydrogen bond to $C_{13^1}{=}O$ raises by ~ 36 meV the energy of the P^+Pheo^- state relative to the same state where D1-130 is a Glu as in higher plants (74). Thus in wild-type *Synechocystis*, P^+Pheo^- is actually ~ 21 meV above the excited state energy at 60 ps. A consequence of this uphill process for radical pair formation is that the electron transfer from $Pheo^-$ to Q_A (250–300 ps) should appear as a component in the excited state relaxation.

DONOR-SIDE SECONDARY ELECTRON TRANSFER

The redox active tyrosines, Y_Z and Y_D (D1-Tyr161 and D2-Tyr160, respectively) [for reviews see (23, 26, 27)], act as secondary electron donors and are oxidized to form the neutral radicals, releasing the phenolic proton. Site-directed mutagenesis has implicated D1-His190 and D2-His189 as the immediate proton acceptors for Y_Z and Y_D, respectively [see (23, 26) and references therein]. Pulsed ENDOR measurements have indicated direct hydrogen bonding of the proton on the τ-imidazole nitrogen of D2-His189 to Y_D· (16). FTIR measurements indicate that this same His is hydrogen bonded by the phenolic proton of Y_D (53). However, there is no clear evidence to indicate the formation of the imidazolium form of histidine (53, 82) upon tyrosine oxidation, implying that the proton on the π-imidazole nitrogen of D2-His189 is likely transferred to a proton acceptor. The proton may, however, remain associated with the acceptor and back hydrogen bond. If the hydrogen-bonded chain were to extend to the bulk phase, then it is possible that the positive charge of the proton could be dissipated by proton release at the surface. In the case of Y_Z, the situation is even less clear. Site-directed replacement of D1-His190 causes a substantial slowing of Y_Z oxidation (29, 52), consistent with a similar role as the acceptor of the phenolic proton of Y_Z. However, pulsed ENDOR measurements do not indicate a direct hydrogen bond between the two, as in the case of Y_D (K. Campbell & R. David Britt, personal communication). Either the two are indirectly linked by intermediary waters or amino acid residue side chains or the hydrogen-bonded interaction between the two is intermittent (gated). Thus for both tyrosines, either the released proton is distributed among a number of likely hydrogen-bonded residues or the charge, if not the actual proton, is released to the bulk phase. There has been considerable debate over this issue, as it has repercussions for the hydrogen atom abstraction model for water oxidation that implicates Y_Z directly in this mechanism (13, 120). This model is a consequence of recent demonstrations of the close proximity between Y_Z and the Mn cluster (113, 116), the lowered activation energy that results from having a strong base prepositioned to accept a proton from bound water as the Mn cluster is being oxidized (67, 145), and the difference in bond dissociation energies that favor hydrogen atom transfer from the Mn cluster to Y_Z• (6, 83). In this model, the phenolic proton of Y_Z has been proposed to be released to the lumen upon Y_Z oxidation. The proton-coupled reduction of Y_Z• with the proton coming from a water molecule ligated to the Mn cluster would then decrease the energetic levels of the reaction intermediates and increase the driving force for the catalysis for water oxidation.

RETENTION OF TYROSINE PHENOLIC PROTON AND ELECTROSTATICS

A number of authors have argued, however, that the Y_Z phenolic proton is retained within the reaction center following oxidation (1, 30, 31, 86, 87), giving rise to a chlorophyll band shift of electrochromic origin in the Soret and Qy spectral regions. A similar band shift is observed accompanying the oxidation of Y_D but shifted by 2–3 nm to the red (31) (see above). The very slow exchange of the phenolic Y_D proton upon replacement of H_2O with D_2O [$t_{1/2} \sim 9$ h (28, 94)] also argues for proton retention upon oxidation of this tyrosine. If instead the phenolic proton were released to the thylakoid lumenal space, then there would be no retained charge, and another explanation would need to be sought to explain the band shift [e.g., a structural change or a hydrogen-bond displacement (122)]. Arguments for proton retention include a pH dependence for proton release coupled to Y_Z oxidation that turns off below pH 5 even though the pK_a of the tyrosyl cation radical is approximately -2 (87). In addition, Rappaport & Lavergne (86) have observed an oscillation of period four of a very similar band shift in the Soret that appears upon S-state advances that do not give rise to proton release (e.g., S1 –> S2, generation of a plus charge).

   The ability to shift the position of the P^+ cation between P_A and P_B by ligand replacement and the difference in the absorbance maxima of P_A and P_B (30) mean that the P^+-P difference spectrum can be used as an electrostatic indicator of nearby charges. Boerner et al. (7) reported and Diner and coworkers (unpublished) later confirmed that the rate of charge recombination between P^+ and Q_A^- is influenced by the presence of Y_D•. D2-Tyr160Phe, a mutant lacking Y_D, showed a rate of charge recombination nearly twice that observed in wild type. This observation is consistent with the positive charge generation associated with Y_D•, which in wild type would increase the reduction potential of P^+/P, thereby decreasing both the concentration of P^+ at the expense of Y_Z and the rate of charge recombination between P^+ and Q_A^- (Y_Z• and Q_A^- are too far apart to allow a competitive direct electron transfer). The positive charge would likely also influence the position of the P^+ cation between P_A and P_B. If we place D2-His190 at roughly 13 and 20 Å, respectively, from P_B and P_A and assume that most of the positive charge is in the immediate environment of the histidine upon Y_D oxidation, then one can do a rough electrostatic calculation on the potential opposing the movement of a plus charge from P_A^+ to P_B. This calculation gives a potential difference of 147 and 59 mV for a dielectric constant of 4 and 10, respectively, which is larger than the estimated 40 mV difference in reduction potential of P_A^+/P_A and P_A^+/P_B (30). The charge that accompanies Y_D oxidation would therefore be expected to have a substantial influence on the localization of the P^+ cation. Indeed, Faller et al. (37) have recently suggested that the P^+ reduction rate by Y_Z is slower in the presence of Y_D than $Y_{D^\bullet}$. It was proposed that, owing to electrostatic interaction, the propensity of the cation to reside on P_A (the closest Chl to Y_Z) would be larger in the presence of $Y_{D^\bullet}$. J. Bautista & B. Diner (unpublished) have examined the P^+-P difference spectrum in the presence and absence of Y_D and have found that, indeed, the mutant lacking Y_D shows a difference spectrum shifted by approximately 0.5 nm to the red relative to that of the wild type. This finding is consistent with the probability that the cation residing on P_B is larger in the absence of $Y_{D^\bullet}$. These observations are consistent with charge retention and the electrochromic interpretation of the very similar $Y_{Z^\bullet}$-Y_Z and the $Y_{D^\bullet}$-Y_D band shifts. In addition, Ananyev et al. (2) have found that the loss of Y_D greatly slows the photoactivated assembly of the Mn cluster relative to wild type. This observation has also been interpreted in terms of an enhancement by $Y_{D^\bullet}$ of the P^+/P reduction potential, thereby increasing the driving force for Mn oxidation. These observations provide a rationale for the evolutionary retention of Y_D even though it is on a side path to the main electron transport pathway that leads to water oxidation.

ALTERNATE ELECTRON DONORS

Illumination under conditions in which the normal donor side electron transfer reactions (Y_Z/Mn oxidation) are blocked (e.g., low temperature) results in the oxidation of alternate electron donors. These include the formation of a β-carotene cation radical [Car^+, $\varepsilon_{990\,nm} = 160$ mM^{-1} cm^{-1} (115)] [(79, 104); for review see (124)], a chlorophyll cation radical [Chl_Z^+, $\varepsilon_{820\,nm} = 7.0$ mM^{-1} cm^{-1} (9)] (140), and an oxidized cytochrome b$_{559}$ [cyt b$_{559}$, $\varepsilon_{560-570\,nm} = 17.5$ mM^{-1} cm^{-1} (19)] [for review see (19, 110)]. The Chl_Z^+ has been attributed to one of the two chlorophylls coordinated by D1-His118 and D2-His117 (Chl_{ZD1} and Chl_{ZD2}, respectively) (Figure 1). This designation is based on several factors: (*a*) the location of these histidines in the second transmembrane helix of D1 and D2 (75), (*b*) the kinetics of energy transfer from peripheral chlorophylls to the central pigments in PSII reaction centers (103), (*c*) the distance (39.5 ± 2.5 nm) of Chl_Z^+ from the PSII non-heme iron, measured using saturation-recovery EPR (65), and (*d*) the perturbation of the Chl_Z^+ Resonance Raman spectrum by site-directed mutations at D1-His118 in *Synechocystis* 6803 (111).

When cyt b$_{559}$ is reduced, it is the only cytochrome that is stably oxidized at temperatures ≤100 K. If the cytochrome is oxidized prior to illumination, then Chl_Z^+ and Car are photooxidized, with the relative proportions of the two dependent on the temperature and on the nature of the PSII preparation. In addition, there appear to be two components to Chl_Z^+ absorbing at 814 nm (in *Synechocystis* 6803; 817 nm in spinach) and at 850 nm (only observed in spinach) (123). Tracewell et al. (123) have suggested that these signals arise from Chl_{ZD1}^+ and Chl_{ZD2}^+, respectively. The oxidation of these chlorophylls is accompanied by very strong signals arising from Car^+ at 987 nm (*Synechocystis*) and 994 nm (spinach) and by a vibronic band at 878 nm and 895 nm, respectively. As the temperature is raised from 20 K to 120 K, the component at 850 nm increases at the expense of the Car^+ signal (123). Additionally, if Car^+ is generated at 20 K then warmed to 120 K, an increase is observed in the 850 nm Chl_Z^+ signal at the expense of the Car^+ (51). These latter observations have been interpreted in terms of an oxidation of Chl_Z by Car^+ in a thermally activated process. Hanley et al. (51) have argued that because cyt b_{559} can be oxidized at 20 K, whereas Chl_Z cannot (in the presence of cyt b_{559}^+), then cyt b_{559} and Chl_Z are located on different branches of the oxidation pathway and compete for the oxidizing equivalent on Car^+ at higher temperatures (Figure 6*B*). Tracewell et al. (123) were, however, able to observe some oxidation of Chl_Z at 20 K and thus consider the question of a parallel (Figure 6*B*) or series (Figure 6*A*) connection for Chl_Z and cyt b_{559} to Car^+ to be open.

There has been some controversy regarding the identity of Chl_Z. Perturbations of Resonance Raman spectra of Chl_Z^+ in site-directed mutations at D1-His118 in *Synechocystis* 6803 led to the conclusion that this residue coordinated Chl_Z (111). In contrast, observations by Ruffle et al. (96) of perturbations in the chlorophyll fluorescence yield caused by mutations at D2-His117 in *Chlamydomonas* and the spin relaxation measurements by Shigemori et al. (107) that yield a 29-Å distance between $Y_{D^\bullet}$ and Chl_Z both argue instead for coordination of Chl_Z by D2-His117. The demonstration by Tracewell et al. (123) of Chl_Z oxidation at 814 nm only in *Synechocystis* led these authors to assign this oxidized chlorophyll to Chl_{ZD1}^+ (coordinated by D1-His118). The observation of Chl^+ absorption at 817 nm and at 850 nm in spinach led to the assignment of the latter to Chl_{ZD2}^+ (coordinated by D2-His117).

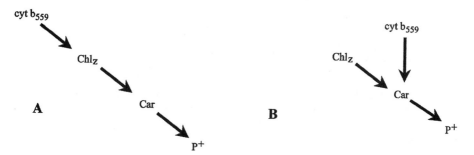

Figure 6   Two alternative models for the donor-side oxidative pathway that include Car, Chl_Z, and cyt b_{559}. In model A (123), Chl_Z and cyt b_{559} are connected in series to Car^+. In model B (51), Chl_Z and cyt b_{559} are parallel donors to Car^+.

The X-ray crystal structure of PSII has placed cyt b_{559} much closer to D2-His117 (Chl_{ZD2}) than to D1-His118 (Chl_{ZD1}) (Figure 1). Regardless of whether Chl_Z and cyt b_{559} form a linear or branched pathway (Figure 6), it would still appear that Chl_{ZD2} is far more apt to compete with or to oxidize the cyt b_{559} present in the crystal structure. The reduction of cyt b_{559} by plastoquinol (146) and, more particularly, by Q_BH_2 (or Q_B^-) (14) would also be more consistent with the interaction of this cytochrome with the D2 side components.

There has been considerable controversy over the years as to whether there are one or two cyt b_{559} associated with PSII [for review see (110)]. One possible solution to this debate is that there are two cyt b_{559} in thylakoid membranes but one is readily lost following detergent treatment. What remains then in the X-ray structure is the more tightly bound of the two. Consequently it may be that there are two pathways symmetrically arranged within the reaction center, both of which are capable of oxiding cyt b_{559}. Alternatively, it may be that only the D2 side pathway truly does cyclic electron transfer when the main donor side pathway is blocked. This would leave the D1 side to generate Chl_{ZD1}^+, which would quench excitation energy, consistent with the observations of Schweitzer & Brudvig [(106), also see (56)].

REEVALUATION OF ENERGETICS

Although a fairly clear and solid picture of the relative position and role of the different PSII cofactors has recently emerged, the energetic picture of PSII is more tenuous. This is, at least partly, attributed to the fact that, at variance with its bacterial homologues in which most of the component reduction potentials are accessible to direct redox titration, in PSII only three among the nine potentials have been directly measured (the Pheo/Pheo$^-$, Q_A/Q_A^-, and $Y_D^•/Y_D$ couples). The determination of the free-energy changes associated with the successive electron transfer reactions has, however, allowed the estimation of reduction potentials that could not be directly determined by titration [see (27, 89, 121) for reviews of the currently accepted values of the P$^+$/P, $Y_Z^•/Y_Z$, S_1/S_0, S_2/S_1, S_3/S_2, and S_0/S_3 couples]. In fact, all of the energetic relations within PSII rely on the determination of the reduction potential of the Pheo/Pheo$^-$ couple [-640 mV (59)] and the assumption that the energy level of the P$^+$Pheo$^-$ state is defined relative to this reduction potential. This latter hypothesis led Klimov et al. (59) to estimate the reduction potential of the P$^+$/P couple to be approximately 1.12 V ($-0.64 - 0.07 + 1.83$) from the free-energy change associated with charge separation (-70 meV) and the singlet-singlet difference energy between P and P* (1.83 eV). The midpoint potential of the P$^+$/P couple has then been used as a reference to estimate the other reduction potentials from measured equilibrium constants. There is growing evidence that our current view of PSII energetics could be erroneous and may require revision. As the reduction potential of the Pheo$_A$/Pheo$_A^-$ couple is more negative than that of the Q_A/Q_A^- couple, the equilibrium redox titration of Pheo$_A$ is inevitably performed in the presence of Q_A^-. Above, we discuss the consequences of the redox state of Q_A on the rate of P^+Pheo^- radical pair decay, i.e., the significantly increased rate in the presence of Q_A^- with respect to Q_AH_2. Gibasiewicz et al. (41) recently estimated the electrostatic interaction between Q_A^- and $Pheo^-$ to be $\sim$90 mV. Thus the energy level of the P^+Pheo^- should be at least 90 mV more positive than predicted by the equilibrium midpoint potential of the $Pheo/Pheo^-$ couple. Similarly, an electrostatic interaction between P^+ and $Pheo^-$ could stabilize the radical pair and further decrease its energy level. Indeed, van Gorkom (132) estimated the free energy difference between $P^+Q_A^-$ and P^+Pheo^- to be 340 mV, i.e., approximately 260 mV smaller than expected from the difference in midpoint potentials. If indeed the energy of the P^+Pheo^- radical pair is closer to that of $P^+Q_A^-$ than previously estimated, then an immediate consequence of this conclusion would be the upward revision of all of the reduction potentials (except for Q_A/Q_A^- and $Y_D^\bullet/Y_D$), including that of the P^+/P couple, which determines the available driving force for water splitting.

CONCLUSION

There are a number of fundamental ways in which the view of PSII function is being altered by (*a*) the newly developing three-dimensional structure of the PSII core complexes, (*b*) biochemical chromophore replacement, (*c*) analysis of site-directed mutations, and (*d*) new kinetic and spectroscopic methods. In this review, we mention that energy transfer and the electron transfer pathways involving Car, Chl_Z, and cyt b_{559} are undergoing intense reevaluation in light of new structural and kinetic information. As in the bacterial reaction centers, the three-dimensional structure likely will allow a better appreciation of how electrostatic interactions contribute to the energetics of the charge-separated states. Although not presently the case, the PSII X-ray structure will ultimately provide a great deal more assurance as to the location of the amino acid side chains of the component polypeptides, critical for the construction and interpretation of site-directed mutations. The intersection of all of these methods will provide a solid foundation for examining questions regarding the role of the protein in the function and assembly of the Mn cluster and other cofactors and the mechanisms of proton-coupled electron transfer; investigations whose impact extends well beyond the domain of photosynthesis.

ACKNOWLEDGMENTS

Support by the Central Research and Development Department of E. I. du Pont de Nemours & Co., the NRICGP/USDA (97-35306-4882) (B.A.D.), and the Centre National de la Recherche Scientifique (F.R.) is gratefully acknowledged. We are also grateful to James Bautista, R. David Britt, Kristy Campbell, Dee Force, and Andy Pascal whose unpublished work is mentioned in the text. We also thank James Bautista for his help with the figures. This paper is contribution No. 8261 of the Central Research and Development Department, E. I. du Pont de Nemours & Co.

Visit the Annual Reviews home page at www.annualreviews.org

LITERATURE CITED

1. Ahlbrink R, Haumann M, Cherepanov D, Boegershausen O, Mulkidjanian A, Junge W. 1998. Function of tyrosine Z in water oxidation by photosystem II: electrostatical promotor instead of hydrogen abstractor. *Biochemistry* 37:1131–42

2. Ananyev GM, Sakiyan I, Diner BA, Dismukes GC. 2002. A functional role for tyrosine-D in assembly of the inorganic core of the water oxidase complex of photosytem II and the kinetics of water oxidation. *Biochemistry* 41:974–80

3. Barter LMC, Bianchietti M, Jeans C, Schilstra MJ, Hankamer B, et al. 2001. Relationship between excitation energy transfer, trapping, and antenna size in photosystem II. *Biochemistry* 40:4026–34

4. Beekman LMP, van Mourik F, Jones MR, Visser HM, Hunter CN, van Grondelle R. 1994. Trapping kinetics in mutants of the photosynthetic purple bacterium *Rhodobacter sphaeroides*: influence of the charge separation rate and consequences for the rate-limiting step in the light-harvesting process. *Biochemistry* 33:3143–47

5. Berthold DA, Babcock GT, Yocum CF. 1981. A highly resolved, oxygen-evolving photosystem II preparation from spinach thylakoid membranes. EPR and electron-transport properties. *FEBS Lett.* 134:231–34

6. Blomberg MRA, Siegbahn PEM, Styring S, Babcock GT, Aakermark B, Korall P. 1997. A quantum chemical study of hydrogen abstraction from manganese-coordinated water by a tyrosyl radical: a model for water oxidation in photosystem II. *J. Am. Chem. Soc.* 119:8285–92

7. Boerner RJ, Bixby KA, Nguyen AP, Noren GH, Debus RJ, Barry BA. 1993. Removal of stable tyrosine radical D^+ affects the structure or redox properties of tyrosine Z in manganese-depleted photosystem II particles from *Synechocystis 6803*. *J. Biol. Chem.* 268:1817–23

8. Booth PJ, Crystall B, Giorgi LB, Barber J, Klug DR, Porter G. 1990. Thermodynamic properties of D1/D2/cytochrome b_{559} reaction centers investigated by time-resolved fluorescence measurements. *Biochim. Biophys. Acta* 1016:141–52

9. Borg DC, Fajer J, Felton RH, Dolphin D. 1970. Pi-cation radical of chlorophyll-a. *Proc. Natl. Acad. Sci. USA* 67:813–20

10. Bosch MK, Proskuryakov II, Gast P, Hoff AJ. 1996. Time-resolved EPR study of the primary donor triplet in D1-D2-cyt b_{559} complexes of photosystem II: temperature dependence of spin-lattice relaxation. *J. Phys. Chem.* 100:2384–90

11. Braun P, Greenberg BM, Scherz A. 1990. D1-D2-cytochrome b_{559} complex from the aquatic plant *Spirodela oligorrhiza*: correlation between complex integrity, spectroscopic properties, photochemical activity, and pigment composition. *Biochemistry* 29:10376–87

12. Briantais J-M, Dacosta J, Goulas Y, Ducruet J-M, Moya I. 1996. Heat stress induces in leaves an increase of the minimum level of chlorophyll fluorescence, Fo: a time-resolved analysis. *Photosynth. Res.* 48:189–96

13. Britt RD. 1996. Oxygen evolution. In *Oxygenic Photosynthesis: The Light Reactions*, ed. DR Ort, CF Yocum, pp. 137–64. Dordrecht: Kluwer Acad.

14. Buser CA, Diner BA, Brudvig GW. 1992. Photooxidation of cytochrome b_{559} in oxygen-evolving photosystem II. *Biochemistry* 31:11449–59

14a. Bylina EJ, Youvan DC. 1988. Directed mutations affecting spectroscopic and electron transfer properties of the primary donor in the photosynthetic reaction center. *Proc. Natl. Acad. Sci. USA* 85:7226–30

15. Bylina EJ, Kirmaier C, McDowell L, Holten D, Youvan DC. 1988. Influence of an amino-acid residue on the optical properties and electron transfer dynamics of a photosynthetic reaction center complex. *Nature* 336:182–84

16. Campbell KA, Peloquin JM, Diner BA, Tang X-S, Chisholm DA, Britt RD. 1997. The tau-nitrogen of D2 histidine 189 is the hydrogen bond donor to the tyrosine radical Y_D• of photosystem II. *J. Am. Chem. Soc.* 119:4787–88

17. Chirino AJ, Lous EJ, Huber M, Allen JP, Schenck CC, et al. 1994. Crystallographic analyses of site-directed mutants of the photosynthetic reaction center from *Rhodobacter sphaeroides*. *Biochemistry* 33:4584–93

18. Clayton RK, Yamamoto T. 1976. Photochemical quantum efficiency and absorption spectra of reaction centers from *Rhodopseudomonas sphaeroides* at low temperature. *Photochem. Photobiol.* 24:67–70

19. Cramer WA, Theg SM, Widger WR. 1986. On the structure and function of cytochrome b_{559}. *Photosynth. Res.* 10:393–403

20. Danielius RV, Satoh K, van Kan PJM, Plijter JJ, Nuijs AM, van Gorkom HJ. 1987. The primary reaction of photosystem II in the D1-D2-cytochrome b_{559} complex. *FEBS Lett.* 213:241–44

21. Dau H, Sauer K. 1996. Exciton equilibration and photosystem II exciton dynamics—a fluorescence study on photosystem II membrane particles of spinach. *Biochim. Biophys. Acta* 1273:175–90

22. Debus RJ. 2000. The polypeptides of photosystem II and their influence on manganotyrosyl-based oxygen evolution. In *Metal Ions in Biological Systems*, ed. A Sigel, H Sigel, pp. 657–711. New York: Dekker

23. Debus RJ. 2001. Amino acid residues that modulate the properties of tyrosine Y_Z and the manganese cluster in the water oxidizing complex of photosystem II. *Biochim. Biophys. Acta* 1503:164–86

24. Dekker JP, van Grondelle R. 2000. Primary charge separation in photosystem II. *Photosynth. Res.* 63:195–208

25. Den Blanken HJ, Hoff AJ, Jongenelis APJM, Diner BA. 1983. High-resolution triplet-minus-singlet absorbance difference spectrum of photosystem II particles. *FEBS Lett.* 157:21–27

26. Diner BA. 2001. Amino acid residues involved in the coordination and assembly of the manganese cluster of photosystem II. Proton-coupled electron transport of the redox-active tyrosines and its relationship to water oxidation. *Biochim. Biophys. Acta* 1503:147–63

27. Diner BA, Babcock GT. 1996. Structure, dynamics, and energy conversion efficiency in photosystem II. In *Oxygenic Photosynthesis: The Light Reactions*, ed. D Ort, C Yocum, pp. 213–47. Dordrecht: Kluwer Acad.

28. Diner BA, Force DA, Randall DW, Britt RD. 1998. Hydrogen bonding, solvent exchange, and coupled proton and electron transfer in the oxidation and reduction of redox-active tyrosine Y_Z in Mn-depleted core complexes of Photosystem II. *Biochemistry* 37:17931–43

29. Diner BA, Nixon PJ. 1998. Evidence for D1-His190 as the proton acceptor implicated in the oxidation of redox-active tyrosine Y_Z of PSII. See Ref. 38a, pp. 1177–80

30. Diner BA, Schlodder E, Nixon PJ, Coleman WJ, Rappaport F, et al. 2001. Site-directed mutations at D1-His198 and D2-His197 of photosystem II in *Synechocystis PCC 6803*: sites of primary charge separation and cation and triplet stabilization. *Biochemistry* 40:9265–81

31. Diner BA, Tang X-S, Zheng M, Dismukes GC, Force DA, et al. 1995. Environment and function of the redox active tyrosines of photosystem II. In *Photosynthesis: From Light to Biosphere*, ed. P Mathis, pp. 229–34. Dordrecht: Kluwer Acad.

32. Donovan B, Walker LA II, Kaplan D, Bouvier M, Yocum CF, Sension RJ. 1997. Structure and function in the isolated

reaction center complex of photosystem II. 1. Ultrafast fluorescence measurements of PSII. *J. Phys. Chem. B* 101:5232–38

33. Dorlet P, Xiong L, Sayre RT, Un S. 2001. High field EPR study of the pheophytin anion radical in wild type and D1-E130 mutants of photosystem II in *Chlamydomonas reinhardtii*. *J. Biol. Chem.* 276: 22313–16

34. Durrant JR, Giorgi LB, Barber J, Klug DR, Porter G. 1990. Characterization of triplet states in isolated photosystem II reaction centers: oxygen quenching as a mechanism for photodamage. *Biochim. Biophys. Acta* 1017:167–75

35. Durrant JR, Hastings G, Joseph DM, Barber J, Porter G, Klug DR. 1992. Subpicosecond equilibration of excitation energy in isolated photosystem II reaction centers. *Proc. Natl. Acad. Sci. USA* 89: 11632–36

36. Ermler U, Fritzsch G, Buchanan SK, Michel H. 1994. Structure of the photosynthetic reaction center from *Rhodobacter sphaeroides* at 2.65 Å resolution: cofactors and protein-cofactor interactions. *Structure* 2:925–36

37. Faller P, Debus RJ, Brettel K, Sugiura M, Rutherford AW, Boussac A. 2001. Rapid formation of the stable tyrosyl radical in Photosystem II. *Proc. Natl. Acad. Sci. USA* 98:14368–73

38. Fleming GR, Martin JL, Breton J. 1988. Rates of primary electron transfer in photosynthetic reaction centers and their mechanistic implications. *Nature* 333: 190–92

38a. Garab G, ed. 1998. *Photosynthesis: Mechanism and Effects*. Dordrecht: Kluwer Acad.

39. Germano M, Shkuropatov AY, Permentier H, de Wijn R, Hoff AJ, et al. 2001. Pigment organization and their interactions in reaction centers of photosystem II: optical spectroscopy at 6 K of reaction centers with modified pheophytin composition. *Biochemistry* 40:11472–82

40. Germano M, Shkuropatov AY, Permentier H, Khatypov RA, Shuvalov VA, et al. 2000. Selective replacement of the active and inactive pheophytin in reaction centers of photosystem II by 13^1-deoxo-13^1-hydroxy-pheophytin *a* and comparison of their 6-K absorption spectra. *Photosynth. Res.* 64:189–98

41. Gibasiewicz K, Dobek A, Breton J, Leibl W. 2001. Modulation of primary radical pair kinetics and energetics in photosystem II by the redox state of the quinone electron acceptor Q_A. *Biophys. J.* 80: 1617–30

42. Giorgi LB, Nixon PJ, Merry SA, Joseph DM, Durrant JR, et al. 1996. Comparison of primary charge separation in the photosystem II reaction center complex isolated from wild-type and D1-130 mutants of the cyanobacterium *Synechocystis PCC 6803*. *J. Biol. Chem.* 271:2093–101

43. Goldsmith JO, King B, Boxer SG. 1996. Mg coordination by amino acid side chains is not required for assembly and function of the special pair in bacterial photosynthetic reaction centers. *Biochemistry* 35:2421–28

44. Greenfield SR, Seibert M, Govindjee, Wasielewski MR. 1997. Direct measurement of the effective rate constant for primary charge separation in isolated photosystem II reaction centers. *J. Phys. Chem. B* 101:2251–55

45. Greenfield SR, Seibert M, Wasielewski MR. 1999. Time-resolved absorption changes of the pheophytin Q_X band in isolated photosystem II reaction centers at 7 K: energy transfer and charge separation. *J. Phys. Chem. B* 103:8364–74

46. Greenfield SR, Wasielewski MR. 1996. Excitation energy transfer and charge separation in the isolated photosystem II reaction center. *Photosynth. Res.* 48:83–97

47. Groot M-L, van Mourik F, Eijckelhoff C, van Stokkum IHM, Dekker JP, van Grondelle R. 1997. Charge separation in the reaction center of photosystem II studied as a function of temperature. *Proc. Natl. Acad. Sci. USA* 94:4389–94

48. Groot M-L, Peterman EJG, van Kan PJM, van Stokkum IHM, Dekker JP, van Grondelle R. 1994. Temperature-dependent triplet and fluorescence quantum yields of the photosystem II reaction center described in a thermodynamic model. *Biophys. J.* 67:318–30

49. Hales BJ, Das Gupta A. 1979. Orientation of the bacteriochlorophyll triplet and the primary ubiquinone acceptor of *Rhodospirillum rubrum* in membrane multilayers determined by ESR spectroscopy. I. *Biochim. Biophys. Acta* 548:276–86

50. Hankamer B, Barber J, Boekema EJ. 1997. Structure and membrane organization of photosystem II in green plants. *Annu. Rev. Plant Physiol. Plant Mol. Biol.* 48: 641–71

51. Hanley J, Deligiannakis Y, Pascal A, Faller P, Rutherford AW. 1999. Carotenoid oxidation in photosystem II. *Biochemistry* 38:8189–95

52. Hays AM, Vassiliev IR, Golbeck JH, Debus RJ. 1998. Role of D1-His190 in proton-coupled electron transfer reactions in photosystem II: a chemical complementation study. *Biochemistry* 37:11352–65

53. Hienerwadel R, Boussac A, Breton J, Diner BA, Berthomieu C. 1997. Fourier transform infrared difference spectroscopy of photosystem II tyrosine D using site-directed mutagenesis and specific isotope labeling. *Biochemistry* 36:14712–23

54. Hillmann B, Brettel K, van Mieghem F, Kamlowski A, Rutherford AW, Schlodder E. 1995. Charge recombination reactions in photosystem II. 2. Transient absorbance difference spectra and their temperature dependence. *Biochemistry* 34:4814–27

55. Hillmann B, Schlodder E. 1995. Electron transfer reactions in Photosystem II core complexes from *Synechococcus* at low temperature—difference spectrum of $P680^+ Q_A^-/P680 Q_A$ at 77 K. *Biochim. Biophys. Acta* 1231:76–88

56. Horton P, Ruban AV, Walters RG. 1996. Regulation of light harvesting in green plants. *Annu. Rev. Plant Physiol. Plant Mol. Biol.* 47:655–84

56a. Ikeuchi M, Inoue Y. 1988. A new photosystem II reaction center component (4.8 kDa protein) encoded by chloroplast genome. *FEBS Lett.* 241:99–104

57. Jankowiak R, Tang D, Small GJ, Seibert M. 1989. Transient and persistent hole burning of the reaction center of photosystem II. *J. Phys. Chem.* 93:1649–54

58. Kamlowski A, Frankemöller L, van der Est A, Stehlik D, Holzwarth A. 1996. Evidence for the delocalization of the triplet state $^3P680$ in the D1D2cyt b_{559}-complex of photosystem II. *Ber. Bunsenges. Phys. Chem.* 100:2045–51

58a. Kirmaier C, Gaul D, DeBey R, Holten D, Schenck CC. 1991. Charge separation in a reaction center incorporating bacteriochlorophyll for photoactive bacteriopheophytin. *Science* 251:922–27

59. Klimov VV, Krasnovskii AA. 1981. Pheophytin as the primary electron acceptor in photosystem II reaction centers. *Photosynthetica* 15:592–609

60. Klug DR, Durrant JR, Barber J. 1998. The entanglement of excitation energy transfer and electron transfer in the reaction center of photosystem II. *Philos. Trans. R. Soc. London Ser. A* 356:449–64

61. Klug DR, Rech T, Joseph DM, Barber J, Durrant JR, Porter G. 1995. Primary processes in isolated Photosystem II reaction centers probed by magic angle transient absorption spectroscopy. *Chem. Phys.* 194:433–42

62. Konermann L, Gatzen G, Holzwarth AR. 1997. Primary processes and structure of the photosystem II reaction center. Part V. Modeling of the fluorescence kinetics of the D1-D2-cyt-b_{559} complex at 77 K. *J. Phys. Chem. B* 101:2933–44

63. Konermann L, Holzwarth AR. 1996. Analysis of the absorption spectrum of photosystem II reaction centers: temperature dependence, pigment assignment, and inhomogeneous broadening. *Biochemistry* 35:829–42

64. Konermann L, Yruela I, Holzwarth AR. 1997. Pigment assignment in the absorption spectrum of the photosystem II reaction center by site-selection fluorescence spectroscopy. *Biochemistry* 36:7498–502

65. Koulougliotis D, Innes JB, Brudvig GW. 1994. Location of chlorophyll$_Z$ in photosystem II. *Biochemistry* 33:11814–22

66. Krabben L, Schlodder E, Jordan R, Carbonera D, Giacometti G, et al. 2000. Influence of the axial ligands on the spectral properties of P700 of photosystem I: a study of site-directed mutants. *Biochemistry* 39:13012–25

67. Krishtalik LI. 1990. Activation energy of photosynthetic oxygen evolution: an attempt at theoretical analysis. *Bioelectrochem. Bioenerg.* 23:249–63

68. Kwa SLS, Eijckelhoff C, van Grondelle R, Dekker JP. 1994. Site-selection spectroscopy of the reaction center complex of photosystem II. 1. Triplet-minus-singlet absorption difference: search for a second exciton band of P680. *J. Phys. Chem.* 98:7702–11

69. Lin S, Taguchi AKW, Woodbury NW. 1996. Excitation wavelength dependence of energy transfer and charge separation in reaction centers from *Rhodobacter sphaeroides*: evidence for adiabatic electron transfer. *J. Phys. Chem.* 100:17067–78

70. Lince MT, Vermaas W. 1998. Association of His117 in the D2 protein of photosystem II with a chlorophyll that affects excitation-energy transfer efficiency to the reaction center. *Eur. J. Biochem.* 256:595–602

71. Mathis P, Breton J, Vermeglio A, Yates M. 1976. Orientation of the primary donor chlorophyll of Photosystem II in chloroplast membranes. *FEBS Lett.* 63:171–73

72. McGlynn SP, Azumi T, Kinoshita M. 1969. *Molecular Spectroscopy of the Triplet State*. Englewood Cliffs, NJ: Prentice Hall

73. Merry SAP, Kumazaki S, Tachibana Y, Joseph DM, Porter G, et al. 1996. Sub-picosecond equilibration of excitation energy in isolated photosystem II reaction centers revisited: time-dependent anisotropy. *J. Phys. Chem.* 100:10469–78

74. Merry SAP, Nixon PJ, Barter LMC, Schilstra M, Porter G, et al. 1998. Modulation of quantum yield of primary radical pair formation in photosystem II by site-directed mutagenesis affecting radical cations and anions. *Biochemistry* 37:17439–47

75. Michel H, Deisenhofer J. 1988. Relevance of the photosynthetic reaction center from purple bacteria to the structure of photosystem II. *Biochemistry* 27:1–7

76. Moënne-Loccoz P, Robert B, Lutz M. 1989. A resonance Raman characterization of the primary electron acceptor in photosystem II. *Biochemistry* 28:3641–45

77. Mueller MG, Hucke M, Reus M, Holzwarth AR. 1996. Primary processes and structure of the photosystem II reaction center 4. Low-intensity femtosecond transient absorption spectra of D1-D2-cyt-b$_{559}$ reaction centers. *J. Phys. Chem.* 100: 9527–36

77a. Nanba O, Satoh K. 1987. Isolation of a Photosystem II reaction center consisting of D-1 and D-2 polypeptides and cytochrome b-559. *Proc. Natl. Acad. Sci. USA* 84:109–12

78. Noguchi T, Inoue Y, Satoh K. 1993. FT-IR studies on the triplet state of P680 in the photosystem II reaction center: triplet equilibrium within a chlorophyll dimer. *Biochemistry* 32:7186–95

79. Noguchi T, Mitsuka T, Inoue Y. 1994. Fourier transform infrared spectrum of the radical cation of beta-carotene photoinduced in photosystem II. *FEBS Lett.* 356:179–82

80. Noguchi T, Tomo T, Inoue Y. 1998. Fourier transform infrared study of the cation radical of P680 in the photosystem II reaction center: evidence for charge delocalization on the chlorophyll dimer. *Biochemistry* 37:13614–25

81. Noguchi T, Tomo T, Kato C. 2001. Triplet

formation on a monomeric chlorophyll in the photosystem II reaction center as studied by time-resolved infrared spectroscopy. *Biochemistry* 40:2176–85

82. O'Malley PJ. 1998. Hybrid density functional studies of the oxidation of phenol-imidazole hydrogen-bonded complexes: a model for tyrosine oxidation in oxygenic photosynthesis. *J. Am. Chem. Soc.* 120:11732–37

83. Pecoraro VL, Baldwin MJ, Caudle MT, Hsieh W-Y, Law NA. 1998. A proposal for water oxidation in photosystem II. *Pure Appl. Chem.* 70:925–29

84. Peloquin JM, Williams JC, Lin X, Alden RG, Taguchi AKW, et al. 1994. Time-dependent thermodynamics during early electron transfer in reaction centers from *Rhodobacter sphaeroides*. *Biochemistry* 33:8089–100

85. Peterman EJG, Van Amerongen H, van Grondelle R, Dekker JP. 1998. The nature of the excited state of the reaction center of photosystem II of green plants: a high-resolution fluorescence spectroscopy study. *Proc. Natl. Acad. Sci. USA* 95:6128–33

86. Rappaport F, Lavergne J. 1991. Proton release during successive oxidation steps of the photosynthetic water oxidation process: stoichiometries and pH dependence. *Biochemistry* 30:10004–12

87. Rappaport F, Lavergne J. 1997. Charge recombination and proton transfer in manganese-depleted photosystem II. *Biochemistry* 36:15294–302

88. Renger G. 1992. Energy transfer and trapping in photosystem II. In *The Photosystems: Structure, Function and Molecular Biology*, ed. J Barber, pp. 45–99. Amsterdam: Elsevier

89. Renger G. 2001. Photosynthetic water oxidation to molecular oxygen: apparatus and mechanism. *Biochim. Biophys. Acta* 1503:210–28

90. Rhee K-H. 2001. Photosystem II: the solid structural era. *Annu. Rev. Biophys. Biomol. Struct.* 30:307–28

91. Rhee K-H, Morris EP, Barber J, Kuhlbrandt W. 1998. Three-dimensional structure of the plant photosystem II reaction centre at 8-Å resolution. *Nature* 396:283–86

92. Rhee K-H, Morris EP, Zheleva D, Hankamer B, Kuhlbrandt W, Barber J. 1997. Two-dimensional structure of plant photosystem II at 8-Å resolution. *Nature* 389:522–26

93. Rigby SEJ, Nugent JHA, O'Malley PJ. 1994. ENDOR and Special Triple Resonance studies of chlorophyll cation radicals in photosystem II. *Biochemistry* 33:10043–50

94. Rodriguez ID, Chandrashekar TK, Babcock GT. 1987. ENDOR characterization of H_2O/D_2O exchange in the D^+Z^+ radical in photosynthesis. In *Progress in Photosynthesis Research*, ed. J Biggins, pp. 471–74. Dordrecht: Martinus Nijhoff

95. Roelofs TA, Lee CH, Holzwarth AR. 1992. Global target analysis of picosecond chlorophyll fluorescence kinetics from pea chloroplasts: a new approach to the characterization of the primary processes in photosystem II a and b units. *Biophys. J.* 61:1147–63

96. Ruffle S, Hutchison R, Sayre RT. 1998. Mutagenesis of the symmetry related H117 residue in the photosystem II D2 protein of *Chlamydomonas*: implications for energy transfer from accessory chlorophylls. *Photosynth.: Mech. Eff., Proc. Int. Congr. Photosynth., 11th*, 2:1013–16

97. Ruffle SV, Donnelly D, Blundell TL, Nugent JHA. 1992. A three-dimensional model of the photosystem II reaction center of *Pisum sativum*. *Photosynth. Res.* 34:287–300

98. Rutherford AW. 1988. Photosystem II, the oxygen evolving photosystem. In *Light Energy Transduction in Photosynthesis: Higher Plant and Bacterial Models*, ed. SE Stevens, DA Bryant, pp. 163–77. Rockville, MD: Am. Soc. Plant Physiol.

99. Rutherford AW, Nitschke W. 1996. Photosystem II and the quinone-iron-containing

reaction centers: comparisons and evolutionary perspectives. In *Origin and Evolution of Biological Energy Conversion*, ed. H Baltscheffsky, pp. 143–75. New York: VCH

100. Rutherford AW, Paterson DR, Mullet JE. 1981. A light-induced spin-polarized triplet detected by EPR in photosystem II reaction centers. *Biochim. Biophys. Acta* 635:205–14

101. Sarcina M, Breton J, Nabedryk E, Diner BA, Nixon PJ. 1998. FTIR studies on the P680 cation and triplet states in WT and mutant PSII reaction centers of *Synechocystis 6803*. See Ref. 38a, pp. 1053–56

102. Schatz GH, Brock H, Holzwarth AR. 1988. Kinetic and energetic model for the primary processes in photosystem II. *Biophys. J.* 54:397–405

103. Schelvis JPM, van Noort PI, Aartsma TJ, van Gorkom HJ. 1994. Energy transfer, charge separation and pigment arrangement in the reaction center of photosystem II. *Biochim. Biophys. Acta* 1184:242–50

104. Schenck CC, Diner B, Mathis P, Satoh K. 1982. Flash-induced carotenoid radical cation formation in photosystem II. *Biochim. Biophys. Acta* 680:216–27

105. Schulz C, Muh F, Beyer A, Jordan R, Schlodder E, Lubitz W. 1998. Investigation of *Rhodobacter sphaeroides* reaction center mutants with changed ligands to the primary donor. See Ref. 38a, pp. 767–70

106. Schweitzer RH, Brudvig GW. 1997. Fluorescence quenching by chlorophyll cations in photosystem II. *Biochemistry* 36:11351–59

107. Shigemori K, Hara H, Kawamori A, Akabori K. 1998. Determination of distances from tyrosine D to Q_A and chlorophyll$_Z$ in photosystem II studied by $2'+1'$ pulsed EPR. *Biochim. Biophys. Acta* 1363:187–98

108. Shkuropatov AY, Khatypov RA, Shkuropatova VA, Zvereva MG, Owens TG, Shuvalov VA. 1999. Reaction centers of photosystem II with a chemically modified pigment composition: exchange of pheophytins with 13^1-deoxo-13^1-hydroxy-pheophytin *a*. *FEBS Lett.* 450:163–67

109. Shkuropatov AY, Khatypov RA, Volshchukova TS, Shkuropatova VA, Owens TG, Shuvalov VA. 1997. Spectral and photochemical properties of borohydride-treated D1-D2-cytochrome b_{559} complex of photosystem II. *FEBS Lett.* 420:171–74

110. Stewart DH, Brudvig GW. 1998. Cytochrome b_{559} of photosystem II. *Biochim. Biophys. Acta* 1367:63–87

111. Stewart DH, Cua A, Chisholm DA, Diner BA, Bocian DF, Brudvig GW. 1998. Identification of Histidine 118 in the D1 polypeptide of photosystem II as the axial ligand to chlorophyll$_Z$. *Biochemistry* 37:10040–46

112. Svensson B, Etchebest C, Tuffery P, van Kan P, Smith J, Styring S. 1996. A model for the photosystem II reaction center core including the structure of the primary donor P680. *Biochemistry* 35:14486–502

113. Szalai VA, Kuhne H, Lakshmi KV, Brudvig GW. 1998. Characterization of the interaction between manganese and tyrosine Z in acetate-inhibited photosystem II. *Biochemistry* 37:13594–603

114. Takahashi Y, Hansson O, Mathis P, Satoh K. 1987. Primary radical pair in the photosystem II reaction center. *Biochim. Biophys. Acta* 893:49–59

115. Tan Q, Kuciauskas D, Lin S, Stone S, Moore AL, Moore TA, Gust D. 1997. Dynamics of photoinduced electron transfer in a carotenoid-porphyrin-dinitronaphthalenedicarboximide molecular triad. *J. Phys. Chem. B* 101:5214–23

116. Tang X-S, Randall DW, Force DA, Diner BA, Britt RD. 1996. Manganese-tyrosine interaction in the photosystem II oxygen-evolving complex. *J. Am. Chem. Soc.* 118:7638–39

117. Telfer A, Lendzian F, Schlodder E, Barber J, Lubitz W. 1998. ENDOR and transient absorption studies of P680$^+$ and other cation radicals in PSII reaction centers before and after inactivation of secondary

electron donors. See Ref. 38a, pp. 1061–64

118. Tetenkin VL, Gulyaev BA, Seibert M, Rubin AB. 1989. Spectral properties of stabilized D1/D2/cytochrome b_{559} photosystem II reaction center complex. Effects of Triton X-100, the redox state of pheophytin, and beta-carotene. *FEBS Lett.* 250:459–63

119. Tiede DM, Dutton PL. 1981. Orientation of the primary quinone of bacterial photosynthetic reaction centers contained in chromatophore and reconstituted membranes. *Biochim. Biophys. Acta* 637:278–90

120. Tommos C, Babcock GT. 1998. Oxygen production in nature: a light-driven metalloradical enzyme process. *Acc. Chem. Res.* 31:18–25

121. Tommos C, Babcock GT. 2000. Proton and hydrogen currents in photosynthetic water oxidation. *Biochim. Biophys. Acta* 1458:199–219

122. Tommos C, McCracken J, Styring S, Babcock GT. 1998. Stepwise disintegration of the photosynthetic oxygen-evolving complex. *J. Am. Chem. Soc.* 120:10441–52

123. Tracewell CA, Cua A, Stewart DH, Bocian DF, Brudvig GW. 2001. Characterization of carotenoid and chlorophyll photooxidation in photosystem II. *Biochemistry* 40:193–203

124. Tracewell CA, Vrettos JS, Bautista JA, Frank HA, Brudvig GW. 2001. Carotenoid photooxidation in photosystem II. *Arch. Biochem. Biophys.* 385:61–69

125. Trebst A. 1986. The topology of plastoquinone and herbicide binding peptides of photosystem II in the thylakoid membrane. *Z. Naturforsch. Teil C.* 41:240–45

126. Vacha F, Joseph DM, Durrant JR, Telfer A, Klug DR, et al. 1995. Photochemistry and spectroscopy of a five-chlorophyll reaction center of photosystem II isolated by using a Cu affinity column. *Proc. Natl. Acad. Sci. USA* 92:2929–33

127. Valkunas L, Liuolia V, Dekker JP, van Grondelle R. 1995. Description of energy migration and trapping in photosystem I by a model with two distance scaling parameters. *Photosynth. Res.* 43:149–54

128. van Brederode ME, Jones MR, van Mourik F, van Stokkum IHM, van Grondelle R. 1997. A new pathway for transmembrane electron transfer in photosynthetic reaction centers of *Rhodobacter sphaeroides* not involving the excited special pair. *Biochemistry* 36:6855–61

129. van Brederode ME, van Mourik F, van Stokkum IHM, Jones MR, van Grondelle R. 1999. Multiple pathways for ultrafast transduction of light energy in the photosynthetic reaction center of *Rhodobacter sphaeroides*. *Proc. Natl. Acad. Sci. USA* 96:2054–59

130. van Brederode ME, van Stokkum IHM, Katilius E, van Mourik F, Jones MR, van Grondelle R. 1999. Primary charge separation routes in the B_{Chl}:B_{Phe} heterodimer reaction centers of *Rhodobacter sphaeroides*. *Biochemistry* 38:7545–55

131. van der Vos R, van Leeuwen PJ, Braun P, Hoff AJ. 1992. Analysis of the optical absorbance spectra of D1-D2-cytochrome b_{559} complexes by absorbance-detected magnetic resonance. Structural properties of P680. *Biochim. Biophys. Acta* 1140:184–98

132. van Gorkom HJ. 1985. Electron transfer in photosystem II. *Photosynth. Res.* 6:97–112

133. van Grondelle R, Dekker JP, Gillbro T, Sundstrom V. 1994. Energy transfer and trapping in photosynthesis. *Biochim. Biophys. Acta* 1187:1–65

134. van Kan PJM, Otte SCM, Kleinherenbrink FAM, Nieveen MC, Aartsma TJ, van Gorkom HJ. 1990. Time-resolved spectroscopy at 10 K of the photosystem II reaction center; deconvolution of the red absorption band. *Biochim. Biophys. Acta* 1020:146–52

135. van Mieghem FJE, Satoh K, Rutherford AW. 1991. A chlorophyll tilted 30° relative to the membrane in the photosystem

II reaction center. *Biochim. Biophys. Acta* 1058:379–85

136. Vasil'ev S, Diner B. 2000. Picosecond time-resolved fluorescence studies on excitation energy transfer in a Histidine 117 mutant of the D2 protein of photosystem II in *Synechocystis 6803. Biochemistry* 39:14211–18

137. Vasil'ev S, Orth P, Zouni A, Owens TG, Diner B. 2001. Excited-state dynamics in photosystem II: insights from the X-ray crystal structure. *Proc. Natl. Acad. Sci. USA* 98:8602–7

138. Vermaas WFJ, Ikeuchi M, Inoue Y. 1988. Protein composition of the photosystem II core complex in genetically engineered mutants of the cyanobacterium *Synechocystis* sp. *PCC 6803. Photosynth. Res.* 17:97–113

139. Visser HM, Groot M-L, van Mourik F, van Stokkum IHM, Dekker JP, van Grondelle R. 1995. Subpicosecond transient absorption difference spectroscopy on the reaction center of photosystem II: radical pair formation at 77 K. *J. Phys. Chem.* 99: 15304–9

140. Visser JWM, Rijgersberg CP, Gast P. 1977. Photooxidation of chlorophyll in spinach chloroplasts between 10 and 180° K. *Biochim. Biophys. Acta* 460:36–46

141. Volk M, Gilbert M, Rousseau G, Richter M, Ogrodnik A, Michel-Beyerle M-E. 1993. Similarity of primary radical pair recombination in photosystem II and bacterial reaction centers. *FEBS Lett.* 336:357–62

142. Vos MH, Breton J, Martin J-L. 1997. Electronic energy transfer within the hexamer

cofactor system of bacterial reaction centers. *J. Phys. Chem.* B 101:9820–32

143. Wasielewski MR, Johnson DG, Govindjee, Preston C, Seibert M. 1989. Determination of the primary charge separation rate in photosystem II reaction centers at 15 K. *Photosynth. Res.* 22:89–99

144. Webber AN, Su H, Bingham SE, Kaess H, Krabben L, et al. 1996. Site-directed mutations affecting the spectroscopic characteristics and midpoint potential of the primary donor in photosystem I. *Biochemistry* 35:12857–63

145. Westphal KL, Tommos C, Cukier RI, Babcock GT. 2000. Concerted hydrogen-atom abstraction in photosynthetic water oxidation. *Curr. Opin. Plant Biol.* 3:236–42

146. Whitmarsh J, Cramer WA. 1978. A pathway for the reduction of cytochrome b_{559} by photosystem II in chloroplasts. *Biochim. Biophys. Acta* 501:83–93

147. Woodbury NW, Allen JP. 1995. The pathway, kinetics and thermodynamics of electron transfer in wild-type and mutant reaction centers of purple nonsulfur bacteria. In *Anoxygenic Photosynthesis*, ed. RE Blankenship, MT Madigan, CE Bauer, pp. 527–57. Dordrecht: Kluwer Acad.

148. Xiong J, Subramaniam S, Govindjee. 1998. A knowledge-based three-dimensional model of the photosystem II reaction center of *Chlamydomonas reinhardtii. Photosynth. Res.* 56:229–54

149. Zouni A, Witt H-T, Kern J, Fromme P, Krauss N, et al. 2001. Crystal structure of photosystem II from *Synechococcus elongatus* at 3.8-Å resolution. *Nature* 409: 739–43

Subject Index

581

CUMULATIVE INDEXES

CONTRIBUTING AUTHORS, VOLUMES 43–53

CHAPTER TITLES, VOLUMES 43–53

Prefatory Chapters

Biochemistry and Biosynthesis

Genetics and Molecular Biology

Cell Differentiation

Tissue, Organ, and Whole Plant Events

Acclimation and Adaptation

Methods